APPLIED MODERN ALGEBRA

APPLIED MODERN ALGEBRA

LARRY L. DORNHOFF
Associate Professor of Mathematics,
University of Illinois, Urbana-Champaign

FRANZ E. HOHN
Late Professor of Mathematics,
University of Illinois, Urbana-Champaign

MACMILLAN PUBLISHING CO., INC.
New York

COLLIER MACMILLAN PUBLISHERS
London

COPYRIGHT © 1978, LARRY L. DORNHOFF
AND ESTATE OF FRANZ E. HOHN

PRINTED IN THE UNITED STATES OF AMERICA

All rights reserved. No part of this book may be reproduced or transmitted in any form or by any means, electronic or mechanical, including photocopying, recording, or any information storage and retrieval system, without permission in writing from the Publisher.

MACMILLAN PUBLISHING CO., INC.
866 Third Avenue, New York, New York 10022

COLLIER MACMILLAN CANADA, LTD.

Library of Congress Cataloging in Publication Data

Dornhoff, Larry.
  Applied modern algebra.

  Bibliography: p.
  Includes index.
  1. Algebra, Abstract. I. Hohn, Franz Edward, 1915–1977, joint author. II. Title.
QA162.D67   512'.02   76-30825
ISBN 0-02-329980-0

Printing: 1 2 3 4 5 6 7 8   Year: 8 9 0 1 2 3 4

PREFACE

This book presents introductions to a variety of algebraic structures useful to students in curricula in mathematics, applied mathematics, electrical engineering, and computer science. No volume of modest size could treat all useful topics in even a limited way, so our list is by no means exhaustive. Our choices are based in part on our own experience in teaching courses in switching theory, automata theory, and coding theory, but even more on the advice of colleagues in electrical engineering and computer science.

The first six chapters provide an introduction to the more elementary aspects of sets, functions, relations, graphs, semigroups, groups, Pólya counting theory, rings, finite fields, posets, lattices, and Boolean algebras. There is more than enough material here for a semester's work. Chapters 4 and 5 are independent of Chapter 3; after Chapters 1 and 2, which are basic, an instructor may choose either Boolean algebras and rings, or semigroups and groups. The exposition is detailed, since our book is intended for a mathematically unsophisticated audience. Indeed, for many of our students, this is their first mathematics course beyond the calculus. We introduce computer-oriented applications of the many abstract notions early and often. We believe strongly that most students find the algebraic concepts easier when specific concrete examples are available, and that applications-oriented examples are best because they provide both motivation and illustration.

The remaining three chapters provide sufficient material to complete a full-year course in the applications of modern algebra to computer engineering. Chapter 7 is a complete, but concise, presentation of the linear algebra and field theory needed in Chapters 8 and 9. Students who have had a course in linear algebra will need only to survey Chapter 7 and study those topics, such as rational canonical forms and the theory of finite fields, which they have not previously encountered. Everything in Chapter 7 is used in Chapters 8 and 9, where linear machines and algebraic codes are discussed thoroughly. Since knowledge of linear algebra is so necessary for the study of these topics, a student with no previous exposure to linear algebra will have to work hard to master Chapters 7 to 9, and perhaps will need to use a linear algebra textbook as a reference. (Some are recommended at the beginning of Chapter 7.)

In order to make abstract ideas as comprehensible as possible, we have included many illustrative examples. There are extensive sets of exercises, some computational, some abstract, and of varying degrees of difficulty, to help the student attain full mastery of the material. The flavor of the book is strongly algorithmic, as is appropriate for the areas in which these topics are applied. We have starred those exercises which are specifically referred to elsewhere in the text or exercises, and we have also starred a number of other exercises that represent important basic concepts. "Exercise x.y.z" denotes Exercise z in Section x.y.

L. L. D.
F. E. H.

June 1977

Franz E. Hohn died in July 1977, and is sorely missed by his co-workers at the University of Illinois and his many friends elsewhere. My own debt to him is large; I benefited greatly from his advice and encouragement for several years preceding the writing of this book. Professor Hohn saw only a few of the page proofs, so responsibility for errors remaining in this book must lie with me.

Many of our colleagues at the University of Illinois have been helpful to us during the years of development of this book, either with advice concerning the selection of topics, with constructive criticism of the exposition, or by the discovery of errors. I wish particularly to thank Professors J. Armstrong, F. Bateman, J. Brown, E. Davidson, M. Day, C. Jockusch, G. Metze, D. Muller, F. Preparata, and J. Walter. I also wish to express my grateful appreciation for the quality workmanship provided by Everett Smethurst and Elaine Wetterau and the staff at Macmillan.

L. L. D.

August 1977

CONTENTS

CHAPTER 1
Sets and Functions — 1
1.1 Sets — 1
1.2 The Indexing of Sets — 2
1.3 Sets Derived from Other Sets — 2
1.4 The Order of a Set — 4
1.5 Functions — 4
1.6 Exercises — 6
1.7 More Notation — 8
1.8 One-to-One and Onto — 8
1.9 Composition and Inversion of Functions — 9
1.10 Exercises — 13
1.11 Finite-State Machines — 15
1.12 Construction of Machines — 20
1.13 Exercises — 23
1.14 Union, Intersection, and Inclusion — 26
1.15 Complementation — 27
1.16 Venn Diagrams — 28
1.17 More Algebra of Subsets — 29
1.18 The Principle of Duality — 30
1.19 Boolean Algebras; Summary of Basic Laws — 32
1.20 Mathematical Induction — 33
1.21 The Characteristic Function of a Subset of a Set — 35
1.22 Exercises — 37

CHAPTER 2
Relations and Graphs — 41
2.1 Relations — 41
2.2 A Matrix Model of Finite Relations — 42
2.3 The Composition of Relations — 45
2.4 Matrix Equivalent of Composition; Digraphs — 46
2.5 Relations on a Set A — 48
2.6 Functions as Relations — 49
2.7 Exercises — 50
2.8 Partial Orderings and Posets — 53
2.9 Special Elements of Posets — 57
2.10 Direct Products of Posets — 60
2.11 Exercises — 61
2.12 Equivalence Relations and Partitions — 63
2.13 Intersection and Union of Equivalence Relations — 66
2.14 Intersection and Union of Partitions — 68

2.15	Exercises	70
2.16	Identification of Equivalent States	71
2.17	Minimal-State Machines	75
2.18	Distinguishing Sequences for States	77
2.19	Exercises	78
2.20	Undirected Graphs	81
2.21	Trees	85
2.22	Exercises	89
2.23	Spanning Trees	91
2.24	Fundamental Circuits and Cut-Sets	94
2.25	Applications and Famous Problems	95
2.26	Exercises	99

CHAPTER 3
Rings and Boolean Algebras — 102

3.1	Algebras	102
3.2	Rings	103
3.3	Congruences	106
3.4	The Ring of Integers Modulo p	110
3.5	Binary Arithmetic Modulo 2^n	111
3.6	Exercises	114
3.7	Boolean Rings and Boolean Algebras	116
3.8	Independent Postulates for a Boolean Algebra	119
3.9	Exercises	121
3.10	The Algebraic Description of Logic Circuits	124
3.11	Switching Functions and Their Basic Properties	127
3.12	Exercises	132
3.13	Disjunctive and Conjunctive Normal Forms	136
3.14	The Exclusive-OR and the Ring Normal Form of f	140
3.15	Inequalities in Switching Algebra	143
3.16	Exercises	144
3.17	Prime Implicants and Minimal Sums of Products	149
3.18	Reusch's Method for Finding All Prime Implicants	151
3.19	The Minimal Sums of a Function	155
3.20	Reusch's Method for Finding All Minimal Sums	158
3.21	Exercises	163
3.22	Conclusion	164

CHAPTER 4
Semigroups and Groups — 165

4.1	Definitions	165
4.2	Zeros and Identities	168
4.3	Cyclic Semigroups and Cyclic Groups	170

4.4	Subsemigroups and Subgroups	172
4.5	Exercises	174
4.6	Congruence Relations on Semigroups	179
4.7	Morphisms	181
4.8	The Semigroup of a Machine	183
4.9	The Machine of a Semigroup	189
4.10	Exercises	190
4.11	Equations in Semigroups	192
4.12	Lagrange's Theorem	193
4.13	Direct Products	196
4.14	Normal Subgroups	196
4.15	Exercises	199
4.16	Structure of Cyclic Groups	201
4.17	Permutation Groups	202
4.18	Dihedral Groups	205
4.19	Additive Abelian Groups	207
4.20	Exercises	208

CHAPTER 5

Applications of Group Theory — 211

5.1	The Binary Symmetric Channel	211
5.2	Block Codes	213
5.3	Weight and Distance	216
5.4	Generator and Parity-Check Matrices	218
5.5	Exercises	224
5.6	Group Codes	226
5.7	Coset Decoding	227
5.8	Hamming Codes	230
5.9	Extended Hamming Codes	233
5.10	Exercises	234
5.11	Fast Adders: Winograd's Theory	235
5.12	Fast Adders: Procedures	238
5.13	Exercises	240
5.14	Pólya Enumeration Theory	242
5.15	Exercises	249
5.16	An Extension of Pólya Enumeration Theory	251
5.17	Equivalence Classes of Switching Functions	255
5.18	Exercises	263

CHAPTER 6

Lattices — 265

6.1	Definition of a Lattice	265
6.2	A Basic Theorem	266

6.3	The Operations "Cup" and "Cap"	267
6.4	Another Description of a Lattice	268
6.5	The Modular and Distributive Laws	271
6.6	Complements in Lattices	276
6.7	Exercises	277
6.8	Free Distributive Lattices	281
6.9	Compatibility Relations and Covers	282
6.10	Atomic Lattices and Boolean Algebras	285
6.11	Exercises	288
6.12	Closed Partitions in a Finite-State Machine	290
6.13	Series and Parallel Decomposition of Machines	295
6.14	Exercises	298

CHAPTER 7
Linear Algebra and Field Theory 300

7.1	Matrices	300
7.2	Elementary Row Operations	301
7.3	The Inverse of a Matrix	303
7.4	Exercises	305
7.5	Vector Spaces	307
7.6	Linear Independence and Dimension	310
7.7	Exercises	314
7.8	Linear Transformations	316
7.9	Matrices of Linear Transformations	320
7.10	Rank	325
7.11	Exercises	327
7.12	Determinants	329
7.13	Similarity	334
7.14	Exercises	335
7.15	Ideals and Homomorphisms in Rings	337
7.16	Polynomial Rings	339
7.17	Rational Canonical Form	341
7.18	Exercises	357
7.19	Field Extensions	360
7.20	Finite Fields	366
7.21	Computation in Finite Fields	369
7.22	Exercises	373
7.23	Automorphisms of Finite Fields	375
7.24	Number of Irreducibles	377
7.25	Exercises	379

CHAPTER 8
Linear Machines 381

8.1	Definition	381

8.2	Shift Registers	382
8.3	Characterizing Matrices	387
8.4	Exercises	390
8.5	The Distinguishing Matrix	392
8.6	Minimization of Linear Machines	394
8.7	Exercises	397
8.8	Rational Transfer Functions	399
8.9	Impulse Response	408
8.10	Exercises	411
8.11	Autonomous Linear Machines	413
8.12	Cycle Structure of Feedback Shift Registers	417
8.13	Recursive Equations	422
8.14	Exercises	425
8.15	Null Sequences	426
8.16	Circulating Shift Registers	429
8.17	Section 5.17 Revisited	430
8.18	Exercises	433

CHAPTER 9
Algebraic Coding Theory 435

9.1	Codes over $GF(q)$	435
9.2	Exercises	439
9.3	Cyclic Codes	440
9.4	BCH Codes	442
9.5	Reed–Solomon Codes; Burst Error Correction	444
9.6	Encoding Cyclic Codes	445
9.7	Exercises	447
9.8	BCH Decoding as an FSR Problem	449
9.9	Shortest FSR with Given Output	453
9.10	Algorithmic BCH Decoding	458
9.11	Exercises	464
9.12	Binary BCH Decoding	465
9.13	The Cyclic Redundancy Check	469
9.14	Fire Codes	475
9.15	Some Coding Tricks	478
9.16	How Good Can a Code Be?	481
9.17	Exercises	483

Bibliography 486
Index 491

CHAPTER 1
Sets and Functions

1.1 Sets

The concept of a **set** as an arbitrary collection of objects of interest is probably the most basic concept of modern mathematics. There are no constraints on what objects one may include in a set except that we do not consider it possible for a set to be one of its own objects. A set could consist of precisely these three objects: the real number π, the Taj Mahal, and the ball Hank Aaron hit when he broke the home-run record. Useful sets are, by contrast, ordinarily defined by obviously useful properties. Some examples are the set of positive integers, the set of teams in the National Football League, the set of FORTRAN instructions for a given computer, a complete collection of Zeppelin airmail stamps, and so on.

In this book we shall make use of the following infinite sets (among others):

$\mathbb{R}$: the set of real numbers.
$[a, b]$: the set of real numbers x such that $a \leq x \leq b$.
$\mathbb{Q}$: the set of rational numbers.
$\mathbb{Z}$: the set of integers (positive, negative, or zero).
$\mathbb{N}$: the set of nonnegative integers.
$\mathbb{P}$: the set of positive integers.
$k\mathbb{Z}$: the set of all integral multiples of the positive integer k.
$k\mathbb{N}$: the set of nonnegative integer multiples of the positive integer k.

We shall also make use of the following finite sets (among others):

$\mathbb{Z}_n$: the set of integers from 0 to $n-1$ inclusive.
$\mathbb{P}_n$: the set of positive integers from 1 to n inclusive.
$\mathbb{B}$: the set consisting of the integers 0 and 1.

In computer science applications, a finite set is often called an **alphabet**, and its elements are called **letters**.

The objects in a set are often called its **elements** or its **members**. If x is an element of the set S, we write $x \in S$, which is read "x belongs to S" or "x belonging to S" as the context requires, as in the phrases "if $x \in S$" and "for all $x \in S$," respectively. If x is not an element of the set S, we write $x \notin S$. Thus $2 \in \mathbb{Z}$ but $\frac{1}{2} \notin \mathbb{Z}$, Chicago Bears $\in$ NFL, J. William Fulbright $\notin$ United States Senate.

Two sets are **equal** if they contain precisely the same elements. Thus the set of all real numbers whose remainders on division by 2 are 0 is equal to the set of all integers of the form $2n$ where $n \in \mathbb{Z}$.

1.2 The Indexing of Sets

It is often useful to name the elements of a set S in some appropriate manner with the aid of another set I, called an **indexing set**. The elements of the indexing set are commonly used as subscripts; each element of S is assigned a subscript name and each element of I is used exactly once as such a name. In the case of a finite set S consisting of n elements, the elements are numbered in some convenient (perhaps random) order, the ith element being denoted by s_i. Then $S = \{s_1, s_2, \ldots, s_n\}$ and $\mathbb{P}_n$ is the indexing set.

Infinite sets may be indexed in a similar way. For example, a **complete polygon** with n distinct vertices consists of n vertices (points in the plane), each pair of vertices joined by a straight line segment (edge). The number E_n of edges of such a polygon is given by the formula $E_n = n(n-1)/2$. The infinite set of numbers $\{E_1, E_2, \ldots, E_n, \ldots\}$ has the indexing set $\mathbb{P}$. Any infinite set whose elements can be indexed by $\mathbb{P}$ is called **enumerable** or **denumerable** or **countably infinite**.

One might think offhand that every finite set can be effectively indexed, but this is not the case. For example, the set of all cloudy days at the site of Tombstone, Arizona, from January 1, A.D. 1700, through December 31, 1974, is a well-defined, reasonably small finite set although, for lack of the proper records, it is not possible to index it. On the other hand, many nondenumerable infinite sets can be indexed. The most familiar example is the indexing of the points P_x of a line by their coordinates x. Here the indexing set is $\mathbb{R}$.

Ordinarily, there is a natural choice for the indexing set and often for the manner of indexing as well. However, many of the finite sets we use in this book have no natural ordering and may be indexed in random order.

1.3 Sets Derived from Other Sets

If U is a set and P is a property (P may in fact be a combination of several properties) which elements x of U may or may not possess, we can define a new set with the "set-builder" notation

$$\{x \in U | P(x)\}.$$

This denotes "the set of all elements x that belong to U and have property P." For example,

$$\{x \in \mathbb{Z} | x > 0\} = \mathbb{P}$$

and

$$\left\{ r \in \mathbb{R} \left| \frac{r}{2} \in \mathbb{Z} \right. \right\} = \{z \in \mathbb{Z} | z = 2n, \quad n \in \mathbb{Z}\}.$$

Sec. 1.3 Sets Derived from Other Sets

A **subset** of a set U is a set S all of whose elements belong to U. The set-builder notation is the basic tool for describing subsets. In a discussion dealing exclusively with subsets of a fixed set U, U is often called the **universal set** or the **universe of discourse**. If S is a subset of U, we say that S is **included** (or **contained**) in U and write $S \subseteq U$. We also say that U **includes** or **contains** S and write $U \supseteq S$. If $U \supseteq S$, then U is called a **superset** of S. The set of *all* subsets of U is called the **power set** of U and is denoted by $\mathscr{P}(U)$.

If the property P is so restrictive that no elements of U have that property, we say that $\{x \in U | P(x)\}$ defines the **empty set** or **null set** $\varnothing$. It is proper to refer to *the* empty set here because all empty sets, regardless of the properties that define them, contain exactly the same elements, namely none, and hence are equal. If S is a subset of U and $S \neq U$, we write $S \subset U$ and say that S is a **proper subset** of U.

It is often useful to denote the elements of a set A that are not in a set B by $A - B$, the **set-theoretic difference** of A and B, in that order.

From two sets S and T, not necessarily distinct, new sets may be derived in a variety of useful ways. Many such sets are based on the concept of an **ordered pair** of elements (s, t), where $s \in S$ and is listed first and $t \in T$ and is listed second. Two ordered pairs (s_1, t_1) and (s_2, t_2) are defined to be **equal** if and only if $s_1 = s_2$ and $t_1 = t_2$. For example, if $S = T = \mathbb{Z}$, then $(2, -3) \neq (-3, 2)$.

We may now define the **Cartesian product** of S and T, denoted by $S \times T$, to be the set of all ordered pairs (s, t) such that $s \in S$ and $t \in T$. That is,

$$S \times T = \{(s, t) | s \in S, t \in T\}.$$

For example, if $S = T = \mathbb{R}$, then the Cartesian product is

$$\mathbb{R} \times \mathbb{R} = \{(x, y) | x \in \mathbb{R}, y \in \mathbb{R}\},$$

which is denoted by $\mathbb{R}^2$ and which is interpreted as the set of points of the familiar Cartesian coordinate plane. This plane and its generalization, the Cartesian product of arbitrary sets S and T, are named after the French philosopher and mathematician René Descartes (1596–1650), who invented analytic geometry.

An indexed set $\{a_1, a_2, \ldots, a_n\}$ is called an **ordered n-tuple** and is written $(a_1, a_2, \ldots, a_n)$ if it matters which element is listed first, which is listed second, and so on; that is, if $(a_1, a_2, \ldots, a_n) \neq (b_1, b_2, \ldots, b_n)$ unless $a_i = b_i$, $i = 1, 2, \ldots, n$. We can now define the **Cartesian product** of n sets $S_1, S_2, \ldots, S_n$ as a set of ordered n-tuples thus:

$$S_1 \times S_2 \times \cdots \times S_n = \{(s_1, s_2, \ldots, s_n) | s_i \in S_i, \quad i = 1, 2, \ldots, n\}.$$

If $S_1 = S_2 = \cdots = S_n = S$, this product is denoted by S^n. For example,

$$\mathbb{R}^3 = \mathbb{R} \times \mathbb{R} \times \mathbb{R} = \{(x, y, z) | x \in \mathbb{R}, \quad y \in \mathbb{R}, \quad z \in \mathbb{R}\}$$

is a set-theoretic description of coordinatized 3-space. A subset of this space (recall that $\mathbb{B} = \{0, 1\}$) is the set of vertices of a unit cube defined by

$$\mathbb{B}^3 = \mathbb{B} \times \mathbb{B} \times \mathbb{B} = \{(b_1, b_2, b_3) | b_i \in \mathbb{B}, \quad i = 1, 2, 3\}.$$

Other examples of Cartesian products will appear naturally in what follows.

1.4 The Order of a Set

The **order** or **cardinality** of a finite set S is the number of elements in S and is denoted by $|S|$. For example, the order of the set $\{1, 3, 5, \ldots, 2n - 1\}$ is n. If a set can be indexed by the set $\mathbb{P}_n$, then its order is n.

It is important, when determining the order of a set, to observe precisely what the elements of the set are. Thus $|\{(1, 1), (2, 3), (3, 2)\}| = 3$ because this is a set of three ordered pairs. We shall have frequent occasion to employ sets whose elements are themselves sets or ordered sets.

Here are some more examples:

$$|\text{set of columns on an IBM card}| = 80,$$

$$|\text{Los Angeles Dodgers traveling squad}| = 25,$$

$$|\{x \in \mathbb{Z} | 0 < x < 100, x = 3n - 2, n \in \mathbb{P}\}| = 33.$$

A set of order 1 is called a **singleton** and a set of order 0 is the empty set.

If $|S| = m$ and $|T| = n$, then in forming an element (s, t) of $S \times T$, we have m choices for s and n choices for t, so $|S \times T| = mn$. This proves that for finite sets S and T,

(1.4.1) $$|S \times T| = |S| \cdot |T|.$$

In the application of algebra to discrete systems, one has frequent occasion to determine the order of a finite set, so counting problems will often be treated in this book, in both text and exercises. In some cases, even though the order of S cannot be precisely determined, one can determine a number μ_S such that $|S| \leq \mu_S$. Such a number is called an **upper bound** for the order of S.

1.5 Functions

If S and T are sets, a **function** f from S to T, denoted by the symbolism $f: S \to T$, may be defined as a rule assigning to each $s \in S$ a unique element $f(s) \in T$. Although for each $s \in S$ there must be exactly one $f(s)$, it is not required that each $t \in T$ be $f(s)$ for some s. Nor is it required that $f(s_1) \neq f(s_2)$ whenever $s_1 \neq s_2$.

Most mathematics through calculus deals with functions from $\mathbb{R}$ to $\mathbb{R}$. Some examples are: if $f(x) = x^2 - 3x + 2$, then $f(3) = 2$ and $f(1.5) = -0.25$; if $g(x) = \sin x$, then $-1 \leq g(x) \leq 1$ for all $x \in \mathbb{R}$; if $h(x) = e^x$, then $0 < h(x)$ for all

Sec. 1.5 *Functions*

$x \in \mathbb{R}$. In this book we shall encounter functions $f: S \to T$ for many different sets S and T.

Given a function $f: S \to T$, S is called the **domain** of f and T is called the **codomain** of f. The element $f(s)$ of T is called the **image** of s by f and s is a **counterimage** or **pre-image** (there may be others) of $f(s)$. A function $f: S \to T$ is also called a **mapping** of S into T and f is said to **map** s onto its image $f(s)$. The **range** or **image** of a function $f: S \to T$ is the subset $f(S)$ of T defined by

$$f(S) = \{t \in T \mid t = f(s) \text{ for some } s \in S\}.$$

If S_0 is a subset of S, we define $f(S_0) \subseteq f(S)$ by

$$f(S_0) = \{f(s) \in T \mid s \in S_0\}.$$

Here are some more examples of functions:

1. $f_1: \mathbb{R} \to \mathbb{R}$, defined by $f_1(x) = x^2 + 4x + 1$. Is every $y \in \mathbb{R}$ the image by f_1 of at least one $x \in \mathbb{R}$? If not, precisely which $y \in \mathbb{R}$ are images and which are not?

2. $f_2: \mathbb{Z} \to \mathbb{Q}$, defined by $f_2(n) = 1/(n + \tfrac{1}{2})$. Precisely which integers have integers as images by f_2?

3. f_3: {nonbigamous married American men} $\to$ {women} defined by $f_3(\text{man}) = \text{man's wife}$. Explain why the conditions "nonbigamous" and "American" are included here.

The third example illustrates the fact that there is no restriction on the nature of the sets S and T appearing in the definition of a function. In the following pages we often make use of functions in which S or T or both are not sets of numbers.

The functions

4. $f_4: \mathbb{Z} \to \{0, 1\}$ defined by $f_4(2n) = 0$, $f_4(2n-1) = 1$, $n \in \mathbb{Z}$,
5. $f_5: \mathbb{Z} \to \{0, 1, 2\}$ defined by $f_5(2n) = 0$, $f_5(2n-1) = 1$, $n \in \mathbb{Z}$,

are distinct functions even though their rules of assignment are the same. This illustrates the fact that a function involves three things: the *domain* S, the *codomain* T, and the *rule* that assigns to each $s \in S$ a unique $t \in T$.

Distinct rules at times effect the same mapping of S into T. We therefore make the following definition: The functions $f: S \to T$ and $g: S \to T$ are **equal**, written $f = g$, if and only if $f(s) = g(s)$ for all s in S. For example, f_1 defined above is equal to f_6, where

6. $f_6: \mathbb{R} \to \mathbb{R}$ is defined by $f_6(x) = (x+2)^2 - 3$.

The ordered pairs $(s, f(s))$ determined by a function $f: S \to T$ constitute a special kind of subset of $S \times T$: Each $s \in S$ appears in exactly one pair (s, t) of the subset and a given pair (s, t) of $S \times T$ belongs to the subset if and only if $t = f(s)$. As a consequence, a function $f: S \to T$ may alternatively be *defined* as any subset of $S \times T$ such that each $s \in S$ appears as the first element of precisely one pair of the subset. Then, given such a subset, the rule of assignment is simply this: "$f(s)$ is the element t of T associated with s in the pair (s, t) of the subset."

To illustrate, if $S = \{0, 1, 2, 3, 4, 5, 6\}$ and $T = \{0, 1, 2\}$, the subset

$$\{(0, 0), (1, 1), (2, 2), (3, 0), (4, 1), (5, 2), (6, 0)\}$$

defines a function $f: S \to T$. This function may also be described by the rule: "If $s = 3q + r$, $0 \leq r \leq 2$, then $f(s) = r$."

As we have pointed out before, one often has to answer the question "How many?" Thus if $|S| = m$, $|T| = n$, how many functions f are there from S to T? For each of the m elements $s \in S$, we may choose any one of the n elements of T as $f(s)$, independently of the choices made for the other elements of S. Hence there are altogether $n \cdot n \cdot \cdots \cdot n = n^m$ functions from S to T. Because of this result, the set of all functions from S to T is often denoted by T^S, and we have proved the following theorem.

Theorem 1.5.1. *For finite sets S and T,*

$$|T^S| = |T|^{|S|}.$$

The basic counting principle used here is that if one of two independent tasks can be performed in p ways and the other in q ways, then the pair of tasks can be performed in pq ways.

1.6 Exercises

1. If $S = \{n \in \mathbb{Z} \mid 1 \leq n \leq 3\}$ and $T = \{n \in \mathbb{Z} \mid 2 \leq n \leq 5\}$, diagram $S \times T$ as a subset of the Cartesian plane.

2. For S and T as in Exercise 1, give a subset of $S \times T$ that does not represent a function.

3. Determine two sets such that the rectangle of Figure 1.6.1(a), including its boundaries, represents their Cartesian product. Then do the same for the rectangle of Figure 1.6.1(b), including only the two boldface boundaries.

4. If $S = \{3, 7, 21\}$, $T = \{11, 111, 10101\}$, under what conditions, if any, is $S = T$? Is indexing involved?

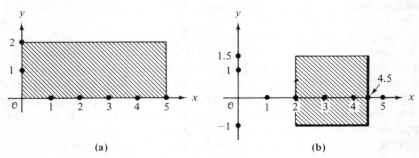

FIGURE 1.6.1. *Cartesian Products*

Sec. 1.6 Exercises

5. How would you index the seats in a large rectangular concert hall divided by a central aisle so that the index would readily identify the location of the seat? What is the indexing set in this case?

6. If $S = \{s_1, s_2, s_3\}$ and $T = \{t_1, t_2, t_3\}$, tabulate all the functions from S to T. Invent a way to index these functions such that the index identifies the function completely (that is, the index reveals the image of every $s \in S$).

7. Invent a function whose domain is a set of pairs of people and whose codomain is most appropriately $\mathbb{N}$.

8. Given $f: \mathbb{R} - \{0\} \to \mathbb{R}$ where $f(x) = \frac{1}{2}(x + 2/x)$, what is the range of f? For what values of x is $f(x) = x$? (For any $f: A \to B$, the elements x such that $x \in A$, $x \in B$, and $f(x) = x$ are called the **fixed points** of f.)

*9. The function $f: S \to S$, where $f(s) = s$ for all $s \in S$, is called the **identity function** on S. Describe the corresponding subset of $S \times S$. Show that for all functions $g: S \to S$, and for all $s \in S$, $f(g(s)) = g(f(s)) = g(s)$.

10. Let $A = \varnothing$, $B = \{\varnothing\}$, $C = \mathcal{P}(\varnothing)$, and $D = \mathcal{P}(\{\varnothing\})$. Are any of these sets equal? What are $|A|$, $|B|$, and $|C|$?

11. No two of $R \times S \times T$, $(R \times S) \times T$, and $R \times (S \times T)$ are equal. Explain.

*12. Let U be a set with n elements. Determine, for each $k \in \mathbb{N}$ such that $0 \leq k \leq n$, the number of subsets S of U such that $|S| = k$. Then find the total number of subsets of U, including the empty set $\varnothing$ and the set U itself.

13. Let A be a finite set (alphabet) of m elements. Find the number of ordered n-tuples (code words) in the Cartesian product $A^n = A \times A \times \cdots \times A$. Use this result to show that the 26 letters of the English alphabet can be encoded using code words of length ≤ 4 formed from the alphabet $\{\cdot, -\}$. Then look up Morse Code and the International Code in an encyclopedia to see the encodings that are actually used.

14. Consider a standard 8×8 chess board, regarded as a set of vertices in nine horizontal rows and nine vertical columns, joined by lines forming the edges of the 64 small squares. Let P denote the set of paths along the *lines* of the chess board from one vertex (the "lower left" vertex) to the diagonally opposite ("upper right") vertex. A path proceeds one step at a time, either one square to the right (H) or one square upward (V), and consists of eight horizontal moves H and eight vertical moves V. Determine $|P|$.

15. An ordered set of three positive integers (a, b, c) is a **Pythagorean triad** if and only if $a^2 + b^2 = c^2$. The most familiar such triad is $(3, 4, 5)$.

(a) A **primitive Pythagorean triad** is an ordered triple $(2mn, m^2 - n^2, m^2 + n^2)$, $m \in \mathbb{P}$, $n \in \mathbb{P}$, $0 < n < m$. Show that any such triple is a Pythagorean triad.

(b) It is shown in number theory that any Pythagorean triad has form (ra_0, rb_0, rc_0) or (rb_0, ra_0, rc_0), where $r \in \mathbb{P}$ and (a_0, b_0, c_0) is a primitive Pythagorean triad. Write a program, in whatever computer language is available to you, that will produce all Pythagorean triads (a, b, c) with $c \leq 150$.

1.7 More Notation

In the following sections, we shall often have occasion to deal with **equivalent statements**, that is, statements which are both true or both false. We use the abbreviation "iff" for "if and only if." Then the theorem that statements P and Q are equivalent may be written "P iff Q." Proof of such a theorem always involves two parts: proof that P implies Q (written "$P \Rightarrow Q$") and proof that $Q \Rightarrow P$. This is reflected in the notation "$P \Leftrightarrow Q$," which means the same as "P iff Q." Informally, proof that $P \Rightarrow Q$ is called the "only if" part of the proof of "$P \Leftrightarrow Q$," and proof that $Q \Rightarrow P$ is called the "if" part.

A statement "P iff Q" may also be read "Q is a *necessary and sufficient condition* (abbreviated n.a.s.c. or n.s.c.) for P." Proof that $P \Rightarrow Q$ is proof of the necessity of the condition; in other words, for P to be true it is necessary that Q be true. Proof that $Q \Rightarrow P$ is proof of the sufficiency of the condition; that is, in order to conclude that P is true, it is sufficient to know that Q is true.

It is not uncommon that one of the two parts of an if-and-only-if theorem is obvious. In such a case, a simple remark may dispense with this half of the argument. The reader is warned against the common error of assuming that half of the proof may *always* be ignored. To the contrary, in many cases both parts of the proof require substantial arguments.

Many theorems apply to all the elements of a certain set. We use "$\forall$" to mean "for all." Thus "$\forall x \in X$" is read "for all x belonging to X" or "for each x belonging to X."

Some theorems, called **existence theorems**, assert the existence of certain objects. We use "$\exists$" to mean "there exists" and "$\exists!$" to mean "there exists a unique." The symbol "$\ni$" is used to mean "such that." Thus

$$\text{"}\exists! \, x \in \mathbb{R} \ni x^3 - 1 = 0\text{"}$$

means "there exists a unique real number x such that $x^3 - 1 = 0$."

1.8 One-to-One and Onto

A function $f: S \to T$ is said to be **one-to-one** (abbreviated one-one) iff for each $t \in T$ there exists at most one $s \in S$ such that $f(s) = t$; in other words, iff a given element of T has at most one counterimage in S. Equivalent definitions are that $f: S \to T$ is one-one iff $f(s_1) = f(s_2)$ implies that $s_1 = s_2$, or iff $s_1 \neq s_2$ implies that $f(s_1) \neq f(s_2)$.

For example, let $S = T = \mathbb{R}$ and consider $f: S \to T$ defined by $f(s) = t = as + b$, $a \neq 0$, a and b fixed real numbers. For each t, the unique s whose image is t is given by $s = (t-b)/a$. Also, if $as_1 + b = as_2 + b$, then $s_1 = s_2$ since $a \neq 0$. This example also illustrates the next definition.

A function $f: S \to T$ is **onto** iff for each $t \in T$ there exists at least one $s \in S$ such that $f(s) = t$; that is, f is onto iff every element of T has at least one counterimage in S. An equivalent definition is that $f: S \to T$ is onto iff the range of f is T, that is, if $f(S) = T$. We also say that f **maps** S **onto** T when the range of f is T. When the range of f is a proper subset of T, we say that f **maps** S **strictly into** T. If we do not require f to be onto or if we do not know whether or not it is onto, we say that f **maps** S **into** T. Thus "into" means "onto or strictly into."

A function $f: S \to T$ that is one-one is called an **injection**. A function $f: S \to T$ that is onto is called a **surjection**. A function $f: S \to T$ that is both one-one and onto is called a **bijection** or a **one-one correspondence**. In this case, each s has precisely one image t; because f is onto and one-one, each t has one and only one counterimage s; hence the name "correspondence."

In each of the following examples, the reader should verify that the function has the stated characteristics:

1. $f_1 : \mathbb{R}^2 \to \mathbb{R}^2$, defined by $f_1((x, y)) = (x+y, x-y)$, is both one-one and onto, so it is a bijection.
2. $f_2: \mathbb{Z} \to \mathbb{Z}$, defined by $f_2(z) = z^2$, is neither one-one nor onto.
3. $f_3: \mathbb{P} \to \mathbb{P}$, defined by $f_3(n) = n^2$, is one-one but not onto (recall that $\mathbb{P} = \{\text{positive integers}\}$), so it is an injection.
4. $f_4: \mathbb{N} \to \{0, 1\}$, defined by $f_4(n) = 0$ when n is even, $f_4(n) = 1$ when n is odd, is onto but not one-one, o it is a surjection.

The several possibilities, for finite sets S and T, are illustrated in Figure 1.8.1.

If one considers functions $f: S \to T$, where S and T are finite, a variety of counting questions may be asked. For example, if $|S| = m$, $|T| = n$, how many one-one functions are there from S to T? For a one-one function to exist, we must have $n \geq m$ since $s_1 \neq s_2$ implies that $f(s_1) \neq f(s_2)$. If $n = m$, each $t \in T$ must be the image of precisely one $s \in S$. If $S = \{s_1, s_2, \ldots, s_m\}$, then there are $m = |T|$ choices for $f(s_1)$, $m - 1 = |T - \{f(s_1)\}|$ choices for $f(s_2)$, $m - 2 = |T - \{f(s_1), f(s_2)\}|$ choices for $f(s_3)$, $\ldots$, so the number of one-one functions is, in this case, just $m! = m(m-1)(m-2) \cdots 1$. If $n > m$, we may select any m distinct elements from T, in any order, as images for the elements of S. Thus there are $n(n-1) \cdots (n-m+1) = n!/(n-m)!$ one-one functions in this case.

1.9 Composition and Inversion of Functions

Let S, T, and U be sets and let $f: S \to T$ and $g: T \to U$ be functions. (Note that the codomain of f is the domain of g.) We define the **composite function** $g \circ f: S \to U$ (the **composite of f and g**, in that order) thus:

$$\forall s \in S, \quad (g \circ f)(s) = g(f(s)).$$

Since this rule assigns to each $s \in S$ a unique $u \in U$, $g \circ f$ is indeed a function from S to U.

For example, if $S = T = U = \mathbb{R}$, $f(x) = x^2$, and $g(x) = x + 1$, then

$$(g \circ f)(x) = g(f(x)) = g(x^2) = x^2 + 1.$$

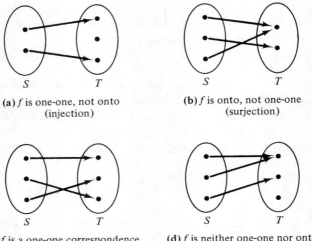

(a) f is one-one, not onto (injection)

(b) f is onto, not one-one (surjection)

(c) f is a one-one correspondence (bijection)

(d) f is neither one-one nor onto

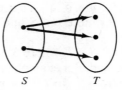

(e) Not a function

FIGURE 1.8.1. *Functions and a Nonfunction*

In general, $f \circ g$ need not be defined just because $g \circ f$ is. In this example, both are defined because S and U are the same set. We have, in fact,

$$(f \circ g)(x) = f(g(x)) = f(x+1) = (x+1)^2 = x^2 + 2x + 1,$$

which illustrates the fact that we need not have $f \circ g = g \circ f$, even though both are defined. Although the composition of functions is therefore not a commutative operation, we have the following theorem.

Theorem 1.9.1. *The composition of functions is associative.*

Proof. Let S, T, U, and V be sets and let $f: S \to T$, $g: T \to U$, and $h: U \to V$ be functions. Then the theorem says that

(1.9.1) $$h \circ (g \circ f) = (h \circ g) \circ f.$$

Figure 1.9.1 shows what is happening: If $s \in S$, $f(s) = t$, $g(t) = u$, and $h(u) = v$, then $g \circ f$ maps s onto u via t and $h \circ g$ maps t onto v via u.

The proof follows from the definition of the operation of composition.

Sec. 1.9 Composition and Inversion of Functions

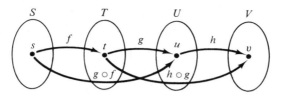

FIGURE 1.9.1. *Associative Law of Composition of Functions*

Since

$$(h \circ (g \circ f))(s) = h((g \circ f)(s)) = h(g(f(s)))$$

and

$$((h \circ g) \circ f)(s) = (h \circ g)(f(s)) = h(g(f(s))),$$

the two functions produce the same image for all $s \in S$ and hence are equal. ∎ (The symbol "∎" means "end of proof" and will be used throughout this book.)

This is a very powerful theorem because it does not depend on the specific nature of the sets and functions involved.

If $f: S \to S$ is a function, we define $f^2 = f \circ f$, $f^3 = f^2 \circ f, \ldots, f^{k+1} = f^k \circ f, \ldots$, any $k \in \mathbb{P}$.

For any set S, the **identity function** 1_S is the function $1_S: S \to S$ such that for all $s \in S$, $1_S(s) = s$. For every S, 1_S is a bijection such that for all $f: S \to S$,

$$1_S \circ f = f \circ 1_S = f.$$

(Proof of this fact was Exercise 1.6.9.)

A function $f: S \to T$ is **invertible** iff there exists a function $g: T \to S$ such that $g \circ f = 1_S$ and $f \circ g = 1_T$. We then say g is the **inverse** of f and f is the inverse of g, and we write $g = f^{-1}$, $f = g^{-1}$. We think of f^{-1} as a function that reverses the effect of f. The notation "f^{-1}" in effect assumes that the inverse of f is unique when it exists. This is justified by the following theorem.

Theorem 1.9.2. *A function $f: S \to T$ has at most one inverse.*

Proof. Suppose that, in fact, $g: T \to S$ and $h: T \to S$ are inverses of f, so

$$g \circ f = 1_S, \quad f \circ g = 1_T, \quad h \circ f = 1_S, \quad f \circ h = 1_T.$$

Then, by Theorem 1.9.1,

$$h = 1_S \circ h = (g \circ f) \circ h = g \circ (f \circ h) = g \circ 1_T = g. \blacksquare$$

Next we prove the basic theorem concerning invertibility.

Theorem 1.9.3. *A function $f: S \to T$ is invertible iff f is a bijection.*

Proof. Suppose that f is invertible with inverse g, so $g \circ f = 1_S$ and $f \circ g = 1_T$. If s_1 and $s_2 \in S$ and $f(s_1) = f(s_2)$, then
$$g(f(s_1)) = g(f(s_2)),$$
$$(g \circ f)(s_1) = (g \circ f)(s_2),$$
$$1_S(s_1) = 1_S(s_2),$$
$$s_1 = s_2,$$
so f is one-one. If $t \in T$, then $t = 1_T(t) = (f \circ g)(t) = f(g(t))$. If we put $s = g(t)$, then $t = f(s)$, and this proves that f is onto. So f is a bijection.

Now suppose that f is a bijection. For any $t \in T$, then, there is a unique $s \in S$ such that $f(s) = t$. We define $g(t)$ to be this s. Then $s = g(t) = g(f(s)) = (g \circ f)(s)$. Since every s is the counterimage of some t, this equation holds for all $s \in S$, so $g \circ f = 1_S$. Also, for all $t \in T$, $(f \circ g)(t) = f(g(t)) = f(s) = t$, so $f \circ g = 1_T$. Thus f is invertible. ∎

If $f: S \to T$ is a bijection, we frequently write $f: S \leftrightarrow T$ even though the leftward arrow actually denotes f^{-1}.

For us the special case in which S and T are finite sets of the same order is especially important and is the subject of the next theorem.

Theorem 1.9.4 (Pigeonhole Principle). *Let S and T be finite sets with $|S| = |T| = n$, and let $f: S \to T$ be a function. Then the following statements are equivalent (that is, all three are true or all three are false):*

(a) *f is one-one.*
(b) *f is onto.*
(c) *f is invertible.*

Proof. We must prove that each of these implies the other two. We already know by Theorem 1.9.3 that if (c) is true, both (a) and (b) are true. Also, by the same theorem, if both (a) and (b) are true, (c) is true. Hence all that we have to prove is that (a) implies (b) and that (b) implies (a), that is, that (a) and (b) are equivalent.

While proving (a) and (b) equivalent, we assume (in accord with the name of the theorem) that set S consists of n pigeons and set T consists of n pigeonholes, with f a function sending pigeons into pigeonholes. If (a) holds, no two pigeons go into the same pigeonhole, so the n pigeons must go into n different pigeonholes and hence each pigeonhole gets a pigeon. Thus (b) holds. On the other hand, if (b) holds so that each of the n pigeonholes gets a pigeon, then, since there were only n pigeons to start with, we cannot have two pigeons in the same pigeonhole. Thus (a) holds. ∎

The importance of Theorem 1.9.4 is that it shows we can prove that a function from a finite set S to a finite set T of the same order is invertible by proving that it is either one-one or onto; we do not need both.

1.10 Exercises

1. Tell whether the following functions are one-one or onto, and when they are not onto, determine the range.
(a) $f_1: \mathbb{R} \to \mathbb{R}$ defined by $f_1(x) = e^x$.
(b) $f_2: \mathbb{Z} \to \mathbb{Z}$ defined by $f_2(n) = n^2 + 1$.
(c) $f_3: \mathbb{R} \to \mathbb{R}$ defined by $f_3(x) = x^2$.
(d) $f_4: \mathbb{R} \to \mathbb{R}$ defined by $f_4(x) = x^3$.
(e) $f_5: \mathbb{R} \to \mathbb{R}$ defined by $f_5(x) = \cos x$.

*__2.__ Let $f: S \to S$ and $g: S \to S$ be arbitrary functions from S to S. Prove:
(a) If g is not onto, then $g \circ f$ is not onto.
(b) If f is not one-one, then $g \circ f$ is not one-one.
(c) If f and g are both onto, so is $g \circ f$.
(d) If f and g are both one-one, so is $g \circ f$.
(e) If f and g are both bijections, so are $f \circ g$ and $g \circ f$.

3. If $S = \{a, b, c\}$ and $T = \{0, 1\}$, which subsets of $S \times T$ are functions from S to T?

4. Let $f: \mathbb{R} \to \mathbb{R}$ be defined by $f(x) = x^n$, where $n \in \mathbb{P}$. For what integers n does f have an inverse? Explain.

5. Find $f \circ g$ and $g \circ f$, where $S = \{a, b, c, d, e\}$ and $f: S \to S$ and $g: S \to S$ are defined by this table:

s	$f(s)$	$g(s)$
a	b	e
b	c	a
c	d	b
d	e	c
e	a	d

6. Compute $g \circ f$, where $f: \mathbb{B}^3 \to \mathbb{B}^2$ and $g: \mathbb{B}^2 \to \mathbb{B}$ are defined by the following tables and the rules $f((x, y, z)) = (a, b)$, $g((a, b)) = \alpha$.

$f:$	x	y	z	a	b
	0	0	0	0	0
	0	0	1	0	1
	0	1	0	0	1
	0	1	1	1	0
	1	0	0	0	1
	1	0	1	1	0
	1	1	0	1	0
	1	1	1	1	1

$g:$	a	b	α
	0	0	0
	0	1	1
	1	0	1
	1	1	0

7. Show that the function $g: \mathbb{B}^4 \to \mathbb{B}$ defined by the following table is independent of x [that is, knowing w, y, and z is sufficient to determine $g((w, x, y, z))$].

w	x	y	z	$g((w, x, y, z))$
0	0	0	0	1
0	0	0	1	1
0	0	1	0	0
0	0	1	1	0
0	1	0	0	1
0	1	0	1	1
0	1	1	0	0
0	1	1	1	0
1	0	0	0	0
1	0	0	1	1
1	0	1	0	1
1	0	1	1	0
1	1	0	0	0
1	1	0	1	1
1	1	1	0	1
1	1	1	1	0

8. Express the area of an equilateral triangle as a function of its edge, then identify the domain and the range of the function.

*9. Show that if S and T are finite and $f: S \to T$ is one-one, then $|S| \leq |T|$.

*10. Let the function $f: X \to Y$ be a surjection and let $|Y| = n$. For all $y \in Y$, let $f^{(-1)}(y)$ denote the set of all counterimages of Y. Prove that if $|f^{(-1)}(y)| = m$ for all $y \in Y$, then $|X| = mn$. (This illustrates the **sheep-pen principle**; if you have m sheep in each of n pens, you have mn sheep altogether.)

11. Let $f: X \to Y$ and $g: Y \to X$ be functions such that whenever $f(x) = y$, then $g(y) = x$. Also, let g be injective. Prove that whenever $g(y) = x$, then $f(x) = y$. Is $g = f^{-1}$, necessarily?

*12. Consider the function $f: \mathbb{P} \to \mathbb{P}$, where $f(n) = kn$, $k > 1$ a fixed integer. The range of f is the proper subset $k\mathbb{P}$ of $\mathbb{P}$ and f is one-one. What does this imply about $\mathbb{P}$ in view of the pigeonhole principle? Using the terms "one-one" and "onto," can you now define when a set is (a) finite; (b) infinite?

13. (For those who know matrix theory over $\mathbb{R}$.) (a) Let S be the set of all $n \times n$ real matrices and let $f: S \to \mathbb{R}$ be defined by $f(\mathbf{A}) = \det \mathbf{A}$. What is the range of f?

(b) Let T be the set of all real $n \times n$ matrices $\mathbf{A} = [a_{ij}]$ such that $|a_{ij}| \leq 1$ for $1 \leq i, j \leq n$. Let $g: T \to \mathbb{R}$ be defined by $g(\mathbf{A}) = \det \mathbf{A}$. What can you say about the range of g?

Sec. 1.11 Finite-State Machines

14. Recall that $\mathbb{R}^n$ denotes the set of all ordered n-tuples of real numbers. Determine the range of the function $f: \mathbb{R}^3 \to \mathbb{R}^2$ where $f((x_1, x_2, x_3)) = (x_1 + x_2 + x_3, x_1 - x_2 - x_3)$. Is the function one-one? If not, which points of $\mathbb{R}^3$ map onto a given point of $\mathbb{R}^2$?

*15. Let $S = \mathbb{R} - \{0, 1\}$. Define $f_1: S \to S$ and $f_2: S \to S$ by $f_1(r) = 1/(1-r)$ and $f_2(r) = 1/r$. Determine the formulas that represent $f_3 = f_2 \circ f_1$, $f_4 = f_1 \circ f_2$, $f_5 = f_2 \circ f_4$, and $f_6 = f_2 \circ f_2$. Show that no other functions arise by composition from these six.

(HINT: Make a 6×6 table of all composites $f_i \circ f_j$.)

16. Given that f and g are functions from S to S such that $f^2 = 1_S$, $g^2 = 1_S$, and $f \circ g = g \circ f$, show that the set $\{1_S, f, g, f \circ g\}$ is closed under composition, that is, that no new functions can be generated from these four by composition. Find such a set S and functions f and g, with $f \neq g$, $f \neq 1_S$, and $g \neq 1_S$.

17. Let $f: S \to T$ and $g: T \to S$ (where S and T are subsets of $\mathbb{R}$) be defined by $f(x) = x^2 - 1$, $g(x) = \sqrt{x + 1}$. What are the largest sets S and T such that f and g are defined and inverse to each other?

18. The function $\sigma: \mathbb{P} \to \mathbb{P}$, where $\sigma(n) = n + 1$, is known as the **successor function**. Is it one-one? Onto? An injection? A surjection? A bijection? What is the function $\sigma^k: \mathbb{P} \to \mathbb{P}$, where k is any positive integer?

19. Earth certainly has a finite number of grains of sand. How could you determine an upper bound for this number?

20. If there are precisely 60 injections from a set S of three elements to a set T of n elements, what is the value of n?

*21. If $|S| = m$ and $|T| = n$, how many onto functions are there from S to T?

*22. If $f: S \to T$ is a function and V a subset of T, denote by $f^*(V)$ the pre-image $\{s \in S | f(s) \in V\}$ of V. Prove that $f(f^*(V)) \subseteq V$. When does equality hold?

*23. Assume that $f: S \to T$ and $g: T \to U$ are invertible functions. Prove that $g \circ f: S \to U$ is also invertible and $(g \circ f)^{-1} = f^{-1} \circ g^{-1}$.

1.11 Finite-State Machines

Digital computers and other systems, such as telephone central offices, automatic milling machines, and elevators, for example, contain two basic types of control circuits. Each type has one or more **input leads** (wires) and one or more **output leads** along which electrical signals are respectively provided to and received from the circuit. A **combinational circuit** is one whose output at any point in time depends only on the input present at that time. For mathematical purposes, one commonly assumes that the response to a change in input is instantaneous. (One can alternatively take account of the small but nontrivial delays in the response of the components of the circuit and treat that case mathematically also.) A **sequential circuit** has a set of internal components the combined state of which, called the **internal**

state, determines the response (output) of the circuit to the signals present on the input leads, the so-called **present input**. The internal state and the present input also determine the next internal state. Since the response of a sequential circuit to a given combination of input signals depends not only on that combination but also on the internal state, the circuit is capable of taking account of its past history (stored in the internal state) in determining its output response.

We now introduce an abstract model, the finite-state machine, which is an important tool in the design of physical sequential circuits. This model will serve as a continuing source of substantial examples of applications of sets, functions, and a variety of algebraic concepts that will appear in following chapters.

A **finite-state machine**, **sequential machine**, or **finite automaton** is a 5-tuple (ordered set of five objects) $[S, X, Z, \tau, \omega]$, where S, X, and Z are finite sets and $\tau: S \times X \to S$ and $\omega: S \times X \to Z$ are functions. S is the set of **internal states**, X is the **input alphabet** or set of **input symbols**, and Z is the **output alphabet** or set of **output symbols**. If $S = \{\sigma_1, \sigma_2, \ldots, \sigma_n\}$, $X = \{\xi_1, \xi_2, \ldots, \xi_m\}$, and $Z = \{\zeta_1, \zeta_2, \ldots, \zeta_p\}$, the machine is called an **(n, m, p)-machine**. Variable elements of S, X, and Z will be denoted by s, x, and z, respectively. States will often be denoted by other symbols: $\{A, B, C, D\}$, $\{0, 1, 2, 3, 4\}$, $\{00, 01, 10\}$, $\{\sigma_0, \sigma_1\}$, and so on, as may be convenient at the moment; similarly for inputs and outputs. Sets S, X, and Z are nonempty.

The function τ is the **next-state function** or **transition function** of the machine. We write

$$s' = \tau(s, x)$$

where s' represents the **next state** and s and x represent the **present state** and the **present input**, respectively. The function ω is the **output function** and expresses the present output in terms of the present state and the present input:

$$z = \omega(s, x).$$

The expressions "present state," "present input," "present output," and "next state" reflect the fact that in some physical implementations of this model, inputs appear successively at equally spaced points in time, and outputs are delivered before the next input appears so they appear in the "present" time interval. The next internal state $s' = \tau(s, x)$, however, does not affect the output until the next time interval.

A machine of this kind is commonly called a **Mealy machine** after G. H. Mealy, who invented this model (1955). An earlier, related model is called a **Moore machine** after E. F. Moore, who originated the notion of an abstract sequential machine (1956). The paper by Huffman (1954) was also important in development of the notion of machine.

Sec. 1.11 Finite-State Machines

A specific finite-state machine $M = [S, X, Z, \tau, \omega]$ can be described in various ways. First there is the **defining table** (or **transition table** or **state table**) of M, which simply lists the values of τ and ω for each pair (s, x). The elements of S identify the rows, and the elements of X identify the columns of two adjacent parts of the table in which the values of $\tau(s, x)$ and $\omega(s, x)$ are listed. We give a simple example, with $X = \{0, 1\}$, $Z = \{00, 01, 10\}$, and $S = \{\sigma_0, \sigma_1, \sigma_2\}$, in Table 1.11.1.

In this example, $\tau(\sigma_0, 1) = \sigma_1$, $\tau(\sigma_2, 0) = \sigma_2$, $\omega(\sigma_1, 0) = 01$, and so on. The machine works as follows: Suppose that the machine is initially in state σ_0 and we feed in the sequence of inputs 1, 0, 1, 1, 0, 0, 1. The first output is $\omega(\sigma_0, 1) = 01$ and the next state is $\tau(\sigma_0, 1) = \sigma_1$. The second output is $\omega(\sigma_1, 0) = 01$ and the machine goes to state $\tau(\sigma_1, 0) = \sigma_1$, and so on. The response to the given input sequence is summarized in Table 1.11.2. In this table, the input and state in any one column, that is, the present state and the present input, determine the present output, which appears in the same column, and the next state, which appears in the following column.

Because 00, 01, and 10 are the binary equivalents of 0, 1, and 2, respectively, and because of the manner in which the output cycles, the machine is an example of a **binary ring counter**, that is, a machine which counts incoming ones thus: $1, 2, 3, \ldots, n-1, 0, 1, \ldots$ in a binary code. In our example, $n = 3$ and the output at any time is r, where the total number of 1's received by that time is $m = 3k + r$, $0 \leq r \leq 2$; that is, the output is m "reduced modulo 3."

TABLE 1.11.1

Present State s \ Input x	Next State $\tau(s, x)$		Present Output $\omega(s, x)$	
	0	1	0	1
σ_0	σ_0	σ_1	00	01
σ_1	σ_1	σ_2	01	10
σ_2	σ_2	σ_0	10	00

TABLE 1.11.2

Input sequence	1	0	1	1	0	0	1	—
State sequence	σ_0	σ_1	σ_1	σ_2	σ_0	σ_0	σ_0	σ_1
Output sequence	01	01	10	00	00	00	01	—

Since, as in the preceding example, we are often concerned with the responses of machines to input sequences, it is useful to extend the definitions of τ and ω accordingly. First, we define X^* to be the set of all finite input sequences that can be formed from the letters (elements) of X, Z^* to be the set of all finite output sequences that can be formed from the letters of Z. Both X^* and Z^* include the **empty sequence** λ of no symbols at all. Let $\mathbf{x} = x_1 x_2 \cdots x_r$ be an arbitrary member of X^* and let $s_1 \in S$. Then if

$$\tau(s_1, x_1) = s_2,$$

we define

(1.11.1)
$$\tau(s_1, x_1 x_2) = \tau(\tau(s_1, x_1), x_2) = \tau(s_2, x_2) = s_3,$$
$$\tau(s_1, x_1 x_2 x_3) = \tau(s_3, x_3) = s_4,$$
$$\vdots$$
$$\tau(s_1, \mathbf{x}) = \tau(s_1, x_1 x_2 \cdots x_r) = \tau(s_r, x_r) = s_{r+1}.$$

With the same meanings for the s_j, if

$$\omega(s_j, x_j) = z_j,$$

we define

(1.11.2)
$$\omega(s_1, x_1 x_2) = \omega(s_1, x_1)\omega(s_2, x_2) = z_1 z_2,$$
$$\omega(s_1, x_1 x_2 x_3) = \omega(s_1, x_1)\omega(s_2, x_2)\omega(s_3, x_3) = z_1 z_2 z_3,$$
$$\vdots$$
$$\omega(s_1, \mathbf{x}) = \omega(s_1, x_1 x_2 \cdots x_r) = \omega(s_1, x_1)\omega(s_2, x_2) \cdots \omega(s_r, x_r)$$
$$= z_1 z_2 \cdots z_r.$$

This is the symbolic representation of the fact that the response of M to a finite **input sequence** (also called an **input string**, **word**, or **tape**) is a **final state** and an **output sequence** (or **string**, **word**, or **tape**). The usefulness of this extension of the domains of τ and ω to $S \times X^*$ is apparent from the following definition: Given a finite-state machine $M = [S, X, Z, \tau, \omega]$, two states s_i and s_j of M are **r-equivalent** iff for all input sequences $\mathbf{x} = x_1 x_2 \cdots x_r$, we have $\omega(s_i, \mathbf{x}) = \omega(s_j, \mathbf{x})$; in words, iff, given any input sequence of length r, the output sequence will be the same whether M is started in s_i or s_j. If s_i and s_j are r-equivalent for all positive integers r, we say that s_i and s_j are **equivalent**. We shall make extensive use of these concepts in following sections.

We can also describe a machine with the aid of a **state diagram**, which is a set of nodes (vertices) interconnected by arrows such that there is precisely one node for each state and such that the notation

$$\underset{}{(s)} \xrightarrow{x,z} \underset{}{(t)}$$

means that $\tau(s, x) = t$ and $\omega(s, x) = z$. Thus the arrow (directed branch) represents a transition from state s to state t, with associated output z, in

Sec. 1.11 Finite-State Machines

response to the input x. The complete set of nodes and labeled arrows duplicates exactly the information in the defining table of the machine. For example, the binary ring counter defined by Table 1.11.1 has the state diagram shown in Figure 1.11.1.

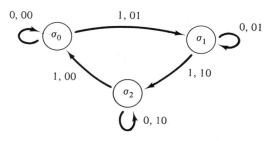

FIGURE 1.11.1. *Modulo 3 Ring Counter*

The state diagram makes it easier to see which states are reachable or accessible from other states and also reveals other important properties at a glance. We define some of these concepts as follows: State t of a machine $M = [S, X, Z, \tau, \omega]$ is **reachable (accessible)** from state s if either $s = t$ or there exists a nonempty input sequence **x** that takes the machine from initial state s to final state t. State s is **transient** if no nonempty sequence of inputs takes M from s back to s. State s is **conditionally transient with respect to** t if t is reachable from s but no nonempty input sequence takes M from t back to s. M is **strongly connected** if every state is reachable from every other state. Many machines are strongly connected, but a machine need not be strongly connected to be useful. If $\tau(s, x) = s$ for all $x \in X$, s is called a **sink state**. If $M = [S, X, Z, \tau, \omega]$ is a machine and if S_0 and X_0 are subsets of S and X such that $\tau(s_0, x_0) \in S_0$ for all $s_0 \in S_0$ and for all $x_0 \in X_0$, then $M_0 = [S_0, X_0, Z, \tau, \omega]$, with τ and ω now defined on (restricted to) $S_0 \times X_0$, is a **submachine** of M.

To illustrate these concepts, consider the machine with $X = Z = \{0, 1\}$ defined by the state diagram of Figure 1.11.2. Here σ_0 and σ_1 are equivalent states. States σ_1 and σ_5 are reachable from σ_3, but σ_2 is not reachable from σ_6. The states σ_0 and σ_4 are transient and σ_1 is conditionally transient with respect to σ_4. The subsets $S_0 = \{\sigma_4, \sigma_5, \sigma_6, \sigma_7\}$ and $X_0 = X$ determine a submachine of M. The subsets $S_0 = \{\sigma_5, \sigma_6\}$ and $X_0 = X$ determine a strongly connected submachine of M. State σ_7 is a sink state.

The notions just defined come from the branch of mathematics known as *graph theory*. They have applications in several aspects of computer science, in biology, in communication theory, and so on. Graph-theoretic concepts will be introduced in this book whenever they help to clarify the subject matter being considered.

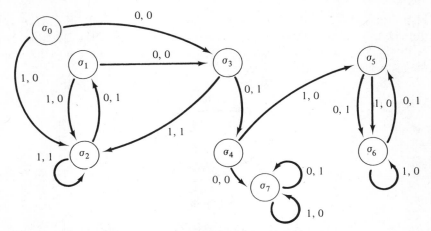

FIGURE 1.11.2. *A Sequential Machine*

1.12 Construction of Machines

To **construct a machine** means to determine the entries of its defining table or, equivalently, to determine its state diagram. We illustrate by constructing two basic types of machines: those that simply delay each input signal a certain number of units of time and those that respond with a specified output whenever a specified input sequence is completed.

Assume the input and output alphabets are $X = Z = \{0, 1\}$, and let $x_0 x_1 x_2 \cdots$ denote an arbitrary input sequence formed from X. The problem will be to construct a machine M that will output this same sequence but delayed one unit in time, the initial output being 0. That is, the output sequence is to be $0 x_0 x_1 x_2 \cdots$.

Let the starting state of the machine be σ_0. The states σ_1 and σ_2, which result from initial inputs 0 and 1, respectively, must be distinct because the machine must always be able to identify what the previous input was. The output resulting from either of these initial inputs is to be 0. State σ_1 stores the fact that the immediately preceding input was 0 and state σ_2 stores the fact that the immediately preceding input was 1.

If an input 0 arrives while the machine is in state σ_1, the output is 0 because the previous input was 0. The next state is σ_1, because in the next time interval the immediately previous input is again 0. If a 1 arrives while the machine is in state σ_1, the output is again 0 because the previous input was 0, but the next state must be σ_2 because now in the next time interval the immediately previous input is 1. Similarly, a 1 received while the machine is in state σ_2 results in the output 1 and a return to state σ_2, while an input 0 received while the machine is in state σ_2 results in the output 1 and a transition to σ_1. The state diagram is now complete (Figure 1.12.1) and the defining table can also be completed.

Sec. 1.12 Construction of Machines

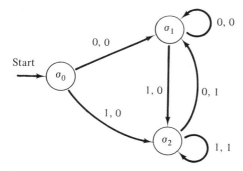

FIGURE 1.12.1. *Unit-Delay Machine*

The defining table (also in Figure 1.12.1) shows that the next-state and output information for σ_0 and σ_1 are identical. Consequently, σ_0 is not needed; we can combine σ_0 with σ_1 and obtain a simpler, two-state machine with σ_1 as the initial state, which responds in exactly the same way as the original machine to any input sequence (Figure 1.12.2).

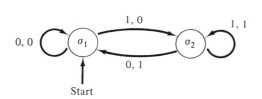

FIGURE 1.12.2. *Reduced Unit-Delay Machine*

Machines—like states—that cannot be distinguished by their responses to any finite input sequences are called **equivalent**. In the construction of state diagrams, equivalent states often appear naturally, as in this example. In Sections 2.16 and 2.17 we show how to detect and combine equivalent states, even when the equivalence is not immediately obvious. This will enable us to obtain minimal-state machines that have specified input–output behavior.

As another example, let us construct a machine that regards an input sequence as a sequence of nonoverlapping triads of inputs and that outputs a 1 whenever the triad 101 is completed. We say a machine **recognizes** or **accepts** any input sequence at the completion of which it outputs a specified signal. The present machine recognizes with output 1 exactly those input sequences of length $3n$ that end with the triad 101. Here is a sample input–output response:

```
in:  0 1 0 | 1 1 0 | 1 0 1 | 1 0 1 | 0 1 1 |
out: 0 0 0 | 0 0 0 | 0 0 1 | 0 0 1 | 0 0 0 |
```

We begin the state diagram with a sequence of states corresponding to the acceptable triad, then direct every other sequence through states that take the machine back to the initial state (reset it) at the count of three. The result is shown in Figure 1.12.3.

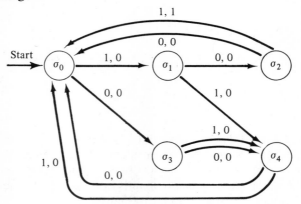

FIGURE 1.12.3. *Finite Sequence Recognizer*

If a machine is to output a 1 whenever the subsequence 101 appears, regardless of its position in the sequence (that is, if overlapping is permitted), a sample input–output response is this:

in: 0 1 0 1 1 0 1 0 1 1 0 1 0 1 1
out: 0 0 0 1 0 0 1 0 1 0 0 1 0 1 0

The state diagram appears in Figure 1.12.4. Here state σ_1 stores the fact that a 1, potentially the beginning of a triad 101, has appeared and σ_2 stores the fact that a 0 preceded by a 1 has appeared.

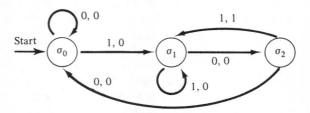

FIGURE 1.12.4. *Finite Sequence Recognizer*

The machine will stay in state σ_0 until an input 1 appears. Once in state σ_1, the machine will stay there until an input 0 appears. When the machine is in state σ_2, an input 1 elicits the output 1 and returns the machine to state σ_1 while an input 0 elicits the output 0 and a return to state σ_0.

1.13 Exercises

1. Draw the state diagram of the following machine and state in words what the machine does:

	τ		ω	
	0	1	0	1
σ_0	σ_0	σ_1	0	0
σ_1	σ_1	σ_2	0	0
σ_2	σ_2	σ_3	0	0
σ_3	σ_3	σ_0	0	1

*__2.__ Assume that the machine of Figure 1.13.1 starts in state σ_0. Observe that no matter what sequence of input symbols is fed into the machine, it is afterward in state σ_0 if the number of 1's in the sequence is even, in state σ_1 if the number of 1's in the sequence is odd. Determine the output response of this machine to the input sequence 100101100011101. Then describe the output behavior in words. This machine is called a **binary counter** or a **parity-check machine**. If the arrow labeled "0, 1" from σ_1 back to σ_1 is relabeled "0, 0," the resulting machine is a **binary pulse divider**. Describe its output behavior in words. Draw the state diagram of a binary pulse divider that outputs a 1 precisely on receipt of every even-numbered input 1.

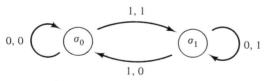

FIGURE 1.13.1. *Binary Counter*

3. Construct the defining table of the machine whose state diagram is given in Figure 1.11.2. Assume that the machine starts in state σ_1 and find the next-state and output sequences corresponding to the input sequence 0100000101.

In Exercises 4 to 7, state diagrams may be used to describe the machines being constructed.

4. Construct a machine with $X = Z = \{0, 1\}$ which recognizes every sequence ending in 0101. (Overlapping is allowed by this statement.)

5. Construct a machine with $X = Z = \{0, 1\}$ which recognizes every sequence of length $4n$ which ends in 0101. (Now overlapping is *not* allowed.)

6. Construct a two-unit delay machine. Note that the second state attained after the initial state may be regarded as identifying the fact that the immediately preceding input pair is 00, 01, 10, or 11 (one state for each pair). Can you find any equivalent states in your diagram? What will be the general nature of the state diagram of an n-unit delay machine?

7. Construct an "up–down counter" with $X = \{0, 1\}$, $Z = \{00, 01, 10, 11\}$, which counts forward one step in the cycle

$$\circlearrowright 00 \to 01 \to 10 \to 11 \circlearrowleft$$

when an input 1 arrives and backward one step when an input 0 arrives. The present state may be thought of as storing the immediately preceding output so that the proper response to the present input can be given. Assume that the machine starts in the state that stores 00 as the immediately preceding response. (Counters of various sorts constitute a third basic class of machines, after delay machines and sequence recognizers.)

8. A **JK-flipflop** is a two-state machine with two input leads labeled J and K and two output leads labeled y and y'. Input and output signals are 0 or 1, so the input alphabet is $X = \{00, 01, 10, 11\}$ (input 01 means $J = 0$, $K = 1$); since y' is always opposite to y, the output alphabet is $Z = \{01, 10\}$. The defining table for the machine is as follows:

	τ				ω			
	00	01	10	11	00	01	10	11
σ_0	σ_0	σ_0	σ_1	σ_1	01	01	10	10
σ_1	σ_1	σ_0	σ_1	σ_0	10	01	10	01

Describe in words the input–output behavior of the flipflop. (Flipflops, used for temporary storage, are a fourth basic class of machines.)

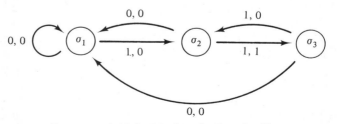

FIGURE 1.13.2. *Machine for Exercise 12*

Sec. 1.13 Exercises

9. Examine the following machines for transient states, conditionally transient states, and strongly connected submachines:

	τ		ω	
	0	1	0	1
σ_1	σ_4	σ_2	1	0
σ_2	σ_3	σ_2	1	1
σ_3	σ_2	σ_4	0	0
σ_4	σ_4	σ_6	0	1
σ_5	σ_4	σ_3	0	1
σ_6	σ_4	σ_5	1	0

	τ		ω	
	0	1	0	1
1	1	6	1	0
2	2	5	1	0
3	7	3	0	0
4	6	4	0	1
5	2	2	1	1
6	4	4	0	1
7	5	2	0	1

10. Show by an example that a submachine of a strongly connected machine need not be strongly connected.

11. Show that every finite-state machine contains at least one strongly connected submachine.

(HINT: Consider any one input symbol ξ_j. If there is a state σ_i such that $\tau(\sigma_i, \xi_j) = \sigma_i$, and if $\omega(\sigma_i, \xi_j) = \zeta_k$, then the submachine defined by the state diagram

$$\sigma_i \circlearrowright \quad \xi_j, \zeta_k$$

is a strongly connected submachine. If no such single-state submachine exists for the input ξ_j, what else must happen?)

12. Prepare a defining table for the machine whose state diagram appears in Figure 1.13.2 and tell in words what it does.

13. Fix an integer $r \in \mathbb{P}$. Draw the state diagram of a machine that recognizes precisely the input strings of r 1's, overlapping being allowed.

14. Is it possible to construct a finite-state machine which, for every input sequence $a_1 a_2 \cdots a_{2n-1} a_{2n}$, produces the output sequence $0 a_1 0 a_2 0 a_3 \cdots 0 a_n$? Explain your answer fully.

*15. How many distinct finite-state machines $M = [S, X, Z, \tau, \omega]$ are there with $|S| = n$, $|X| = m$, $|Z| = p$? [Apply (1.4.1) and Theorem 1.5.1.]

16. A finite-state machine is **simply minimal** if no two rows of outputs in the table defining the machine are identical. For given n, m, and p as in Exercise 15, how many simply minimal machines are there?

*17. (a) Show that the output response of every finite-state machine M to a constant input sequence $aaa \cdots$ is ultimately periodic.

(b) Show that the output response to any periodic input sequence $a_1 a_2 \cdots a_k a_1 a_2 \cdots a_k \cdots$ is ultimately periodic. What is the period of the output sequence?

18. Let $\{0, 1\}$ be a subset of the input and output alphabets of a finite-state machine M and let M receive the constant input sequence $111\cdots$. Show that it cannot output a 1 on receipt of the inputs numbered 2, 6, 12, 20, 30, ..., $k(k+1)$, ... and output a 0 otherwise.

19. Draw the state diagram of the following machine and invent a penny-tossing game that it describes:

	τ		ω	
	H	T	H	T
a	b	c	0	0
b	b	a	1	0
c	a	c	0	1

*20. A **serial binary adder** is a two-state machine capable of adding arbitrarily long binary numbers. It has two input leads a and b and one output lead s. The two states, σ_0 and σ_1, indicate whether 0 or 1 is carried after each step in the addition. Binary numbers $a_k a_{k-1} \cdots a_2 a_1 = a_k \cdot 2^{k-1} + \cdots + a_2 \cdot 2 + a_1$ and $b_k b_{k-1} \cdots b_2 b_1$ are added to yield sum $s_k s_{k-1} \cdots s_2 s_1$; the first input is $a_1 b_1$, the first output is s_1, the second input is $a_2 b_2$, the second output is s_2, and so forth. Draw the state diagram and defining table for the binary adder, with $S = \{\sigma_0, \sigma_1\}$, $X = \{00, 01, 10, 11\}$, $Z = \{0, 1\}$, and starting state σ_0. You may assume that 00 is the last input, so the entire sum is obtained. (Adders are a fifth important class of machines.)

1.14 Union, Intersection, and Inclusion

If S and T are sets, we define the **union** $S \cup T$ of S and T to be the set of all elements in either S or T. We define the **intersection** $S \cap T$ of S and T to be the set of all elements that are members of both S and T. For example,

$$\{001, 011, 100\} \cup \{010, 011, 100, 110\} = \{001, 010, 011, 100, 110\}$$

and

$$\{001, 011, 100\} \cap \{010, 011, 100, 110\} = \{011, 100\}.$$

If $S \cap T = \emptyset$, we say that S and T are **disjoint sets**.

When a set S is a subset of the set T, we write "$S \subseteq T$," which is read "S is **included** in T" or "S is **contained** in T." We also say that "T **includes** (or **contains**) S," and write "$T \supseteq S$." For example, we have at once, for all sets R,

S, and T:

(1.14.1) $\quad S \cap T \subseteq S \subseteq S \cup T \quad$ and $\quad S \cap T \subseteq T \subseteq S \cup T.$

(1.14.2) $\quad\quad\quad\quad\quad \varnothing \subseteq S.$

(1.14.3) $\quad S \subseteq S$ (the **reflexive property** of the inclusion relation).

(1.14.4) $\quad\quad\quad$ If $R \subseteq S$ and $S \subseteq T$, then $R \subseteq T$

(the **transitive property** of the inclusion relation).

If sets S and T are equal, every member of either is also a member of the other, so $S \subseteq T$ and $T \subseteq S$. Conversely, if $S \subseteq T$ and $T \subseteq S$, neither set contains any element not also in the other, so $S = T$. We can express all this compactly, thus:

(1.14.5) $\quad\quad\quad S = T \quad$ iff $\quad S \subseteq T \quad$ and $\quad T \subseteq S.$

This fact (1.14.5) is the **antisymmetric property** of the inclusion relation: one cannot have inclusion holding symmetrically between *distinct* sets S and T.

We have, finally, the two forms of the **consistency principle**:

(1.14.6) $\quad\quad\quad\quad S \subseteq T \quad$ iff $\quad S \cap T = S;$

(1.14.7) $\quad\quad\quad\quad S \subseteq T \quad$ iff $\quad S \cup T = T.$

Proof of (1.14.6). First, suppose that $S \subseteq T$. Then every element of S belongs also to T, so $S \cap T$ is precisely S. Suppose next that $S \cap T = S$. This implies that every element of S is also in T, so $S \subseteq T$ and (1.14.6) follows. ∎

The proof of (1.14.7) is similar and is left to the reader.

Laws (1.14.6) and (1.14.7) permit the translation of inequalities into equalities, and vice versa. This makes them particularly valuable in constructing proofs, as later sections will reveal.

1.15 Complementation

In many kinds of applications, the sets of concern are all subsets of a fixed set U called the **universal set**. In addition to the binary operations "$\cup$" and "$\cap$" (binary because they operate on pairs of sets), there is the unary operation of **complementation** (unary because it operates on individual sets). The complement is denoted by a prime "$'$" or by an overbar "$^-$" and is defined as follows:

$$\forall A \subseteq U, \quad A' = \{x \in U \mid x \notin A\}.$$

For example, if $U = \{00, 01, 10, 11\}$, then $\{00, 01, 11\}' = \{10\}$.

We have at once from the definition that for all subsets A and B of U:

(1.15.1) $\quad U' = \emptyset, \quad \emptyset' = U.$

(1.15.2) $\quad A = B$ iff $A' = B'$.

(1.15.3) $\quad (A')' = A.$ **(law of involution)**

(1.15.4a) $\quad A \cup A' = U.$

(1.15.4b) $\quad A \cap A' = \emptyset.$ **(laws of complementarity)**

We also have **De Morgan's laws**: For all subsets A and B of U,

(1.15.5a) $\quad (A \cup B)' = A' \cap B'$

and

(1.15.5b) $\quad (A \cap B)' = A' \cup B'.$

In words, *the complement of a union is the intersection of the separate complements and the complement of an intersection is the union of the separate complements.*

Proof of (1.15.5a). We note that if $x \in (A \cup B)'$, then $x \notin A$ and $x \notin B$, so $x \in A'$ and $x \in B'$, hence $x \in A' \cap B'$. Thus $(A \cup B)' \subseteq A' \cap B'$. If $x \in A' \cap B'$, then $x \in A'$ and $x \in B'$, so $x \notin A$ and $x \notin B$. Hence $x \notin A \cup B$, so $x \in (A \cup B)'$ and hence $A' \cap B' \subseteq (A \cup B)'$. The result now follows by the antisymmetric property of inclusion. ∎

Proof of (1.15.5b). One possible proof is similar to that of (1.15.5a), but another type of argument is instructive. Since (1.15.5a) applies to *all* subsets of U, we have, with the aid of (1.15.3),

$$(A' \cup B')' = (A')' \cap (B')' = A \cap B.$$

Then, by (1.15.2),

$$((A' \cup B')')' = (A \cap B)'$$

and, by (1.15.3),

$$A' \cup B' = (A \cap B)'. \quad \blacksquare$$

1.16 Venn Diagrams

Relations among subsets of U may be pictured using what are called **Venn diagrams** (more properly, **Euler–Venn diagrams**). In the examples of Figure 1.16.1, the shaded area in the picture illustrates the set named below the picture.

The last two Venn diagrams of Figure 1.16.1 suggest the theorem that $A \cap (B \cup C) = (A \cap B) \cup (A \cap C)$. Although Venn diagrams are valuable as an aid to intuition, they are not in general to be regarded as providing proofs of identities like this one, for not all sets are representable as convex regions

Sec. 1.17   *More Algebra of Subsets*   29

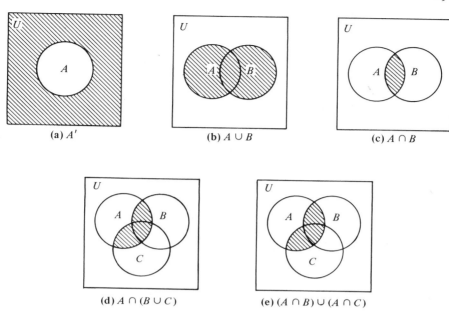

FIGURE 1.16.1. *Venn Diagrams*

in a plane; in cases involving more than three sets, the Venn diagram is often very difficult to draw correctly.

As preceding sections illustrate, proofs of the most basic identities involve consideration of what elements belong to what sets. Once a few laws are established in this way, others follow by algebraic arguments. We turn to these in the next section.

1.17 More Algebra of Subsets

For all subsets A, B, and C of U, we have:

(1.17.1a) $A \cup B = B \cup A$,
(1.17.1b) $A \cap B = B \cap A$.   **(commutative laws)**
(1.17.2a) $A \cup (B \cup C) = (A \cup B) \cup C$,
(1.17.2b) $A \cap (B \cap C) = (A \cap B) \cap C$.   **(associative laws)**
(1.17.3a) $A \cap (B \cup C) = (A \cap B) \cup (A \cap C)$,
(1.17.3b) $A \cup (B \cap C) = (A \cup B) \cap (A \cup C)$.   **(distributive laws)**

Proof of (1.17.1a). We note that every member of $A \cup B$ belongs to at least one of A and B, and hence to $B \cup A$, so $A \cup B \subseteq B \cup A$. Also, every member of $B \cup A$ belongs to at least one of B and A, hence to $A \cup B$, so $B \cup A \subseteq A \cup B$. Therefore, by (1.14.5), $A \cup B = B \cup A$. ∎

Similar arguments prove (1.17.1b), (1.17.2a), and (1.17.2b) and are left to the reader as exercises.

To prove (1.17.3b), note that

$$x \in A \cup (B \cap C) \quad \text{iff} \quad x \in A \quad \text{or} \quad x \in B \text{ and } x \in C$$
$$\text{iff} \quad x \in A \cup B \quad \text{and} \quad x \in A \cup C$$
$$\text{iff} \quad x \in (A \cup B) \cap (A \cup C).$$

This completely proves (1.17.3b), since all steps are iff (reversible) steps. The reader should provide a similar proof for (1.17.3a).

Note that if we let "$\cup$" correspond to "+" and "$\cap$" to "·," the first five of these laws are analogous to the commutative, associative, and distributive laws of the algebra of real numbers. However, (1.17.3b) would correspond to the rule $A + BC = (A + B)(A + C)$, which is not in general valid for real numbers. This warns us against letting prior algebraic experience dominate our intuitions in the present context. Similarities and contrasts of this sort also appear in the next six laws, which are immediate from the definitions of $\cup$, $\cap$, $\varnothing$, and U:

(1.17.4a) $\qquad A \cup \varnothing = A,$

(1.17.4b) $\qquad A \cap U = A,$

in which $\varnothing$ and U are **identity elements** for $\cup$ and $\cap$, respectively;

(1.17.5a) $\qquad A \cup U = U,$

(1.17.5b) $\qquad A \cap \varnothing = \varnothing,$

in which U and $\varnothing$ are **dominant elements** or **zero elements** for $\cup$ and $\cap$, respectively; and finally the **laws of idempotency**,

(1.17.6a) $\qquad A \cup A = A,$

(1.17.6b) $\qquad A \cap A = A.$

1.18 The Principle of Duality

Basic laws (1.15.4), (1.15.5), and (1.17.1) to (1.17.6), which govern union, intersection, and complementation, appear in pairs such that either member of a pair may be obtained from the other, called its **dual**, by interchanging $\cap$ and $\cup$, $\varnothing$ and U throughout. This applies even to (1.15.3), in which none of $\cap$, $\cup$, $\varnothing$, and U appear, so (1.15.3) may be regarded as its own dual.

Sec. 1.18 The Principle of Duality

Consider any theorem derived from the nine basic pairs of laws just listed. We can parallel the steps of the proof, using dual steps throughout, and thus derive a theorem dual to the first. (In this new **dual theorem**, the hypothesis is the dual of the old hypothesis and the conclusion is the dual of the old conclusion.) Once this fact is known, there is no need to write the dual proof, and we thus get two theorems for the proof of one. This fact is known as the **principle of duality**.

We illustrate this principle with the **absorption laws**:

(1.18.1a) $\qquad A \cup (A \cap B) = A,$

(1.18.1b) $\qquad A \cap (A \cup B) = A.$

Proof of (1.18.1a).

$$\begin{aligned}
A \cup (A \cap B) &= (A \cap U) \cup (A \cap B) & &\text{by (1.17.4b)} \\
&= A \cap (U \cup B) & &\text{by (1.17.3a)} \\
&= A \cap (B \cup U) & &\text{by (1.17.1a)} \\
&= A \cap U & &\text{by (1.17.5a)} \\
&= A & &\text{by (1.17.4b).} \blacksquare
\end{aligned}$$

Proof of (1.18.1b) is not needed, by the principle of duality, but the dual proof parallels the given one step by step and should be written as an exercise.

Similarly, we have the **laws of redundancy**:

(1.18.2a) $\qquad A \cup (A' \cap B) = A \cup B,$

(1.18.2b) $\qquad A \cap (A' \cup B) = A \cap B.$

Proof of (1.18.2a).

$$\begin{aligned}
A \cup (A' \cap B) &= (A \cup A') \cap (A \cup B) & &\text{by (1.17.3b)} \\
&= U \cap (A \cup B) & &\text{by (1.15.4a)} \\
&= (A \cup B) \cap U & &\text{by (1.17.1b)} \\
&= A \cup B & &\text{by (1.17.4b).} \blacksquare
\end{aligned}$$

Again, the proof of the dual result should be written for practice.

As a final example, consider the **laws of consensus**:

(1.18.3a) $\quad (A \cup B) \cap (A' \cup C) \cap (B \cup C) = (A \cup B) \cap (A' \cup C),$

(1.18.3b) $\quad (A \cap B) \cup (A' \cap C) \cup (B \cap C) = (A \cap B) \cup (A' \cap C).$

Proof of (1.18.3a). Here a special trick is helpful. We have

$(A \cup B) \cap (A' \cup C) \cap (B \cup C)$
$= (A \cup B) \cap (A' \cup C) \cap [\varnothing \cup (B \cup C)]$
$= (A \cup B) \cap (A' \cup C) \cap [(A \cap A') \cup (B \cup C)]$
$= (A \cup B) \cap (A' \cup C) \cap \{[A \cup (B \cup C)] \cap [A' \cup (B \cup C)]\}$
$= \{(A \cup B) \cap [(A \cup B) \cup C]\} \cap \{(A' \cup C) \cap [(A' \cup C) \cup B]\}$
$= (A \cup B) \cap (A' \cup C)$. ∎

The last equality uses (1.18.1b) twice. The reasons justifying the other steps should be determined carefully by the reader.

The three pairs of laws just established are frequently useful in proofs and in effecting transformations or reductions of complex expressions.

1.19 Boolean Algebras; Summary of Basic Laws

The basic laws of the last four sections are not unique to the algebra of the subsets of a set. Analogous sets of laws arise, for example, in switching theory, in the theory of relations, and in the algebra of propositions. The latter, which represents the original application, was the creation of George Boole in his book, *An Investigation into the Laws of Thought*, first published in 1854. Accordingly, we make the following definition: A **Boolean algebra** is any set B with at least two distinct special elements 0 and 1, called the **zero element** and the **universal element**, respectively; two binary operations, "$\cup$" and "$\cap$"; and a unary operation, "$'$" (that is, functions $\cup : B \times B \to B$, $\cap : B \times B \to B$, and $' : B \to B$) such that, for all a, b, and c in B, we have:

(B1) $a \cup b = b \cup a$, $\quad a \cap b = b \cap a$, $\qquad$ **(commutative laws)**

(B2) $a \cup (b \cup c) = (a \cup b) \cup c$, $\quad a \cap (b \cap c) = (a \cap b) \cap c$,
$\qquad\qquad\qquad\qquad\qquad\qquad\qquad\qquad\qquad$ **(associative laws)**

(B3) $a \cap (b \cup c) = (a \cap b) \cup (a \cap c)$, $\quad a \cup (b \cap c) = (a \cup b) \cap (a \cup c)$,
$\qquad\qquad\qquad\qquad\qquad\qquad\qquad\qquad\qquad$ **(distributive laws)**

(B4) $a \cup a = a$, $\quad a \cap a = a$, $\qquad$ **(laws of idempotency)**

(B5) $a \cup 0 = a$, $\quad a \cap 1 = a$, $\qquad$ **(identity elements)**

(B6) $a \cup 1 = 1$, $\quad a \cap 0 = 0$, $\qquad$ **(dominant elements)**

(B7) $a \cup a' = 1$, $\quad a \cap a' = 0$, $\qquad$ **(laws of complementarity)**

(B8) $\quad (a')' = a$, $\qquad$ **(law of involution)**

(B9) $(a \cup b)' = a' \cap b'$, $\quad (a \cap b)' = a' \cup b'$. $\qquad$ **(De Morgan's laws)**

Actually, a short subset of these laws would suffice to define a Boolean algebra, for the remaining laws could be derived from the subset. We shall return to this point later.

We have seen that for every set U, the set $\mathcal{P}(U)$ forms a Boolean algebra with zero element $\emptyset$, universal element U, and operations of union, intersection, and complementation.

In every Boolean algebra, we have also the useful laws

(B10) $a \cup (a \cap b) = a, \quad a \cap (a \cup b) = a,$ (absorption laws)

(B11) $a \cup (a' \cap b) = a \cup b, \quad a \cap (a' \cup b) = a \cap b,$ (laws of redundancy)

(B12) $(a \cup b) \cap (a' \cup c) \cap (b \cup c) = (a \cup b) \cap (a' \cup c),$
$(a \cap b) \cup (a' \cap c) \cup (b \cap c) = (a \cap b) \cup (a' \cap c);$ (laws of consensus)

for these are algebraic consequences of (B1) to (B9), as seen in Section 1.18. They are independent of any particular interpretation of the elements of B.

The **principle of duality** holds in every Boolean algebra, and in every such algebra **inclusion** is defined thus:

(I1) $\forall\, a, b \in B, \quad a \subseteq b \quad \text{iff} \quad a \cap b = a.$

Using this, we can prove, for all $a, b, c \in B$:

(I2) $a \subseteq b \quad \text{iff} \quad a \cup b = b,$

(I3) $0 \subseteq a \subseteq 1,$

(I4) $a \subseteq a,$

(I5) if $a \subseteq b$ and $b \subseteq a$, then $a = b$,

(I6) if $a \subseteq b$ and $b \subseteq c$, then $a \subseteq c$.

The elements of a Boolean algebra B are subject to varying interpretations, depending on the application. The binary operations "$\cup$" and "$\cap$" may be denoted by "$+$" and "$\cdot$" or by "$\vee$" and "$\wedge$," respectively, and the unary operation "$'$" may be denoted by "$\bar{}$" (overbar) or by the prefix "$\sim$." Inclusion "$\subseteq$" may be denoted "$\leq$."

1.20 Mathematical Induction

There are many mathematical statements that are functions of an integer variable. A familiar example is this general statement:

$$\forall\, n \in \mathbb{P}, \quad 1 + 3 + 5 + \cdots + (2n - 1) = n^2,$$

which, for $n = 1, 2, 3, \ldots$, yields the particular statements

$$1 = 1^2, \quad 1 + 3 = 2^2, \quad 1 + 3 + 5 = 3^2, \ldots.$$

For the proof of such general statements, we use a property of the positive integers known as the **principle of finite** (or **mathematical**) **induction**: *If $S(n)$ is a statement that is a function of the positive integer n, if $S(1)$ is true, and if $S(k+1)$ is true whenever $S(k)$ is true, then $S(n)$ is true for all positive integers n.*

Returning to the example just cited, we see that the formula holds when $n = 1$. Now let k be any positive integer for which the formula holds, so
$$1 + 3 + 5 + \cdots + (2k - 1) = k^2.$$
This equality then implies that
$$1 + 3 + 5 + \cdots + (2k - 1) + (2k + 1) = k^2 + (2k + 1)$$
$$= (k + 1)^2.$$
Thus the formula holds for $n = k + 1$ if it holds for $n = k$. Because the formula holds for $n = 1$, it therefore holds for $n = 2$, hence for $n = 3$, and so on. That is, by the principle of finite induction, it holds for all positive integers n.

Note that it must be shown that both conditions stated in the principle are satisfied: The truth of $S(1)$ allows the process of concluding the truth of successive particular statements to get started. The fact that the truth of $S(k)$ implies the truth of $S(k+1)$ allows the process to continue. The two together justify the general conclusion.

As a further example of the use of this principle, we prove a generalized law of De Morgan: If $n \in \mathbb{P}$, B is a Boolean algebra, and $b_i \in B$, $i = 1, 2, \ldots, n$, then
$$S(b): \quad (b_1 \cup b_2 \cup \cdots \cup b_n)' = b_1' \cap b_2' \cap \cdots \cap b_n'.$$

For $n = 1$, we have just $(b_1)' = b_1'$; and for $n = 2$, we have the usual form of the law: $(b_1 \cup b_2)' = b_1' \cap b_2'$, both of which are true, the latter because B is a Boolean algebra.

Now let k be any integer ≥ 2 for which the formula holds, so that
$$S(k): \quad (b_1 \cup b_2 \cup \cdots \cup b_k)' = b_1' \cap b_2' \cap \cdots \cap b_k'.$$

(This assumption is the **induction hypothesis**.) Then, assuming that the associative law may be employed in the usual way,

$$(b_1 \cup b_2 \cup \cdots \cup b_k \cup b_{k+1})' = ((b_1 \cup b_2 \cup \cdots \cup b_k) \cup b_{k+1})'$$
$$= (b_1 \cup b_2 \cup \cdots \cup b_k)' \cap b_{k+1}' \qquad \text{by (B9)}$$
$$= (b_1' \cap b_2' \cap \cdots \cap b_k') \cap b_{k+1}' \qquad \text{by } S(k)$$
$$= b_1' \cap b_2' \cap \cdots \cap b_k' \cap b_{k+1}'.$$

Thus $S(k)$ implies $S(k+1)$ and hence, by the principle of finite induction, $S(n)$ holds for all positive integers n.

We shall make frequent use of this method of proof in following sections.

1.21 The Characteristic Function of a Subset of a Set

One often has occasion to map a set S onto the set $\{0, 1\}$, which we have denoted by the symbol $\mathbb{B}$. The most familiar example is the function $f: \mathbb{Z} \to \mathbb{B}$ where, for all $z \in \mathbb{Z}$, $f(z) = 0$ if z is even and $f(z) = 1$ if z is odd. The set of all functions from any set S to the set $\mathbb{B}$ is denoted by $\mathbb{B}^S$.

Now consider again the set of all subsets of a universal set U, that is, the power set $\mathcal{P}(U)$ of U. A mapping from $\mathcal{P}(U)$ to $\mathbb{B}^U$ will assign to each element S of $\mathcal{P}(U)$ (that is, to each subset of U) a function from U to $\mathbb{B}$. A useful mapping $S \to f_S$ of this sort is defined in this way:

$$\forall S \in \mathcal{P}(U), \quad \forall x \in U, \quad f_S(x) = \begin{cases} 0 & \text{if } x \notin S, \\ 1 & \text{if } x \in S. \end{cases}$$

For each $S \in \mathcal{P}(U)$, the function $f_S: U \to \mathbb{B}$ is called the **characteristic function** of S.

For example, if $U = \{u_1, u_2, \ldots, u_n\}$ and $S = \{u_2, u_3\}$, we have $f_S(u_1) = 0$, $f_S(u_2) = 1$, $f_S(u_3) = 1$, $f_S(u_4) = 0$, ..., $f_S(u_n) = 0$. These values are conveniently written as the ordered n-tuple $(0, 1, 1, 0, \ldots, 0)$, where the ith entry is $f_S(u_i)$.

This example makes it clear that, given an indexed finite set $U = \{u_1, u_2, \ldots, u_n\}$, each subset $S \subseteq U$ defines a unique ordered n-tuple or n-vector $\mathbf{E}_S = (e_1, e_2, \ldots, e_n)$ of $\mathbb{B}^n$, where $e_i = 0$ if $u_i \notin S$ and $e_i = 1$ if $u_i \in S$. We call $\mathbf{E}_S$ the **characteristic vector** of S. Moreover, given a vector $(e_1, e_2, \ldots, e_n)$, its components 1 identify the elements of the unique subset that determines this vector. Thus we have a one-one correspondence between subsets of U and n-vectors of 0's and 1's. In particular, $\varnothing$ and $(0, 0, \ldots, 0)$, U and $(1, 1, \ldots, 1)$ correspond.

Given a subset S of $U = \{u_1, u_2, \ldots, u_n\}$ with characteristic vector $\mathbf{E}_S = (e_1, e_2, \ldots, e_n)$, the subset S' of U determines the vector $\mathbf{E}_{S'} = (e'_1, e'_2, \ldots, e'_n)$, where we define

$$0' = 1, \quad 1' = 0.$$

Given any vector $\mathbf{E} = (e_1, e_2, \ldots, e_n)$ of $\mathbb{B}^n$, we define its **complement** thus:

$$\mathbf{E}' = (e'_1, e'_2, \ldots, e'_n).$$

From this it follows that

$$\mathbf{E}_{S'} = (\mathbf{E}_S)',$$

so the characteristic vector of the complement of a subset is the complement of the characteristic vector of the subset.

If T is also a subset of U with characteristic vector $\mathbf{E}_T = (\eta_1, \eta_2, \ldots, \eta_n)$, the characteristic vectors of $S \cup T$ and $S \cap T$ are, respectively,

$$\mathbf{E}_{S \cup T} = (\alpha_1, \alpha_2, \ldots, \alpha_n)$$

and
$$E_{S \cap T} = (\beta_1, \beta_2, \ldots, \beta_n),$$
where $\alpha_i = 1$ iff $u_i \in S$ *or* $u_i \in T$ and where $\beta_i = 1$ iff $u_i \in S$ *and* $u_i \in T$.
If we define
$$0 \cup 0 = 0, \quad 0 \cup 1 = 1 \cup 0 = 1 \cup 1 = 1,$$
$$0 \cap 0 = 0 \cap 1 = 1 \cap 0 = 0, \quad 1 \cap 1 = 1,$$
and define **union** and **intersection** of characteristic vectors thus:
$$E_S \cup E_T = (e_1 \cup \eta_1, e_2 \cup \eta_2, \ldots, e_n \cup \eta_n),$$
$$E_S \cap E_T = (e_1 \cap \eta_1, e_2 \cap \eta_2, \ldots, e_n \cap \eta_n),$$
it now follows that
$$E_{S \cup T} = E_S \cup E_T,$$
$$E_{S \cap T} = E_S \cap E_T.$$

Thus the characteristic vector of the union of two subsets of U is the union of their characteristic vectors, and the characteristic vector of the intersection of two subsets of U is the intersection of their characteristic vectors.

In summary, we have corresponding operations ', $\cup$, and $\cap$ in each of $\mathscr{P}(U)$ and $\mathbb{B}^n$, and a one-one correspondence between their elements such that the vector corresponding to a complement, union, or intersection of sets in $\mathscr{P}(U)$ is the complement, union, or intersection, respectively, of the vectors corresponding to these sets. From this it follows (by mathematical induction on the number of operations) that the outcomes of any finite sequence of these three operations performed on corresponding elements of the two sets also correspond. For example, the subsets $A \cup (A' \cap B)$ and $A \cup B$ of U correspond, respectively, to the vectors $E_A \cup (E_A' \cap E_B)$ and $E_A \cup E_B$. Then, since $A \cup (A' \cap B) = A \cup B$, we must also have $E_A \cup (E_A' \cap E_B) = E_A \cup E_B$. As this example suggests, it now follows that the laws (B1) to (B12) of Section 1.19 that characterize a Boolean algebra [and in particular the algebra of $\mathscr{P}(U)$] apply to the vectors of $\mathbb{B}^n$. That is, $\mathbb{B}^n$ with $\cup$, $\cap$, and ' as defined above, with zero element $(0, 0, \ldots, 0)$ and universal element $(1, 1, \ldots, 1)$, is a Boolean algebra.

The parallelism just outlined is described in brief as a *one-one correspondence between the elements of $\mathscr{P}(U)$ and those of $\mathbb{B}^n$ which "preserves operations"* (to a union corresponds a union, and so on). The correspondence is called an **isomorphism** and the algebras of $\mathscr{P}(U)$ and $\mathbb{B}^n$ are said to be **isomorphic** (iso-morphic = same-form).

This isomorphism provides a good way to determine the number of subsets of a finite set, that is, $|\mathscr{P}(U)|$, where $|U| = n$. To each $S \in \mathscr{P}(U)$ there corresponds exactly one vector $(e_1, e_2, \ldots, e_n)$ in $\mathbb{B}^n$. Each e_j is 0 or 1

(according as $u_j \notin S$ or $u_j \in S$), so there are 2^n such vectors altogether and
(1.21.1) $$|\mathscr{P}(U)| = |\mathbb{B}^n| = 2^n.$$
Mathematical systems with corresponding operations governed by the same set of laws (except possibly for purely notational differences) are of frequent occurrence. The concept of *isomorphism* and its generalizations are thus of major importance. These matters will be treated more formally after an intuitive basis for the general definitions has been developed. The determination of $|\mathscr{P}(U)|$ by examination of the elements of $\mathbb{B}^n$ is a small illustration of the fact that what one knows about one system may be useful in answering questions about an isomorphic system.

1.22 Exercises

1. Prove by considering elements that
$$A \subseteq B \quad \text{iff} \quad A \cap B' = \varnothing.$$
In Exercises 2 to 4, do as many of the problems as you can using the laws derived in Sections 1.14 to 1.18, without considering the individual elements in the sets. Do not use Venn diagrams except as a supplementary aid to intuition.

*2. Given that A, B, and C are subsets of a universal set U, prove each of the following, and state the dual theorem in each case:
  (a) If $A \cup B = \varnothing$, then $A = \varnothing$ and $B = \varnothing$.
  (b) If $\exists C \ni A \cup C = B \cup C$ and $A \cap C = B \cap C$, then $A = B$.
  (c) If $A \cap B' = \varnothing$ and $B \cap C' = \varnothing$, then $A \cap C' = \varnothing$.
  (d) If $A \cup B = B$ and $A' \cap B = \varnothing$, then $A = B$.
  (e) If $A \cup B = U$ and $A \cap B = \varnothing$, then $B = A'$. (This shows that only one subset has the properties of the complement.)
  (f) If $(A \cap C) \cup (B \cap C') = \varnothing$, then $A \cap B = \varnothing$.
  (g) If $(A \cap C) \cup (B \cap C') = U$, then $A \cup B = U$, $B \cup C = U$, and $A \cup C' = U$.
  (h) If $A \cap B = A$ and $B \cup C = C$, then $A \cap C' = \varnothing$.
  (i) If $A \subseteq B$ and $B \subseteq C$, then $A \cup B \cup C = C$.
  (j) If $A \cap B = A$ and $B \cup C = B$, then $(A \cup C) \cap B = A \cup C$.
  (k) If $A \cap B' = \varnothing$ and $A \neq \varnothing$, then $B \neq \varnothing$.
  (l) If $A \cap B' = \varnothing$ and $A \cap B = \varnothing$, then $A = \varnothing$.
  (m) $A = \varnothing$ iff $\exists B \subseteq U \ni (A \cap B') \cup (A' \cap B) = B$.
  (n) If $A \cup B = A$ for all A, then $B = \varnothing$.

*3. Prove each of the following, where A, B, C, D, and X are subsets of U:
  (a) If $A \subseteq B$ and $C \subseteq D$, then $A \cup C \subseteq B \cup D$ and $A \cap C \subseteq B \cap D$.
  (b) If $A \subseteq B$ and $C \subseteq B$, then $A \cup C \subseteq B$ and $A \cap C \subseteq B$.
  (c) If $A \cup B \subseteq C$, then $A \subseteq C$ and $B \subseteq C$.

(d) If $A \subseteq B$ and $A \subseteq C$, then $A \subseteq B \cup C$ and $A \subseteq B \cap C$.
(e) If $A \supseteq B \supseteq C$, then $(A \cap B') \cup (B \cap C') = A \cap C'$.
(f) If $A \subseteq A'$, then $A = \varnothing$, and if $A' \subseteq A$, then $A = U$.
(g) If $A \subseteq B$ and $A' \subseteq B$, then $B = U$.
(h) $A \subseteq B$ iff $A' \cup B = U$.
(i) $A \cap B \subseteq (A \cap C) \cup (B \cap C') \subseteq A \cup B$.
(j) $A \subseteq B$ iff $B' \subseteq A'$.
(k) If $A \cup X = A \cup B$ and $A \cap X = \varnothing$, then $X = A' \cap B$.

*4. Using the definition

$$A \oplus B = (A \cap B') \cup (A' \cap B),$$

prove each of the following:
(a) $A \oplus B = B \oplus A$.
(b) $A \oplus (B \oplus C) = (A \oplus B) \oplus C$.
(c) $A \cap (B \oplus C) = (A \cap B) \oplus (A \cap C)$.
(d) $A \cup B = A \oplus B \oplus (A \cap B)$.
(e) $A \oplus A = \varnothing$.
(f) $A \oplus \varnothing = A$.
(g) $A \oplus U = A'$.
(h) $A \oplus A' = U$.
(i) $A \oplus B = A \cup B$ iff $A \cap B = \varnothing$.
(j) If $A \oplus B = C$, then $A = B \oplus C$ and $B = A \oplus C$.

(The operation "$\oplus$" is known as the **exclusive-or** or as the **symmetric difference**. It has important applications in computer science.)

5. Use Venn diagrams to illustrate laws (1.15.5), (1.17.1), (1.17.2), (1.17.3), (1.18.1), (1.18.2), (1.18.3), and the first four parts of Exercise 4.

6. The following is a portion of a report submitted by an investigator for a well-known market analysis agency with standards so high that it boasts that an employee's first mistake is his last:
Number of consumers interviewed: 100.
Number of consumers using brand X: 78.
Number of consumers using brand Y: 71.
Number of consumers using both brands: 48.
Why was the interviewer fired?

7. A famous and often-quoted problem of Lewis Carroll reads thus: "In a very hotly fought battle, 70 per cent, at least, of the combatants lost an eye, 75 per cent, at least, lost an ear, 80 per cent, at least, lost an arm, and 85 per cent, at least, lost a leg. How many, at least, lost all four?" (*A Tangled Tale*.) First solve the problem, then generalize it and derive a formula.

8. If A_1 and A_2 are subsets of U such that

$$\frac{|A_1 \cap A_2|}{|A_2|} > \frac{|A_1 \cap A_2'|}{|A_2'|},$$

Sec. 1.22 *Exercises*

prove that also

$$\frac{|A_1 \cap A_2|}{|A_1|} > \frac{|A_1' \cap A_2|}{|A_1'|};$$

then illustrate with a Venn diagram.

9. Prove that for all subsets A_1, A_2, and A_3 of U, $|A_1 \cap A_2' \cap A_3'| = |A_1| - |A_1 \cap A_2| - |A_1 \cap A_3| + |A_1 \cap A_2 \cap A_3|$.

10. If A_1 and A_2 are subsets of U such that $|A_1| = |A_1'|$ and $|A_2| = |A_2'|$, what relations exist among $|A_1 \cap A_2|$, $|A_1 \cap A_2'|$, $|A_1' \cap A_2|$, and $|A_1' \cap A_2'|$?

11. Prove that for all sets A, B, and C,

$$(A - B) \cap (A - C) = A - (B \cup C).$$

12. Prove that

$$((A \cap B) \cup (A' \cap C))' = (A \cap B') \cup (A' \cap C').$$

13. Given that A and B are fixed subsets of U, define $f: \mathcal{P}(U) \to \mathcal{P}(U)$ by $f(X) = A \cup (B \cap X)$. Find and simplify an expression for $(f \circ f)(X)$.

14. Assume that in a set of 16 defective integrated circuit chips, 10 possess defect d_1, 6 possess defect d_2, 8 possess defect d_3, 3 possess defects d_1 and d_2, 3 possess defects d_2 and d_3, and 4 possess defects d_1 and d_3. How many possess all three defects?

*15. Verify in detail that laws (B1) to (B9) of Section 1.19 hold for the vectors of $\mathbb{B}^n$.

*16. If A_i, $i = 1, 2, \ldots, k$, are subsets of $U = \{u_1, u_2, \ldots, u_n\}$ and if $\mathbf{E}_{A_i} = (e_{i1}, e_{i2}, \ldots, e_{in})$, compute $\mathbf{E}_{(A_1 \cup A_2) \cap (A_1' \cup A_2')}$ and $\mathbf{E}_{f(A_1, A_2, \ldots, A_k)}$, where $f(A_1, A_2, \ldots, A_k)$ is any function of the A_i involving a finite number of uses of union, intersection, and complementation only.

17. Let $U = \{u_1, u_2, \ldots, u_n\}$. Prove by induction on the number p of operations in an expression that if any finite sequence of operations $\cup$, $\cap$, and $'$ is performed on elements of $\mathcal{P}(U)$ and similarly on their corresponding vectors in $\mathbb{B}^n$, then the outcomes also correspond.

*18. Prove that

$$|A \cup B| = |A| + |B| - |A \cap B|;$$

then prove by induction that

$$\left| \bigcup_{i=1}^{n} A_i \right| = \sum_{i=1}^{n} |A_i| - \sum_{1 \leq i_1 < i_2 \leq n} |A_{i_1} \cap A_{i_2}|$$
$$+ \sum_{1 \leq i_1 < i_2 < i_3 \leq n} |A_{i_1} \cap A_{i_2} \cap A_{i_3}| - \cdots + (-1)^{n-1} |A_1 \cap A_2 \cap \cdots \cap A_n|.$$

19. Prove by mathematical induction the following generalized distributive law:

$$A \cap \left(\bigcup_{i=1}^{k} B_i\right) = \bigcup_{i=1}^{k} (A \cap B_i)$$

and then state the dual law.

20. Assuming that every nonempty set of positive integers has a least member, prove the principle of finite induction.

21. An alternative form of the principle of finite induction is as follows: "If $S(n)$ is a statement that is a function of the positive integer n, if $S(1)$ is true, and if the truth of $S(1), \ldots, S(k)$ implies the truth of $S(k+1)$, then $S(n)$ is true for all positive integers n." Show how each form of the principle may be used to prove the other.

22. Find the fault in the following "proof" by induction that $1+2+\cdots+n = (2n+1)^2/8$ for all integers n.

"If $1+2+\cdots+n = (2n+1)^2/8$, then

$$1+2+\cdots+n+(n+1) = \frac{(2n+1)^2}{8} + (n+1) = \frac{4n^2+4n+1+8n+8}{8}$$

$$= \frac{4n^2+12n+9}{8} = \frac{(2n+3)^2}{8} = \frac{[2(n+1)+1]^2}{8},$$

so truth of the statement for n implies its truth for $n+1$."

CHAPTER 2
Relations and Graphs

2.1 Relations

Let A and B be arbitrary sets, not necessarily distinct. Then any subset R of $A \times B$ is called a **binary relation from A to B**, or simply a **relation from A to B**. A binary relation R from A to A is called a **relation on A**. If the pair $(a, b) \in R$, we write $a\, R\, b$, which may be read, "a bears the relation R to b". If $(a, b) \notin R$, we write $a\, \not{R}\, b$. Not all relations from a given A to a given B are ordinarily useful, but in a general theory of relations, all possibilities must be included. In most cases the property defining a useful subset of $A \times B$ is clear and natural.

The concept of a binary relation provides the mathematician with a means for making precise various familiar notions of relations among mathematical objects. For example,

1. $\{(x, y) \in \mathbb{R} \times \mathbb{R} | x \text{ is less than or equal to } y\}$ is the usual "$\leq$" relation on $\mathbb{R}$, since $(x, y) \in $ "$\leq$" iff $x \leq y$.

2. Let U be a set. Then $\{(S, T) \in \mathcal{P}(U) \times \mathcal{P}(U) | S \text{ is a subset of } T\}$ is the usual inclusion relation "$\subseteq$" on $\mathcal{P}(U)$.

3. Let S be a set. Then $\{(x, x) \in S \times S\}$ is the usual relation of equality of elements of the set S.

4. Let $f: S \to T$ be a function. Then $G_f = \{(x, f(x)) \in S \times T | x \in S\}$ is a relation from S to T that is often called the **graph** of f. For functions $f: \mathbb{R} \to \mathbb{R}$ this "graph" yields the usual picture in the xy-plane.

5. $\{(x, y) \in \mathbb{R} \times \mathbb{R} | x^2 + y^2 = 25\}$ is a relation on $\mathbb{R}$ which can be pictured as a circle in the xy-plane. It is *not* the graph of a function. Why?

6. Let $A = B = S$ be the set of states of a finite-state machine $M = [S, X, Z, \tau, \omega]$. Let C denote the set of all pairs (s_i, s_j) such that for some $x_k \in X$, $\tau(s_i, x_k) = s_j$. This subset of $S \times S$ is the **simple connection relation** C for M. Let R denote the set of all pairs (s_i, s_j) such that s_j is reachable from s_i. This subset of $S \times S$ is the **reachability relation** R for M. If M is strongly connected, $R = S \times S$.

Given two sets A and B, let $U = A \times B$. Then the set $\mathcal{R}(A, B)$ of *all* relations from A to B is simply $\mathcal{P}(U)$. The empty subset of U contains no ordered pairs at all and is called the **empty relation** $\varnothing$ from A to B. In this relation, no member of A is related to any member of B. U itself consists of all possible pairs and is called the **universal relation** from A to B. In this relation, every member of A is related to every member of B.

For example, let A and B be sets of finite-state machines, each with a specified initial state. (Such machines are called **initialized machines**, or **initial machines**.) All machines of both sets are assumed to have common input and output alphabets. Let M_A of A and M_B of B be paired iff they respond in the same way to every input sequence of length k, assuming each starts in the designated initial state. The machines of such a pair may be called **k-equivalent**. If no M_A is k-equivalent to any M_B, the k-equivalence relation from A to B is the empty relation. If every M_A is k-equivalent to every M_B, the k-equivalence relation from A to B is the universal relation from A to B.

Since the set $\mathcal{R}(A, B)$ of all relations from A to B is the set $\mathcal{P}(U)$ of all subsets of $U = A \times B$, $\mathcal{R}(A, B)$ *is a Boolean algebra* with the empty relation $\emptyset$ as its zero element and with the universal relation U as its unit element. The *complementary relation* R' of a relation R of $\mathcal{R}(A, B)$ is the subset of $A \times B$ consisting of all pairs *not* in R. The *union* of two relations R_1 and R_2 from A to B is the relation $R_1 \cup R_2$ consisting of the union of the sets of pairs of R_1 and R_2, and the *intersection* of R_1 and R_2 is the relation $R_1 \cap R_2$ consisting of the intersection of the sets of pairs of R_1 and R_2. The **converse** of the relation R from A to B is the relation $\check{R}$ from B to A such that $(b, a) \in \check{R}$ iff $(a, b) \in R$.

For example, in a group of people, the union of the relations "is the mother of" and "is the father of" is the relation "is a parent of." The converse of the relation "is a parent of" is the relation "is a child of." The intersection of the relations "is of the same sex as" and "is a parent of" is the union of the "mother–daughter" and "father–son" relations.

2.2 A Matrix Model of Finite Relations

When A and B are finite sets, the relations from A to B are called **finite relations**. If $A = \{a_1, a_2, \ldots, a_m\}$ and $B = \{b_1, b_2, \ldots, b_n\}$, then there are mn distinct ordered pairs $(a, b) \in A \times B$, so we have

$$|\mathcal{R}(A, B)| = |\mathcal{P}(A \times B)| = 2^{mn}.$$

An $m \times n$ **matrix** over a set S is a rectangular array of mn elements of S with m rows and n columns. It is denoted

$$[s_{ij}]_{m \times n} = [s_{ij}] = \begin{bmatrix} s_{11} & s_{12} & \cdots & s_{1n} \\ s_{21} & s_{22} & \cdots & s_{2n} \\ \cdots & \cdots & \cdots & \cdots \\ s_{m1} & s_{m2} & \cdots & s_{mn} \end{bmatrix},$$

where s_{ij} is the element, or **entry**, in the ith row and jth column. The **transpose** of $[s_{ij}]_{m \times n}$ is $[s_{ji}]_{n \times m}$, the $n \times m$ matrix whose i,j-entry (entry in the ith row and jth column) is the same as the j,i-entry in $[s_{ij}]_{m \times n}$.

Each $R \in \mathcal{R}(A, B)$ may be represented uniquely by an $m \times n$ matrix $\mathbf{M}_R = [r_{ij}]$ over $\mathbb{B}$, called a **relation matrix**. The entries r_{ij} satisfy

(2.2.1) $$r_{ij} = \begin{cases} 0 & \text{if } (a_i, b_j) \notin R, \\ 1 & \text{if } (a_i, b_j) \in R. \end{cases}$$

Sec. 2.2 A Matrix Model of Finite Relations

Conversely, any given $m \times n$ matrix over $\mathbb{B}$, with entries interpreted according to (2.2.1), determines a unique $R \in \mathcal{R}(A, B)$, so we have a bijection between $\mathcal{R}(A, B)$ and the set of all $m \times n$ matrices of 0's and 1's.

The empty relation from A to B corresponds to the $m \times n$ matrix with all entries 0, and the universal relation $\mathcal{U}$ from A to B corresponds to the $m \times n$ matrix with all entries 1. For example, if $A = \{a_1, a_2, a_3\}$, $B = \{b_1, b_2, b_3, b_4\}$, and

$$R = \{(a_1, b_3), (a_2, b_1), (a_2, b_4), (a_3, b_3)\},$$

we have

$$\mathbf{M}_R = \begin{bmatrix} 0 & 0 & 1 & 0 \\ 1 & 0 & 0 & 1 \\ 0 & 0 & 1 & 0 \end{bmatrix}, \quad \mathbf{M}_\varnothing = \begin{bmatrix} 0 & 0 & 0 & 0 \\ 0 & 0 & 0 & 0 \\ 0 & 0 & 0 & 0 \end{bmatrix}, \quad \mathbf{M}_\mathcal{U} = \begin{bmatrix} 1 & 1 & 1 & 1 \\ 1 & 1 & 1 & 1 \\ 1 & 1 & 1 & 1 \end{bmatrix}.$$

Now define, as in Section 1.21,

(2.2.2)
$$0' = 1, \quad 1' = 0,$$
$$0 \cup 0 = 0, \quad 0 \cup 1 = 1 \cup 0 = 1 \cup 1 = 1,$$
$$0 \cap 0 = 0 \cap 1 = 1 \cap 0 = 0, \quad 1 \cap 1 = 1,$$

and for $m \times n$ relation matrices define, for $\mathbf{M}_R = [r_{ij}]$ and $\mathbf{M}_S = [s_{ij}]$,

(2.2.3)
$$(\mathbf{M}_R)' = [(r_{ij})']_{m \times n},$$
$$\mathbf{M}_R \cup \mathbf{M}_S = [(r_{ij} \cup s_{ij})]_{m \times n},$$
$$\mathbf{M}_R \cap \mathbf{M}_S = [(r_{ij} \cap s_{ij})]_{m \times n}.$$

Then

(2.2.4)
$$\mathbf{M}_{R'} = (\mathbf{M}_R)',$$
$$\mathbf{M}_{R \cup S} = \mathbf{M}_R \cup \mathbf{M}_S,$$
$$\mathbf{M}_{R \cap S} = \mathbf{M}_R \cap \mathbf{M}_S.$$

We now have two sets, $\mathcal{R}(A, B)$ and $\{\mathbf{M}_R\}$, whose elements are in one-one correspondence, and which have corresponding operations that are preserved by the correspondence between elements. The matrix of the complement of a relation is the complement of the matrix of the relation, the matrix of the union of two relations is the union of their relation matrices, and the matrix of the intersection of two relations is the intersection of their relation matrices. Once again we have an example of an isomorphism: the algebra of $\mathcal{R}(A, B)$, where $|A| = m$, $|B| = n$, and the algebra of $m \times n$ matrices of 0's and 1's, with respect to the operations complement, union, and intersection, are isomorphic. This isomorphism expedites computations and the proofs of theorems concerning finite relations.

As an example, consider a finite-state machine $M = [S, X, Z, \tau, \omega]$, where $S = \{\sigma_1, \sigma_2, \ldots, \sigma_n\}$ and $X = \{\xi_1, \xi_2, \ldots, \xi_m\}$. With each input symbol ξ_k we associate a relation matrix

$$\mathbf{T}_{\xi_k} = [t_{ij,k}]_{n \times n},$$

where

$$t_{ij,k} = \begin{cases} 0 & \text{if } \tau(\sigma_i, \xi_k) \neq \sigma_j, \\ 1 & \text{if } \tau(\sigma_i, \xi_k) = \sigma_j. \end{cases}$$

Then the matrix $\mathbf{M}_C$ of the simple connection relation C defined in Section 2.1 is related to the matrices $\mathbf{T}_{\xi_k}$ by the equation

$$\mathbf{M}_C = \bigcup_{k=1}^{m} \mathbf{T}_{\xi_k}.$$

For the ternary counter of Figure 1.11.1, we have

$$\mathbf{M}_C = \mathbf{T}_0 \cup \mathbf{T}_1 = \begin{bmatrix} 1 & 0 & 0 \\ 0 & 1 & 0 \\ 0 & 0 & 1 \end{bmatrix} \cup \begin{bmatrix} 0 & 1 & 0 \\ 0 & 0 & 1 \\ 1 & 0 & 0 \end{bmatrix} = \begin{bmatrix} 1 & 1 & 0 \\ 0 & 1 & 1 \\ 1 & 0 & 1 \end{bmatrix}.$$

Let R and S be relations from A to B. Because R and S are subsets of $A \times B$, $R \subseteq S$ iff every ordered pair (a, b) that belongs to R also belongs to S. For example, if $A = B =$ a group of people, then the relation "is a son of" is included in the relation "is a child of."

Again suppose that $|A| = m$ and $|B| = n$ and consider $\mathbf{M}_R = [r_{ij}]$ and $\mathbf{M}_S = [s_{ij}]$, R and S relations from A to B. If $R \subseteq S$, then whenever $r_{ij} = 1$, $s_{ij} = 1$ also. This suggests the following set of definitions:

(2.2.5) $\qquad 0 \leq 0, \quad 0 \leq 1, \quad 1 \leq 1, \quad 1 \not\leq 0,$

and, for arbitrary $m \times n$ matrices of 0's and 1's,

(2.2.6) $\qquad [r_{ij}]_{m \times n} \leq [s_{ij}]_{m \times n} \quad \text{iff} \quad r_{ij} \leq s_{ij} \quad \text{for all } i \text{ and } j.$

For example,

$$\begin{bmatrix} 0 & 1 & 0 \\ 1 & 0 & 1 \end{bmatrix} \leq \begin{bmatrix} 0 & 1 & 1 \\ 1 & 1 & 1 \end{bmatrix};$$

also, for all matrices $[r_{ij}]_{m \times n}$,

$$[0]_{m \times n} \leq [r_{ij}]_{m \times n} \leq [1]_{m \times n}.$$

That is, for all relations R from A to B,

$$\mathbf{M}_\varnothing \leq \mathbf{M}_R \leq \mathbf{M}_{A \times B}.$$

It now follows that for all relations R and S from A to B,

$$R \subseteq S \quad \text{iff} \quad \mathbf{M}_R \leq \mathbf{M}_S.$$

2.3 The Composition of Relations

Let A, B, and C be sets and let Q be a relation from A to B, R a relation from B to C. Then the **composition** or **composite** of Q and R is the relation $Q \circ R$ from A to C such that $(a, c) \in Q \circ R$ iff there exists $b \in B$ such that $(a, b) \in Q$ and $(b, c) \in R$.

For example, let C be the simple connection relation (see Section 2.1) of a finite-state machine $M = [S, X, Z, \tau, \omega]$ and consider the relation $C^2 = C \circ C$. A pair $(s_i, s_j) \in C \circ C$ iff there exists an s_k such that $(s_i, s_k) \in C$ and $(s_k, s_j) \in C$, that is, iff there exists an input string $x_1 x_2$ such that $\tau(s_i, x_1 x_2) = s_j$. Thus $C \circ C$ consists of those pairs of states the second member of which is reachable from the first via at least one input string of length two (Figure 2.3.1).

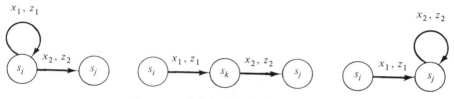

FIGURE 2.3.1. *The Relation* $C \circ C$

If we define $C^r = C^{r-1} \circ C$, it is not hard to see that $(s_i, s_j) \in C^r$ iff there exists an input string $x_1 x_2 \cdots x_r$ of length r such that $\tau(s_i, x_1 x_2 \cdots x_r) = s_j$.

It can happen that the composite $Q \circ R$ of two relations is empty even though neither Q nor R is empty. An example involving sets of small orders is not hard to invent and the reader should do this.

Now consider four sets A, B, C, and D and relations Q from A to B, R from B to C, and S from C to D. Consider any pair $(a, d) \in (Q \circ R) \circ S$, that is, such that $a\, (Q \circ R) \circ S\, d$. The relation $Q \circ R$ is a relation from A to C. We have $a\, (Q \circ R) \circ S\, d$ because there exists $c \in C$ such that $a\, Q \circ R\, c$ and $c\, S\, d$. But $a\, Q \circ R\, c$ because there exists $b \in B$ such that $a\, Q\, b$ and $b\, R\, c$ (Figure 2.3.2). Now because $b\, R\, c$ and $c\, S\, d$, we have $b\, R \circ S\, d$. Hence, because $a\, Q\, b$, $a\, Q \circ (R \circ S)\, d$. This shows that $(Q \circ R) \circ S \subseteq Q \circ (R \circ S)$. A similar argument shows that $Q \circ (R \circ S) \subseteq (Q \circ R) \circ S$, so $(Q \circ R) \circ S = Q \circ (R \circ S)$; that is, we have the following theorem.

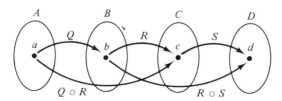

FIGURE 2.3.2. *The Composition of Relations*

Theorem 2.3.1. *The composition of relations is associative.*

The conclusion here is independent of the nature of the sets and relations involved. As a result, this one theorem has many associative laws of algebra, including Theorem 1.9.1, as special cases. (We shall discuss functions as special cases of relations in Section 2.6.)

2.4 Matrix Equivalent of Composition; Digraphs

In the case of finite relations, the operation of composition also has a matrix interpretation. Let R and S be relations from A to B and from B to C, respectively, where $A = \{a_1, a_2, \ldots, a_m\}$, $B = \{b_1, b_2, \ldots, b_n\}$, and $C = \{c_1, c_2, \ldots, c_p\}$. Then $\mathbf{M}_R$ is an $m \times n$ matrix and $\mathbf{M}_S$ is an $n \times p$ matrix. We define the **composite** $\mathbf{M}_R \circ \mathbf{M}_S$ of these two matrices as follows:

$$\mathbf{M}_R \circ \mathbf{M}_S = \left[\bigcup_{j=1}^{n} (r_{ik} \cap s_{kj}) \right]_{m \times p},$$

where the standard $\cup, \cap$-arithmetic of 0 and 1 [defined in (2.2.2)] is to be used. The i,j-entry, expanded, is

$$(r_{i1} \cap s_{1j}) \cup (r_{i2} \cap s_{2j}) \cup \cdots \cup (r_{in} \cap s_{nj}),$$

which has the value 1 iff, for at least one value of k, $r_{ik} = s_{kj} = 1$. Hence the i,j-entry is 1 iff there exists $b_k \in B$ such that $a_i R b_k$ and $b_k S c_j$, that is, iff $a_i R \circ S b_j$. Thus we have the following theorem.

Theorem 2.4.1. *For all finite relations R from A to B and S from B to C,*

$$\mathbf{M}_{R \circ S} = \mathbf{M}_R \circ \mathbf{M}_S.$$

A natural application of this result is to the theory of **directed graphs** or **digraphs**, which are finite sets of vertices (nodes), represented as points or small circles, with arrows from one vertex to another and possibly arrows from a vertex to itself (**loops**). Sometimes digraphs are required to be loop-free, but *not in this book*. In the simplest case, a **simple digraph**, there are no labels on the arrows, and no two arrows can have the same initial vertex and terminal vertex. If the arrows are labeled, as in the case of the state diagram of a finite-state machine, the digraph is said to be **labeled** or **weighted**. In every case it is of interest to determine which vertices are reachable from which other vertices.

For each relation R on a set $A = \{a_1, a_2, \ldots, a_n\}$, we define the (**simple**) **digraph** G_R of R to have one vertex v_i for each $a_i \in A$, and an arrow from v_i to v_j iff $(a_i, a_j) \in R$. Conversely, any simple digraph G on n vertices $v_1, v_2, \ldots, v_n$ defines a relation R_G on A by $(a_i, a_j) \in R_G$ iff G has an arrow

Sec. 2.4 *Matrix Equivalent of Composition; Digraphs*   47

from v_i to v_j. The functions $R \to G_R$ and $G \to R_G$ determine a one-one correspondence between $\mathscr{R}(A, A)$ and the set of all simple digraphs with vertices $v_1, v_2, \ldots, v_n$.

Let G be a directed graph with vertices $v_1, v_2, \ldots, v_n$. A sequence of arrows beginning at vertex v_i and terminating at vertex v_j will be called a **walk** from v_i to v_j, if each arrow ends where the next one begins. If no vertex is entered more than once, the walk will be called a **path** from v_i to v_j.

We define the **simple connection relation** C on the digraph G as follows: $(v_i, v_j) \in C$ iff there exists an arrow from v_i to v_j. Let $\mathbf{M}_C$ be the matrix of C. Then by Theorem 2.4.1 we have $\mathbf{M}_C \circ \mathbf{M}_C = \mathbf{M}_{C \circ C} = \mathbf{M}_{C^2}$, and $\mathbf{M}_{C^2}$ has i,j-entry 1 iff there exists at least one two-step walk from v_i to v_j, so C^2 is the two-step connection relation on G (recall Figure 2.3.1 here). Then $\mathbf{M}_{C^2} \circ \mathbf{M}_C$ has i,j-entry 1 iff there exists v_k such that there is a two-step walk from v_i to v_k and an arrow from v_k to v_j, that is, iff there exists a three-step walk from v_i to v_j. Hence $\mathbf{M}_{C^2} \circ \mathbf{M}_C = \mathbf{M}_{C^3}$, where C^3 is the three-step connection relation of G. In general, $\mathbf{M}_C \circ \mathbf{M}_C \circ \cdots \circ \mathbf{M}_C$ to k factors is $\mathbf{M}_{C^k}$, the matrix of the **k-step connection relation**. If $j \neq i$ and v_j is reachable from v_i, then it is reachable by a walk of length no greater than $n-1$ steps; a longer walk would necessarily involve entering at least one vertex twice, so a shorter path would exist (Figure 2.4.1). Any v_i is reachable from itself by the "empty walk" of 0 steps. Hence $(v_i, v_j) \in R$, the reachability relation for G, iff there exists at least one walk of length $\leq n-1$ from v_i to v_j. Therefore,

$$\mathbf{M}_R = \mathbf{I} \cup \mathbf{M}_C \cup \mathbf{M}_{C^2} \cup \cdots \cup \mathbf{M}_{C^{n-1}}.$$

(Here $\mathbf{I}$ is the identity matrix, that is, the matrix of the identity or 0-step connection relation, with 1's on the main diagonal, 0's elsewhere.)

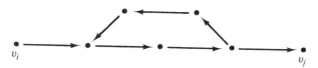

FIGURE 2.4.1. *Example of Walk and Path*

Some of the terms in this union may not be needed, for it may well occur that for some $p < k < n$, $\mathbf{M}_{C^p} = \mathbf{M}_{C^k}$. In that event, $\mathbf{M}_{C^{k+1}} = \mathbf{M}_{C^k} \circ \mathbf{M}_C = \mathbf{M}_{C^p} \circ \mathbf{M}_C = \mathbf{M}_{C^{p+1}}$, then $\mathbf{M}_{C^{k+2}} = \mathbf{M}_{C^{p+2}}$, and so on for higher powers. That is, the powers repeat cyclically and account for no new 1 entries in $\mathbf{M}_R$, which in this case is given by

(2.4.1) $$\mathbf{M}_R = \mathbf{I} \cup \mathbf{M}_C \cup \mathbf{M}_{C^2} \cup \cdots \cup \mathbf{M}_{C^{k-1}},$$

where k is the smallest integer for which the repetition in question occurs.

2.5 Relations on a Set A

Relations on a set A are of particular usefulness, so we now examine some of their special properties. A relation R on a set A is **reflexive** iff for all $a \in A$, $a\,R\,a$; R is **symmetric** iff for all $a, b \in A$, whenever $a\,R\,b$, then $b\,R\,a$ also; R is **antisymmetric** iff for all $a, b \in A$, whenever $a\,R\,b$ and $b\,R\,a$, then $a = b$; R is **transitive** iff for all $a, b, c \in A$, whenever $a\,R\,b$ and $b\,R\,c$, then $a\,R\,c$. Note that "antisymmetric" does not mean the same as "not symmetric." If a relation R is not symmetric, there may be instances in which $a\,R\,b$ and $b\,R\,a$, $a \neq b$. This *never* happens if R is antisymmetric. All four of the properties just defined hold whenever there is no violation of them. Thus if there exist no $a, b \in A$ such that $a\,R\,b$ and $b\,R\,a$, then R is antisymmetric.

For example, consider the inclusion relation "$\subseteq$" on the set $\mathcal{P}(U)$ for given U. The relation is, as we have already seen in Section 1.14, reflexive, antisymmetric, and transitive; it is not symmetric if $U \neq \emptyset$.

Consider the set $\mathbb{P}$ of positive integers and write $a|b$ to mean "a divides b"; that is, $b = aq$ for some positive integer q. The relation "is a divisor of" consists of all ordered pairs (a, b) of positive integers such that $a|b$. This relation is reflexive, is not symmetric, is antisymmetric, and is transitive.

The relation of reachability, defined on the set S of states of a finite machine M, is not necessarily symmetric or antisymmetric, but is reflexive and transitive in every case.

For a given set A, the set of all pairs (a, a), where $a \in A$, is the **identity relation** I on A. I is reflexive, symmetric, antisymmetric, and transitive.

The empty relation on A is not reflexive, but is symmetric, antisymmetric, and transitive.

When A is finite, the properties of relations just defined can be characterized in terms of properties of the corresponding relation matrices. Let $A = \{a_1, a_2, \ldots, a_n\}$ and let $\mathbf{M}_R = [r_{ij}]$ denote the matrix corresponding to a relation R on A. Then we have (with $\mathbf{M}^T$ denoting the transpose of the matrix $\mathbf{M}$) the following theorem.

Theorem 2.5.1. *For all relations R on a finite set A:*
(a) *R is reflexive iff $\mathbf{M}_I \leq \mathbf{M}_R$.*
(b) *R is symmetric iff $\mathbf{M}_R^T = \mathbf{M}_R$, that is, iff $\mathbf{M}_R$ is symmetric.*
(c) *R is transitive iff $\mathbf{M}_{R^2} \leq \mathbf{M}_R$.*
(d) *R is antisymmetric iff $\mathbf{M}_R \cap \mathbf{M}_R^T \leq \mathbf{M}_I$.*

Proof. Conclusions (a) and (b) follow at once from the definitions and (d) is left to the reader as an exercise. To prove (c), assume first that R is transitive. Then whenever $(a_i, a_k) \in R$ and $(a_k, a_j) \in R$, $(a_i, a_j) \in R$ also. Since the i,j-entry of $\mathbf{M}_{R^2}$ is $\bigcup_{k=1}^{n} (r_{ik} \cap r_{kj})$, this means that whenever the i,j-entry of $\mathbf{M}_{R^2}$ is 1 (because some $r_{ik} \cap r_{kj} = 1$), the i,j-entry of $\mathbf{M}_R$ is also 1, so $\mathbf{M}_{R^2} \leq \mathbf{M}_R$.

Now assume that $\mathbf{M}_{R^2} \leq \mathbf{M}_R$. Then whenever the i,j-entry of $\mathbf{M}_{R^2}$ is 1, so is the i,j-entry of $\mathbf{M}_R$; that is, whenever some $r_{ik} \cap r_{kj} = 1$, $r_{ij} = 1$ also, which implies that R is transitive. ∎

2.6 Functions as Relations

We have already pointed out in Section 1.5 that a function $f: A \to B$ may be regarded as a subset of $A \times B$ such that every $a \in A$ appears in exactly one pair of the subset, so a function is a particular kind of relation. For example, let $A = \mathbb{B}^2$ and $B = \mathbb{B}$ and define a function f from A to B by the following set of pairs (a, b):

$$f = \{((0, 0), 0), ((0, 1), 1), ((1, 0), 1), ((1, 1), 0)\}.$$

This function is called the **exclusive-OR function** (more briefly, **XOR**) and is usually represented by Table 2.6.1, where (x_1, x_2) represents the variable pair of $\mathbb{B}^2$. The standard symbol for this function is "$\oplus$," which is often called the **ring sum**. This example also illustrates the next definition.

TABLE 2.6.1. The exclusive-OR function.

x_1	x_2	$f((x_1, x_2)) = x_1 \oplus x_2$
0	0	0
0	1	1
1	0	1
1	1	0

Let A and B be sets and let $f: A^2 \to B$ be a function. (Recall that $A^2 = A \times A$.) This particular kind of function is called a **binary operation** because it assigns to each ordered pair (a_1, a_2) of elements of A a unique element of B, so f is a set of pairs of the form $((a_1, a_2), b)$ in which every possible pair (a_1, a_2) appears exactly once. If $A = B$, f is called a **binary operation on** A. We write $f(a_1, a_2) = b$ instead of $f((a_1, a_2)) = b$. Some familiar examples are these:

1. $A = B = \mathbb{R}, f(a_1, a_2) = a_1 + a_2$.
2. $A = B = \mathcal{P}(U), g(A_1, A_2) = A_1 \cap A_2$.
3. $A = B = \{\text{vectors in 3-space}\}, h(V_1, V_2) = V_1 \times V_2$ (cross product of two vectors).
4. $A = \{\text{vectors in 3-space}\}, B = \mathbb{R}, d(V_1, V_2) = V_1 \cdot V_2$ (dot or scalar product of two vectors).

In the case of frequently used binary operations, the functional notation is replaced by more convenient special symbols, such as $+, \times, \cup, \cap, \oplus$, and so on.

Recall now that A^n denotes the set of all ordered n-tuples $(a_1, a_2, \ldots, a_n)$ of elements of A. A function $f: A^n \to B$ is called an **n-ary operation** or an **n-ary operator**. Such a function is a set of pairs $((a_1, a_2, \ldots, a_n), b)$ in which every n-tuple $(a_1, a_2, \ldots, a_n)$ appears exactly once. It will often be defined by a *table of values* in this book and will be represented by an equation of the form $f(x_1, x_2, \ldots, x_n) = y$, where the x_i are variables over A, called the **arguments** of the function, and y is a variable over B. A basic problem will be to transform a tabular definition into an appropriate algebraic expression $f(x_1, x_2, \ldots, x_n)$. When $((a_1, a_2, \ldots, a_n), b) \in f$, we write $f(a_1, a_2, \ldots, a_n) = b$. A function with finitely many arguments is said to be **finitary**.

For an example of an n-ary function, let $A = \{0, 1\}$ and define $f: A^n \to \mathbb{N}$ by

$$f(a_1, a_2, \ldots, a_n) = \sum_{i=1}^{n} a_i,$$

where the summation on the right involves the addition of integers. This function is called the **weight** of the vector $(a_1, a_2, \ldots, a_n)$ and is important in switching and coding theory.

2.7 Exercises

1. Test the following binary relations R on sets S to see if they are reflexive, symmetric, antisymmetric, or transitive:

S	R
$\mathbb{R}$	xRy iff $x \geq y$ (usual $\geq$ of real numbers)
$\mathbb{R} \times \mathbb{R}$	$(x_1, x_2)R(y_1, y_2)$ iff $x_1 \leq y_1$
$\mathbb{R} \times \mathbb{R}$	$(x_1, x_2)R(y_1, y_2)$ iff $x_1 = y_1$
$\mathbb{R} \times \mathbb{R}$	$(x_1, x_2)R(y_1, y_2)$ iff $x_1 \leq y_1$ and $x_2 \leq y_2$
$\mathbb{P}$	xRy iff $x + y$ is even
$\mathbb{P}$	xRy iff $x + y$ is odd
$\mathbb{Z}$	xRy iff $x + y$ is divisible by 7
$\mathbb{Z}$	xRy iff $x - y$ is divisible by 7
$\mathbb{Z} \times (\mathbb{Z} - \{0\})$	$(p, q)R(p', q')$ iff $pq' = p'q$
$(\mathbb{R} \times \mathbb{R}) \times (\mathbb{R} \times \mathbb{R})$	$((x_1, y_1), (x_2, y_2))R((x_3, y_3), (x_4, y_4))$ iff $(y_2 - y_1)(x_4 - x_3) = (y_4 - y_3)(x_2 - x_1)$

2. Using matrices of 0's and 1's, give examples of finite relations with each of the following combinations of properties:
  (a) Reflexive, antisymmetric, not symmetric, not transitive.
  (b) Reflexive, symmetric, not transitive, not antisymmetric.
  (c) Not reflexive, not symmetric, not antisymmetric, not transitive.

3. A digraph with four vertices has the simple connection matrix shown below. Is the relation it represents reflexive? Symmetric? Antisymmetric?

Transitive? Give a reason for each answer.
$$\begin{bmatrix} 1 & 1 & 1 & 0 \\ 1 & 1 & 1 & 0 \\ 0 & 0 & 1 & 1 \\ 0 & 0 & 0 & 0 \end{bmatrix}.$$

4. If $\mathbf{M}_R$ and $\mathbf{M}_S$ are matrices of binary relations on a finite set A, what properties (reflexive, symmetric, and so on) do R and S possess? Compute $\mathbf{M}_{R \circ S}$ and answer the same question for $R \circ S$.

$$\mathbf{M}_R = \begin{bmatrix} 1 & 0 & 1 & 1 \\ 1 & 1 & 0 & 1 \\ 0 & 1 & 1 & 0 \\ 0 & 0 & 0 & 1 \end{bmatrix}, \quad \mathbf{M}_S = \begin{bmatrix} 1 & 0 & 1 & 1 \\ 0 & 1 & 0 & 1 \\ 1 & 0 & 1 & 1 \\ 1 & 1 & 1 & 1 \end{bmatrix}.$$

*5. By means of a matrix (or other) example, show that if R and S on A are both symmetric relations, it does not follow that their composite $R \circ S$ is symmetric.

*6. Let I be the identity relation on a set A. Prove that a given relation R on A is:
(a) Reflexive iff $I \subseteq R$.
(b) Symmetric iff $R = \breve{R}$.
(c) Antisymmetric iff $R \cap \breve{R} \subseteq I$.
(d) Transitive iff $R \circ R \subseteq R$.

*7. Prove that if a relation R on a set A is reflexive, or symmetric, or antisymmetric, or transitive, then so is its converse $\breve{R}$.

*8. Prove that for all relations R and S on a set A:
(a) $\overline{R \cup S} = \breve{R} \cup \breve{S}$.
(b) $\overline{R \cap S} = \breve{R} \cap \breve{S}$.
(c) $\overline{R \circ S} = \breve{S} \circ \breve{R}$.

9. Restate the results of Exercise 8 in matrix form on the assumption that the set A is finite.

10. Prove algebraically, using the results of Exercises 6 and 8, that if a relation R on a set A is symmetric, then so is $R \cup R^2$. Assuming that A is finite, interpret with respect to the associated digraphs.

11. Let R and R_s, $s \in S$, be relations on a set A. Prove that

$$\left(\bigcup_S R_s\right) \circ R = \bigcup_S (R_s \circ R) \quad \text{and} \quad R \circ \left(\bigcup_S R_s\right) = \bigcup_S (R \circ R_s).$$

Then give matrix examples to show that identities do not result if $\cup$ is replaced throughout by $\cap$ in these equations.

12. Let $\mathbf{M}_C$ be the simple connection matrix of a finite-state machine M. What is the condition, expressed in terms of $\mathbf{M}_C$, that M be strongly connected?

13. If ordinary integer arithmetic is used to compute $(\mathbf{M}_C)^k$ (that is, ordinary matrix multiplication over $\mathbb{Z}$), what do the entries of the product signify? (First try examples, $k = 2$ or 3.)

14. Prove Theorem 2.5.1(d).

15. Let R be a relation on $\mathbb{R}$. Describe geometrically the properties of reflexivity, symmetry, antisymmetry, and transitivity of such a relation.

16. If relation R on $\mathbb{R}$ is transitive and includes all points of the lines $y = \frac{1}{2}x$ and $y = 2x$, what other points must it include? Answer the same question for $y = \frac{1}{2}x$ and $y = 3x$.

17. If each entry "1" of the simple connection matrix of a machine M is replaced by some one input symbol that is responsible for the corresponding transition, a **simple transition matrix** $\mathbf{M}_T$ of M results. If concatenation (juxtaposition) of input symbols and union of strings are used to form matrix products, what is signified by the entries of $\mathbf{M}_T^2, \mathbf{M}_T^3, \ldots$?

***18.** Assume that we are given the machine M with inputs $X = \{1, 2, 3\}$, and unspecified outputs, with defining table as follows:

	τ		
	1	2	3
σ_1	σ_1	σ_2	σ_5
σ_2	σ_1	σ_2	σ_4
σ_3	σ_4	σ_3	σ_1
σ_4	σ_1	σ_1	σ_1
σ_5	σ_4	σ_2	σ_3

(a) Obtain the matrix $\mathbf{M}_R$ of the reachability relation.

(b) Determine a shortest input sequence that causes the machine to pass through all the states and return to its starting state.

(c) What other shortest sequences that have this same effect can be obtained from the one you found in part (b)?

19. For the machine M of Exercise 18, let $\mathbf{T}_i$ be the relation matrix associated with the input i (Section 2.2). Compute the composite $\mathbf{T}_1 \circ \mathbf{T}_2 \circ \mathbf{T}_1$ and indicate the meaning of the entries.

***20.** If $|S| = n$, how many binary operations are there on S? How many ternary (3-ary) operations? How many k-ary operations?

21. A k-ary operation f on a set S is **commutative** if $f(x_1, x_2, \ldots, x_k) = f(\pi(x_1), \pi(x_2), \ldots, \pi(x_k))$ for all permutations (rearrangements) π of any $x_1, x_2, \ldots, x_k \in S$. In Exercise 20, how many of each of these types of operations are commutative?

***22.** Let R be a relation on a set S of order n. Among all transitive relations on S that include R, show that there is a unique smallest one $\hat{R}$, and give an algorithm for constructing $\hat{R}$. ($\hat{R}$ is called the **transitive closure** of R.)

23. How many relations are there on a set A, where $|A|=n$, which are
   (a) Both symmetric and antisymmetric?
   (b) Antisymmetric? (Ans.: $2^n \cdot 3^{n(n-1)/2}$.)

2.8 Partial Orderings and Posets

Certain special types of binary relations are particularly useful in applications. Among these are the partial orderings. A binary relation R on a set P is called a **partial ordering** iff it is reflexive, antisymmetric, and transitive. The set P is then called a **partially ordered set** or **poset** with respect to the relation R. We denote the poset by P_R or by $[P, R]$. A given set P may give rise to many different posets, depending on how R is chosen. If there is no doubt about which relation R is intended, we permit ourselves to say simply "the poset P."

Since the most familiar partial ordering is the usual "less than or equal to" relation, denoted by "$\leq$," on $\mathbb{R}$, mathematicians allow themselves to use the symbol "$\leq$" to denote partial orderings in general, even when the set P is not a set of real numbers. The symbol may still be read "is less than or equal to." Other possible readings are "is included in," "includes," "is the same as or precedes," or whatever may be appropriate in a particular case.

The familiar relation "$\leq$" on $\mathbb{R}$ has the additional property that, for all $a, b \in \mathbb{R}$, we have $a \leq b$ or $b \leq a$ (or both). This law of dichotomy does not hold in all posets. For example, the usual relation of inclusion "$\subseteq$" defined on $\mathscr{P}(U)$ is a partial ordering but if $|U| \geq 2$ it is not true that for all $A, B \in \mathscr{P}(U)$, either $A \subseteq B$ or $B \subseteq A$. When the **law of dichotomy** does hold in a poset $P_\leq$, that is, when for all $a, b \in P$ either $a \leq b$ or $b \leq a$, then $P_\leq$ is called a **chain** (or **totally ordered set** or **ordered set**), and the ordering is called a **linear ordering**. Whenever a and b belong to a poset $P_\leq$ and $a \not\leq b, b \not\leq a$, we say that a and b are **not comparable**.

In order to make following definitions easier to grasp intuitively, we present next a simple way of picturing small, finite posets. Let $P_\leq$ be a poset and let $x, y \in P$. We write $x < y$ iff $x \leq y$ and $x \neq y$. (This notation is chosen to agree with the usual "strictly less than" notation on $\mathbb{R}$.) Suppose that $x < y$. We say that y **covers** x iff $x \leq z \leq y$ implies that $z = x$ or $z = y$, that is, iff there is no element of P properly between x and y. The pairs (x, y) such that y covers x form a subset of the pairs of the partial ordering that is called the **covering relation** of the poset. For P finite, the covering relation is pictured as a linear graph called the **Hasse diagram** (pronounced hah'-suh) of the poset. In this diagram, we represent each element of P as a node or vertex of the graph and we draw a rising line from the vertex x to the vertex y iff y covers x. (These lines intersect only at the vertices of the graph.) Subgraphs of the types shown in Figure 2.8.1(a) and (b) do not appear in Hasse diagrams of posets, for in each case the left branch says that a covers b, whereas the right branch contradicts this. Figures 2.8.1(c) and (d), on the other hand, represent distinct posets, each with five elements.

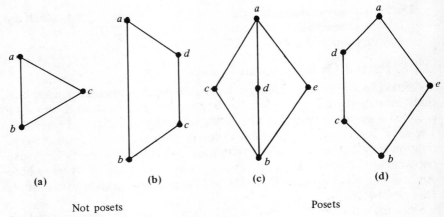

　　　Not posets　　　　　　　　　Posets

FIGURE 2.8.1

In the following examples of posets, the reader should check in each case that the three defining properties hold.

1. $\{1, 2, 3, 4, 5\}$ with the usual relation "$\leq$."

2. The set D_{18} of positive integral divisors of 18 with the relation "$|$" (divides). (Recall that a nonzero integer a **divides** an integer b, and we write $a|b$, iff there exists an integer c such that $b = ac$.)

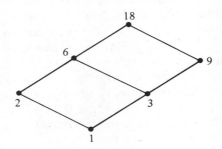

Sec. 2.8 Partial Orderings and Posets

3. $\{3, 5, 7\}$ with the relation "$|$."

4. The set D_{30} of positive integral divisors of 30 with the relation "$|$." (This should be thought of as a three-dimensional picture, since lines intersect only at vertices.)

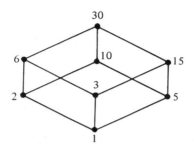

5. The subsets of $S = \{a, b, c\}$ with the relation $\subseteq$.

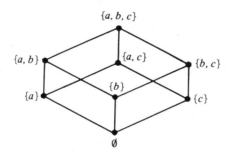

6. $\mathbb{B}^3$ with "$\leq$" as defined for matrices of 0's and 1's in (2.2.6). (Elements of $\mathbb{B}^3$ *are* matrices, with one row and three columns.)

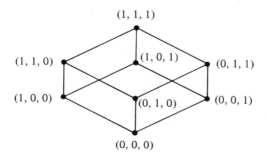

The Hasse diagrams of the last three examples are the same except for the labeling. If we establish the following bijections between elements of these sets:

$$1 \leftrightarrow \emptyset \leftrightarrow (0, 0, 0)$$
$$2 \leftrightarrow \{a\} \leftrightarrow (1, 0, 0)$$
$$3 \leftrightarrow \{b\} \leftrightarrow (0, 1, 0)$$
$$5 \leftrightarrow \{c\} \leftrightarrow (0, 0, 1)$$
$$6 \leftrightarrow \{a, b\} \leftrightarrow (1, 1, 0)$$
$$10 \leftrightarrow \{a, c\} \leftrightarrow (1, 0, 1)$$
$$15 \leftrightarrow \{b, c\} \leftrightarrow (0, 1, 1)$$
$$30 \leftrightarrow \{a, b, c\} \leftrightarrow (1, 1, 1),$$

we see that if p and $q \in D_{30}$, A and B are subsets of S, $\mathbf{x}$ and $\mathbf{y}$ are members of $\mathbb{B}^3$, and $p \leftrightarrow A \leftrightarrow \mathbf{x}$, $q \leftrightarrow B \leftrightarrow \mathbf{y}$, then $p \leq q$ iff $A \subseteq B$ iff $\mathbf{x} \leq \mathbf{y}$; that is, the correspondences preserve the partial ordering relation in each case.

This example suggests the following definition: The posets $P_\leq$ and $Q_\leq$ are **isomorphic posets** iff there exists a bijection $h: P \leftrightarrow Q$ such that for all $a, b \in P$,

$$a \leq b \quad \text{iff} \quad h(a) \leq h(b).$$

In view of this definition, the three posets we have been comparing are isomorphic. They have identical structure; any theorem about the structure of one will be true for all.

The *converse* $\check{R}$ of a partial ordering R on a set P is, as defined in Section 2.1, the set of all pairs (b, a) such that $(a, b) \in R$. We denote the relation $\check{R}$ by "$\geq$," where R is denoted by $\leq$; that is, $b \geq a$ iff $a \leq b$. The reader should show that the relation $\geq$ is also a partial ordering. The poset $P_\geq$ is called the **dual** of the poset $P_\leq$.

If we set up the bijection $a \leftrightarrow a$ between the elements of $P_\leq$ and $P_\geq$, we note that since

$$a \leq b \quad \text{iff} \quad b \geq a,$$

the relation between corresponding elements is not preserved but is, in fact, reversed (Figure 2.8.2).

In general, posets $P_\leq$ and $Q_\leq$ are **anti-isomorphic** iff there exists a bijection $h: P \leftrightarrow Q$ such that for all $a, b \in P$,

$$a \leq b \quad \text{iff} \quad h(b) \leq h(a).$$

The preceding paragraph reveals then that a poset $P_\leq$ and its dual $P_\geq$ are anti-isomorphic under the identity mapping.

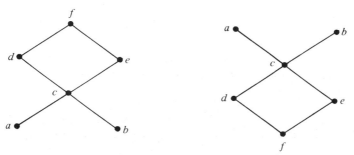

FIGURE 2.8.2. *Dual Posets*

2.9 Special Elements of Posets

A poset $P_\leqslant$ may have a variety of special elements. A **maximal element** of a poset $P_\leqslant$ is any element a of P such that for all $b \neq a$ in P, $a \not\leqslant b$. A **minimal element** of a poset $P_\leqslant$ is any element c of P such that for all $d \neq c$ in P, $d \not\leqslant c$. In words, an element is maximal iff there is no element strictly greater than it, and an element is minimal iff there is no element strictly less than it. A poset may have more than one maximal or minimal element. Hasse diagrams of some examples appear in Figure 2.9.1. Figures 2.9.1(a) and (b) illustrate the following theorem.

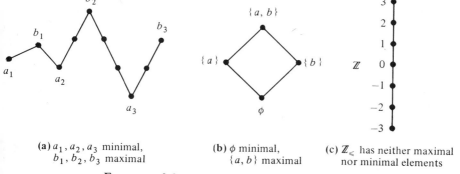

(a) a_1, a_2, a_3 minimal, b_1, b_2, b_3 maximal

(b) ϕ minimal, $\{a, b\}$ maximal

(c) $\mathbb{Z}_\leqslant$ has neither maximal nor minimal elements

FIGURE 2.9.1. *Maximal and Minimal Elements*

Theorem 2.9.1. *A finite nonempty poset $P_\leqslant$ has at least one maximal element and at least one minimal element.*

Proof. To prove this, let p_1 be any element of P. If p_1 is not maximal, then there exists $p_2 \in P$ such that $p_1 < p_2$. If p_2 is not maximal, there exists $p_3 \in P$ such that $p_2 < p_3$, and so on. Because P is finite, this chain must end with some element p_k of P. Since there is no element q of P such that $p_k < q$, p_k is maximal. A similar argument shows there exists at least one minimal element. ∎

Certain special maximal and minimal elements are useful. Given a poset $P_\leq$, an element z of P is called a **zero element** (or **least element**) of $P_\leq$ iff for all $p \in P$, $z \leq p$. An element u of P is called a **unit element** (or **greatest element**) of $P_\leq$ iff for all $p \in P$, $p \leq u$. For example, in the poset $\mathcal{P}(U)_\subseteq$, the empty set $\varnothing$ is a zero element and the universal set U is a unit element. In the poset determined by the positive integral divisors of 30, the integer 1 is a zero element and the integer 30 is a unit element. In Figure 2.9.1, (b) has zero element $\varnothing$ and unit element $\{a, b\}$, but (a) and (c) do not have zero or unit elements. Concerning zero and unit elements, we have the following theorems.

Theorem 2.9.2. *A poset has at most one zero element and at most one unit element.*

Proof. Let z_1 and z_2 be zero elements of a poset $P_\leq$. Then $z_1 \leq z_2$, since z_1 is a zero element; and $z_2 \leq z_1$, since z_2 is a zero element. Therefore, $z_1 = z_2$ and a zero element is unique if it exists. The uniqueness of a unit element is proved similarly. ∎

Theorem 2.9.3. *If a finite poset has a unique maximal element, that element is a unit element; if it has a unique minimal element, that element is a zero element.*

Proof. Let m be a unique maximal element of $P_\leq$. Then any $a_1 \in P$, $a_1 \neq m$, is not maximal, and hence we can construct a chain $a_1 < a_2 < \cdots$ which, because P is finite, necessarily terminates in a maximal element. Since this maximal element must be m, it follows that for all $a_1 \in P$, $a_1 \leq m$, so m is a unit element. The other half of the theorem is proved analogously. ∎

A poset may have other special elements. Thus if a poset $P_\leq$ has the zero element z and if a covers z, the element a is called an **atom** of $P_\leq$. If $P_\leq$ has the unit element u, and if u covers the element b, then b is called an **antiatom** (or **co-atom**) of $P_\leq$.

For example, if $P_\leq = \mathcal{P}(U)_\subseteq$, where $U = \{a_1, a_2, \ldots, a_n\}$, the n singletons $\{a_1\}, \{a_2\}, \ldots, \{a_n\}$ are atoms and the n subsets $\{a_1, a_2, \ldots, a_{n-1}\}, \ldots, \{a_2, a_3, \ldots, a_n\}$ of order $n-1$ are antiatoms. Moreover, there are no other atoms or antiatoms. (Why?) Note that every proper subset of U, that is, every member of $\mathcal{P}(U)_\subseteq$, may be represented as a union of atoms and as an intersection of antiatoms. (Explain in detail.)

Another important pair of concepts is defined as follows: Given a poset $P_\leq$ and a subset Q of P, an element a of P is called an **upper bound** for Q iff, for all $b \in Q$, $b \leq a$. An element c of P is called a **lower bound** for Q iff, for all $b \in Q$, $c \leq b$.

For example, consider the poset of Figure 2.9.2. For the subset $Q = \{a_5, a_6, a_7\}$, a_9 and a_{10} are upper bounds, and a_1, a_2, and a_3 are lower bounds. The subset $S = \{a_5, a_6, a_8\}$ has a_9 and a_{10} as upper bounds but has

Sec. 2.9 *Special Elements of Posets* 59

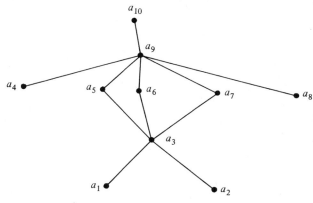

FIGURE 2.9.2. *Poset*

no lower bounds. The subset $T = \{a_3, a_5, a_7, a_9\}$ has a_9 and a_{10} as upper bounds and a_1, a_2, and a_3 as lower bounds. These examples show that upper or lower bounds of a subset are sometimes members of the subset, sometimes not.

An upper bound a for a subset Q of P is called a **least upper bound** (l.u.b.) for Q iff for every upper bound b of Q, $a \leq b$. A lower bound c for a subset Q of P is called a **greatest lower bound** (g.l.b.) for Q iff for every lower bound d of Q, $d \leq c$.

In the preceding example, the subsets Q and T both have least upper bound a_9 and greatest lower bound a_3. Again, such elements in some cases belong to the sets of which they are bounds and in other cases do not.

Theorem 2.9.4. *The least upper bound (or greatest lower bound) of a subset Q of elements of a poset is unique if it exists.*

Proof. The proof is like that of Theorem 2.9.2 and should be provided by the reader. ∎

Earlier, we defined a poset $P_\leq$ to be a *chain* or a *totally ordered set* iff for all $a, b \in P$, $a \leq b$ or $b \leq a$. An example is the set of all positive integral divisors of p^n, where p is a prime and n is a fixed positive integer, with respect to the divisor relation (Figure 2.9.3). (A **prime** is a positive integer $p > 1$ whose only positive integer divisors are p and 1.)

Whether or not $P_\leq$ is a chain, it may have subsets that are totally ordered. In fact, any subset $\{a_1, a_2, \ldots, a_k\}$ of P such that $a_1 < a_2 < \cdots < a_k$ is called a **chain in $P_\leq$**. If there is no $b \in P$ such that $a_i < b < a_{i+1}$ for any i, the chain is called a **maximal chain** with **endpoints** a_1 and a_k. A maximal chain $a_1 < a_2 < \cdots < a_k$ in $P_\leq$ is said to have **length** $k - 1$.

For example, in the poset $\mathscr{P}(U)_\subseteq$, where $U = \{a_1, a_2, \ldots, a_n\}$, the chain $\varnothing \subset \{a_1\} \subset \{a_1, a_2\} \subset \cdots \subset \{a_1, a_2, \ldots, a_n\}$ is a maximal chain of length n.

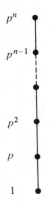

FIGURE 2.9.3. *A Chain*

A poset $P_\leqslant$ may or may not satisfy what is called the **Jordan–Dedekind chain condition**: For each fixed pair of elements of P, all maximal chains with the elements of the pair as end elements are of the same length. Many posets important in applications satisfy this condition.

Let a and b be elements of a finite poset $P_\leqslant$ and suppose that $a < b$. Then if b does not cover a, there exists $c \in P$ such that $a < c < b$. Now if c does not cover a, there exists $d \in P$ such that $a < d < c < b$; and similarly, if b does not cover c, there exists $d \in P$ such that $a < c < d < b$. The process of inserting distinct elements wherever possible into the sequence must come to a halt because P is finite. The end result is a maximal chain $a = a_1 < a_2 < \cdots < a_k = b$ in which each element other than b is covered by its successor. This shows that in the Hasse diagram of $P_\leqslant$, whenever $a < b$ there is an upward path from a to b along the lines of the graph. Conversely, if there is an upward path from a to b in the diagram, $a < b$ by the transitivity of the relation $\leqslant$. Thus the Hasse diagram displays by implication *all* pairs (a, b) such that $a \leqslant b$, not just the instances of covering.

2.10 Direct Products of Posets

In applications, one is often concerned with representing a given system as some sort of "product" of systems that are in some sense simpler. For example, one might want to decompose an abstract machine into an interconnected system of simpler machines. This is important both for studying the properties of machines and for determining how to assemble them economically in terms of suitably simple components.

In the case of posets, we make the following definition: given posets $P_\leqslant$ and $Q_\leqslant$, their **direct product** is defined to be the system $(P \times Q)_\leqslant$ where for the pairs comprising $P \times Q$ we define $(p_1, q_1) \leqslant (p_2, q_2)$ iff $p_1 \leqslant p_2$ in $P_\leqslant$ and $q_1 \leqslant q_2$ in $Q_\leqslant$. (Note that there are three distinct interpretations of "$\leqslant$" involved here.)

Theorem 2.10.1. *The direct product* $(P \times Q)_\leqslant$ *of posets* $P_\leqslant$ *and* $Q_\leqslant$ *is also a poset.*

The reflexive, antisymmetric, and transitive properties follow immediately from those of $P_\leqslant$ and $Q_\leqslant$. (Check this.)

As an example, let $U = \{a_1, a_2\}$ and let $P_\subseteq = \{\varnothing, \{a_1\}\}_\subseteq$, $Q_\subseteq = \{\varnothing, \{a_2\}\}_\subseteq$. Then

$$(P \times Q)_\subseteq = \{(\varnothing, \varnothing), (\varnothing, \{a_2\}), (\{a_1\}, \varnothing), (\{a_1\}, \{a_2\})\}_\leqslant.$$

For this poset, the Hasse diagram is as shown in Figure 2.10.1. From the diagram it is clear that $(P \times Q)_\leqslant$ is isomorphic to $\mathscr{P}(U)_\subseteq$, the natural correspondence being

$$\varnothing \leftrightarrow (\varnothing, \varnothing),$$
$$\{a_1\} \leftrightarrow (\{a_1\}, \varnothing),$$
$$\{a_2\} \leftrightarrow (\varnothing, \{a_2\}),$$
$$\{a_1, a_2\} \leftrightarrow (\{a_1\}, \{a_2\}).$$

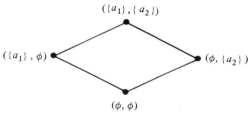

FIGURE 2.10.1. *Poset of a Direct Product*

Note that the direct products $(P \times Q)_\leqslant$ and $(Q \times P)_\leqslant$ are distinct but are isomorphic. If $P_i = \{\varnothing, \{a_i\}\}$ for $i = 1, 2, \ldots, n$ and $U = \{a_1, a_2, \ldots, a_n\}$, $\mathscr{P}(U)_\subseteq$ is isomorphic to

$$((\cdots ((P_1 \times P_2)_\leqslant \times P_3)_\leqslant \times \cdots)_\leqslant \times P_n)_\leqslant,$$

where $\leqslant$ is defined as before for ordered pairs. As an exercise, the reader should provide the details for the case $n = 3$.

2.11 Exercises

1. Draw as many distinct Hasse diagrams of posets with one, two, three, or four elements as you can. Label all zero and all unit elements which appear.

2. The set of all ordered sequences of 0's and 1's with n elements is denoted by $\mathbb{B}^n$. Draw the Hasse diagram for $\mathbb{B}^4_\leqslant$, where $\leqslant$ is the usual partial ordering for matrices of 0's and 1's.

3. Let C denote the set of all real functions continuous on the closed interval $0 \leq x \leq 1$. Define, for $f, g \in C$,

$$f \leq g \quad \text{iff} \quad \forall x \ni 0 \leq x \leq 1, \quad f(x) \leq g(x).$$

Show that C is a poset that has neither zero nor unit element.

4. Prove that if $P_\leq$ is a poset and if $Q \subseteq P$, then $Q_\leq$, where $b_1 \leq b_2$ in $Q_\leq$ iff $b_1 \leq b_2$ in $P_\leq$, is also a poset. If $Q_\leq$ has a l.u.b. in $P_\leq$, does $Q_\leq$ necessarily have a l.u.b. in $Q_\leq$? If $Q_\leq$ has a g.l.b. in $P_\leq$, does $Q_\leq$ necessarily have a g.l.b. in $Q_\leq$?

5. A chain will be called **nontrivial** if it has at least two elements (that is, length ≥ 1). What posets do not contain subsets which are nontrivial chains with respect to the partial ordering of the poset?

6. Show that in the poset of real numbers with the usual inequality relation $\leq$, no element covers any other.

7. How many distinct maximal chains connect $\emptyset$ and U in $\mathcal{P}(U)_\subseteq$, where $U = \{a_1, a_2, \ldots, a_n\}$?

8. How many distinct maximal chains connect the elements $\{a_{i_1}, a_{i_2}, \ldots, a_{i_k}\}$ and $\{a_{j_1}, a_{j_2}, \ldots, a_{j_p}\}$, $k < p$, of the poset $\mathcal{P}(U)$, where $U = \{a_1, a_2, \ldots, a_n\}$?

*9. Show that the bijection $h: \mathcal{P}(U) \leftrightarrow \mathcal{P}(U)$ defined by $h(A) = A'$, for all $A \in \mathcal{P}(U)$, determines an anti-isomorphism of $\mathcal{P}(U)_\subseteq$ with itself.

10. Show that for the posets $\mathcal{P}(U)_\subseteq$ and $\mathcal{P}(U)_\supseteq$, the bijection $h: \mathcal{P}(U) \leftrightarrow \mathcal{P}(U)$ defined as in Exercise 9 is an isomorphism. Illustrate with the Hasse diagrams for the case $|U| = 3$.

11. Give an example other than $\mathcal{P}(U)_\subseteq$ and $\mathcal{P}(U)_\supseteq$ of a poset that is isomorphic to its dual. (Think of the Hasse diagram.)

12. If $P_\leq$ and $Q_\leq$ are anti-isomorphic, $P_\geq$ is the dual of $P_\leq$, and $Q_\geq$ is the dual of $Q_\leq$, prove that $P_\leq$ and $Q_\geq$ are isomorphic, $P_\geq$ and $Q_\leq$ are isomorphic, and $P_\geq$ and $Q_\geq$ are anti-isomorphic.

*13. Prove that if a and b are atoms of a poset $P_\leq$ and if $a \leq b$, then $a = b$.

14. Show that the antiatoms of a poset are precisely the atoms of its dual poset. Conclude that Exercise 13 also holds for antiatoms a and b satisfying $a \leq b$.

15. Given a poset $P_\leq$ and given that $p_1 \leq p_2 \leq \cdots \leq p_k \leq p_1$, where $p_j \in P$, prove that $p_1 = p_2 = \cdots = p_k$; that is, a poset cannot contain any closed chains.

16. Give an example of a poset that does not satisfy the Jordan–Dedekind chain condition.

17. In the finite poset of Figure 2.11.1, find l.u.b. $\{d, f\}$, l.u.b. $\{a, i\}$, g.l.b. $\{b, i\}$, and g.l.b. $\{b, f\}$, if they exist. What are the maximal and minimal elements of this poset?

18. Draw the Hasse diagram of the poset $[P, |]$, where P is the set of positive integral divisors of 72. What will be the nature of the Hasse

Sec. 2.12 *Equivalence Relations and Partitions*

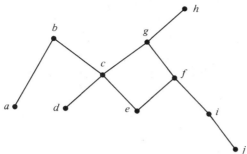

FIGURE 2.11.1. *Poset for Exercise 17*

diagrams of the sets of positive integral divisors of (a) $p_1^{k_1} p_2^{k_2}$ and (b) $p_1^{k_1} p_2^{k_2} p_3^{k_3}$, where p_1, p_2, and p_3 are distinct primes?

*19. The Hasse diagram of the poset of all sequences of n 0's and 1's ($\mathbb{B}^n_\leq$) is called an **n-cube**. A **k-dimensional face** or **subcube** of an n-cube is obtained by fixing some $n-k$ of the components as 0's or 1's, the other components remaining variable. How many k-dimensional subcubes does an n-cube have? How many nonempty subcubes does it have altogether?

$$\left[\text{ANS.: } \frac{2^k (n!)}{(k!)[(n-k)!]};\ 3^n.\right]$$

2.12 Equivalence Relations and Partitions

Two more tools of basic importance in both pure and applied mathematics are equivalence relations and their associated partitions.

A relation E on a set S is called an **equivalence relation** on S iff it is reflexive, symmetric, and transitive. Some examples are these:

1. The *identity relation* on set S (each element related to itself and no other).
2. The *universal relation* on set S (each element related to every element).
3. The similarity relation for geometric figures in a plane.
4. The equality of complex numbers.
5. The equality relation on the set B^A of all functions from a set A to a set B.
6. The relation "$\equiv$" on $\mathbb{Z}$ defined thus:

    $\forall a, b \in \mathbb{Z},\ a \equiv b$ iff $\exists q \in \mathbb{Z} \ni a - b = 3q$

    (congruence modulo 3).
7. The relation E on the state set S of a machine M such that $s_1 E s_2$ iff $\omega(s_1, \mathbf{x}) = \omega(s_2, \mathbf{x})$ for every finite input sequence $\mathbf{x}$.

The reader should verify for each of these examples that the relation in question is indeed reflexive, symmetric, and transitive. (Note that examples 4 and 5 are special cases of example 1.)

With each given equivalence relation E on a set S we associate a collection of subsets of S called equivalence classes. A nonempty subset T of S is an **equivalence class** with respect to E iff

(a) $x, y \in T \Rightarrow x E y$; and
(b) $x \in T, y \in S - T \Rightarrow x \cancel{E} y$.

We will see in Theorem 2.12.1 that such classes exist, indeed that each element of S is in exactly one equivalence class.

For the study of equivalence classes, we need the following definition: A **partition** (or **partitioning**) of a set S is a collection $\pi = \{S_i | i \in$ an indexing set$\}$ of nonempty subsets of S such that each element of S is in exactly one of the S_i, that is, such that $S_i \cap S_j = \emptyset$ if $i \neq j$ and $\bigcup_i S_i = S$. The subsets S_i are called the **blocks** of the partition π of S. The S_i are **pairwise-disjoint** subsets whose union is S.

The Venn diagram of a partition of S with finitely many blocks $S_1, S_2, \ldots, S_k$ is shown in Figure 2.12.1. The regions do not overlap, but together they exhaust S.

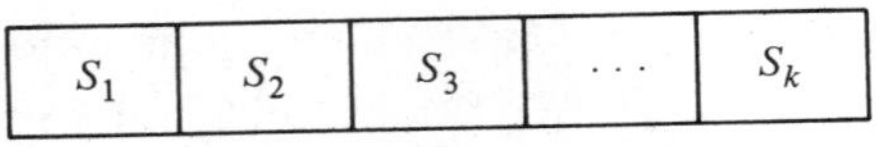

S_1	S_2	S_3	$\cdots$	S_k

FIGURE 2.12.1. *Venn Diagram of a Finite Partition of S*

Some typical examples of partitions are these:
1. S arbitrary; $\pi = \{S\}$. π has just one block. This is called the **unit partition** of S.
2. S arbitrary; $\pi = \{\{s\} | s \in S\}$. π has one block for each element of S. This is called the **zero partition** of S.
3. $S = \{1, 2, 3, 4, 5, 6\}$; $\pi = \{\{1, 2, 4\}, \{5\}, \{3, 6\}\}$. π has three blocks. A commonly used alternative notation designates the blocks with overbars: $\pi = \{\overline{1, 2, 4}; \overline{5}; \overline{3, 6}\}$.
4. $S = \mathbb{Z}$; $S_0 = \{3k | k \in \mathbb{Z}\}$, $S_1 = \{3k + 1 | k \in \mathbb{Z}\}$, $S_2 = \{3k + 2 | k \in \mathbb{Z}\}$; $\pi = \{S_0, S_1, S_2\}$. The S_i are the equivalence classes of congruence modulo 3.
5. Consider the output portion of the defining table of a machine $M = [S, X, Z, \tau, \omega]$, where $S = \{\sigma_1, \ldots, \sigma_n\}$, $X = \{\xi_1, \xi_2, \ldots, \xi_m\}$, and $Z = \{\zeta_1, \zeta_2, \ldots, \zeta_p\}$. Define the contents of the ith row, namely

   $$\omega(\sigma_i, X) = (\omega(\sigma_i, \xi_1), \omega(\sigma_i, \xi_2), \ldots, \omega(\sigma_i, \xi_m)),$$

   to be the **output vector** associated with state σ_i. There are $p^m = |Z^m|$ distinct output vectors possible. Ordinarily, not all of them appear in the defining table of a given machine M. It is convenient to name these output vectors $\mathbf{Z}_0, \mathbf{Z}_1, \ldots, \mathbf{Z}_{p^m-1}$.

   Now form subsets T_i of S where T_i consists of all the states whose output vector is $\mathbf{Z}_i$. Some of the T_i are probably empty; let $\{S_0, S_1, \ldots, S_k\}$ be the set of all nonempty T_i's. Then $\bigcup_i S_i = S$, $S_i \cap S_j = \emptyset$ when $i \neq j$, so $\pi = \{S_0, S_1, \ldots, S_k\}$ is a partition of S. This

Sec. 2.12   Equivalence Relations and Partitions

partition is used in the construction of minimal state machines, which we shall treat in Sections 2.16 and 2.17.

If, in this last example, we define $\sigma_\alpha E_1 \sigma_\beta$ iff σ_α and σ_β are in the same subset S_i, then E_1 is readily seen to be an equivalence relation and the S_i are the corresponding equivalence classes. This illustrates the following theorem.

Theorem 2.12.1. (a) *Every equivalence relation E on a set S determines a unique partition Π_E of S, namely the set of equivalence classes of E on S.*

(b) *Every partition Π of S determines a unique equivalence relation E_Π on S: two elements of S are E_Π-equivalent iff they belong to the same block of Π.*

(c) $E_{(\Pi_E)} = E$ *and* $\Pi_{(E_\Pi)} = \Pi$. *That is, if $\{E\}_S$ denotes the set of all equivalence relations on S and if $\{\Pi\}_S$ denotes the set of all partitions on S, then the function*

$$f: \{E\}_S \to \{\Pi\}_S, \qquad f(E) = \Pi_E,$$

is a bijection with inverse

$$g: \{\Pi\}_S \to \{E\}_S, \qquad g(\Pi) = E_\Pi.$$

Proof. To prove (a), let E be an equivalence relation on S and for each $s \in S$ consider the set $E(s)$ of all elements of S equivalent to s. Since E is reflexive, $s \in E(s)$. If $a, b \in E(s)$, then $a E s$ and $s E b$ (recall that E is symmetric). Hence $a E b$ by the transitivity of E. Thus all elements of $E(s)$ are mutually equivalent. Transitivity shows also that no element of $S - E(s)$ can be equivalent to any element of $E(s)$. These two properties show that the sets $E(s)$ are the equivalence classes of E on S.

Now consider any $E(s_1)$ and $E(s_2)$ where $s_1, s_2 \in S$. If there exists s such that $s \in E(s_1)$ and $s \in E(s_2)$, and if $a \in E(s_1)$, $b \in E(s_2)$, then $a E s_1 E s E s_2 E b$, so $a E s_2$ and $b E s_1$. Thus every a in $E(s_1)$ is also in $E(s_2)$ and every b in $E(s_2)$ is also in $E(s_1)$. This implies that either $E(s_1) \cap E(s_2) = \emptyset$ or $E(s_1) = E(s_2)$. We have already noted that for all $s \in S$, $s \in E(s)$. Hence $\bigcup_s E(s) = S$. Thus the set of distinct equivalence classes $E(s)$ constitutes a partition Π_E of S.

For (b), let $\Pi = \{S_\alpha\}$ be a partition of S and for all $s_1, s_2 \in S$ define $s_1 E_\Pi s_2$ iff both s_1 and s_2 belong to a common block S_α of Π. Then because $\bigcup_\alpha S_\alpha = S$, every $s \in S$ is in some S_α, so $s E_\Pi s$ and E_Π is reflexive. If s_1 and s_2 belong to a common block S_α, so do s_2 and s_1. Hence $s_1 E_\Pi s_2$ iff $s_2 E_\Pi s_1$ and E_Π is symmetric. Finally, because $S_\alpha \cap S_\beta = \emptyset$ if $\alpha \neq \beta$, whenever $s_1 E_\Pi s_2$ and $s_2 E_\Pi s_3$, all three of s_1, s_2, and s_3 must be in a common block S_α of Π, so $s_1 E_\Pi s_3$. Thus E_Π is also transitive and is an equivalence relation.

For (c), note that $E_{(\Pi_E)}$ is the equivalence relation on S whose equivalence classes are the blocks of Π_E. By the definition of Π_E, these are just the equivalence classes of E. That is, $E_{(\Pi_E)} = E$. Similarly, $\Pi_{(E_\Pi)}$ is the partition on S whose blocks are the equivalence classes of E_Π, which are, by the definition of E_Π, just the blocks of Π. That is, $\Pi_{(E_\Pi)} = \Pi$.

Now $(g \circ f)(E) = g(\Pi_E) = E_{(\Pi_E)} = E$, so $g \circ f = 1_{\{E\}_S}$. Similarly, $(f \circ g)(\Pi) = f(E_\Pi) = \Pi_{(E_\Pi)} = \Pi$, so $f \circ g = 1_{\{\Pi\}_S}$ and $g = f^{-1}$. ∎

For an example, let p be a fixed integer >1. We say that x and y in $\mathbb{Z}$ are **congruent modulo** p and write

(2.12.1) $\qquad\qquad\qquad x \equiv y \pmod{p}$

iff $p|(x-y)$, that is, iff there exists $k \in \mathbb{Z}$ such that $x - y = pk$. It is not hard to show that this is an equivalence relation on $\mathbb{Z}$ with equivalence classes

$$E_0 = \{\ldots, -2p, -p, 0, p, 2p, \ldots\},$$
$$E_1 = \{\ldots, -2p+1, -p+1, 1, p+1, 2p+1, \ldots\},$$
$$E_2 = \{\ldots, -2p+2, -p+2, 2, p+2, 2p+2, \ldots\},$$
$$\vdots$$
$$E_{p-1} = \{\ldots, -p-1, -1, p-1, 2p-1, 3p-1, \ldots\}.$$

When $p = 2$, the two resulting classes are the even integers and the odd integers. In later chapters we shall make extensive use of the case where p is a prime.

For another example, let $f: A \to B$ be a function. Define $a_1, a_2 \in A$ to be **f-equivalent** iff $f(a_1) = f(a_2)$. Then f-equivalence is easily shown to be an equivalence relation on A. In particular, let $M = [S, X, Z, \tau, \omega]$; let $A = X^*$ be the set of all finite strings, including the empty string, formed from X; and let $B = S^S$, the set of all functions from S to S. Each input symbol, and hence, by composition, each input string, induces a unique mapping of S into S. The set of all such mappings constitutes a function $f: X^* \to S^S$. Two input sequences are f-equivalent iff they induce the same mapping of S into S. This equivalence relation is basic in the decomposition theory of finite-state machines.

2.13 Intersection and Union of Equivalence Relations

Let E_1 and E_2 be equivalence relations on a set S. It is natural to consider the set $E_1 \cap E_2$ of all pairs (s_1, s_2) that belong to both E_1 and E_2. For example, if E_1 and E_2, defined on $\mathbb{Z}$, are congruence mod 2 and congruence mod 3, respectively, then $E_1 \cap E_2$ is congruence mod 6, which is also an equivalence relation. (Check this.)

The example just given leads to the question: Is $E_1 \cap E_2$ always an equivalence relation? For all $s \in S$, (s, s) is in both E_1 and E_2, so $(s, s) \in E_1 \cap E_2$, which is therefore reflexive. Similarly, if $(s_1, s_2) \in E_1 \cap E_2$, then (s_1, s_2) is in both E_1 and E_2, so (s_2, s_1) is in both E_1 and E_2, hence $(s_2, s_1) \in E_1 \cap E_2$, which is therefore symmetric. Finally, if (s_1, s_2) and (s_2, s_3) are in $E_1 \cap E_2$, then both pairs are in E_1 and E_2, so (s_1, s_3) is in both E_1 and

Sec. 2.13 Intersection and Union of Equivalence Relations

E_2, so $(s_1, s_3) \in E_1 \cap E_2$, which is therefore transitive. *Thus the intersection of two equivalence relations is again an equivalence relation.*

We now determine the equivalence classes of $E_1 \cap E_2$, where E_1 and E_2 are equivalence relations on a set S. Since $s_1(E_1 \cap E_2)s_2$ iff $s_1 E_1 s_2$ and $s_1 E_2 s_2$, the equivalence classes of $E_1 \cap E_2$ are just the intersections of each of the equivalence classes of E_1 with each of the equivalence classes of E_2. Here is an example; if

$$S = \{s_1, s_2, s_3, s_4, s_5, s_6, s_7\}$$

and if

$$\Pi_{E_1} = \{\overline{s_1, s_2, s_6, s_7}; \overline{s_3, s_4, s_5}\},$$
$$\Pi_{E_2} = \{\overline{s_1, s_2}; \overline{s_3, s_4, s_6, s_7}; \overline{s_5}\},$$

then

$$\Pi_{E_1 \cap E_2} = \{\overline{s_1, s_2}; \overline{s_6, s_7}; \overline{s_3, s_4}; \overline{s_5}\}.$$

Now consider the set-theoretic union of E_1 and E_2. This union includes all pairs (s, s), where $s \in S$, and so is reflexive. Also, the union contains (s_2, s_1) whenever it contains (s_1, s_2) and so is symmetric. However, if $(s_1, s_2) \in E_1$ and $(s_2, s_3) \in E_2$, both pairs belong to the union of the two sets, which may fail to include (s_1, s_3) and therefore fail to be transitive. For example, if E_1 is congruence mod 2 on $\mathbb{Z}$ and E_2 is congruence mod 3 on $\mathbb{Z}$, then $(2, 4) \in E_1$ and $(4, 7) \in E_2$, but $(2, 7)$ is in neither E_1 nor E_2, so in this case the union is not transitive.

In view of these considerations, we define the **transitive union**, denoted by $E_1 \cup E_2$, of two equivalence relations E_1 and E_2 on a set S to be the set-theoretic union of E_1 and E_2 augmented by precisely those pairs required to guarantee transitivity. [Note that if we include (s_1, s_3) because $s_1 E_1 s_2$ and $s_2 E_2 s_3$, we also include (s_3, s_1) because $s_3 E_2 s_2$ and $s_2 E_1 s_1$, so symmetry is preserved by this operation.] If, say, $s_1 E_1 s_2$, $s_2 E_2 s_3$, $s_3 E_1 s_4$, and $s_4 E_2 s_5$, we must include pairs (s_1, s_3), (s_1, s_4), (s_1, s_5), (s_2, s_4), (s_2, s_5), (s_3, s_5), (s_3, s_1), (s_4, s_1), (s_5, s_1), (s_4, s_2), (s_5, s_2), and (s_5, s_3).

We now determine the equivalence classes of $E_1 \cup E_2$. By the definition of the transitive union, if $s_1 E_1 s_2$ and $s_2 E_2 s_3$, then $s_1(E_1 \cup E_2)s_3$. This implies that whenever equivalence classes of E_1 and E_2 overlap (have a nonempty intersection), their set-theoretic union is contained in an equivalence class of $E_1 \cup E_2$. Hence, to obtain the equivalence classes of $E_1 \cup E_2$, we must, at the very least, unite all overlapping classes of E_1 and E_2. (As seen in the last paragraph, this may require several steps.) Since every element of S appears in exactly one of the resulting subsets, the result is a partition of S, so the operation is sufficient to guarantee transitivity by Theorem 2.12.1. The associated equivalence relation is therefore $E_1 \cup E_2$.

Here is an example. Let
$$S = \{s_1, s_2, s_3, s_4, s_5, s_6, s_7, s_8, s_9\}$$
and let
$$\Pi_{E_1} = \{\overline{s_1, s_3}; \overline{s_2, s_4}; \overline{s_5, s_6}; \overline{s_7, s_8, s_9}\},$$
$$\Pi_{E_2} = \{\overline{s_1, s_2}; \overline{s_3, s_4}; \overline{s_5, s_7}; \overline{s_6, s_8}; \overline{s_9}\}.$$
Then
$$\Pi_{E_1 \cup E_2} = \{\overline{s_1, s_2, s_3, s_4}; \overline{s_5, s_6, s_7, s_8, s_9}\}.$$

2.14 Intersection and Union of Partitions

Because the equivalence classes of an equivalence relation E on S are the blocks of the corresponding partition Π_E of S, we now can make useful definitions of operations on partitions. Let π_1 and π_2 be partitions on a set S. Then their **intersection** $\pi_1 \cap \pi_2$ is defined to be the partition whose blocks are the intersections of the blocks of π_1 and π_2; and their **union** $\pi_1 \cup \pi_2$ is defined to be the partition whose blocks are the unions of overlapping blocks of π_1 and π_2.

For example, let
$$S = \{1, 2, 3, 4, 5, 6, 7, 8, 9\}$$
and let
$$\pi_1 = \{\overline{1, 2, 4}; \overline{3, 8}; \overline{5, 7, 9}; \overline{6}\},$$
$$\pi_2 = \{\overline{2, 4, 5}; \overline{3, 6, 8}; \overline{1, 7, 9}\}.$$
Then
$$\pi_1 \cap \pi_2 = \{\overline{1}; \overline{2, 4}; \overline{3, 8}; \overline{5}; \overline{6}; \overline{7, 9}\},$$
$$\pi_1 \cup \pi_2 = \{\overline{1, 2, 4, 5, 7, 9}; \overline{3, 6, 8}\}.$$

From the definitions in this and the preceding section, it follows that

(2.14.1)
$$\Pi_{E_1 \cap E_2} = \Pi_{E_1} \cap \Pi_{E_2},$$
$$\Pi_{E_1 \cup E_2} = \Pi_{E_1} \cup \Pi_{E_2};$$

that is, the bijection $f: \{E\}_S \to \{\Pi\}_S$, $f(E) = \Pi_E$ is an *isomorphism* because it preserves the operations.

We say that π_1 is **included** in π_2 (or π_2 **includes** π_1) and write $\pi_1 \leq \pi_2$ (or $\pi_2 \geq \pi_1$) iff every block of π_1 is contained in a block of π_2. When $\pi_1 \leq \pi_2$, we

Sec. 2.14 Intersection and Union of Partitions

call π_1 a **refinement** of π_2. For example, it is always true that

$$\pi \leq \pi,$$

(2.14.2)
$$\pi_1 \cap \pi_2 \leq \pi_1, \quad \pi_1 \cap \pi_2 \leq \pi_2,$$
$$\pi_1 \leq \pi_1 \cup \pi_2, \quad \pi_2 \leq \pi_1 \cup \pi_2.$$

It is now not hard to show that the relation "$\leq$" on $\{\Pi\}_S$ is a partial ordering. The reader should do this and then show that

(2.14.3) $\quad\quad\quad \pi_1 \leq \pi_2 \quad \text{iff} \quad \pi_1 \cap \pi_2 = \pi_1;$

(2.14.4) $\quad\quad\quad \pi_1 \leq \pi_2 \quad \text{iff} \quad \pi_1 \cup \pi_2 = \pi_2.$

Note that the definitions of union and intersection imply that $\pi_1 \cup \pi_2$ is the *smallest* partition containing both π_1 and π_2 and that $\pi_1 \cap \pi_2$ is the *largest* partition included in each of π_1 and π_2. ("Smallest" and "largest" here refer to the ordering on $\{\Pi\}_S$, definitely not to the number of blocks in the partitions.) This means that in the poset $[\{\Pi\}_S, \leq]$,

$$\pi_1 \cup \pi_2 = \text{l.u.b.} \{\pi_1, \pi_2\}, \quad \pi_1 \cap \pi_2 = \text{g.l.b.} \{\pi_1, \pi_2\}.$$

The *smallest* partition of a set S is the **zero partition** "0," in which each block consists of a single element of S. The *largest* partition of a set S is the **unit partition** "1," in which there is precisely one block containing all elements of S. For all partitions π of S,

(2.14.5) $\quad\quad\quad\quad\quad\quad\quad 0 \leq \pi \leq 1.$

From the definitions of $\cup$ and $\cap$ and from (2.14.3) and (2.14.4) it now follows readily that for all partitions π_1, π_2, and π_3 of a set S, we have

$$\pi_1 \cap \pi_2 = \pi_2 \cap \pi_1, \quad \pi_1 \cup \pi_2 = \pi_2 \cup \pi_1, \quad \text{(commutativity)}$$

$$\pi_1 \cap (\pi_2 \cap \pi_3) = (\pi_1 \cap \pi_2) \cap \pi_3, \quad \pi_1 \cup (\pi_2 \cup \pi_3) = (\pi_1 \cup \pi_2) \cup \pi_3,$$

(2.14.6) $\quad\quad\quad\quad\quad\quad\quad\quad\quad\quad\quad\quad\quad\quad\quad\quad$ (associativity)

$$\pi_1 \cap \pi_1 = \pi_1, \quad \pi_1 \cup \pi_1 = \pi_1, \quad \text{(idempotency)}$$

$$\pi_1 \cap (\pi_1 \cup \pi_2) = \pi_1, \quad \pi_1 \cup (\pi_1 \cap \pi_2) = \pi_1. \quad \text{(absorption)}$$

The proofs are left as exercises for the reader.

Consider the set of all partitions of $\{a, b, c\}$. The Hasse diagram of this poset appears in Figure 2.14.1. We have in this case

$$\alpha \cup (\beta \cap \gamma) = \alpha \cup 0 = \alpha; \quad (\alpha \cup \beta) \cap (\alpha \cup \gamma) = 1 \cap 1 = 1.$$

Since this figure appears in the Hasse diagram of $\{\Pi\}_S$ whenever $|S| \geq 3$, we conclude that the algebra of $\{\Pi\}_S$ does not satisfy the distributive law and hence is *not* a Boolean algebra (Section 1.19) if $|S| \geq 3$. Because it does satisfy (2.14.6), it is an example of a system called a *lattice*, which we shall study in a later chapter.

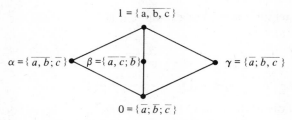

FIGURE 2.14.1. *Partitions of $\{a, b, c\}$*

2.15 Exercises

1. Can a binary relation on a set S be both a partial ordering and an equivalence relation? If so, identify it.

2. Consider the following binary relations α on the set of triangles in the Euclidean plane. Which are equivalence relations? Which are partial orderings?
 (a) $T_1 \alpha T_2$ iff T_1 is congruent to T_2.
 (b) $T_1 \alpha T_2$ iff T_1 is similar to T_2.
 (c) $T_1 \alpha T_2$ iff some angle of T_1 equals some angle of T_2.
 (d) $T_1 \alpha T_2$ iff the area of T_1 equals the area of T_2.
 (e) $T_1 \alpha T_2$ iff all points of the boundary of T_1 lie on or within the boundary of T_2.

3. List at least four distinct kinds of equivalence relation not already given in the text.

4. If R and S are relations from X to Y and if $(x, y) \in R$ implies that $(x, y) \in S$, R is called a **refinement** of S. Give examples of refinements of three distinct relations.

*5. Let R and S be equivalence relations on a set X. Show that if R is a refinement of S, the equivalence classes determined by R are subsets of the equivalence classes determined by S. Illustrate with an example.

6. What are the intersection and the union of all the equivalence relations on a set A?

7. What are the intersections of the equivalence classes of congruence mod 2 on $\mathbb{Z}$ and congruence mod 6 on $\mathbb{Z}$? Mod 2 and mod 5? Mod 4 and mod 6? Can you give a general rule?

8. What are the equivalence classes of the (transitive) union of congruence mod 2 and congruence mod 3 on $\mathbb{Z}$? Mod 4 and mod 6? Can you give a general rule?

9. If $f: A \to B$ is a function and $R = \{(a_1, a_2) | f(a_1) = f(a_2)\}$, show that R is an equivalence relation on A. Show that every equivalence relation on A arises this way, for some choice of B and $f: A \to B$.

10. Show that the set of all equivalence relations on the set S, $|S| \geq 3$, with $\cup$ and $\cap$ as defined in Section 2.13, is not a Boolean algebra, no matter how the complement is defined.

11. Compute the intersection and the union of the partitions
(a) $\pi_1 = \{\overline{1,2,8};\ \overline{3,5,9};\ \overline{4,6,7}\}$, $\pi_2 = \{\overline{1,5,8};\ \overline{2,4,9};\ \overline{3,6,7}\}$.
(b) $\pi_1 = \{\overline{1,3,5};\ \overline{2,7};\ \overline{4,6};\ \overline{8};\ \overline{9}\}$, $\pi_2 = \{\overline{1,3,7};\ \overline{2,5,9};\ \overline{4,6,8}\}$.

12. In the set of all partitions of $\{a,b,c\}$, show that the partition $\pi = \{a,b;\bar{c}\}$ has two complements; that is, there exist two distinct partitions α such that $\pi \cup \alpha = 1$, $\pi \cap \alpha = 0$.

13. Draw the Hasse diagram of the set of all partitions of $\{a,b,c,d\}$ and mark in separate colors several examples of Figure 2.14.1 within your diagram.

14. Prove the results stated in (2.14.3) and (2.14.4). Then use these results to prove the absorption laws of partition algebra.

2.16 Identification of Equivalent States

Let $M = [S, X, Z, \tau, \omega]$ be a finite-state machine, r a positive integer. Denote by X^r the set of all input strings $\mathbf{x} = x_1 x_2 \cdots x_r$ of length r and by Z^r the set of all output strings of length r. When the machine starts in state s and we input the string $\mathbf{x}$ of length r, the resulting output string of length r may be denoted by $\omega_{(r)}(s, \mathbf{x})$. We have then a function

$$\omega_{(r)} \colon S \times X^r \to Z^r.$$

In terms of this notation, two states s_i and s_j of M are **r-equivalent**, written $s_i \, E_r \, s_j$, iff

$$\forall\, \mathbf{x} \in X^r, \qquad \omega_{(r)}(s_i, \mathbf{x}) = \omega_{(r)}(s_j, \mathbf{x}),$$

and **equivalent**, written $s_i \, E \, s_j$, iff

$$\forall\, r \in \mathbb{P}, \qquad s_i \, E_r \, s_j.$$

If the states s_i and s_j are r-inequivalent, there must be some input sequence $\mathbf{x} \in X^r$ such that $\omega_{(r)}(s_i, \mathbf{x}) \neq \omega_{(r)}(s_j, \mathbf{x})$. Then for all $x_k \in X$, $\omega_{(r+1)}(s_i, \mathbf{x}x_k) \neq \omega_{(r+1)}(s_j, \mathbf{x}x_k)$, so s_i and s_j are $(r+1)$-inequivalent also. *Hence we need examine for $(r+1)$-equivalence only those states already known to be r-equivalent.*

If the states s_i and s_j are r-equivalent and for all $x_k \in X$ their x_k-successors $\tau(s_i, x_k)$ and $\tau(s_j, x_k)$ are r-equivalent, then for all $\mathbf{x} \in X^r$,

$$\omega_{(r)}(\tau(s_i, x_k), \mathbf{x}) = \omega_{(r)}(\tau(s_j, x_k), \mathbf{x});$$

that is, for all input sequences $x_k \mathbf{x}$ of length $r+1$,

$$\omega_{(r+1)}(s_i, x_k \mathbf{x}) = \omega_{(r+1)}(s_j, x_k \mathbf{x}),$$

so s_i and s_j are $(r+1)$-equivalent. [Recall (1.11.1) and (1.11.2) here.] On the other hand, if for any $x_k \in X$, the x_k-successors of s_i and s_j are not r-equivalent,

$$\exists\, \mathbf{x} \in X^r \ni \omega_{(r)}(\tau(s_i, x_k), \mathbf{x}) \neq \omega_{(r)}(\tau(s_j, x_k), \mathbf{x}),$$

and hence

$$\omega_{(r+1)}(s_i, x_k\mathbf{x}) \neq \omega_{(r+1)}(s_j, x_k\mathbf{x}),$$

so s_i and s_j are not $(r+1)$-equivalent. In summary, we have the following theorem.

Theorem 2.16.1. *Two states s_i and s_j of $M = [S, X, Z, \tau, \omega]$ are $(r+1)$-equivalent iff s_i and s_j are r-equivalent and for all $x_k \in X$, $\tau(s_i, x_k)$ and $\tau(s_j, x_k)$ are r-equivalent.*

Let the partition of S determined by E_r be denoted by P_r. In particular, P_1 identifies states of M which have identical output vectors (see Section 2.12, Example 5) and may be written by inspection of the output portion of the defining table of M. We shall assume that P_1 has at least two blocks, since otherwise the states of M would all have precisely the same output behavior and hence all be equivalent. The partition of S determined by the relation of equivalence is denoted by P_E.

Theorem 2.16.1 shows that P_{r+1} is a *refinement* (Section 2.14) of P_r. Since a nontrivial refinement of P_r has at least one more block than P_r does, and since P_1 is assumed to have at least two blocks, at most $n-1$ distinct partitions $P_1, P_2, \ldots, P_{n-1}$ can appear, for if each P_{r+1} is a nontrivial refinement of P_r, we must have $P_{n-1} = 0$, the partition whose blocks are singletons (the only partition with as many as n blocks). If any $P_r = 0$, no two states of M are equivalent.

Now suppose that $P_r = P_{r+1}$; that is, $s_i E_r s_j$ iff $s_i E_{r+1} s_j$. By Theorem 2.16.1, $s_i E_{r+1} s_j$ implies that $s_i E_r s_j$ and

$$\forall x_k \in X, \quad \tau(s_i, x_k) E_r \tau(s_j, x_k).$$

Then, because $E_r = E_{r+1}$, we have also that

$$\forall x_k \in X, \quad \tau(s_i, x_k) E_{r+1} \tau(s_j, x_k),$$

so $s_i E_{r+2} s_j$ whenever $s_i E_{r+1} s_j$. That is, $P_{r+1} \leq P_{r+2}$. But $P_{r+1} \geq P_{r+2}$ always, so $P_{r+1} = P_{r+2}$. Similarly, $P_{r+2} = P_{r+3}$, and so on. We can now prove by induction that $P_r = P_{r+k}$ for all positive integers k.

If no $P_r = 0$, then refinements must cease before the partition 0 is reached, so for some $r < n-1$ we must have $P_r = P_{r+1} = P_{r+2} = \cdots > 0$. Then the states in a block of P_r are k-equivalent for all k, so P_r is, in fact, P_E, the equivalence partition on S. This implies the following theorem.

Theorem 2.16.2. *If $P_r = P_{r+1}$, then $P_r = P_E$.*

The two preceding theorems are the basis for a simple *algorithm for identifying equivalent states* of a given finite-state machine M:

Sec. 2.16 Identification of Equivalent States

STEP 1. By inspecting the output table of M, partition the set of states of M into blocks as large as possible, such that all the states in any one block have the same output vector. The resulting partition is P_1, whose blocks are the equivalence classes of E_1.

STEP 2. If the most recently constructed partition is P_r, construct by inspection of the transition table of M the refinement P_{r+1} of P_r. To do this, consider pairs of states s_i, s_j in the same block of P_r and place them in the same block of P_{r+1} iff for all $x_k \in X$, $\tau(s_i, x_k) \, E_r \, \tau(s_j, x_k)$; that is, iff for each $x_k \in X$, the x_k-successors of s_i and s_j are in a common block of P_r. (Be sure to list *every* s_i in the appropriate block of P_{r+1}.)

STEP 3. If $P_{r+1} \neq P_r$, repeat step 2. If $P_{r+1} = P_r$, put $P_r = P_E$ and stop.

Note that if $P_r = P_{r+1} = P_E$, then for each $x_k \in X$, all s_i of a given block of P_E have x_k-successors in the same block of P_E. That is, each x_k defines a function from P_E (a set of blocks) to itself.

For example, consider the machine defined by Table 2.16.1. The 1-equivalence classes are found by inspection of the output portion of the table:

$$P_1 = \{\overline{1, 2, 3, 4}; \overline{5, 6}\}.$$

TABLE 2.16.1

	τ		ω	
	0	1	0	1
1	2	3	0	0
2	2	2	0	0
3	5	6	0	0
4	4	2	0	0
5	4	2	0	1
6	1	4	0	1

Next we determine the 2-equivalence classes. We have $1 \, E_1 \, 2$. The 0-successors of states 1 and 2 are both state 2, and $2 \, E_1 \, 2$. The 1-successors of states 1 and 2 are states 3 and 2, respectively, and $3 \, E_1 \, 2$. Hence $1 \, E_2 \, 2$. Now consider $1 \, E_1 \, 3$. The 0-successors of 1 and 3 are 2 and 5 and $2 \, \not{E_1} \, 5$, so $1 \, \not{E_2} \, 3$. We have next $1 \, E_1 \, 4$, $2 \, E_1 \, 4$, and $3 \, E_1 \, 2$, so $1 \, E_2 \, 4$. Also, $5 \, E_1 \, 6$, $4 \, E_1 \, 1$, and $2 \, E_1 \, 4$, so $5 \, E_2 \, 6$. Hence

$$P_2 = \{\overline{1, 2, 4}; \overline{3}; \overline{5, 6}\}.$$

To find P_3, we note that $1 \, E_2 \, 2$, but for the 1-successors of 1 and 2, respectively, we have $3 \, \not{E_2} \, 2$, so $1 \, \not{E_3} \, 2$. However, since $2 \, E_2 \, 4$, $2 \, E_2 \, 4$, and

2 E_2 2, we have 2 E_3 4. Also 5 E_2 6, 4 E_2 1, and 2 E_2 4, so 5 E_3 6. Hence

$$P_3 = \{\bar{1}; \overline{2,4}; \bar{3}; \overline{5,6}\}.$$

In the same way we find that

$$P_4 = \{\bar{1}; \overline{2,4}; \bar{3}; \bar{5}; \bar{6}\} = P_5 = P_E.$$

Hence states 2 and 4 are the only equivalent states.

If the number of states or inputs is large or if successive refinements involve separating blocks into several smaller blocks, an alternative second step [Gill (1962)] makes it easy to keep track of details:

STEP 2'. For the most recently determined partition P_r, construct the P_r**-table** of M by listing the states (rows) in the same order as that in which they appear in P_r, listing the next states as usual and listing in place of outputs the names B_{rj} of the blocks of P_r to which the recorded next states belong. Partition each block of P_r into maximal subblocks whose states all have the same **next-block vector**. The resulting partition is P_{r+1}.

In the case of our example, the P_2-table is Table 2.16.2. From this table, we have at once that

$$P_3 = \{\bar{1}; \overline{2,4}; \bar{3}; \overline{5,6}\}.$$

TABLE 2.16.2. P_2-table for example of Table 2.16.1.

		τ		Next Block		
		0	1	0	1	
B_{21}	1	2	3	B_{21}	B_{22}	$\}B_{31}$
	2	2	2	B_{21}	B_{21}	$\}B_{32}$
	4	4	2	B_{21}	B_{21}	
B_{22}	3	5	6	B_{23}	B_{23}	$\}B_{33}$
B_{23}	5	4	2	B_{21}	B_{21}	$\}B_{34}$
	6	1	4	B_{21}	B_{21}	

Actually, singleton blocks of P_r need not be listed in the P_r-table, but they must of course be recorded in P_E. Also, simpler names may be used for the blocks to save writing, as in the P_E-table for M, which is given in Table 2.16.3. Here, for later reference, it is useful to list all states, whether or not they appear in singleton blocks, and the outputs as well.

Sec. 2.17 Minimal-State Machines

TABLE 2.16.3. P_E-table for example of Table 2.16.1.

		τ		Next Block		Output	
		0	1	0	1	0	1
B_1	1	2	3	B_2	B_3	0	0
B_2	$\begin{cases} 2 \\ 4 \end{cases}$	2 4	2 2	B_2 B_2	B_2 B_2	0 0	0 0
B_3	3	5	6	B_4	B_5	0	0
B_4	5	4	2	B_2	B_2	0	1
B_5	6	1	4	B_1	B_2	0	1

2.17 Minimal-State Machines

The discussion in this section will further emphasize the importance of the preceding algorithm. Loosely, we want to say machine M_1 *covers* machine M_2, written $M_1 \geqslant M_2$, iff M_1 can do everything M_2 can do. The precise definition is as follows. We say that $M_1 = [S_1, X, Z, \tau_1, \omega_1]$ **covers** $M_2 = [S_2, X, Z, \tau_2, \omega_2]$ iff there exists a function $\varphi: S_2 \to S_1$ such that for all $s \in S_2$ and for all $\mathbf{x} \in X^*$, $\omega_1(\varphi(s), \mathbf{x}) = \omega_2(s, \mathbf{x})$.

The machines M_1 and M_2 are called **equivalent**, written $M_1 \equiv M_2$, iff $M_1 \geqslant M_2$ and $M_2 \geqslant M_1$. As we shall see, two machines need not be identical in order to have precisely the same capabilities.

Finally, we say that a finite-state machine $M = [S, X, Z, \tau, \omega]$ is a **minimal-state machine** iff for all $M' = [S', X, Z, \tau', \omega']$ such that $M' \geqslant M$, we have $|S'| \geqslant |S|$. Informally, a machine is a minimal-state machine iff no machine with fewer states can do the same job. A minimal-state machine equivalent to a given machine M may be found by the following *algorithm*.

Let $P_E = \{B_1, B_2, \ldots, B_q\} > 0$ be the equivalence partition of a machine $M = [S, X, Z, \tau, \omega]$, obtained by the algorithm of the preceding section. Form the P_E-table of M, easily done even if the intermediate P_r-tables were not constructed. From the P_E-table of M we construct the defining table of a machine $M' = [S', X, Z, \tau', \omega']$, called the **reduced form** of M, whose state set is $S' = \{B_1, B_2, \ldots, B_q\}$. The next-state vector in the B_j-row of M' is the next-block vector that is common to all the states in the block B_j of the P_E-table. The output vector in the B_j-row is the output vector common to all the states of the block B_j. A consequence of this construction is that the functions $\tau': S' \times X \to S'$ and $\omega': S' \times X \to Z$, defined by the table of M', are related to τ and ω as follows. If $s_i \in B_{\alpha_i}$, $s_j \in B_{\alpha_j}$, $\tau(s_i, x_k) = s_j$ and $\omega(s_i, x_k) = z_j$, then $\tau'(B_{\alpha_i}, x_k) = B_{\alpha_j}$ and $\omega'(B_{\alpha_i}, x_k) = z_j$.

This algorithm actually may yield any one of $q!$ reduced forms of M, depending on the order in which the states are numbered. It is customary to call the one at hand "the" reduced form of M.

In the case of the example of the preceding section, the algorithm yields Table 2.17.1 for M'.

TABLE 2.17.1 Reduced form for example of Table 2.16.1.

	τ'		ω'	
	0	1	0	1
B_1	B_2	B_3	0	0
B_2	B_2	B_2	0	0
B_3	B_4	B_5	0	0
B_4	B_2	B_2	0	1
B_5	B_1	B_2	0	1

To establish the significance of the algorithm, we first prove the following theorem.

Theorem 2.17.1. *The reduced form M' of M is equivalent to M.*

Proof. We show that $M' \geq M$ and that $M \geq M'$. Consider first this function:

$$\varphi: S \to S' \quad \text{defined by} \quad \varphi(s_j) = B_{\alpha_j}, \quad \text{where } s_j \in B_{\alpha_j}.$$

Thus φ maps each state s_j onto the block B_{α_j} of P_E to which it belongs.
Now let $\mathbf{x} = x_1 x_2 \cdots x_r \in X^*$, let $s_1 \in S$, and let

$$\tau(s_1, x_1) = s_2, \tau(s_2, x_2) = s_3, \ldots, \tau(s_r, x_r) = s_{r+1}.$$

Then, by the definition of M',

$$\tau'(\varphi(s_1), x_1) = \varphi(s_2), \tau'(\varphi(s_2), x_2) = \varphi(s_3), \ldots, \tau'(\varphi(s_r), x_r) = \varphi(s_{r+1}).$$

Hence, by (1.11.1), (1.11.2), and the definition of M',

$$\omega'(\varphi(s_1), \mathbf{x}) = \omega'(\varphi(s_1), x_1) \omega'(\varphi(s_2), x_2) \cdots \omega'(\varphi(s_r), x_r)$$
$$= \omega(s_1, x_1) \omega(s_2, x_2) \cdots \omega(s_r, x_r)$$
$$= \omega(s_1, \mathbf{x}),$$

and this holds for all $s_1 \in S$, all $\mathbf{x} \in X^*$. Hence $M' \geq M$.

Now consider the function $\varphi': S' \to S$ defined by $\varphi'(B_j) = s_{\beta_j}$, where s_{β_j} is any particular member of B_j. By an argument that closely parallels the one just completed, we conclude that $M \geq M'$. That is, $M' \equiv M$. ∎

Sec. 2.18 *Distinguishing Sequences for States*   77

Theorem 2.17.2. *A finite-state machine* $M = [S, X, Z, \tau, \omega]$ *is minimal-state iff it has no distinct equivalent states.*

Proof. If M has two or more states that are equivalent, by the preceding algorithm we can replace them by a single state and have an equivalent machine with fewer states, so M is not a minimal-state machine and the condition is necessary.

Suppose now that M has no two states equivalent and that $M' = [S', X, Z, \tau', \omega']$ covers M. By the definition of covering, there is a function $\varphi: S \to S'$ such that for all $s \in S$ and for all $\mathbf{x} \in X^*$, $\omega(s, \mathbf{x}) = \omega'(\varphi(s), \mathbf{x})$. We prove that φ is one-one. If s_1 and s_2 in S satisfy $\varphi(s_1) = \varphi(s_2)$, then for all $\mathbf{x} \in X^*$ we have

$$\omega(s_1, \mathbf{x}) = \omega'(\varphi(s_1), \mathbf{x}) = \omega'(\varphi(s_2), \mathbf{x}) = \omega(s_2, \mathbf{x}),$$

so s_1 and s_2 are equivalent. Since M has no distinct equivalent states, $s_1 = s_2$ and φ is one-one. The fact that there is a one-one function from S to S' implies that $|S| \leq |S'|$ for any M' that covers M (Exercise 1.10.9). Hence M is a minimal-state machine. ∎

Since the reduced form of a machine has no distinct equivalent states, we have reached our goal:

Theorem 2.17.3. *The reduced form of a finite-state machine M is a minimal-state machine.*

Since all $q!$ reduced forms of M are equivalent to M, they are all equivalent to each other, and all have the minimal number of states, namely q.

The procedure for finding the reduced form of M is called the **minimization procedure**.

2.18 Distinguishing Sequences for States

If two states s_1 and s_2 of $M = [S, X, Z, \tau, \omega]$ appear in distinct blocks of P_{r+1}, they are not $(r+1)$-equivalent. Hence there exists $\mathbf{x} \in X^{r+1}$ such that $\omega_{(r+1)}(s_i, \mathbf{x}) \neq \omega_{(r+1)}(s_j, \mathbf{x})$. Any such sequence $\mathbf{x}$ is a **distinguishing sequence** for s_1 and s_2 and is said to **distinguish** s_1 and s_2.

Let s_1 and s_2 belong to distinct blocks of P_{r+1} but to a common block of P_r. Then $s_1 E_r s_2$, so $r + 1$ is the length of a *shortest* distinguishing sequence for s_1 and s_2 (there may be more than one of this length). Such a sequence may be found by the following *algorithm*, which uses Theorem 2.16.1 repeatedly.

STEP ·1. Select any x_1 such that the x_1-successors $\tau(s_1, x_1) = s_{11}$ and $\tau(s_2, x_1) = s_{21}$ are in distinct blocks of P_r. [Since $s_1 E_r s_2$, if no such x_1 were to exist, s_1 and s_2 would be $(r+1)$-equivalent, contrary to fact.]

STEP 2. Select any x_2 such that the x_2-successors of s_{11} and s_{21} are in distinct blocks of P_{r-1}. (Absence of any such x_2 would mean $s_{11} E_r s_{21}$, a contradiction, since $s_1 E_r s_2$ implies that $s_{11} E_{r-1} s_{21}$.)

STEP 3. Complete a sequence $x_1 x_2 \cdots x_r$ of inputs that map the states s_1 and s_2, s_{11} and s_{21}, and their successors, respectively, into distinct blocks of the partitions $P_r, P_{r-1}, \ldots, P_2, P_1$. (The reasoning used in step 2 also applies at each following step.)

STEP 4. Select a final input x_{r+1} that will distinguish the two states, obtained in step 3, that appear in distinct blocks of P_1. (Such an x_{r+1} exists by the definition of P_1.) Then $x_1 x_2 \cdots x_r x_{r+1}$ is the required distinguishing sequence for x_1 and x_2.

In the case of the example of Section 2.16, states 5 and 6 may be distinguished by an input sequence of length 4. The work may be arranged as follows.

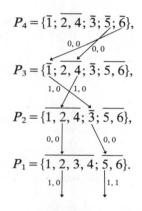

$$\begin{array}{rcccc}
\text{input:} & 0 & 1 & 0 & 1 \\
\text{output} \begin{cases} \text{initial state 5:} & 0 & 0 & 0 & 0 \\ \text{initial state 6:} & 0 & 0 & 0 & 1 \end{cases}
\end{array}$$

2.19 Exercises

1. Show that both r-equivalence and equivalence of states, as well as equivalence of machines, are indeed equivalence relations.

2. In the definition of a partial ordering on a set P, equality may be replaced by any equivalence relation on P. Show that the covering relation for machines is a partial ordering in this larger sense.

*3. For each of the following machines, find the minimum-state equivalent machine.

Sec. 2.19 Exercises

(a)

	τ		ω	
	0	1	0	1
s_0	s_3	s_0	1	0
s_1	s_2	s_2	0	1
s_2	s_0	s_3	0	1
s_3	s_0	s_2	1	0

(b)

	τ		ω	
	0	1	0	1
σ_0	σ_5	σ_2	0	0
σ_1	σ_4	σ_6	1	0
σ_2	σ_5	σ_1	1	1
σ_3	σ_6	σ_4	0	1
σ_4	σ_1	σ_3	1	0
σ_5	σ_6	σ_5	0	0
σ_6	σ_3	σ_1	0	1

(c)

	τ			ω		
	a	b	c	a	b	c
1	1	2	5	0	1	1
2	2	1	6	1	0	0
3	7	2	5	0	1	1
4	4	7	6	1	0	0
5	2	5	1	1	1	1
6	4	6	3	1	0	0
7	3	4	5	0	1	1

(d)

	τ		ω	
	0	1	0	1
1	8	2	0	0
2	9	4	0	0
3	9	1	0	0
4	8	3	0	0
5	8	7	0	0
6	9	6	0	0
7	10	5	0	1
8	2	10	0	1
9	4	10	0	1
10	9	1	0	1

*4. In Exercise 3(d) find shortest distinguishing sequences for (a) states 9 and 10; (b) states 6 and 7; (c) states 4 and 5.

5. Minimize the machine whose state diagram appears in Figure 2.19.1.

6. Devise an algorithm that will give all shortest distinguishing sequences for two specified states of a given machine M; then apply it to Exercise 4.

7. A finite-state machine is to examine serially pairs (p_i, q_i) of digits of two given three-digit binary numbers $p_2p_1p_0$ and $q_2q_1q_0$, then output a 1 if $p_2p_1p_0 > q_2q_1q_0$. Otherwise, it is to output 0. The machine is to reset after each pair of three-digit numbers has been compared. Draw the state diagram of the machine and minimize it. Repeat for the case in which the output is to be 1 iff $p_2p_1p_0 \geq q_2q_1q_0$. These machines are examples of a class of machines called **comparators**.

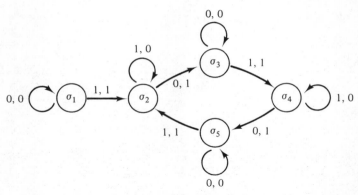

FIGURE 2.19.1. *Machine for Exercise 5*

*8. States s of $M = [S, X, Z, \tau, \omega]$ and s' of $M' = [S', X, Z, \tau', \omega']$ are said to be **equivalent**, written $s\,E\,s'$, iff for all $\mathbf{x} \in X^*$, $\omega(s, \mathbf{x}) = \omega'(s', \mathbf{x})$, that is, iff they agree in their responses to any given input sequence. Prove that $M \equiv M'$ iff (1) $\forall\, s \in S\ \exists\, s' \in S' \ni s'\,E\,s$, and (2) $\forall\, s' \in S'\ \exists\, s \in S \ni s\,E\,s'$. What conclusion about equivalent states may be drawn if $M' \geqslant M$?

*9. Two machines $M = [S, X, Z, \tau, \omega]$ and $M' = [S', X, Z, \tau', \omega']$ are said to be **isomorphic** (or **state-isomorphic**) iff there exists a bijection $\varphi: S \to S'$ such that for all $s \in S$, $\varphi(s)\,E\,s$. Assume that M and M' are minimal-state machines, and prove that the following statements are equivalent (all are true or all are false).

(a) M and M' are isomorphic.
(b) M and M' are equivalent.
(c) There is a bijection $\varphi: S \to S'$ such that for all $x_k \in X$ and all $s \in S$,
$$\tau'(\varphi(s), x_k) = \varphi(\tau(s, x_k)), \qquad \omega'(\varphi(s), x_k) = \omega(s, x_k).$$

10. Under what conditions will two minimal-state initialized machines (Section 2.1) be isomorphic? How can you use this to determine whether or not two given minimal-state machines are isomorphic?

11. Show that the following machines are isomorphic and specify the bijection that exhibits the pairs of equivalent states.

	τ		ω	
	0	1	0	1
A	B	D	0	0
B	D	C	0	1
C	A	B	1	0
D	A	C	1	0

	τ'		ω'	
	0	1	0	1
α	γ	β	0	1
β	δ	α	1	0
γ	δ	β	1	0
δ	α	γ	0	0

*12. The function $\varphi: S \to S'$ is a **homomorphism** from machine $M = [S, X, Z, \tau, \omega]$ to machine $M' = [S', X, Z, \tau', \omega']$ if, for all $x_k \in X$ and all $s \in S$,

$$\tau'(\varphi(s), x_k) = \varphi(\tau(s, x_k)) \quad \text{and} \quad \omega'(\varphi(s), x_k) = \omega(s, x_k).$$

Show that if φ is such a homomorphism, then:
 (a) State s of M is equivalent to state $\varphi(s)$ of M'.
 (b) M' covers M.

13. It is desired to drive a machine from its initial state to a specified terminal state. Devise an algorithm that will produce all input sequences that will set the machine in the specified final state, regardless of the initial state. (Such a sequence is called a synchronizing sequence; for a given machine, there may exist no such sequence.) Here the output sequence is of no interest and is not observed.

For a treatment of various specialized forms of input sequences, see F. C. Hennie (1968, Chapter 3).

2.20 Undirected Graphs

We have defined directed graphs (*digraphs*), and we saw in Section 2.4 how the simple directed graphs on a set of vertices A are equivalent to the set of binary relations on A. We also discussed connectivity and reachability, and we saw in Section 2.4 how these properties of graphs can be tested by using the relation matrices of the corresponding relations.

Hasse diagrams of posets (Section 2.8) are also directed graphs, of the corresponding covering relations; arrowheads are not specifically drawn on the edges, but all edges are directed downward. State diagrams of finite-state machines (Section 1.11) are also directed graphs; they are labeled (and therefore not *simple*). Notions of connectivity and reachability are important for finite-state machines also.

An **undirected graph** (or simply **graph**) consists of a nonempty set V, whose elements are called **vertices** (or **nodes**), and a set E of (unordered) pairs of elements of V, called **edges** (or **lines** or **branches**). We can picture a finite graph by representing each vertex with a point, and each edge $\{v_1, v_2\}$ with a line between points v_1 and v_2. Figures 2.20.1(a) and (b) both picture

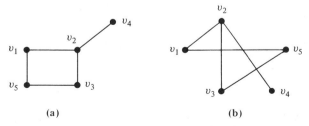

FIGURE 2.20.1. *Representations of the Same Graph*

the graph $\{\{v_1, v_2\}, \{v_1, v_5\}, \{v_2, v_3\}, \{v_2, v_4\}, \{v_3, v_5\}\}$. If $\{v_i, v_j\}$ is an edge, we say that vertices v_i and v_j are **adjacent**; in Figure 2.20.1, v_1 is adjacent to v_2 but not to v_3.

In general, we may allow a graph to have more than one edge $\{v_i, v_j\}$ (*parallel edges*), or a *loop* $\{v_i, v_i\}$ from vertex v_i to itself. (Some authors use the term *multigraph* for such graphs and reserve the term *graph* for multigraphs without loops or parallel edges. We shall call a graph without loops or parallel edges a *simple graph*.) Figure 2.20.2 illustrates parallel edges $\{v_3, v_4\}$, a loop at v_5, and two loops at v_2. Vertex v_6, which is not a **terminal** of any edge, is called an **isolated vertex**; v_7, which is a terminal of only one edge, is a **pendant vertex**. In general, the **degree** $d(v)$ of vertex v is the number of times v is a terminal of an edge, so $d(v_6) = 0$, $d(v_7) = 1$, $d(v_3) = 4$, $d(v_2) = 7$. [In a digraph, each vertex v has an *in-degree* $d^+(v)$ and an *out-degree* $d^-(v)$.]

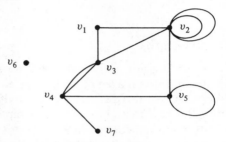

FIGURE 2.20.2. *Special Edges and Vertices*

A *directed* graph is **symmetric** if it is the digraph of a symmetric relation; that is, whenever there is an arrow from v_i to v_j, there is also an arrow from v_j to v_i. The set of all undirected graphs on a set V of vertices may be made to correspond to the set of all symmetric directed graphs on V, by replacing each edge in each graph by two oppositely directed arrows with the same terminals. Figure 2.20.3 illustrates the symmetric digraph corresponding to

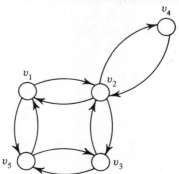

FIGURE 2.20.3. *Symmetric Digraph*

Sec. 2.20 Undirected Graphs

Figure 2.20.1(a). In this way, the theory of undirected graphs is a special case of the theory of directed graphs; but many results and applications are, as we shall see, peculiar to the undirected case, so each should be treated separately.

Given any digraph, we obtain the *associated* undirected graph by changing all the arrows to (undirected) edges. On the other hand, any given undirected graph G may be made a digraph by changing the edges to arrows; the digraph is then an **orientation** of G. Each arrow has two possible directions, so a simple undirected graph with e edges has 2^e different orientations. (What if the graph is not simple?)

From now on, "graph" will mean *undirected graph*. All graphs considered are *finite* (that is, have finitely many vertices and finitely many edges). Graph $G = [V, E]$ has set V of vertices and set E of edges. If $e_i \in E$, we write $t(e_i) = \{v_j, v_k\}$, where v_j and v_k are the terminals of e_i; t is the **terminal function** of G. Graphs $G_1 = [V_1, E_1]$ and $G_2 = [V_2, E_2]$ are **isomorphic** if there are one-one correspondences $\nu: V_1 \to V_2$ and $\varepsilon: E_1 \to E_2$, such that if $e_i \in E_1$, $v_j, v_k \in V_1$, then

(2.20.1) $t(e_i) = \{v_j, v_k\}$ iff $t(\varepsilon(e_i)) = \{\nu(v_j), \nu(v_k)\}$.

That is, G_1 and G_2 are isomorphic iff vertices and edges correspond one-one and the vertex–edge incidence relationship is preserved. Pictorially, isomorphic graphs amount to different ways of drawing the same graph. The graphs of pairs (a) and (b) in Figure 2.20.4 are isomorphic, but the graphs of pair (c) are not (in the first graph of this pair, the degree-3 vertex is adjacent to two pendant vertices, and in the second, the degree-3 vertex is adjacent to only one pendant vertex).

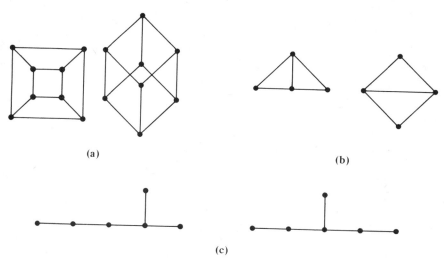

FIGURE 2.20.4. *Isomorphic and Nonisomorphic Graphs*

In general, it is an important and difficult problem to determine when two given graphs are isomorphic. Efficient computer algorithms for some special classes of graphs have been found; see Section 11-7 of Deo (1974) for discussion and references.

A **subgraph** of a graph G is any graph obtained from G by a deletion of edges and/or vertices from G (when a vertex v is deleted, all edges terminating at v must also be deleted). In Figure 2.20.5, all graphs (a) to (i) are subgraphs of (a).

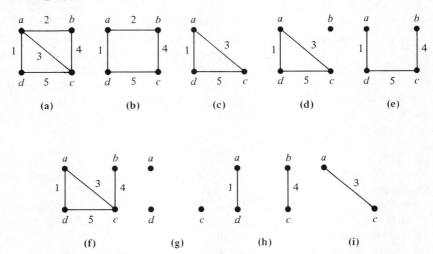

FIGURE 2.20.5. *Subgraphs of a Graph*

Many special kinds of subgraphs are important; unfortunately, terminology in the literature is not standard. By a **walk** we shall mean a finite sequence of edges of form

(2.20.2) $\quad \{v_0, v_1\}, \{v_1, v_2\}, \{v_2, v_3\}, \{v_3, v_4\}, \ldots, \{v_{n-1}, v_n\}$

(successive edges are incident at a vertex). The **length** of a walk is the number of edges it contains. The walk (2.20.2) is **closed** if $v_0 = v_n$; otherwise, it is **open**. In either case, it has length n. A **simple walk** is a walk in which no edge appears more than once. In Figure 2.20.5(a), walks 1, 2, 4, 5, 1, 3 and 1, 2, 4 are both open walks from d to c, but only the second is simple. Graphs (a), (c), (e), (f), and (i) of Figure 2.20.5 can be interpreted as simple, open walks.

Assume that $n \geq 1$ and that vertices $v_0, v_1, \ldots, v_{n-1}, v_n$ of walk (2.20.2) are distinct; then (2.20.2) is called a **path** from v_0 to v_n. If $v_0, v_1, \ldots, v_{n-1}$ are distinct but $v_0 = v_n$, then (2.20.2) is called a **circuit**. Graphs (e) and (i) of Figure 2.20.5 are paths, and (b) and (c) are circuits. Loops are circuits of length 1.

We say that vertices v_i and v_j are **connected** if $v_i = v_j$ or there is a path from v_i to v_j. Connectedness is obviously an equivalence relation on the set V of vertices; the equivalence classes are called **components**, and the graph is **connected** iff there is only one component. In Figure 2.20.5, graph (g) has three components, graphs (d) and (h) have two components each, and the other graphs are connected. The components of a graph G may be considered subgraphs of G.

A vertex v of a connected graph G is a **cut point** if removal of v (together with all edges terminating at v) results in a disconnected subgraph of G. Circuits never have cut points, but paths of length at least 2 always do. In Figure 2.20.5, graph (e) has cut points d and c, while graph (f) has cut point c; no other (connected) graph has a cut point in Figure 2.20.5. If the vertices of a graph represent cities and the edges represent communications links (highways or telephone lines, for example), then cut points are undesirable; trouble at that one point will mean that some cities cannot communicate with others.

A graph $G = [V, E]$ is **bipartite** if $V = V_1 \cup V_2$, $V_1 \neq \emptyset$, $V_2 \neq \emptyset$, $V_1 \cap V_2 = \emptyset$, such that every edge has one terminal in V_1 and one in V_2. (There may be isolated vertices in V_1, or V_2, or both.) The marriage relationship in a set of men and women ($\{v_i, v_j\}$ is an edge iff v_i and v_j are married) is a simple example.

Two (or more) subgraphs of a graph are **edge-disjoint** if they have no edge in common; subgraphs (h) and (i) are edge-disjoint subgraphs of graph (a) in Figure 2.20.5. Distinct components of a disconnected graph are not only edge-disjoint, but even **vertex-disjoint**; that is, they have no vertices in common.

2.21 Trees

A **tree** is a connected graph without any circuits. All the trees (up to isomorphism) with at most five vertices are shown in Figure 2.21.1.

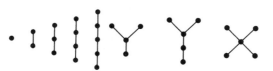

FIGURE 2.21.1. *Trees with at Most Five Vertices*

There are several equivalent conditions for a graph to be a tree, as follows.

Theorem 2.21.1. *The following are equivalent, for a graph $G = [V, E]$:*
(a) *G is a tree.*
(b) *There is one and only one path between any pair of vertices of G.*

(c) *G is connected, but removal of any edge leaves a disconnected subgraph (consisting of two components, each of which is a tree).*
(d) *G has no circuits and $|V| = |E| + 1$.*
(e) *G is connected and $|V| = |E| + 1$.*
(f) *G has no circuits, and if any two nonadjacent vertices are joined by an edge to form a graph $\bar{G}$, then $\bar{G}$ has exactly one circuit.*

Proof. (a)$\Rightarrow$(b). Since G is connected, there is at least one path; two distinct paths would yield a circuit.

(b)$\Rightarrow$(c). Clearly, G is connected; if edge $\{v_i, v_j\}$ is removed, then the resulting subgraph cannot be connected, as otherwise there would be two paths from v_i to v_j. From every vertex of G, there is a path to v_i in G. Removal of the edge $\{v_i, v_j\}$ breaks only the paths that contain this edge. This removal of $\{v_i, v_j\}$ creates a subgraph G such that every vertex is connected by a path to v_i or to v_j, but not both. Thus, G consists of two components, each of which is connected. Neither component has any circuits, since such circuits would also be circuits of G, thus contradicting (b).

(c)$\Rightarrow$(d). Clearly, G has no circuits. The proof that $|V| = |E| + 1$ is by induction on $|V|$; true because $|E| = 0$ (no circuits!) if $|V| = 1$, and clearly true if $|V| = 2$. Now remove one edge from G. By (c), we obtain two components which are connected subgraphs $G_1 = [V_1, E_1]$ and $G_2 = [V_2, E_2]$; deletion of an edge from a component disconnects the component, as it would have disconnected G, so the components therefore satisfy (c). Hence by the induction hypothesis, $|V_1| = |E_1| + 1$ and $|V_2| = |E_2| + 1$, so

$$|V| = |V_1| + |V_2| = |E_1| + 1 + |E_2| + 1 = |E| + 1.$$

(d)$\Rightarrow$(e). If G were disconnected, with components (necessarily trees) $G_1, \ldots, G_m$, we could choose vertices $v_1, \ldots, v_m$ in components $G_1, \ldots, G_m$, respectively; join them with new edges $\{v_1, v_2\}, \{v_2, v_3\}, \ldots, \{v_{m-1}, v_m\}$; and obtain a tree $\bar{G} = [V, \bar{E}]$, $|V| = |E| + 1 = (|\bar{E}| - (m-1)) + 1 = |\bar{E}| - (m-2) \leq |\bar{E}|$, contradicting the fact that (a)$\Rightarrow$(d) for $\bar{G}$.

(e)$\Rightarrow$(f). If G has a circuit C containing k vertices, say, then that circuit contains k edges. Since G is connected, we can join the other vertices to C, one by one, using edges in G; each vertex requires at least one additional edge, giving $|E| \geq |V|$, a contradiction, so G has no circuit.

Now we know that G is a tree, so (a) to (e) hold. If we add edge $\{v_i, v_j\} = \{v_j, v_i\}$ to form $\bar{G}$, and P is the unique path from v_i to v_j in G, then $P, \{v_j, v_i\}$ is a circuit in $\bar{G}$. G is circuitless, so any second circuit in $\bar{G}$ would have to use $\{v_j, v_i\}$ also; denote such a circuit by $Q, \{v_j, v_i\}$. Then Q must be a path from v_i to v_j, and the existence of paths P and Q from v_i to v_j contradicts (b).

(f)$\Rightarrow$(a). If G were not connected, joining vertices in separate components would not yield a circuit. ∎

Sec. 2.21 Trees

A tree with only one vertex is necessarily an isolated vertex, and is called a **degenerate tree**. Other trees are **nondegenerate**.

Lemma 2.21.1. *Any nondegenerate tree has at least two pendant vertices.*
Proof. Each edge has two terminals, so any graph $G = [V, E]$ satisfies

(2.21.1) $$2|E| = \sum_{v \in V} d(v),$$

where $d(v)$ is the degree of $v \in V$. In the case of a tree, all $d(v) \geq 1$ and $2|E| = 2|V| - 2$, so the equation (2.21.1) implies that at least two v's satisfy $d(v) = 1$. ∎

A pendant vertex in a tree is often called a **leaf**. Other picturesque terminology is common; edges are called **branches**, and a graph whose components are trees is called a **forest**.

In any connected graph $G = [V, E]$, the **distance** $d(v_i, v_j)$ between vertices v_i and v_j is the length of the shortest path from v_i to v_j; $d(v_i, v_i) = 0$. The usual properties of a distance function hold:

(2.21.2) $$d(v_i, v_j) = 0 \quad \text{iff} \quad v_i = v_j,$$
$$d(v_i, v_j) = d(v_j, v_i), \quad d(v_i, v_k) \leq d(v_i, v_j) + d(v_j, v_k)$$

(the easy proof is Exercise 2.22.20). Distances are easy to find in a tree, because of Theorem 2.21.1(b). For example, in the tree of Figure 2.21.2, $d(c, l) = 3$ and $d(i, p) = 7$.

The **eccentricity** of the vertex v_i in the graph $G = [V, E]$ is $\mathscr{E}(v_i) = \max_{v_j \in V} d(v_i, v_j)$; in Figure 2.21.2, $\mathscr{E}(l) = 5$ and $\mathscr{E}(b) = 6$. It is natural to say that v_i is a **center** of a connected graph, if its eccentricity is a minimum. A fast method for finding the center(s) of a tree is suggested by the proof of the following interesting theorem.

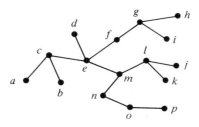

FIGURE 2.21.2. *A Tree*

Theorem 2.21.2. *Every tree $T = [V, E]$ has one or two centers.*
Proof. The result is trivial if $|V| = 1$ or 2; so we use induction on $|V|$ and assume that $|V| > 2$. By Lemma 2.21.1, T has at least two leaves, which certainly cannot be centers (if v_i is a leaf and $\{v_i, v_j\}$ an edge, then $\mathscr{E}(v_i) =$

$\mathscr{E}(v_j)+1$, since $d(v_i, v_k) = d(v_j, v_k)+1$ for all $v_k \neq v_i$). Let T' be the subtree of T obtained by deleting all leaves (and all edges terminating in them). All vertices of T' have eccentricity in T' one less than their eccentricity in T (why?), so centers of T' are precisely the centers of T; by induction, we know that T' has one or two centers. ∎

When finding the center(s) of a tree, we can quickly find T', then $T'' = (T')'$, and so on, until we reach a tree with at most two vertices; its vertices are the centers. For the tree of Figure 2.21.2, the steps to find the centers are shown in Figure 2.21.3. The two centers are e and m.

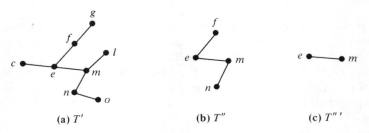

FIGURE 2.21.3. *Finding the Centers of Figure 2.21.2*

A **rooted tree** is a tree with a distinguished vertex, called the **root**. Any vertex of a tree may be designated the root, but we shall customarily draw a rooted tree with the root at the top, and circled. We see in Figure 2.21.1 that there are only *three* nonrooted trees with five vertices, but we see in Figure 2.21.4 that there are *nine* 5-vertex rooted trees.

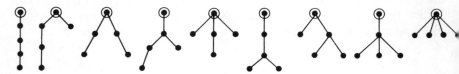

FIGURE 2.21.4. *Rooted Trees with Five Vertices*

Decision trees (or *sorting trees*) are labeled rooted trees which occur often in applications, especially in computer programming and computer algorithms. The root represents a starting point (that's why it is drawn at the top), later vertices represent later decision points, and one proceeds downward through the tree, choosing an edge at each step according to observed data. A particular case is the *binary tree*, in which the root has degree 2 and all other vertices have degree 1 (are leaves) or 3 (are decision points). For example, if a student must pass two of three tests to pass a course, the (binary) decision tree appears in Figure 2.21.5, with label P denoting "pass" and label F denoting "fail." If the student passes both of the first two tests, or

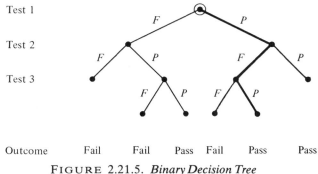

Outcome   Fail   Fail   Pass   Fail   Pass   Pass

FIGURE 2.21.5. *Binary Decision Tree*

fails both, there is no decision needed from test 3. We have emphasized the path to the outcome for a student who passes the first test, fails the second, and passes the third: the student passes the course.

In a binary tree, each decision has only two possible outcomes (YES or NO, TRUE or FALSE, 0 or 1, etc.). A decision tree, or any rooted tree, has a natural orientation as a digraph, all edges becoming arrows directed away from the root; but the arrowheads are customarily not drawn.

2.22 Exercises

1. Find all isolated vertices, pendant vertices, parallel edges, loops, degrees of vertices, and circuits in Figure 2.22.1.

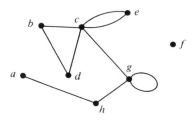

FIGURE 2.22.1. *Graph for Exercise 1*

2. Which pairs of graphs in Figure 2.22.2 are isomorphic?

3. Prove that in a connected graph with at least three vertices at least two of the vertices are not cut points.

4. Prove that if a connected graph has no cut points, then every edge is part of a circuit.

5. Prove that a graph is a circuit iff it is connected and every vertex has degree 2.

6. Show that a graph is connected iff the corresponding symmetric digraph is strongly connected.

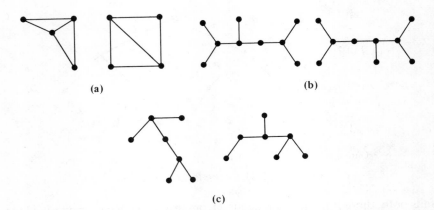

FIGURE 2.22.2. *Graphs for Exercises 2 and 15*

7. Show that if $G = [V, E]$ is connected, then $|V| \leq |E| + 1$.

*8. Show that in any graph, the number of vertices of odd degree is even.

9. Prove that a connected graph G remains connected after removal of an edge e iff e is in a circuit in G.

10. Show that a vertex v in a connected graph G is a cut point iff there are vertices v_i and v_j such that every path from v_i to v_j passes through v.

11. State an algorithm for finding a shortest path from vertex v_1 to vertex v_2 in a connected graph G.

12. Draw all nonisomorphic simple graphs with exactly four vertices.

13. Find two nonisomorphic graphs $G_1 = [V_1, E_1]$ and $G_2 = [V_2, E_2]$ such that $|V_1| = |V_2|$, $|E_1| = |E_2|$, and for each $d \in \mathbb{N}$, G_1 and G_2 have the same number of vertices of degree d.

14. (a) Find all distinct (that is, pairwise nonisomorphic) trees with exactly six vertices.

(b) Repeat part (a), assuming that the trees are rooted.

15. Find the centers of the trees in Figure 2.22.2(b) and (c) and Figure 2.22.3.

16. Describe two conceivably practical uses of decision trees not specifically discussed in the text.

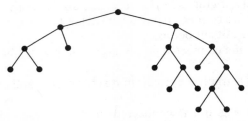

FIGURE 2.22.3. *Graph for Exercise 15*

17. Draw your family tree. (One strategy: Use yourself as root and go back as far as you can.)

***18.** (a) Show that if n is the number of vertices in a binary tree, then n is odd. (Use Exercise 8.)

(b) If the binary tree T has n vertices, find the number of leaves in T and the number of vertices of degree 3 in T.

19. (A. Cayley, 1857) In chemistry, connected graphs can be used to represent saturated hydrocarbons of the paraffin series $C_k H_{2k+2}$. A carbon atom is represented by a vertex of degree 4, and a hydrogen atom is represented by a vertex of degree 1. Questions (a) to (d) deal with the graph of $C_k H_{2k+2}$.

(a) Find the total number of vertices.

(b) Find the total number of edges.

(c) Conclude that the graph must be a tree T.

(d) Show that the carbon atoms alone constitute the vertices of the tree T', which is the first step when finding the center(s) of T. Show that T is determined by T'.

(e) Draw graphs of the possible isomers (different graph-isomorphism types) of pentane, $C_5 H_{12}$.

***20.** Prove (2.21.2).

2.23 Spanning Trees

If $G = [V, E]$ is any connected graph, a **spanning tree** (or **skeleton**) in G is a subgraph $T = [V, E']$ of G which is a tree. Note that we have *three* requirements:

1. T has the same set V of vertices as does G.
2. T is a tree.
3. T is a subgraph (so $E' \subseteq E$).

Nevertheless, spanning trees within a given graph are easy to find; a connected graph that is not itself a tree will generally have many spanning trees. In Figure 2.23.1 we picture the eight spanning trees of Figure 2.20.5(a). (Why is this the complete set of such trees?)

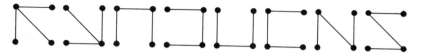

FIGURE 2.23.1. *Spanning Trees*

We can always obtain a spanning tree from a connected graph G, by the following simple algorithm:

1. Choose an edge $\{v_i, v_j\}$ which is not a loop, together with its two vertices.

2. Since G is connected, find a new vertex v_k joined by an edge e_l to the tree subgraph obtained so far, and choose v_k and e_l.
3. Repeat step 2 until all $|V|$ vertices (and $|V|-1$ edges) have been chosen. (Why is the end result necessarily a tree?)

If $T = [V, E']$ is a spanning tree in $G = [V, E]$ and $E'' = E - E'$, then $T^c = [V, E'']$ is a subgraph of G called the **cotree** of T; edges in E'' are called **chords**. (May cotrees be disconnected or contain circuits? Illustrate with examples.)

A graph G is a **weighted graph** if we associate a real number with each edge. For example, in Figure 2.23.2 suppose that the vertices a, b, c, d, e, f, and g represent towns, the edges represent existing roads, and the integer labels (weights) of the edges represent distance in miles. We wish to lay underground communication lines, parallel to existing roads, so that any two towns can communicate. We wish to use a minimal number of miles of cable. The solution will certainly be a spanning tree, since any solution not a tree would contain a spanning tree (hence a better solution) within it. (We are allowing cut points in our solution.) But some spanning tree solutions will be better than others, and there are too many to try them all (try it). We need the following definitions and theorem.

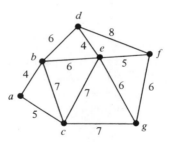

FIGURE 2.23.2. *Weighted Graph*

The **weight** of a spanning tree $T = [V, E']$ in a weighted graph $G = [V, E]$ is $w(T) = \sum_{e \in E'} w(e)$, $w(e)$ the weight of edge e. Tree T is a **minimal spanning tree** in G if T is a spanning tree such that $w(T) \leq w(T_1)$, all spanning trees T_1 in G.

Theorem 2.23.1. *We can obtain a minimal spanning tree in a connected weighted graph $G = [V, E]$ as follows:*

(a) *Choose an initial edge $e_1 = \{v_1, v_2\}$ of minimal weight and its vertices v_1, v_2, and let $T_2 = [\{v_1, v_2\}, e_1]$ be the initial subtree.*

(b) *Assume that $T_i = [V_i, E_i]$ is the current subtree. Among all pairs (v, v_k) with $v \in V - V_i$, $v_k \in V_i$, and v adjacent to v_k, choose one with edge $\{v, v_k\}$ of minimal weight and set $v = v_{i+1}$, $e_i = \{v, v_k\}$, $T_{i+1} = [V_i \cup \{v_{i+1}\}, E_i \cup \{e_i\}]$.*

(c) *Repeat step b until a spanning tree $T_{|V|}$ is obtained; $T_{|V|}$ is minimal.*

Sec. 2.23 Spanning Trees

For example, in Figure 2.23.2 we may obtain successively (abbreviating edge $\{x, y\}$ by $\overline{xy}$):

$T_2 = [\{a, b\}, \{\overline{ab}\}]$,

$T_3 = [\{a, b, c\}, \{\overline{ab}, \overline{ac}\}]$,

$T_4 = [\{a, b, c, e\}, \{\overline{ab}, \overline{ac}, \overline{be}\}]$,

$T_5 = [\{a, b, c, e, d\}, \{\overline{ab}, \overline{ac}, \overline{be}, \overline{de}\}]$,

$T_6 = [\{a, b, c, e, d, f\}, \{\overline{ab}, \overline{ac}, \overline{be}, \overline{de}, \overline{ef}\}]$,

$T_7 = [\{a, b, c, e, d, f, g\}, \{\overline{ab}, \overline{ac}, \overline{be}, \overline{de}, \overline{ef}, \overline{fg}\}]$.

The minimal spanning tree T_7 obtained is emphasized in Figure 2.23.3; it has weight $4+5+6+4+5+6=30$. Are there other minimal spanning trees? If so, find them.

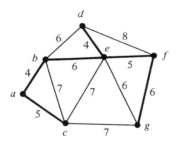

FIGURE 2.23.3. *Minimal Spanning Tree*

Proof of Theorem 2.23.1. If $|V| = n$, then we certainly obtain a spanning tree T_n (T_n has n vertices, $n-1$ edges, and is connected). Let $e_1, e_2, \ldots, e_{n-1}$ be the sequence of edges obtained for T_n by the algorithm. Let T be a *minimal* spanning tree, and label the edges $f_1, f_2, \ldots, f_{n-1}$ of T in such a way that each initial subset $\{f_1, f_2, \ldots, f_k\}$ is a subtree. By minimality of T, $w(T) \leq w(T_n)$; we must show that $w(T_n) \leq w(T)$.

Choose the first integer i such that $e_i \neq f_i$. By Theorem 2.21.1(f), the graph $T\langle e_i \rangle$ obtained by adding edge e_i to T has exactly one circuit C. Since T_n is a tree, some edge in C is not in T_n. In fact, since $e_i \in T_n$ is an edge in C, and since i is the first integer such that $e_i \neq f_i$, if $i > 1$ there is an edge f_j of C that is not in the set $e_1, \ldots, e_{i-1}, e_i$, but is incident with a vertex of e_1 or e_2 or $\cdots$ or e_{i-1} since otherwise C could not include e_i. If $i = 1$, f_j may be chosen as *any* edge other than e_1 in C. Let $\bar{T}$ be the graph obtained by deleting f_j from $T\langle e_i \rangle$. Since $\bar{T}$ is a spanning tree, $w(T) \leq w(\bar{T})$; T and $\bar{T}$ differ only in that T

has edge f_j and $\bar{T}$ has edge e_i, so $w(f_j) \leq w(e_i)$. However, when T_n was constructed, e_i was chosen instead of the still available edge f_j, so $w(e_i) \leq w(f_j)$, proving that $\bar{T} = \{e_1 = f_1, \ldots, e_{i-1} = f_{i-1}, e_i, \ldots\}$ is also a minimal spanning tree.

Now $\bar{T}$ and T_n agree in the first i edges, one more than before. We can repeat the construction in the last paragraph for larger and larger i, until a minimal spanning tree equals T_n. ∎

Spanning trees and minimal spanning trees are important in many applications; computer algorithms for generating them (even *all* of them) quickly in large graphs are described in Section 11.4 of Deo (1974).

2.24 Fundamental Circuits and Cut-Sets

In this section, $G = [V, E]$ is a connected undirected graph and $T = [V, E']$ is a fixed spanning tree in G. If $e_i \in E - E'$ and $T\langle e_i \rangle$ is the graph obtained by adding edge e_i to T, then we know by Theorem 2.21.1(f) that $T\langle e_i \rangle$ contains exactly one circuit C; C is a **fundamental circuit** of G (with respect to T). If $|V| = n$ and $|E| = e$, then $|E'| = n - 1$ and there are $|E - E'| = e - n + 1$ fundamental circuits with respect to T.

In the graph G of Figure 2.24.1, with spanning tree T emphasized, there are $e - n + 1 = 10 - 6 + 1 = 5$ fundamental circuits with respect to T, namely, $\{e_2, e_9, e_3\}$. $\{e_4, e_1, e_3\}$, $\{e_5, e_1, e_3, e_6\}$, $\{e_8, e_9, e_{10}\}$, and $\{e_7, e_6, e_9, e_{10}\}$.

A **cut-set** in our connected $G = [V, E]$ is a subset E_1 of E such that $[V, E - E_1]$ is disconnected, but if E_0 is any proper subset of E_1, then $[V, E - E_0]$ is connected. In Figure 2.24.1, $\{e_1, e_2, e_3\}$ is a cut-set, since deletion of these three edges disconnects the graph, but it is again connected if any one of the three edges is restored. The set $\{e_1, e_4, e_6, e_8, e_{10}\}$ is also a cut-set in Figure 2.24.1. A cut-set, together with the vertices terminating its edges, may be considered a subgraph in a graph.

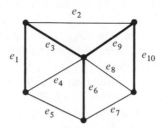

FIGURE 2.24.1. *Fundamental Circuits*

Again let $T = [V, E']$ be a fixed spanning tree in the connected graph $G = [V, E]$. If e is an edge of T, then $\{e\}$ is a cut-set of T, so let $V = V_1 \cup V_2$, $V_1 \cap V_2 = \varnothing$, the V_i the sets of vertices of the two components of T. If S is the set of all edges of G joining vertices of V_1 and V_2, then S is a cut-set of G satisfying $S \cap E' = \{e\}$; S is called a **fundamental cut-set** of G with respect to T. There will be $|E'|$ such fundamental cut-sets. In Figure 2.24.1, the five fundamental cut-sets are $\{e_1, e_4, e_5\}$, $\{e_3, e_2, e_4, e_5\}$, $\{e_6, e_5, e_7\}$, $\{e_9, e_2, e_7, e_8\}$, and $\{e_{10}, e_7, e_8\}$.

If $G_1 = [V_1, E_1]$ and $G_2 = [V_2, E_2]$ are subgraphs of $G = [V, E]$, then the **ring sum** of G_1 and G_2 is the subgraph

$$G_1 \oplus G_2 = [V_1 \cup V_2, E_1 \oplus E_2],$$

where $E_1 \oplus E_2 = (E_1 - E_2) \cup (E_2 - E_1)$. The reader should show that the ring sum of subgraphs is commutative and associative.

The proof of the following remarkable theorem may be found in Deo (1974) and many other texts. The proof involves much of the algebra that appears later in this book: abelian groups, matrices, vector spaces, finite fields, and so on.

Theorem 2.24.1. *Let $T = [V, E']$ be a spanning tree in the connected undirected graph $G = [V, E]$. Then:*

(a) *Any ring sum of circuits in G is a circuit or an edge-disjoint union of circuits in G.*

(b) *Any ring sum of cut-sets in G is a cut-set or an edge-disjoint union of cut-sets in G.*

(c) *Any circuit in G is a ring sum of some of the fundamental circuits of G with respect to T.*

(d) *Any cut-set in G is a ring sum of some of the fundamental cut-sets of G with respect to T.*

Fundamental circuits, introduced by Kirchhoff, are a basic tool in the analysis of electrical networks. *Kirchhoff's current law* asserts that the sum of the voltage drops around any circuit (often called a *loop* in engineering) within a network is zero. (Dually, the sum of the currents in any cut-set is zero.) A complex network may contain very many circuits, but it is a consequence of Theorem 2.24.1(c) that we need only consider the fundamental circuits with respect to one fixed spanning tree. Cut-sets are important in the analysis of transportation and communication networks; see subsection 5 in the next section.

2.25 Applications and Famous Problems

1. *Königsberg bridge problem.* This famous solved problem in graph theory was formulated and solved by L. Euler in 1736. The city of Königsberg (Figure 2.25.1), situated on a river containing two islands, included seven bridges. Euler asked if one could find a closed walk traversing each bridge exactly once. (The answer is *no*; see Exercise 2.26.12.)

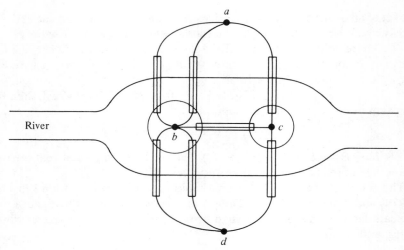

FIGURE 2.25.1. *Königsberg Bridge Problem*

2. Planarity. Much research has been done on graphs which are **planar**, that is, which can be drawn in the Euclidean plane without intersecting edges. The graph in Figure 2.25.2(a) is planar, since it can be redrawn as Figure 2.25.2(b). Nonplanar graphs are discussed in the next paragraph.

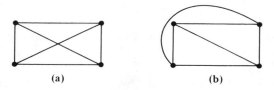

FIGURE 2.25.2. *Planar Graph*

Two graphs are **homeomorphic** if one can be obtained from the other by operations of the following kind:

(a) Replace an edge $\{v_i, v_j\}$ by a new vertex v_x and two edges $\{v_i, v_x\}$ and $\{v_x, v_j\}$, so that

(b) For a vertex v_k of degree 2, replace v_k and edges $\{v_i, v_k\}$ and $\{v_k, v_j\}$ by a single edge $\{v_i, v_j\}$, so that

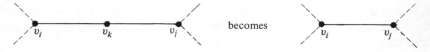

Sec. 2.25 *Applications and Famous Problems*

The graphs in Figure 2.25.3 are homeomorphic; the degree-2 vertex v_3 is deleted and the degree-2 vertex v_6 is added, to convert graph (a) to graph (b).

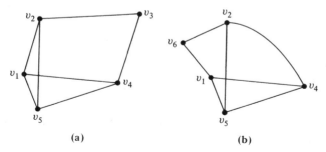

FIGURE 2.25.3. *Homeomorphic Graphs*

A remarkable theorem of Kuratowski states that *the two graphs in Figure 2.25.4 are nonplanar, and any graph is nonplanar iff it has a subgraph homeomorphic to one of these two.* For a proof, see p. 108 of Harary (1969). The graph of Figure 2.25.4(a) is called the *complete graph on five vertices*; a graph is **complete** if any two vertices are joined by an edge.

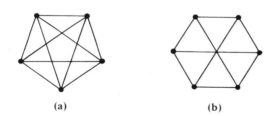

FIGURE 2.25.4. *Kuratowski's Nonplanar Graphs*

Planar graphs are important, because an electrical network whose graph is planar can be etched on a printed circuit board or integrated circuit chip without intersecting lines. If a network is nonplanar, we can ask if it is *two-planar*, that is, if it can be placed on two parallel printed circuit boards, or the two sides of one board of an insulating material.

3. *Four-color problem*. In cartography, a map is normally colored in such a way that no countries with a common boundary have the same color. In the map of Figure 2.25.5, the four colors R = red, Y = yellow, G = green, B = blue suffice. Cartographers soon formulated a question surprisingly difficult to answer: Can every map be colored by four colors?

If we replace each country by a vertex and join vertices iff the corresponding countries have a common boundary, we obtain the following equivalent graph-theoretic problem.

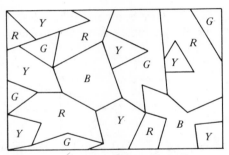

FIGURE 2.25.5. *Coloring of a Map*

Four-color problem. *Can we color (label) the vertices of any planar graph with four colors so that no two adjacent vertices have the same color?*

After a century of mathematical effort, recent computer-aided research has yielded an affirmative answer to this question [Appel and Haken (1976)]. Special properties of graphs for which two or three colors suffice have also been studied. (Any tree is two-colorable.)

4. *Traveling salesperson problem.* In this problem, a salesperson is required to visit n cities on a trip. The problem is to find a route by which the salesperson can visit each city exactly once and return home, with minimum distance traveled.

The relevant graph is the complete graph on n vertices, each vertex representing a city. The graph is weighted, each edge being labeled with the distance between the two cities represented by its terminals.

The problem is important in operations research, and in its full generality is very difficult; it is believed that the computer time required to solve the n-city problem is an *exponential* function of n. If a near-best, rather than a best, solution is satisfactory, the problem is more tractable. Algorithms for this and other graph-theoretic problems are discussed in Aho, Hopcroft, and Ullman (1974).

5. *Transportation problem.* In the weighted directed graph of Figure 2.25.6, vertex s is called a *source* and t a *sink*, for obvious reasons. The weights of the arrows are to be interpreted as *capacities*; the weight of an arrow is the maximum amount of some commodity that can be transported along that arrow in unit time. We can ask for the maximum amount that can be transported from s to t in unit time.

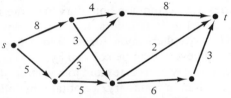

FIGURE 2.25.6. *Max-flow = Min-cut*

A **cut** in this graph is a cut-set that leaves s and t in different components. The *capacity* of a cut is the sum of the capacities of its arrows. The *max-flow min-cut theorem* [Section 14-2 of Deo (1974)] asserts that the maximum possible flow from s to t equals the minimum of the capacities of all cuts between s and t. (What is the maximum flow in Figure 2.25.6?)

We can also associate with each arrow a *cost*, the cost of unit flow through that arrow, and then ask for the flow pattern which transports commodities from s to t at minimum unit cost. This is known as the *transportation problem*, is often solved with linear programming, and is important in operations research.

The preceding examples should convince the reader of the vastness and importance of graph theory. Section 15-5 of Deo (1974) describes many more, and provides references to (lists of) over 3000 research papers. Good reference texts on graph theory include Mayeda (1972), Deo (1974), Harary (1969), and Busacker and Saaty (1965).

2.26 Exercises

1. Draw the complete graph on four vertices, and find all spanning trees within it. Are any of the cotrees actually trees?

*2. If graph G has n vertices and the subgraph T of G has $n-1$ edges and no circuits, show that G is connected and that T is a spanning tree in G.

3. Assume that G_0 is a subgraph of the connected graph G. Show that G_0 is a subgraph of a spanning tree in G iff G_0 contains no circuit.

4. If e is an edge in the connected graph G, show that some spanning tree T_1 in G contains e. If e is part of some circuit in G, show that some spanning tree T_2 in G does not contain e.

5. If v is a vertex in the connected graph G, show that there is a spanning tree T in G such that for all vertices v_i, the distance $d(v, v_i)$ is the same in T and in G.

*6. Find minimal spanning trees for the weighted graphs in Figure 2.26.1. What are their weights?

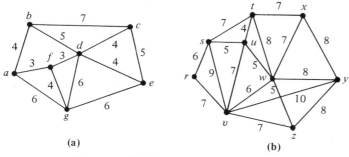

FIGURE 2.26.1. *Graphs for Exercise 6*

7. Find all fundamental circuits and fundamental cut-sets in Figure 2.26.1(b), with respect to the spanning tree found in Exercise 6. Also find several other circuits and cut-sets, by taking ring sums of fundamental ones.

8. Show that if graph $G = [V, E]$ has k components, then G contains a forest F of k spanning trees, and we can define $|E| - |V| + k$ fundamental circuits with respect to F.

9. Let C be a circuit in the connected graph G, with e_1 and e_2 edges in C. Show that there is a cut-set S in G such that $\{e_1, e_2\}$ is the set of all edges in both C and S.

*__10.__ Assume that $T = [V, E']$ is a spanning tree in the connected graph $G = [V, E]$, C a circuit in G, S a cut-set in G. Prove:
 (a) $C \cap (E - E') \neq \emptyset$.
 (b) $S \cap E' \neq \emptyset$.
 (c) $|C \cap S|$ is even.
(HINT: S partitions V into two subsets, V_1 and V_2. If C passes from V_1 to V_2, it must return to V_1.)

11. With G and T as in Exercise 10, assume that C is a fundamental circuit with respect to T containing $e_1 \in E - E'$ and S is a fundamental cut-set with respect to T containing $e_2 \in E'$. Prove that $e_2 \in C$ iff $e_1 \in S$. [Use Exercise 10(c).]

*__12.__ An **Euler walk** in an undirected graph G is a walk that traverses each edge exactly once.
 (a) Show that if G has an Euler walk, then G is connected and contains either zero or two vertices of odd degree.
(HINT: Intermediate vertices on the walk are left each time they are entered.)
 (b) Show the converse: if G is connected with zero or two vertices of odd degree, then G has an Euler walk.
(HINT: If we try to draw an Euler walk in a graph with all vertices even, we can continue until we reach the starting point. If we start at one of two odd vertices, we must reach the other.)
 (c) Conclude that the answer to the Königsberg bridge problem is *no*.

13. Prove that a graph has a closed Euler walk iff it is the edge–disjoint union of circuits.

14. Prove that if G is connected and has $2k$ vertices of odd degree ($k > 0$), then there are k edge-disjoint walks, which together include all edges in the graph. But $k - 1$ walks are not sufficient.

15. (a) Draw a graph with 64 vertices corresponding to the squares of a chessboard, with vertices v_i and v_j joined by an edge iff a knight can move from v_i to v_j.
 (b) Can a knight start anywhere and find a walk that makes every permissible move exactly once? (Use Exercise 13.)

*__16.__ A **Hamiltonian path** is a path that enters each vertex exactly once. No general criterion is known for the existence of Hamiltonian paths.

(a) Which of the graphs in Figures 2.20.1(a), 2.20.2, 2.20.4, 2.21.2, 2.23.2, and 2.24.1 possess Hamiltonian paths?
(b) Show that the graph in Figure 2.26.2 possesses no Hamiltonian path.
(c) Which trees possess Hamiltonian paths?

FIGURE 2.26.2. *Graph for Exercise 16*

17. Show that a subgraph of a connected graph is a Hamiltonian path iff it is a spanning tree with all vertices of degree at most 2.

18. (a) Show that if the connected planar graph G can be colored with two colors, then any circuit in G has even length.
(HINT: The colors alternate.)
(b) Prove the converse; if all circuits have even length, then G is two-colorable.

19. Show that the planar graph of Figure 2.26.3 cannot be colored with three colors.

FIGURE 2.26.3. *Graph for Exercise 19*

20. Try the traveling salesperson problem by choosing some number n of cities and finding the distance between any two in a mileage chart. Do this for $n = 3, 4, 5, \ldots$ and comment on the increased difficulty for larger n.

CHAPTER 3
Rings and Boolean Algebras

3.1 Algebras

Although some sets are intrinsically of interest—the set of all prime numbers, for example—most sets are interesting because of operations that are performed on their elements. A familiar example is $[\mathcal{P}(U), \cup, \cap, ']$, the set of all subsets of a universal set U subject to the operations union, intersection, and complementation. In general, we have a set S and a set of finitary operations (see Section 2.6) with respect to each of which S is closed; that is, an operation applied to elements of S yields an element of S. The set S together with these operations is called an **algebra**. In this book, an algebra will be assumed to have only a finite number of operations, but some algebras used in generalizations of finite-state machine theory have infinitely many operations. If $\theta_1, \theta_2, \ldots, \theta_n$ denote the operations on the elements of S, we denote the algebra by $[S, \theta_1, \theta_2, \ldots, \theta_n]$. Thus the algebra of complex numbers can be denoted by $[\mathbb{C}, +, \cdot, ^-]$, where "$^-$" indicates the unary operation of forming the complex conjugate.

Consider an arbitrary algebra $\mathcal{S} = [S, \theta_1, \theta_2, \ldots, \theta_n]$ and suppose that θ_i is a binary operation. Then θ_i is **commutative** iff $\forall s_1, s_2 \in S$,

$$\theta_i(s_1, s_2) = \theta_i(s_2, s_1),$$

and is **associative** iff $\forall s_1, s_2, s_3 \in S$,

$$\theta_i(s_1, \theta_i(s_2, s_3)) = \theta_i(\theta_i(s_1, s_2), s_3).$$

[Read θ_i as "the sum of" (or "the product of"), commas as "and," to get a feel for the notation.]

If θ_i and θ_j are both binary operations of the algebra S, then θ_i is **left-** or **right-distributive** over θ_j iff $\forall s_1, s_2, s_3 \in S$,

$\theta_i(s_1, \theta_j(s_2, s_3)) = \theta_j(\theta_i(s_1, s_2), \theta_i(s_1, s_3)),$ (left distributive)

$\theta_i(\theta_j(s_1, s_2), s_3) = \theta_j(\theta_i(s_1, s_3), \theta_i(s_2, s_3)).$ (right distributive)

(In each of these, read θ_i as "the product of," θ_j as "the sum of," commas as "and," for an example of what these mean.)

An element e_i of S is called an **identity element** or **unity element** for the binary operation θ_i of S iff $\forall s \in S$,

$$\theta_i(s, e_i) = \theta_i(e_i, s) = s.$$

Without even knowing what the elements or operations of S may be, we can prove the next theorem.

Theorem 3.1.1. *An identity element e_i for a binary operation θ_i of an algebra $\mathscr{S} = [S, \theta_1, \theta_2, \ldots, \theta_n]$ is unique; that is, a binary operation θ_i has at most one identity element.*

Proof. Suppose that e_i and $\tilde{e}_i$ are both identity elements for θ_i. Then

$$e_i = \theta_i(e_i, \tilde{e}_i) = \tilde{e}_i,$$

the first equality because $\tilde{e}_i$ is an identity element, the second because e_i is. ∎

If e_i is an identity element for θ_i and if $\theta_i(s_1, s_2) = \theta_i(s_2, s_1) = e_i$, then each of s_1 and s_2 is said to be θ_i-**invertible** and each is called the θ_i-**inverse** of the other. We write $s_1 = s_2^{-1}$. With regard to inverses, we have the following theorem.

Theorem 3.1.2. *If an associative, binary operation θ_i of $\mathscr{S} = [S, \theta_1, \theta_2, \ldots, \theta_n]$ has the identity e_i and if $s \in S$ has a θ_i-inverse $\tilde{s}$, then $\tilde{s}$ is unique; that is, an element $s \in S$ has at most one θ_i-inverse.*

Proof. Suppose that

$$\theta_i(s, \tilde{s}) = \theta_i(\tilde{s}, s) = e_i, \qquad \theta_i(s, s') = \theta_i(s', s) = e_i.$$

Then

$$s' = \theta_i(e_i, s') = \theta_i(\theta_i(\tilde{s}, s), s') = \theta_i(\tilde{s}, \theta_i(s, s')) = \theta_i(\tilde{s}, e_i) = \tilde{s}. \quad ∎$$

Theorems like the two just proved, which hold for all algebras, are theorems in "universal algebra." We could continue to develop such universal theorems, but these two will serve our immediate needs. In the remainder of this chapter, we study several examples of the class of algebras called *rings*.

3.2 Rings

Consider a nonempty set R which is closed under two binary operations: "addition," denoted by "$+$," and "multiplication," denoted by "$\cdot$" or simply by juxtaposition. The algebra $\mathscr{R} = [R, +, \cdot]$ is called a **ring** (we say "the ring R" when no confusion can result) iff the following postulates are satisfied.

With respect to addition:

A_1. $\forall\, r_1, r_2 \in R,\ r_1 + r_2 = r_2 + r_1$ *(commutativity)*.
A_2. $\forall\, r_1, r_2, r_3 \in R,\ r_1 + (r_2 + r_3) = (r_1 + r_2) + r_3$ *(associativity)*.
A_3. $\exists\, 0 \in R \ni \forall\, r \in R, r + 0 = r$ *(zero element or additive identity element)*.
A_4. $\forall\, r \in R\ \exists\, -r \in R \ni r + (-r) = 0$ *(negatives or additive inverses)*.

With respect to multiplication:

M_1. $\forall\, r_1, r_2, r_3 \in R,\ r_1(r_2 r_3) = (r_1 r_2) r_3$ *(associativity)*.

Relating the two operations are the left- and right-distributive laws of multiplication with respect to addition:

D_1. $\forall\, r_1, r_2, r_3 \in R$, $r_1(r_2+r_3) = r_1r_2 + r_1r_3$ (*left distributivity*).

D_2. $\forall\, r_1, r_2, r_3 \in R$, $(r_1+r_2)r_3 = r_1r_3 + r_2r_3$ (*right distributivity*).

This completes the list of required postulates.

Note that by Theorems 3.1.1 and 3.1.2, *the zero element and negatives are unique.* Note also that *multiplication precedes addition* here, in the familiar way, so the right members of the equalities in D_1 and D_2 are unambiguous. Two distributive laws are needed, since multiplication is not in general assumed to be commutative.

The zero element of a ring behaves as one would expect in multiplication:

Theorem 3.2.1. *In every ring* $[R, +, \cdot\,]$, *for all r in R,* $r \cdot 0 = 0 \cdot r = 0$.

Proof. We have

$$r \cdot 0 = r \cdot 0 + 0 = r \cdot 0 + (r \cdot r + [-(r \cdot r)])$$
$$= (r \cdot 0 + r \cdot r) + [-(r \cdot r)] = r(0+r) + [-(r \cdot r)]$$
$$= r \cdot r + [-(r \cdot r)] = 0.$$

In the same way, one proves that $0 \cdot r = 0$. ∎

Some optional postulates concerning multiplication, one or more of which are satisfied by most rings of interest, are as follows:

M_2. $\exists\, e \in R \ni \forall\, r \in R$, $re = er = r$. (Such a multiplicative identity element is called the *unity* or *unit element* of the ring. By Theorem 3.1.1, it is unique if it exists. The unity is denoted by 1, I, U, or whatever symbol may be appropriate in a given case.)

M_3. $\forall\, r \in R - \{0\}$, $\exists\, r^{-1} \in R \ni rr^{-1} = r^{-1}r = e$. (Note that M_3 cannot be satisfied unless M_2 is also satisfied, and that if r^{-1} exists, it is unique by Theorem 3.1.2.)

M_4. $\forall\, r_1, r_2 \in R$, $r_1r_2 = r_2r_1$ (*commutativity of multiplication*).

M_5. $\forall\, r \in R$, $r^2 = r$ (*idempotency of multiplication*).

M_6. $\forall\, r_1, r_2 \in R$, $r_1r_2 = 0$ implies that $r_1 = 0$ or $r_2 = 0$. (If $r_1r_2 = 0$ and $r_1 \neq 0$, $r_2 \neq 0$, the elements r_1 and r_2 are called **divisors of zero**. The postulate M_6 may therefore be restated thus: The ring $[R, +, \cdot\,]$ has no divisors of zero.)

Rings with various combinations of the properties M_2 to M_6 have special names:

(a) A ring with property M_2 is called a **ring with unity**.

(b) A ring with property M_4 is called a **commutative ring**.

(c) A ring with property M_5 is called an **idempotent ring**.

(d) A ring with properties M_2 and M_5 is called a **Boolean ring**.

(e) A ring with properties M_2, M_4, and M_6 is called an **integral domain**.

(f) A ring with properties M_2 and M_3 is called a **division ring**.

(g) A ring with properties M_2, M_3, and M_4 is called a **field**.

There are many familiar examples of these various kinds of rings. For each of the examples in the following list, the reader should verify that all requisite postulates are satisfied. The natural operations of addition and multiplication are to be assumed in each case unless otherwise specified.

1. The sets of all rational numbers, all real numbers, and all complex numbers, are fields.

2. The set of all integers is an integral domain. So is every field.

3. The set of all polynomials in a single variable x and with integer coefficients is an integral domain. This ring is important in combinatorial analysis.

4. The set of all rational functions $P(x)/Q(x)$, where $P(x)$ and $Q(x)$ belong to the integral domain of example 3 and where $Q(x)$ is not the zero polynomial, is a field.

5. (For those who know matrix algebra.) The set of all $n \times n$ matrices with complex entries is a ring with unity. (The unity is the identity matrix $\mathbf{I}_n$. For $n > 1$, this ring is not commutative and contains divisors of zero. The reader should illustrate these latter assertions with examples.)

6. The set of $n \times n$ relation matrices, where "+" means addition modulo 2 so $1 + 1 = 0$ and multiplication is defined as usual, is a ring with unity.

7. The set of $n \times n$ relation matrices, where "+" means "OR" so $1 + 1 = 1$ and multiplication is defined as usual, is also a ring with unity. (This example and the previous one show that the same set R can appear in distinct rings.)

8. The set of all continuous real functions on a closed interval $[a, b]$ is a commutative ring with unity. (This ring has divisors of zero. Illustrate.)

9. The set of all even integers is a commutative ring. (This ring does not have a unity. Explain.)

10. The set $[\mathcal{P}(U), \oplus, \cap]$, where $\mathcal{P}(U)$ is the power set of U and $\oplus$ denotes the symmetric difference (Exercise 1.22.4), is a commutative ring with unit. (For most sets U, this ring has divisors of zero. What are the exceptions?)

Other examples will be introduced in following sections.

In every ring $[R, +, \cdot]$, because multiplication is associative, the usual laws of exponents hold for every element r of R and for all positive integers m and n:

(3.2.1)
$$r^m \cdot r^n = r^{m+n},$$
$$(r^m)^n = r^{mn}.$$

If the ring is commutative, we have also, $\forall\, r, s \in R$, and for all positive integers n:

(3.2.2)
$$(rs)^n = r^n s^n.$$

If R is a ring with unit e, we define $r^0 = e$ and in (3.2.1) and (3.2.2), m and n may be 0. If r^{-1} exists, m and n in (3.2.1) are arbitrary members of $\mathbb{Z}$, as is n in (3.2.2) if r^{-1} and s^{-1} exist.

For the integers n of $\mathbb{Z}$ we define, for every element $r \in R$:

$0 \cdot r = 0$ (on the left, the integer 0; on the right, the ring element 0),

(3.2.3) $nr = r + r + \cdots + r$ to n terms,

$(-n)r = (-r) + (-r) + \cdots + (-r)$ to n terms.

Then for all n and m in $\mathbb{Z}$ and all r and s in R, we have, from (3.2.3),

(3.2.4)
$$mr + nr = (m+n)r,$$
$$n(mr) = (nm)r,$$
$$(-n)r = -(nr),$$
$$n(rs) = (nr)s = r(ns).$$

Note that in neither (3.2.3) nor (3.2.4) is it a requirement that $\mathbb{Z} \subseteq R$. The integers act like scalars here.

3.3 Congruences

Recall from arithmetic that every integer $n > 1$ that is not prime can be factored into a product of primes, uniquely except for the order of the factors. If $m = p_1^{a_1} p_2^{a_2} \cdots p_k^{a_k}$ and $n = p_1^{b_1} p_2^{b_2} \cdots p_k^{b_k}$, where the exponents are all nonnegative integers and the p_i represent distinct primes, then the least common multiple and the greatest common divisor of m and n are given by

l.c.m. $(m, n) = p_1^{e_1} p_2^{e_2} \cdots p_k^{e_k}$, $\quad e_i = \max(a_i, b_i), \quad i = 1, \ldots, k,$

g.c.d. $(m, n) = p_1^{\varepsilon_1} p_2^{\varepsilon_2} \cdots p_k^{\varepsilon_k}$, $\quad \varepsilon_i = \min(a_i, b_i), \quad i = 1, \ldots, k.$

For example, $204 = 2^2 \cdot 3^1 \cdot 17^1$, $108 = 2^2 \cdot 3^3 \cdot 17^0$, l.c.m. $(204, 108) = 2^2 \cdot 3^3 \cdot 17^1 = 1836$, g.c.d. $(204, 108) = 2^2 \cdot 3^1 \cdot 17^0 = 12$.

We say two positive integers m and n are **relatively prime** if g.c.d. $(m, n) = 1$. For example, $24 = 2^3 \cdot 3^1 \cdot 5^0 \cdot 7^0$ and $35 = 2^0 \cdot 3^0 \cdot 5^1 \cdot 7^1$ are relatively prime.

Recall also from arithmetic that when any integer z is divided by a positive integer p, there results a quotient q and a remainder r such that $z = qp + r$, $0 \le r \le p - 1$. If $r = 0$, we say p **divides** z and write $p|z$. For example, if $p = 7$,

$$0 = 0 \cdot 7 + 0; \qquad q = 0, r = 0, 7|0;$$
$$-35 = (-5) \cdot 7 + 0; \qquad q = -5, r = 0, 7|-35;$$
$$45 = 6 \cdot 7 + 3; \qquad q = 6, r = 3;$$
$$-15 = (-3) \cdot 7 + 6; \qquad q = -3, r = 6.$$

Sec. 3.3 Congruences

For each integer $p > 1$ and for all a, b in $\mathbb{Z}$, we define a to be **congruent modulo p** to b, written $a \equiv b \pmod{p}$, iff $a - b$ is divisible by p, that is, iff there exists q in $\mathbb{Z}$ such that $a = qp + b$. In particular, $a \equiv 0 \pmod{p}$ iff for some q in $\mathbb{Z}$, $a = qp$. For example,

$$45 \equiv 3 \pmod{7} \quad \text{because } 45 - 3 = 6 \cdot 7,$$
$$35 \equiv 0 \pmod{7} \quad \text{because } 35 = 5 \cdot 7,$$
$$-15 \equiv 6 \pmod{7} \quad \text{because } -15 - 6 = (-3) \cdot 7.$$

Computation with congruences proceeds according to the rules summarized in the next five theorems.

Since equal integers are congruent modulo p, we have at once the following theorem.

Theorem 3.3.1. *For all integers $p > 1$, any identity in the ring $[\mathbb{Z}, +, \cdot]$ holds also modulo p. In particular,*
(a) $a + b \equiv b + a \pmod{p}$.
(b) $a + (b + c) \equiv (a + b) + c \pmod{p}$.
(c) $a(b + c) \equiv ab + ac \pmod{p}$. ∎

The next result shows that we can add and multiply congruences.

Theorem 3.3.2. *For all a, b, c, and d in $\mathbb{Z}$ and any $p > 1$ in $\mathbb{Z}$, if $a \equiv b \pmod{p}$ and $c \equiv d \pmod{p}$, we have*
(a) $a + c \equiv b + d \pmod{p}$.
(b) $ac \equiv bd \pmod{p}$.
(c) $-a \equiv -b \pmod{p}$.

Proof. If $a \equiv b \pmod{p}$ and $c \equiv d \pmod{p}$, there exist q_1 and q_2 in $\mathbb{Z}$ such that

$$a = q_1 p + b, \quad c = q_2 p + d.$$

Then

$$a + c = (q_1 + q_2)p + b + d,$$
$$ac = (q_1 q_2 p + q_1 d + q_2 b)p + bd,$$
$$-a = (-q_1)p + (-b),$$

and these imply (a), (b), and (c), respectively. ∎

Theorem 3.3.3 (The Substitution Principle). *If $a \equiv b \pmod{p}$, then for all z in $\mathbb{Z}$,*
(a) $a + z \equiv b + z \pmod{p}$,
(b) $az \equiv bz \pmod{p}$.

Proof. Since $z \equiv z$ (mod p), these are special cases of Theorem 3.3.2(a) and (b). ∎

This result is helpful in computations, for it shows that one can replace an expression by a congruent expression at any point in a computation. For example, $2^4 \equiv 1$ (mod 5), so for all integers $n \geq 1$, $2^{4n} - 2^{4(n-1)} = (2^4)^n - (2^4)^{n-1} \equiv 1^n - 1^{n-1} \equiv 0$ (mod 5); that is, for all n in $\mathbb{P}$, $2^{4n} - 2^{4(n-1)}$ is divisible by 5.

The **laws of cancellation** are summarized in the following theorem.

Theorem 3.3.4. *For all a, b, c, and z in $\mathbb{Z}$ and for any $p > 1$,*
(a) $a + z \equiv b + z$ (mod p) *implies that* $a \equiv b$ (mod p).
(b) z, p *relatively prime and* $az \equiv bz$ (mod p) *imply that* $a \equiv b$ (mod p).
(c) *g.c.d.* $(z, p) = d < p$ *and* $az \equiv bz$ (mod p) *imply that* $a \cdot (z/d) \equiv b \cdot (z/d)$ (mod p/d).

Note that cancellation is unrestricted in the case of addition but that cancellation from products is possible only under special conditions.

Proof. The proof of (a) follows from Theorem 3.3.2(a) and the fact that $-z \equiv -z$ (mod p). To prove (b), observe that $(a - b)z = qp$ implies that z divides q, since g.c.d. $(z, p) = 1$. Hence $a - b = (q/z)p$, where q/z is an integer, and (b) follows. For (c), observe that by hypothesis z/d and p/d are integers. From $az = qp + bz$ then follows $a(z/d) = q(p/d) + b(z/d)$; that is, $a \cdot (z/d) \equiv b \cdot (z/d)$ (mod p/d). ∎

Two frequently used identities are due to Fermat:

Theorem 3.3.5. *If p is prime, then for all $a \in \mathbb{N}$,*

(3.3.1) $$a^p \equiv a \pmod{p},$$

and if $a \not\equiv 0$ (mod p),

(3.3.2) $$a^{p-1} \equiv 1 \pmod{p}.$$

For example, $243 = 3^5 \equiv 3$ (mod 5) and $81 = 3^4 \equiv 1$ (mod 5).

Proof. Let p be any given prime number. The result (3.3.1) holds for $a = 0$ and $a = 1$:

$$0^p \equiv 0 \pmod{p}, \qquad 1^p \equiv 1 \pmod{p}.$$

Now let k be any positive integer such that

$$k^p \equiv k \pmod{p}.$$

Then, by the binomial theorem,

$$(k+1)^p = k^p + pk^{p-1} + \frac{p(p-1)}{1 \cdot 2} k^{p-2} + \cdots + pk + 1.$$

Because p is prime, every binomial coefficient in this expansion, except for the first and the last, contains a factor p and hence

$$(k+1)^p \equiv k^p + 1 \pmod{p}.$$

By the induction hypothesis and Theorem 3.3.3, we therefore have

$$(k+1)^p \equiv k+1 \pmod{p},$$

so the stated result holds for $k+1$ if it holds for k. Hence it holds for all integers $a \geq 0$, by the principle of finite induction.

For the second part of the theorem, note that since p is prime, g.c.d. (a, p) is 1 or p. If $a \not\equiv 0 \pmod{p}$, $a \neq qp$, so g.c.d. $(a, p) = 1$. Then, by Theorem 3.3.4(b), (3.3.1) implies (3.3.2). ∎

Congruences can help one draw interesting conclusions that may be hard to obtain otherwise. For example, if $a \not\equiv 0 \pmod{p}$, where p is prime, the congruence $a^{p-1} \equiv 1 \pmod{p}$ implies that for all m and n in $\mathbb{N}$, $a^{(p-1)m} - a^{(p-1)n}$ is divisible by p. In particular, since $5 \not\equiv 0 \pmod{7}$, $5^{30} - 5^6$ is divisible by 7.

If $a_n a_{n-1} \cdots a_1 a_0$ is the base-b representation of a nonnegative integer a, then

$$a = a_n b^n + a_{n-1} b^{n-1} + \cdots + a_1 b + a_0.$$

If $1 < k < b$ and $b \equiv 1 \pmod{k}$, then

$$a \equiv a_n + a_{n-1} + \cdots + a_1 + a_0 \pmod{k}$$

so a is divisible by k iff $a_n + a_{n-1} + \cdots + a_1 + a_0$ is divisible by k. In particular, this holds for $k = b-1$. When $b = 10$, this yields the familiar tests for divisibility by 9 and by 3 since $10 \equiv 1 \pmod{9}$ and $10 \equiv 1 \pmod{3}$.

Again, if $1 < k < b$ and there is a smallest integer s such that $b^s \equiv 0 \pmod{k}$, then

$$a \equiv a_{s-1} b^{s-1} + \cdots + a_1 b + a_0 \pmod{k},$$

so a is divisible by k iff the truncated base-b integer $a' = a_{s-1} a_{s-2} \cdots a_1 a_0$ is divisible by k. Thus if $b = 16$, for $k = 2, 4,$ or 8 we have $s = 1$, so a is divisible by 2, 4, or 8 iff a_0 is divisible by 2, 4, or 8, respectively. If $b = 10$, for $k = 4$ or 25 we have $s = 2$, so a is divisible by 4 or 25 iff $a_1 a_0$ is divisible by 4 or 25, respectively.

3.4 The Ring of Integers Modulo p

As in the preceding section, let p be any positive integer greater than 1. Let addition and multiplication on integers of the set $\mathbb{Z}_p = \{0, 1, \ldots, p-1\}$ be performed as usual, then reduced modulo p to a member of $\mathbb{Z}_p$, so $\mathbb{Z}_p$ is closed under these two operations. For example, if $p = 12$, $8 + 4 \equiv 0$, $9 + 11 \equiv 8$, $3 \cdot 4 \equiv 0$, and $3 \cdot 9 \equiv 3$, all modulo 12. By Theorem 3.3.1, these operations are commutative and associative and the distributive laws hold as well. For all k in $\mathbb{Z}_p$, we have $0 + k \equiv k$, $k + (p-k) \equiv 0$, so $-k \equiv p - k$, and $1 \cdot k \equiv k$, all modulo p. These facts imply the next theorem.

Theorem 3.4.1. *The system $[\mathbb{Z}_p, +, \cdot]$, where the arithmetic is executed modulo p, is a commutative ring with unity.* ∎

We call this ring simply the ring $\mathbb{Z}_p$ and drop the notation "mod p" when no confusion can result.

As an example, we give the addition and multiplication tables for $\mathbb{Z}_5$:

+	0	1	2	3	4		·	0	1	2	3	4
0	0	1	2	3	4		0	0	0	0	0	0
1	1	2	3	4	0		1	0	1	2	3	4
2	2	3	4	0	1		2	0	2	4	1	3
3	3	4	0	1	2		3	0	3	1	4	2
4	4	0	1	2	3		4	0	4	3	2	1

The multiplication table for $p = 5$ shows that each nonzero element of $\mathbb{Z}_5$ has a multiplicative inverse in $\mathbb{Z}_5$; in fact, $1^{-1} = 1$, $2^{-1} = 3$, $3^{-1} = 2$, $4^{-1} = 4$. This illustrates the next theorem.

Theorem 3.4.2. *The ring $\mathbb{Z}_p$ is a field iff p is prime.*

A field $\mathbb{Z}_p$ is called the **Galois field** GF(p).

Proof. Suppose that p is prime. Since all other properties of a field have already been established, we have only to show that every member other than 0 of $\mathbb{Z}_p$ has an inverse. By Theorem 3.3.5 we have, for every element other than 0 of $\mathbb{Z}_p$,

$$a^{p-1} \equiv 1 \quad \text{or} \quad a \cdot a^{p-2} \equiv 1,$$

so

(3.4.1) $$a^{-1} = a^{p-2}.$$

Hence $\mathbb{Z}_p$ is a field.

Sec. 3.5 Binary Arithmetic Modulo 2^n

Now suppose that $\mathbb{Z}_p$ is a field. If p is not prime, then $p = ab$, where $1 < a < p$ and $1 < b < p$. Then $ab \equiv 0$ and, since a^{-1} exists because $\mathbb{Z}_p$ is a field,

$$b \equiv 1 \cdot b \equiv (a^{-1}a)b \equiv a^{-1}(ab) \equiv a^{-1} \cdot 0 \equiv 0,$$

which is a contradiction. Hence p must be prime. ∎

In the field $GF(p)$, one can always solve a linear congruence $ax \equiv b$, where $a \not\equiv 0$. In fact, since $a^{-1} = a^{p-2}$, the solution is $x = a^{p-2}b$. (The set of all integers in $\mathbb{Z}$ that satisfy $ax \equiv b$ is then $\{a^{p-2}b + qp | q \in \mathbb{Z}\}$.) For example, in $GF(11)$, $2^{-1} \equiv 2^9 \equiv (2^4)^2 \cdot 2 \equiv 5^2 \cdot 2 \equiv 3 \cdot 2 \equiv 6$. Hence $2x \equiv 5$ has the solution $x \equiv 2^{-1} \cdot 5 \equiv 6 \cdot 5 \equiv 8 \pmod{11}$.

By checking all the possibilities, we find that in $GF(5)$, $x^2 + 4 \equiv 0$ has the solutions 1 and 4, so $x^2 + 4 \equiv (x+4)(x+1)$. On the other hand, the congruence $x^2 + x + 1 \equiv 0$ has no solutions in $GF(5)$, so $x^2 + x + 1$ cannot be represented as a product of nonconstant factors with coefficients in $GF(5)$. A polynomial $a_n x^n + a_{n-1} x^{n-1} + \cdots + a_1 x + a_0$ with coefficients in $GF(p)$ is called a **polynomial over** $GF(p)$. When a polynomial $f(x)$ over $GF(p)$ cannot be factored into a product of at least two nonconstant polynomials over $GF(p)$, it is said to be **irreducible**. Such polynomials are of basic importance in coding theory.

For each prime p and each positive integer n, there exists a finite field with p^n elements, denoted $GF(p^n)$. In later chapters, we shall see how these fields may be constructed with the aid of irreducible polynomials and how they are used in coding theory and in the theory of linear machines.

3.5 Binary Arithmetic Modulo 2^n

In this section we discuss the arithmetic of binary (base 2) numbers which is actually used for addition, subtraction, and multiplication within many large computers. See Kapur (1970, Chap. 6) and Vickers (1971, Chap. 4) for further discussion.

Consider the ring $\mathbb{Z}_{2^n}$ but with the integers $0, 1, 2, \ldots, 2^n - 1$ written in n-digit binary form: $00 \cdots 0, 00 \cdots 1, \ldots, 11 \cdots 1$. We denote the resulting ring of sequences of n 0's and 1's by $\mathbb{B}^n$. The unique negative of an element b of $\mathbb{B}^n$ is the n-digit binary expansion of $2^n - b$ and is called the **two's complement** of b. Since $2^n - b = (2^n - 1) - b + 1$, and $2^n - 1$ is the binary number $11 \cdots 1$, the two's complement of b may be found by complementing each digit of the n-digit binary representation of b, that is, by first finding the **one's complement** $(2^n - 1) - b$ of b, adding 1 to the result, and finally reducing mod 2^n. For example, if $n = 5$, $(2^5 - 1)_{\text{dec}} = 11111_{\text{bin}}$, so the two's complement of 01101 is

$$11111 - 01101 + 00001 \equiv 10010 + 00001 \equiv 10011.$$

Subtraction in $\mathbb{B}^n$ is accomplished by adding the two's complement of the subtrahend, then reducing mod 2^n.

Note that $00\cdots 0$ and $10\cdots 0$ are their own negatives in $\mathbb{B}^n$ and that the two's complement of the two's complement of b is again b.

When $n > 1$, 2^n is not prime, so $\mathbb{B}^n$ has divisors of zero and the cancellation law for multiplication holds only in restricted form (Theorem 3.3.4). For example, if $n = 4$, $0100 \cdot 1100 \equiv 0000$, so 0100 and 1100 are divisors of zero, and $0100 \cdot 0101 \equiv 0100 \cdot 1001$ but $0101 \neq 1001$.

The ring $\mathbb{B}^n$ is made useful for practical computation, despite the presence of divisors of zero, by the introduction of a certain mapping and a restriction.

First we observe that the set $\mathbb{B}^n$ may be partitioned into the subset of n-digit binary representations of the integers $0, 1, 2, \ldots, 2^{n-1}-1$, all of which have initial digit 0, and the subset of two's complements of the n-digit binary representations of the integers $1, 2, \ldots, 2^{n-1}$, all of which have initial digit 1. (We agree always to regard $100\cdots 0$ as the two's complement of 2^{n-1} and never as 2^{n-1} in order to eliminate the ambiguity that arises from the fact that $100\cdots 0$ is its own two's complement.)

Now let $\tilde{a}$ denote the binary representation of a. Then the mapping referred to earlier is the following:

$$h: \{-2^{n-1}, \ldots, -1, 0, 1, \ldots, 2^{n-1}-1\} \to \mathbb{B}^n,$$
$$h(a) = \tilde{a} \quad \text{if} \quad a \in \{0, 1, \ldots, 2^{n-1}-1\},$$
$$h(-a) = \widetilde{2^n - a} \quad \text{if} \quad -a \in \{-2^{n-1}, \ldots, -1\}.$$

As a result of this mapping, the first digit of a member of $\mathbb{B}^n$ becomes in effect a sign digit: 0 for zero and positive numbers, 1 for negative numbers. For example, when $n = 3$, we have these sequences and their decimal interpretations:

$$0 \leftrightarrow 000, \qquad -1 \leftrightarrow 111,$$
$$1 \leftrightarrow 001, \qquad -2 \leftrightarrow 110,$$
$$2 \leftrightarrow 010, \qquad -3 \leftrightarrow 101,$$
$$2^{n-1}-1 = 3 \leftrightarrow 011, \qquad -2^{n-1} = -4 \leftrightarrow 100.$$

To exploit this interpretation for arithmetic purposes, we now admit only sums and products whose inputs and outcomes are integers that lie in the interval $[-2^{n-1}, 2^{n-1}-1]$. With this restriction, the ring operations in $\mathbb{B}^n$ yield results corresponding to the usual sum and product of decimal or binary integers, as we shall show. For example, if $n = 5$, then $2^{n-1} = 16$, and we have corresponding computations like the following (additions and

Sec. 3.5 Binary Arithmetic Modulo 2^n

multiplications are done as usual for n-digit binary numbers):

$$6+9=15, \quad 00110+01001 \equiv 01111,$$
$$3+5=8, \quad 00011+00101 \equiv 01000,$$
$$-7-9=-16, \quad 11001+10111 \equiv 10000,$$
$$15-9=6, \quad 01111+10111 \equiv 00110,$$
$$3 \cdot 5 = 15, \quad 00011 \cdot 00101 \equiv 01111,$$
$$-2 \cdot 8 = -16, \quad 11110 \cdot 01000 \equiv 10000,$$
$$3 \cdot -4 = -12, \quad 00011 \cdot 11100 \equiv 10100,$$
$$-3 \cdot -5 = 15, \quad 11101 \cdot 11011 \equiv 01111.$$

We now prove that the ring operations in $\mathbb{B}^n$, with the mapping and restriction defined above, faithfully represent the results of the corresponding operations in $\mathbb{Z}$. We do this by considering the following seven cases, in each of which a and b denote nonnegative integers.

1. $a+b$; $0 \le a \le 2^{n-1}-1$, $0 \le b \le 2^{n-1}-1$, $0 \le a+b \le 2^{n-1}-1$. The stated restrictions and the definition of h imply that

$$h(a) = \tilde{a} \quad \text{and} \quad h(b) = \tilde{b}, \quad \text{so } h(a)+h(b) \equiv \tilde{a}+\tilde{b} \equiv \widetilde{a+b} \equiv h(a+b).$$

But $h(a+b) \equiv h(a)+h(b)$ says that the image of a sum is the sum of the images, that is, that the sum in $\mathbb{Z}$ is represented correctly by the sum in $\mathbb{B}^n$, under the stated conditions.

2. $a-b = a+(-b)$; $0 \le b \le a \le 2^{n-1}-1$. The restrictions on a and b imply that $0 \le a+(-b) \le 2^{n-1}-1$. We have now $h(a) = \tilde{a}$ and $h(-b) = \widetilde{2^n - b}$, so that

$$h(a)+h(-b) \equiv \tilde{a} + (\widetilde{2^n - b}) \equiv \widetilde{2^n} + (\tilde{a}-\tilde{b}) \equiv \tilde{a}-\tilde{b} \equiv h(a+(-b))$$
$$\equiv h(a-b).$$

3. $a-b = a+(-b)$; $0 \le a \le 2^{n-1}-1$, $-2^{n-1} \le -b \le 0$, $a \le b$. The restrictions on a and b imply that $-2^{n-1} \le a-b \le 0$. We have $h(a) = \tilde{a}$, $h(-b) = \widetilde{2^n-b}$.

$$h(a)+h(-b) \equiv \tilde{a}+\widetilde{2^n-b} \equiv \widetilde{2^n}-(\tilde{b}-\tilde{a}) \equiv h(-(b-a)) \equiv h(a-b).$$

4. $-a-b = (-a)+(-b)$; $-2^{n-1} \le -a \le 0$, $-2^{n-1} \le -b \le 0$, $-2^{n-1} \le -a-b \le 0$. Now $h(-a) = \widetilde{2^n-a}$, $h(-b) = \widetilde{2^n-b}$, so

$$h(-a)+h(-b) \equiv (\widetilde{2^n-a})+(\widetilde{2^n-b}) \equiv \widetilde{2^n-(a+b)} \equiv h(-(a+b))$$
$$\equiv h(-a-b).$$

5. ab; $0 \le a \le 2^{n-1}-1$, $0 \le b \le 2^{n-1}-1$, $0 \le ab \le 2^{n-1}-1$. Here $h(a) \cdot h(b) \equiv \tilde{a}\tilde{b} \equiv \widetilde{ab} \equiv h(ab)$ because of the stated restrictions.

6. $a(-b)$; $0 \leq a \leq 2^{n-1}-1$, $-2^{n-1} \leq -b \leq 0$, $-2^{n-1} \leq a(-b) \leq 0$. We now have

$$h(a) \cdot h(-b) \equiv \tilde{a}\widetilde{(2^n-b)} \equiv \tilde{a}\widetilde{2^n} - \tilde{a}\tilde{b} \equiv \widetilde{2^n} - \tilde{a}\tilde{b} \equiv \widetilde{2^n - ab} \equiv h(-ab)$$
$$\equiv h(a(-b)).$$

7. $(-a)(-b)$; $-2^{n-1} \leq -a \leq 0$, $-2^{n-1} \leq -b \leq 0$, $0 \leq (-a)(-b) \leq 2^{n-1}-1$. We have

$$h(-a)h(-b) \equiv \widetilde{(2^n-a)}\widetilde{(2^n-b)} \equiv \tilde{a}\tilde{b} \equiv \widetilde{ab} \equiv h(ab).$$

Thus in every case the image of a sum (product) is the sum (product) of the images, and these cases include all possibilities. Hence the modular arithmetic unambiguously represents the arithmetic of $\mathbb{Z}$ in the specified range.

3.6 Exercises

1. Use congruences to show that for all $n \geq 1$, 35 is a divisor of $6^{2n} - 6^{2(n-1)}$ and 70 is a divisor of $101^{6n} - 1$.

2. Compute 7^{-1} (mod 11), 4^{-1} (mod 43), and 2^{-1} (mod 101).

3. Is the symbol a/b, $b \neq 0$, unambiguous in $\mathrm{GF}(p)$? If so, what is $\frac{25}{37}$ (mod 41)?

4. Show that if $p_1, p_2, \ldots, p_k$ are distinct primes $\neq 2$, then $p_1 p_2 \cdots p_k$ is a divisor of

$$2^{(p_1-1)(p_2-1)\cdots(p_k-1)} - 1.$$

5. Prove that in every ring $[R, +, \cdot]$, for all a, b in R,

$$(-a)(-b) = ab$$

by simplifying

$$(-a)(-b) + a(-b) + ab$$

in two different ways.

6. Prove that no divisor of zero in a ring $[R, +, \cdot]$ has an inverse in $[R, +, \cdot]$. Conclude that every field is an integral domain.

*7. Show that if $[R, +, \cdot]$ is a ring, then for each $r \in R$, so is $[\{nr \mid n \in \mathbb{Z}\}, +, \cdot]$. (This ring is a *subring* of $[R, +, \cdot]$; that is, it is a subset of R and is a ring with respect to the same operations.)

*8. Prove that if $[R, +, \cdot]$ is a finite ring with unity e, then there exists a smallest positive integer n such that

$$0 = e + e + \cdots + e \quad \text{(to } n \text{ terms)}.$$

Then show that for all r in R,

$$0 = r + r + \cdots + r \quad \text{(to } n \text{ terms)}.$$

[This integer n is called the **characteristic** of the ring. In particular, for any prime p, GF(p) is a finite field of characteristic p.]

9. How many distinct rings are there with precisely three distinct elements: $[R, +, \cdot]$, where $R = \{0, 1, a\}$ and 1 is a unity element?

10. Show that if a ring $[R, +, \cdot]$ has unity e and if for some $r \in R$ there exist a and b in R such that $ar = e$ and $rb = e$, then $a = b$, so r has an inverse in $[R, +, \cdot]$. (Simplify arb in two ways.)

*__11.__ If a ring $[R, +, \cdot]$ contains an element a such that $a^2 = a$, then a is called an **idempotent element** of R. Find all idempotent elements in the ring $\mathbb{Z}_{36}$.

12. Let $[R_1, +, \cdot]$ and $[R_2, +, \cdot]$ be rings such that if r and s each belong to both rings, then $r + s$ has the same value in both rings, and similarly for rs. Prove that the intersection of these two rings is either a ring or is empty.

13. In the ring of real functions continuous on the closed interval $[0, 1]$, there exist divisors of zero. Give examples.

*__14.__ Prove that if $[R, +, \cdot]$ is a ring with unity e and if $a, b \in R$, $a \neq 0$, $b \neq 0$, $ab = 0$, then there exists no c in R such that $ca = e$ and no d in R such that $bd = e$. [If $ab = 0$ and $a \neq 0$, $b \neq 0$, a is called a **left divisor of zero** and b is called a **right divisor of zero**. If $ca = e$, c is a **left inverse** of a, and if $bd = e$, d is a **right inverse** of b. Thus the problem here is to prove that a left (right) divisor of zero has no left (right) inverse.]

*__15.__ Prove that if a is an element of a ring $[R, +, \cdot]$ and a is not a left divisor of zero, then $ar = as$ implies that $r = s$. If a is not a right divisor of zero, then $ra = sa$ implies that $r = s$. (This is the *cancellation law for multiplication* in rings.)

16. Prove that the set C of all elements of a ring $[R, +, \cdot]$ which commute with every element of R constitutes a ring $[C, +, \cdot]$. (This ring is a subring of $[R, +, \cdot]$ and is called the **center** of $[R, +, \cdot]$.)

17. Determine the center of the ring of $n \times n$ real matrices.

18. Determine $[2\mathbb{Z}, +, \cdot] \cap [3\mathbb{Z}, +, \cdot]$ and $[12\mathbb{Z}, +, \cdot] \cap [18\mathbb{Z}, +, \cdot]$. (All are subrings of $[\mathbb{Z}, +, \cdot]$.) Then give a rule for computing $[m\mathbb{Z}, +, \cdot] \cap [n\mathbb{Z}, +, \cdot]$, where m and n are arbitrary positive integers.

19. Prove by induction that if $[R, +, \cdot]$ has no divisors of zero, then an equation $r_1 r_2 \cdots r_n = 0$, where the $r_i \in R$, implies that some $r_i = 0$.

*__20.__ Prove that if $[R, +, \cdot]$ is a finite, commutative ring without divisors of zero, then $[R, +, \cdot]$ is a field. (First prove that $[R, +, \cdot]$ has a unity; then prove that every nonzero element is invertible. This one is fairly difficult.)

*__21.__ Let Q denote a set of subsets of a universal set U; that is, let $Q \subseteq \mathcal{P}(U)$. Show that if Q is closed under intersection and complementation, then $[Q, \oplus, \cap]$ is an idempotent ring with unity.

22. In combinatorial mathematics, formal power series such as

$$S_1 = a_0 + a_1 t + a_2 t^2 + \cdots + a_n t^n + \cdots,$$
$$S_2 = b_0 + b_1 t + b_2 t^2 + \cdots + b_n t^n + \cdots,$$

where the coefficients are integers, are a basic tool of analysis. These series are added term by term:

$$S_1 + S_2 = (a_0 + b_0) + (a_1 + b_1)t + (a_2 + b_2)t^2 + \cdots + (a_n + b_n)t^n + \cdots,$$

and multiplied according to the rule

$$S_1 S_2 = a_0 b_0 + (a_1 b_0 + a_0 b_1)t + (a_2 b_0 + a_1 b_1 + a_0 b_2)t^2 + \cdots$$
$$+ (a_n b_0 + a_{n-1} b_1 + \cdots + a_i b_{n-i} + \cdots + a_0 b_n)t^n + \cdots.$$

Show that the algebra $[\{S\}, +, \cdot]$, where $\{S\}$ is the set of all such series, is a commutative ring with unity.

23. Given that $n = 6$, show how the following decimal computations would be represented in $\mathbb{B}^6$:

(a) $18 + 11$ (b) $18 - 11$ (c) $-18 - 11$
(d) $4 \cdot 7$ (e) $4 \cdot (-7)$ (f) $(-4) \cdot (-7)$

*24. Prove by induction that the extended distributive laws

$$(a_1 + a_2 + \cdots + a_n)b = a_1 b + a_2 b + \cdots + a_n b,$$
$$a(b_1 + b_2 + \cdots + b_n) = ab_1 + ab_2 + \cdots + ab_n,$$
$$(a_1 + a_2 + \cdots + a_n)(b_1 + b_2 + \cdots + b_m) = \sum_{i=1}^{n} \sum_{j=1}^{m} a_i b_j$$

hold in every ring.

3.7 Boolean Rings and Boolean Algebras

If the law of idempotency $r^2 = r$ holds for all elements r of a ring with unity, the ring is called a **Boolean ring**. The basic example is the set $\mathscr{P}(U)$ of all subsets of a universal set U for which addition, denoted by "$\oplus$," is defined to be the symmetric difference:

$$A \oplus B = (A \cap B') \cup (A' \cap B),$$

and multiplication is defined to be intersection:

$$AB = A \cap B.$$

Easy computations show that for all A, B, and C in $\mathscr{P}(U)$,

$A \oplus B = B \oplus A$ (addition is commutative),
$A \oplus (B \oplus C) = (A \oplus B) \oplus C$ (addition is associative),
$A \oplus \varnothing = A$ ($\varnothing$ is the zero element),
$A \oplus A = \varnothing$ (every element is its own negative).

Sec. 3.7 Boolean Rings and Boolean Algebras

We already know that for all A, B, and C in $\mathcal{P}(U)$,

$AU = A$ (U is the unit element),

$AB = BA$ (multiplication is commutative),

$A(BC) = (AB)C$ (multiplication is associative).

Because multiplication is commutative, we need demonstrate only one distributive law:

$$A(B \oplus C) = AB \oplus AC.$$

This is easy to do and is left to the reader. Since also

$$AA = A,$$

these computations prove the following theorem.

Theorem 3.7.1. $[\mathcal{P}(U), \oplus, \cdot]$ *is a Boolean ring.* ∎

In the case of Boolean rings, we shall always denote addition by "$\oplus$" and call it the "ring sum" in accordance with the most common usage in switching algebra, which is the major practical application of this theory.

The law of idempotency has two important first consequences:

Theorem 3.7.2. *If $[B, \oplus, \cdot]$ is a Boolean ring, then for all a and b in B, $-a = a$ and $ab = ba$.*

Proof. We have, for all a and b in B,

$$a \oplus b = (a \oplus b)^2 = a^2 \oplus ab \oplus ba \oplus b^2 = a \oplus ab \oplus ba \oplus b,$$

so, by the cancellation law for addition,

$$0 = ab \oplus ba,$$

and hence

$$-ab = ba.$$

If we put $b = a$, then we have, for all a in B,

(3.7.1) $\qquad -a = a, \qquad a \oplus a = 0.$

Since $ab \in B$, we have

$$-ab = ab,$$

but since $-ab = ba$, we have, for all a and b in B,

(3.7.2) $\qquad ab = ba.$ ∎

In a Boolean ring $[B, \oplus, \cdot]$ we define the **complement** $\bar{a}$ of an element a of B thus:

(3.7.3) $$\bar{a} = a \oplus 1,$$

and the **sum** or **alternation** or **join** $a + b$ of a and b by

(3.7.4) $$a + b = a \oplus b \oplus ab.$$

It is often convenient to read "$a + b$" as "a or b." It is essential to remember that the operation "$\oplus$," and *not* "$+$," represents the addition operation that appears in the general definition of a ring.

The operations $+$ and $^-$ exhibit a long list of familiar properties, which we assemble here for ease of reference.

Theorem 3.7.3. *In every Boolean ring* $[B, \oplus, \cdot]$, *for all* a, b, *and* c *in* B, *the operations* "$+$" *and* "$^-$" *have these basic properties*:
 (a) $0 + 0 = 0, 0 + 1 = 1 + 0 = 1 + 1 = 1$.
 (b) $a + b = b + a$.
 (c) $a + (b + c) = (a + b) + c$.
 (d) $a(b + c) = ab + ac$, $a + bc = (a + b)(a + c)$.
 (e) $a + a = a$.
 (f) $a + 0 = a$.
 (g) $a + 1 = 1$.
 (h) $a + \bar{a} = 1$, $a\bar{a} = 0$.
 (i) $\overline{(a + b)} = \bar{a}\bar{b}$, $\overline{(ab)} = \bar{a} + \bar{b}$.
 (j) $\overline{(\bar{a})} = a$.
 (k) $a \oplus b = a\bar{b} + \bar{a}b$.
 (l) *The complement* $\bar{a}$ *of* a *is unique*.

Proof. The proofs involve only straightforward computations of types used earlier and are left to the reader. ∎

We may now state

Theorem 3.7.4. *Every Boolean ring* $[B, \oplus, \cdot]$, *with* "$+$" *and* "$^-$" *defined as in* (3.7.3) *and* (3.7.4), *is a Boolean algebra*.

Proof. The laws of rings combined with those of Theorems 3.7.2 and 3.7.3 include all the defining properties of a Boolean algebra. ∎

In every Boolean ring, we may therefore use all the laws of Boolean algebra enumerated in Section 1.19.

The operations $\oplus$ and $+$ yield the same result under one condition, as the equation $a + b = (a \oplus b) \oplus ab$ reveals.

Theorem 3.7.5. *In a Boolean ring* $[B, \oplus, \cdot]$, $a + b = a \oplus b$ *iff* $ab = 0$. ∎

It is often useful to note that the condition $a_i a_j = 0$, $i \neq j$, $i, j = 1, 2, \ldots, k$, implies that
(3.7.5) $$a_1 + a_2 + \cdots + a_k = a_1 \oplus a_2 \oplus \cdots \oplus a_k,$$
as the reader may prove by induction. For $k > 2$, the condition is not a necessary one, however, as the following example shows. By the absorption law and the definition of "$+$," we have
$$x + y + xy = x + y = x \oplus y \oplus xy$$
regardless of whether or not $xy = 0$.

In an arbitrary Boolean algebra B, we define **inclusion** just as in Section 1.19, in the following way: For all a and b in B, $a \leq b$ ("a **is included in** b"; "b **includes** a") iff $ab = a$. Then, as the reader may prove using familiar techniques, we have results (I1) to (I6) of Section 1.19, and others, as shown in the following theorem.

Theorem 3.7.6. *For all a, b, and c of an arbitrary Boolean algebra B:*
(a) $a \leq a$ (*the reflexive property*).
(b) If $a \leq b$ and $b \leq a$, then $a = b$ (*the antisymmetric property*).
(c) If $a \leq b$ and $b \leq c$, then $a \leq c$ (*the transitive property*).
(d) B is a poset with respect to inclusion.
(e) $0 \leq a \leq 1$ (*universal bounds*).
(f) $a \leq b$ iff $a + b = b$.
(g) $a \leq b$ iff $a\bar{b} = 0$.
(h) $ab \leq a \leq a + b$. ∎

Many properties of inclusion were derived earlier in this chapter and in Chapter 1 from the list of properties above. These are consequently independent of the particular application and may be employed in any Boolean algebra. The properties of Boolean algebras in general will be developed further in Chapter 6, for these algebras are of considerable importance in multivalued switching theory.

3.8 Independent Postulates for a Boolean Algebra

In certain cases the usual list of postulates for an algebraic system is not, in fact, an independent set (that is, some of the postulates may be derived from others). For example, in the case of a ring R with unit 1, we have, for all a and b in R, the two expansions

$$(a+b)(1+1) = a(1+1) + b(1+1) = a + a + b + b,$$
$$(a+b)(1+1) = (a+b)1 + (a+b)1 = a + b + a + b,$$

so

$$a + a + b + b = a + b + a + b,$$

which, by the cancellation law for addition, implies that

$$a + b = b + a.$$

Thus postulating a unit element (M₂) renders the postulate of commutativity of addition redundant.

Similarly, in Section 1.19 we gave a natural and convenient list of nine postulates for a Boolean algebra, eight of which are, in fact, dual pairs. As was pointed out at the time, these postulates, although consistent, are not an independent set.

There exist several independent sets of postulates for a Boolean algebra. The best known of these appears in E. V. Huntington's classic paper [Huntington (1904)]. In our notation, his definition may be stated thus: A *Boolean algebra* $\mathscr{B}$ is a set on which are defined an equivalence relation " = " and binary operations " + " and " · " such that the following postulates hold:

P1. For all a and b of $\mathscr{B}$, $a + b$ and ab are uniquely defined elements of $\mathscr{B}$. (Closure.)

P2. There are elements 0 and 1 in $\mathscr{B}$ such that for all a in $\mathscr{B}$, $a + 0 = a$ and $a \cdot 1 = a$. (Identity elements.)

P3. For all a and b of $\mathscr{B}$, $a + b = b + a$ and $ab = ba$. (Commutativity.)

P4. For all a, b, and c of $\mathscr{B}$, $a + bc = (a + b)(a + c)$ and $a(b + c) = ab + ac$. (Distributivity.)

P5. For every element a of $\mathscr{B}$, there is an element $\bar{a}$ of $\mathscr{B}$ such that $a + \bar{a} = 1$ and $a\bar{a} = 0$. (Laws of complementarity.)

P6. There are at least two inequivalent elements in $\mathscr{B}$.

P7. If $a = b$, then for all c in $\mathscr{B}$, $a + c = b + c$ and $ac = bc$. (Principle of substitution.)

This last postulate is not explicitly in Huntington's list. It appears informally in his preliminary discussion of the properties of equivalence, just as it did in ours. We also did not include closure (P1) in our formal list, nor the assumption that $\mathscr{B}$ has at least two distinct elements (P6). Thus Huntington's four postulates P2 to P5 replace our list of nine.

Huntington proves the independence and the consistency of P1 to P6 and derives the other most familiar laws of Boolean algebra from these six in the following order:

H1. The elements 0 and 1 are unique.

H2. $a + a = a$ and $a \cdot a = a$. (Idempotency.)

H3. $a + 1 = 1$ and $a \cdot 0 = 0$. (Dominance of 0 and 1.)

H4. $a + ab = a$ and $a(a + b) = a$. (Absorption laws.)

H5. $\bar{a}$ is uniquely determined by a; $\bar{\bar{a}} = a$.

H6. $\overline{a + b} = \bar{a}\bar{b}$ and $\overline{ab} = \bar{a} + \bar{b}$. (De Morgan's laws.)

H7. $(a + b) + c = a + (b + c)$ and $(ab)c = a(bc)$. (Associativity.)

The relation " ≤ " is now defined thus: $a \leq b$ iff $a + b = b$. One has then

H8. $a \leq b$ iff $ab = a$ iff $\bar{a} + b = 1$ iff $a\bar{b} = 0$.

Proofs follow familiar patterns and are left to the reader.

Brevity and the fact that laws come in dual pairs make Huntington's list a mathematically elegant list. In the case of a given finite system, it may be necessary to establish distributivity by an examination of cases, which can be tedious. An advantage of the Huntington postulates is that one does not have the same chore for associativity since it is not in the postulate list but is rather implied by the others.

3.9 Exercises

*1. Prove that in every Boolean algebra

$$\overline{b_1 \oplus b_2 \oplus \cdots \oplus b_k} = b_{i_1} \oplus b_{i_2} \oplus \cdots \oplus b_{i_r} \oplus \overline{(b_{i_{r+1}} \oplus \cdots \oplus b_{i_k})}$$
$$= \bar{b}_{i_1} \oplus \bar{b}_{i_2} \oplus \cdots \oplus \bar{b}_{i_{2s-1}} \oplus b_{i_{2s}} \oplus \cdots \oplus b_{i_k}$$

and

$$b_1 \oplus b_2 \oplus \cdots \oplus b_k = \bar{b}_{i_1} \oplus \bar{b}_{i_2} \oplus \cdots \oplus \bar{b}_{i_{2s}} \oplus b_{i_{2s+1}} \oplus \cdots \oplus b_{i_k}$$
$$= \overline{(b_{i_1} \oplus b_{i_2} \oplus \cdots \oplus \bar{b}_{i_{2s}})} \oplus b_{i_{2s+1}} \oplus \cdots \oplus b_{i_k}$$

where $i_1, i_2, \ldots, i_k$ is any permutation of $1, 2, \ldots, k$ and $1 \leq r < k$, $2 \leq 2s \leq k$.

2. Simplify $(a+b+c)(a+b)(a+c)(b+c)$, where a, b, and c are elements of an arbitrary Boolean algebra $\mathcal{B}$.

*3. Let a, b, and c belong to a Boolean algebra $\mathcal{B}$. Prove that:
(a) $a \leq b$ implies that $(a+c) \leq (b+c)$ for all c.
(b) $a+b \leq c$ iff $a \leq c$ and $b \leq c$.
(c) $ab \geq c$ iff $a \geq c$ and $b \geq c$.
(d) $a \leq b$ iff $a\bar{b} = 0$ iff $\bar{a}+b = 1$.
(e) $a = b$ iff $a\bar{b} + \bar{a}b = 0$.
(f) If $a \geq b \geq c$, then $a\bar{b} + b\bar{c} = a\bar{c}$.
(g) If $a \leq c$ and $b \leq c$, then $a \oplus b \leq c$.
(h) $a+c = b+c$ iff $a \oplus b \leq c$.
(i) $ac = bc$ iff $c \leq \overline{a \oplus b}$.

4. Solve each of $ab = 0$ and $a+b = 1$ for b; then find the simultaneous solution. (This demonstrates the uniqueness of the complement.)

5. For the sets A and B shown in Figure 3.9.1, find $A \oplus B$ if (a) the boundaries of A and B are included in these sets, (b) the boundaries are not part of the sets.

6. Prove theorems H1 to H8 of Section 3.8.

7. Prove that if $a+b = \text{l.c.m.}\,(a,b)$, $ab = \text{g.c.d.}\,(a,b)$, and $\bar{a} = 30/a$, then the set of positive integer divisors of 30 constitutes a Boolean algebra. What does $a \leq b$ mean in this algebra?

8. Prove Theorem 3.7.6 in detail.

*9. A nonsymmetrical but very compact set of postulates for a Boolean algebra is the following. Not counting closure and the substitution principle,

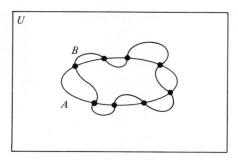

FIGURE 3.9.1. *Venn Diagram for Exercise 5*

there are in this list five individual operational laws versus Huntington's four *pairs*.

We define a *Boolean algebra* to be a set $\mathscr{B}$ of elements on which are defined a binary operation "·" (denoted by juxtaposition) and a unary operation "¯" such that the following postulates hold:

A1. For all a and b in $\mathscr{B}$, ab and $\bar{a}$ are uniquely defined elements of $\mathscr{B}$.
A2. For all a and b in $\mathscr{B}$, $ab = ba$.
A3. For all a, b, and c in $\mathscr{B}$, $(ab)c = a(bc)$.
A4. There is an element 0 in $\mathscr{B}$ such that for all a in $\mathscr{B}$, $a\bar{a} = 0$.
A5. For all a and b in $\mathscr{B}$, $a\bar{b} = 0$ implies that $ab = a$.
A6. For all a and b in $\mathscr{B}$, $ab = a$ implies that $a\bar{b} = 0$.
A7. If a, b, and c are in $\mathscr{B}$ and $a = b$, then $ac = bc$ and $\bar{a} = \bar{b}$.

The element "1," the binary operation "+," and the relation "$\leq$" are now introduced by means of these definitions:

D1. "1" means the unique element $\bar{0}$ of $\mathscr{B}$.
D2. $a + b$ means the unique element $\overline{\bar{a}\bar{b}}$ of $\mathscr{B}$.
D3. $a \leq b$ means that $ab = a$.

Postulate A4 implies that $\mathscr{B}$ is nonempty, since it contains at least the element "0." This single element with the rules $0 \cdot 0 = 0$, $0 + 0 = 0$, and $\bar{0} = 0$ satisfies all the postulates and is consistent with the definitions D1 to D3. A further postulate excludes this special algebra and makes the two definitions of a Boolean algebra equivalent:

A8. There are at least two elements a and b in $\mathscr{B}$ such that $a \neq b$.

With this postulate added to the list, one can show that $0 \neq 1$.

That these postulates define a Boolean algebra in the sense of Section 1.19 is now a consequence of the following sequence of theorems, each of which is needed to prove one of the later theorems in the list or is one of the basic laws of Boolean algebra:

For all a, b, and c in $\mathscr{B}$,
T1. $a \cdot a = a$.
T2. $ab \leq a$.
T3. $0 \leq a$.

T4. $a \cdot 0 = 0$.
T5. $a \leq b$ iff $a\bar{b} = 0$.
T6. $a \leq b$ iff $a + b = b$.
T7. $a \leq b$ and $b \leq a$ imply that $a = b$.
T8. $(\bar{a}) = a$.
T9. $a \leq b$ iff $\bar{b} \leq \bar{a}$.
T10. $a + a = a$.
T11. $(a+b)+c = a+(b+c)$.
T12. $\overline{ab} = \bar{a} + \bar{b}$.
T13. $\overline{a+b} = \bar{a}\bar{b}$.
T14. $a + \bar{a} = 1$.
T15. $a + 1 = 1$.
T16. $a \cdot 1 = a$.
T17. $a + 0 = a$.
T18. $a + b = b + a$.
T19. $a \leq a + b$.
T20. $a(a+b) = a$.
T21. $a + ab = a$.
T22. $a(\bar{a} + b) = ab$.
T23. $a \leq b$ implies that $ac \leq bc$ and $a + c \leq b + c$.
T24. $a \leq c$ and $b \leq c$ imply that $ab \leq c$ and $a + b \leq c$.
T25. $a(b+c) = ab + ac$.
T26. $a + bc = (a+b)(a+c)$.

For a challenging exercise, prove this list of theorems. [Some help with the harder ones can be found in Rosenbloom (1950, pp. 10–12).]

10. Which of the postulates A1 to A7 of Exercise 9 are not satisfied if $\mathcal{B}$ is the set of positive integral divisors of 24, $\bar{a} = 24/a$, and $ab = $ g.c.d. (a, b)?

11. Let U be the set of all binary expansions

$$a = 0.a_1 a_2 a_3 \cdots$$

of real numbers in the closed interval [0, 1]. If

$$b = 0.b_1 b_2 b_3 \cdots,$$

define $a = b$ to mean that a and b represent the same real number,

$$ab = 0.(a_1 \cdot b_1)(a_2 \cdot b_2)(a_3 \cdot b_3) \cdots,$$

and

$$\bar{a} = 0.\bar{a}_1 \bar{a}_2 \bar{a}_3 \cdots,$$

where, as usual, $\bar{0} = 1$, $\bar{1} = 0$, $0 \cdot 0 = 0 \cdot 1 = 1 \cdot 0 = 0$, and $1 \cdot 1 = 1$. Show that postulates A1 to A6 of Exercise 9 hold for $[U, \cdot, ^-]$ but that A7 does not. This proves that postulate A7 is independent of the other six. By constructing other such systems, one can prove the mutual independence of the postulates of this set.

3.10 The Algebraic Description of Logic Circuits

The laws of Boolean algebra listed earlier provide mathematical models of certain classes of electronic or electromechanical circuits known as **switching circuits**. We consider here the class commonly known as **logic circuits** to illustrate how a mathematical model may be constructed from simple physical assumptions. For those already familiar with this application, a quick reading of Sections 3.10 and 3.11 should suffice. However, the exercises in Section 3.12 are designed to provide a deeper and broader than average understanding of the algebra and should not be neglected.

In electronic logic circuits, **inputs**, typically in the form of voltages, are applied to component circuits, called **logic elements** or **gates** (constructed from transistors, diodes, resistors, capacitors, and so on) by means of wires that are called **input leads** [Figure 3.10.1(a)]. The logic elements process these voltages and produce **outputs**, again in the form of voltages, that appear on wires known as **output leads**.

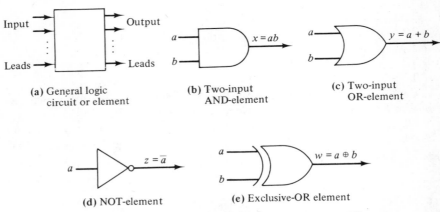

FIGURE 3.10.1. *Standard Symbols for Common Logic Elements*

In Figure 3.10.1, the output of what is called an **AND-gate** or **AND-element** is denoted by the product of the inputs, the output of what is called an **OR-gate** or **OR-element** is denoted by the sum (+) of the inputs, the output of a **NOT-element** or **inverter** is denoted by the complement (¯) of the input, and the output of an **exclusive-OR element** is denoted by the symbol $\oplus$, called the **ring sum**, all symbols being chosen to agree with standard usage in the integrated circuits industry. Since the voltages that appear on the several leads of a logic circuit are ordinarily not fixed, we call a and b in these diagrams **input variables** and call w, x, y, and z **output variables**. The operations addition, multiplication, complementation, and ring sum (as well as other binary operations) are often called **connectives**.

Sec. 3.10 The Algebraic Description of Logic Circuits

Logic circuits can be highly complex, but unless memory elements must be included (which we exclude for now) they can all be designed as suitably interconnected assemblages of three basic kinds of logic elements: the *two-input AND-element*, the *two-input OR-element*, and the *NOT-element* or *inverter*. Actually, as we shall see, the NOT-element and just one of the other two types would suffice. Other logic elements, such as the exclusive-OR or *half-adder*, multiple-input AND- and OR-elements, and NAND- and NOR-elements, for example, often reduce complexity and costs, and will be discussed later.

In logic circuits, the output of one element may serve as the input to one or more following elements. Recognizing this, we can describe the diagrams of logic circuits not employing memory elements, which we call **combinational circuits**, by algebraic **output expressions** involving the ultimate input variables, the connectives $+$, $\cdot$, and $^{-}$, and appropriately distributed parentheses. Examples appear in Figure 3.10.2. The ultimate output expressions are constructed by representing outputs of logic elements, beginning with those nearest the original inputs, proceeding to those next nearest, and so on, writing the sum, product, or complement of the inputs for each OR-, AND-, and NOT-element, and employing sufficient parentheses to make the order of application of the connectives fully clear.

Note that the output expression of either of the circuits in Figure 3.10.2 fully implies the circuit diagram. In general, an expression defining a finite *sequence* of OR's, AND's, and NOT's, applied to a given set of input

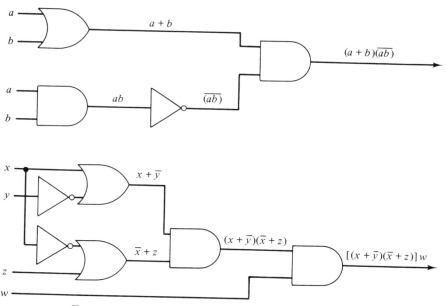

FIGURE 3.10.2. *Output Expressions for Logic Circuits*

variables, defines uniquely the diagram of a combinational circuit. However, as will later become clear, it is not necessary or useful in practice to write expressions that are so painfully explicit, for it is not essential that an expression define a unique circuit.

Two-valued logic circuits, the kind we treat here, employ voltages of two levels that are reliably distinguishable by the components being used. These are applied, from external sources or internally by logic elements, to the various leads of a logic circuit. For convenience, one of these voltages is given the name "0," the other the name "1." (The actual voltages could be, respectively, 0 V and 5 V, or -0.7 V and -1.6 V, or the reverse, and so on, depending on the physical components being used.) Then input and output variables have values belonging to the set $\{0, 1\}$. (Inputs and outputs may also be constants, of course.)

A voltage 1 appears on the output lead of a two-input AND-element (we say it "outputs" 1) iff the voltages on both of its input leads are 1 (briefly, "both inputs are 1"). An OR-element outputs a 1 iff at least one input is 1. A NOT-element outputs 0 when the input is 1 and outputs 1 when the input is 0. These behaviors explain the names of these elements. Graphical representations of these behaviors appear in Figure 3.10.3.

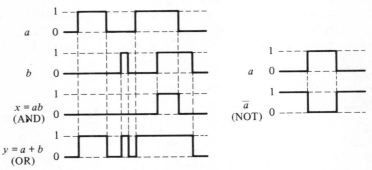

FIGURE 3.10.3. *Input–Output Behavior of Basic Logic Elements*

Figure 3.10.3 reflects the fact that to keep the model simple, we assume that the voltages "0" are all precisely the same, and similarly for the voltages "1." We assume also that changes in input or output voltages are instantaneous rather than gradual, hence the square corners on the voltage graphs in Figure 3.10.3. We assume that the output responds instantaneously to any change in the input. Of course, none of these assumptions hold precisely true for physical devices. Voltage levels vary somewhat and tend to deteriorate as various logic elements process them. Changes in level are always to some degree gradual rather than sharp, and logic elements delay voltage changes to some extent, so output does not really respond instantaneously to input. These departures from the ideal are in some cases of no consequence because the circuits in question can tolerate some variation. In other cases they must be dealt with by special engineering techniques, such as the introduction of amplifiers, signal reshaping circuits, delay elements, and so on. Sometimes more sophisticated logical designs must be employed to avoid excessive input or output requirements. In all cases, however, the ideal model we present here is the basic design tool.

3.11 Switching Functions and Their Basic Properties

A function-table description of the behavior of the three basic logic elements appears in Table 3.11.1.

TABLE 3.11.1. The basic logic functions.

a	b	$x = ab$	$y = a + b$	$z = \bar{a}$
0	0	0	0	1
0	1	0	1	1
1	0	0	1	0
1	1	1	1	0

If we make once more the familiar definitions:

$$0 \cdot 0 = 0 \cdot 1 = 1 \cdot 0 = 0, \quad 1 \cdot 1 = 1,$$
$$0 + 0 = 0, \quad 0 + 1 = 1 + 0 = 1 + 1 = 1,$$
$$\bar{0} = 1, \quad \bar{1} = 0,$$

the symbols "$\cdot$," "$+$," and "$^-$" now represent not only special circuit elements but arithmetic operations on 0 and 1 as well. Using this special arithmetic, we can compute the values of arbitrary output expressions at arbitrary combinations of values of the input variables. For example, at $a = 0$, $b = 1$, $(a+b)(\overline{ab})$ has the value $(0+1)(\overline{0 \cdot 1}) = 1 \cdot \bar{0} = 1 \cdot 1 = 1$.

The question that now arises is this: If the input voltages are those corresponding to the values assigned to the input variables, is the corresponding physically determined output voltage correctly represented by the computed value of the output expression? Certainly, the answer is "yes" for our three basic logic elements, for we have defined the arithmetic of 0 and 1 so as to guarantee it. The general answer is given by the following theorem.

Theorem 3.11.1. *Given the output expression $f(X)$ of a combinational logic circuit with input variables $x_1, x_2, \ldots, x_n$, the output voltage of the circuit resulting from input voltages corresponding to any combination X_0 of values of the input variables is correctly represented by $f(X_0)$.*

Proof. The output expression represents a finite sequence of applications of the connectives OR, AND, and NOT to the input variables $x_1, x_2, \ldots, x_n$.

For expressions involving only one of these connectives, the theorem holds by definition of the arithmetic of 0 and 1. Now let k denote any integer such that the theorem holds for any output expression involving a sequence of k or fewer of these connectives. Let $f(X)$ denote any output expression involving $k+1$ of these connectives. Then, according as "+," "·," or "¯" is the last connective employed, $f(X)$ has the form $g(X)+h(X)$, $g(X) \cdot h(X)$, or $\overline{(g(X))}$, where $g(X)$ and $h(X)$ represent output expressions involving k or fewer connectives. Applying the induction hypothesis first to $g(X)$ and $h(X)$, and then to $f(X)$ regarded as an expression operating on the values of $g(X)$ and $h(X)$ by one of the three basic connectives, we conclude that for every combination X_0 of values of the input variables, $f(X_0)$ correctly represents the output of the circuit of which $f(X)$ is the output expression. ∎

In view of the theorem just proved, an output expression is not merely a device for representing a certain interconnection of logic elements; it also defines the output values of a combinational logic circuit as a function of the input values. We call the function $f: \mathbb{B}^n \to \mathbb{B}$ defined by the output expression $f(X)$ the **output function** or the **switching function** of the circuit, and the circuit is said to **implement** the function. Switching functions are also called **Boolean functions** or **transmissions**.

Let $\mathbb{F}_n$ denote the set of all switching functions of two-valued logic circuits with n input variables. Then the only definitions for the sum, product, and complement of functions f and g of $\mathbb{F}_n$ that are consistent with the already defined arithmetic of 0 and 1 are the following:

$$\forall X_0 \in \mathbb{B}^n, \quad (f+g)(X_0) = f(X_0) + g(X_0),$$
$$(fg)(X_0) = f(X_0)g(X_0),$$
$$\bar{f}(X_0) = \overline{(f(X_0))}.$$

In view of these definitions, we may treat the basic logic elements as processors of functions, not just of zeros and ones (Figure 3.11.1).

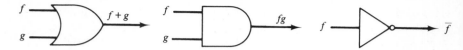

FIGURE 3.11.1. *Logic Elements as Function Processors*

Another consequence of these definitions is stated in the following theorem.

Theorem 3.11.2. *With respect to the operations $+$, $\cdot$, and $^-$ just defined, the set $\mathbb{F}_n$ of all switching functions of n variables constitutes a Boolean algebra.*

Sec. 3.11 Switching Functions and Their Basic Properties

That is, for all f, g, and h in $\mathbb{F}_n$,

$$f + g = g + f, \quad fg = gf \quad (commutativity),$$
$$(f + g) + h = f + (g + h), \quad (fg)h = f(gh) \quad (associativity),$$
$$f(g + h) = fg + fh, \quad f + gh = (f + g)(f + h) \quad (distributivity),$$
$$f + 0 = f, \quad f \cdot 1 = f \quad (identity\ elements),$$
$$f + 1 = 1, \quad f \cdot 0 = 0 \quad (dominant\ elements),$$
$$f + f = f, \quad f \cdot f = f \quad (idempotency),$$
$$f + \bar{f} = 1, \quad f \cdot \bar{f} = 0 \quad (complementarity),$$
$$\overline{(f + g)} = \bar{f} \cdot \bar{g}, \quad \overline{(fg)} = \bar{f} + \bar{g} \quad (De\ Morgan's\ laws),$$
$$\overline{(\bar{f})} = f \quad (involution),$$
$$f + fg = f, \quad f(f + g) = f \quad (absorption),$$
$$f + \bar{f}g = f + g, \quad f(\bar{f} + g) = fg \quad (redundancy),$$
$$fg + \bar{f}h + gh = fg + \bar{f}h, \quad (f + g)(\bar{f} + h)(g + h) = (f + g)(\bar{f} + h)$$
$$(consensus).$$

(In these rules, multiplication has priority over addition, as usual. Also, "0" represents the *function* 0, that is, the function that has the value 0 for all combinations of values of the input variables. Similarly, "1" represents the function 1.)

Proof. The values of f, g, and h are 0's and 1's and these identities all hold for every combination of values of these functions. This may be verified in detail by means of function tables that are sometimes called *truth tables*. For example, consider the distributive law $f + gh = (f + g)(f + h)$. Table 3.11.2 lists the eight possible combinations of values of f, g, and h and the

TABLE 3.11.2. Function values for all $X_0 \in \mathbb{B}^n$.

f g h	gh	$f + gh$	$f + g$	$f + h$	$(f + g)(f + h)$
0 0 0	0	0	0	0	0
0 0 1	0	0	0	1	0
0 1 0	0	0	1	0	0
0 1 1	1	1	1	1	1
1 0 0	0	1	1	1	1
1 0 1	0	1	1	1	1
1 1 0	0	1	1	1	1
1 1 1	1	1	1	1	1

corresponding values of $f+gh$ and $(f+g)(f+h)$. Since these two expressions have the same values under all conditions, the functions $f+gh$ and $(f+g)(f+h)$ are identically equal. [Note that row 1 of the table compares the values of $f+gh$ and $(f+g)(f+h)$ for all $X_0 \in \mathbb{B}^n$ such that $f(X_0) = g(X_0) = h(X_0) = 0$, and similarly for the other rows. Thus Table 3.11.2 compares function values for all $X_0 \in \mathbb{B}^n$, regardless of how large n may be.] The other laws may be established similarly. ∎

The function table approach just illustrated examines all possible cases and hence is a valid general procedure for proving identities in $\mathbb{F}_n$. However, algebraic procedures are often much more economical. For example, to prove that $\forall f, g, h$, and k in $\mathbb{F}_n$,

$$fg + \bar{f}h + \bar{h}k + gk = fg + \bar{f}h + \bar{h}k + gh,$$

would require a 16-row table, but if we apply the law of consensus twice, once to introduce gh and once to remove gk, we have

$$\overline{fg + \bar{f}h} + \bar{h}k + gk = fg + \bar{f}h + \overline{\bar{h}k + gk + gh} = fg + \bar{f}h + \bar{h}k + gh.$$

Actually, many of the laws in Theorem 3.11.2 could be derived from a small subset thereof by algebraic means. (See Section 3.8.)

In view of the associative laws, the expressions $f+g+h$ and fgh are unambiguous, and since these laws generalize in the usual way, so are $f_1+f_2+\cdots+f_k$ and $f_1f_2\cdots f_k$, where the f_i are arbitrary members of $\mathbb{F}_n$ and k is any positive integer. Thus switching functions need not employ, in addition to NOT-elements, *only* two-input AND's and OR's. Indeed, multiple-input AND- and OR-elements are available for the representation of these extended sums, products, and their complements, the latter being called NOR- and NAND-elements, respectively (Figure 3.11.2).

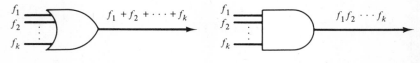

(a) Multiple-input OR-element   (b) Multiple-input AND-element

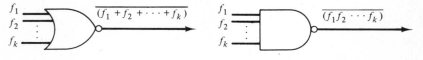

(c) Multiple-input NOR-element   (d) Multiple-input NAND-element

FIGURE 3.11.2. *Useful Multiple-Input Logic Elements*

Sec. 3.11 Switching Functions and Their Basic Properties

One writes switching functions for circuits containing multiple-input elements just as before; also, circuits corresponding to given switching functions and employing multiple-input logic elements may be drawn just as before (Figure 3.11.3). However, a given switching function no longer defines a unique circuit, since it is always possible to implement a given function with different combinations of logic elements, as will be illustrated presently.

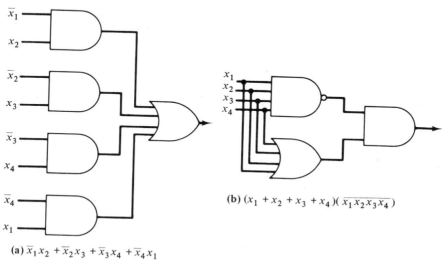

(a) $\bar{x}_1 x_2 + \bar{x}_2 x_3 + \bar{x}_3 x_4 + \bar{x}_4 x_1$

(b) $(x_1 + x_2 + x_3 + x_4)(\overline{x_1 x_2 x_3 x_4})$

FIGURE 3.11.3. *Example of Equivalent Circuits*

Given an identity in $\mathbb{F}_n$, we can draw a circuit for each of its members. Ordinarily, these circuits are distinct, but they determine the same function from $\mathbb{B}^n$ to $\mathbb{B}$. Two such circuits are called **equivalent**. In particular, the output expression for one circuit may often be transformed algebraically into an identically equal expression that leads to a more useful or more economical circuit.

For example, let

(a) $$f(x_1, x_2, x_3, x_4) = \bar{x}_1 x_2 + \bar{x}_2 x_3 + \bar{x}_3 x_4 + \bar{x}_4 x_1.$$

Then, since De Morgan's laws generalize thus:

$$\overline{(f_1 + f_2 + \cdots + f_k)} = \bar{f}_1 \bar{f}_2 \cdots \bar{f}_k,$$
$$\overline{(f_1 f_2 \cdots f_k)} = \bar{f}_1 + \bar{f}_2 + \cdots + \bar{f}_k,$$

as the reader may prove by induction, we have

$$\bar{f}(x_1, x_2, x_3, x_4) = (x_1 + \bar{x}_2)(x_2 + \bar{x}_3)(x_3 + \bar{x}_4)(x_4 + \bar{x}_1)$$
$$= x_1 x_2 x_3 x_4 + \bar{x}_1 \bar{x}_2 \bar{x}_3 \bar{x}_4,$$

so
(b) $\quad f(x_1, x_2, x_3, x_4) = \overline{(x_1 x_2 x_3 x_4)}(x_1 + x_2 + x_3 + x_4).$

Circuits implementing (a) and (b) appear in Figure 3.11.3. If these two circuits were being fabricated from logic elements, the second would be more economical, for there are two fewer logic elements in the latter and two fewer input wires to be connected to gates. The second circuit also has the sometimes helpful property of not requiring complements as inputs; the NAND-element is what is responsible for this special circumstance.

There are many tricks of algebraic manipulation other than those illustrated in this section. The only way to learn them is by lots of practice. The exercises that follow in the next section are designed to provide some of this practice.

When working the exercises, the reader should not forget that the *principle of duality* holds here, just as it does in all Boolean algebras.

3.12 Exercises

1. Write each of the following as a product of sums of the functions f, g, h and, as needed, their complements:
(a) $f + ghk$.
(b) $f\bar{g} + \bar{f}g$.
(c) $fg + \bar{f}\bar{g}$.
(d) $fg + \bar{f}h$.
(e) $fg + hk$.
(f) $f\bar{g} + g\bar{h} + h\bar{f}$.
(g) $f\bar{g}h + \bar{f}gh$.
(h) $fg + gh + fh$.

2. Prove each of the following Boolean identities:
(a) $(x + \bar{y})(y + \bar{z})(z + \bar{x}) = (\bar{x} + y)(\bar{y} + z)(\bar{z} + x).$
(b) $(x + \bar{y})(\bar{x} + y)(x + z) = (x + \bar{y})(\bar{x} + y)(y + z).$
(c) $\overline{\bar{x}y + z} \cdot \overline{\bar{x} + z} + (\bar{x}y + z)(\bar{x} + z) = x + y + z.$
(d) $(f + g + h)(f + h + \bar{k})(g + k) = (f + h + \bar{k})(g + k).$
(e) $\bar{f}g + fh + \bar{g}h = gh + \bar{g}h + fh + \bar{f}h.$
(f) $\bar{f}\bar{g} + \bar{f}g + f\bar{g} + fg = 1.$

*__3.__ Prove that if $f = gh + \bar{g}k$, then $\bar{f} = g\bar{h} + \bar{g}\bar{k}$; in particular, $\overline{gh + \bar{g}h} = gh + \bar{g}\bar{h}.$

4. Prove that if $x + y + z = xyz$, then $x = y = z$.

5. Is this statement correct?
$$\overline{x_1 x_2 + x_2 x_3 + x_3 x_1} = \bar{x}_1 \bar{x}_2 + \bar{x}_2 \bar{x}_3 + \bar{x}_3 \bar{x}_1.$$

*__6.__ Prove that $ab(ax + by) = ab(x + y)$, and use this result to express $abx + aby + acxz + bcyz$ as a product of two factors involving altogether eight literals. (A **literal** is a symbol for a Boolean function or variable, such as $f, \bar{a},$ and x_i, for example.)

7. Show that the expression
$$x_1 x_2 x_3 + x_1 x_3 x_4 x_5 + x_1 x_2 x_4 x_6 + x_4 x_5 x_6$$
can be factored into a form requiring only eight literals.

8. Simplify these output expressions:
(a) $(f+g)(\bar{f}+h)(\bar{g}+h)$.
(b) $(f+g+h)(f+\bar{g})\bar{h}$.
(c) $fg + \bar{f}h + \bar{g}h + h$.
(d) $(x_1+x_2+x_3+x_4)(x_1+x_2+x_3+\bar{x}_4)(x_1+x_2+\bar{x}_3+x_4)$.
(e) $x + y(zw + \bar{z}\bar{w}) + \bar{x}y\bar{w}$.
(f) $\bar{x}_1\bar{x}_2 + \bar{x}_1 x_2 + x_1\bar{x}_2$.

The object in each case is to obtain an equivalent expression with the smallest possible number of literals.

9. Prove that if $f+g=g$ and $\bar{f}g=0$, then $f=g$. Then state the dual result. Use these results to prove the laws of consensus.

10. Simplify each of the following, using exactly one identity from Theorem 3.11.2 in each case:
(a) $(fg+h)+(fg+h)(r+st)$.
(b) $(f+g)+\bar{f}\bar{g}h$.
(c) $fg + \bar{f}(\bar{g}+h) + gh$.
(d) $(f+gh)(f+gh+rs)$.

11. Prove that for all $n \geq 2$,

$$x_1\bar{x}_2 + x_2\bar{x}_3 + \cdots + x_{n-1}\bar{x}_n + x_n\bar{x}_1 = \left(\sum_{i=1}^{n} x_i\right)\overline{\left(\prod_{i=1}^{n} x_i\right)}.$$

***12.** Prove the generalized De Morgan laws by induction.

***13.** Generalize the distributive laws and prove your generalizations by induction.

14. Given that

$$gh\bar{f} + \bar{g}\bar{h}f = 0,$$
$$hk\bar{f} + \bar{h}\bar{k}f = 0,$$
$$gk\bar{f} + \bar{g}\bar{k}f = 0,$$

prove that

$$f = gh + hk + kg.$$

(This one is rather difficult.)

15. Write the output expression for each circuit in Figure 3.12.1, and then simplify and draw corresponding equivalent circuits, with the object of minimizing the number of inputs to gates.

***16.** Prove that every Boolean function can be expressed in terms of the operators " $+$ " and " $^-$ " only, also in terms of " $\cdot$ " and " $^-$ " only.

***17.** Given that $f(x_1, x_2, \ldots, x_n)x_j = g(x_1, x_2, \ldots, x_n)x_j$ for all values of j and x_j, show by an example that it does not follow that $f = g$.

18. Write the switching function of the circuit in Figure 3.12.2. Then redesign it as economically as possible, using exactly one NOT-element and some AND- and OR-elements.

19. Prove that if $fg + h\bar{g} = 1$, then $f + h = 1$. (Remember that 1 here denotes the function 1, which has the value 1 for all combinations of values of the input variables.)

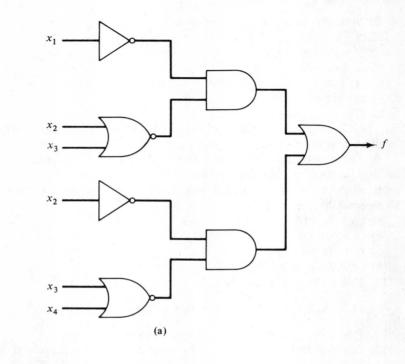

(a)

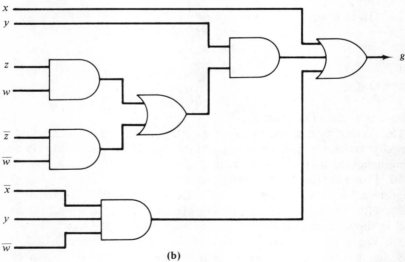

(b)

FIGURE 3.12.1. *Circuits for Exercise 15*

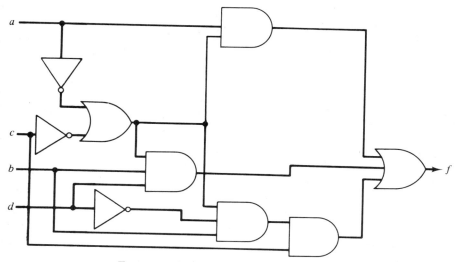

FIGURE 3.12.2. *Circuit for Exercise 18*

20. Determine all combinations of values of x_1, x_2, x_3, and x_4 that simultaneously satisfy

$$(x_1 x_2 + \bar{x}_3)\bar{x}_4 = 1 \quad \text{and} \quad \overline{(x_1 + x_4 + x_2)x_3} = 0.$$

21. Prove the identity

$$(f + hk\bar{g})(g + hk) = fg + hk\bar{g} = (fg + hk)(f + \bar{g}).$$

22. Write the switching function of the circuit in Figure 3.12.3. Then redesign it to require only AND-elements, OR-elements, and a single NOT-element.

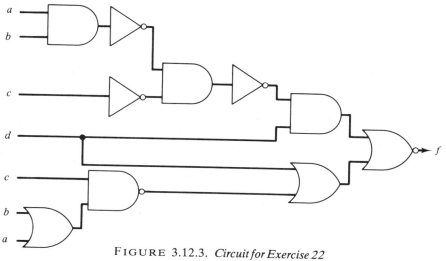

FIGURE 3.12.3. *Circuit for Exercise 22*

3.13 Disjunctive and Conjunctive Normal Forms

In practice, a switching function f of n variables $x_1, x_2, \ldots, x_n$ is often defined initially by a table that lists the value of f at each of the 2^n combinations of values of the x's, or by a verbal statement that implies such a table. Since each of the 2^n values of f is 0 or 1, we have this result.

Theorem 3.13.1. *There are $2^{(2^n)}$ switching functions of n variables.* ∎

For $n = 1, 2, 3$, and 4 we have $2^{(2^n)} = 4, 16, 256$, and 65,536, respectively. (Note the pattern here. Prove that it holds in general and determine the count for $n = 5$.)

When a switching function is defined by a table, we may wish to write the function in one of several standard forms or in a reduced form of some specific type. For the discussion of these forms, it is helpful to introduce some additional notation.

Let f denote a switching function from $\mathbb{B}^n$ to $\mathbb{B}$, let $f(X)$ denote some algebraic representation of f in terms of the variables $x_1, x_2, \ldots, x_n$, and let $(e_1, e_2, \ldots, e_n)$ be a particular combination of values of the x_i. Then if

$$j_{\text{dec}} = (e_1 e_2 \cdots e_n)_{\text{bin}},$$

that is, if $(e_1, e_2, \ldots, e_n)$ is the jth *combination* of values of the x_i, we write as a convenient abbreviation

$$f(e_1, e_2, \ldots, e_n) = f(j).$$

For example, if $n = 6$, $23_{\text{dec}} = 010111_{\text{bin}}$, so $f(0, 1, 0, 1, 1, 1) = f(23)$. (What happens when there is only one variable, x_1?)

We also define

$$0^0 = \bar{0} = 1, \qquad 0^1 = 0; \qquad 1^0 = \bar{1} = 0, \qquad 1^1 = 1,$$

so if $e = 0$ or 1,

$$e^{\bar{e}} = 0, \qquad e^e = 1,$$

and for each x_k,

$$x_k^0 = \bar{x}_k, \qquad x_k^1 = x_k.$$

For each n there are two sets of 2^n functions from $\mathbb{B}^n$ to $\mathbb{B}$ known as the **minterms** or **minimal polynomials** represented by

$$p_i(X) = x_1^{e_1} x_2^{e_2} \cdots x_n^{e_n},$$

and the **maxterms** or **maximal polynomials** represented by

$$s_i(X) = x_1^{\bar{e}_1} + x_2^{\bar{e}_2} + \cdots + x_n^{\bar{e}_n} = \overline{p_i(X)},$$

Sec. 3.13 Disjunctive and Conjunctive Normal Forms

where each e_j is 0 or 1 and where again $i_{dec} = (e_1 e_2 \cdots e_n)_{bin}$, so $i = 0, 1, 2, \ldots, 2^n - 1$. We often abbreviate $p_i(X)$ and $s_i(X)$ to p_i and s_i, respectively, when this causes no confusion. Also, as with any switching function f, we denote the value of p_i at the jth combination of values of the x's by $p_i(j)$.

For example, if $n = 3$, the minterms and maxterms are as shown in Table 3.13.1. In this table, note that 0's in the ith combination (e_1, e_2, e_3) imply complements of the corresponding variables of the minterm p_i, whereas the 1's imply complements of the corresponding variables of the maxterm s_i. This makes these products and sums easy to write. For another example, if $n = 5$, then $11_{dec} = 01011_{bin}$, so $p_{11} = \bar{x}_1 x_2 \bar{x}_3 x_4 x_5$ and $s_{11} = x_1 + \bar{x}_2 + x_3 + \bar{x}_4 + \bar{x}_5$.

TABLE 3.13.1. Minterms and maxterms for $n = 3$.

i	e_1	e_2	e_3	p_i	s_i
0	0	0	0	$\bar{x}_1 \bar{x}_2 \bar{x}_3$	$x_1 + x_2 + x_3$
1	0	0	1	$\bar{x}_1 \bar{x}_2 x_3$	$x_1 + x_2 + \bar{x}_3$
2	0	1	0	$\bar{x}_1 x_2 \bar{x}_3$	$x_1 + \bar{x}_2 + x_3$
3	0	1	1	$\bar{x}_1 x_2 x_3$	$x_1 + \bar{x}_2 + \bar{x}_3$
4	1	0	0	$x_1 \bar{x}_2 \bar{x}_3$	$\bar{x}_1 + x_2 + x_3$
5	1	0	1	$x_1 \bar{x}_2 x_3$	$\bar{x}_1 + x_2 + \bar{x}_3$
6	1	1	0	$x_1 x_2 \bar{x}_3$	$\bar{x}_1 + \bar{x}_2 + x_3$
7	1	1	1	$x_1 x_2 x_3$	$\bar{x}_1 + \bar{x}_2 + \bar{x}_3$

The basic property of minterms and maxterms is given in the following theorem.

Theorem 3.13.2. *For $i_{dec} = (e_1 e_2 \cdots e_n)_{bin}$ and $j_{dec} = (\varepsilon_1 \varepsilon_2 \cdots \varepsilon_n)_{bin}$, and for all i and j in $\{0, 1, 2, \ldots, 2^n - 1\}$,*

$$p_i(j) = \delta_{ij}, \qquad s_i(j) = \bar{\delta}_{ij},$$

where $\delta_{ij} = 1$ if $i = j$ but $\delta_{ij} = 0$ if $i \neq j$.

That is, p_i has the value 1 at the ith combination and has the value 0 everywhere else, while s_i has the value 0 at the ith combination and has the value 1 everywhere else. Thus, *among nonconstant functions*, each p_i has the value 1 at the minimum possible number of combinations while each s_i has the value 1 at the maximum possible number of combinations; this explains the names given these functions.

Proof. We have $p_i(j) = \varepsilon_1^{e_1} \varepsilon_2^{e_2} \cdots \varepsilon_n^{e_n}$, which is 1 iff each factor is 1, that is, iff $\varepsilon_k = e_k, k = 1, 2, \ldots, n$. Similarly, $s_i(j) = \varepsilon_1^{\bar{e}_1} + \varepsilon_2^{\bar{e}_2} + \cdots + \varepsilon_n^{\bar{e}_n}$, which is 0 iff each term is 0, that is, iff $\varepsilon_k = e_k, k = 1, 2, \ldots, n$. The stated results follow. ∎

The preceding theorem now implies the following result.

Theorem 3.13.3. *Every switching function $f: \mathbb{B}^n \to \mathbb{B}$ can be represented uniquely as a sum of minterms, called the **disjunctive normal form (d.n.f.)** of f:*

$$f(X) = \sum_{i=0}^{2^n-1} f(i)p_i(X),$$

*and uniquely as a product of maxterms, called the **conjunctive normal form (c.n.f.)** of f:*

$$f(X) = \prod_{i=0}^{2^n-1} [f(i) + s_i(X)].$$

Proof. For all j in $\{0, 1, 2, \ldots, 2^n - 1\}$, we have, by Theorem 3.13.2,

$$\sum_{i=0}^{2^n-1} f(i)p_i(j) = \sum_{i=0}^{2^n-1} f(i)\delta_{ij} = f(j),$$

$$\prod_{i=0}^{2^n-1} [f(i) + s_i(j)] = \prod_{i=0}^{2^n-1} [f(i) + \bar{\delta}_{ij}] = f(j).$$

Thus the values of the d.n.f. and c.n.f. coincide with the value of f at every combination of values of the x_i; hence the d.n.f. and c.n.f. each equal the function f as claimed. Since f is uniquely determined by the values assigned to $f(0), f(1), \ldots, f(2^n - 1)$, these normal forms are necessarily unique. ∎

Note that whenever $f(i) = 0$, the corresponding term $f(i)p_i$ in the d.n.f. is zero, and so p_i is eliminated from the sum. Similarly, whenever $f(i) = 1$, the corresponding factor $f(i) + s_i$ in the c.n.f. is 1, so the factor s_i is eliminated from the product. Thus $f(X)$ *is the sum of the minterms p_i such that $f(i) = 1$ and is the product of the maxterms s_i such that $f(i) = 0$*. [If $f(i) = 1$, we call p_i a **minterm of f**; if $f(i) = 0$, we call s_i a **maxterm of f**.] These observations make it possible to write the d.n.f. and the c.n.f. by inspection from a table for f. For example, consider f and $\bar{f}$ as defined in Table 3.13.2. We have for f and $\bar{f}$ the d.n.f.'s

$$f(X) = p_1 + p_2 + p_7 = \bar{x}_1\bar{x}_2 x_3 + \bar{x}_1 x_2 \bar{x}_3 + x_1 x_2 x_3,$$

$$\bar{f}(X) = p_0 + p_3 + p_4 + p_5 + p_6 = \bar{x}_1\bar{x}_2\bar{x}_3 + \bar{x}_1 x_2 x_3 + x_1\bar{x}_2\bar{x}_3 + x_1\bar{x}_2 x_3 + x_1 x_2 \bar{x}_3$$

and the c.n.f.'s

$$f(X) = s_0 s_3 s_4 s_5 s_6$$

$$= (x_1 + x_2 + x_3)(x_1 + \bar{x}_2 + \bar{x}_3)(\bar{x}_1 + x_2 + x_3)(\bar{x}_1 + x_2 + \bar{x}_3)(\bar{x}_1 + \bar{x}_2 + x_3),$$

$$\bar{f}(X) = s_1 s_2 s_7 = (x_1 + x_2 + \bar{x}_3)(x_1 + \bar{x}_2 + x_3)(\bar{x}_1 + \bar{x}_2 + \bar{x}_3).$$

Sec. 3.13   Disjunctive and Conjunctive Normal Forms

TABLE 3.13.2. A function and its complement.

i	x_1	x_2	x_3	f	$\bar{f}$
0	0	0	0	0	1
1	0	0	1	1	0
2	0	1	0	1	0
3	0	1	1	0	1
4	1	0	0	0	1
5	1	0	1	0	1
6	1	1	0	0	1
7	1	1	1	1	0

Note that the sets of minterms in the d.n.f.'s of f and $\bar{f}$ are complementary, as are the sets of maxterms in the c.n.f.'s of these functions.

A further abbreviation is standard. Instead of writing $f(X) = p_{i_1} + p_{i_2} + \cdots + p_{i_k}$ and $g(X) = s_{j_1} s_{j_2} \cdots s_{j_m}$ one writes simply $f = \sum(i_1, i_2, \ldots, i_k)$ and $g = \prod(j_1, j_2, \ldots, j_m)$, so minterms and maxterms are replaced by their indices. Thus in the preceding example we have $f = \sum(1, 2, 7) = \prod(0, 3, 4, 5, 6)$ and $\bar{f} = \sum(0, 3, 4, 5, 6) = \prod(1, 2, 7)$.

When f is given in algebraic rather than in tabular form, we can, of course, compute and tabulate the $f(i)$, then write the d.n.f. and c.n.f. as before. Alternatively, we can use De Morgan's laws and the distributive laws to expand the given expression into a sum of products or a product of sums of the x_i and their complements. If the d.n.f. is desired, we examine each product of the sum-of-products expansion and for each i such that neither x_i nor $\bar{x}_i$ appears in the product, we insert the factor $x_i + \bar{x}_i$, then expand again into a sum of products and reduce by idempotency only. The result is the d.n.f. of f. The c.n.f. is obtained in dual fashion. For example, if $n = 3$ and $f = \overline{(x_1 + \bar{x}_2)\bar{x}_2 x_3}$, then $f = \overline{(x_1 + \bar{x}_2)} (\bar{x}_2 + \bar{x}_3) = x_1 \bar{x}_3 + \bar{x}_2$. Hence

$$f = x_1(x_2 + \bar{x}_2)\bar{x}_3 + (x_1 + \bar{x}_1)\bar{x}_2(x_3 + \bar{x}_3)$$
$$= x_1 x_2 \bar{x}_3 + x_1 \bar{x}_2 \bar{x}_3 + x_1 \bar{x}_2 x_3 + \bar{x}_1 \bar{x}_2 x_3 + \bar{x}_1 \bar{x}_2 \bar{x}_3$$

and

$$f = (x_1 + \bar{x}_2 + x_3 \bar{x}_3)(x_1 \bar{x}_1 + \bar{x}_2 + \bar{x}_3)$$
$$= (x_1 + \bar{x}_2 + x_3)(x_1 + \bar{x}_2 + \bar{x}_3)(\bar{x}_1 + \bar{x}_2 + \bar{x}_3).$$

If only the decimal indices of the minterms are needed, another procedure may be used. Each product of a sum-of-products representation of f is represented by an ordered sequence of 0's, 1's, and dashes, thus: If x_i appears in the product, we write 1 in the ith position of the sequence; if $\bar{x}_i$ appears, we write 0; if neither appears, we write a dash. Thus if $n = 5$, $\bar{x}_1 x_2 \bar{x}_3 x_5$ is represented by the sequence 010–1.

From these sequences we can determine by inspection the indices of the minterms of which a product is the sum, simply by replacing each dash by 0 and 1 in all possible combinations and translating the resulting n-digit binary numbers into decimal form. The union of the sets of indices associated with the product terms of f is the set of indices of the minterms of f. For example, if $n = 4$ and $f = x_1\bar{x}_2 x_3 + \bar{x}_1\bar{x}_2\bar{x}_3 x_4 + x_3\bar{x}_4$, we have these computations:

Product	Binary Representation	Minterms
$x_1\bar{x}_2 x_3$	101–	10, 11
$\bar{x}_1\bar{x}_2\bar{x}_3 x_4$	0001	1
$x_3\bar{x}_4$	– –10	2, 6, 10, 14

$$f = \sum(1, 2, 6, 10, 11, 14).$$

The reader should describe a dual procedure for determining the indices of the maxterms of f.

3.14 The Exclusive-OR and the Ring Normal Form of f

Since the set of all switching functions of n variables is a Boolean algebra, it is also a Boolean ring. Hence, given two switching functions f and g of n variables, we can compute their *ring sum* $f \oplus g = \bar{f}g + f\bar{g}$. This sum is also called the **exclusive-OR** of f and g. [This name is explained by the table of Figure 3.14.1(b).]

The ring sum is extensively used in the design of logic circuits. Its principal algebraic properties have appeared earlier in this chapter but are reassembled here for the sake of convenience. For all switching functions f, g, and h from $\mathbb{B}^n$ to $\mathbb{B}$, we have

(3.14.1)
  (a) $f \oplus g = g \oplus f$ (commutativity).
  (b) $f \oplus (g \oplus h) = (f \oplus g) \oplus h$ (associativity).
  (c) $f(g \oplus h) = fg \oplus fh$ ("$\cdot$" distributes over "$\oplus$").
  (d) $f \oplus 0 = f$ (0 is the identity for $\oplus$).
  (e) $f \oplus f = 0$ (f is its own negative).
  (f) $f \oplus g = f \oplus h$ iff $g = h$ (cancellation law for $\oplus$).
  (g) $f \oplus g = h$ iff $f = g \oplus h$ (solution of a "linear" equation).
  (h) $f \oplus 1 = \bar{f}$, $\bar{f} \oplus 1 = f$ (laws of complementation).
  (i) $\overline{(f \oplus g)} = f \oplus \bar{g} = \bar{f} \oplus g$
  (j) $f + g = f \oplus g \oplus fg$
  (k) $f + g = f \oplus g$ iff $fg = 0$ (relations between $+$ and $\oplus$).
  (l) $\sum_{i=1}^{k} f_i = \bigoplus_{i=1}^{k} f_i$ if $f_i f_j = 0$, $1 \leq i < j \leq k$

Sec. 3.14 The Exclusive-OR and the Ring Normal Form of f

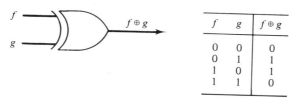

(a) Exclusive-OR element   (b) Values of $f \oplus g$

FIGURE 3.14.1

We say that f and g are **orthogonal** iff $fg = 0$. Thus (k) says that $f + g = f \oplus g$ iff f and g are orthogonal. In (l), Σ denotes a continued ring sum. This rule gives a sufficient but not a necessary condition for the replacement of "+" by "$\oplus$" throughout an expression. For example:

$$\sum_{i=0}^{2^n-1} f(i)p_i = \sum_{i=0}^{2^n-1} f(i)p_i \quad \text{because} \quad p_i p_j = \delta_{ij} p_j,$$

so the terms are mutually orthogonal.

Generalizations of some of these identities appear in Exercise 3.9.1. Their applicability here should not be overlooked.

Just as for AND and OR, there exist packaged circuits which generalize the exclusive-OR. An r-bit parity checker (Figure 3.14.2) outputs a 1 on the "Σ-odd" lead iff the total number of 1's on the input leads is odd. Since $0 \oplus 0 = 0$, $0 \oplus 1 = 1 \oplus 0 = 1$, $1 \oplus 1 = 0$, the output is simply the ring sum, $\Sigma_{i=1}^{r} f_i$, of the inputs. Such a circuit customarily has also a "Σ-even" output lead, for which the output function is $(\Sigma_{i=1}^{r} f_i) \oplus 1$, which is 1 iff the total number of input 1's is even.

One way to implement an AND-OR-NOT switching function with the aid of exclusive-OR elements or parity checkers is to transform it into a ring sum of products. This is accomplished by appropriate use of the rules (j), (k), and (l) in (3.14.1). The resulting ring sum may then be transformed into complement-free form, if that is desired, by use of the substitution $\bar{f} = f \oplus 1$.

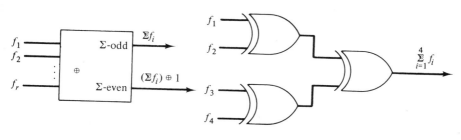

(a) r-bit parity checker   (b) 4-bit parity checker

FIGURE 3.14.2. *Parity Checkers*

For example, since the terms are mutually orthogonal, we have

$$x_1 x_2 x_3 + x_1 x_2 \bar{x}_3 + x_1 \bar{x}_2 x_3 + \bar{x}_1 x_2 x_3$$
$$= x_1 x_2 x_3 \oplus x_1 x_2 \bar{x}_3 \oplus x_1 \bar{x}_2 x_3 \oplus \bar{x}_1 x_2 x_3$$
$$= x_1 x_2 x_3 \oplus x_1 x_2 (x_3 \oplus 1) \oplus x_1 (x_2 \oplus 1) x_3 \oplus (x_1 \oplus 1) x_2 x_3$$
$$= x_1 x_2 x_3 \oplus x_1 x_2 x_3 \oplus x_1 x_2 \oplus x_1 x_2 x_3 \oplus x_1 x_3 \oplus x_1 x_2 x_3 \oplus x_2 x_3$$
$$= x_1 x_2 \oplus x_1 x_3 \oplus x_2 x_3.$$

The circuit is shown in Figure 3.14.3.

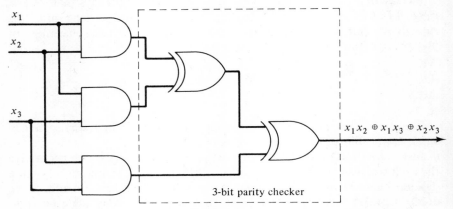

FIGURE 3.14.3. *Implementation Using a Parity Checker*

There are altogether 2^n complement-free products of the variables $x_1, x_2, \ldots, x_n$, for we must accept or reject each variable for inclusion in a product. If we reject all the variables, the product is by definition 1. There is a natural scheme for indexing these products: We write $\pi_0 = 1$ and when $i \geq 1$, let π_i denote the product obtained by deleting all complemented variables from the minterm p_i. In particular, if $n = 1$, $p_0 = \bar{x}_1$ and $p_1 = x_1$, so $\pi_0 = 1$, $\pi_1 = x_1$. If $n = 2$, $p_0 = \bar{x}_1 \bar{x}_2$, $p_1 = \bar{x}_1 x_2$, $p_2 = x_1 \bar{x}_2$, $p_3 = x_1 x_2$, so $\pi_0 = 1$, $\pi_1 = x_2$, $\pi_2 = x_1$, and $\pi_3 = x_1 x_2$.

Theorem 3.14.1. *A given switching function f from $\mathbb{B}^n$ to $\mathbb{B}$ has a unique representation as a ring sum of products π_i:*

$$f(X) = \sum_{j=0}^{2^n - 1} g_j \pi_j, \quad \text{all } g_j \in \{0, 1\}.$$

*This is the **ring normal form** of f.*

Proof. The preceding example of transformation to ring normal form makes it clear that such an expression always exists; one can (if necessary) write $f(X)$ in d.n.f., next replace all additions by ring sums, and then replace each $\bar{x}_i$ by $x_i \oplus 1$, expand, and collect. The uniqueness follows from the facts that there are exactly $2^{(2^n)}$ distinct expressions $\sum g_j \pi_j$, and exactly $2^{(2^n)}$ n-variable switching functions. ∎

The theorem implies that every switching function can be implemented as a circuit involving only AND's and parity checkers. (This assumes that a constant input 1 is available.)

3.15 Inequalities in Switching Algebra

Since $\mathbb{F}_n$ is a Boolean algebra, we have, for any two switching functions f and g, $f \leq g$ iff $fg = f$. If $f \not\leq g$ and $g \not\leq f$, f and g are **not comparable**. The identity $fg = f$ holds at any combination X_0 of values of the variables such that $f(X_0) = 0$. If, however, $f(X_0) = 1$, the identity holds iff $g(X_0) = 1$ also. We also have (Section 3.7) $f \leq g$ iff $f + g = g$, and this identity holds iff $f(X_0) = 0$ whenever $g(X_0) = 0$. These observations prove the following theorem.

Theorem 3.15.1. *For all switching functions f and g of $\mathbb{F}_n$, $f \leq g$ iff g has the value 1 at least at every combination at which f has the value 1. Dually, $f \leq g$ iff f has the value 0 at least at every combination at which g has the value 0.* ∎

If f and g are defined by a table, it is easy to see whether $f \leq g$, or $g \leq f$, or f and g are not comparable. For example, let f, g, and h be defined by the following table:

x_1	x_2	f	g	h
0	0	0	1	1
0	1	1	0	1
1	0	1	0	1
1	1	0	1	0

Then f and g are not comparable and the same holds for g and h, but $f \leq h$.

The tabular representation makes apparent the truth of the following theorem about the d.n.f. and c.n.f.

Theorem 3.15.2. *For all f and g in $\mathbb{F}_n$, $f \leq g$ iff every minterm of f is also a minterm of g; dually, $f \leq g$ iff every maxterm of g is also a maxterm of f.* ∎

If $f \leq g$, we say that f **implies** g and write $f \Rightarrow g$, for $f(X_0) = 1$ implies that $g(X_0) = 1$ in this case. In particular, if p_i is a minterm of f, then $p_i \leq f$ and p_i implies f. If p_i is not a minterm of f, then p_i implies $\bar{f}$.

The variables $x_1, x_2, \ldots, x_n$ and their complements are called **literals**. Let P_1 and P_2 be products of literals in which no x_i or $\bar{x}_i$ appears more than once and in which x_i and $\bar{x}_i$ do not both appear for any i. Only such products, called *terms* or *product terms*, will be used in the remainder of this chapter. If either x_i or $\bar{x}_i$ appears in such a product P, we say P **involves** the variable x_i. If $P_1 = x_{i_1}^{e_1} x_{i_2}^{e_2} \cdots x_{i_k}^{e_k}$, where $k \le n$, $e_i \in \{0, 1\}$, and the i_j are distinct, then P_1 has **length** k and we write $|P_1| = k$. If $|P_2| < |P_1|$, we say that P_2 is **shorter** than P_1. The **empty product** 1 results when all literals are rejected; it has length 0. There are altogether 3^n distinct products of the kind just described, determined by n variables.

If every literal of a product P_2 is also a literal of P_1, we say that P_1 **subsumes** P_2 and P_2 is a **shortening** of P_1. In particular, P_1 subsumes itself and is also a trivial shortening of itself. If P_1 subsumes P_2, $|P_2| \le |P_1|$. Also, $P_1 P_2 = P_1$, so $P_1 \le P_2$; that is, P_1 implies P_2. For example, $x_1 \bar{x}_2 x_3 \le x_1 \bar{x}_2$. If $n = 4$, the left member has the value 1 at the combinations 1010 and 1011, while the right member has the value 1 not only at 1010 and 1011 but also at 1000 and 1001. The example illustrates the fact that introducing an additional literal in a product P results in a lesser, though longer, product.

The following immediate results are often useful:

Lemma 3.15.1. *Let P and Q be product terms. Then*:
(a) P *subsumes* Q *iff* $P \Rightarrow Q$.
(b) *If P subsumes Q, then* $P + Q = Q$.
(c) *If* $P \Rightarrow Q$ *and* $|P| = |Q|$, *then* $P = Q$. ∎

By repeated application of the basic laws of Boolean algebra, one can expand any given Boolean function into a sum of products of literals, $f = Q_1 + Q_2 + \cdots + Q_r$. Such a sum is called a **sum-of-products representation** of f. Each Q_i implies f and if P_i subsumes Q_i, then P_i also implies f. In particular, every minterm that subsumes any Q_i implies f.

Any product of literals that implies f is called an **implicant** of f. If Q is an implicant of f and if Q subsumes no shorter product that implies f, Q is called a **prime implicant** of f. In following sections, we shall see how prime implicants are used to express switching functions in minimal form.

3.16 Exercises

1. Construct the tabular representation of the function

$$g(x_1, x_2, x_3, x_4) = x_1 + x_2 \bar{x}_3 + x_3 \bar{x}_4.$$

2. There are 16 functions from $\mathbb{B}^2$ to $\mathbb{B}$. Make a table that lists them all and devise a natural scheme for numbering them. Also express each in both d.n.f. and c.n.f.

3. Write the d.n.f.'s and the c.n.f.'s of the functions f, g, and h defined in Table 3.16.1.

Sec. 3.16 Exercises

TABLE 3.16.1

x_1	x_2	x_3	f	g	h
0	0	0	0	1	0
0	0	1	1	1	1
0	1	0	0	1	1
0	1	1	1	1	0
1	0	0	1	0	1
1	0	1	0	0	0
1	1	0	1	1	0
1	1	1	0	1	1

4. Given the d.n.f.

$$f(X) = \bar{x}_1\bar{x}_2\bar{x}_3 + \bar{x}_1\bar{x}_2 x_3 + x_1 x_2 \bar{x}_3,$$

write the d.n.f. of $\bar{f}$ and the c.n.f.'s of f and $\bar{f}$.

5. Given the c.n.f.

$$f(X) = (\bar{x}_1 + \bar{x}_2 + \bar{x}_3)(\bar{x}_1 + \bar{x}_2 + x_3)(x_1 + x_2 + \bar{x}_3),$$

write the c.n.f. of $\bar{f}$ and the d.n.f.'s of f and $\bar{f}$.

6. Write the d.n.f. for the function defined in Table 3.16.2, and simplify the function. Why should the result have been expected from the table?

TABLE 3.16.2

x_1	x_2	x_3	x_4	f
0	0	1	0	1
0	0	1	1	1
0	1	1	0	1
0	1	1	1	1
All others				0

7. Obtain the d.n.f. and c.n.f. of f, where

$$f(X) = \overline{x_1(\bar{x}_2 + x_3)} \quad (n = 3),$$

by expanding the given expression algebraically using the distributive and other laws appropriately.

*8. Prove that if $i_0, i_1, \ldots, i_k, i_{k+1}, \ldots, i_{2^n-1}$ is any permutation of $0, 1, 2, \ldots, 2^n - 1$, then

$$\left(\sum_{j=0}^{k} p_{i_j}\right) = \sum_{j=k+1}^{2^n-1} p_{i_j}.$$

*9. Prove:

(a) $p_i p_j = \delta_{ij} p_i = \delta_{ij} p_j$.

(b) $\sum_{i=0}^{2^n-1} p_i = 1$.

(c) $s_i + s_j = \bar{\delta}_{ij} + s_i = \bar{\delta}_{ij} + s_j$.

(d) $\prod_{i=0}^{2^n-1} s_i = 0$.

*10. Prove:

(a) $f(X) + g(X) = \sum_{j=0}^{2^n-1} [f(j) + g(j)] p_j(X)$.

(b) $f(X) \cdot g(X) = \sum_{j=0}^{2^n-1} [f(j) \cdot g(j)] p_j(X)$.

(c) $\overline{f(X)} = \sum_{j=0}^{2^n-1} \overline{f(j)} \cdot p_j(X)$.

*11. Prove:

(a) $f(X) + g(X) = \prod_{j=0}^{2^n-1} [f(j) + g(j) + s_j(X)]$.

(b) $f(X) \cdot g(X) = \prod_{j=0}^{2^n-1} [f(j) \cdot g(j) + s_j(X)]$.

(c) $\overline{f(X)} = \prod_{j=0}^{2^n-1} [\overline{f(j)} + s_j(X)]$.

*12. Prove that if f and g are arbitrary functions of $\mathbb{F}_n$ and if φ is an arbitrary function of $\mathbb{F}_2$, then

$$\varphi(f, g) = \sum_{j=0}^{2^n-1} \varphi(f(j), g(j)) p_j = \prod_{j=0}^{2^n-1} (\varphi(f(j), g(j)) + s_j).$$

(This requires only a few well-chosen words, *no* messy computation.)

13. Prove that if $f \in \mathbb{F}_n$,

$$f(X) = f(0, x_2, \ldots, x_n)\bar{x}_1 + f(1, x_2, \ldots, x_n)x_1$$
$$= f(0, 0, x_3, \ldots, x_n)\bar{x}_1\bar{x}_2 + f(0, 1, x_3, \ldots, x_n)\bar{x}_1 x_2$$
$$+ f(1, 0, x_3, \ldots, x_n)x_1\bar{x}_2 + f(1, 1, x_3, \ldots, x_n)x_1 x_2.$$

What is the ultimate outcome if this expansion process is continued?

14. Write the conjunctive normal form for the switching function defined by Table 3.16.3, then simplify using only the distributive law $(a+b)(a+c) = a+bc$.

TABLE 3.16.3

x_1	x_2	x_3	x_4	f
0	0	0	1	0
0	0	1	1	0
1	0	1	1	0
1	0	1	0	0
1	1	1	0	0
All others				1

15. Are the following equations correct?

$$\overline{ab \oplus bc \oplus ca} = \bar{a}\bar{b} \oplus \bar{b}\bar{c} \oplus \bar{c}\bar{a} = \overline{ab} \oplus \overline{bc} \oplus \overline{ca}.$$

16. Use the properties of "$\oplus$" appropriately to express

$$f(X) = x_1x_2 \oplus x_1x_3 \oplus x_2x_3 \oplus x_1x_2x_3$$

in terms of AND, OR, and NOT only.

17. Obtain d.n.f.'s and c.n.f.'s for f and $\bar{f}$, where $n = 4$ and

$$f(X) = \overline{x_1x_2x_3x_4} \oplus x_1x_2 \oplus x_3x_4.$$

18. Obtain the ring normal form for f, where

$$f(X) = s_0s_1s_5s_6 \qquad (n = 3).$$

19. Write in ring normal form:

$$f(X) = x_1\bar{x}_2 + x_2\bar{x}_3 + x_3\bar{x}_1 \qquad (n = 3).$$

20. Prove that these identities hold for any function $f \in \mathbb{F}_1$:

$$f(X) = f(1)x \oplus f(0)(x \oplus 1)$$
$$= f(0) \oplus [f(0) \oplus f(1)]x$$
$$= f(1) \oplus [f(0) \oplus f(1)]\bar{x}$$
$$= f(0) \oplus f(1) \oplus f(0)x \oplus f(1)\bar{x}.$$

21. Express the function defined in Table 3.16.4 as a product of exclusive-OR's.

TABLE 3.16.4

x_1	x_2	x_3	x_4	f
0	1	0	1	1
0	1	1	0	1
1	0	0	1	1
1	0	1	0	1
All others				0

22. (Requires a knowledge of matrix algebra.) Let $f(X) = \sum_{i=0}^{2^n-1} f(i)p_i = \sum_{i=0}^{2^n-1} g_i \pi_i$, $\mathbf{F} = [f(0), f(1), \ldots, f(2^n-1)]^T$, and $\mathbf{G} = [g_0, g_1, \ldots, g_{2^n-1}]^T$. Show that the relation between the $f(i)$ and the g_i can be expressed as $\mathbf{G} = \mathbf{M}_n \mathbf{F}$, where $\mathbf{M}_n$ is a $2^n \times 2^n$ matrix of 0's and 1's, $\mathbf{M}_1 = \begin{bmatrix} 1 & 0 \\ 1 & 1 \end{bmatrix}$, and $\forall n \in \mathbb{P}, \mathbf{M}_{n+1} = \begin{bmatrix} \mathbf{M}_n & 0 \\ \mathbf{M}_n & \mathbf{M}_n \end{bmatrix}$. Moreover, if the ring sum is used in matrix multiplication, $\mathbf{M}_n^2 = \mathbf{I}_{2^n}$, so $\mathbf{M}_n$ is its own inverse.

23. Prove that $\sum_{i=0}^{2^n-1} \pi_i = \bar{x}_1 \bar{x}_2 \cdots \bar{x}_n$.

24. Let $\varphi(a, b) = ab + \bar{a}\bar{b} = a \oplus b$. Suppose that you have lots of φ-elements. Show how to design a circuit, using *only* φ-elements, that will implement the function $f(X) = x_1 \oplus x_2 \oplus x_3$. Can you do the same if $f(X) = x_1 \oplus x_2 \oplus x_3 \oplus x_4$?

25. If $n = 4$ and outputs $f_1(X) = p_0 + p_1 + p_4 + p_5$ and $f_2(X) = p_{10} + p_{12} + p_{14} + p_{15}$ are available for use as inputs of following circuits, show that it is possible to implement the function $f(X) = p_0 + p_4 + p_{10} + p_{14}$ in the form $f = g(f_1 + f_2)$. Find the simplest function g that will serve here.

26. Prove that when $n = 3$,
$$p_0 + p_2 + p_5 = s_1 s_3 s_4 s_6 s_7.$$

27. Determine all minterms which imply the function
$$f(X) = (x_1 + x_2 + x_3)(x_1 + \bar{x}_3 + x_4)(x_2 + \bar{x}_3 + \bar{x}_4) \quad (n = 4).$$

28. If $n = 4$ and if we need a function f such that
$$(p_0 + p_1 + p_3 + p_5 + p_{11} + p_{15}) \cdot f \leq (p_4 + p_5 + p_{11} + p_{13}),$$
what minterms:
  (a) Must be included in the d.n.f. of f?
  (b) May not be included in the d.n.f. of f?
  (c) Are optional in the d.n.f. of f?

29. Prove that for all f, g, and h in $\mathbb{F}_n$,
  (a) $fh = gh$ iff $h \leq fg + \bar{f}\bar{g}$.
  (b) $f + h = g + h$ iff $\bar{f}g + f\bar{g} \leq h$.
Then show that there exists h that satisfies both equations iff $f = g$.

Sec. 3.17 Prime Implicants and Minimal Sums of Products

30. Prove Lemma 3.15.1.

*31. Prove that the constant function 0 has no prime implicant, that the constant function 1 has the single prime implicant 1, and that 1 is not a prime implicant of any other function.

*32. Prove that if $f \Rightarrow gh$, then $f \Rightarrow g$ and $f \Rightarrow h$. What is the dual result? Explain what these results mean in terms of minterms and maxterms, respectively.

*33. Prove that if $f \Rightarrow g$, then also $fh \Rightarrow g$, for all $h \in \mathbb{F}_n$.

*34. Prove that if $fg = fh$, $f \neq 0$, and no variable of f appears in g or h, then $g = h$.

35. Prove that a function of the form

$$f(x_1, x_2, \ldots, x_n) = \bar{x}_1 \bar{x}_2 \cdots \bar{x}_k f(0, 0, \ldots, 0, x_{k+1}, \ldots, x_n)$$
$$+ \cdots + x_1^{e_1} x_2^{e_2} \cdots x_k^{e_k} f(e_1, e_2, \ldots, e_k, x_{k+1}, \ldots, x_n)$$
$$+ \cdots + x_1 x_2 \cdots x_k f(1, 1, \ldots, 1, x_{k+1}, \ldots, x_n)$$

is identically 1 iff each of the functions $f(e_1, e_2, \ldots, e_k, x_{k+1}, \ldots, x_n)$ is identically 1. Prove that every Boolean function can be written in the indicated form.

*36. It is at times necessary to solve simultaneously a system of equations $f_i = g_i$, $i = 1, 2, \ldots, k$, where the f_i and g_i belong to $\mathbb{F}_n$, that is, to find all combinations $(e_1, e_2, \ldots, e_n)$ for which all k equations are simultaneously satisfied. Show how a truth table of the f_i and the g_i may be used for this purpose. Then apply your method to the following systems:

(a) $\begin{cases} x_1 x_2 = x_1 + x_3 x_4 \\ x_1 + x_2 = x_3 + x_2 x_4 \end{cases} \quad (n = 4)$.

(b) $\begin{cases} (x_1 + x_2)\bar{x}_3 = (x_3 + x_4)\bar{x}_5 \\ \bar{x}_1 \bar{x}_2 + x_3 = \bar{x}_3 x_4 + \bar{x}_5 \end{cases} \quad (n = 5)$.

37. Compute for each of the systems in Exercise 36 the function $\prod_i (f_i g_i + \bar{f}_i \bar{g}_i)$ and represent it in d.n.f. What can you conclude?

3.17 Prime Implicants and Minimal Sums of Products

The products of which a given sum-of-products representation of a switching function f is the sum are called its **terms** or **product terms**. If either of the literals x_i or $\bar{x}_i$ appears in a product, we say that the variable x_i is **involved** in the product. The given sum-of-products representation is called **irredundant** iff no term and no literal appearing in any term may be deleted from the representation without rendering the resulting sum unequal to f. A sum-of-products representation f_1 of f is **shorter** than another such representation f_2 iff

1. f_1 has no more terms than f_2.
2. f_1 has fewer total appearances of literals than f_2 does.

Because a given function f has only finitely many irredundant sum-of-products representations, it necessarily has at least one that is shortest. All shortest or **minimal** sum-of-products representations of f are necessarily irredundant but are not necessarily unique. For example, $\bar{x}_1 x_2 + \bar{x}_2 x_3 + \bar{x}_3 x_1$ and $x_1 \bar{x}_2 + x_2 \bar{x}_3 + x_3 \bar{x}_1$ are minimal sum-of-products representations of the same function of three variables, as is easily shown with the law of consensus or a table of function values.

In the design of logic circuits, minimal sum-of-products forms do not necessarily lead to the simplest or most useful circuits. Nevertheless, because of their usefulness as starting points in logical design, they are of major interest. We begin our study of minimal sum-of-products forms by establishing some basic properties of prime implicants.

Lemma 3.17.1. *If $P \Rightarrow f$ but P is not a prime implicant of f, then P subsumes at least one prime implicant of f.*

Proof. By the definition of a prime implicant, P subsumes at least one shorter product that implies f. If this product is not a prime implicant of f, it, too, subsumes a shorter product that implies f. This process must stop with a product $Q < 1$ if $f \neq 1$, with $Q = 1$ if $f = 1$. Then Q is a prime implicant of f. ∎

Thus, among all implicants of f, the prime implicants are the most economical of literals. The next lemma shows that they have a certain independence.

Lemma 3.17.2. *If P and Q are prime implicants of f and if $P \Rightarrow Q$, then $P = Q$; that is, no prime implicant of f implies another.*

Proof. P subsumes Q by Lemma 3.15.1. If Q is shorter than P, then P is not a prime implicant of f. Hence $|P|=|Q|$, so $P = Q$, again by Lemma 3.15.1. ∎

With respect to the relation "$\leqslant$" defined for Boolean functions in Section 3.15, the implicants of f constitute a poset, for "$\leqslant$" is a partial ordering. In this poset, the prime implicants of f are the maximal elements and the minterms of f are the minimal elements. For example, in the case of the function $f(X) = x_2 x_3 + x_1 x_2 x_4$ ($n = 4$), the Hasse diagram of the poset of implicants is given in Figure 3.17.1.

The next result shows the relation of prime implicants to minimality; it implies that all terms of any minimal sum-of-products representation of f are prime implicants of f.

Theorem 3.17.1. *All terms of an irredundant sum-of-products representation of f are prime implicants of f.*

Proof. If $f = 1$, its only prime implicant is 1, and 1 is the shortest expression of this function, so the theorem holds in this case. Assume that

Sec. 3.18 Reusch's Method for Finding All Prime Implicants

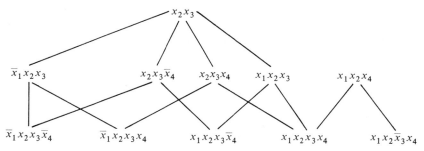

FIGURE 3.17.1. *Poset of Implicants*

$f \neq 1$. If P is a term of an irredundant sum-of-products representation of f and if P is not a prime implicant of f, then P subsumes a prime implicant Q of f, Q shorter than P, by Lemma 3.17.1. If $f = g + P$, where g represents the sum of all terms of f other than P, then because $P \Rightarrow Q \Rightarrow f$, we have

$$f = f + Q = (g + P) + Q = g + (P + Q) = g + Q,$$

contradicting irredundancy of the expression $f = g + P$ for f. ∎

By the preceding theorem, any switching function f may be written as a sum of prime implicants of f. Adding the remaining prime implicants of f to this sum leaves it still equal to f, for if p implies f, $p \leq f$, so $f + p = f$. Hence f is equal to the sum of all its prime implicants. This particular sum-of-products representation is called the **complete sum** of f. We see in the next section how it may be computed.

3.18 Reusch's Method for Finding All Prime Implicants

The procedures here and in Section 3.20 are based on B. Reusch (1975).

Lemma 3.18.1. *Given a Boolean function* $f(x_1, x_2, \ldots, x_n)$, *write*

$$f = \bar{x}_1 f_0 + x_1 f_1,$$

where

$$f_0 = f(0, x_2, \ldots, x_n), \qquad f_1 = f(1, x_2, \ldots, x_n).$$

Then

(a) *If* P_0 *and* P_1 *are products involving only* $x_2, \ldots, x_n$ *and if* P_0 *implies* f_0, P_1 *implies* f_1, *it follows that* $\bar{x}_1 P_0$, $x_1 P_1$, *and* $P_0 P_1$ *each imply* f.

(b) *If* P *is a prime implicant of* f, *one of the following holds:*
   (1) $P = \bar{x}_1 P_0$, P_0 *a prime implicant of* f_0.
   (2) $P = x_1 P_1$, P_1 *a prime implicant of* f_1.
   (3) $P = P_0 P_1$, P_0 *a prime implicant of* f_0, P_1 *a prime implicant of* f_1.

Proof. (a) Suppose that $P_0 = f_0 P_0$ and $P_1 = f_1 P_1$. Then

$$P_0 P_1 f = \bar{x}_1 \cdot P_0 f_0 \cdot P_1 + x_1 P_0 \cdot P_1 f_1 = \bar{x}_1 P_0 P_1 + x_1 P_0 P_1 = P_0 P_1,$$

so P_0P_1 implies f. The other two cases are treated similarly.

(b) Let P be any prime implicant of f. Then P has the factor $\bar{x}_1$, the factor x_1, or neither of these.

(1) If $P = \bar{x}_1P_0$, P_0 free of $\bar{x}_1$ and x_1, we have the identity $\bar{x}_1P_0 = \bar{x}_1P_0 \cdot f = \bar{x}_1P_0 \cdot f_0$. When $x_1 = 0$, this reduces to the identity in $x_2, \ldots, x_n$: $P_0 = P_0f_0$, so P_0 implies f_0. If P_0 is not prime, some shortening P_{00} of P_0 is a prime implicant of f_0. Then, by (a), $\bar{x}_1P_{00}$, a shortening of P, implies f, so P is not prime. The contradiction shows that P_0 is a prime implicant of f_0.

(2) The proof is similar to that of (1).

(3) If P is free of x_1 and $\bar{x}_1$, then the identity $P = Pf = \bar{x}_1f_0P + x_1f_1P$ implies the identities in $x_2, \ldots, x_n$: $P = f_0P$ (when $x_1 = 0$), and $P = f_1P$ (when $x_1 = 1$), so P implies each of f_0 and f_1. Hence some shortening P_0 of P implies f_0 and some shortening P_1 of P implies f_1. Then, by (a), P_0P_1 implies f. Now P_0P_1 is P or a shortening of P. Since P is prime, we must have $P = P_0P_1$. ∎

Theorem 3.18.1. *We can find all prime implicants of f by finding all prime implicants P_0 of f_0 and all prime implicants P_1 of f_1, listing all products $\bar{x}_1P_0$, x_1P_1, and P_0P_1, then deleting every term a shortening of which appears in the list and also deleting every duplicate term.*

Proof. By Lemma 3.18.1(a) every product in the list is an implicant of f; by (b) the list includes all prime implicants, and every nonprime implicant in the list may be shortened to at least one of the prime implicants, hence will be deleted. (No simplifications other than the two types of deletion are permissible.) ∎

If the prime implicants of f_0 and f_1 are not at once apparent, we decompose each of f_0 and f_1 with respect to the next variable x_2 (or any other convenient one) and continue the process until the resulting functions are recognized as sums of their prime implicants. In particular, $x_{i1}^{e_1} + x_{i2}^{e_2} + \cdots + x_{ik}^{e_k}$, $x_i\bar{x}_j + \bar{x}_ix_j$, and $x_ix_j + \bar{x}_i\bar{x}_j$ are sums whose terms are prime implicants of those sums. The pattern is that of an expanding tree:

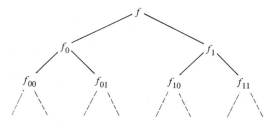

Obvious simplifications using absorption and redundancy laws may be made in the $f_{e_1e_2\cdots e_k}$ as one goes along; this tends to reduce the computation

Sec. 3.18 Reusch's Method for Finding All Prime Implicants

needed, but it does not affect the final outcome. The development stops at each function $f_{e_1 e_2 \cdots e_k}$ which is recognized as a sum of its prime implicants, and stops finally when all such functions are so recognized.

One now applies the theorem iteratively to work back, finding the prime implicants of the $f_{e_1 e_2 \cdots e_k}$ at successively higher levels until those of f are finally determined. This work can be arranged in the form of a converging tree, as the example in Figure 3.18.1 illustrates. When developing the initial tree, it is useful to recall that if $f = \bar{x}_1 \tilde{f}_0 + \tilde{f}_2 + x_1 \tilde{f}_1$, where $\tilde{f}_0, \tilde{f}_1$, and $\tilde{f}_2$ are free of $\bar{x}_1$ and x_1, then $f = \bar{x}_1(\tilde{f}_0 + \tilde{f}_2) + x_1(\tilde{f}_1 + \tilde{f}_2)$, so $f_0 = \tilde{f}_0 + \tilde{f}_2$, $f_1 = \tilde{f}_1 + \tilde{f}_2$. The last expression obtained in the algorithm represents f as the sum of *all* its prime implicants, and hence is what we have called the **complete sum** of f.

To make it easier to check the work, we recorded in Figure 3.18.1 all distinct terms $P_0 P_1$ at each step of the process. At any stage, one could, of

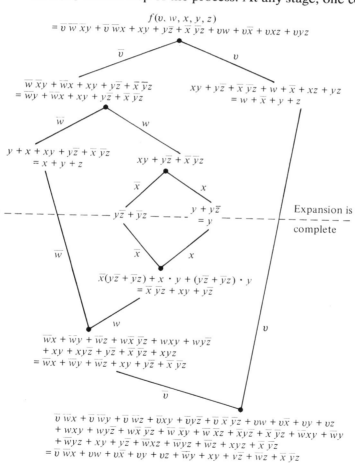

FIGURE 3.18.1. *Finding the Complete Sum*

course, refrain from recording any term that would obviously be eliminated later, thus saving a considerable amount of writing in some cases.

Often a switching function is defined by listing the indices of its minterms. For example, if $n = 5$, $f = \sum(0, 7, 18, 30)$ denotes the function $f = p_0 + p_7 + p_{18} + p_{30}$. In such a case, there is a compact way of implementing Reusch's procedure that avoids writing the minterms as products. In this case, an abbreviated truth table, with the indices of the minterms actually included in f listed in increasing order, is in effect already the diverging tree we constructed before. The table for the function

$$f = \sum(0, 1, 2, 4, 7, 8, 9, 14, 17, 18, 26, 28, 29, 30, 31), \qquad n = 5,$$

is given in Table 3.18.1.

TABLE 3.18.1. $f = \sum(0, 1, 2, 4, 7, 8, 9, 14, 17, 18, 26, 28, 29, 30, 31)$ $(n = 5)$

	x_1	x_2	x_3	x_4	x_5	A	x_3	x_2	x_1
0	0	0	0	0	0	$\bar{x}_4 + \bar{x}_5$	$\bar{x}_3\bar{x}_4$ $+\bar{x}_3\bar{x}_5$ $+x_3x_4x_5$ $+\bar{x}_4\bar{x}_5$	$\bar{x}_2\bar{x}_3\bar{x}_5$ $+\bar{x}_2x_3x_4x_5$ $+\bar{x}_2\bar{x}_4\bar{x}_5$ $+x_2x_3x_4\bar{x}_5$ $+\bar{x}_3\bar{x}_4$	$\bar{x}_1\bar{x}_2\bar{x}_3\bar{x}_5$ $+\bar{x}_1\bar{x}_2x_3x_4x_5$ $+\bar{x}_1\bar{x}_2\bar{x}_4\bar{x}_5$ $+\bar{x}_1\bar{x}_3\bar{x}_4$ $+x_1x_2x_3$ $+x_1x_2x_4\bar{x}_5$ $+x_1\bar{x}_3x_4\bar{x}_5$ $+\bar{x}_2\bar{x}_3x_4\bar{x}_5$ $+x_2x_3\bar{x}_4\bar{x}_5$ $+\bar{x}_2\bar{x}_3\bar{x}_4\bar{x}_5$
1	0	0	0	0	1				
2	0	0	0	1	0				
4	0	0	1	0	0	$\bar{x}_4\bar{x}_5$ $+x_4x_5$			
7	0	0	1	1	1				
8	0	1	0	0	0	$\bar{x}_4$	$\bar{x}_3\bar{x}_4$ $+x_3x_4\bar{x}_5$		
9	0	1	0	0	1				
14	0	1	1	1	0	$x_4\bar{x}_5$			
17	1	0	0	0	1	$\bar{x}_3\bar{x}_4x_5$ $+\bar{x}_3x_4\bar{x}_5$	$\bar{x}_3\bar{x}_4x_5$ $+\bar{x}_3x_4\bar{x}_5(*)$		
18	1	0	0	1	0				
26	1	1	0	1	0	$x_4\bar{x}_5$		$\bar{x}_2\bar{x}_3\bar{x}_4x_5$ $+x_2x_3$ $+x_2x_4\bar{x}_5$ $+\bar{x}_3x_4\bar{x}_5$	
28	1	1	1	0	0	1	$x_3 + x_4\bar{x}_5$		
29	1	1	1	0	1				
30	1	1	1	1	0				
31	1	1	1	1	1				

Sec. 3.19   *The Minimal Sums of a Function*   155

In Table 3.18.1 the horizontal dividing lines beginning in column x_i correspond to branching associated with x_i in the tree. Entries above such a line correspond to $\bar{x}_i$-coefficients, entries below to x_i-coefficients. A vertical line between the columns x_i and x_{i+1} and spanning the rows between two successive horizontal lines implies that the remaining entries in these rows define a function of $x_{i+1}, \ldots, x_n$ that is recognized as the sum of all its prime implicants. (Recall that 1 is a prime implicant of itself; Exercise 3.16.31.) These sums of prime implicants are recorded in column "A" of the table, regardless of how many variables they involve.

We now construct what is in essence the converging tree, beginning just as before with the last variable x_i with respect to which partitioning of the rows was effected and working back to x_1. (In our example, this variable is x_3.) If we are combining with respect to x_i, and if P_0 is listed above a horizontal x_i-dividing line and P_1 below it, we compute and record in the next column, also labeled "x_i," the appropriately reduced form of $\bar{x}_i P_0 + x_i P_1 + P_0 P_1$. All possible x_i-combinings are effected before the x_{i-1}-combinings are begun.

When an entry in one column duplicates that of the preceding column [see (*) in Table 3.18.1], it need not be recorded a second time. One simply looks back to its initial appearance when the entry is needed.

There are other methods of finding all the prime implicants, but we chose to discuss Reusch's method because it is simple, easy to execute, and has not yet appeared in a textbook. Other methods are briefly discussed in Exercise 3.21.2 and Section 3.22.

3.19 The Minimal Sums of a Function

It is not hard in principle to determine all minimal sum-of-products representations of a given switching function f, but the computations may be tedious. We first discuss a procedure that is easy, but perhaps lengthy, when we know both the prime implicants and all the minterms of a given function. Since every minimal sum-of-products representation is a sum of prime implicants, we first find all the prime implicants of f, perhaps by the method of the last section. For each prime implicant we next list the minterms of which it is the sum (see the algorithm in Section 3.13 and Table 3.19.1), and list all the minterms of f (the union of the preceding lists) if the latter list is not already at hand. Next we make a table with a column for each minterm of f and a row for each prime implicant of f, the prime implicants commonly being listed in order of increasing length. If P is a prime implicant of f and if p_i is a minterm of P, we record an "×" in row P and column i of the table; otherwise, there is no entry at all. In the case of the example of Table 3.18.1, this procedure results in the entries of Table 3.19.2 (other marks in the table will be explained in the next two pages).

TABLE 3.19.1

	Prime Implicant	Binary Representation	Minterms
A	$x_1 x_2 x_3$	111--	28, 29, 30, 31
B	$\bar{x}_1 \bar{x}_3 \bar{x}_4$	0-00-	0, 1, 8, 9
C	$\bar{x}_1 \bar{x}_2 \bar{x}_3 \bar{x}_5$	000-0	0, 2
D	$\bar{x}_1 \bar{x}_2 \bar{x}_4 \bar{x}_5$	00-00	0, 4
E	$x_1 x_2 x_4 \bar{x}_5$	11-10	26, 30
F	$x_1 \bar{x}_3 x_4 \bar{x}_5$	1-010	18, 26
G	$\bar{x}_2 \bar{x}_3 x_4 \bar{x}_5$	-0010	2, 18
H	$x_2 x_3 x_4 \bar{x}_5$	-1110	14, 30
I	$\bar{x}_2 \bar{x}_3 \bar{x}_4 x_5$	-0001	1, 17
J	$\bar{x}_1 \bar{x}_2 x_3 x_4 x_5$	00111	7

If a sum of prime implicants is equal to f, that sum must include ("inclusion" is the relation "$\geqslant$") every minterm of f. Hence if any column of the table contains a *single* "×," there is only one prime implicant that includes the corresponding minterm, and that prime implicant must be a term of every minimal sum-of-products representation of f. Such prime implicants are called **essential prime implicants** and the first task is to identify all of them. The single ×'s referred to are encircled, and the corresponding essential prime implicants are marked with asterisks. In our example, A, B, D, H, I, and J are essential.

TABLE 3.19.2

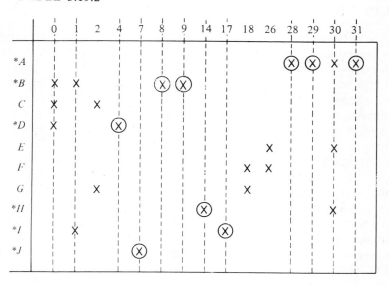

Sec. 3.19 The Minimal Sums of a Function

Now we delete every column that contains an × in a row of an essential prime implicant, for the minterms of f associated with such columns will all be included, by virtue of the presence of the essential prime implicants, in each minimal sum. Only the undeleted minterms are of further concern. They must be included in each minimal sum by a suitable choice of nonessential prime implicants that include them. For ease of reading, we now write a reduced table, using the rows of the nonessential prime implicants and the columns of minterms still to be accounted for. In the case of our example, this results in Table 3.19.3.

TABLE 3.19.3

	2	18	26
C	×		
E			×
F		×	×
G	×	×	

At this point, each remaining minterm can be included in a sum by choice of at least two prime implicants. We include the first by choosing one *or* another of the prime implicants that include it, *and* then include the second by choosing one *or* another of the prime implicants that include it, *and* so on. These choices can be translated into a Boolean product of sums, the **Petrick function** Π of this table. In the case of our example, we have

$$\Pi = (C+G)(F+G)(E+F).$$

One choice must be made from each factor of Π, and we expand the product Π by the rules of Boolean algebra to determine the various choices that exist. The commutative, associative, and distributive laws are consistent with the operation of choosing one term from each factor. Since choosing X or X, like choosing X and X, amounts to choosing X, the laws of idempotency $X+X=X$ and $XX=X$ apply. Also, since we always want the minimum number of prime implicants in a sum, the absorption laws $X+XY=X$ and $X(X+Y)=X$ are used. Indeed, if a choice "X or $(X$ and $Y)$" or a choice "X and $(X$ or $Y)$" presents itself, the simplest resolution of the choice is X. In the case of our example, we have

$$\Pi = CF + EG + FG.$$

The terms of the sum-of-products representation of Π represent the alternative means of including the remaining minterms in a sum of prime implicants. We discard all but the shortest products, for only the latter can lead to minimal sums. From the shortest products, we select those whose corresponding prime implicants include the minimum total number of literals. Then each minimal sum-of-products representation of f is the sum of the essential prime implicants and the prime implicants represented by the factors of one of the just identified terms of Π. In our example, choosing C and F, or E and G, or F and G involves a total of eight literals in each case, so there are three minimal forms for f:

$$f = x_1x_2x_3 + \bar{x}_1\bar{x}_3\bar{x}_4 + \bar{x}_1\bar{x}_2\bar{x}_4\bar{x}_5 + x_2x_3x_4\bar{x}_5 + \bar{x}_2\bar{x}_3\bar{x}_4x_5$$

$$+ \bar{x}_1\bar{x}_2x_3x_4x_5 + \begin{cases} \bar{x}_1\bar{x}_2\bar{x}_3\bar{x}_5 + x_1\bar{x}_3x_4\bar{x}_5 \\ x_1x_2x_4\bar{x}_5 + \bar{x}_2\bar{x}_3x_4\bar{x}_5 \\ x_1\bar{x}_3x_4\bar{x}_5 + \bar{x}_2\bar{x}_3x_4\bar{x}_5. \end{cases}$$

3.20 Reusch's Method for Finding All Minimal Sums

Now we discuss a procedure for finding the irredundant forms for f when given the complete sum of f, due to Reusch (1975) but based on previous work by Ghazala (1956) and others. As we proceed, we shall illustrate the process with the following example:

(3.20.1)
$$f = x_1\bar{x}_2x_5 + x_2x_3x_4 + x_1\bar{x}_4x_5 + \bar{x}_1x_2x_4 + x_1x_3x_5 + x_2x_4\bar{x}_5$$
$$+ x_1\bar{x}_2x_3 + x_1x_3x_4 \quad \text{(complete sum)}.$$

After stating the procedure and presenting a second example, we shall prove that the procedure is correct.

STEP 1. Let $f = P_1 + P_2 + \cdots + P_n$ be the complete sum of f and $f = Q_1 + Q_2 + \cdots + Q_m$ any sum-of-products representation of f; we make a chart with n rows and m columns. In the ith row and jth column, indexed by P_i and Q_j, respectively, the entry is
 1. A dash (—) if $P_iQ_j = 0$.
 2. Otherwise, the product of all literals appearing in P_i but not in Q_j (if there are none, this entry is 1).

For (3.20.1), we take $P_i = Q_i$, $m = n = 8$; if we knew a shorter sum of products for f we would use it, as the chart would be smaller. The chart for

TABLE 3.20.1

	$x_1\bar{x}_2x_5$	$x_2x_3x_4$	$x_1\bar{x}_4x_5$	$\bar{x}_1x_2x_4$	$x_1x_3x_5$	$x_2x_4\bar{x}_5$	$x_1\bar{x}_2x_3$	$x_1x_3x_4$
1 $x_1\bar{x}_2x_5$	1	—	~~$\bar{x}_2$~~	—	$\bar{x}_2$	—	~~$\bar{x}_5$~~	$\bar{x}_2x_5$
2 $x_2x_3x_4$	—	1	—	~~x_3~~	x_2x_4	~~x_3~~	—	x_2
3 $x_1\bar{x}_4x_5$	~~$\bar{x}_4$~~	—	1	—	$\bar{x}_4$	—	~~$\bar{x}_4x_5$~~	—
4 $\bar{x}_1x_2x_4$	—	$\bar{x}_1$	—	1	—	~~$\bar{x}_4$~~	—	—
5 $x_1x_3x_5$	~~x_3~~	x_1x_5	~~x_3~~	—	1	—	~~x_5~~	x_5
6 $x_2x_4\bar{x}_5$	—	$\bar{x}_5$	—	~~$\bar{x}_5$~~	—	1	—	$x_2\bar{x}_5$
7 $x_1\bar{x}_2x_3$	~~x_3~~	—	~~$\bar{x}_2x_3$~~	—	$\bar{x}_2$	—	1	$\bar{x}_2$
8 $x_1x_3x_4$	~~x_3x_4~~	x_1	—	—	x_4	~~x_1x_3~~	~~$\bar{x}_4$~~	1

Sec. 3.20 Reusch's Method for Finding All Minimal Sums

(3.20.1) is Table 3.20.1. In the first column, third row, only $\bar{x}_4$ appears in $x_1\bar{x}_4x_5$ but not in $x_1\bar{x}_2x_5$, so the entry is $\bar{x}_4$. Note that we have given each prime implicant one of the indices 1, 2, 3, ..., 8. (The reason some entries are crossed out is explained in the next step.)

STEP 2. In each column of entries, determine which variables are *monoform*: that is, appear only uncomplemented or only complemented, not both. Every column entry containing a variable monoform in that column may be deleted (replaced by a dash).

In the first column of Table 3.20.1, x_3 is monoform, so entries x_3, x_3, and x_3x_4 are crossed out. Then x_4 in the first column becomes monoform, as only $\bar{x}_4$ appears; it is also crossed out. In the same way, many other table entries become dashes; those which were crossed out because a variable not initially monoform became monoform as a result of a prior deletion are indicated by double dashes.

STEP 3. For each column, determine all the *minimal sets of entries that sum to (logical) 1*, and for each such set form the symbolic product of the indices of the rows in which those entries appear. The switching-algebra sum of these products will be called the **column factor** of that column.

In the example, columns 1, 3, 4, 6, and 7 contain only the entry 1, so the only such minimal set in any of these columns is {1}; the 1's appear in rows 1, 3, 4, 6, and 7, respectively, so the column factors are 1, 3, 4, 6, and 7.

In column 2 of Table 3.20.1 the minimal-length sums equal to 1 are 1, $\bar{x}_1 + x_1 = 1$, and $\bar{x}_1 + x_1x_5 + \bar{x}_5 = 1$, yielding symbolic products 2, 48, and 456, and column factor $2+48+456$. In column 5 we see that $1=1$, $\bar{x}_4 + x_4 = 1$, $\bar{x}_2 + x_2x_4 + \bar{x}_4 = 1$, and $x_2x_4 + \bar{x}_4 + \bar{x}_2 = 1$, so the column factor is $5+38+123+237$. The column factor for column 8 is $8+27+567$. (We shall discuss later how to find column factors in difficult cases.)

STEP 4. Form the product of the column factors, and treat it as we did the Petrick function: in the example, treating the integers as Boolean variables, we have

$$p = (1)(3)(4)(6)(7)(2+48+456)(5+38+123+237)(8+27+567)$$
(3.20.2)
$$= (1)(3)(4)(6)(7)(2+8+5)(5+8+2+2)(8+2+5)$$
$$= (1)(3)(4)(6)(7)(2+5+8).$$

The three irredundant forms are thus

(3.20.3) $x_1\bar{x}_2x_5 + x_1\bar{x}_4x_5 + \bar{x}_1x_2x_4 + x_2x_4\bar{x}_5 + x_1\bar{x}_2x_3 + \begin{cases} x_2x_3x_4 & \text{or} \\ x_1x_3x_5 & \text{or} \\ x_1x_3x_4. \end{cases}$

All are minimal. [When simplifying (3.20.2), we used the relation $x(y+xz) = x(y+z)$ of switching algebra to eliminate unneeded literals, and then used idempotency.]

When column factors are hard to find, we can use the following lemma. Let $T_1+T_2+\cdots+T_r$ be a Boolean sum of indexed terms. Denote by $\mathcal{R}(T_1+T_2+\cdots+T_r)$ the switching function

$$\mathcal{R}(T_1+T_2+\cdots+T_r) = \sum_{T_{u_1}+\cdots+T_{u_s}=1} u_1 u_2 \cdots u_s$$

of the r indices $1, 2, \ldots, r$. (The sum runs over all subsets $\{u_1, u_2, \ldots, u_s\}$ of $\{1, 2, \ldots, r\}$ such that $T_{u_1}+T_{u_2}+\cdots+T_{u_s}=1$.) This sum is the same switching function as the column factor of a column with entries $T_1, T_2, \ldots, T_r$ but in general has additional terms, which are absorbed upon simplification.

Lemma 3.20.1. *If f_0, f_1, and f_2 are sums of terms not involving x, then* $\mathcal{R}(\bar{x}f_0+xf_1+f_2) = \mathcal{R}(f_0+f_2)\mathcal{R}(f_1+f_2)$.

Proof. The function

$$\bar{x}f_0+xf_1+f_2 = \bar{x}(f_0+f_2)+x(f_1+f_2)$$

is 1 identically iff the quantities in parentheses (coefficients of $\bar{x}$ and of x) are both 1, so terms of $\mathcal{R}(\bar{x}f_0+xf_1+f_2)$ are products of terms of $\mathcal{R}(f_0+f_2)$ and $\mathcal{R}(f_1+f_2)$, and conversely. ∎

The variable x in this proof is called the **pivotal variable** of the expansion.

For example, we shall compute the column factors for columns 5 and 8 of Table 3.20.1, using superscripts to keep track of the row-indices of terms and using x_2 as the pivotal variable in both cases:

$$\mathcal{R}(\bar{x}_2^1+x_2x_4^2+\bar{x}_4^3+1^5+\bar{x}_2^7+x_4^8)$$
$$=\mathcal{R}(1^1+\bar{x}_4^3+1^5+1^7+x_4^8)\mathcal{R}(x_4^2+\bar{x}_4^3+1^5+x_4^8)$$
$$=(1+5+7+38)(5+23+38) = 5+38+(1+7)23 = 5+38+123+237.$$

Note how each term not involving literals of x_2 appears in both $\mathcal{R}$-factors, but a term $\bar{x}_2 w$ becomes a term w in the first $\mathcal{R}$-factor only, and so on.

$$\mathcal{R}(\bar{x}_2 x_5^1+x_2^2+x_5^5+x_2\bar{x}_5^6+\bar{x}_2^7+1^8)$$
$$=\mathcal{R}(x_5^1+x_5^5+1^7+1^8)\mathcal{R}(1^2+x_5^5+\bar{x}_5^6+1^8)$$
$$=(7+8)(2+8+56) = 8+7(2+56) = 8+27+567.$$

In a more complex case, the lemma might need to be applied to each of f_0+f_2 and f_1+f_2, and so on.

As a second example of the complete procedure, consider

(3.20.4) $\quad f = x_1x_2x_6+x_3x_4x_6+\bar{x}_1x_3x_4+\bar{x}_2\bar{x}_4x_6+x_4x_5x_7.$

We must first get the complete sum of f. Using Reusch's method, we get

$$f = \bar{x}_4(x_1x_2x_6+\bar{x}_2x_6)+x_4(x_1x_2x_6+x_3x_6+\bar{x}_1x_3+x_5x_7)$$
$$= \bar{x}_4(x_1x_6+\bar{x}_2x_6)+x_4(x_1x_2x_6+x_3x_6+\bar{x}_1x_3+x_5x_7),$$

Sec. 3.20 Reusch's Method for Finding All Minimal Sums

with complete sums in the latter parentheses, so the complete sum of f is

(3.20.5) $$f = x_1x_2x_6 + x_1x_3x_6 + x_1x_5x_6x_7 + \bar{x}_2x_3x_6 + \bar{x}_2x_5x_6x_7$$
$$+ x_1\bar{x}_4x_6 + \bar{x}_2\bar{x}_4x_6 + x_3x_4x_6 + \bar{x}_1x_3x_4 + x_4x_5x_7.$$

The chart for this example is Table 3.20.2. (It has 10 rows and 5 columns. Since f has 61 minterms, the chart using the procedure in Section 3.19 would have 10 rows and 61 columns!) We use (3.20.4) as $Q_1 + \cdots + Q_{\hat{m}}$, as it has only half as many terms as (3.20.5).

TABLE 3.20.2

		$x_1x_2x_6$	$x_3x_4x_6$	$\bar{x}_1x_3x_4$	$\bar{x}_2\bar{x}_4x_6$	$x_4x_5x_7$
1	$x_1x_2x_6$	1	x_1x_2	—	—	$x_1x_2x_6$
2	$x_1x_3x_6$	x_3	x_1	—	x_1x_3	$x_1x_3x_6$
3	$x_1x_5x_6x_7$	x_5x_7	$x_1x_5x_7$	—	$x_1x_5x_7$	x_1x_6
4	$\bar{x}_2x_3x_6$	—	$\bar{x}_2$	$\bar{x}_2x_6$	x_3	$\bar{x}_2x_3x_6$
5	$\bar{x}_2x_5x_6x_7$	—	$\bar{x}_2x_5x_7$	$\bar{x}_2x_5x_6x_7$	x_5x_7	$\bar{x}_2x_6$
6	$x_1\bar{x}_4x_6$	$\bar{x}_4$	—	—	x_1	—
7	$\bar{x}_2\bar{x}_4x_6$	—	—	—	1	—
8	$x_3x_4x_6$	x_3x_4	1	x_6	—	x_3x_6
9	$\bar{x}_1x_3x_4$	—	$\bar{x}_1$	1	—	$\bar{x}_1x_3$
A	$x_4x_5x_7$	$x_4x_5x_7$	x_5x_7	x_5x_7	—	1

In the second column, using x_1 as the pivotal variable, and noting that since x_5 and x_7 are monoform in this column, they cannot appear in any minimal sum equal to 1, so the entries of rows 3, 5, and A may be ignored, we get

$$\mathcal{R}(x_1x_2^1 + x_1^2 + \bar{x}_2^4 + 1^8 + \bar{x}_1^9)$$
$$= \mathcal{R}(\bar{x}_2^4 + 1^8 + 1^9)\mathcal{R}(x_2^1 + 1^2 + \bar{x}_2^4 + 1^8)$$
$$= (8+9)(2+8+14) = 8 + 9(2+14) = 8 + 29 + 149.$$

Now we simplify the column factor product

$$(1)(8+29+149)(9)(7)(A) = (1)(7)(9)(A)(8+2+4),$$

and the three irredundant forms of f are

(3.20.6) $$x_1x_2x_6 + \bar{x}_2\bar{x}_4x_6 + \bar{x}_1x_3x_4 + x_4x_5x_7 + \begin{cases} x_1x_3x_6 & \text{or} \\ \bar{x}_2x_3x_6 & \text{or} \\ x_3x_4x_6. \end{cases}$$

All are minimal.

We remark that if the Q_i's are the minterms of f, then the Ghazala–Reusch procedure is exactly that used in Section 3.19, with 1's instead of ×'s.

The proof of this procedure constitutes the remaining four results in this section; it is moderately difficult, but we present it here briefly, for the interested reader.

Lemma 3.20.2. *Let $Q_k, P_1, P_2, \ldots, P_n$ be product terms. Denote $\mathcal{U}_k = \{P_j | Q_k P_j \neq 0\}$. If $P_j \in \mathcal{U}_k$, let $P_j = P_j^{(1)} P_j^{(2)}$, where $P_j^{(1)}$ consists of literals in Q_k and $P_j^{(2)}$ consists of literals not in Q_k, and denote $\tilde{\mathcal{U}}_k = \{P_j^{(2)} | P_j \in \mathcal{U}_k\}$. Then Q_k is an implicant of $P_1 + P_2 + \cdots + P_n$ iff $\sum_{T \in \tilde{\mathcal{U}}_k} T = 1$.*

(Note that $\tilde{\mathcal{U}}_k$ is exactly the set of entries we put in the chart, in the column headed Q_k, and $P_j^{(2)} = 1$ if $P_j Q_k = Q_k$.)

Proof. Consider the equation

$$(P_1 + \cdots + P_n) Q_k = \sum_{P \in \mathcal{U}_k} P Q_k = \sum_{P \in \mathcal{U}_k} Q_k P^{(1)} P^{(2)} = \sum_{P \in \mathcal{U}_k} Q_k P^{(2)}$$
$$= \sum_{T \in \tilde{\mathcal{U}}_k} Q_k T = Q_k \sum_{T \in \tilde{\mathcal{U}}_k} T.$$

Q_k is an implicant of $P_1 + \cdots + P_n$ iff the left side equals Q_k, iff the right side equals Q_k, and iff $\sum_{T \in \tilde{\mathcal{U}}_k} T = 1$, since $\sum_{T \in \tilde{\mathcal{U}}_k} T$ does not involve variables in Q_k. ∎

Lemma 3.20.3. *A sum $\sum_{i=1}^{k} P_i$ of product terms is 1 iff it is also 1 when all terms containing monoform variables are deleted.*

Proof. Assume that $P_1, \ldots, P_l$ involve x (say $P_i = x P_{i0}$, $1 \leq i \leq l$), and $P_{l+1}, \ldots, P_k$ do not involve x or $\bar{x}$. Then

$$\sum_{i=1}^{k} P_i = x \sum_{i=1}^{l} P_{i0} + \sum_{i=l+1}^{k} P_i = x \left(\sum_{i=1}^{l} P_{i0} + \sum_{i=l+1}^{k} P_i \right) + \bar{x} \left(\sum_{i=l+1}^{k} P_i \right).$$

This is 1 iff $\sum_{i=l+1}^{k} P_i = 1$. ∎

Theorem 3.20.1. *Assume that $\tilde{\mathcal{U}}_k$ is defined from Q_k and $P_1, \ldots, P_n$ as in Lemma 3.20.2, and $\{j_1, j_2, \ldots, j_l\} \subseteq \{1, 2, \ldots, n\}$. Then $Q_1 + Q_2 + \cdots + Q_m$ implies $P_{j_1} + P_{j_2} + \cdots + P_{j_l}$ iff $j_1 j_2 \cdots j_l$ is an implicant of $\prod_{k=1}^{m} \mathcal{R}(\sum_{T \in \tilde{\mathcal{U}}_k} T)$.*

(Note that this last product is exactly the product of column factors that we used in step 4.)

Proof. $Q_1 + \cdots + Q_m$ implies $P_{j_1} + \cdots + P_{j_l}$ iff Q_k implies $\sum_{\{P_{j_i} | Q_k P_{j_i} \neq 0\}} P_{j_i}$ for every k, iff $\sum_{\{P_{j_i} | Q_k P_{j_i} \neq 0\}} P_{j_i}^{(2)} = 1$ for every k, iff $j_1 j_2 \cdots j_l$ is an implicant of $\mathcal{R}(\sum_{T \in \tilde{\mathcal{U}}_k} T)$ for every k, and iff $j_1 j_2 \cdots j_l$ is an implicant of $\prod_{k=1}^{m} \mathcal{R}(\sum_{T \in \tilde{\mathcal{U}}_k} T)$. ∎

Corollary 3.20.1. *Let $P_1 + P_2 + \cdots + P_n$ be the complete sum of $f = Q_1 + Q_2 + \cdots + Q_m$. Then $P_{j_1} + P_{j_2} + \cdots + P_{j_l}$ is an irredundant sum for f iff $j_1 j_2 \cdots j_l$ is a prime implicant of $\prod_{k=1}^{m} \mathcal{R}(\sum_{T \in \tilde{\mathcal{U}}_k} T)$.*

Sec. 3.21 *Exercises*   163

Proof. If $P_{j_1}+P_{j_2}+\cdots+P_{j_l}$ is an irredundant sum for f, then Theorem 3.20.1 implies that $j_1 j_2 \cdots j_l$ is an implicant of $\prod_{k=1}^{m} \mathcal{R}(\sum_{T \in \tilde{u}_k} T)$. If $j_1 j_2 \cdots j_l$ is not a prime implicant, then some shortening of $j_1 j_2 \cdots j_l$ is still an implicant. Therefore, $Q_1+Q_2+\cdots+Q_m$ implies, and hence equals, a shortening of the sum $P_{j_1}+P_{j_2}+\cdots+P_{j_l}$, a contradiction to its irredundancy; we conclude that $j_1 j_2 \cdots j_l$ is a prime implicant.

Similarly, if $j_1 j_2 \cdots j_l$ is a prime implicant, we can see that $P_{j_1}+P_{j_2}+\cdots+P_{j_l}$ is an irredundant sum. ∎

It is pointed out by Reusch (1975) that the procedure above may also be used to simplify "incompletely specified functions": "functions" that are only defined for some, not all, of the possible combinations of the switching variables. These are frequently encountered in practical problems.

3.21 Exercises

*1. Find the prime implicants and all the minimal sum-of-products representations of the functions:

(a) $f_1 = \Sigma(0, 1, 3, 7, 15)$ $(n=4)$.
(b) $f_2 = x_1 x_2 x_3 + x_1 \bar{x}_3 + x_1 x_2 \bar{x}_4 + \bar{x}_1 x_3 + \bar{x}_1 \bar{x}_2 \bar{x}_3 \bar{x}_4$ $(n=4)$.
(c) $f_3 = \Sigma(0, 1, 2, 3, 4, 5, 16, 20)$ $(n=5)$.
(d) $f_4 = x_1 x_2 x_3 x_5 + x_1 \bar{x}_3 + x_1 x_2 \bar{x}_4 + \bar{x}_1 x_3 + \bar{x}_1 \bar{x}_2 \bar{x}_3 \bar{x}_4 \bar{x}_5$ $(n=5)$.
(e) $f_5 = \Sigma(0, 1, 4, 5, 16, 17, 20, 21, 32, 33, 36, 37, 52, 53)$ $(n=6)$.
(f) $f_6 = x_1 x_2 + \bar{x}_1 x_3 + \bar{x}_2 x_3 + \bar{x}_3 x_4 + \bar{x}_4 x_5 + \bar{x}_5 x_6$ $(n=6)$.
(g) $f_7 = \Sigma(8, 9, 10, 11, 12, 13, 15, 28, 29, 58, 75, 79, 92)$ $(n=7)$.

*2. The method of **iterated consensus**: Given any sum-of-products representation of a nonconstant function f, the complete sum of f may be obtained by performing the following two operations in any order as often as possible:

(a) *Reduction by absorption.* If term P of a sum of products equal to f subsumes (implies) any other term Q of that sum, delete the term P from the sum.

(b) *Expansion by consensus.* If terms P and Q of a sum of products equal to f have consensus C (i.e., $P = xP_0$, $Q = \bar{x}Q_0$, $C = P_0 Q_0$, so that $P+Q+C = P+Q$ by consensus, for some literal x) and if C does not subsume any other term of that sum, add term C to the sum.

Apply this procedure to the following:
(1) $f = \bar{x}_1 \bar{x}_2 + x_2 x_3 + x_1 \bar{x}_3$ $(n=3)$.
(2) $g = x_1 \bar{x}_2 x_3 + x_2 x_3 x_4 + \bar{x}_1 x_3 x_4 + x_1 \bar{x}_3 \bar{x}_4$ $(n=4)$.
(3) Functions (a), (b), and (c) of Exercise 1.

The consensus method was first discussed in Blake (1937) and later emphasized by Quine (1955). Tison (1967) showed that if we do all reductions (a) possible, then add all consensus terms (b) possible with respect to x_1, then all reductions (a) possible, then all consensus terms (b) possible with respect to x_2, and so on, we shall reach the complete sum after considering all switching variables x_i once each. So he reduced the consensus method to a "one-pass" procedure. Tison gave a somewhat similar algebraic procedure for finding all irredundant sums, given all prime implicants.

3. Show by examples that a prime implicant of a switching function f need not necessarily imply or be a shortening of any term of any one given sum-of-products representation of f.

*4. A switching function that has a sum-of-products representation that is free of complemented variables is called a **frontal** or **positive switching function**. Show that if $f = \sum \pi_i$ is a positive sum of products and if π is a positive product term, then $\pi \Rightarrow f$ iff π subsumes some π_i of the sum.

*5. Show that every prime implicant of a positive switching function:

(a) Is itself positive.

(b) Implies some term of each positive sum-of-products representation of f.

(c) Is subsumed by some term of the sum.

*6. Prove that if a positive switching function expressed in sum-of-products form has the property that no term is a shortening of any other, then the terms of the sum are precisely the prime implicants of f.

7. Prove that a prime implicant of f involves only variables that appear in every sum-of-products representation of f. (See Exercise 3.16.34.)

8. Prove that if a prime implicant P of f is not a term of a given sum-of-products representation of f, then P implies no term of the sum.

9. Prove that if a product P implies f and contains all the variables present in a given sum-of-products representation of f, then P implies some term Q of the given representation of f.

10. Prove that if a prime implicant P of f is not a term of a given sum-of-products representation of f, then P does not contain all the variables appearing in the given sum.

3.22 Conclusion

More of the general theory of Boolean algebras and additional applications will appear in Chapter 6. The application of switching algebra to relay switching circuits is treated in Chapter 1 of Hohn (1966). The application to propositional logic is treated in Chapter 3 of Hohn (1966).

The map method and the Quine–McCluskey procedures for minimizing switching functions may be found in almost any book on switching circuits; in particular, they are treated in Chapter 4 of Hohn (1966).

CHAPTER 4
Semigroups and Groups

4.1 Definitions

In this section we define groups and semigroups and give some basic examples. In Sections 4.8 and 4.9 we shall see how semigroups are applied in the theory of finite-state machines. Both topics are important in Chapters 5 and 6. We have written Chapter 4 so that it (and Chapter 5) are independent of Chapter 3. Some of the notions here are special cases of more general concepts in the early part of Chapter 3, but are repeated here in full.

Given sets S and T, a function $\circ$ from $S \times S$ to T is called a **binary operation**. We commonly abbreviate $\circ((s_1, s_2))$ to $s_1 \circ s_2$ or simply to $s_1 s_2$ if there is no confusion. A binary operation $\circ: S \times S \to T$ is **commutative** iff $s_1 \circ s_2 = s_2 \circ s_1$ for all s_1 and s_2 in S. We have seen examples where the set T is different from S (the dot product of two vectors is a scalar), but in very many important cases, $T = S$. When this is the case, that is, when $s_1 \circ s_2 \in S$ for all s_1 and s_2 in S, we say that S is **closed** with respect to $\circ$ and that $\circ: S \times S \to S$ is a *binary operation on S*. A binary operation $\circ$ on S is **associative** iff $s_1 \circ (s_2 \circ s_3) = (s_1 \circ s_2) \circ s_3$ for all s_1, s_2, and s_3 in S.

A **binary algebra** $[S, \circ]$ is defined to be a nonempty set S with a binary operation $\circ$ on S. When there is no confusion as to which operation is intended, we may abbreviate $[S, \circ]$ by S.

Let $[S, \circ]$ be a binary algebra. If $\circ$ is associative, $[S, \circ]$ is called a **semigroup**. If S is a finite set, then the **order** of the binary algebra $[S, \circ]$ is defined to be the order of the set S.

If $+$ and $\cdot$ denote ordinary addition and multiplication of integers, then $[\mathbb{Z}, +]$, $[\mathbb{Z}, \cdot]$, $[\mathbb{P}, +]$, and $[\mathbb{P}, \cdot]$ are all semigroups.

A semigroup $[M, \circ]$ is a **monoid** if there is a special element "1" $\in M$ such that $1 \circ m = m \circ 1 = m$, for all $m \in M$. The element 1 is then called the **identity element** (or simply the **identity**) of M. The article "the" is appropriate because of the following lemma.

Lemma 4.1.1. *A binary algebra $[M, \circ]$ has at most one identity element; that is, any identity element is unique.*

Proof. If 1 and $1'$ are identity elements of $[M, \circ]$, then

$$1' = 1' \circ 1 = 1,$$

the first equality since 1 is an identity and the second since $1'$ is an identity. ∎

$[\mathbb{Z}, +]$ and $[\mathbb{P}, \cdot]$ are monoids, but $[2\mathbb{Z}, \cdot]$ and $[\mathbb{P}, +]$ are not. (Why?)

A monoid $[M, \circ]$ with identity element 1 is a **group** iff for each $m \in M$ there is an inverse element $m^{-1} \in M$ such that

$$m^{-1} \circ m = m \circ m^{-1} = 1.$$

Any such $m \in M$ is called an **invertible element** of M (with respect to the operation $\circ$).

Lemma 4.1.2. *If $[M, \circ]$ is a monoid and $m \in M$ has an inverse element m^{-1}, then m^{-1} is unique.*

Proof. If $x \in M$ is also an inverse of m so that $x \circ m = m \circ x = 1$, then

$$x = x \circ 1 = x \circ (m \circ m^{-1}) = (x \circ m) \circ m^{-1} = 1 \circ m^{-1} = m^{-1}. \blacksquare$$

Note that in a monoid, some elements may be invertible while others are not, but in a group, every element is invertible. Thus in the monoid of positive integers with operation multiplication, only the element 1 is invertible; but in the group of positive rationals with operation multiplication, every element is invertible because $(p/q) \cdot (q/p) = 1$.

Note also that if S is a set and $\circ$ is a function with domain $S \times S$, then $[S, \circ]$ is a semigroup iff $\circ$ is *closed* and *associative*. $[S, \circ]$ is a group iff the following four properties (sometimes called the *group properties*) hold: $\circ$ is *closed* and *associative*, S includes an *identity element* with respect to $\circ$, and every element of S has an *inverse* with respect to $\circ$.

Finally, any binary algebra $[S, \circ]$ is **commutative** (or **abelian**) iff the operation $\circ$ is commutative.

All of the following examples will be important in later chapters of this book.

1. Let X be a finite set, X^* the set of all finite sequences (or "strings" or "words") $\mathbf{x} = x_1 x_2 \cdots x_n$ of elements of X, including the empty word λ. The binary operation $\circ$ on X^* called **concatenation** is defined by "writing the two sequences together"; for example,

$$x_1 x_5 x_1 x_2 \circ x_3 x_1 x_4 x_1 = x_1 x_5 x_1 x_2 x_3 x_1 x_4 x_1,$$

$$x_1 x_7 x_2 x_3 x_2 \circ \lambda = x_1 x_7 x_2 x_3 x_2.$$

$[X^*, \circ]$ is a semigroup with identity element λ, so it is a monoid. (Why is it not a group?) The set $X^+ = X^* - \{\lambda\}$ also forms a semigroup with operation $\circ$, but it is not a monoid. $[X^+, \circ]$ is called the **free semigroup** generated by X, $[X^*, \circ]$ the **free monoid** generated by X.

If formally distinct elements of such a monoid are identified as equal by an equivalence relation, the equivalence classes may form a monoid, but it will no longer be free. We shall later see this for the free monoid of all possible input sequences to a given finite-state machine.

2. Let S be a nonempty set, S^S the set of all functions from S to S, and the composition of functions. By Theorem 1.9.1, $[S^S, \circ]$ is a semigroup; and

Exercise 1.6.9 shows that the *identity function* $1_S: S \to S$ defined by $1_S(s) = s$ is the identity element of $[S^S, \circ]$, making $[S^S, \circ]$ a monoid. The inverse of a function f, defined in Section 1.9, is the same as the inverse element f^{-1} of the present section. If S has more than one element, some functions in S^S do not have inverses (are not bijections), so $[S^S, \circ]$ is not a group.

Let Sym $(S) = \{f \in S^S | f \text{ is invertible}\}$. Exercise 1.10.2(e) and Theorem 1.9.3 show that the set Sym (S) is closed with respect to the operation $\circ$. The reader should complete the verification that [Sym (S), $\circ$] is a group, called the **symmetric group on** S. If S is finite, elements of Sym (S) are called **permutations**; we saw in Section 1.8 that if $|S| = n$, then $|\text{Sym}(S)| = n!$. If $|S| > 2$, Sym (S) is not commutative.

3. If B is a Boolean algebra with binary operations $\cup$ and $\cap$, zero element 0 and unit element 1, then it is clear from Section 1.19 that $[B, \cup]$ is a commutative monoid with identity element 0 and $[B, \cap]$ is a commutative monoid with identity element 1.

4. Let $+$ and $\cdot$ be the usual addition and multiplication operations on $\mathbb{R}$. Then the reader can easily verify the following statements.
   (a) $[\mathbb{Z}, +]$ is a commutative group.
   (b) $[\mathbb{Z}, \cdot]$ and $[\mathbb{Z} - \{0\}, \cdot]$ are commutative monoids, but not groups.
   (c) $[\mathbb{P}, +]$ is a commutative semigroup, but not a monoid; $[\mathbb{N}, +]$ and $[\mathbb{P}, \cdot]$ are commutative monoids, but not groups.
   (d) $[\mathbb{R}, +]$ and $[\mathbb{R} - \{0\}, \cdot]$ are commutative groups; $[\mathbb{R}, \cdot]$ is a commutative monoid but not a group.

5. Let $\mathcal{R}(S, S)$ be the set of all binary relations from S to S, $\circ$ the composition of relations. Then $[\mathcal{R}(S, S), \circ]$ is a monoid with identity element the identity relation $\{(s, s) \in S \times S | s \in S\}$.

6. Define the binary operation $+$ on $\mathbb{B} = \{0, 1\}$ by

$$0 + 0 = 1 + 1 = 0, \qquad 0 + 1 = 1 + 0 = 1,$$

and then define $+$ on $\mathbb{B}^n$ by

$$(a_1, a_2, \ldots, a_n) + (b_1, b_2, \ldots, b_n) = (a_1 + b_1, a_2 + b_2, \ldots, a_n + b_n),$$

all $a_i, b_j \in \mathbb{B}$. Then $\mathbb{B}^n$ is a commutative group in which $(0, 0, \ldots, 0)$ is the identity element and each element is its own inverse. (Note that here "+" denotes addition modulo 2, not the logical sum of Boolean algebra; $\mathbb{B}^n$ is not a group with respect to the latter operation.)

Any binary algebra $[S, \circ]$ with a small finite number m of elements can be specified by giving its *multiplication table*. Index the rows and columns of an $m \times m$ table (matrix) with the elements of S in some order. The entry in the row indexed x and column indexed y is the product $x \circ y$. All group properties except associativity are easy to check in a multiplication table.

Let $S = \{1, 2\}$, so $S^S = \{f_1, f_2, f_3, f_4\}$, where $f_1: 1 \to 1, 2 \to 2$; $f_2: 1 \to 1, 2 \to 1$; $f_3: 1 \to 2, 2 \to 2$; and $f_4: 1 \to 2, 2 \to 1$. By Theorem 1.9.1, any subset of S^S that is a binary algebra under composition is also a semigroup. Multiplication tables of four such semigroups are

(a)

∘	f_1
f_1	f_1

(b)

∘	f_1	f_2
f_1	f_1	f_2
f_2	f_2	f_2

(c)

∘	f_1	f_4
f_1	f_1	f_4
f_4	f_4	f_1

(d)

∘	f_2	f_3
f_2	f_2	f_2
f_3	f_3	f_3

For example, in table (d), $(f_2 \circ f_3)(s) = f_2(f_3(s)) = f_2(2) = 1 = f_2(s)$, all $s \in S$, so $f_2 \circ f_3 = f_2$. The reader should verify similarly all other entries of these tables. Tables (a), (b), and (c) represent monoids with identity element f_1. Tables (a) and (c) represent groups, with $f_4^{-1} = f_4$ in (c).

The table hinted at in Exercise 1.10.15 is the multiplication table of a group.

4.2 Zeros and Identities

Let S be a semigroup. If $1_l \in S$ satisfies $1_l x = x$ for all $x \in S$, then 1_l is a **left identity** in S. If $1_r \in S$ satisfies $x 1_r = x$ for all $x \in S$, then 1_r is a **right identity**. Note that an element of S is an *identity element* (or **two-sided identity** or **unit element**) iff it is both a left identity and a right identity. If $0_l \in S$ satisfies $0_l x = 0_l$ for all $x \in S$, then 0_l is a **left zero**. If $0_r \in S$ satisfies $x 0_r = 0_r$ for all $x \in S$, then 0_r is a **right zero**. An element of S that is both a left zero and a right zero is called a **zero element** (or **two-sided zero**) in S.

Lemma 4.2.1. *If 1_l is a left identity and 1_r a right identity in the semigroup S, then*
  (a) $1_l = 1_r$, *so we denote the identity element $1_l = 1_r$ simply by 1.*
  (b) *1 is the only left or right identity in S.*

Proof. For (a), $1_l = 1_l 1_r = 1_r$, the first equality since 1_r is a right identity and the second since 1_l is a left identity. Part (b) also must hold, since any other left identity 1_l^* or right identity 1_r^* can be treated in the same way as 1_l or 1_r. ∎

Lemma 4.2.2. *If 0_l is a left zero and 0_r a right zero in the semigroup S, then*
  (a) $0_l = 0_r$, *so we denote the zero element $0_l = 0_r$ simply by 0.*
  (b) *0 is the only left or right zero in S.*

Proof. For (a), $0_l = 0_l 0_r = 0_r$, the first equality since 0_l is a left zero and the second since 0_r is a right zero. Part (b) follows as does Lemma 4.2.1(b). ∎

Sec. 4.2 *Zeros and Identities* 169

Here are some examples of these concepts:
1. If X is a finite set, X^* and $\circ$ as in example 1 of the previous section, then X^* has the two-sided identity λ but no left or right zero.
2. Let S be a set containing more than one element. We saw in example 2 of the previous section that $[S^S, \circ]$ is a monoid with two-sided identity 1_S. By Lemma 4.2.1, no other element of S^S can be either a left or a right identity.

For each $x \in S$, define $f_x : S \to S$ by $f_x(s) = x$, all $s \in S$. Here f_x is called a **constant function**, or a **reset**. For any $g \in S^S$ we see that

$$(f_x \circ g)(s) = f_x(g(s)) = x = f_x(s) \; \forall \, s \in S,$$

so $f_x \circ g = f_x$. This proves that each f_x is a left zero. Lemma 4.2.2 shows that since there is more than one left zero, there cannot be any right zero.

3. If B is a Boolean algebra with binary operations $\cup$ and $\cap$, zero element 0 and unit element 1, then $[B, \cap]$ has zero element 0 and identity element 1, while $[B, \cup]$ has identity element 0 and zero element 1. We do not really want to call 1 a "zero element," so we used the term *dominant element* instead in (1.17.5) and Section 1.19.

4. A group G with more than one element has no zeros, either left or right. For if $x \neq y$ in G and z is, say, a left zero in G, then $zx = z = zy$, $z^{-1}zx = z^{-1}zy$, $1x = 1y$, $x = y$, which is a contradiction.

5. Let S be any set, and define the binary operation $\circ$ on S by $x \circ y = y$, all $x, y \in S$. Then

$$x \circ (y \circ z) = y \circ z = z = (x \circ y) \circ z,$$

so S is a semigroup. Clearly, every element of S is a left identity and a right zero.

We can always make a semigroup into a monoid by adjoining an identity element, as follows.

Theorem 4.2.1. *Let $[S, \circ]$ be a semigroup and let e be some element satisfying $e \notin S$. Form the set $\bar{S} = S \cup \{e\}$, and define the binary operation $\diamond$ on $\bar{S}$ by:*

$$s_1 \diamond s_2 = s_1 \circ s_2 \quad \text{if } s_1, s_2 \in S,$$

$$s \diamond e = e \diamond s = s \quad \text{if } s \in S,$$

$$e \diamond e = e.$$

Then $[\bar{S}, \diamond]$ is a monoid, obviously containing $[S, \circ]$.

Proof. By the definition of $\diamond$, it will be a binary operation on $\bar{S}$ and e will be an identity for $[\bar{S}, \diamond]$, provided we check that $\diamond$ is associative. We must show $(x \diamond y) \diamond z = x \diamond (y \diamond z)$ always. This is true if $x, y, z \in S$, since $\circ$ is associative. We check the separate cases $x = e$, $y = e$, and $z = e$, the other two variables representing arbitrary elements of $\bar{S}$:

$$(e \circ y) \circ z = y \circ z = e \circ (y \circ z),$$
$$(x \circ e) \circ z = x \circ z = x \circ (e \circ z),$$
$$(x \circ y) \circ e = x \circ y = x \circ (y \circ e). \blacksquare$$

4.3 Cyclic Semigroups and Cyclic Groups

We define the **powers** of an element x of a semigroup S as follows:

$$x^1 = x, \qquad x^2 = x \cdot x, \qquad x^3 = x^2 \cdot x, \ldots, x^{n+1} = x^n \cdot x, \ldots$$

This defines x^m for positive integers m. If S is a monoid so that it has an identity element e, then we also define

$$x^0 = e.$$

Finally, if x has an inverse x^{-1} with $x \cdot x^{-1} = x^{-1} \cdot x = e$ (this will be true if S is a group), then we define negative integer powers of x by

$$x^{-1} = x^{-1}, \qquad x^{-2} = (x^{-1})^2, \ldots, x^{-n} = (x^{-1})^n, \ldots$$

A **cyclic semigroup** is a semigroup consisting of the powers $\{x, x^2, \ldots, x^n, \ldots\}$ of a single element x. The element x is then called a **generator** of the cyclic semigroup.

A **cyclic group** is a group consisting of the powers $\{\ldots, x^{-n}, \ldots, x^{-2}, x^{-1}, x^0 = e, x, x^2, \ldots, x^n, \ldots\}$ of a single element x. The element x is then called a **generator** of the cyclic group.

The above sets appear infinite, but this will not be the case if the terms repeat. Suppose that for positive integers $l < m$, $x^l = x^m$. Then $x^{l+1} = x^l \cdot x = x^m \cdot x = x^{m+1}$, $x^{l+2} = x^{m+2}, \ldots, x^{l+k} = x^{m+k}$ for all $k \in \mathbb{P}$. This means that the set $\{x, x^2, \ldots, x^n, \ldots\}$ is a finite set $\{x, x^2, \ldots, x^l, \ldots, x^{m-1}\}$, whose elements are distinct if l and m are the smallest positive integers $l < m$ with $x^l = x^m$ (prove this). In this case, we can draw a picture of the cyclic semigroup as in Figure 4.3.1, the arrows denoting the result of multiplication by x. If no two distinct powers of x are equal, the (infinite) cyclic semigroup can be pictured as in Figure 4.3.2. As an exercise, the reader should convince himself that Figures 4.3.1 and 4.3.2 are the only possible "pictures" of cyclic semigroups.

Suppose now that such a repetition $x^l = x^m$, $l < m$, occurs in a cyclic group. Then

$$x^{-l} \cdot x^l = x^{-l} \cdot x^m,$$
$$(x^{-1})^l \cdot x^l = (x^{-1})^l \cdot x^l \cdot x^{m-l},$$
$$e = x^{m-l}$$

Sec. 4.3 Cyclic Semigroups and Cyclic Groups

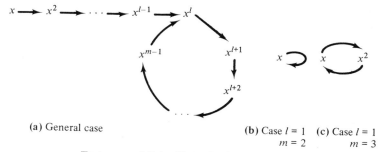

(a) General case   (b) Case $l = 1$ (c) Case $l = 1$
                         $m = 2$ $m = 3$

FIGURE 4.3.1. *Finite Cyclic Semigroups*

$$x \longrightarrow x^2 \longrightarrow x^3 \longrightarrow \cdots \longrightarrow x^n \longrightarrow \cdots$$

FIGURE 4.3.2. *Infinite Cyclic Semigroup*

for the integer $m - l > 0$. (Exercise 4.5.6 will further discuss laws of exponents inside a group.) Whenever there is a positive integer t such that $x^t = e$, the least such integer is called the **order** of x, and the cyclic group generated by x is the set

$$\{x^0 = e, x^1 = x, x^2, \ldots, x^{t-1}\}$$

of order t. This set contains not only the higher positive powers $x^t = e$, $x^{t+1} = x, \ldots$ of x but also the negative powers; for $x^{t-1} \cdot x = x^t = e = x \cdot x^{t-1}$ implies that $x^{t-1} = x^{-1}$ (Lemma 4.1.2), so x^{-1} is in the set, and similarly $x^{-2} = x^{t-2}, x^{-3} = x^{t-3}, \ldots, x^{-t} = e, x^{-(t+1)} = x^{2t-(t+1)}, \ldots$ are all in the set.

Now we discuss some examples.

1. Consider the rotations of the square shown here (i.e., rotations in the plane of the square and about its center). Let R be the 90° clockwise rotation

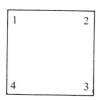

sending 1 to 2, 2 to 3, 3 to 4, and 4 to 1. Then R^2 is a 180° rotation, R^3 a 270° rotation, and $R^4 = I$ the identity map. We see that R^3 is R^{-1}, the inverse of R; $R^2 = R^{-2}$, $R = R^{-3}$, and so on. So $\{R, R^2, R^3, R^4 = I\}$ is both a cyclic semigroup and a cyclic group, with generator R. (R^3 is also a generator.)

Rotations of the regular n-gon (again, in its plane and about its center) provide us with an example of a cyclic semigroup (which is also a cyclic group) of order n, any $n \geq 3$. The "picture" looks like the n-gon itself.

2. We defined powers in a semigroup S in terms of a binary operation of multiplication, in a way that looks natural. The same definition applies when the binary operation is denoted $+$ and called addition, but it looks less natural. Just as the second power of x is $x \cdot x$ in the case of multiplication, so must it be $x + x$ in the case of addition; we usually denote $x + x$ by $2x$. So the positive powers of x are

$$x, x + x = 2x, x + x + x = 3x, \ldots, nx + x = (n+1)x, \ldots.$$

If S is a monoid, the zeroth power of x is the identity for addition, usually denoted 0. We define $0x = 0$ for all $x \in S$. On the left, "0" denotes the integer 0. On the right, it denotes the additive identity in S. If $x \in S$ is invertible, the inverse of x is usually denoted $-x$, so it can be denoted $(-1)x$; the negative powers of x are

$$-x, (-x) + (-x) = (-2)x, \ldots, (-n)x + (-x) = [-(n+1)]x, \ldots.$$

Note that for all m and n in $\mathbb{Z}$,

$$mx + nx = (m+n)x$$

and

$$(mx) + (-mx) = (m)x + (-m)x = (m + (-m))x = 0x = 0.$$

These notations do not imply that the integers necessarily belong to S.

For example, $[\mathbb{Z}, +]$ is a cyclic group with generator 1 but is not a *cyclic semigroup*. $[\mathbb{P}, +]$ is a cyclic semigroup with generator 1 but is not a group.

4.4 Subsemigroups and Subgroups

Suppose that $[S, \circ]$ is a semigroup and that T is a nonempty subset of S such that whenever $t_1, t_2 \in T$, then $t_1 \circ t_2 \in T$. Then $\circ$ defines a new binary operation $T \times T \to T$, called the *restriction* of $\circ$ to $T \times T$ and denoted also by $\circ$. For any $x, y, z \in T$, the associative law $x \circ (y \circ z) = (x \circ y) \circ z$ holds since it holds in S. Therefore, $[T, \circ]$ is a semigroup, called a **subsemigroup** of $[S, \circ]$. We emphasize that for a semigroup $[A, \square]$ to be a subsemigroup of another semigroup $[B, \circ]$, we require not only that A be a subset of B but also that $\square$ be the restriction of $\circ$ to $A \times A$, that is, $a_1 \square a_2 = a_1 \circ a_2$, all $a_1, a_2 \in A$.

If $[G, \circ]$ is a group and the subsemigroup $[H, \circ]$ of $[G, \circ]$ is also a group, we say that $[H, \circ]$ is a **subgroup** of $[G, \circ]$.

Lemma 4.4.1. *If H is a subgroup of a group G, then the identity element 1 of G is in H and is the identity element of H.*

Proof. Let e be the identity element of H. Then $e \cdot e = e$. If e^{-1} is the inverse of e in G, then $e = 1e = (e^{-1}e)e = e^{-1}(ee) = e^{-1}e = 1$ is in H. ∎

Lemma 4.4.2. *Let M be a monoid, $x, x_1, x_2, \ldots, x_{n-1}, x_n$ invertible elements of M. Then*

(a) *x^{-1} is invertible and $(x^{-1})^{-1} = x$.*

(b) *$x_1 x_2 \cdots x_{n-1} x_n$ is invertible, and $(x_1 x_2 \cdots x_{n-1} x_n)^{-1} = x_n^{-1} x_{n-1}^{-1} \cdots x_2^{-1} x_1^{-1}$.*

Proof. (a) $x \cdot x^{-1} = x^{-1} \cdot x = 1$, so x satisfies all requirements to be the inverse of x^{-1}.

(b) Using the associative law repeatedly, we have $x_1 x_2 \cdots x_{n-1} x_n \cdot x_n^{-1} x_{n-1}^{-1} \cdots x_2^{-1} x_1^{-1} = x_1 x_2 \cdots x_{n-1} x_{n-1}^{-1} \cdots x_2^{-1} x_1^{-1} = \cdots = x_1 x_1^{-1} = 1$. Similarly, $x_n^{-1} x_{n-1}^{-1} \cdots x_2^{-1} x_1^{-1} \cdot x_1 x_2 \cdots x_n = 1$. ∎

Theorem 4.4.1. *A nonempty subset H of a group G is a subgroup of G if and only if*

(a) *$x \in H$ implies that $x^{-1} \in H$, and*

(b) *$x \in H$, $y \in H$ imply that $xy \in H$.*

Proof. Suppose that H is a subgroup. By Lemma 4.4.1, the inverse x^* of x in H must also be an inverse of x in G. Lemma 4.1.2 then shows that $x^* = x^{-1}$, so (a) holds. The operation in H is the restriction of the operation in G, so (b) holds.

Suppose that (a) and (b) hold. Part (b) shows that multiplication in G defines a binary operation on H, so H is at least a semigroup. If $x \in H$, then (a) shows $x^{-1} \in H$, so (b) shows $1 = xx^{-1} \in H$; H is a monoid. Now (a) shows that H is a group. ∎

Another form of Theorem 4.4.1 is the next result.

Theorem 4.4.2. *A nonempty subset H of a group G is a subgroup of G iff $x \in H$ and $y \in H$ imply that $xy^{-1} \in H$.*

Proof. If H is a subgroup and $x, y \in H$, then Theorem 4.4.1(a) shows that $y^{-1} \in H$ and Theorem 4.4.1(b) shows that $xy^{-1} \in H$. Conversely, if $x \in H$ and $y \in H$ imply that $xy^{-1} \in H$, then if $x \in H$, the hypothesis implies that $1 = xx^{-1} \in H$. Then, since 1 and x are in H, the hypothesis implies that $x^{-1} = 1x^{-1} \in H$. Also, if $y \in H$, then $y^{-1} = 1y^{-1} \in H$, so, again by the hypothesis, $x(y^{-1})^{-1} \in H$. But $(y^{-1})^{-1} = y$, so $xy = x(y^{-1})^{-1} \in H$. We have proved that (a) and (b) of Theorem 4.4.1 hold here, so H is a subgroup. ∎

If G is finite (has finitely many elements), we can improve Theorem 4.4.1.

Theorem 4.4.3. *A nonempty subset H of a finite group G is a subgroup of G iff $x \in H$ and $y \in H$ imply that $xy \in H$.*

Proof. By Theorem 4.4.1, it suffices to show that if H is closed under the multiplication of G and if $x \in H$, then $x^{-1} \in H$. Consider the subset $S = \{x, x^2, x^3, \ldots, x^n, \ldots\}$ of G of positive powers of $x \in H$. Since H is closed under multiplication, $S \subseteq H$. Since G is finite, S is finite and $x^k = x^l$, for some positive integers k and l, $k < l$. Multiplying by x^{-k}, we have $1 = x^{-k}x^k = x^{-k}x^l = x^{l-k} \in S \subseteq H$. If $l - k = 1$, then $x = 1 = 1^{-1} = x^{-1} \in H$; done. If $l - k > 1$, then multiply $1 = x^{l-k}$ by x^{-1} to get $x^{-1} = x^{l-k-1} \in S \subseteq H$. ∎

The idea used in the last proof is an important general method for finding subgroups. If S is any semigroup and $x \in S$, then $\{x, x^2, \ldots, x^n, \ldots\}$ is a cyclic subsemigroup of S, called the subsemigroup **generated** by x. If G is any group and $x \in G$, then

$$\langle x \rangle = \{\ldots, x^{-n}, \ldots, x^{-2}, x^{-1}, 1, x, x^2, \ldots, x^n, \ldots\}$$

is a cyclic subgroup of G, called the subgroup **generated** by x. As was explained in Section 4.3, these sets may be finite. If $\langle x \rangle$ is finite, and if t is the smallest positive integer such that $1 = x^t$, the number t of elements in $\langle x \rangle = \{1, x, x^2, \ldots, x^{t-1}\}$ will equal the order of x.

We discuss some examples.

1. If X^* is the monoid of all finite sequences of elements of the input alphabet X of a finite-state machine M, then a cyclic subsemigroup $\{\mathbf{x}, \mathbf{x}^2, \ldots, \mathbf{x}^n, \ldots\}$ is the set of all repetitions

$$x_1x_2 \cdots x_n, x_1x_2 \cdots x_nx_1x_2 \cdots x_n,$$
$$\ldots, x_1x_2 \cdots x_nx_1x_2 \cdots x_n \cdots x_1x_2 \cdots x_n, \ldots$$

of a particular input sequence $\mathbf{x} = x_1x_2 \cdots x_n$. Such input tapes are called **periodic**.

2. $[\mathbb{Z}, +]$ and $[\mathbb{Q}, +]$ are subgroups of $[\mathbb{R}, +]$, but $[\mathbb{Q} - \{0\}, \cdot]$ is not; it has a different operation.

3. Any group G with more than one element has two distinct trivial subgroups, itself and the subgroup $\{1\}$ consisting of the identity element alone. Any other subgroup of G is called a **proper subgroup** of G.

4. Let $S = \{1, 2, \ldots, n\}$, $n > 2$, and $G = \text{Sym}(S)$, the set of all invertible functions $S \to S$. We know $|G| = n!$. Denote $G_n = \{f \in G | f(n) = n\}$. Using Theorem 4.4.3, we see easily that G_n is a subgroup of G. G_n is called the **stabilizer** of n in G. (What is $|G_n|$? How is G_n related to $\text{Sym}(T)$, where $T = \{1, 2, \ldots, n-1\}$?)

4.5 Exercises

*1. Explain why the following binary algebras are *not* groups.
(a) The rational numbers under multiplication.
(b) The nonzero integers under multiplication.
(c) Functions from $\{1, 2, 3\}$ to $\{1, 2, 3\}$ under function composition ∘.
(d) The set $\{a, b, c, d\}$, with the multiplication table shown.

Sec. 4.5 *Exercises* 175

	a	b	c	d
a	a	b	c	d
b	b	a	b	c
c	c	d	d	a
d	d	c	a	c

(e) All real numbers under subtraction.

(f) X a set with more than one element, and with binary operation $\circ$ satisfying $x_1 \circ x_2 = x_1$, all $x_1, x_2 \in X$.

(g) Nonnegative real numbers under addition.

2. Show that if S is a semigroup and $x \in S$ is both a left zero and a left identity, then $|S| = 1$.

*__3.__ We say the **left cancellation law** holds in the semigroup S if

$$xy = xz \text{ implies that } y = z \quad \forall\ x, y, z \in S.$$

The **right cancellation law** holds in S if

$$xz = yz \text{ implies that } x = y \quad \forall\ x, y, z \in S.$$

(a) Show that both cancellation laws hold in any group.

(b) Show that if the left cancellation law holds in the finite semigroup S, then in the multiplication table for S, each element appears exactly once in each row. Make a similar statement for columns and the right cancellation law.

(c) Use (a) and (b) to do Exercise 1(d).

(d) Show that both cancellation laws hold in the semigroup X^* of example 1, Section 4.1, although that semigroup is not a group.

*__4.__ An element x in a binary algebra S is called an **idempotent** if $xx = x$.

(a) Show that the set of all idempotent elements in a commutative semigroup C forms a subsemigroup of C.

(b) Show that if both cancellation laws hold in the semigroup S, then any idempotent in S is a two-sided identity element.

5. (a) Show that the only idempotent in a group is the identity element.

(b) If B is a Boolean algebra with binary operations $\cup$ and $\cap$, which elements of the semigroups $[B, \cup]$ and $[B, \cap]$ are idempotents?

(c) Consider the picture of a finite cyclic semigroup S (Figure 4.3.1), where $0 < l < m$, $x^l = x^m$, and the arrows denote the result of multiplication by x. Let $s = x^i \in S$, some $i > 0$. Show that if $i < l$, then s is *not* an idempotent. If $i \geq l$, show that s is an idempotent iff $m - l$ divides i. Show that the subset $x^l, x^{l+1}, \ldots, x^{m-1}$ is a subgroup of S, and find its identity. Finally, show that $m - l$ divides exactly one of the $m - l$ integers $l, l+1, l+2, \ldots, m-1$, and conclude that S contains exactly one idempotent.

(d) Use part (c) to show that *any* finite semigroup contains an idempotent.

(e) Assume that M is a finite monoid with 1 as its only idempotent. If $x \in M$, use part (c) to conclude that some power x^n of x is an idempotent, hence $x^n = 1$. Conclude that x^{n-1} is x^{-1}, so M is a group.

(f) Note that parts (a) and (e) together prove the following theorem: *A finite monoid is a group iff* 1 *is its only idempotent.* How would this be useful in checking whether a given multiplication table represents a group? Why, then, does the table of Exercise 1(d) not represent a group, even though the identity element a is its only idempotent?

*6. (a) Let S be any semigroup, $x \in S$, $m, n \in \mathbb{P}$. Prove the *laws of exponents*:

$$x^m x^n = x^{m+n}, \quad (x^m)^n = x^{mn}.$$

(b) If S is a commutative semigroup, $x, y \in S$ and $n \in \mathbb{P}$, prove that

$$(xy)^n = x^n y^n.$$

Why is "commutative" needed here, but not in (a)? Must all of S be commutative, or is $xy = yx$ enough to imply $(xy)^n = x^n y^n$?

(c) Do parts (a) and (b) if S is a monoid and $m, n \in \mathbb{N}$.

(d) Do parts (a) and (b) if S is a group and $m, n \in \mathbb{Z}$.

(e) What do the laws in parts (a) and (b) look like if S is commutative and the binary operation in S is denoted "+" and called addition? (See example 2, Section 4.3.)

*7. (a) Let S be a semigroup, $x_1, \ldots, x_n \in S$, $T = \{x \in S | x$ has an expression as a finite product of x_i's$\}$. Show that T is a subsemigroup of S. This subsemigroup is called the subsemigroup **generated** by $x_1, \ldots, x_n$.

(b) Let G be a group, $x_1, \ldots, x_n \in G$, $H = \{x \in G | x$ has an expression as a finite product of x_i's and x_i^{-1}'s$\}$. Show that H is a subgroup of G. This subgroup is called the subgroup **generated** by $x_1, \ldots, x_n$, and we write $H = \langle x_1, \ldots, x_n \rangle$.

*8. Let M be any monoid and let $G = \{m \in M | m$ is invertible in $M\}$. Use Lemma 4.4.2 to show that G is a group.

9. Let M be a finite monoid, $a \in M$. Prove that the following are equivalent.

(1) a has a *left inverse*. ($\exists b \in M \ni ba = 1$.)

(2) The function $x \to ax$ is one-one from M to M.

(3) a has a *right inverse*. ($\exists c \in M \ni ac = 1$.)

(4) The function $x \to xa$ is one-one from M to M.

[HINT: Remember the pigeonhole principle, and prove that $(1) \Rightarrow (2)$, $(2) \Rightarrow (3)$, $(3) \Rightarrow (4)$, and $(4) \Rightarrow (1)$.]

Sec. 4.5 Exercises

10. (For those who know matrix theory over $\mathbb{R}$) Let S_0 be the set of all 2×2 matrices over $\mathbb{R}$, and let

$$S_1 = \left\{ \begin{bmatrix} a & b \\ c & 0 \end{bmatrix} \middle| a, b, c \in \mathbb{R} \right\}, \quad S_2 = \left\{ \begin{bmatrix} a & 0 \\ c & d \end{bmatrix} \middle| a, c, d \in \mathbb{R} \right\},$$

$$S_3 = \left\{ \begin{bmatrix} a & 0 \\ c & 0 \end{bmatrix} \middle| a, c \in \mathbb{R} \right\}, \quad S_4 = \left\{ \begin{bmatrix} a & b \\ 0 & 0 \end{bmatrix} \middle| a, b \in \mathbb{R} \right\},$$

$$S_5 = \left\{ \begin{bmatrix} a & 0 \\ 0 & d \end{bmatrix} \middle| a, d \in \mathbb{R} \right\}, \quad S_6 = \left\{ \begin{bmatrix} 0 & b \\ c & 0 \end{bmatrix} \middle| b, c \in \mathbb{R} \right\},$$

$$S_7 = \left\{ \begin{bmatrix} a & 0 \\ 0 & 0 \end{bmatrix} \middle| a \in \mathbb{R} \right\}, \quad S_8 = \left\{ \begin{bmatrix} 0 & b \\ 0 & 0 \end{bmatrix} \middle| b \in \mathbb{R} \right\}.$$

The operation on each S_i is ordinary matrix multiplication. For each S_i, answer the following questions:

(a) Is S_i a semigroup? A monoid? A group?

(b) If S_i is a semigroup, what are the identity elements in S_i? Left identities? Right identities?

(c) If S_i is a semigroup, what are the zero elements in S_i? Left zeros? Right zeros?

(d) If S_i is a monoid, what are the invertible elements of S_i?

11. (a) Let $G_3 = \{1 = x^3, x, x^2\}$ be a cyclic group of order 3 with generator x. Show that x^2 is also a generator of G_3 (i.e., show that every element of G_3 is a power of x^2).

(b) Which elements of a cyclic group $G_5 = \{1, y, y^2, y^3, y^4\}$ of order 5 are generators of G_5?

(c) Which elements of a cyclic group $G_6 = \{1, z, z^2, z^3, z^4, z^5\}$ of order 6 are generators of G_6? Can you find any nontrivial subgroups of G_6? Are they cyclic?

(d) Try to make some general conjectures about generators and subgroups of finite cyclic groups. (Cyclic groups are discussed further in Section 4.16.)

12. Can a cyclic semigroup have more than one generator? If so, give an example.

*13. If H and K are subgroups of the group G, show that $H \cap K$ is a subgroup of G.

*14. Consider the six symmetries of the equilateral triangle shown. Denote by (1 2 3) the 120° rotation sending $1 \to 2, 2 \to 3, 3 \to 1$; by (1 3) the reflection in the vertical axis sending $1 \to 3, 2 \to 2, 3 \to 1$. The other four symmetries are $e =$ identity, (1 3 2) = 240° (clockwise) rotation, and the reflections (1 2) and (2 3). The six symmetries form a group G under

composition ∘; for example,

$$(1 \ 2 \ 3) \circ (1 \ 3) = (2 \ 3),$$
$$(1 \ 3 \ 2) \circ (2 \ 3) = (1 \ 3).$$

(As always with composition, we apply the right function first. This notation is developed further in Section 4.17.)

(a) Write the multiplication table of G, and find the order of each element. Is G cyclic? Is G abelian?

(b) Find as many subgroups of G as you can. Use these subgroups to show that Exercise 13 fails if ∩ is replaced by ∪.

15. Let S be a semigroup, $Z(S) = \{z \in S | zs = sz \ \forall \ s \in S\}$. Show that $Z(S)$ is a commutative subsemigroup of S. If S is a group, show that $Z(S)$ is also a group. [$Z(S)$ is called the **center** of S.]

***16.** If $\langle x \rangle$ is a cyclic group of order n, $t \in \mathbb{Z}$, and $x^t = 1$, prove that n divides t.

17. Let $S = \{s_1, s_2, \ldots, s_n\}$ be a finite set with binary operation ∘, and let $\mathbf{M} = [s_i \circ s_j]_{1 \leq i,j \leq n}$, the multiplication table of $[S, \circ]$. We shall describe a matrix procedure for determining whether ∘ is associative (whether S is a semigroup). Denote by $\mathbf{M} \circ s_k$ the matrix whose i,jth entry is $(s_i \circ s_j) \circ s_k$. Let $\mathbf{R}_i$ be the matrix whose kth row is the ith row of $\mathbf{M} \circ s_k$, $1 \leq k \leq n$, and let $\mathbf{R}_i^T$ be the transpose of $\mathbf{R}_i$ (interchange rows and columns).

(a) Show that ∘ is associative iff
$$s_i \circ \mathbf{M} = \mathbf{R}_i^T, \qquad i = 1, 2, \ldots, n.$$

(b) Show that if S contains identity or zero elements, we may delete their corresponding rows and columns from $\mathbf{M}$ when performing this test.

(c) Use parts (a) and (b) to do Exercise 1(d).

(d) Find x and y such that

	1	2	3
1	3	2	1
2	x	y	2
3	1	2	3

is the multiplication table of a semigroup $[\{1, 2, 3\}, \circ]$.

18. [R. E. Moore (1966)] For $a, b \in \mathbb{R}$, let $[a, b] = \{x \in \mathbb{R} | a \leq x \leq b\}$ denote a closed interval on the real number line. Let $I(\mathbb{R})$ denote the set of all such intervals, including the "points" $[a, a] = a$.

Let ∗ denote any one of the operations addition (+), subtraction (−), multiplication (·), or division (÷), and define, for each interpretation of ∗ and for all $A, B \in I(\mathbb{R})$,

$$A * B = \{a * b | a \in A, \ b \in B\}$$

except that $A \div B$ is not defined if $0 \in B$.

(a) Prove that each operation leads again to a member of $I(\mathbb{R})$.
(b) Prove that the system $[I(\mathbb{R}), +]$ is a commutative semigroup with identity and with a cancellation law but that most intervals have no additive inverses. What intervals do have additive inverses?
(c) Investigate the properties of $[I(\mathbb{R}), -]$, $[I(\mathbb{R}), \cdot]$, and $[I(\mathbb{R}), \div]$.

4.6 Congruence Relations on Semigroups

Let $[S, \circ]$ be a semigroup, and suppose that we also have an equivalence relation R on the set S. Denote by S/R the set of equivalence classes of S under R (that is, the *partition* of S induced by R); S/R is a set whose elements are subsets of S. If R satisfies the condition

$$s R s', \quad t R t' \quad \text{imply that} \quad s \circ t R s' \circ t'$$

for all $s, s', t, t' \in S$, then R is called a **congruence relation** on $[S, \circ]$.

Theorem 4.6.1. *Assume that R is a congruence relation on the semigroup $[S, \circ]$. If $s \in S$, denote*

$$[s] = [s]_R = \{x \in S | x R s\} \in S/R.$$

Then we can define a binary operation (also denoted $\circ$) on the set S/R by

$$[s] \circ [t] = [s \circ t],$$

*and $[S/R, \circ]$ is a semigroup, called the **factor semigroup** of S by R.*

Proof. Each equivalence class in S/R may have several names; if $s R s'$ in S, then $[s] = [s']$. To check that $\circ$ is a binary operation on S/R, we must show that if $[s] = [s']$ and $[t] = [t']$, then $[s \circ t] = [s' \circ t']$. That is, if one reader uses names $[s]$ and $[t]$ for two classes and gets the product $[s] \circ [t] = [s \circ t]$, while another reader uses the names $[s']$ and $[t']$ for the same classes and gets the product $[s'] \circ [t'] = [s' \circ t']$, we must know they get the same answer. (We are showing that $\circ$ is **well-defined**.) Now $[s] = [s']$ and $[t] = [t']$ imply that $s R s', t R t'$. Since R is a congruence relation, $s \circ t R s' \circ t'$, so $[s \circ t] = [s' \circ t']$ as desired.

To check the associative law in $[S/R, \circ]$, we note that if $[s], [t], [u] \in S/R$, then

$$[s] \circ ([t] \circ [u]) = [s] \circ [t \circ u] = [s \circ (t \circ u)] = [(s \circ t) \circ u] = [s \circ t] \circ [u]$$
$$= ([s] \circ [t]) \circ [u]. \ \blacksquare$$

For our first example, we use the semigroup $[\mathbb{Z}, +]$ (which is actually a group). Fix an integer $p > 1$, and let R be the equivalence relation on $\mathbb{Z}$ called *congruence modulo p* (Section 2.12). Thus

$$a\, R\, b \quad \text{iff} \quad p|(b-a).$$

If $a\, R\, b$ and $c\, R\, d$, then $p|(b-a)$ and $p|(d-c)$, say $b-a = pk$, $d-c = pl$, $k, l \in \mathbb{Z}$. Then

$$(b+d) - (a+c) = (b-a) + (d-c) = p(k+l),$$

so $p|[(b+d)-(a+c)]$, $(a+c)\, R\, (b+d)$. That is, R is a congruence relation on $[\mathbb{Z}, +]$. By Theorem 4.6.1, $\mathbb{Z}/R$ is a semigroup under an operation denoted $+$, called **addition modulo p**. We saw in Section 2.12 that the classes in $\mathbb{Z}/R$ are $[0], [1], [2], \ldots, [p-1]$. Some examples are

if $p = 5$: $[3]+[4]=[7]=[2]$;

if $p = 8$: $[5]+[6]=[11]=[3]$,

$\phantom{\text{if } p = 8:\ }[5]+[3]=[8]=[0]$,

$\phantom{\text{if } p = 8:\ }[2]+[4]=[6]$.

Since there is one class for each integer in $\mathbb{Z}_p = \{0, 1, 2, \ldots, p-1\}$, addition modulo p is also considered an operation on $\mathbb{Z}_p$. Examples are

$$3+4 = 2 \quad \text{modulo } 5;$$

$$5+6 = 3, \quad 5+3 = 0, \quad 2+4 = 6 \quad \text{modulo } 8.$$

$[\mathbb{Z}_p, +]$ is actually a group, not just a semigroup. It has identity 0; 0 is its own inverse (negative), since $0 + 0 = 0$; if $i \in \mathbb{Z}_p - \{0\}$, then the inverse of i is $p - i$, since $i + (p-i) = (p-i) + i = 0$ (modulo p).

We can try the same equivalence relation R on the semigroup $[\mathbb{Z}, \cdot]$. If $a\, R\, b$ and $c\, R\, d$, say $b-a = pk$ and $d-c = pl$, then $bd - ac = (pk+a)(pl+c) - ac = p(klp + al + ck)$, so $ac\, R\, bd$. Thus R is a congruence relation on $[\mathbb{Z}, \cdot]$. This means that $\mathbb{Z}/R$, or equivalently $\mathbb{Z}_p$, is a semigroup under **multiplication modulo p**. Examples of this in $\mathbb{Z}/R$ are

if $p = 5$: $[3] \cdot [4] = [12] = [2]$;

if $p = 8$: $[5] \cdot [6] = [30] = [6]$,

$\phantom{\text{if } p = 8:\ }[5] \cdot [3] = [15] = [7]$,

$\phantom{\text{if } p = 8:\ }[2] \cdot [4] = [8] = [0]$.

Hence in $\mathbb{Z}_p$ we have

if $p = 5$: $\quad 3 \cdot 4 = 2$;

if $p = 8$: $\quad 5 \cdot 6 = 6, \quad 5 \cdot 3 = 7, \quad 2 \cdot 4 = 0.$

$[\mathbb{Z}_p, \cdot]$ is a monoid with identity 1 but is not a group. ($\mathbb{Z}_p$, with these operations $+$ and $\cdot$, was discussed as a ring in Section 3.4.)

For a second example of a congruence relation, let X be the input alphabet of a finite-state machine $M = [S, X, Z, \tau, \omega]$, X^* the monoid of all finite sequences from X (all *input tapes*). With each input sequence $\mathbf{x} \in X^*$ we associate a **state transition map** $T_\mathbf{x}: S \to S$, defined by $T_\mathbf{x}(\sigma_i) = \tau(\sigma_i, \mathbf{x})$, all $\sigma_i \in S$. For example, for the empty sequence λ we have $T_\lambda = 1_S$. Thus $\mathbf{x} \to T_\mathbf{x}$ is a function from the infinite monoid X^* to the finite set S^S. (If $|S| = n$, then $|S^S| = n^n$ by Theorem 1.5.1.) We define a relation E on X^* by $\mathbf{x} E \mathbf{y}$ iff $T_\mathbf{x} = T_\mathbf{y}$. This is an equivalence relation. (Why?)

Before showing that E is a congruence relation, we must identify $T_{\mathbf{x} \circ \mathbf{y}}$, $\circ$ the (concatenation) operation in X^*. If $\sigma_i \in S$,

$$T_{\mathbf{x} \circ \mathbf{y}}(\sigma_i) = \tau(\sigma_i, \mathbf{x} \circ \mathbf{y}) = \tau(\tau(\sigma_i, \mathbf{x}), \mathbf{y}) = T_\mathbf{y}(\tau(\sigma_i, \mathbf{x})) = T_\mathbf{y}(T_\mathbf{x}(\sigma_i))$$

$$= (T_\mathbf{y} \circ T_\mathbf{x})(\sigma_i),$$

$\circ$ also denoting composition of functions in S^S. Hence $T_{\mathbf{x} \circ \mathbf{y}} = T_\mathbf{y} \circ T_\mathbf{x}$. If $\mathbf{u}, \mathbf{v}, \mathbf{x}, \mathbf{y} \in X^*$ and $\mathbf{u} E \mathbf{v}$, $\mathbf{x} E \mathbf{y}$, then $T_\mathbf{u} = T_\mathbf{v}$ and $T_\mathbf{x} = T_\mathbf{y}$, so $T_\mathbf{x} \circ T_\mathbf{u} = T_\mathbf{y} \circ T_\mathbf{v}$, $T_{\mathbf{u} \circ \mathbf{x}} = T_{\mathbf{v} \circ \mathbf{y}}$, $\mathbf{u} \circ \mathbf{x} E \mathbf{v} \circ \mathbf{y}$, E is a congruence relation on $[X^*, \circ]$. We conclude that X^*/E is a semigroup, by Theorem 4.6.1. Since S^S is finite, there can be only finitely many equivalence classes in X^*/E. Also, X^*/E is a monoid (why?), called the **input monoid** of M.

4.7 Morphisms

Assume that $[S, \cdot]$ and $[T, \circ]$ are binary algebras. A function $f: S \to T$ is a **homomorphism** (or **morphism**) if

(4.7.1) $\qquad\qquad\qquad f(s_1 \cdot s_2) = f(s_1) \circ f(s_2),$

all $s_1, s_2 \in S$. The morphism f is an **epimorphism** if it is onto, a **monomorphism** if it is one-one, an **isomorphism** if it is both one-one and onto. If f is an isomorphism, $[S, \cdot]$ and $[T, \circ]$ are **isomorphic**. If f is an epimorphism, $[T, \circ]$ is a **homomorphic image** of $[S, \cdot]$.

As in the case of posets (Section 2.8), an isomorphism implies identical structure. If (4.7.1) represents a known isomorphism f and operation $\cdot$ is known but $\circ$ is not, then for any $t_1, t_2 \in T$ we can find $t_1 \circ t_2$ by finding $s_1, s_2 \in S$ with $f(s_1) = t_1$, $f(s_2) = t_2$; then $t_1 \circ t_2 = f(s_1 \cdot s_2)$. Thus the structure of $[S, \cdot]$ reveals the structure of $[T, \circ]$ completely; the converse is also true.

The presence of a homomorphism indicates that the structure is somewhat the same. As a first example, suppose that $[S, \circ]$ is a semigroup and R is a congruence relation on S. If $[s] \in S/R$ is defined by $[s] = \{x \in S | x R s\}$ and we define $f: S \to S/R$ by $f(s) = [s]$, then for $s, t \in S$ we have $f(s \circ t) = [s \circ t] = [s] \circ [t] = f(s) \circ f(t)$ by the very definition of $\circ$ on S/R. Thus f is an epimorphism from S onto S/R.

In particular, if the semigroup is $[\mathbb{Z}, +]$ or $[\mathbb{Z}, \cdot]$ and R is congruence modulo p ($p > 1$), then, denoting addition and multiplication modulo p by $\underset{p}{+}$ and $\underset{p}{\cdot}$ for emphasis, we obtain homomorphisms from $[\mathbb{Z}, +]$ onto $[\mathbb{Z}_p, \underset{p}{+}]$ and from $[\mathbb{Z}, \cdot]$ onto $[\mathbb{Z}_p, \underset{p}{\cdot}]$. Now (4.7.1) becomes

$$f(m+n) = f(m) \underset{p}{+} f(n), \quad f(mn) = f(m) \underset{p}{\cdot} f(n),$$

where $f(z)$, $z \in \mathbb{Z}$, is that integer in $\mathbb{Z}_p$ which is congruent to z modulo p.

Now we see how a homomorphism "preserves structure."

Theorem 4.7.1. *Assume that $f: S \to T$ is a binary algebra homomorphism, and as usual let $f(S) = \{f(s) \in T | s \in S\}$ denote the image of f.*

(a) *If S is a semigroup, so is $f(S)$.*

(b) *If S is a monoid with identity element 1, then $f(S)$ is a monoid with identity element $f(1)$.*

(c) *If S is a group, then $f(S)$ is a group and $f(x)^{-1} = f(x^{-1})$ for all $x \in S$.*

Proof. (a) All we have to show is that the operation of the binary algebra T, restricted to $f(S)$, is associative. If $x, y, z \in S$, then because f is a homomorphism,

$$f(x)[f(y)f(z)] = f(x)f(yz) = f(x(yz)) = f((xy)z) = f(xy)f(z)$$
$$= [f(x)f(y)]f(z).$$

(b) Each element of $f(S)$ is some $f(x)$, $x \in S$. But for any $x \in S$, $f(x)f(1) = f(x1) = f(x)$ and $f(1)f(x) = f(1x) = f(x)$, so $f(1)$ is the identity element of $f(S)$, which is already a semigroup by (a).

(c) For any $x \in S$, $f(x)f(x^{-1}) = f(xx^{-1}) = f(1)$ and $f(x^{-1})f(x) = f(x^{-1}x) = f(1)$. Thus each element of the monoid $f(S)$ is invertible, so $f(S)$ is a group. ∎

Exponents and logarithms provide natural examples of isomorphisms. Denote by $[\mathbb{R}, +]$ the additive group of real numbers and by $[\mathbb{R}_+, \cdot]$ the multiplicative group of positive real numbers. Then $\exp: \mathbb{R} \to \mathbb{R}_+$ and $\ln: \mathbb{R}_+ \to \mathbb{R}$ defined by $\exp(x) = e^x$ and $\ln(x) = \ln x$ are homomorphisms because

$$\exp(x+y) = e^{x+y} = e^x e^y = \exp(x) \cdot \exp(y)$$

and

$$\ln(xy) = \ln x + \ln y.$$

As exp and ln are inverses of one another, they are indeed isomorphisms (Why?)

Now suppose that G is a group, and define $f: G \to G$ by $f(x) = x^{-1}$. Then $f(xy) = (xy)^{-1} = y^{-1}x^{-1} = f(y)f(x)$, by Lemma 4.4.2(b). For any $x \in G$, $f(x^{-1}) = (x^{-1})^{-1} = x$, so f is onto. If $f(x) = f(y)$, then $x^{-1} = y^{-1}$, so $x = (x^{-1})^{-1} = (y^{-1})^{-1} = y$ and f is one-one. So if f is a homomorphism, f is indeed

Sec. 4.8 *The Semigroup of a Machine* 183

an isomorphism. To be a homomorphism, f must satisfy $f(xy) = f(x)f(y)$; but $f(xy) = f(y)f(x)$, so we get a homomorphism only if $f(x)f(y) = f(y)f(x)$, all $f(x), f(y) \in G$. Since f is onto, f is a homomorphism (and an isomorphism) iff G is abelian. In any case, f satisfies the following definition.

A function $f: S \to T$ from one binary algebra to another is an **anti-homomorphism** if $f(xy) = f(y)f(x)$, all $x, y \in S$. It is an **anti-isomorphism** if it is also one-one and onto; then we say that S and T are **anti-isomorphic**. As we have just seen, *anti-isomorphic* need not mean *different structure*; indeed, every group G is anti-isomorphic to itself via the anti-isomorphism $x \to x^{-1}$.

For another natural example of an anti-isomorphism, return to the second example of the previous section, X^*, the monoid of all input tapes for the machine $M = [S, X, Z, \tau, \omega]$. As before, for $\mathbf{x} \in X^*$, $T_\mathbf{x} \in S^S$ is defined by $T_\mathbf{x}(\sigma_i) = \tau(\sigma_i, \mathbf{x})$, all $\sigma_i \in S$, and the congruence relation E on X^* is defined by $\mathbf{x} E \mathbf{y}$ iff $T_\mathbf{x} = T_\mathbf{y}$. The function $\varphi: X^* \to S^S$ defined by $\varphi(\mathbf{x}) = T_\mathbf{x}$ is an anti-homomorphism because we showed that $\varphi(\mathbf{x} \circ \mathbf{y}) = T_{\mathbf{x} \circ \mathbf{y}} = T_\mathbf{y} \circ T_\mathbf{x} = \varphi(\mathbf{y}) \circ \varphi(\mathbf{x})$. This same equation shows that the image $\varphi(X^*)$ of φ is a binary algebra; it is a semigroup because

$$\varphi(\mathbf{x}) \circ [\varphi(\mathbf{y}) \circ \varphi(\mathbf{z})] = \varphi(\mathbf{x}) \circ \varphi(\mathbf{z} \circ \mathbf{y}) = \varphi((\mathbf{z} \circ \mathbf{y}) \circ \mathbf{x}) = \varphi(\mathbf{z} \circ (\mathbf{y} \circ \mathbf{x}))$$
$$= \varphi(\mathbf{y} \circ \mathbf{x}) \circ \varphi(\mathbf{z}) = [\varphi(\mathbf{x}) \circ \varphi(\mathbf{y})] \circ \varphi(\mathbf{z}).$$

(It is even a monoid. Why?)

Denoting $[\mathbf{x}] = \{\mathbf{y} \in X^* | \mathbf{x} E \mathbf{y}\} \in X^*/E$, we can also define a function $\bar{\varphi}: X^*/E \to \varphi(X^*)$ by $\bar{\varphi}([\mathbf{x}]) = \varphi(\mathbf{x})$; this is well defined because if $[\mathbf{x}] = [\mathbf{y}]$, then $\mathbf{x} E \mathbf{y}$, so $T_\mathbf{x} = T_\mathbf{y}$ and $\bar{\varphi}([\mathbf{x}]) = \varphi(\mathbf{x}) = T_\mathbf{x} = T_\mathbf{y} = \varphi(\mathbf{y}) = \bar{\varphi}([\mathbf{y}])$. The function $\bar{\varphi}$ is an anti-homomorphism, since $\bar{\varphi}([\mathbf{x} \circ \mathbf{y}]) = \varphi(\mathbf{x} \circ \mathbf{y}) = \varphi(\mathbf{y}) \circ \varphi(\mathbf{x}) = \bar{\varphi}([\mathbf{y}]) \circ \bar{\varphi}([\mathbf{x}])$; is onto, since φ is; and is one-one, since $\bar{\varphi}([\mathbf{x}]) = \bar{\varphi}([\mathbf{y}])$ implies that $\varphi(\mathbf{x}) = \varphi(\mathbf{y})$, $T_\mathbf{x} = T_\mathbf{y}$, $\mathbf{x} E \mathbf{y}$, $[\mathbf{x}] = [\mathbf{y}]$. So $\bar{\varphi}$ is an anti-isomorphism from the monoid X^*/E onto the submonoid $\varphi(X^*) = \{T_\mathbf{x} | \mathbf{x} \in X^*\}$ of S^S.

4.8 The Semigroup of a Machine

By now, the reader is probably uncomfortable with the fact that φ and $\bar{\varphi}$ in the example above are anti-homomorphisms rather than homomorphisms. The difficulty arises because we naturally read input tapes under concatenation from left to right; $\mathbf{x} \circ \mathbf{y}$ means first input sequence $\mathbf{x}$, then sequence $\mathbf{y}$. On the other hand, function composition as we know it proceeds from right to left; $(T_\mathbf{x} \circ T_\mathbf{y})(\sigma_i) = T_\mathbf{x}(T_\mathbf{y}(\sigma_i))$ means first apply input sequence $\mathbf{y}$, then input sequence $\mathbf{x}$.

To overcome this difficulty, we change our policy on function notation and composition. Whenever it is desirable for specific reasons, but only then, we shall write a function on the right of its argument; xf instead of $f(x)$, for a function $f: A \to B$ and $x \in A$. The composite (called **right composite**) of two functions written on the right is denoted $f \diamond g$ or simply fg; the expression $x(fg) = (xf)g$ means the same as $(g \circ f)(x) = g(f(x))$. The associative law (Theorem 1.9.1) still holds for right composition: $(fg)h = f(gh)$.

The set S^S and its subset $T = \{T_x | x \in X^*\}$ are now monoids under the binary operation $\diamond$, and the functions φ and $\bar{\varphi}$ are now a homomorphism and an isomorphism, respectively, because for all $\sigma_i \in S$,

$$\sigma_i \varphi(\mathbf{x} \circ \mathbf{y}) = \sigma_i T_{\mathbf{x} \circ \mathbf{y}} = \tau(\sigma_i, \mathbf{x} \circ \mathbf{y}) = \tau(\tau(\sigma_i, \mathbf{x}), \mathbf{y}) = \tau(\sigma_i T_\mathbf{x}, \mathbf{y})$$
$$= (\sigma_i T_\mathbf{x}) T_\mathbf{y} = \sigma_i (T_\mathbf{x} \diamond T_\mathbf{y}) = \sigma_i (\varphi(\mathbf{x}) \diamond \varphi(\mathbf{y})),$$

which proves that $\varphi(\mathbf{x} \circ \mathbf{y}) = T_{\mathbf{x} \circ \mathbf{y}} = T_\mathbf{x} \diamond T_\mathbf{y} = \varphi(\mathbf{x}) \diamond \varphi(\mathbf{y})$.

The semigroup $[T, \diamond]$ is isomorphic to the input monoid X^*/E and may be called the **semigroup of the machine** $M = [S, X, Z, \tau, \omega]$; the elements of T reveal all the different shifts of internal state possible, when input tapes are applied to M. However, since another function $\omega: S \times X \to Z$ is involved, T alone does not usually reveal the output behavior of M.

For this reason, semigroups are more commonly associated with the state-output or Moore machine, mentioned in Section 1.11. A **state-output machine** (**Moore machine**), defined by E. F. Moore in Moore (1956), is a 5-tuple $M = [S, X, Z, \nu, \delta]$ consisting of three finite sets, S, X, and Z, and two functions, $\nu: S \times X \to S$ and $\delta: S \to Z$. S is the set of *states*, X the set of *inputs*, Z the set of *outputs*, ν the *next-state function*, and δ the *output function*. Now the output depends only on the present state σ_i via the function δ, and the semigroup of M, revealing the transitions of states within M, can reveal the entire behavior of M. (The input monoid X^*/E of M and the semigroup of M are defined just as before.) For a concrete example, see Figure 4.8.2 later in this section.

Sometimes even more specialized machines are considered, such as the **state machine**, in which $Z = S$ and $\delta = 1_S$ so the output of M in state σ_i is just σ_i, and the **autonomous machine**, which has no outputs at all. The definition of the input monoid and the semigroup remain unchanged even in these cases, and they are always finite.

To contrast with the Moore machine, the finite-state machine defined in Section 1.11 will often be called a **Mealy machine**; it was defined by G. H. Mealy in Mealy (1955). Certainly, given any Moore machine $M = [S, X, Z, \nu, \delta]$, we can construct a Mealy machine $M' = [S, X, Z, \tau, \omega]$ with the same output behavior by simply defining $\tau = \nu$, $\omega(\sigma_i, x_j) = \delta(\sigma_i)$, all $\sigma_i \in S$, all $x_j \in X$. More surprisingly, we can construct a Moore machine to imitate the behavior of any given Mealy machine with a specified initial state, provided that we are willing to enlarge the set of states. Recall that for each state of a Mealy machine, each input produces an output and a next state. In a Moore machine, for each state, each input needs to produce just a next state, since that state has an output associated with it. If we let the states of a Moore machine be pairs: (Mealy state, output), then a present input

Sec. 4.8 The Semigroup of a Machine

can elicit the same output response as in the Mealy machine (see Figure 4.8.1) and also generate a next state that will respond properly to following inputs. In the resulting Moore machine, the output associated with the initial state is not determined by the Mealy machine being imitated, hence is arbitrary. All this is made precise in the following theorem. [Note that the **total state** of a machine = ((internal state), (output state)) = *Moore state* is important in the design of asynchronous circuits.]

FIGURE 4.8.1. *Corresponding Mealy and Moore Machines*

Theorem 4.8.1. *Let* $M = [S, X, Z, \tau, \omega]$ *be a Mealy machine, with a specified starting state* s_1. *Then the Moore machine* $M' = [S \times Z, X, Z, \nu, \delta]$ *defined by*

$$\nu(((\sigma_i, \zeta_j), \xi_k)) = (\tau(\sigma_i, \xi_k), \omega(\sigma_i, \xi_k)),$$

$$\delta((\sigma_i, \zeta_j)) = \zeta_j$$

produces the same string of outputs as does M for every string of inputs, with a suitable choice of starting state in M' *and with the first output of* M' *ignored.*

Proof. Let the sequence of inputs be $x_1, x_2, \ldots, x_n$. Table 4.8.1 specifies the behavior of M. Then start M' in state (s_1, z_0) ($z_0 \in Z$ arbitrary). Table 4.8.2 specifies the output behavior of M', given the same inputs; the outputs

TABLE 4.8.1. The Mealy Machine M.

Input sequence	x_1	x_2		x_n
State sequence	s_1	$s_2 = \tau(s_1, x_1)$	...	$s_n = \tau(s_{n-1}, x_{n-1})$
Output sequence	$z_1 = \omega(s_1, x_1)$	$z_2 = \omega(s_2, x_2)$		$z_n = \omega(s_n, x_n)$

are the same as those of M, proving the theorem. [Note that to get the nth output of M from a string of n inputs, it is necessary to feed some arbitrary $(n+1)$th input x_{n+1} into M'. ∎

As with Mealy machines, Moore machines are specified either by means of a defining table or by a state diagram; in a state diagram, the output $\delta(\sigma_i)$ of a state σ_i is specified within the node representing σ_i, usually below a diagonal or horizontal line.

For example, consider the machine M_1 whose table and diagram are given in Figure 4.8.2. We shall specify the action of the state transmission maps by showing what they do to the ordered vertical column of states. In this

TABLE 4.8.2. The Moore Machine M'.

Input sequence	x_1	x_2	x_3
State sequence	(s_1, z_0)	$\nu(((s_1, z_0), x_1)) = (\tau(s_1, x_1), \omega(s_1, x_1)) = (s_2, z_1)$	$\nu(((s_2, z_1), x_2)) = (\tau(s_2, x_2), \omega(s_2, x_2)) = (s_3, z_2)$
Output sequence	$\delta((s_1, z_0)) = z_0$	$\delta((s_2, z_1)) = z_1$	$\delta((s_3, z_2)) = z_2$
		x_n	x_{n+1}
	$\cdots$	$\nu(((s_{n-1}, z_{n-2}), x_{n-1})) = (\tau(s_{n-1}, x_{n-1}), \omega(s_{n-1}, x_{n-1})) = (s_n, z_{n-1})$	$\nu(((s_n, z_{n-1}), x_n)) = (\tau(s_n, x_n), \omega(s_n, x_n)) = (s_{n+1}, z_n)$
		$\delta((s_n, z_{n-1})) = z_{n-1}$	$\delta((s_{n+1}, z_n)) = z_n$

example we have

$$\begin{bmatrix} \sigma_1 \\ \sigma_2 \\ \sigma_3 \\ \sigma_4 \end{bmatrix} T_0 = \begin{bmatrix} \sigma_3 \\ \sigma_4 \\ \sigma_1 \\ \sigma_2 \end{bmatrix}, \quad \begin{bmatrix} \sigma_1 \\ \sigma_2 \\ \sigma_3 \\ \sigma_4 \end{bmatrix} T_1 = \begin{bmatrix} \sigma_4 \\ \sigma_3 \\ \sigma_2 \\ \sigma_1 \end{bmatrix};$$

that is, $\sigma_1 T_0 = \sigma_3$, $\sigma_2 T_0 = \sigma_4$, $\sigma_2 T_1 = \sigma_3$, and so on. $T_\lambda = 1_S$, so

$$\begin{bmatrix} \sigma_1 \\ \sigma_2 \\ \sigma_3 \\ \sigma_4 \end{bmatrix} T_\lambda = \begin{bmatrix} \sigma_1 \\ \sigma_2 \\ \sigma_3 \\ \sigma_4 \end{bmatrix}.$$

We get another element of the semigroup $T = \{T_\mathbf{x} | \mathbf{x} \in X^*\}$ when we compute T_{01}, using right composition of T_0 and T_1:

$$\begin{bmatrix} \sigma_1 \\ \sigma_2 \\ \sigma_3 \\ \sigma_4 \end{bmatrix} T_{01} = \begin{bmatrix} \sigma_1 \\ \sigma_2 \\ \sigma_3 \\ \sigma_4 \end{bmatrix} T_0 T_1 = \begin{bmatrix} \sigma_3 \\ \sigma_4 \\ \sigma_1 \\ \sigma_2 \end{bmatrix} T_1 = \begin{bmatrix} \sigma_2 \\ \sigma_1 \\ \sigma_4 \\ \sigma_3 \end{bmatrix}.$$

When we form a multiplication table of T, starting with T_λ, T_0, T_1, T_{01}, we find that we complete the table without finding any new elements. Table 4.8.3 is the complete multiplication table of the semigroup of M_1. There are $|S^S| = 4^4 = 256$ possible maps of the state set into itself, but the semigroup of M_1 has only four of these elements. The reader should verify that Table 4.8.3 actually represents a group of order 4 which is not cyclic.

Sec. 4.8 The Semigroup of a Machine

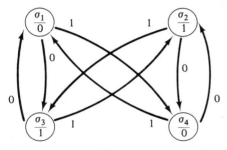

	ν		δ
	0	1	
σ_1	σ_3	σ_4	0
σ_2	σ_4	σ_3	1
σ_3	σ_1	σ_2	1
σ_4	σ_2	σ_1	0

FIGURE 4.8.2. *The Machine* M_1

TABLE 4.8.3. The semigroup of M_1.

	T_λ	T_0	T_1	T_{01}
T_λ	T_λ	T_0	T_1	T_{01}
T_0	T_0	T_λ	T_{01}	T_1
T_1	T_1	T_{01}	T_λ	T_0
T_{01}	T_{01}	T_1	T_0	T_λ

The semigroup of a Moore machine is, in principle, not hard to compute. However, the execution may be very long since there may be as many as $|S|^{|S|}$ elements, S the set of states. An algorithm for obtaining the list of elements of the semigroup is as follows. Construct an extension of the next-state table which lists the next states resulting from input sequences of length 2, then those resulting from input sequences of length 3,..., always deleting columns that duplicate earlier columns. The process stops when no new columns are generated. The finally remaining columns identify the elements of the semigroup.

For an example, consider machine M_2 in Figure 4.8.3. The column deletions occurred because $T_{00} = T_0$, $T_{000} = T_0$, $T_{001} = T_{01}$, and so on. So the semigroup of M_2 is

$$\{T_\lambda, T_0, T_1, T_{01}, T_{10}, T_{11}\}.$$

Sometimes it is important to know in detail which input sequences lead to which state transformations. This amounts to knowing the elements of each equivalence class $[\mathbf{x}] = \{\mathbf{y} \in X^* | T_\mathbf{x} = T_\mathbf{y}\}$ in the input monoid X^*/E. For example, the tabular computation in Figure 4.8.3 for the machine M_2 showed $T_{00} = T_{000} = T_{010} = T_0$, $T_{001} = T_{01}$, $T_{011} = T_{110} = T_{111} = T_{11}$, $T_{100} = T_{10}$, $T_{101} = T_1$, so 00, 000, $010 \in [0]$, $001 \in [01]$, 011, 110, $111 \in [11]$, $100 \in [10]$, and $101 \in [1]$.

This information enables us to describe the classes $[\mathbf{x}]$ completely. Since $T_0 T_{11} = T_{011} = T_{11}$, $T_1 T_{11} = T_{111} = T_{11}$, $T_{11} T_0 = T_{110} = T_{11}$, and $T_{11} T_1 =$

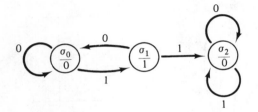

FIGURE 4.8.3. *The Machine* M_2

$T_{111} = T_{11}$, every sequence containing two consecutive 1's is in $[11]$. $T_{10}T_0 = T_{100} = T_{10}$, so $T_{100\cdots 0} = T_{10}T_0 \cdots T_0 = T_{10}$; hence $T_{10\cdots 01} = T_{10}T_1 = T_{101} = T_1$. For example, $T_{10101} = T_{101}T_{01} = T_1 T_{01} = T_{101} = T_1$; in general, any sequence **x** starting in 1 and ending in 1 but without consecutive 1's is in $[1]$. If **x** is such a sequence, then $T_{\mathbf{x}0} = T_{\mathbf{x}00} = \cdots = T_{10}$, so x0, x00, ... are in $[10]$. Similarly, 0x, 00x, ... are in $[01]$ and 0x0, 00x0, 0x00, ... are in $[010] = [0]$.

Let P be the set of all $\mathbf{x} \in X^*$ starting in 1 and ending in 1 but without consecutive 1's. We now describe the input monoid of M_2 completely:

$$[\lambda] = \{\lambda\},$$
$$[11] = \{\mathbf{y}11\mathbf{z} | \mathbf{y}, \mathbf{z} \in X^*\},$$
$$[0] = \{00^*\} \cup \{00^*10^*0\} \cup \{00^*\mathbf{x}0^*0 | \mathbf{x} \in P\},$$
$$[1] = \{1\} \cup P,$$
$$[10] = \{10\} \cup \{\mathbf{x}0^*0 | \mathbf{x} \in P\},$$
$$[01] = \{01\} \cup \{00^*\mathbf{x} | \mathbf{x} \in P\}.$$

Here 0* denotes any sequence (including λ) of 0's. The reader should show that every input sequence appears in one of these classes, so we have described X^*/E completely.

The classes in the input monoid of M_1 are easily described, because the semigroup of M_1 is a commutative group in which $T_\mathbf{x}^2 = T_\mathbf{x} T_\mathbf{x} = T_\lambda$, all $\mathbf{x} \in X^*$. For example,

$$T_{011010001} = T_0 T_1^2 T_0 T_1 T_0^2 T_0 T_1 = T_0 T_0 T_1 T_0 T_1 = T_1 T_0 T_1 = T_1^2 T_0 = T_0,$$

Sec. 4.9 The Machine of a Semigroup

so $011010001 \in [0]$. In general,

$[\lambda] = \{x \in X^* | x$ has an even number of 0's and an even number of 1's$\}$,
$[0] = \{x \in X^* | x$ has an odd number of 0's and an even number of 1's$\}$,
$[1] = \{x \in X^* | x$ has an even number of 0's and an odd number of 1's$\}$,
$[01] = \{x \in X^* | x$ has an odd number of 0's and an odd number of 1's$\}$.

4.9 The Machine of a Semigroup

Given any finite semigroup $[S, \circ]$, we can define a Moore machine M from S as follows: $M = [S, S, S, \circ, 1_S]$. This is a state machine in which inputs, outputs, and states all coincide with the set S, and the next-state function $\nu: S \times S \to S$ is simply the operation $\circ$ in S, so $\nu(s_1, s_2) = s_1 \circ s_2$.

For example, if S is the cyclic group $\{1, x, x^2\}$ of order 3, then the multiplication table of S is in Table 4.9.1 and the state diagram of the corresponding machine is Figure 4.9.1.

TABLE 4.9.1

	1	x	x^2
1	1	x	x^2
x	x	x^2	1
x^2	x^2	1	x

Now consider the semigroup of Table 4.8.3. From it, we can define a machine M_1'; in Figure 4.9.2 we draw only the part of the state diagram of M_1' corresponding to inputs T_0 and T_1. We find that Figure 4.9.2 is exactly the state diagram of the machine M_1 (Figure 4.8.2), whose semigroup was Table 4.8.3. (We are not comparing outputs here.)

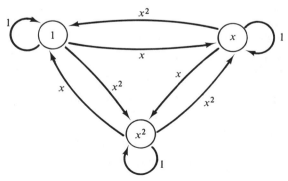

FIGURE 4.9.1. *Machine of a Cyclic Group*

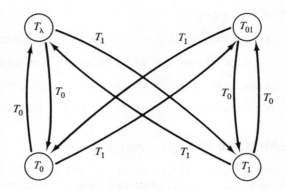

FIGURE 4.9.2. *Machine of a Semigroup*

Since we were able to recover machine M_1 by first taking the semigroup T of M_1 and then the machine of T, this graphically illustrates the fact (not discussed further in this book) that all basic properties of a Moore machine are reflected in its semigroup. Good references for semigroups in machine theory include Arbib (1968), Ginzburg (1968), and Hartmanis and Stearns (1966).

4.10 Exercises

1. Show that $[\mathbb{Z}_p, +]$ is a cyclic group of order p with generator 1.

2. In the monoid $M = [\mathbb{Z}_p, \cdot]$, let G be the group of invertible elements of M (Exercise 4.5.8). Identify the elements of G in the cases $p = 2, 3, 4, 5, 6, 7$. Can you make a general statement?

*3. When we formed the factor semigroup $[\mathbb{Z}_p, +]$ of the group $[\mathbb{Z}, +]$ with respect to the congruence relation "congruence modulo p," we found that $[\mathbb{Z}_p, +]$ is a group. Show that the factor semigroup G/R of a group G with respect to a congruence relation R is always a group.

*4. (a) Show that every cyclic group of order p, $p > 1$ an integer, is isomorphic to $[\mathbb{Z}_p, +]$.

(b) Show that every infinite cyclic group is isomorphic to $[\mathbb{Z}, +]$.

5. Show that if $m \in \mathbb{Z}$, $m > 1$, then $f: z \to mz$ is a homomorphism from $[\mathbb{Z}, +]$ into itself. Describe $f(\mathbb{Z})$. Conclude from Theorem 4.7.1 that $[f(\mathbb{Z}), +]$ is a group. Show that f is an isomorphism from $[\mathbb{Z}, +]$ onto $[f(\mathbb{Z}), +]$. (Thus an infinite group can be isomorphic to a proper subgroup of itself.)

6. (a) Show that if $f: S \to T$ and $g: T \to U$ are homomorphisms of binary algebras, so is $g \circ f: S \to U$.

(b) In (a), show that if f is an isomorphism, so is f^{-1}.

(c) Show that isomorphism is an equivalence relation on any set of binary algebras.

(d) If $f: S \to T$ and $g: T \to U$ are anti-homomorphisms, what is $g \circ f$?

Sec. 4.10 *Exercises* 191

(e) Show that if groups G and H are anti-isomorphic, then they are also isomorphic.

(f) Find a monoid that is not anti-isomorphic to itself.

*7. (a) A morphism from a binary algebra S to itself is an **endomorphism** of S. Show that the set of all endomorphisms of S is a monoid under function composition.

(b) An isomorphism from a binary algebra S onto itself is an **automorphism** of S. Show that the set of all automorphisms of S is a group under function composition (the **automorphism group** of S).

8. Let $[\mathbb{B}^n, +]$ be the group of example 6, Section 4.1. For example, $(0, 1, 1, 0) + (0, 0, 1, 1) = (0, 1, 0, 1)$ in $\mathbb{B}^4$.

(a) Show that $f: \mathbb{B}^4 \to \mathbb{B}^3$ defined by

$$f: (b_1, b_2, b_3, b_4) \to (b_1, b_2, b_3)$$

is an epimorphism.

(b) Show that $g: \mathbb{B}^3 \to \mathbb{B}^4$ defined by

$$g: (b_1, b_2, b_3) \to (b_1, b_2, b_3, 0)$$

is a monomorphism. Is $(b_1, b_2, b_3) \to (b_1, b_2, b_3, 1)$ a monomorphism? Show that $g(\mathbb{B}^3)$ is isomorphic to $\mathbb{B}^3$.

(c) Show that whenever $h: S \to T$ is a binary algebra monomorphism, S is isomorphic to $h(S)$.

9. Show that a homomorphic image of an abelian group is abelian, and a homomorphic image of a cyclic group is cyclic.

10. If $f: G \to H$ is a group isomorphism and $x \in G$, show that x and $f(x)$ have the same order.

11. Find the multiplication tables of the semigroups of the Moore machines diagrammed in Figure 4.10.1. Also, describe the elements in each equivalence class of the input monoids.

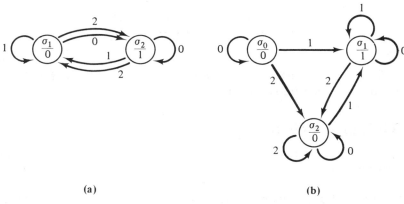

(a)   (b)

FIGURE 4.10.1. *Moore Machines for Exercise 11*

12. Show that the Mealy machine diagrammed in Figure 4.10.2 can produce output sequences 11, 00, and 1100, but not 0011. Conclude that the output sequences of a machine need not form a semigroup under the binary operation of concatenation.

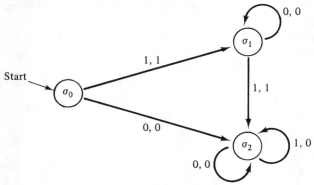

FIGURE 4.10.2. *Mealy Machine for Exercise 12*

***13.** Let $S = \{0, 1\}$ and let T be the semigroup $[S^S, \circ]$, so that $|T| = 4$. Write the multiplication table of T. Draw the state diagram of the state machine M defined by T, and then find the multiplication table of the semigroup T' of the machine M. Are T and T' isomorphic?

14. Use what you learned in Exercise 13 to state and prove a general theorem.

15. Suppose that M is a Moore machine, T the semigroup of M, and M' the state machine constructed from T. How much can you say in general about the relationship between M and M'?

16. Let M be the Mealy machine with diagram

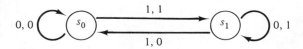

(a) Find the Moore machine M' corresponding to M as in Theorem 4.8.1.
(b) Find the semigroup of M'. Is it a monoid? A group?

4.11 Equations in Semigroups

We collect in this section some occasionally useful sufficient conditions for a semigroup to be a group.

Theorem 4.11.1. *A semigroup S is a group iff for all $a, b \in S$, the equations $ax = b$ and $ya = b$ have unique solutions x and y in S.*

Proof. If S is a group, then
$$ax = b \quad \text{iff} \quad a^{-1}ax = a^{-1}b \quad \text{iff} \quad x = a^{-1}b$$
and
$$ya = b \quad \text{iff} \quad yaa^{-1} = ba^{-1} \quad \text{iff} \quad y = ba^{-1},$$
so unique solutions exist.

Now assume that the condition holds in the semigroup S, and fix $a \in S$; let $e_1, e_2 \in S$ solve $ae_1 = a$, $e_2 a = a$. For any $b \in S$, let $c_b, d_b \in S$ satisfy $c_b a = b$, $ad_b = b$; then
$$be_1 = (c_b a)e_1 = c_b(ae_1) = c_b a = b, \qquad e_2 b = e_2(ad_b) = (e_2 a)d_b = ad_b = b.$$
This proves that e_1 is a right identity and e_2 is a left identity in S; by Lemma 4.2.1, $e_1 = e_2 = e$ is an identity element.

Now let $a_1, a_2 \in S$ solve $aa_1 = e$, $a_1 a_2 = e$. Then
$$a_1 a = (a_1 a)e = (a_1 a)(a_1 a_2) = a_1(aa_1)a_2 = a_1 e a_2 = a_1 a_2 = e,$$
so $a_1 = a^{-1}$. ∎

Theorem 4.11.2. *A finite semigroup S is a group iff both cancellation laws hold in S.*

Proof. Definitions and part of the proof appear in Exercise 4.5.3. Suppose that the cancellation laws hold. For any $a \in S$, part (b) of that exercise implies that the functions $a_L, a_R: S \to S$ defined by $a_L: x \to ax$, $a_R: y \to ya$ are bijections. Hence, for each $b \in S$, there are unique $x, y \in S$ such that $ax = b$ and $ya = b$. By Theorem 4.11.1, S is a group. ∎

Theorem 4.11.3. *If a semigroup S has a left identity e_l and each $a \in S$ has a left inverse a_l^{-1} with respect to e_l, then S is a group.*

Proof. For any $a \in S$,
$$aa_l^{-1} = e_l(aa_l^{-1}) = ((a_l^{-1})_l^{-1} a_l^{-1})(aa_l^{-1}) = (a_l^{-1})_l^{-1}(a_l^{-1}a)a_l^{-1}$$
$$= (a_l^{-1})_l^{-1} e_l a_l^{-1} = (a_l^{-1})_l^{-1} a_l^{-1} = e_l.$$
Hence
$$ae_l = a(a_l^{-1}a) = (aa_l^{-1})a = e_l a = a.$$
The second equation shows that e_l is a right identity, hence an identity element. Then the first equation shows that a_l^{-1} is a right inverse, hence an inverse, of a. ∎

4.12 Lagrange's Theorem

In the remainder of this chapter we develop results from group theory that will be applied in future chapters, especially Chapter 5.

Let G be a group, H a subgroup of G. Define a binary relation L on G by $x L y$ iff $x^{-1}y \in H$. Then, using Lemma 4.4.2:
1. $x L x$ for all x, since $x^{-1}x = 1 \in H$.
2. If $x L y$, then $y L x$. For if $x L y$, then $x^{-1}y \in H$. Hence H contains $(x^{-1}y)^{-1} = y^{-1}(x^{-1})^{-1} = y^{-1}x$, so $y L x$.
3. If $x L y$ and $y L z$, then $x L z$. For if $x^{-1}y$ and $y^{-1}z$ are in H, then $x^{-1}y \cdot y^{-1}z = x^{-1}z$ is in H.

Thus L is an equivalence relation and partitions G into equivalence classes. The equivalence class of $x \in G$ is

$$\{y | x L y\} = \{y | x^{-1}y \in H\} = \{y | x^{-1}y = h, \text{ some } h \in H\}$$
$$= \{y | y = xh, \text{ some } h \in H\}$$
$$= \{xh | h \in H\};$$

it is denoted xH, the set of all xh for $h \in H$, and is called the **left coset** of H containing x.

Lemma 4.12.1. *If H is a finite subgroup of G and $x \in G$, then $|H| = |xH|$; that is, all left cosets of H have the same order.*

Proof. It is enough to find a one-one correspondence $f: H \to xH$. Define f by $f(h) = xh$, all $h \in H$; by definition of xH, f is onto. If $f(h_1) = f(h_2)$, then $xh_1 = xh_2$, $x^{-1}xh_1 = x^{-1}xh_2$, $h_1 = h_2$, f is one-one. ∎

If G is a group, H a subgroup, and there are only finitely many equivalence classes of G under the relation L above, then the **index** of H in G, denoted $|G : H|$, is the number of equivalence classes; that is,

$$|G : H| = \text{number of distinct left cosets of } H \text{ in } G.$$

Theorem 4.12.1 (Lagrange's Theorem). *If G is a finite group and H a subgroup of G, then*

$$|G| = |G : H| \, |H|;$$

in particular, $|H|$ divides $|G|$.

Proof. Each element of G is in exactly one left coset since L is an equivalence relation. There are $|G : H|$ left cosets, and by Lemma 4.12.1 exactly $|H|$ elements in each coset, so $|G : H| \, |H|$ elements of G. (This uses the *sheep-pen principle* of Exercise 1.10.10.) ∎

Corollary 4.12.1. *The order of every element x of a finite group G divides $|G|$.*

Proof. The order of x is the order of the cyclic subgroup $\langle x \rangle$ generated by x, and this order divides $|G|$ by Lagrange's theorem. ∎

Sec. 4.12 Lagrange's Theorem

Corollary 4.12.2. *If G is a group of order p, p a prime, then $G = \langle x \rangle$ for any nonidentity element x of G. In particular, the only subgroups of G are $\{1\}$ and G, 1 the identity element of G.*

Proof. The only positive integer divisors of p are 1 and p. Now $\langle x \rangle$ has order dividing p; since $1, x \in \langle x \rangle$, the order of $\langle x \rangle$ exceeds 1; hence it must be p, so $\langle x \rangle = G$. Any subgroup of G other than $\{1\}$ must contain some such x, hence contain $\langle x \rangle = G$, hence equal G. ■

Lagrange's theorem and its corollaries are powerful tools in any work with finite groups. For example, consider the group G (see Exercise 4.5.14) of six symmetries of the equilateral triangle shown. Since $|G| = 6$, Lagrange's

theorem says that any subgroup of G has order 1, 2, 3, or 6. The only subgroups of orders 1 and 6 are $\{e\}$ and G, respectively. By Corollary 4.12.2, subgroups of order 2 or 3 must be cyclic, generated by some nonidentity element of G. The three reflections (1 2), (1 3), and (2 3) along the medians yield the identity e when performed twice, so there are three (cyclic) subgroups of order 2:

$$H_1 = \{e, (1\ 2)\}, \quad H_2 = \{e, (1\ 3)\}, \quad H_3 = \{e, (2\ 3)\}.$$

The rotations (1 2 3) and $(1\ 2\ 3)^2 = (1\ 3\ 2)$ yield the identity when performed three times. The only subgroup of order 3 is

$$H_4 = \{e, (1\ 2\ 3), (1\ 2\ 3)^2 = (1\ 3\ 2)\}.$$

Sometimes (as we will see in coding theory, Chapter 5) it is important to actually list the elements of the left cosets. For example, one left coset of H_1 above is

$$H_1 = eH_1 = \{e, (1\ 2)\}.$$

(1 3) is not in this coset, so a new coset is

$$(1\ 3)H_1 = \{(1\ 3)e, (1\ 3)(1\ 2)\} = \{(1\ 3), (1\ 2\ 3)\}.$$

(These symmetries are functions, and multiply like functions under composition; apply the right one first.) The reflection (2 3) is not in either of the first two cosets, so a third coset is

$$(2\ 3)H_1 = \{(2\ 3)e, (2\ 3)(1\ 2)\} = \{(2\ 3), (1\ 3\ 2)\}.$$

This exhausts the elements of G, so $|G : H_1|$, the number of left cosets, is 3 and $|G : H_1||H_1| = 3 \cdot 2 = 6 = |G|$, in accordance with Lagrange's theorem.

4.13 Direct Products

The **direct product** $S \times T$ of two binary algebras S and T is the Cartesian product $S \times T = \{(s, t) | s \in S, t \in T\}$ of sets S and T, with binary operation

$$(s, t)(s', t') = (ss', tt').$$

Note that there are three perhaps different operations here, in S, T, and $S \times T$, all denoted by juxtaposition.

Theorem 4.13.1. (a) *If S and T are semigroups, so is $S \times T$.*
(b) *If S and T are monoids with respective identities 1_S and 1_T, then $S \times T$ is a monoid with identity $(1_S, 1_T)$.*
(c) *If S and T are groups, so is $S \times T$, with $(s, t)^{-1} = (s^{-1}, t^{-1})$.*
(d) *If S and T are commutative, so is $S \times T$.*
Proof. An easy exercise. ∎

We can also form the direct product of three or more binary algebras (semigroups, groups, etc.) by using the Cartesian product of three or more sets and a componentwise binary operation. Results similar to Theorem 4.13.1 hold in this case, also. For example, the group $[\mathbb{B}^n, +]$ of example 6, Section 4.1, is the direct product of n copies of $[\mathbb{Z}_2, +]$; it is abelian because $[\mathbb{Z}_2, +]$ is abelian.

If S and T are finite, we know by (1.4.1) that $|S \times T| = |S| |T|$. For example, we can construct a group of $24 = 6 \cdot 4$ elements by forming $G \times H$, G the group of Exercise 4.5.14 and H the group of Table 4.8.3.

If S and T are binary algebras, the functions $\pi_S: S \times T \to S$ and $\pi_T: S \times T \to T$ defined by

$$\pi_S((s, t)) = s, \qquad \pi_T((s, t)) = t$$

are called **projections**. If S has identity element 1_S and T has identity element 1_T, then the functions $i_S: S \to S \times T$ and $i_T: T \to S \times T$ defined by

$$i_S(s) = (s, 1_T), \qquad i_T(t) = (1_S, t)$$

are called **injections**. The projections are epimorphisms. The injections are monomorphisms; in particular, i_S is an isomorphism from S to the subalgebra $\{(s, 1_T) | s \in S\}$ of $S \times T$ and i_T is an isomorphism from T onto the subalgebra $\{(1_S, t) | t \in T\}$ of $S \times T$. (As an exercise, the reader should verify these statements.)

4.14 Normal Subgroups

Returning to the situation in Section 4.12, we have a group G, a subgroup H, and an equivalence relation L on G defined by $x L y$ iff $x^{-1}y \in H$. We now ask when this equivalence relation is a congruence relation on G. Before stating the main theorem, we first present some notation.

Sec. 4.14 *Normal Subgroups*

$Hx = \{hx \mid h \in H\}$ is the **right coset** of H determined by x. (These are treated in Exercise 4.15.3.)

$x^{-1}Hx = \{x^{-1}hx \mid h \in H\}$ is the **conjugate of H by x**; also $x^{-1}hx \in G$ is the **conjugate of h by x**. (These are treated in Exercise 4.15.9.)

Theorem 4.14.1. *Let G be a group with subgroup H, and let L be the equivalence relation on G defined by $x L y$ iff $x^{-1}y \in H$. Then the following are equivalent.*
  (a) *L is a congruence relation on G.*
  (b) *$x^{-1}Hx \subseteq H$ for all $x \in G$.*
  (c) *$x^{-1}Hx = H$ for all $x \in G$.*
  (d) *$Hx = xH$ for all $x \in G$.*

Note that (c) says that every conjugate of H is equal to H. Part (d) says that every left coset is also a right coset, and every right coset is also a left coset. These conditions do not hold for all subgroups H of groups G, but the theorem says that conditions (a) to (d) are all true or all false. A subgroup H of a group G satisfying one (hence all) of conditions (a) to (d) is called a **normal subgroup** (or **self-conjugate subgroup**) of G. We then write $H \triangleleft G$.

Proof of Theorem 4.14.1. (a)$\Rightarrow$(b). If $x \in G$, $h \in H$, we must show that $x^{-1}hx \in H$. But $x^{-1}hx \in H$ iff $x L hx$. Since $1 L h, x L x$, and L is a congruence relation, $1 \cdot x L h \cdot x$; done.

(b)$\Rightarrow$(c). We need only show that if $x \in G$, then $H \subseteq x^{-1}Hx$; that is, if $x \in G, h \in H$, we must show that $h = x^{-1}h_1 x$, some $h_1 \in H$. Since (b) holds for $x^{-1} \in G$, $(x^{-1})^{-1}Hx^{-1} \subseteq H$, $xHx^{-1} \subseteq H$, $xhx^{-1} = h_1$ for some $h_1 \in H$, $x^{-1}xhx^{-1}x = x^{-1}h_1 x$, $h = x^{-1}h_1 x$; done.

(c)$\Rightarrow$(d). Let $hx \in Hx$. By (c), $x^{-1}hx = h_1$ for $h_1 \in H$. Thus $hx = xh_1 \in xH$, so $Hx \subseteq xH$. Now let $xh_2 \in xH$. By (c), $h_2 = x^{-1}h_3 x$ for some $h_3 \in H$. Thus $xh_2 = xx^{-1}h_3 x = h_3 x \in Hx$, so $xH \subseteq Hx$.

(d)$\Rightarrow$(a). If $x_1, x_2, y_1, y_2 \in G$ and $x_1 L y_1$, $x_2 L y_2$, we must show that $x_1 x_2 L y_1 y_2$. That is, we must prove that $(x_1 x_2)^{-1} y_1 y_2 \in H$ whenever $x_1^{-1} y_1 = h_1$ and $x_2^{-1} y_2 = h_2$ are in H. We have

$$(x_1 x_2)^{-1} y_1 y_2 = x_2^{-1} x_1^{-1} y_1 y_2 = x_2^{-1} h_1 y_2 = x_2^{-1} y_2 y_2^{-1} h_1 y_2$$
$$= h_2 y_2^{-1} h_1 y_2,$$

so it is enough to show that $y_2^{-1} h_1 y_2 \in H$. By (d), $h_1 y_2 \in H y_2 = y_2 H$, so $h_1 y_2 = y_2 h_3$ for some $h_3 \in H$, and so $y_2^{-1} h_1 y_2 = h_3 \in H$. ∎

Corollary 4.14.1. *If N is a normal subgroup of the group G, then the set of left cosets xN, $x \in G$, is a group G/N with the binary operation*

$$(xN)(yN) = (xy)N, \quad \text{all } x, y \in G.$$

Proof. Since L is a congruence relation on G, Theorem 4.6.1 shows that G/N is a semigroup. We observed in Exercise 4.10.3 that it must actually be a group. ∎

This group G/N is called the **factor group** or **quotient group** of G by N. If G is finite, Lagrange's theorem shows that

(4.14.1) $$|G/N| = |G:N| = |G|/|N|.$$

G/N has identity $1N = N$, since $xN \cdot 1N = xN = 1N \cdot xN$. Also, $(xN)^{-1} = x^{-1}N$.

Any subgroup of an abelian group is normal; in fact, $x^{-1}hx = x^{-1}xh = h$, so certainly $x^{-1}Hx = H$. In the group G of symmetries of the equilateral triangle discussed in Section 4.12, the subgroup H_1 is not normal, since

$$(1\ 3)H_1 = (1\ 3)\{e, (1\ 2)\} = \{(1\ 3), (1\ 2\ 3)\}$$

does not equal

$$H_1(1\ 3) = \{e, (1\ 2)\}(1\ 3) = \{(1\ 3), (1\ 3\ 2)\}.$$

The other subgroups H_2 and H_3 of order 2 are also not normal, but the subgroup H_4 of order 3 is normal. (Check this.) The factor group G/H_4 has order $|G|/|H_4| = \frac{6}{3} = 2$, so by Corollary 4.12.2 it is cyclic of order 2.

There is a close relation between normal subgroups and homomorphisms, which we shall now determine. If $f: G \to H$ is a group homomorphism and 1_H is the identity of H, the **kernel** of f is defined to be

$$\ker f = \{g \in G | f(g) = 1_H\}.$$

Theorem 4.14.2. (a) *If N is a normal subgroup of G, then the function $f: G \to G/N$ defined by $f(x) = xN$ is an epimorphism with kernel N.*

(b) *If $\varphi: G \to H$ is a group homomorphism, then $N = \ker \varphi$ is a normal subgroup of G and we obtain an isomorphism $\bar{\varphi}: G/N \to \varphi(G)$ by defining $\bar{\varphi}(xN) = \varphi(x)$.*

Part (a) shows that any normal subgroup naturally induces a homomorphism, and part (b) shows that any homomorphism yields a normal subgroup. Part (b) is sometimes called the **First Homomorphism Theorem**.

Proof. (a) Note that f maps each element x of G onto the left coset xN to which it belongs. We have $f(xy) = (xy)N = (xN)(yN) = f(x)f(y)$, as seen in Corollary 4.14.1, and f is certainly onto, so f is an epimorphism. Also, $\ker f = \{x \in G | xN = N\} = \{x \in G | x \in N\} = N$.

(b) Assume that $n_1, n_2 \in N$, $x \in G$. Then by Theorem 4.7.1(c),

$$\varphi(n_1 n_2^{-1}) = \varphi(n_1)\varphi(n_2^{-1}) = \varphi(n_1)\varphi(n_2)^{-1} = 1_H 1_H^{-1} = 1_H,$$

so $n_1 n_2^{-1} \in N$; by Theorem 4.4.2, N is therefore a subgroup.

Next, for all $x \in G$,

$$\varphi(x^{-1}n_1x) = \varphi(x^{-1})\varphi(n_1x) = \varphi(x)^{-1}\varphi(n_1)\varphi(x) = \varphi(x)^{-1}1_H\varphi(x) = 1_H,$$

so $x^{-1}n_1x \in N$, $N \triangleleft G$. Note now that $\bar{\varphi}$ maps the coset containing $x \in G$ onto the image of x in H. If $xN = yN$, then $y = xn$, $n \in N$, so $\varphi(y) = \varphi(x)\varphi(n) = \varphi(x)1_H = \varphi(x)$; this proves that $\bar{\varphi}$ is well defined (does not give different values for two names of the same coset). Next, $\bar{\varphi}$ is a homomorphism because

$$\bar{\varphi}((xN)(yN)) = \bar{\varphi}((xy)N) = \varphi(xy) = \varphi(x)\varphi(y) = \bar{\varphi}(xN)\bar{\varphi}(yN).$$

By its definition, $\bar{\varphi}$ is onto $\varphi(G)$. If $\bar{\varphi}(xN) = \bar{\varphi}(yN)$, then $\varphi(x) = \varphi(y)$, $1_H = \varphi(x)^{-1}\varphi(y) = \varphi(x^{-1})\varphi(y) = \varphi(x^{-1}y)$, $x^{-1}y \in N$. If $x^{-1}y = n \in N$, then $y = xn \in xN$, $yN = xN$, which proves that $\bar{\varphi}$ is one-one. Thus $\bar{\varphi}$ is a bijection and hence an isomorphism. ∎

4.15 Exercises

*1. State and prove the analogue of Theorem 4.11.3 for right identity and right inverse.

2. Find a semigroup S with left identity e_l, such that every element has a right inverse with respect to e_l but S is not a group.

*3. If H is a subgroup of G, define a binary relation R on G by $x R y$ iff $xy^{-1} \in H$. Prove:
  (a) R is an equivalence relation on G.
  (b) The equivalence classes under R are the right cosets of H in G.
  (c) R is a congruence relation on G iff $H \triangleleft G$.
  (d) The assignment $xH \to Hx^{-1}$ provides a well-defined one-one correspondence between left cosets and right cosets of H in G.
  (e) If H is finite and $x \in G$, $|H| = |Hx|$.

4. What are the possible orders of elements in $[\mathbb{Z}_{24}, +]$? Find at least one element of each possible order.

5. The group G of symmetries of the square has eight elements; four rotations (0, 90, 180, and 270°) and four reflections (with axes the horizontal and vertical medians and the two diagonals).
  (a) List the eight elements, with their orders.
  (b) Verify that the group is not abelian.
  (c) Find all the subgroups of G, and determine which ones are normal. (G will be discussed in Section 4.18.)

6. Prove Theorem 4.13.1.

*7. Define the direct product of n binary algebras, and prove results similar to Theorem 4.13.1.

8. Verify the statements about projections and injections in Section 4.13, and extend the results to the direct products discussed in Exercise 7.

*9. (a) If H is a subgroup of the group G and $x \in G$, show that the *conjugate* $x^{-1}Hx$ is also a subgroup of G.

(b) The relation of **conjugacy** of elements of G is the binary relation C on G defined by $x_1 C x_2$ iff $x_2 = x^{-1} x_1 x$, some $x \in G$. Show that C is an equivalence relation on G.

(c) Show that conjugacy of subgroups of G, as defined in (a), is an equivalence relation on the set of all subgroups of G. Which equivalence classes (**conjugacy classes**) contain only one subgroup?

*10. Show that a group homomorphism $f: G \to H$ is one-one iff $\ker f = \{1_G\}$.

*11. Show that if G is a finite group, H a subgroup of G, and K a subgroup of H, then K is a subgroup of G and

$$|G:K| = |G:H||H:K|.$$

12. Fix $m \in \mathbb{Z}$, $m > 1$, and let (m) be the subgroup of $[\mathbb{Z}, +]$ consisting of all multiples of m. Show that the factor group $[\mathbb{Z}/(m), +]$ is isomorphic to $[\mathbb{Z}_m, +]$.

*13. (a) Show that every cyclic group is abelian.

(b) Show that if $n > 1$, then $[\mathbb{B}^n, +]$ is abelian but not cyclic.

*14. (a) Let $G = \{e, x, y, z\}$ be a group of order 4, with identity element e, which is not cyclic. With the aid of Section 4.12, write the multiplication table of G and show that G is isomorphic to $[\mathbb{B}^2, +]$. (Any noncyclic group of order 4 is called a **Klein 4-group**.)

(b) Conclude that any group of order 4 is abelian and is isomorphic to either $[\mathbb{B}^2, +]$ or $[\mathbb{Z}_4, +]$.

*15. If H is a subgroup of G and $|G:H| = 2$, show that $H \triangleleft G$.

*16. If $H \triangleleft G$ and $K \triangleleft G$, show that $(H \cap K) \triangleleft G$.

17. If G and H are groups with identities 1_G and 1_H, respectively, and

$$\bar{G} = \{(g, 1_H) | g \in G\} \subseteq G \times H \quad \text{and} \quad \bar{H} = \{(1_G, h) | h \in H\} \subseteq G \times H,$$

show that $\bar{G} \triangleleft G \times H$, $\bar{H} \triangleleft G \times H$, and $\bar{G} \cap \bar{H} = \{1_{G \times H}\}$.

18. Let A and B be subgroups of G, and assume that $A \triangleleft G$, $B \triangleleft G$, and $A \cap B = 1$. Show that

(a) If $a \in A$ and $b \in B$, then $ab = ba$.

(b) $\{ab | a \in A, b \in B\}$ is a subgroup of G isomorphic to $A \times B$.

[HINT for (a): $ab = ba$ iff $1 = a^{-1} \cdot b^{-1}ab$.]

19. If G is a group and $g \in G$ is fixed, define $I_g: G \to G$ by $I_g(x) = g^{-1}xg$. Show that I_g is an automorphism (Exercise 4.10.7) of G. Show that $\{I_g | g \in G\}$ is a subgroup of the automorphism group $\text{Aut}(G)$ of G. Is it normal in $\text{Aut}(G)$? (This subgroup is called the group of **inner automorphisms** of G.)

4.16 Structure of Cyclic Groups

In elementary number theory, we say that positive integers a and b are *relatively prime* and write $(a, b) = 1$ if 1 is the only positive integer that divides both. [In general, if their greatest common divisor is d, we write $(a, b) = d$.] For $m \in \mathbb{Z}, m > 1$, the *Euler φ-function* $\varphi(m)$ denotes the number of integers in

$$\mathbb{Z}_m - \{0\} = \{1, 2, \ldots, m-1\}$$

which are relatively prime to m. Also $\varphi(1) = 1$. It is proved in number theory that if $m = p_1^{a_1} p_2^{a_2} \cdots p_k^{a_k}$, the p_i distinct primes and the a_i positive integers, then

(4.16.1) $$\varphi(m) = \prod_{i=1}^{k} p_i^{a_i - 1}(p_i - 1).$$

For example, $\varphi(12) = \varphi(2^2 \cdot 3^1) = 2^1 \cdot 1 \cdot 3^0 \cdot 2 = 4$. Here 1, 5, 7, and 11 are the four positive integers less than 12 and relatively prime to 12.

Another result from number theory, which we shall not prove, is the familiar

Division Algorithm. *If $a \in \mathbb{Z}$ and $b \in \mathbb{P}$, then there are uniquely determined integers q and r such that*

$$a = qb + r, \quad 0 \leq r < b.$$

We first determine which elements of a cyclic group $\langle x \rangle$ generate $\langle x \rangle$.

Theorem 4.16.1. (a) *If $\langle x \rangle = \{1, x, \ldots, x^{n-1}\}$ is a finite cyclic group of order $n > 1$, then x^i generates $\langle x \rangle$ iff $(i, n) = 1$. In this case $\langle x \rangle$ has exactly $\varphi(n)$ generators.*

(b) *If $\langle x \rangle$ is an infinite cyclic group, then x^i generates $\langle x \rangle$ iff $i = \pm 1$. In this case $\langle x \rangle$ has exactly two generators.*

Proof. (a) Suppose that $(i, n) \neq 1$, so there is $d \in \mathbb{P}, d > 1, d | i$, and $d | n$; say $i = di_0, n = dn_0, i_0, n_0 \in \mathbb{P}$. Then

$$(x^i)^{n_0} = x^{di_0(n/d)} = x^{ni_0} = 1^{i_0} = 1,$$

so $|\langle x^i \rangle| \leq n_0 < n$ and x^i does not generate $\langle x \rangle$.

Now suppose that x^i does not generate $\langle x \rangle$; then the set

$$\{1, x^i, x^{2i}, \ldots, x^{(n-1)i}\}$$

contains fewer than n distinct elements, so $x^{ai} = x^{bi}$, $0 \leq a < b \leq n - 1$; therefore, $x^{(b-a)i} = 1$, $0 < b - a < n$. Hence $n | (b-a)i$ (Exercise 4.5.16) but $n \nmid (b-a)$; n and i must therefore have a common divisor greater than 1, that is, $(i, n) \neq 1$.

The observations of the last two paragraphs imply (a).

(b) For all integers $m \neq 0$,
$$\langle x^m \rangle = \langle x^{-m} \rangle = \{\ldots, x^{-3m}, x^{-2m}, x^{-m}, 1, x^m, x^{2m}, x^{3m}, \ldots\},$$
and this is $\langle x \rangle$ iff $m = \pm 1$. ∎

Theorem 4.16.2. *Every subgroup H of a cyclic group G is cyclic.*

Proof. Let $G = \langle x \rangle$. If $H = \{1\}$, $H = \langle 1 \rangle$, done. If not, then H contains a positive power of x; let m be the least positive integer such that $x^m \in H$. Then
$$H_0 = \langle x^m \rangle = \{x^{mn} | n \in \mathbb{Z}\}$$
is a cyclic group such that $H_0 \subseteq H$. It is now enough to show $H \subseteq H_0$. If $x^t \in H$, then by the Division Algorithm we can write $t = qm + r$, $0 \leq r < m$. We have $r = t - qm$, so $x^r = x^t (x^m)^{-q}$; since $x^t \in H$ and $x^m \in H$, $x^r \in H$ also. By the minimality of m this forces $r = 0$, $t = qm$, $x^t = (x^m)^q \in H_0$. ∎

Corollary 4.16.1. *Assume that $d, n \in \mathbb{P}$ and $d | n$. Then a cyclic group of order n has exactly one subgroup of order d and exactly $\varphi(d)$ elements of order d.*

Proof. If $G = \langle x \rangle$, then $\langle x^{n/d} \rangle$ is certainly one such subgroup. If H is any such subgroup, Theorem 4.16.2 says that $H = \langle x^i \rangle$ for some i. Since $(x^i)^d = 1$, $x^{id} = 1$, we must have $n | id$, $(n/d) | i$, $x^i \in \langle x^{n/d} \rangle$, $H = \langle x^{n/d} \rangle$. The count $\varphi(d)$ comes from Theorem 4.16.1(a). ∎

4.17 Permutation Groups

In this section S is the set $\{1, 2, \ldots, n\}$ and S_n is the set of all permutations of S (bijections from S to S). The set S_n is a group under function composition; S_n is called the **symmetric group on n letters**. We saw in example 2 of Section 4.1 that $S_n = \text{Sym}(S)$ is a group of order $n!$. If $f \in S_n$, then $\langle f \rangle$ is a cyclic subgroup of S_n and $f^t = 1_S$, some $t \in \mathbb{P}$.

Declare two elements i and j of S to be f-*equivalent* if $f^s(i) = j$, some $s \in \mathbb{Z}$. This is an equivalence relation; it is reflexive since $f^0(i) = i$, symmetric since $f^s(i) = j$ implies that $f^{-s}(j) = i$, and transitive since $f^s(i) = j$, $f^r(j) = k$ imply that $f^{s+r}(i) = k$. The equivalence classes are called the **cycles** (or **orbits**) of f on S. The usual notation for a cycle is
$$(i \quad f(i) \quad f^2(i) \cdots f^{s-1}(i)),$$
where $f^s(i) = i$; just pick any i in the cycle and list the f-images in order, as long as they are different.

For example, if $n = 14$, then the $f \in S_n$ defined by

i	1	2	3	4	5	6	7	8	9	10	11	12	13	14
$f(i)$	3	14	7	11	10	6	9	13	2	5	8	12	4	1

Sec. 4.17 *Permutation Groups*

is denoted

$$f = (1\ 3\ 7\ 9\ 2\ 14)(4\ 11\ 8\ 13)(5\ 10)(6)(12)$$

or, more briefly,

$$f = (1\ 3\ 7\ 9\ 2\ 14)(4\ 11\ 8\ 13)(5\ 10),$$

since one-element cycles are usually not written. This notation for permutations is called **disjoint cycle notation**; since the sets of elements in the cycles are equivalence classes, the cycles are pairwise disjoint.

In this notation, $f_1 = (1\ 3\ 7\ 9\ 2\ 14)$, $f_2 = (4\ 11\ 8\ 13)$, and $f_3 = (5\ 10)$ are also in S_n; these permutations are often called *cycles*, since they each have only one cycle of length > 1.

When we see

$$(1\ 3\ 7\ 9\ 2\ 14)(4\ 11\ 8\ 13)(5\ 10),$$

do we mean f or $f_1 \circ f_2 \circ f_3$? The answer is that we mean both; they are both the same function. In fact, the cycles can appear in any order; $f = f_2 \circ f_1 \circ f_3 = f_3 \circ f_2 \circ f_1 = \cdots$. That is, *disjoint cycles commute*.

As always (unless otherwise specified), we write a function on the left of its argument, and function composition proceeds from right to left. For example, $(1\ 2\ 3)(2) = 3$, $(1\ 2\ 3)(3) = 1$,

$$(1\ 2\ 3)(5\ 6\ 8) \circ (2\ 5\ 7\ 8)(1\ 4\ 3) = (1\ 4)(2\ 6\ 8\ 3)(5\ 7),$$

$$(2\ 5\ 7\ 8)(1\ 4\ 3) \circ (1\ 2\ 3)(5\ 6\ 8) = (1\ 5\ 6\ 2)(3\ 4)(7\ 8).$$

In the product $(1\ 2\ 3)(5\ 6\ 8) \circ (2\ 5\ 7\ 8)(1\ 4\ 3)$, above, the factor $(2\ 5\ 7\ 8)(1\ 4\ 3)$ maps 1 to 4 and then the factor $(1\ 2\ 3)(5\ 6\ 8)$ maps 4 to 4, so the product maps 1 to 4. Also, $(2\ 5\ 7\ 8)(1\ 4\ 3)$ maps 4 to 3 and then $(1\ 2\ 3)(5\ 6\ 8)$ maps 3 to 1, so the product maps 4 to 1. This accounts for the cycle $(1\ 4)$ in the product, and the other (disjoint) cycles of the product are similarly computed.

A **permutation group** is a subgroup of Sym (U) for some set U. If $|U| = n$, we say that Sym (U) and its subgroups have **degree** n. For n large, S_n has so many subgroups that they are not all classified. In fact, the following theorem shows that any group (finite or infinite) is isomorphic to a subgroup of some Sym (U).

Theorem 4.17.1 (Cayley's Theorem). *Any group G is isomorphic to a permutation group on the set G.*

Proof. For any $g \in G$, define $f_g: G \to G$ by $f_g(x) = gx$. If $gx_1 = gx_2$, then $x_1 = x_2$, so f_g is one-one. If $y \in G$, then $gx = y$ for $x = g^{-1}y$, so f_g is onto, $f_g \in$ Sym (G).

Let $\{f_g | g \in G\} = C \subseteq \text{Sym}(G)$. Define $\psi: G \to C$ by $\psi(g) = f_g$. For any $h, x \in G$,

$$(f_g \circ f_h)(x) = f_g(f_h(x)) = f_g(hx) = ghx = f_{gh}(x),$$

so $\psi(g) \circ \psi(h) = f_g \circ f_h = f_{gh} = \psi(gh)$ and ψ is a homomorphism. Since ψ is readily seen to be one-one and onto, ψ is the required isomorphism from G to a permutation group. ∎

For a finite group G, the rows of the multiplication table enable us specifically to construct the permutation group $C(G)$ isomorphic to G. Consider the noncyclic group of order 4 with the multiplication table shown.

	e	x	y	z
e	e	x	y	z
x	x	e	z	y
y	y	z	e	x
z	z	y	x	e

The functions f_g amount to left multiplication by g. The first row of the table shows that multiplication by e fixes everything, so f_e = identity function on the set G = identity element of group $C(G)$. The second row shows that left multiplication by x sends $e \to x$, $x \to e$, $y \to z$, and $z \to y$, so $f_x = (e\ x)(y\ z)$. Similarly, $f_y = (e\ y)(x\ z)$ and $f_z = (e\ z)(x\ y)$, so the isomorphism ψ is $e \to 1_G$, $x \to (e\ x)(y\ z)$, $y \to (e\ y)(x\ z)$, and $z \to (e\ z)(x\ y)$.

The following notion of action of a group on a set generalizes the notion of permutation group. It will be used in Sections 5.14 and 5.16.

Let G be a group and S be a set. A **left action** of G on S is a function $G \times S \to S$, usually denoted by juxtaposition $(g, s) \to gs \in S$, such that

$$g(hs) = (gh)s \quad \text{and} \quad 1s = s, \quad \text{all } g, h \in G, s \in S.$$

(Here 1 is the identity of G.) We also say that G **acts on** S **on the left**. A **right action** of G on S is a function $S \times G \to S$, again denoted by juxtaposition $(s, g) \to sg \in S$, such that

$$(sg)h = s(gh) \quad \text{and} \quad s1 = s, \quad \text{all } g, h \in G, s \in S.$$

A permutation group G on a set S is an example of a group G acting on a set S, with $gs = g(s)$, all $g \in G$, $s \in S$, because $1s = 1(s) = s$ and $(gh)s = (gh)(s) = g(h(s)) = g(hs)$. But the notion of action is more general; for example, any group G acts on a one-element set $S = \{s\}$ if we define $gs = s$ for all $g \in G$.

If C and R are finite sets and G is a (left) permutation group on C, so that G acts on C on the left, then G also acts on R^C on the right; if $f \in R^C$, so $f: C \to R$, and $\pi \in G$, then $f \circ \pi: C \to R$, $f \circ \pi \in R^C$. $(f, \pi) \to f \circ \pi$ is a right action of G on R^C because $f \circ 1 = f$ and $f \circ (\pi_1 \circ \pi_2) = (f \circ \pi_1) \circ \pi_2$, all $f \in R^C$, $\pi_1, \pi_2 \in G$. This is the situation we deal with in Pólya enumeration theory in Chapter 5.

If group G acts on set S on the left (respectively, on the right), we say that $s_1, s_2 \in S$ are **G-equivalent** if $gs_1 = s_2$ (respectively, $s_1 g = s_2$), some $g \in G$. We will show that G-equivalence is an equivalence relation. It is reflexive because $1s = s$, all $s \in S$. If $gs_1 = s_2$, then $g^{-1}s_2 = g^{-1}(gs_1) = (g^{-1}g)s_1 = 1s_1 = s_1$, proving that G-equivalence is symmetric. If $gs_1 = s_2$ and $hs_2 = s_3$ for $g, h \in G$ and $s_1, s_2, s_3 \in S$, then $(hg)s_1 = h(gs_1) = hs_2 = s_3$, proving that G-equivalence is transitive. (The proof is similar if G acts on S on the right.) The G-equivalence classes in S are called **orbits of G on S**. Group G is said to be **transitive** on each orbit; if there is only one orbit, G is **transitive on S**. (If G is a cyclic permutation group on the set S, generated by an element f, then the *orbits* of G are the same as the *orbits* of f defined earlier.)

Lemma 4.17.1. *If the finite group G acts on the finite set S and $\mathcal{O}$ is an orbit, $s \in \mathcal{O}$, then $G_s = \{g \in G | gs = s\}$ is a subgroup of G and $|G : G_s| = |\mathcal{O}|$. In particular, if G is transitive on S, then $|G : G_s| = |S|$ for all $s \in S$ (G_s is called the **stabilizer** of $s \in S$).*

Proof. We use left notation in the statement and proof; the proof is similar if G acts on the right. To see that G_s is a subgroup, note that $1s = s$, so $1 \in G_s$; if $g \in G_s$, then $g^{-1}s = g^{-1}(gs) = (g^{-1}g)s = 1s = s$, so $g^{-1} \in G_s$; if $g, h \in G_s$, then $(gh)s = g(hs) = gs = s$.

Now let $\mathcal{O} = \{s_1, s_2, \ldots, s_n\}$, where $s = s_1$. Since $\mathcal{O}$ is an orbit, there are elements $g_1, g_2, \ldots, g_n \in G$ with

$$g_i s = s_i, \quad 1 \leq i \leq n.$$

We claim that $g_1 G_s, g_2 G_s, \ldots, g_n G_s$ are the distinct left cosets of G_s in G; this will show that $|G : G_s| = n = |\mathcal{O}|$ and complete the proof. First, if $g \in G$, then $gs = s_i$ for some i; $g_i^{-1}s_i = s$, so $g_i^{-1}gs = g_i^{-1}s_i = s$, $g_i^{-1}g \in G_s$, $g \in g_i G_s$. This proves that the cosets include all elements of G. They are distinct, for if $g \in g_i G_s \cap g_j G_s$, say $g = g_i h = g_j h'$, $h, h' \in G_s$, then $s_i = g_i h s = gs = g_j h' s = s_j$, forcing $i = j$. ∎

For example, consider the group G of six symmetries of the equilateral triangle. Group G acts transitively on the set $S = \{1, 2, 3\}$ of three vertices, since (identity)$\cdot 1 = 1$, $(1 \; 2 \; 3) \cdot 1 = 2$, $(1 \; 3 \; 2) \cdot 1 = 3$. The stabilizer G_1 of $1 \in S$ is $G_1 = \{\text{identity}, (2 \; 3)\}$, and $|G : G_1| = \frac{6}{2} = 3 = |S|$, in agreement with Lemma 4.17.1.

4.18 Dihedral Groups

We have used the symmetry groups of the equilateral triangle and square as examples before, but now we consider symmetries of the regular n-gon.

The cases $n = 5$ and $n = 6$ (pentagon and hexagon) are pictured in Figure 4.18.1. The full set of these symmetries constitutes a group for each n because the inverse of a symmetry and the composite of two symmetries are again symmetries.

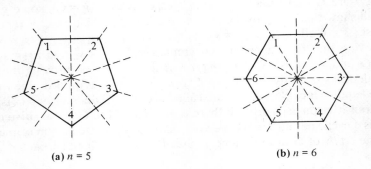

FIGURE 4.18.1. *Regular n-gons*

For any $n \geq 3$, let R represent a clockwise rotation of the regular n-gon through an angle $(360/n)°$ about its center. Then the n rotations are $R, R^2, \ldots, R^n = I$ (identity).

If n is odd, there are n axes of symmetry, each through a vertex and a midpoint of the opposite edge. Thinking of the n-gon in 3-space, we can rotate it $180°$ about such an axis, so the final and initial figures are identical. Note that no such reflection can be accomplished by any rotation, for a reflection reverses the order of the vertices while a rotation does not. If n is even, there are $n/2$ axes of symmetry through pairs of opposite vertices and $n/2$ axes of symmetry through pairs of midpoints of opposite edges.

In any case, the full group D_n of symmetries (or *rigid motions*) of the n-gon has order $2n$. The set $\langle R \rangle = \{I, R, R^2, \ldots, R^{n-1}\}$ of rotations is a subgroup of order n, normal in D_n by Exercise 4.15.15. The other n elements of D_n all have order 2; let F be one of them. If $FR \in \langle R \rangle$, say $FR = R^i$, then $F = R^{i-1} \in \langle R \rangle$, a contradiction; therefore, $FR \in D_n - \langle R \rangle$, which forces FR to have order 2. The equation $I = (FR)^2$ implies that $I = FRFR$, $R^{-1} = FRF$. Because $F = F^{-1}$, we can also write $F^{-1}RF = R^{-1}$.

Since it has index 2, the subgroup $\langle R \rangle$ has only two cosets, $\langle R \rangle$ and $F\langle R \rangle$, in D_n; therefore,

$$D_n = \{I, R, R^2, \ldots, R^{n-1}, F, FR, FR^2, \ldots, FR^{n-1}\}.$$

The equations

(4.18.1) $\qquad R^n = I, \qquad F^2 = I, \qquad F^{-1}RF = R^{-1}$

are often referred to as *defining equations* of the group D_n, since any multiplication in D_n can be performed using these equations. For example,

Sec. 4.19 Additive Abelian Groups

$F^{-1}RF = R^{-1}$ implies that $RF = FR^{-1}$, so we can multiply

$$FR \cdot FR^2 = F \cdot RF \cdot R^2 = F \cdot FR^{-1} \cdot R^2 = F^2 R = R,$$

$$\begin{aligned}FR^3 \cdot F = FR^2 \cdot RF = FR^2 \cdot FR^{-1} = FR \cdot RF \cdot R^{-1} &= FR \cdot FR^{-1} \cdot R^{-1}\\ &= F \cdot RF \cdot R^{-2}\\ &= F \cdot FR^{-1} \cdot R^{-2}\\ &= F^2 R^{-3} = R^{-3},\end{aligned}$$

and so on.

Any group of order $2n$ with elements satisfying (4.18.1) is isomorphic to D_n and is called a **dihedral group** of order $2n$.

4.19 Additive Abelian Groups

The groups we discuss in this section are all *additive*; that is, the binary operation is denoted by $+$ and is called addition, the identity element is denoted 0, and the inverse of x is $-x$. All groups in this section are abelian; they satisfy $x + y = y + x$. (Additive notation is rarely used for nonabelian groups.) The nth power of x is

$$\underbrace{x + x + \cdots + x}_{n\ x\text{'s}},$$

denoted by nx instead of x^n.

One example we have mentioned before is $[\mathbb{B}^n, +]$, the direct product of n copies of $[\mathbb{Z}_2, +]$. When we study codes, we will often write elements of $[\mathbb{B}^n, +]$ without parentheses or commas; for example, instead of

$$(1, 0, 1, 1, 0, 0) + (0, 1, 1, 0, 1, 0) = (1, 1, 0, 1, 1, 0)$$

we write

$$101100 + 011010 = 110110.$$

More generally, let p be a *prime* and form the direct product $[\mathbb{Z}_p \times \cdots \times \mathbb{Z}_p, +] = [\mathbb{Z}_p^n, +]$ of n copies of $[\mathbb{Z}_p, +]$. (In additive notation, we often say *direct sum* instead of direct product.) For example,

$$(1, 0, 3, 4) + (4, 2, 3, 2) = (0, 2, 1, 1)$$

in $\mathbb{Z}_5 \times \mathbb{Z}_5 \times \mathbb{Z}_5 \times \mathbb{Z}_5$. $[\mathbb{Z}_p^n, +]$ is an additive abelian group of order p^n, with identity element $(0, 0, \ldots, 0)$. Every element has order 1 or p, since

$$p(x_1, x_2, \ldots, x_n) = (px_1, px_2, \ldots, px_n) = (0, 0, \ldots, 0)$$

for all $(x_1, x_2, \ldots, x_n) \in \mathbb{Z}_p^n$. $[\mathbb{Z}_p^n, +]$ (or any group isomorphic to it) is called the **elementary abelian group** of order p^n.

By a **prime power** we mean a positive integer power of a prime. The 10 smallest prime powers are 2, 3, 4, 5, 7, 8, 9, 11, 13, and 16. The most important theorem on abelian groups follows.

Fundamental Theorem on Finite Abelian Groups. *Any finite abelian group A is isomorphic to a direct product of cyclic groups of prime power order. The set of prime powers is uniquely determined by A.*

Of course, for the (abelian) group $\{1\}$ of order 1, we have to interpret $\{1\}$ as an empty direct product, to fit the theorem. The order of any direct product is the product of the orders of the factors, so for any A the associated prime powers have product $|A|$. We shall not prove this Fundamental Theorem here; it is proved in many basic texts on abstract algebra.

When considering examples, it is convenient to denote by C_t a cyclic group of order t. (If we like, we may assume that $C_t = [\mathbb{Z}_t, +]$.) Here are some examples.

1. What are the abelian groups of order 4? The number 4 has two expressions as a product of prime powers, 4 and $2 \cdot 2$, so any abelian group of order 4 is isomorphic to C_4 or $C_2 \times C_2$. (We saw in Exercise 4.15.14 that *any* group of order 4 is isomorphic to one of these.)

2. What are the abelian groups of order 6? The number 6 has only one expression as a product of prime powers: $6 = 2 \cdot 3$. So any abelian group of order 6 is isomorphic to $C_2 \times C_3$. What about C_6? C_6 is abelian of order 6, so we conclude that C_6 is isomorphic to $C_2 \times C_3$. (See Exercise 4.20.17.)

3. What are the abelian groups of order 100? The number 100 has four expressions as a product of prime powers:

$$100 = 4 \cdot 25 = 2 \cdot 2 \cdot 25 = 4 \cdot 5 \cdot 5 = 2 \cdot 2 \cdot 5 \cdot 5.$$

Hence any abelian group of order 100 is isomorphic to one of $C_4 \times C_{25}$, $C_2 \times C_2 \times C_{25}$, $C_4 \times C_5 \times C_5$, and $C_2 \times C_2 \times C_5 \times C_5$. The uniqueness part of the Fundamental Theorem says that no two of these are isomorphic to one another. (Which one is isomorphic to C_{100}?)

4.20 Exercises

1. Verify Corollary 4.16.1 for the cyclic group $[\mathbb{Z}_{36}, +]$ of order 36, by explicitly listing all the subgroups and the orders of all the elements.

2. Using Corollary 4.16.1, verify the basic theorem of number theory:

$$\sum_{d|n} \varphi(d) = n,$$

where $n \in \mathbb{P}$ and the sum runs over all $d \in \mathbb{P}$ which divide n.

3. Write the multiplication table of S_3. Show that S_3 is isomorphic to the group of symmetries of the equilateral triangle. (The notation in Exercise 4.5.14 will help.)

4. Show that any nonabelian group G of order 6 is isomorphic to S_3.
(HINT: If there is an element x of order 3, show that $\langle x \rangle \triangleleft G$ and consider cosets of $\langle x \rangle$. If all nonidentity elements have order 2, show that one cannot finish the multiplication table.)

*5. (a) Show that for any $f \in S_5$,
$$f \circ (1 \quad 2 \quad 3)(4 \quad 5) \circ f^{-1} = (f(1) \quad f(2) \quad f(3))(f(4) \quad f(5)).$$
(b) Find $g \in S_6$ such that $g \circ (1 \quad 2 \quad 4 \quad 6 \quad 3) \circ g^{-1} = (2 \quad 5 \quad 3 \quad 6 \quad 4)$.
(c) Two elements of S_n are said to have the same **cycle structure** if they have the same number of cycles of each length, when written as products of disjoint cycles. Generalize (a) and (b) by showing that any two elements of S_n are conjugate iff they have the same cycle structure.

*6. Explain why disjoint cycles commute.

*7. Write the multiplication table of D_4, and find a permutation group of degree 8 which is isomorphic to D_4.

*8. The quaternion group Q_8 of order 8 contains elements x and y satisfying equations $x^4 = y^4 = 1$, $x^2 = y^2$, $x^{-1}yx = y^{-1}$. Complete the multiplication table of Q_8, and find a permutation group of degree 8 that is isomorphic to Q_8.

*9. Find all abelian groups of order 8. (Exercises 7, 8, and 9 specify, up to isomorphism, *all* groups of order 8; but do not try to prove this.)

10. Show that $(1 \quad 2 \quad 3 \quad 4)$ and $(2 \quad 4)$ lie in a subgroup of S_4 isomorphic to D_4. (Finite groups G are often isomorphic to permutation groups of degree less than $|G|$; Cayley's theorem may be a very inefficient way to represent G.)

11. What group does (4.18.1) define when $n = 2$?

12. By considering rigid motions of a regular n-gon, find permutations $F, R \in S_n$ satisfying (4.18.1).
(HINT: Consider the cases n odd and n even separately.) This shows that there is a permutation group of degree n that is isomorphic to D_n; Cayley's theorem only gives one of degree $2n = |D_n|$.

*13. Show that conjugate elements of a finite group have the same order.

*14. Show that if $x, y \in G$, x has order m, y has order n, $(m, n) = 1$, and $xy = yx$, then xy has order mn.

15. Exercises 5 and 13 imply that we can study the orders of elements of S_n by studying one with each cycle structure.
(a) Show that any cycle $(1 \quad 2 \quad 3 \cdots m)$ has order m.
(b) Show that $(1 \quad 2 \quad 3)(4 \quad 5)$ has order 6. (Use Exercise 14.) Show that no element of S_5 has larger order.
(c) Show that if $f \in S_n$ has, when expressed as a product of disjoint cycles, cycles of lengths $l_1, l_2, \ldots, l_k$, then the order of f is the *least common multiple* of $l_1, l_2, \ldots, l_k$; the smallest $m \in \mathbb{P}$ with $l_i | m$, $i = 1, 2, \ldots, k$.
(d) What are the largest orders of elements in S_7, S_8, S_{10}, and S_{15}?

16. Describe explicitly all abelian groups of orders 16, 72, and 210.

*17. In example 2 of Section 4.19, we remarked that C_6 must be isomorphic to $C_2 \times C_3$. Exhibit explicitly an isomorphism from $[\mathbb{Z}_6, +]$ onto $[\mathbb{Z}_2 \times \mathbb{Z}_3, +]$.

*18. (For those who know matrix theory over $\mathbb{R}$ and $\mathbb{C}$.)

(a) Prove that each of the following sets of matrices is a group with respect to matrix multiplication.

(1) $\left\{ \begin{bmatrix} a & -b \\ b & a \end{bmatrix} \middle| a, b \in \mathbb{R}, a^2 + b^2 \neq 0 \right\}$.

(2) $\left\{ \begin{bmatrix} 1 & a_{12} & a_{13} \\ 0 & 1 & a_{23} \\ 0 & 0 & 1 \end{bmatrix} \middle| a_{12}, a_{13}, a_{23} \in \mathbb{R} \right\}$.

(3) $\left\{ \begin{bmatrix} 1 & 0 \\ 0 & 1 \end{bmatrix}, \begin{bmatrix} -1 & 0 \\ 0 & -1 \end{bmatrix}, \begin{bmatrix} 0 & -1 \\ 1 & 0 \end{bmatrix}, \begin{bmatrix} 0 & 1 \\ -1 & 0 \end{bmatrix} \right\}$.

(4) $\{1, -1, i, -i\} \subseteq \mathbb{C}$, where $i^2 = -1$. (Field elements are 1×1 matrices.)

(5) $\left\{ \begin{bmatrix} 1 & 0 \\ c & 1 \end{bmatrix} \middle| c \in \mathbb{C} \right\}$.

(b) Show that the groups in (3) and (4) of (a) are cyclic, and therefore isomorphic.

(c) What is the smallest multiplicative matrix group that contains $\begin{bmatrix} 0 & -1 \\ 1 & 0 \end{bmatrix}$ and $\begin{bmatrix} 0 & i \\ i & 0 \end{bmatrix}$? To what other group mentioned in this exercise set is it isomorphic?

19. Draw the state diagram of the Moore machine $[G, G, G, \circ, 1_G]$ determined by each of the following groups:

(a) C_6.

(b) The group of Exercise 18(a)(4).

(c) The group D_3 of symmetries of the equilateral triangle.

20. Assume that the group G acts on the set S on the left. For each $g \in G$, define $T_g : S \to S$ by $T_g(s) = gs$, all $s \in S$.

(a) Show that $T_g \in \text{Sym}(S)$ for each $g \in G$.

(b) Show that $\varphi : g \to T_g$ is a homomorphism from G onto the subgroup $\{T_g | g \in G\}$ of Sym (S).

(c) Let $K = \ker \varphi$, and conclude that G/K is isomorphic to a permutation group on S.

CHAPTER 5
Applications of Group Theory

5.1 The Binary Symmetric Channel

In this chapter we examine several ways in which the group theory of Chapter 4 has found practical application.

In coding theory, we speak of signals sent over a "noisy channel." The noisy channel might involve radio signals from a space probe or from a far point on earth, telephone signals on wires or microwave, or transmission of information between the memory and the processing unit of a large computer. In any case, some of the information received may be erroneous, and we want to detect, preferably also to correct, the erroneous information. We try to do this by sending additional signals as a check on the information. These additional signals are themselves subject to error, so there is no way to *guarantee* accuracy; we only try to make the probability of accuracy as high as possible.

Mathematicians measure probability with real numbers, usually fractions, between 0 and 1. In the case of a finite number of possibilities, an event with probability 0 cannot occur, and an event with probability 1 is certain to occur. If a fair coin is flipped, the probabilities are $\frac{1}{2}$ that it will land heads up, $\frac{1}{2}$ that it will land tails up, 0 that it will stay up in the air and 1 that it will come down. If a point is chosen at random within the square $0 \leq x \leq 1, 0 \leq y \leq 1$ of the Cartesian x, y-plane, the probability is $\pi/4$ (not a rational fraction, as π is not rational) that the point will lie below the curve

$$y = \frac{1}{1+x^2}. \quad \text{(Why?)}$$

Two principles about probability will be needed here. *The probability that all of several independent events will occur is the product of their individual probabilities, while the probability that some one of several mutually exclusive events will occur is the sum of their individual probabilities.* For example, if a red die and a green die are tossed, the probability that the total is 12 is $(\frac{1}{6})(\frac{1}{6}) = \frac{1}{36}$ (both must turn up 6), while the probability that the total is 10 is

$$\tfrac{1}{6}\cdot\tfrac{1}{6}+\tfrac{1}{6}\cdot\tfrac{1}{6}+\tfrac{1}{6}\cdot\tfrac{1}{6}=\tfrac{1}{12}$$

(either red reads 4 and green 6, or both 5, or red 6 and green 4).

A **binary symmetric channel** consists of a transmitter sending signals 0 and 1, a receiver, and a probability p, $0 < p < 1$, of incorrect transmission of a single digit. That is, if 0 is transmitted, the probability is p that 1 is received and the probability is $q = 1-p$ that 0 is received; if 1 is transmitted, the probability is p that 0 is received and the probability is q that 1 is received.

(See Figure 5.1.1.) The channel is "binary" because the signals are 0 and 1, and "symmetric" because p is the same whether 0 or 1 is the digit transmitted.

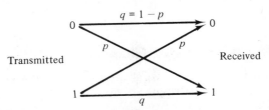

FIGURE 5.1.1. *Binary Symmetric Channel*

The probability that no errors occur when one is transmitting a k-digit message must be q^k. For example, if $p = .001$ and we send a 10,000-digit message, the probability of perfect transmission is

$$(1-.001)^{10,000}, \quad \text{approximately } .00005.$$

We shall illustrate several codes in this chapter by showing how they increase this probability of perfect transmission.

We shall work within the additive group $\mathbb{B}^n$ of n-digit **binary words** $\mathbf{w} = w_1 w_2 \cdots w_n$, $w_i \in \mathbb{B} = \{0, 1\}$. If binary word $\mathbf{w} = w_1 w_2 \cdots w_n$ is sent but binary word $\mathbf{r} = r_1 r_2 \cdots r_n$ is received, the **error pattern** is the binary word $\mathbf{e} = e_1 e_2 \cdots e_n$, where

$$e_i = \begin{cases} 0 & \text{if } w_i = r_i, \\ 1 & \text{if } w_i \neq r_i. \end{cases}$$

Thus the number of 1's in $\mathbf{e}$ is the number of errors made in receiving $\mathbf{w}$, and the locations of the 1's are the locations of the errors. If at the receiver we receive $\mathbf{r} = 010011101$ and we somehow know that the error pattern is $\mathbf{e} = 001000100$, then we find that $\mathbf{w} = 011011001$ by correcting the third and seventh digits. It is obvious from binary arithmetic that the following equations hold in the additive group $\mathbb{B}^n$.

$\mathbf{w} = \mathbf{r} + \mathbf{e}$ (transmitted word = received word + error pattern),

$\mathbf{r} = \mathbf{w} + \mathbf{e}$ (received word = transmitted word + error pattern),

$\mathbf{e} = \mathbf{w} + \mathbf{r}$ (error pattern = transmitted word + received word).

(We remind the reader that the addition in the group $\mathbb{B}^n$ is componentwise addition modulo 2: $0+0 = 1+1 = 0$, $0+1 = 1+0 = 1$.)

We give some more examples.
1. If $\mathbf{w} = 01011010$ and $\mathbf{r} = 01111011$, then $\mathbf{e} = 00100001$.
2. If $\mathbf{r} = 111001001000$ and $\mathbf{e} = 000000001000$, then $\mathbf{w} = 111001000000$.

Theorem 5.1.1. *Assume that we transmit an n-digit message* **w** *through a binary symmetric channel in which the probability of error in each digit is p.*
(a) *If* **e** *is an n-digit error pattern containing k 1's, then the probability of* **e** *occurring is* $p^k(1-p)^{n-k}$.
(b) *There are* $\binom{n}{k} = \frac{n!}{(n-k)!(k!)}$ *error patterns containing k 1's, so the probability that exactly k errors will be made in transmitting* **w** *is* $\binom{n}{k}p^k(1-p)^{n-k}$.

Proof. (a) For the fixed error pattern **e**, the probability of each 1 occurring is p, and the probability of each 0 occurring is $1-p$. To find the probability of each of these n independent events occurring, we multiply their probabilities.

(b) The number of error patterns with exactly k 1's is equal to the number of combinations of n things k at a time, which is $\binom{n}{k} = \frac{n!}{(k!)(n-k)!}$. Each of these $\binom{n}{k}$ error patterns has probability $p^k(1-p)^{n-k}$, and the sum of $\binom{n}{k}$ copies of this number is $\binom{n}{k}p^k(1-p)^{n-k}$. ∎

For example, if $p = .01$ and we transmit a 100-digit message, the probability of no errors is

$$(1-p)^n = (1-.01)^{100}, \quad \text{approximately } .36603;$$

the probability of one error is

$$\binom{n}{1}p(1-p)^{n-1} = 100(.01)(1-.01)^{99}, \quad \text{approximately } .36973;$$

the probability of two errors is

$$\binom{n}{2}p^2(1-p)^{n-2} = \frac{100 \cdot 99}{2}(.01)^2(1-.01)^{98}, \quad \text{approximately } .18486;$$

and the probability of more than two errors is approximately

$$1 - .36603 - .36973 - .18486 = .07938.$$

5.2 Block Codes

An (n, m) **block code** consists of an **encoding function** $E: \mathbb{B}^m \to \mathbb{B}^n$ and a **decoding function** $D: \mathbb{B}^n \to \mathbb{B}^m$. A **code word** is any element of im E. (Here im E denotes the image of the function E. Like most coding topics in this chapter, this definition will be generalized in Chapter 9.)

Generally, we shall have $m < n$, since the function E will be adding "check digits" to the original message. A flow chart for the encoding and decoding process appears in Figure 5.2.1.

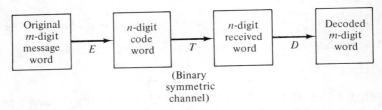

FIGURE 5.2.1. *An (n, m) Block Code Transmission*

Given a long message, we break it into blocks of length m. The function E encodes these into blocks of length n. We assume that E is one-one, so no two message blocks have the same code word. The channel T transmits each digit, with error probability p, and D decodes the received blocks into blocks of length m. We seek to choose E and D so that the probability that a decoded block will equal the original message block will be high. There are two additional requirements. First, we seek an *efficient* code that does not transmit too many extra digits. The fraction $R = m/n$ is called the **rate** of the code; if R is large (close to 1), the code will be efficient. Second, the code will be useless unless the functions E and D can be implemented in some practical fashion, say by digital electronic circuitry. The codes encountered in this chapter are sufficiently simple that the reader with some knowledge of circuit devices will realize that they can be readily implemented. (Implementation is also discussed in Chapter 9.)

Usually, p is quite small, so the greatest likelihood is that a code word will be transmitted without error, and most received words containing errors will contain only one error. Consequently, **single-error-correcting codes**, which decode correctly all received words containing at most one error, are important and are the subject of this chapter. Multiple-error-correcting codes, needed if p is large (e.g., for transmission from space) or if the code is required to have extremely good accuracy, are discussed in Chapter 9.

EXAMPLE 1. The $(m+1, m)$ **parity-check code** is merely an *error-detecting code*. We shall only construct an encoding function and see how it can be used to detect most transmission errors. We define $E: \mathbb{B}^m \to \mathbb{B}^{m+1}$ by

$$E(a_1 a_2 \cdots a_m) = a_1 a_2 \cdots a_m a_{m+1}, \text{ where } a_{m+1} = a_1 + a_2 + \cdots + a_m.$$

Thus a_{m+1} is 0 or 1, depending on whether the number of 1's in $a_1 a_2 \cdots a_m$ (the *weight*) is even or odd. Consequently, *every code word has even weight*. Any single error in transmission will change a code word to a word of odd weight, so the error can be detected by computing the weight of the received word. Of course, if there are two errors (a "double error") in the transmis-

Sec. 5.2 Block Codes

sion of an $(m+1)$-digit block, the errors will be missed. So this type of code is used only when double errors are unlikely; it is frequently used internally in modern computers.

As a specific example, consider the problem of Section 5.1: transmitting a 10,000-digit message, with $p = .001$. We break the message into 1000 10-digit blocks. A single 10-digit block is encoded into an 11-digit block, and the probability that the latter is transmitted without error is $(.999)^{11}$, approximately .989055. The probability of exactly one error is, by Theorem 5.1.1,

$$\binom{11}{1} p(1-p)^{10} = 11(.001)(.999)^{10}, \quad \text{approximately } .01089.$$

So the probability of at most one error in an 11-digit block is about

$$.989055 + .01089 = .999945.$$

If we have some outside method for correcting these single-error 11-digit blocks (perhaps by having the transmitter repeat that block), then our chance of eventually getting the 10,000-digit message through without error is at least $(.999945)^{1000}$, approximately .946.

We are transmitting 11,000 digits instead of 10,000 (rate $\frac{10}{11}$), but now our chances are about 18 of 19 that all errors are detected; before (Section 5.1), our chance of perfect transmission was only 1 in 20,000.

EXAMPLE 2. The $(3m, m)$ **triple-repetition code** is a true *error-correcting code*, although a rather trivial one. The encoding function merely repeats each m-digit block three times:

$$E(a_1 a_2 \cdots a_m) = a_1 a_2 \cdots a_m a_1 a_2 \cdots a_m a_1 a_2 \cdots a_m.$$

During decoding, the function $D: \mathbb{B}^{3m} \to \mathbb{B}^m$ chooses as the ith digit the one that appears as the ith digit at least twice in the three transmissions. For example, if $m = 3$, then $E(010) = 010010010$. If transmission makes an error in the sixth digit, the received word will be 010011010. This will be decoded as 010, the correct message, since the first, fourth, and seventh digits are all 0; the second, fifth, and eighth digits are all 1; and two out of three of the third, sixth, and ninth digits are 0. Clearly, the triple-repetition code automatically corrects all single transmission errors.

Consider again our 10,000-digit message, with $p = .001$. A single digit d is encoded as ddd. The probability that this is transmitted correctly is $(.999)^3$, approximately .997003, and the probability of a single error is

$$\binom{3}{1}(.001)(.999)^2, \quad \text{approximately } .002994,$$

so the probability of correct decoding of the digit d is about

$$.997003 + .002994 = .999997.$$

There are 10,000 digits, so the probability of correctly decoding the entire message is about

$$(.999997)^{10,000}, \quad \text{approximately } .97.$$

Now not only does our code have 33 chances in 34 of complete accuracy, but also no outside error-correcting method is required. But we are paying a high price for this accuracy and error-detection capability; we now have to transmit 30,000 digits to get the 10,000 we want.

EXAMPLE 3. With the $(5m, m)$ **five-times-repetition code** we can correct double errors. In the case of our 10,000-digit message with $p = .001$, the probability of correctly decoding a single digit will be

$$(.999)^5 + \binom{5}{1}(.001)(.999)^4$$

$$+ \binom{5}{2}(.001)^2(.999)^3, \quad \text{approximately } .99999999.$$

The probability of accuracy in decoding the entire message is about

$$(.99999999)^{10,000}, \quad \text{approximately } .9999.$$

So now the chance of incorrect decoding of the message is only 1 in 10,000! We pay dearly for this, having to transmit 50,000 digits instead of 10,000 (rate $\frac{1}{5}$).

5.3 Weight and Distance

In this section we state two general definitions from coding theory and prove two theorems showing when a code will have good error-detecting or error-correcting capability.

The **weight** $w(\mathbf{a})$ of a binary word $\mathbf{a}$ is the number of 1's in it. The **distance** $d(\mathbf{a}, \mathbf{b})$ between binary words $\mathbf{a} = a_1 a_2 \cdots a_n$, $\mathbf{b} = b_1 b_2 \cdots b_n \in \mathbb{B}^n$ is $w(\mathbf{a} + \mathbf{b})$, the number of locations i with $a_i \neq b_i$.

Lemma 5.3.1. *If* $\mathbf{a}, \mathbf{b}, \mathbf{c} \in \mathbb{B}^n$, *then* $d(\mathbf{a}, \mathbf{b}) = d(\mathbf{b}, \mathbf{a})$ *and* $d(\mathbf{a}, \mathbf{c}) \leq d(\mathbf{a}, \mathbf{b}) + d(\mathbf{b}, \mathbf{c})$.

(Mathematically, $\mathbb{B}^n$ is a *metric space* with distance function d.)

Proof. $d(\mathbf{a}, \mathbf{b}) = w(\mathbf{a} + \mathbf{b}) = w(\mathbf{b} + \mathbf{a}) = d(\mathbf{b}, \mathbf{a})$ since $\mathbf{a} + \mathbf{b} = \mathbf{b} + \mathbf{a}$. Now suppose that $\mathbf{a} = a_1 a_2 \cdots a_n$, $\mathbf{b} = b_1 b_2 \cdots b_n$, $\mathbf{c} = c_1 c_2 \cdots c_n$. If we define $d(a_i, b_i)$ by

$$d(a_i, b_i) = \begin{cases} 0 & \text{if } a_i = b_i, \\ 1 & \text{if } a_i \neq b_i, \end{cases}$$

then certainly

$$d(a_i, c_i) \leq d(a_i, b_i) + d(b_i, c_i);$$

Sec. 5.3 Weight and Distance

for this must be true if $a_i = c_i$, and if $a_i \neq c_i$ then $a_i \neq b_i$ or $b_i \neq c_i$. We therefore have

$$d(\mathbf{a}, \mathbf{c}) = \sum_{i=1}^{n} d(a_i, c_i) \leq \sum_{i=1}^{n} d(a_i, b_i) + \sum_{i=1}^{n} d(b_i, c_i) = d(\mathbf{a}, \mathbf{b}) + d(\mathbf{b}, \mathbf{c}). \blacksquare$$

Theorem 5.3.1. *A code can detect all error patterns of weight $\leq k$ iff the minimum distance between code words is at least $k+1$.*

Proof. Here the encoding function $E: \mathbb{B}^m \to \mathbb{B}^n$ is fixed and known, so we know all the code words $E(\mathbf{w})$, $\mathbf{w} \in \mathbb{B}^m$. Suppose that code word $\mathbf{a}$ is transmitted and word $T(\mathbf{a})$ received. The error pattern is $\mathbf{e} = \mathbf{a} + T(\mathbf{a})$, and $w(\mathbf{e}) = d(\mathbf{a}, T(\mathbf{a}))$. The error can be detected iff $T(\mathbf{a})$ is not a code word [just look at the (known) list of code words]. So all $\mathbf{e}$'s of weight $\leq k$ can be detected iff no code word $\mathbf{b} \neq \mathbf{a}$ satisfies $d(\mathbf{a}, \mathbf{b}) \leq k$, that is, iff the minimum distance between code words is at least $k+1$. (See Figure 5.3.1, where small circles represent code words.) $\blacksquare$

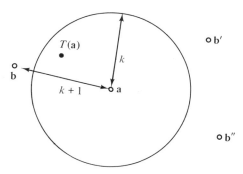

FIGURE 5.3.1. *Distance and Error Detection*

Theorem 5.3.2. (a) *If the minimum distance between code words is at least $2k+1$, we can choose a decoding function D that will correct all error patterns of weight $\leq k$.*

(b) *Conversely, if two different code words $\mathbf{a}$ and $\mathbf{b}$ have $d(\mathbf{a}, \mathbf{b}) \leq 2k$, then some error pattern of weight $\leq k$ cannot be corrected.*

Proof. (a) Assume that code word $\mathbf{c}$ is sent and word $\mathbf{r}$ received, with $\leq k$ errors, so that $d(\mathbf{c}, \mathbf{r}) \leq k$. Inside the "sphere" $S_k(\mathbf{r}) = \{\mathbf{x} \in \mathbb{B}^n | d(\mathbf{x}, \mathbf{r}) \leq k\}$, there can be no code word $\mathbf{c}' \neq \mathbf{c}$ [Figure 5.3.2(a)], since then

$$d(\mathbf{c}, \mathbf{c}') \leq d(\mathbf{c}, \mathbf{r}) + d(\mathbf{r}, \mathbf{c}') \leq k + k < 2k + 1,$$

contradicting the hypothesis. So we can define D on $\mathbb{B}^n$ by

$$D(\mathbf{r}) = \begin{cases} \mathbf{c} & \text{if code word } \mathbf{c} \text{ is in } S_k(\mathbf{r}), \\ \text{arbitrary} & \text{if } S_k(\mathbf{r}) \text{ contains no code word (if such an} \\ & \mathbf{r} \text{ is received, more than } k \text{ errors occurred).} \end{cases}$$

(b) If $d(\mathbf{a},\mathbf{b}) \leq 2k$, then we can find $\mathbf{r} \in \mathbb{B}^n$ with $d(\mathbf{a},\mathbf{r}) \leq k$, $d(\mathbf{b},\mathbf{r}) \leq k$. (Take $r_i = a_i = b_i$ wherever $a_i = b_i$, and choose $r_i = a_i$ about half the time when $a_i \neq b_i$.) Here, $D(\mathbf{r})$ must be something, say $D(\mathbf{r}) = \mathbf{a}$. Then $\mathbf{b}$ could be transmitted, received as $\mathbf{r}$ with error pattern $\mathbf{r}+\mathbf{b}$ of weight $\leq k$, and decoded wrongly as $\mathbf{a}$ [Figure 5.3.2(b)]. ∎

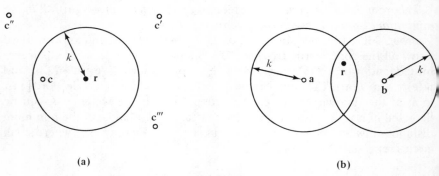

FIGURE 5.3.2. *Distance and Error Correction*

5.4 Generator and Parity-Check Matrices

For the reader who does not know the subject, linear algebra and matrix theory will be developed in Chapter 7, before it is used heavily in Chapters 8 and 9; but we need matrix multiplication in this section and shall define it now.

As defined before, an $m \times n$ *matrix* over a set S is a rectangular array $[a_{ij}]$ of elements of S with m rows and n columns, $a_{ij} \in S$ in the ith row and jth column. In this section, all matrices are over $\mathbb{B} = \{0,1\}$, and addition and multiplication in $\mathbb{B}$ are modulo 2. If $\mathbf{A} = [a_{ij}]$ is an $l \times m$ matrix over $\mathbb{B}$ and $\mathbf{B}$ an $m \times n$ matrix over $\mathbb{B}$, their product $\mathbf{AB}$ is the $l \times n$ matrix $\mathbf{C} = [c_{ij}]$ defined by

(5.4.1) $$c_{ij} = \sum_{k=1}^{m} a_{ik} b_{kj}.$$

In particular, a vector $\mathbf{x} \in \mathbb{B}^m$ is a $1 \times m$ matrix: one row, and m columns with only one entry each. We shall generally write such a vector without brackets, as we have already done when we called it a *word* in $\mathbb{B}^m$.

The $m \times m$ **identity matrix**

$$\mathbf{I}_m = \begin{bmatrix} 1 & 0 & \cdots & 0 \\ 0 & 1 & \cdots & 0 \\ \vdots & & \ddots & \\ 0 & 0 & \cdots & 1 \end{bmatrix}$$

Sec. 5.4 Generator and Parity-Check Matrices

has 1's on the main diagonal, 0's elsewhere; in other words, $\mathbf{I}_m = [\delta_{ij}]$, where

(5.4.2) $$\delta_{ij} = \begin{cases} 0 & \text{if } i \neq j \\ 1 & \text{if } i = j \end{cases} \quad \textbf{(Kronecker delta)}.$$

Any $m \times n$ matrix, $m < n$, over $\mathbb{B}$ will be called a **generator matrix G** if its first m columns form $\mathbf{I}_m$. Given such a matrix $\mathbf{G}$, we can define an encoding function $E: \mathbb{B}^m \to \mathbb{B}^n$ by

(5.4.3) $$E(\mathbf{x}) = \mathbf{x}\mathbf{G}, \quad \text{all } \mathbf{x} \in \mathbb{B}^m.$$

For example, suppose that

(5.4.4) $$\mathbf{G} = \begin{bmatrix} 1 & 0 & 0 & 1 & 1 & 0 \\ 0 & 1 & 0 & 1 & 0 & 1 \\ 0 & 0 & 1 & 0 & 1 & 1 \end{bmatrix}.$$

Then

(5.4.5) $$E(101) = 101 \begin{bmatrix} 1 & 0 & 0 & 1 & 1 & 0 \\ 0 & 1 & 0 & 1 & 0 & 1 \\ 0 & 0 & 1 & 0 & 1 & 1 \end{bmatrix} = 101101;$$

in general, for any $a_1 a_2 a_3 \in \mathbb{B}^3$ we have

(5.4.6) $$E(a_1 a_2 a_3) = a_1 a_2 a_3 \begin{bmatrix} 1 & 0 & 0 & 1 & 1 & 0 \\ 0 & 1 & 0 & 1 & 0 & 1 \\ 0 & 0 & 1 & 0 & 1 & 1 \end{bmatrix}$$
$$= a_1 a_2 a_3 (a_1 + a_2)(a_1 + a_3)(a_2 + a_3) \in \mathbb{B}^6.$$

Thus encoding using the generator matrix is very easy; we just do a matrix multiplication, which is easily programmed for a computer. Decoding is also easy when the transmission is correct; because the first m columns of $\mathbf{G}$ are the identity matrix, the message word is just the first m digits of the code word. Decoding to correct an error cannot be done with the generator matrix alone; the methods in the rest of this section, or Section 5.7, can be used.

The example above is $E: \mathbb{B}^3 \to \mathbb{B}^6$, where

(5.4.7) $$E(a_1 a_2 a_3) = a_1 a_2 a_3 a_4 a_5 a_6,$$
$$a_4 = a_1 + a_2, \quad a_5 = a_1 + a_3, \quad a_6 = a_2 + a_3.$$

These last three equations, called **parity-check equations**, generalize the equation

$$a_{m+1} = a_1 + a_2 + \cdots + a_m$$

of example 1, Section 5.2, which arises from the generator matrix

$$\begin{bmatrix} 1 & 0 & \cdots & 0 & 1 \\ 0 & 1 & \cdots & 0 & 1 \\ \cdot & \cdot & & \cdot & \cdot \\ 0 & 0 & \cdots & 1 & 1 \end{bmatrix}.$$

In the present example, a_4 is a "parity-check" on a_1 and a_2, a_5 a parity-check on a_1 and a_3, a_6 a parity-check on a_2 and a_3. The three equations can be rewritten

(5.4.8) $\quad a_1 + a_2 + a_4 = 0, \qquad a_1 + a_3 + a_5 = 0, \qquad a_2 + a_3 + a_6 = 0.$

We can then express all three equations in one matrix equation which is satisfied by every code word $\mathbf{c} = a_1 a_2 a_3 a_4 a_5 a_6$:

(5.4.9) $\quad \begin{bmatrix} 1 & 1 & 0 & 1 & 0 & 0 \\ 1 & 0 & 1 & 0 & 1 & 0 \\ 0 & 1 & 1 & 0 & 0 & 1 \end{bmatrix} \begin{bmatrix} a_1 \\ a_2 \\ a_3 \\ a_4 \\ a_5 \\ a_6 \end{bmatrix} = \begin{bmatrix} 0 \\ 0 \\ 0 \end{bmatrix}.$

The relation of the coefficient matrix here to the generator matrix will appear following the description of the general process of which this is an example.

In general, if $m < n$, then any $(n-m) \times n$ matrix $\mathbf{H}$ whose last $n-m$ columns are $\mathbf{I}_{n-m}$ will be called a **parity-check matrix**. The parity-check matrix provides an encoding function $E: \mathbb{B}^m \to \mathbb{B}^n$; for any message word $\mathbf{w} \in \mathbb{B}^m$, the code word is the unique word $E(\mathbf{w}) \in \mathbb{B}^n$ whose first m digits are the digits of $\mathbf{w}$ and whose remaining digits are determined by the equation

(5.4.10) $\qquad\qquad \mathbf{H} \cdot (E(\mathbf{w}))^T = \mathbf{0}.$

Indeed, this matrix equation provides $n-m$ equations for the last $n-m$ digits in $E(\mathbf{w})$, the ith equation expressing the $(m+i)$th digit in terms of known digits of $\mathbf{w}$ [just as (5.4.9) leads to the equations for a_4, a_5, and a_6 in (5.4.7)].

It is important to remember that the fact that $E(\mathbf{w})$ is unique implies that if a word $\mathbf{u}$ has $\mathbf{w}$ as its first m digits and if $\mathbf{Hu}^T = \mathbf{0}$, then $\mathbf{u} = E(\mathbf{w})$.

The parity-check matrix $\mathbf{H}$ associated with a generator matrix $\mathbf{G}$ provides a way to correct errors. We start with our example (5.4.6); we saw in (5.4.9) that it has parity-check matrix

(5.4.11) $\qquad\qquad \mathbf{H} = \begin{bmatrix} 1 & 1 & 0 & 1 & 0 & 0 \\ 1 & 0 & 1 & 0 & 1 & 0 \\ 0 & 1 & 1 & 0 & 0 & 1 \end{bmatrix},$

Sec. 5.4 Generator and Parity-Check Matrices

and we saw in (5.4.5) that $E(101) = 101101$. Suppose an error is made in the third digit in the transmission, so at the receiver we see only the received word $\mathbf{r} = 100101$. We compute the *syndrome*

$$(5.4.12) \quad \mathbf{S} = \mathbf{H} \cdot \mathbf{r}^T = \begin{bmatrix} 1 & 1 & 0 & 1 & 0 & 0 \\ 1 & 0 & 1 & 0 & 1 & 0 \\ 0 & 1 & 1 & 0 & 0 & 1 \end{bmatrix} \begin{bmatrix} 1 \\ 0 \\ 0 \\ 1 \\ 0 \\ 1 \end{bmatrix} = \begin{bmatrix} 0 \\ 1 \\ 1 \end{bmatrix}.$$

If the received word were a code word, the result would be 000. However, **S** is the *third* column of **H**, and this is our clue that the error is in the *third* digit of **r**. So we know that the transmitted word is $\mathbf{c} = E(\mathbf{w}) = 101101$, and we decode $\mathbf{w} = 101$ (correctly).

Why did this happen? Recall that for every code word $\mathbf{c}$, $\mathbf{Hc}^T = \mathbf{0}$. Since $\mathbf{r} = \mathbf{c} + \mathbf{e}$, where $\mathbf{e} = 00 \cdots 010 \cdots 00$ (*i*th digit 1) is the error pattern when the *i*th digit of **r** is an error, we have

$$(5.4.13) \quad \mathbf{Hr}^T = \mathbf{H}(\mathbf{c}+\mathbf{e})^T = \mathbf{H}(\mathbf{c}^T + \mathbf{e}^T) = \mathbf{Hc}^T + \mathbf{He}^T = \mathbf{0} + \mathbf{He}^T = \mathbf{He}^T$$

$$= i\text{th column of } \mathbf{H}.$$

Thus any error pattern of weight 1 will be decoded correctly. Single errors occur most often (for p small), so this is good progress.

What about double errors? Suppose that $\mathbf{c} = 101101$ is transmitted as above, but the error pattern is 001010. We receive

$$101101 + 001010 = 100111,$$

and compute the syndrome

$$\begin{bmatrix} 1 & 1 & 0 & 1 & 0 & 0 \\ 1 & 0 & 1 & 0 & 1 & 0 \\ 0 & 1 & 1 & 0 & 0 & 1 \end{bmatrix} \begin{bmatrix} 1 \\ 0 \\ 0 \\ 1 \\ 1 \\ 1 \end{bmatrix} = \begin{bmatrix} 0 \\ 0 \\ 1 \end{bmatrix}.$$

This is the sixth column of **H**, so we would guess an error pattern 000001 and decode incorrectly.

Or suppose that the error pattern is 010010. We receive

$$101101 + 010010 = 111111,$$

with syndrome

$$\begin{bmatrix} 1 & 1 & 0 & 1 & 0 & 0 \\ 1 & 0 & 1 & 0 & 1 & 0 \\ 0 & 1 & 1 & 0 & 0 & 1 \end{bmatrix} \begin{bmatrix} 1 \\ 1 \\ 1 \\ 1 \\ 1 \\ 1 \end{bmatrix} = \begin{bmatrix} 1 \\ 1 \\ 1 \end{bmatrix}.$$

This syndrome is not any column of **H**; it is the first column plus the sixth column, or the second plus the fifth, or the third plus the fourth. Assuming two digits in error, we would be equally likely to guess error pattern 100001 (incorrect), 010010 (correct), or 001100 (incorrect).

Thus a double error will probably be decoded wrongly. Fortunately, when p is close to 0, double errors do not occur very often.

Now we state the parity-check matrix decoding procedure in general. When an n-digit word $\mathbf{r} = r_1 r_2 \cdots r_m r_{m+1} \cdots r_n$ is received, we compute the syndrome

(5.4.14) $$\mathbf{H}\mathbf{r}^T = \mathbf{S}.$$

There are three cases:

1. If $\mathbf{S} = \mathbf{0}$, transmission was probably correct, so we assume that $\mathbf{r}$ is the code word sent and that the original message word is $r_1 r_2 \cdots r_m$.

2. If $\mathbf{S}$ is the ith column of $\mathbf{H}$, there was probably a single error in the ith digit. We assume that the code word sent is $\mathbf{c} = (\mathbf{r}$ with its ith digit changed), and the original message word is the first m digits of $\mathbf{c}$. This conclusion is justified by (5.4.13).

3. If $\mathbf{S}$ is neither $\mathbf{0}$ nor a column of $\mathbf{H}$, then at least two errors occurred in transmission. In this case (which seldom occurs when such decoding is applied), no high-reliability decoding can be done.

The conditions under which this decoding procedure is effective are summarized in the following theorem.

Theorem 5.4.1. *An $(n-m) \times n$ parity-check matrix $\mathbf{H}$ will decode all single errors correctly iff the columns of $\mathbf{H}$ are nonzero and distinct.*

Proof. If some column (say the ith) of $\mathbf{H}$ is $\mathbf{0}$, then if $\mathbf{e}$ is $0 \cdots 010 \cdots 0$ (ith digit 1) and $\mathbf{c}$ is any code word, $\mathbf{H}(\mathbf{c}+\mathbf{e})^T = \mathbf{0}$, so $\mathbf{c}+\mathbf{e}$ appears to be a code word and an error in the ith digit will not be detected at all. Also, if ith column = jth column = $\mathbf{S}$, we cannot tell if the error is in the ith or jth digit. Conversely, if the columns of $\mathbf{H}$ are nonzero and distinct, then by the previously established case 2, $\mathbf{H}$ will correctly decode all single errors. ∎

We conclude this section by pointing out how to go from generator matrix to parity-check matrix, or vice versa, for the same code. The matrices $\mathbf{G}$ and

Sec. 5.4 Generator and Parity-Check Matrices

H of (5.4.4) and (5.4.11) provide an example. When the goal is error correction, it is best to start with the parity-check matrix and then find the corresponding generator matrix, because the parity-check matrix visibly reveals whether the code will correct errors (Theorem 5.4.1). The following proof involves distributive and associative laws for matrix multiplication discussed at the beginning of Chapter 7. Also, recall that $\mathbf{I}_p$ denotes a $p \times p$ identity matrix.

Theorem 5.4.2. (a) *If* $\mathbf{A}$ *is an* $m \times (n-m)$ *matrix over* $\mathbb{B}$, *so that*

$$\mathbf{G} = [\mathbf{I}_m \quad \mathbf{A}]$$

is an $m \times n$ *generator matrix, then*

$$\mathbf{H} = [\mathbf{A}^T \quad \mathbf{I}_{n-m}]$$

is the unique parity-check matrix for the same code.
  (b) *If* $\mathbf{B}$ *is an* $(n-m) \times m$ *matrix over* $\mathbb{B}$, *so that*

$$\mathbf{H} = [\mathbf{B} \quad \mathbf{I}_{n-m}]$$

is an $(n-m) \times n$ *parity-check matrix, then*

$$\mathbf{G} = [\mathbf{I}_m \quad \mathbf{B}^T]$$

is the unique generator matrix for the same code.

Proof. If we assume $\mathbf{G}$ given, the code encodes $\mathbf{w}$ to $\mathbf{wG}$. Also,

$$\mathbf{H} \cdot (\mathbf{wG})^T = \mathbf{H}(\mathbf{G}^T \mathbf{w}^T) = (\mathbf{HG}^T)\mathbf{w}^T = \left([\mathbf{A}^T \quad \mathbf{I}_{n-m}]\begin{bmatrix}\mathbf{I}_m \\ \mathbf{A}^T\end{bmatrix}\right)\mathbf{w}^T$$
$$= (\mathbf{A}^T + \mathbf{A}^T)\mathbf{w}^T = \mathbf{0} \cdot \mathbf{w}^T = \mathbf{0}.$$

Since the product is $\mathbf{0}$ and $\mathbf{wG}$ has $\mathbf{w}$ as its first m digits, $\mathbf{H}$ as parity-check matrix encodes $\mathbf{w}$ to $\mathbf{wG}$; $\mathbf{G}$ and $\mathbf{H}$ give the same code. Conversely, if we assume $\mathbf{H}$ given, then $\mathbf{H}$ determines an encoding of $\mathbf{w}$ to the unique word $E(\mathbf{w})$ with $\mathbf{w}$ as its first m digits, and with remaining $n-m$ digits such that $\mathbf{H} \cdot (E(\mathbf{w}))^T = \mathbf{0}$. But $\mathbf{wG}$ has $\mathbf{w}$ as its first m digits, and

$$\mathbf{H} \cdot (\mathbf{wG})^T = \mathbf{H}(\mathbf{G}^T \mathbf{w}^T) = (\mathbf{HG}^T)\mathbf{w}^T = \left([\mathbf{B} \quad \mathbf{I}_{n-m}]\begin{bmatrix}\mathbf{I}_m \\ \mathbf{B}\end{bmatrix}\right)\mathbf{w}^T$$
$$= (\mathbf{B} + \mathbf{B})\mathbf{w}^T = \mathbf{0} \cdot \mathbf{w}^T = \mathbf{0},$$

so $\mathbf{wG} = E(\mathbf{w})$.

The ith row of $\mathbf{G}$ is the code word $\mathbf{wG}$ for the message $\mathbf{w}$, which has the ith as its only nonzero digit, so changing $\mathbf{G}$ certainly changes the code given by $\mathbf{G}$; $\mathbf{G}$ is unique for a given code. We have seen that $\mathbf{H}$ determines $\mathbf{G}$, so $\mathbf{H}$ is also unique. ∎

5.5 Exercises

1. Assume that we transmit a 14-digit code word through a binary symmetric channel, with $p = .01$.
  (a) What is the probability that exactly five errors will occur?
  (b) What is the probability that five or fewer errors will occur?

2. Assume that we are using the $(11, 10)$ parity-check code, with $p = .01$, and that we transmit one 11-digit code word. (We observed in Sec. 5.2 that single errors are detected and double errors are missed.)
  (a) What about triple errors? Are they detected or are they missed?
  (b) What about quadruple errors?
  (c) What is the probability that an error will go undetected in our 11-digit block?

3. Assume that static is terrible, so that p is very high, say $p = .4$.
  (a) What is the probability of receiving and decoding a digit d correctly, using triple repetition?
  (b) Answer the question of part (a) but use five-times repetition.
  (c) Using a computer or small calculator, try to find an integer $l \geq 3$ such that $(2l + 1)$-times repetition will enable us to transmit and receive any digit d with 99 per cent accuracy.

*4. We give here three examples of encoding functions. For each, find the minimum distance between code words. Tell how many errors the code will detect, and how many errors the code will correct.
  (a) $E_1: \mathbb{B}^3 \to \mathbb{B}^8$, defined by

$$000 \to 00000000, \quad 011 \to 01101111, \quad 110 \to 11011011,$$
$$001 \to 00100101, \quad 100 \to 10010010, \quad 111 \to 11111100.$$
$$010 \to 01001010, \quad 101 \to 10110111,$$

  (b) $E_2: \mathbb{B}^2 \to \mathbb{B}^{10}$, defined by

$$00 \to 0000000000, \quad 10 \to 1111100000,$$
$$01 \to 0000011111, \quad 11 \to 1111111111.$$

  (c) $E_3: \mathbb{B}^3 \to \mathbb{B}^6$, defined by

$$000 \to 000000, \quad 010 \to 010101, \quad 100 \to 100110, \quad 110 \to 110011,$$
$$001 \to 001011, \quad 011 \to 011110, \quad 101 \to 101101, \quad 111 \to 111000.$$

*5. Use the matrix **H** of (5.4.11) to decode the received words
$$101101 \quad 110010 \quad 010000 \quad 111111 \quad 100101.$$

Sec. 5.5 Exercises

*6. Define a code $E: \mathbb{B}^3 \to \mathbb{B}^6$ with the parity-check matrix

$$\begin{bmatrix} 1 & 0 & 0 & 1 & 0 & 0 \\ 1 & 1 & 0 & 0 & 1 & 0 \\ 1 & 1 & 1 & 0 & 0 & 1 \end{bmatrix}.$$

List all eight code words. Does this code correct all single errors?

*7. Show that the rows of a generator matrix $\mathbf{G}$ are the code words $E(100\cdots 0), E(010\cdots 0), E(0010\cdots 0), \ldots, E(00\cdots 01)$. Then show that every code word is a sum of some of the rows of $\mathbf{G}$.

*8. Show that

$$\mathbf{H} = [1 \quad 1 \quad 1 \quad 1 \quad \cdots \quad 1 \quad 1]_{1 \times (m+1)}$$

is the parity-check matrix and

$$\mathbf{G} = \begin{bmatrix} 1 & 0 & 0 & \cdots & 0 & 1 \\ 0 & 1 & 0 & \cdots & 0 & 1 \\ 0 & 0 & 1 & \cdots & 0 & 1 \\ \vdots & & & & & \vdots \\ 0 & 0 & 0 & \cdots & 1 & 1 \end{bmatrix}_{m \times (m+1)}$$

the generator matrix for the $(m+1, m)$ single parity-check code of Section 5.2.

*9. Show that

$$\mathbf{G} = [1 \quad 1 \quad 1 \quad 1 \quad \cdots \quad 1 \quad 1]_{1 \times (m+1)}$$

is the generator matrix and

$$\mathbf{H} = \begin{bmatrix} 1 & 1 & 0 & 0 & \cdots & 0 \\ 1 & 0 & 1 & 0 & \cdots & 0 \\ 1 & 0 & 0 & 1 & \cdots & 0 \\ \vdots & & & & & \vdots \\ 1 & 0 & 0 & 0 & \cdots & 1 \end{bmatrix}_{m \times (m+1)}$$

the parity-check matrix for an $(m+1, 1)$ $(m+1)$-times repetition code. (Note the similarity between the matrices in Exercises 8 and 9. When an (n, m) code has generator matrix $\mathbf{G} = [\mathbf{I}_m \quad \mathbf{A}]$ and parity-check matrix $\mathbf{H} = [\mathbf{A}^T \quad \mathbf{I}_{n-m}]$, the $(n, n-m)$ code with parity-check matrix $[\mathbf{A} \quad \mathbf{I}_m]$ and generator matrix $[\mathbf{I}_{n-m} \quad \mathbf{A}^T]$ is called the *dual code*.)

10. (a) Suppose that we transmit n-digit code words, and our code will correct all error patterns of weight $\leq t$. Find a formula for the probability that some erroneous received word will not be corrected.

(b) Using a computer, evaluate your formula in (a) for a number of sets of values n, p, and t of your choice.

(c) If n, p, and a small number $\varepsilon > 0$ are given, find an expression showing how large t must be, in order that the formula probability be $< \varepsilon$. For which values of ε will this be impossible? (Use Theorem 5.3.2.)

11. Illustrate Theorems 5.3.1 and 5.3.2 with specific nontrivial examples, showing error patterns that can and cannot be detected, can and cannot be corrected.

12. (a) Find the generator matrix of the code of Exercise 6.

(b) Find generator and parity-check matrices for the dual codes (see Exercise 9) to the codes of Exercises 5 and 6. Do they correct all single errors?

*13. Define a code $E: \mathbb{B}^2 \to \mathbb{B}^5$ with the generator matrix

$$\begin{bmatrix} 1 & 0 & 1 & 0 & 1 \\ 0 & 1 & 0 & 1 & 1 \end{bmatrix}.$$

Does this code correct all single errors? Does its dual code correct all single errors? Use the code to decode the received words

       01100   10110   11011   10001   00110   01001   00011.

14. Find a code $E: \mathbb{B}^2 \to \mathbb{B}^6$ with minimum distance 4. If $p = .01$, what is the probability that an erroneous six-digit received word will go undetected? (Use the fact that some error patterns of weight ≥ 4 are detected, some are not.)

15. Describe the generator matrices $\mathbf{G}$ whose associated parity-check matrices $\mathbf{H}$ will correctly decode all words with no more than one error. (See Theorem 5.4.1.)

5.6 Group Codes

In the next sections we shall make powerful use of the fact that $\mathbb{B}^n$ is a *group*, not just a *set*, during our further study of binary block codes.

Lemma 5.6.1. *The encoding function* $E: \mathbb{B}^m \to \mathbb{B}^n$ *given by either a generator matrix* $\mathbf{G}$ *or a parity-check matrix* $\mathbf{H}$ *is a group homomorphism.*

Proof. In the first case E is right multiplication by the matrix $\mathbf{G}$, so $E(\mathbf{w}) = \mathbf{w}\mathbf{G}$ and we have

$$E(\mathbf{w}_1 + \mathbf{w}_2) = (\mathbf{w}_1 + \mathbf{w}_2)\mathbf{G} = \mathbf{w}_1\mathbf{G} + \mathbf{w}_2\mathbf{G} = E(\mathbf{w}_1) + E(\mathbf{w}_2),$$

proving that E is a group homomorphism.

In the second case, $E(\mathbf{w})$ is the unique word whose first m digits are $\mathbf{w}$ and which satisfies $\mathbf{H} \cdot E(\mathbf{w})^T = \mathbf{0}$. Hence the first m digits of $E(\mathbf{w}_1) + E(\mathbf{w}_2)$ are the m digits of $\mathbf{w}_1 + \mathbf{w}_2$, and

$$\mathbf{H} \cdot (E(\mathbf{w}_1) + E(\mathbf{w}_2))^T = \mathbf{H} \cdot (E(\mathbf{w}_1)^T + E(\mathbf{w}_2)^T) = \mathbf{H} \cdot E(\mathbf{w}_1)^T + \mathbf{H} \cdot E(\mathbf{w}_2)^T$$

$$= \mathbf{0} + \mathbf{0} = \mathbf{0},$$

proving that

$$E(\mathbf{w}_1) + E(\mathbf{w}_2) = E(\mathbf{w}_1 + \mathbf{w}_2). \blacksquare$$

Sec. 5.7 Coset Decoding

A code with encoding function $E: \mathbb{B}^m \to \mathbb{B}^n$ is a **group code** (or **linear code**; see Chapter 9) if the set im E of all code words is a subgroup of $\mathbb{B}^n$.

Corollary 5.6.1. *Any code obtained from a generator matrix or parity-check matrix is a group code.*
Proof. Lemma 5.6.1 and Theorem 4.7.1. ∎

We saw in Theorems 5.3.1 and 5.3.2 that the ability of a code to detect or correct errors depends on the minimum distance between two code words. This minimum distance is easier to compute in a group code, because of the next theorem.

Theorem 5.6.1. *In a group code, the minimum distance between code words is the minimum weight of a nonzero code word.*
Proof. Let d be the minimum distance between code words, say $d = d(\mathbf{a}, \mathbf{b})$, where $\mathbf{a}$ and $\mathbf{b}$ are distinct code words. Let w be the minimum weight of a nonzero code word, say $w = w(\mathbf{c})$; $\mathbf{a} \neq \mathbf{b}$ implies that $\mathbf{a} + \mathbf{b} \neq \mathbf{0}$, so $d = d(\mathbf{a}, \mathbf{b}) = w(\mathbf{a} + \mathbf{b}) \geq w$. Also, $\mathbf{0}$ is a code word, since any subgroup of $\mathbb{B}^n$ contains the identity element of $\mathbb{B}^n$; hence $w = w(\mathbf{c}) = w(\mathbf{c} + \mathbf{0}) = d(\mathbf{c}, \mathbf{0}) \geq d$. We proved that $d \geq w$ and $w \geq d$, so $d = w$. ∎

For an example, consider the code $E: \mathbb{B}^3 \to \mathbb{B}^6$ of Section 5.4, with generator and parity-check matrices given by (5.4.4) and (5.4.11), respectively. The code words are actually those of Exercise 5.5.4(c). The weights of nonzero code words are 3, 3, 4, 3, 4, 4, and 3, the minimum of which is 3; so the code has a minimum distance of 3.

5.7 Coset Decoding

Every group code can be decoded by the following tabular procedure. We illustrate the procedure with the example of Section 5.4.

STEP 1. List the words forming the group C of all code words in a row, with $000 \cdots 0$ first.

000000 100110 010101 001011 110011 101101 011110 111000

STEP 2. Choose a word $\mathbf{x}$ in $\mathbb{B}^n$ of least weight among those not yet used, and list the left coset $\mathbf{x} + C$ as the next row, $\mathbf{x} + \mathbf{c}$ appearing below $\mathbf{c}$ for every $\mathbf{c} \in C$.

000000 100110 010101 001011 110011 101101 011110 111000
100000 000110 110101 101011 010011 001101 111110 011000

STEP 3. Repeat step 2 until all elements of $\mathbb{B}^n$ are exhausted. (Since the cosets are equivalence classes on $\mathbb{B}^n$, each element of $\mathbb{B}^n$ will be used exactly once.) This display, in Table 5.7.1, is called a **decoding table**, or **Slepian's standard array**.

STEP 4. Decode each received word as the code word at the top of its column.

TABLE 5.7.1. Decoding table for code of Section 5.4.

000000	100110	010101	001011	110011	101101	011110	111000
100000	000110	110101	101011	010011	001101	111110	011000
010000	110110	000101	011011	100011	111101	001110	101000
001000	101110	011101	000011	111011	100101	010110	110000
000100	100010	010001	001111	110111	101001	011010	111100
000010	100100	010111	001001	110001	101111	011100	111010
000001	100111	010100	001010	110010	101100	011111	111001
100001	000111	110100	101010	010010	001100	111111	011001

For example, the received words

110110   001111   101111   101010   101000   001000   011110

would be decoded as

100110   001011   101101   001011   111000   000000   011110

and we would conclude that the original message was

100   001   101   001   111   000   011

(each code word is the code for its first three digits).

The **x**'s chosen (the ones in the first column) are the **coset leaders** in their respective cosets. We say that we are *decoding by coset leaders*. Sometimes there is a unique choice for the coset leader; in the first through seventh rows of Table 5.7.1, the coset leader is the unique word of smallest weight in its coset. But in the eighth row, there are two other choices we could have made; 010010 and 001100 have the same minimum weight, 2, as does 100001.

This decoding by coset leaders always gives an answer, for every received message. The following theorem shows that the answer this method gives is always as good as any other; it is not necessarily better. It can even be applied for multiple-error-correcting codes; see Chapter 9 for examples.

Theorem 5.7.1. *During decoding by coset leaders, every received word* **r** *is decoded as a code word* **c** *such that* $d(\mathbf{r}, \mathbf{c}) \leq d(\mathbf{r}, \mathbf{b})$, *all code words* **b**. [That is, any other decoding of **r** would assume at least as many errors, and therefore (for p close to 0) be at least as unlikely.]

Proof. Let **b** be another code word. Let **x** be the coset leader in the coset containing **r**. Then

$$d(\mathbf{r}, \mathbf{c}) = w(\mathbf{r} + \mathbf{c}) = w(\mathbf{x}),$$

Sec. 5.7 Coset Decoding

since $\mathbf{r} = \mathbf{x} + \mathbf{c}$ and $\mathbf{x} = \mathbf{r} + \mathbf{c}$. We have $d(\mathbf{r}, \mathbf{b}) = w(\mathbf{r} + \mathbf{b})$, where

$$\mathbf{r} + \mathbf{b} = (\mathbf{x} + \mathbf{c}) + \mathbf{b} = \mathbf{x} + (\mathbf{c} + \mathbf{b}) \in \mathbf{x} + C,$$

C the group of code words. (The decoding table looks like Figure 5.7.1.)

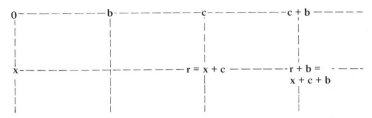

FIGURE 5.7.1. Decoding Table

Since $\mathbf{x}$ was chosen of minimal weight in its coset, $w(\mathbf{x}) \leq w(\mathbf{r} + \mathbf{b})$; therefore,

$$d(\mathbf{r}, \mathbf{c}) = w(\mathbf{x}) \leq w(\mathbf{r} + \mathbf{b}) = d(\mathbf{r}, \mathbf{b}). \blacksquare$$

Lemma 5.7.1. *If, for a parity-check matrix* $\mathbf{H}$, *the group code* C *is* $\{\mathbf{c} \in \mathbb{B}^n | \mathbf{H} \cdot \mathbf{c}^T = \mathbf{0}\}$, *then words* $\mathbf{x}$ *and* $\mathbf{y}$ *in* $\mathbb{B}^n$ *are in the same coset iff they have the same syndrome.*

Proof. $\mathbf{x}$ and $\mathbf{y}$ are in the same coset iff $\mathbf{x} = \mathbf{y} + \mathbf{c}$, some $\mathbf{c} \in C$; iff $\mathbf{x} + \mathbf{y} = \mathbf{c}$, some $\mathbf{c} \in C$; iff $\mathbf{x} + \mathbf{y} \in C$ iff $\mathbf{H}(\mathbf{x} + \mathbf{y})^T = \mathbf{0}$ iff $\mathbf{H}(\mathbf{x}^T + \mathbf{y}^T) = \mathbf{0}$ iff $\mathbf{H}\mathbf{x}^T + \mathbf{H}\mathbf{y}^T = \mathbf{0}$ iff $\mathbf{H}\mathbf{x}^T = \mathbf{H}\mathbf{y}^T$. $\blacksquare$

Note that Theorem 5.7.1 and Lemma 5.7.1 hold for any group code whatever. But the code will correct all single errors iff all error patterns of weight 1 are coset leaders, which will be true iff the columns of H are nonzero and distinct (Theorem 5.4.1).

When we add the syndrome (written as rows instead of columns) at the left, Table 5.7.1 becomes Table 5.7.2. When programming a computer to do this decoding, we need only store the first two columns, syndrome and coset leader, in the computer memory. If the computer computes syndrome 011,

TABLE 5.7.2. Table 5.7.1 with syndromes.

000	000000	100110	010101	001011	110011	101101	011110	111000
110	100000	000110	110101	101011	010011	001101	111110	011000
101	010000	110110	000101	011011	100011	111101	001110	101000
011	001000	101110	011101	000011	111011	100101	010110	110000
100	000100	100010	010001	001111	110111	101001	011010	111100
010	000010	100100	010111	001001	110001	101111	011100	111010
001	000001	100111	010100	001010	110010	101100	011111	111001
111	100001	000111	110100	101010	010010	001100	111111	011001

for example, from received word **r**, it is supposed to decode **r** as **c** = **r**+**x**, where **x** = 001000; so it just adds 001000 to **r** to get **c**, and does not need the rest of the decoding table.

This argument is based on the fact that *when one is decoding by coset leaders, the error patterns corrected are the coset leaders themselves*. In particular: decoding by coset leaders will correct all single errors iff all words of weight 1 are coset leaders, will correct all single and double errors iff all words of weight ≤ 2 are coset leaders, and so on. This method of decoding is often used for double- or triple-error-correcting codes, whenever the two columns (syndrome and coset leader) are not too long for storage.

5.8 Hamming Codes

Recall Theorem 5.4.1: A code $E: \mathbb{B}^m \to \mathbb{B}^n$ defined by a parity-check matrix **H** will correct all single errors iff the columns of **H** are nonzero and distinct. Denoting $n - m = k$, the $k \times n$ matrix **H** amounts to k parity-check equations determining an $(n, n-k)$ code. The number of information digits is $m = n - k$. For a fixed number k of parity-check equations, we want to send as much information as possible, so we make the number n of columns of **H** as large as possible; that is, $n = 2^k - 1$, the number of different nonzero k-digit columns (the binary representations of the numbers $1, 2, \ldots, 2^k - 1$). Such a code is called a **Hamming code**.

EXAMPLE 1: $k = 2$. If $k = 2$, then $n = 2^2 - 1 = 3$ and

$$\mathbf{H} = \begin{bmatrix} 1 & 1 & 0 \\ 1 & 0 & 1 \end{bmatrix}.$$

As seen in Exercise 5.5.9, this is just the (3, 1) triple-repetition code.

EXAMPLE 2: $k = 3$. Here $n = 2^3 - 1 = 7$ and

$$\mathbf{H} = \begin{bmatrix} 1 & 1 & 0 & 1 & 1 & 0 & 0 \\ 1 & 0 & 1 & 1 & 0 & 1 & 0 \\ 0 & 1 & 1 & 1 & 0 & 0 & 1 \end{bmatrix}.$$

This is the Hamming (7, 4) code, single-error-correcting with rate $R = \frac{4}{7}$, considerably better than the rate $\frac{1}{3}$ for the triple-repetition code.

EXAMPLE 3: $k = 4$. Here $n = 2^4 - 1 = 15$ and

$$\mathbf{H} = \begin{bmatrix} 1 & 1 & 1 & 0 & 0 & 0 & 1 & 1 & 1 & 0 & 1 & 1 & 0 & 0 & 0 \\ 1 & 0 & 0 & 1 & 1 & 0 & 1 & 1 & 0 & 1 & 1 & 0 & 1 & 0 & 0 \\ 0 & 1 & 0 & 1 & 0 & 1 & 1 & 0 & 1 & 1 & 1 & 0 & 0 & 1 & 0 \\ 0 & 0 & 1 & 0 & 1 & 1 & 0 & 1 & 1 & 1 & 0 & 0 & 0 & 0 & 1 \end{bmatrix}.$$

This is the Hamming (15, 11) code, with rate $\frac{11}{15}$.

Sec. 5.8 Hamming Codes

In general, for every $k \geq 2$ we get a $(2^k - 1, 2^k - 1 - k)$ Hamming code, with rate

$$\frac{2^k - 1 - k}{2^k - 1} = 1 - \frac{k}{2^k - 1}.$$

In fact, for each k we get a whole family of Hamming codes, depending on how we arrange the nonzero columns within $\mathbf{H}$. In practice, the Hamming code is usually used with the nonzero columns in normal binary order; for $k = 3$, use

(5.8.1) $$\mathbf{H} = \begin{bmatrix} 0 & 0 & 0 & 1 & 1 & 1 & 1 \\ 0 & 1 & 1 & 0 & 0 & 1 & 1 \\ 1 & 0 & 1 & 0 & 1 & 0 & 1 \end{bmatrix}.$$

This is not a parity-check matrix in the sense of our definition (it will be in the sense of the more general Chapter 9 definition), but is used in the same way. The columns of the identity matrix appear in the first, second, and fourth positions instead of the fifth, sixth, and seventh, so we use these components for check digits; given a message $\mathbf{w} = w_1 w_2 w_3 w_4$, the code is $\mathbf{c} = E(\mathbf{w}) = c_1 c_2 w_1 c_3 w_2 w_3 w_4$, where

(5.8.2) $$\mathbf{H} \cdot E(\mathbf{w})^T = \mathbf{0}.$$

For any larger value of k, the check digits again appear in positions $1, 2, 4, \ldots, 2^{k-1}$, the message digits in the other positions of $E(\mathbf{w})$, and (5.8.2) holds; decoding with syndromes proceeds as usual.

We saw in Section 5.7 that since each Hamming code

$$E: \mathbb{B}^{2^k - 1 - k} \to \mathbb{B}^{2^k - 1}$$

is single-error-correcting, all the words of weight 1 are coset leaders in a decoding table. There are $2^k - 1$ such words of weight 1. If $C = E(\mathbb{B}^{2^k - 1 - k})$ is the group of code words, then

$$|\mathbb{B}^{2^k - 1} : C| = \frac{|\mathbb{B}^{2^k - 1}|}{|C|} = \frac{2^{2^k - 1}}{2^{2^k - 1 - k}} = 2^k,$$

so there are 2^k cosets; the only coset leaders are the $2^k - 1$ of weight 1, plus $000 \cdots 0$. Every word is thus at distance ≤ 1 from a code word. Each Hamming code is a *perfect* single-error-correcting code, in the following sense.

An (n, m) block code is a **perfect t-error-correcting code** if it corrects all error patterns of weight t or less, and no others.

Suppose that an (n, m) group code is perfect t-error-correcting; then the number of code words is 2^m, so the number of cosets is $2^n / 2^m = 2^{n-m}$. The number of n-digit error patterns with t or fewer errors is, from Theorem 5.1.1, $\binom{n}{0}+\binom{n}{1}+\binom{n}{2}+\cdots+\binom{n}{t}$, and each error pattern determines a coset. So for a perfect t-error-correcting code, we must have

$$\binom{n}{0}+\binom{n}{1}+\binom{n}{2}+\cdots+\binom{n}{t}=2^{n-m}.$$

For $t>1$ the left side of this equation is very rarely a power of 2. Even if it is, only quite a coincidence will make the set of words of weight $\leq t$ precisely a set of coset leaders for some subgroup of $\mathbb{B}^n$. So we would expect perfect codes to be very rare.

In fact, perfect codes were recently all determined. The only perfect binary group codes are the $(2t+1, 1)$ $(2t+1)$-times repetition codes, the Hamming codes, and *one more*: a $(23, 12)$ perfect three-error-correcting group code discovered by Golay (1949). Its existence is possible because

$$\binom{23}{0}+\binom{23}{1}+\binom{23}{2}+\binom{23}{3}=1+23+253+1771=2048=2^{11}=2^{23-12}.$$

Perfect codes will be discussed a bit more in Chapter 9.

Let us see how effective the Hamming $(7, 4)$ code is at correcting errors in our 10,000-digit message of Section 5.1, with $p=.001$. Each four-digit block becomes a seven-digit block for transmission. The probability of no errors in sending that seven-digit block is $(.999)^7$, and the probability of one error is $\binom{7}{1}(.999)^6(.001)$, so the probability of correctly decoding the four-digit block is

$$(.999)^7+7(.999)^6(.001)=(.999)^6(1.006)=.999979,$$

approximately. There are 2500 such blocks, so the probability of correctly decoding the whole message is about $(.999979)^{2500}$, or approximately $.9489$, about 18 out of 19. So we did not do quite as well as with triple repetition in Section 5.2; but we have a better rate, transmitting only 7500 extra check digits instead of 20,000.

What about the Golay $(23, 12)$ code? In a 23-digit block, the probability of correct decoding is

$$(.999)^{23}+\binom{23}{1}(.999)^{22}(.001)+\binom{23}{2}(.999)^{21}(.001)^2$$

$$+\binom{23}{3}(.999)^{20}(.001)^3=.9999999913, \text{ approximately.}$$

Sec. 5.9 Extended Hamming Codes

We have to send $10^4/12 =$ about 834 23-digit blocks, so the probability of correct decoding is about $(.9999999913)^{834}$, approximately .99999274. The chance of an error in the decoded message is about 1 in 137,000! This is much better than even five-times repetition (Section 5.2), and we need send only about 9170 check digits instead of 40,000. This example illustrates the power of some of the codes to be discussed in Chapter 9.

5.9 Extended Hamming Codes

In practice, the Hamming code is often used in an extended form. We add an "overall parity check" to the **H**-matrix, increasing the minimum distance from 3 to 4. (See Exercise 5.10.10.) For the matrix (5.8.1), the extended matrix is

(5.9.1) $$\bar{\mathbf{H}} = \begin{bmatrix} 0 & 0 & 0 & 1 & 1 & 1 & 1 & 0 \\ 0 & 1 & 1 & 0 & 0 & 1 & 1 & 0 \\ 1 & 0 & 1 & 0 & 1 & 0 & 1 & 0 \\ 1 & 1 & 1 & 1 & 1 & 1 & 1 & 1 \end{bmatrix}.$$

We add a bottom row of 1's to provide the overall parity-check equation $a_1 + a_2 + \cdots + a_n + a_{n+1} = 0$, where the Hamming code word was $a_1 a_2 \cdots a_n$; the last column has 0's above the last row, so the parity-check equations obtained from **H** are also obtained from $\bar{\mathbf{H}}$.

If $a_1 a_2 \cdots a_n$ is a nonzero Hamming code word of minimum weight, its weight is 3, and then $a_{n+1} = 1$, so nonzero extended Hamming code words have weight ≥ 4. This extended Hamming code is not only able to correct all single errors, but also to detect double errors (not erroneously correct them as though they were some other single errors). We can decode an eight-digit received word **r**, using the matrix $\bar{\mathbf{H}}$ of (5.9.1), as follows. Let

$$\bar{\mathbf{H}} \cdot \mathbf{r}^T = \begin{bmatrix} s_1 \\ s_2 \\ s_3 \\ d \end{bmatrix}, \qquad \mathbf{S} = \begin{bmatrix} s_1 \\ s_2 \\ s_3 \end{bmatrix};$$

S is called the **syndrome** (it is the product of **H** and the first seven digits of $\mathbf{r}^T$), d the **overall check symbol**.

1. If $d = 1$, $\bar{\mathbf{H}} \cdot \mathbf{r}^T$ is the ith column of $\bar{\mathbf{H}}$ for some i, so we assume a single error in the ith digit.
2. If $d = 0$ and $\mathbf{S} = \mathbf{0}$, assume that no error occurred.
3. If $d = 0$ and $\mathbf{S} \neq \mathbf{0}$, $\bar{\mathbf{H}} \cdot \mathbf{r}^T$ is not a column of $\bar{\mathbf{H}}$, so an uncorrectable (probably double) error has been detected.

5.10 Exercises

1. Decode the message

010110   110001   101110   000110   101111   011010   101001

using Table 5.7.2.

2. Construct a complete decoding table (with syndromes) for the code of Exercise 5.5.13, and use it to decode the message

10001   11101   01011   00101   10110.

3. Let $E: \mathbb{B}^4 \to \mathbb{B}^7$ be the (7, 4) Hamming code defined by the matrix $\mathbf{H}$ of (5.8.1). Find a 4×7 matrix $\mathbf{G}$ such that $E(\mathbf{w}) = \mathbf{w}\mathbf{G}$, all $\mathbf{w} \in \mathbb{B}^4$.

*4. Use the Hamming (7, 4) code to encode the message

0110   1011   0010   1110   0101.

5. Assume that the received words are

0101101   1101001   1000111.

Decode using the Hamming (7, 4) code.

6. Construct a two-column decoding table (syndrome and coset leader) for the Hamming (7, 4) code.

7. Encode the message in Exercise 4 using the extended Hamming (8, 4) code.

8. Assume that the received words are

11001101   10011101   10001100   00110110   01101101.

Decode using the extended Hamming (8, 4) code.

*9. (a) Show that in a group code, the sum of two code words of even weight is a code word of even weight.

(b) Show that the set E of code words of even weight is a subgroup of the group C of all code words.

(c) Suppose there is a code word $\mathbf{d}$ of odd weight. Show that any other code word $\mathbf{c}$ of odd weight satisfies $\mathbf{c} = \mathbf{d} + \mathbf{e}$, some $\mathbf{e} \in E$.

(d) Use part (c) to show that there are at most two left cosets of E in C.

(e) Conclude that either all code words have even weight, or half have even weight and half have odd weight.

*10. Let $E: \mathbb{B}^m \to \mathbb{B}^n$ be a group code, $E_1: \mathbb{B}^n \to \mathbb{B}^{n+1}$ the $(n+1, n)$ parity-check code. Show that $E_1 \circ E$ is a group code. Show that if the minimum distance for E is an odd number $2l+1$, then $E_1 \circ E$ has minimum distance $2l+2$. What is $E_1 \circ E$ when E is a Hamming code?

11. Prove that the decoding procedure for extended Hamming codes does correct all single errors and identify as uncorrectable all double errors.

12. Let C be *any* proper subgroup of $\mathbb{B}^n$.

(a) Use Lagrange's theorem to show that $|C| = 2^m$ for some m with $0 < m < n$.

(b) Use the Fundamental Theorem on Finite Abelian Groups to show that there exist $c_1, c_2, \ldots, c_m \in C$, such that every element of C has a unique expression as a sum $c_{i_1} + c_{i_2} + \cdots + c_{i_l}$ of some subset $\{c_{i_1}, c_{i_2}, \ldots, c_{i_l}\}$ of $\{c_1, c_2, \ldots, c_m\}$.

(c) Show that the $m \times n$ matrix $\mathbf{G}$ with rows $c_1, c_2, \ldots, c_m$ defines a group code $E: \mathbb{B}^m \to \mathbb{B}^n$ with $E(\mathbf{w}) = \mathbf{wG}$, in which $C = \text{im } E$ is exactly the subgroup of code words.

13. Again consider a 10,000-digit message, with $p = .001$.

(a) Assume that this message is transmitted using the Hamming (15, 11) code, and find the approximate probability that the entire message is decoded without error.

(b) Do the same for several other Hamming $(2^k - 1, 2^k - 1 - k)$ codes, k larger than 4 (use a calculator or computer), and see how this probability decreases. How do you explain this, since all the codes are single-error-correcting?

14. How many bits of computer memory storage are needed to store the syndrome and coset leader columns for the Golay (23, 12) code?

15. Write the matrix $\bar{\mathbf{H}}$ for the extended Hamming (16, 11) code, starting with a matrix $\mathbf{H}$ having its columns in normal binary order. If $E: \mathbb{B}^{11} \to \mathbb{B}^{16}$ is this code, find an 11×16 matrix $\mathbf{G}$ such that $E(\mathbf{w}) = \mathbf{wG}$ for all $\mathbf{w} \in \mathbb{B}^{11}$.

16. Explain why $(2t + 1)$-times repetition is a perfect t-error-correcting code.

17. If $E: \mathbb{B}^m \to \mathbb{B}^n$ and $E_1: \mathbb{B}^n \to \mathbb{B}^r$ are group codes, show that $E_1 \circ E$ is also a group code. If E has generator matrix $\mathbf{G}$ and E_1 has generator matrix $\mathbf{G}_1$, try to find a generator matrix for $E_1 \circ E$ in terms of $\mathbf{G}$ and $\mathbf{G}_1$.

18. Suppose a group code with parity-check matrix $\mathbf{H}$ has minimum weight d. Prove that

(a) The sum of $(d - 1)$ or fewer columns of $\mathbf{H}$ cannot equal zero.

(b) The sum of at least one set of d columns of $\mathbf{H}$ is zero.

19. We assume in this problem that there exists a double-error-correcting group code $E: \mathbb{B}^{14} \to \mathbb{B}^{22}$. (It is true.) In the decoding table for this code, how many coset leaders have weight > 2?

20. Using Exercise 10, show that the (23, 12) three-error-correcting Golay code extends to a (24, 12) code that not only corrects all error patterns of weight ≤ 3 but also detects quadruple errors. What is the probability that an error will go undetected when decoding our 10,000-digit message, $p = .001$, with this code?

5.11 Fast Adders: Winograd's Theory

Modern computers contain numerous circuits for performing addition; computations are necessarily of limited length, so the addition is usually actually modulo n for some large value of n. In this section we discuss the possibility of performing addition faster than with the usual binary expressions; the ideas are due to Winograd (1965).

Adders within computers or calculators are built from devices (usually called gates) with a small number r of inputs and one output. Such devices must have r small for reasons of economy, reliability, and speed; a typical restriction might be $r \leq 4$. Each device needs a certain unit of time (perhaps .0000001 second) to establish its new output, when it receives new inputs.

For a real number x we denote by $\lceil x \rceil$ (read "ceiling of x") the smallest integer $\geq x$; for example, $\lceil -1.5 \rceil = -1$, $\lceil 2 \rceil = 2$, $\lceil \pi \rceil = 4$. The fundamental lemma is as follows:

Lemma 5.11.1. *The time required to compute an m-variable switching function with r-input devices is at least $\lceil \log_r m \rceil$ time units.*

Proof. The final device emitting the output has r inputs, so in two time units, r^2 signals can influence the output. (The case $m = 8, r = 3, \lceil \log_3 8 \rceil = 2$ is pictured in Figure 5.11.1.) In t time units, r^t signals can influence the output, and m signals must, so $r^t \geq m$, $t \geq \log_r m$. ∎

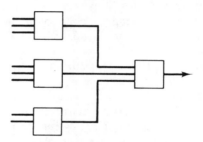

FIGURE 5.11.1. $m = 8, r = 3$ in Lemma 5.11.1

Corollary 5.11.1. *Ordinary binary addition of two m-digit binary numbers requires at least $\lceil \log_r 2m \rceil$ time units.*

Proof. In the sum

(5.11.1)
$$\begin{array}{r} a_{m-1}a_{m-2} \cdots a_1 a_0 \\ +b_{m-1}b_{m-2} \cdots b_1 b_0 \\ \hline s_m s_{m-1} s_{m-2} \cdots s_1 s_0, \end{array}$$

the leftmost digit s_m depends on all $2m$ digits of the numbers being added. (To see this, note that in the sums

$$\begin{array}{r} 1111 \cdots 11 \\ +0000 \cdots 01 \\ \hline 10000 \cdots 00, \end{array} \qquad \begin{array}{r} 0000 \cdots 01 \\ +1111 \cdots 11 \\ \hline 10000 \cdots 00, \end{array}$$

any change in any digit of the first summand of the first sum or the second summand of the second sum changes $s_m = 1$.) Thus s_m is a switching function

Sec. 5.11 Fast Adders: Winograd's Theory

of all $2m$ variables $a_0, a_1, \ldots, a_{m-1}, b_0, b_1, \ldots, b_{m-1}$. Now use Lemma 5.11.1. ∎

Corollary 5.11.2. *Addition modulo n with ordinary binary addition requires at least $\lceil \log_r (2 \lceil \log_2 n \rceil) \rceil$ time units.*
Proof. To write all the integers modulo n as distinct m-digit binary numbers, we need $2^m \geq n$, $m \geq \log_2 n$, $m \geq \lceil \log_2 n \rceil$. Use Corollary 5.11.1. ∎

We now use group theory to prove a theorem of Winograd (1965), which indicates that we might be able to add faster with some kind of addition other than ordinary binary addition. As usual, $\mathbb{Z}_n = \{0, 1, 2, \ldots, n-1\}$ and $[\mathbb{Z}_n, +]$ is the group $\mathbb{Z}_n$ under addition modulo n. By a **representation** of $[\mathbb{Z}_n, +]$ we shall mean a one-one function $\mathbf{h}: \mathbb{Z}_n \to \mathbb{B}^m$, some m, with $\mathbf{h}(0) = 000\cdots00$. (Since $\mathbf{h}$ is one-one, $|\mathbb{Z}_n| \leq |\mathbb{B}^m|$, so $n \leq 2^m$.) An **h-adder** (or, a **modulo-n adder for the representation h**) is a device that computes, for any given $\mathbf{h}(x)$ and $\mathbf{h}(y)$, $x, y \in \mathbb{Z}_n$, the value $\mathbf{h}(x+y)$. [If we have an **h**-adder, we can then hope to find out $x+y$ from $\mathbf{h}(x+y)$; we shall discuss examples later.]

In any group G, a nonidentity element u is **ubiquitous** if it is in every nonidentity subgroup of G.

Theorem 5.11.1 (Winograd). *Assume that $[\mathbb{Z}_n, +]$ has a ubiquitous element u and $\mathbf{h}: \mathbb{Z}_n \to \mathbb{B}^m$ is a representation of $\mathbb{Z}_n$. Then at least one output of any **h**-adder depends on at least $2\lceil \log_2 n \rceil$ distinct inputs.*
Proof. For each $x \in \mathbb{Z}_n$, denote $\mathbf{h}(x) = h(x)_1 h(x)_2 \cdots h(x)_m \in \mathbb{B}^m$. Since $u \neq 0$, $\mathbf{h}(u) \neq 000\cdots 0$, so choose i such that $h(u)_i \neq 0$. We shall show that for $x, y \in \mathbb{Z}_n$, $h(x+y)_i$ depends on at least $\lceil \log_2 n \rceil$ distinct components of $\mathbf{h}(x)$; it will similarly depend on at least $\lceil \log_2 n \rceil$ distinct components of $\mathbf{h}(y)$, as required.

If $h(x+y)_i$ depends on $k < \lceil \log_2 n \rceil$ components of $\mathbf{h}(x)$, then since there are $n > 2^k$ x's there must exist at least two elements $x_1, x_2 \in \mathbb{Z}_n$ such that $\mathbf{h}(x_1)$ and $\mathbf{h}(x_2)$ have these k components identical, and hence such that

(5.11.2) $\qquad h(x_1+y)_i = h(x_2+y)_i, \quad$ all $y \in \mathbb{Z}_n$.

Hence

(5.11.3) $\qquad \begin{aligned} h((x_1-x_2)+y)_i &= h(x_1+(y-x_2))_i = h(x_2+(y-x_2))_i \\ &= h(y)_i, \end{aligned}$

all $y \in \mathbb{Z}_n$, the second equality holding by (5.11.2) with $y - x_2$ in place of y. For $y = 0$, (5.11.3) says that

(5.11.4) $\qquad h(x_1 - x_2)_i = h(0)_i = 0.$

If we set $y = x_1 - x_2$, (5.11.3) also says that

(5.11.5) $\qquad h(2(x_1 - x_2))_i = h(x_1 - x_2)_i = 0.$

Choosing consecutive multiples of $x_1 - x_2$ for y, we find that for the entire subgroup $H = \{k(x_1 - x_2) | k \in \mathbb{Z}_n\}$ of $\mathbb{Z}_n$, any $y \in H$ has $h(y)_i = 0$; but since u is ubiquitous, $u \in H$ and $h(u)_i \neq 0$, which is a contradiction. ∎

Corollary 5.11.3. *If $[\mathbb{Z}_n, +]$ has a ubiquitous element, any **h**-addition of elements of $\mathbb{Z}_n$ requires at least $\lceil \log_r (2 \lceil \log_2 n \rceil) \rceil$ time units.*

Proof. Theorem 5.11.1 and Lemma 5.11.1. ∎

$[\mathbb{Z}_n, +]$ is a cyclic group of order n. We shall use our structure theory of cyclic groups (Section 4.16) to determine the ubiquitous elements in any finite cyclic group. By a **prime power** we mean a positive integer power of a prime integer >1.

Theorem 5.11.2. *A cyclic group C_n of order n has a ubiquitous element iff n is a prime power. If n is a prime power, say $n = p^m$, then the ubiquitous elements in C_n are the $p-1$ nonidentity elements in the unique subgroup of order p.*

Proof. We may assume that $n > 1$. If n is not a prime power, then two different primes p and q divide n; by Corollary 4.16.1, C_n has subgroups C_p of order p and C_q of order q. Any ubiquitous element u satisfies $u \in C_p \cap C_q$, so the order of u must divide both p and q, impossible for a nonidentity element u.

Now suppose that $n = p^m$. If H is any nonidentity subgroup of C_n, then Theorem 4.16.2 shows that H is cyclic. Corollary 4.16.1 shows that both C_n and H have unique subgroups of order p; $H \subseteq C_n$, so these must coincide, proving that nonidentity elements of the unique subgroup of C_n of order p are in H (i.e., are ubiquitous). ∎

Corollary 5.11.4. *Denote by $u(n)$ the order of the largest subgroup of $[\mathbb{Z}_n, +]$ with a ubiquitous element. Then $u(n)$ is the largest prime power dividing n, and **h**-addition of elements of $\mathbb{Z}_n$ requires at least $\lceil \log_r (2 \lceil \log_2 u(n) \rceil) \rceil$ time units.*

Proof. We saw in Section 4.16 that $[\mathbb{Z}_n, +]$ has one subgroup of each order d dividing n, all of them cyclic; therefore, Theorem 5.11.2 implies that $u(n)$ is the largest prime power dividing n. Let H be a subgroup of $[\mathbb{Z}_n, +]$ with order $u(n)$ and a ubiquitous element. The **h**-addition of elements of $\mathbb{Z}_n$ does in particular give a way of adding elements of H; but H is isomorphic to $[\mathbb{Z}_{u(n)}, +]$ by Exercise 4.10.4(a), so that gives a way of adding "in" $\mathbb{Z}_{u(n)}$, bounded in time by Corollary 5.11.3, and addition "in" $\mathbb{Z}_n$ cannot be faster than addition "in" H. ∎

5.12 Fast Adders: Procedures

It is clear from Corollary 5.11.4 that if we hope to do fast addition, we must add modulo n for some n that is not divisible by any large prime

Sec. 5.12 Fast Adders: Procedures

powers. If we restrict ourselves to prime powers <50, we can choose

$$n = 2^5 \cdot 3^3 \cdot 5^2 \cdot 7^2 \cdot 11 \cdot 13 \cdot 17 \cdot 19 \cdot 23 \cdot 29 \cdot 31 \cdot 37 \cdot 41 \cdot 43 \cdot 47$$
$$> 3 \times 10^{21},$$

and with prime powers <100 we can even have

$$n = 2^6 \cdot 3^4 \cdot 5^2 \cdot 7^2 \cdot 11 \cdot \cdots \cdot 89 \cdot 97 > 5 \times 10^{42}.$$

How can we actually carry out such computations? We shall show here that addition modulo these huge numbers n reduces to separate additions modulo the small prime power factors of n. These can then be performed simultaneously, each requiring few time units.

In the rest of this section we assume that $n = n_1 n_2 \cdots n_s$, each n_i a power of a different prime p_i. We denote addition modulo n by $+\limits_n$ and addition modulo n_i by $+\limits_{n_i}$.

Theorem 5.12.1 (Chinese Remainder Theorem). *Define*

$$\theta: \mathbb{Z}_n \to \mathbb{Z}_{n_1} \times \mathbb{Z}_{n_2} \times \cdots \times \mathbb{Z}_{n_s}$$

by $\theta(x) = (x_1, x_2, \ldots, x_s)$, *where x_i is the unique integer $0 \leq x_i < n_i$ such that $x \equiv x_i \pmod{n_i}$. Then θ is an isomorphism of additive groups.*

Proof. Assume that $\theta(x) = (x_1, x_2, \ldots, x_s)$, $\theta(y) = (y_1, y_2, \ldots, y_s)$. To show that θ is a homomorphism, we must show that

(5.12.1) $$\theta(x +\limits_n y) = \theta(x) + \theta(y),$$

where

(5.12.2) $$\theta(x) + \theta(y) = (x_1 +\limits_{n_1} y_1, x_2 +\limits_{n_2} y_2, \ldots, x_s +\limits_{n_s} y_s).$$

Since $x \equiv x_i$ and $y \equiv y_i \pmod{n_i}$, we must have

$$x + y \equiv x_i +\limits_{n_i} y_i \pmod{n_i}.$$

Also, $x + y \equiv x +\limits_n y \pmod{n_i}$, since n divides $(x+y) - (x +\limits_n y)$ and n_i divides n. Therefore,

(5.12.3) $$x +\limits_n y \equiv x_i +\limits_{n_i} y_i \pmod{n_i},$$

proving that (5.12.1) holds.

If $\theta(x) = \theta(y)$, then $x \equiv x_i \equiv y_i \equiv y \pmod{n_i}$ for all i, so $x - y \equiv 0 \pmod{n_i}$ for all i. Therefore, n_i divides $x - y$ for all i, implying that n divides $x - y$; since $x, y \in \mathbb{Z}_n$, $x = y$. This proves that θ is one-one. Since

$$|\mathbb{Z}_n| = n = n_1 n_2 \cdots n_s = |\mathbb{Z}_{n_1} \times \mathbb{Z}_{n_2} \times \cdots \times \mathbb{Z}_{n_s}|,$$

the pigeonhole principle (Theorem 1.9.4) implies that θ is also onto. ∎

The Chinese Remainder Theorem shows that we *can* do our addition modulo the small prime powers n_i, provided that the computer can perform the functions

θ: given x and y, find $\theta(x)$ and $\theta(y)$,

and

θ^{-1}: find $x + y \atop n$ from $\theta(x + y) \atop n$.

[Equations (5.12.1) and (5.12.2) show how to go from $\theta(x)$ and $\theta(y)$ to $\theta(x + y) \atop n$, using the small moduli.]

Both of these tasks can be accomplished if the computer stores a little precalculated information. If x and y are entered as decimal numbers, we can perhaps perform the function θ by having the computer store

$\theta(1), \theta(10), \theta(100), \theta(10^3), \ldots, \theta(10^k)$, where $10^{k+1} > n$.

Then, for example,

$\theta(1,286,439) = \theta(10^6) + 2 \cdot \theta(10^5) + 8 \cdot \theta(10^4) + 6 \cdot \theta(10^3)$
$+ 4 \cdot \theta(100) + 3 \cdot \theta(10) + 9 \cdot \theta(1)$,

the additions and multiplications being performed componentwise modulo the small moduli.

Denote $\mathbf{e}_1 = (1, 0, 0, \ldots, 0), \mathbf{e}_2 = (0, 1, 0, \ldots, 0), \ldots, \mathbf{e}_s = (0, 0, \ldots, 0, 1)$ in $\mathbb{Z}_{n_1} \times \mathbb{Z}_{n_2} \times \cdots \times \mathbb{Z}_{n_s}$. To perform the function θ^{-1}, the computer might prestore the values $\theta^{-1}(\mathbf{e}_1), \theta^{-1}(\mathbf{e}_2), \ldots, \theta^{-1}(\mathbf{e}_s)$; then, for example,

$\theta^{-1}((3, 7, 1, \ldots, 2)) = 3 \cdot \theta^{-1}(\mathbf{e}_1) + 7 \cdot \theta^{-1}(\mathbf{e}_2) + \theta^{-1}(\mathbf{e}_3) + \cdots + 2 \cdot \theta^{-1}(\mathbf{e}_s)$,

the addition being modulo n.

This latter idea ends with several additions to be performed modulo n without the aid of the small moduli, so this general idea of working with small moduli will be most useful when many (hundreds, thousands, etc.) additions are to be performed after coding the given information (using θ) and before decoding (using θ^{-1}) to get the final answer.

5.13 Exercises

1. Does an infinite cyclic group contain any ubiquitous elements?

2. Do any dihedral groups (Section 4.18) contain ubiquitous elements?

3. Which finite abelian groups (Section 4.19) contain ubiquitous elements?

4. Cayley proved that every finite group of order divisible by a prime p has an element of order p. Assuming this, what can you say about finite groups with ubiquitous elements?

Sec. 5.13 Exercises

5. Show that the quaternion group Q_8 of order 8 (Exercise 4.20.8) has a ubiquitous element.

6. Let p be a prime, $G_m = \{\gamma \in \mathbb{C} \mid \gamma^{p^m} = 1\}$, $G = \bigcup_{m=1}^{\infty} G_m$. Show that G is an infinite multiplicative subgroup of $\mathbb{C}$ with a ubiquitous element.

*__7.__ For the following pairs of values of r, the number of inputs to a single device, and t, the number of time units of computation, use Corollary 5.11.4 to estimate the largest n such that one may be able to add modulo n in t time units with r-input devices. [Corollary 5.11.4 estimates $u(n)$, then you estimate the largest n with $u(n)$ no larger.]
 (a) $r = 4, t = 2$.   (b) $r = 2, t = 3$.
 (c) $r = 2, t = 4$.   (d) $r = 3, t = 3$.
 (e) $r = 3, t = 4$.   (f) $r = 4, t = 3$.

8. For the same pairs of values (r, t) as in Exercise 7, what is the largest n such that you can add modulo n with ordinary binary addition?

9. A **full binary adder** is a device with three binary inputs a, b, and c_{in} and two binary outputs s and c_{out}; s is the sum of the inputs modulo 2 and c_{out} is the "carry digit." We can add two m-digit binary numbers $a_{m-1} \cdots a_2 a_1 a_0$ and $b_{m-1} \cdots b_2 b_1 b_0$ by using m full binary adders; the first adds a_0 and b_0 (and an initial 0 carry-in) to output sum s_0 and carry-out c_1, the second adds c_1, a_1, and b_1 to output sum s_1 and carry-out $c_2, \ldots$, the final output being the binary sum $s_m s_{m-1} \cdots s_2 s_1 s_0$. (This is called **ripple-carry** addition.) Assume that each adder is made with two 3-input "gates," each operating separately in the same one time unit, one to produce s and the other to produce c_{out}. Express in terms of t the largest n such that we can add modulo n in t time units, using ripple-carry addition.

10. A somewhat faster binary adder scheme than the ripple-carry is the **carry-lookahead adder**. Each **stage** of the adder has $2v + 1$ inputs for a fixed v. In the case of the first stage the inputs are $a_0, a_1, a_2, \ldots, a_{v-1}$, $b_0, b_1, b_2, \ldots, b_{v-1}$, and $c_{in} = 0$. The first stage produces v sum digits $s_0, s_1, s_2, \ldots, s_{v-1}$ and a carry digit c_{out}. This carry digit becomes c_{in} for the next stage, with other inputs $a_v, a_{v+1}, \ldots, a_{2v-1}, b_v, b_{v+1}, \ldots, b_{2v-1}$ and outputs $s_v, s_{v+1}, \ldots, s_{2v-1}$ and a new carry digit. Thus a carry-lookahead adder with w $(2v+1)$-input stages can add two vw-digit binary numbers. For the following values of v, r, and t, estimate the largest n such that one may be able to add modulo n in t time units, using the carry-lookahead scheme. How does carry-lookahead compare with the theoretical "best" binary addition (Corollary 5.11.1)?
 (a) $r = 3, v = 4, t = 6$.   (b) $r = 3, v = 13, t = 6$.
 (c) $r = 4, v = 7, t = 8$.   (d) $r = 2, v = 7, t = 12$.

11. Take $n = 60 = 4 \cdot 3 \cdot 5$, the largest n with $u(n) = 5$. $\theta: \mathbb{Z}_{60} \to \mathbb{Z}_4 \times \mathbb{Z}_3 \times \mathbb{Z}_5$ is as in Section 5.12.
 (a) Find $\theta(1)$ and $\theta(10)$.
 (b) Find $\theta^{-1}((1, 0, 0))$, $\theta^{-1}((0, 1, 0))$, and $\theta^{-1}((0, 0, 1))$.

(c) Use parts (a) and (b) and the methods of Section 5.12 to perform the addition $17+13+19+7$, modulo 60.

12. If $n = n_1 n_2 \cdots n_s$, the n_i powers of distinct primes, can we also do *multiplication* modulo n by doing it modulo n_i for the different n_i's? Is the θ of Theorem 5.12.1 a ring isomorphism?

13. Use Corollary 4.16.1 to characterize the subgroups of order p of the groups C_n and H of Theorem 5.11.2, and thus show directly that they are indeed the same group.

14. Show that digit s_{m-1} in sum (5.11.1) also depends on all $2m$ of the digits being added.

5.14 Pólya Enumeration Theory

Consider the following problem, illustrated in Figure 5.14.1. We have four spheres fixed into the corners of a square by four rods. We wish to paint each of the four spheres either red or blue. In how many ways can this be done?

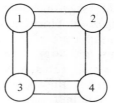

FIGURE 5.14.1. *Example for Section 5.14*

If the square had no symmetries, this would be easy. There are two possibilities for each sphere and hence $2 \cdot 2 \cdot 2 \cdot 2 = 16$ possibilities in all. But painting 1 and 4 red, 2 and 3 blue would not really be different from painting 1 and 4 blue, 2 and 3 red, since by flipping the square over on its vertical axis we have the same figure as before but with the second color arrangement.

The "Pólya enumeration theory" of the next few sections is a twentieth-century discovery used for solving such problems and has found numerous applications to combinatorial problems in computer science and elsewhere. It orginates with Redfield (1927) and Pólya (1937). We shall consider the general theory soon, but first we continue to discuss the example of Figure 5.14.1. We say any two of the possible paintings of the spheres are *equivalent* if a symmetry of the square changes one to the other. We are seeking the number of *equivalence classes* under the *group of symmetries* of the square.

Precisely, let $C = \{1, 2, 3, 4\}$ be the set of four spheres (the "configuration") and let $R = \{\text{red, blue}\}$ be the set of colors (the "range"); the 16 possible paintings are the 16 functions $f: C \to R$. The eight symmetries of the square (studied in Section 4.18) are actually permutations of C, not R^C, the

Sec. 5.14 Pólya Enumeration Theory

set of all functions from C to R. How, then, do they act on R^C? The two paintings we mentioned above were

$$f_1: 1 \to \text{red}, 4 \to \text{red}, 2 \to \text{blue}, 3 \to \text{blue}$$

and

$$f_2: 1 \to \text{blue}, 4 \to \text{blue}, 2 \to \text{red}, 3 \to \text{red},$$

and we said they were equivalent under the permutation $\pi = (1 \; 2)(3 \; 4)$: $1 \to 2, 2 \to 1, 3 \to 4, 4 \to 3$, which results from flipping the square about its vertical axis. It is readily checked that

$$f_1 \circ \pi = f_2, \quad f_2 \circ \pi = f_1;$$

for $(f_1 \circ \pi)(1) = f_1(\pi(1)) = f_1(2) = \text{blue} = f_2(1)$, and so on; that is, we permute in the definition of f_1 the elements of C, leaving the elements of R in place.

This one example suggests that the group G of symmetries of the square, a permutation group on C, acts on R^C on the right; the action $R^C \times G \to R^C$ is defined by $(f, \pi) \to f \circ \pi$, all $f \in R^C$, $\pi \in G$. We saw that this was an action in Section 4.17. (In Exercise 5.15.7 we discuss under what circumstances G can be considered a permutation group on R^C.)

A very important application of Pólya theory will be discussed in Section 5.17. An electronic device with several inputs can produce many different output functions if we permute (and perhaps also complement some signals on) the input wires. We will see that these different functions may be considered equivalent under a certain group of permutations of all functions of the given inputs, and we will study them, using Pólya theory.

Now we start the general theory. G is a permutation group on the finite set C, and R is another finite set. G acts on R^C on the right via the map $(f, \pi) \to f \circ \pi$ from $R^C \times G$ to R^C.

If $f \in R^C$ and $R = \{r_1, r_2, \ldots, r_m\}$, then the **weight** $W(f)$ of f is the expression

$$r_1^{e_1} r_2^{e_2} \cdots r_m^{e_m},$$

e_i the number of elements of C that f sends to r_i. Functions f_1 and f_2 in R^C are **G-equivalent** if there is some $\pi \in G$ such that $f_2 = f_1 \circ \pi$. (We saw in Section 4.17 that G-equivalence is an equivalence relation, whenever a group G acts on a set.)

In the example of Figure 5.14.1, if

$$f_3: 1 \to \text{blue}, 2 \to \text{red}, 3 \to \text{blue}, 4 \to \text{blue}$$

then $W(f_3) = \text{red}^1 \text{blue}^3$. If

$$f_4: 1 \to \text{red}, 2 \to \text{red}, 3 \to \text{red}, 4 \to \text{red}$$

then $W(f_4) = \text{red}^4 \text{blue}^0$ or simply $W(f_4) = \text{red}^4$. Functions f_1 and f_2, defined before, are G-equivalent, with $W(f_1) = W(f_2) = \text{red}^2 \text{blue}^2$. Clearly, f_4 can only be G-equivalent to itself.

Theorem 5.14.1 (Frobenius–Burnside). *Let R and C be finite sets, G a permutation group on C.*

(a) Let $f \in R^C$ and $\pi \in G$, and write π as a product of disjoint cycles on C. Then $f = f \circ \pi$ iff f is constant on each cycle of π.

(b) The number of G-equivalence classes in R^C is

$$(5.14.1) \qquad N(R^C; G) = \frac{1}{|G|} \sum_{\pi \in G} n_\pi,$$

where $n_\pi = |\{f \in R^C | f = f \circ \pi\}|$, the number of elements of R^C not changed by the action of π.

(c) Denote by $l(\pi)$ the number of cycles when $\pi \in G$ is written as a product of disjoint cycles on C. Then

$$(5.14.2) \qquad n_\pi = |R|^{l(\pi)},$$

so

$$(5.14.3) \qquad N(R^C; G) = \frac{1}{|G|} \sum_{\pi \in G} |R|^{l(\pi)}.$$

Proof. (a) Suppose that $f = f \circ \pi$; then $f \circ \pi^2 = f \circ (\pi \circ \pi) = (f \circ \pi) \circ \pi = f \circ \pi = f$, and, in fact, $f \circ \pi^i = f$ for any positive integer i. If $c_1, c_2 \in C$ are in the same cycle B of π, then $c_2 = \pi^i(c_1)$ for some i, so $f(c_2) = f(\pi^i(c_1)) = (f \circ \pi^i)(c_1) = f(c_1)$, proving that f is constant on B. Conversely, if f is constant on every cycle of π, then since any $c \in C$ is in the same cycle as $\pi(c)$, we have $f(c) = f(\pi(c)) = (f \circ \pi)(c)$, all $c \in C$, proving that $f = f \circ \pi$.

(b) Denote $\mathscr{S} = \{(f, \pi) \in R^C \times G | f \circ \pi = f\}$. For each $\pi \in G$ there are n_π pairs $(f, \pi) \in \mathscr{S}$, so $|\mathscr{S}| = \sum_{\pi \in G} n_\pi$.

On the other hand, let $\mathscr{C} = \{f_1, f_2, \ldots, f_e\}$ be one of the G-equivalence classes in R^C. If $G_i = \{\pi \in G | f_i \circ \pi = f_i\}$, then Lemma 4.17.1 says that G_i is a subgroup of G and $|G : G_i| = e$. We have

$$\{(f, \pi) \in \mathscr{S} | f \in \mathscr{C}\} = \bigcup_{i=1}^{e} \{(f_i, \pi) | \pi \in G_i\},$$

so

$$|\{(f, \pi) \in \mathscr{S} | f \in \mathscr{C}\}| = \sum_{i=1}^{e} |G_i| = \sum_{i=1}^{e} \frac{|G|}{e} = |G|$$

for each class $\mathscr{C}$. There are $N(R^C; G)$ classes, so $|\mathscr{S}| = N(R^C; G) \cdot |G|$.

Equating the two expressions for $|\mathscr{S}|$, we get (5.14.1).

(c) We must prove (5.14.2); then substitution in (5.14.1) gives (5.14.3). Fix $\pi \in G$. By (a), n_π is the number of $f \in R^C$ which are constant on each cycle of π on C. There are $l(\pi)$ cycles, and on each cycle the constant value of f may be any one of the $|R|$ elements of R, so the number of f's is

$$n_\pi = \underbrace{|R| \cdot |R| \cdot \cdots \cdot |R|}_{l(\pi) \text{ factors}} = |R|^{l(\pi)}. \blacksquare$$

Sec. 5.14 Pólya Enumeration Theory

In our problem of the four spheres (Figure 5.14.1), we now list the eight symmetries of the square in disjoint cycle notation:

(5.14.4) $G = \{(1)(2)(3)(4),\ (1\ 2\ 3\ 4),\ (1\ 3)(2\ 4),\ (1\ 4\ 3\ 2),$
$(1\ 3)(2)(4),\ (2\ 4)(1)(3),\ (1\ 4)(2\ 3),\ (1\ 2)(3\ 4)\}.$

The numbers of disjoint cycles in these expressions are, respectively,

(5.14.5) $\qquad 4, 1, 2, 1, 3, 3, 2, 2$

and $|G|=8$, $|R|=2$, so by Theorem 5.14.1 the number of equivalence classes of paintings of the four spheres is

$$\frac{1}{|G|}\sum_{\pi \in G} |R|^{l(\pi)} = \tfrac{1}{8}(2^4 + 2^1 + 2^2 + 2^1 + 2^3 + 2^3 + 2^2 + 2^2) = 6.$$

Abbreviate red = r, blue = b. A little thought reveals that 6 inequivalent paintings are as given in Figure 5.14.2. Since G-equivalence is an equivalence relation, any one of the 16 possible paintings is G-equivalent to exactly one of these 6, under some symmetry. (See Exercise 5.15.6.)

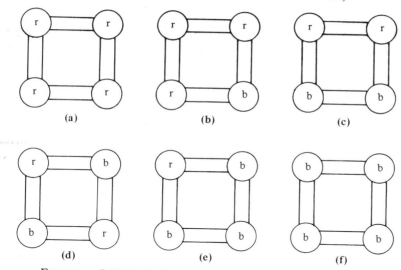

FIGURE 5.14.2. *Six Inequivalent Paintings of Figure 5.14.1*

The six paintings in Figure 5.14.2 have respective weights

(5.14.6) $\qquad r^4, r^3b, r^2b^2, r^2b^2, rb^3, b^4.$

In the general situation of Theorem 5.14.1, it is clear that G-equivalent functions $f, f \circ \pi \in R^C$ have the same weight; if $r \in R$ and e elements $c_1, \ldots, c_e$ of C satisfy $f(c_i) = r$, then $\pi^{-1}(c_1), \ldots, \pi^{-1}(c_e)$ are the e elements of C satisfying $(f \circ \pi)(\pi^{-1}(c_i)) = r$. However, list (5.14.6) shows that functions with the same weight need not be G-equivalent. It is in general desirable to know the number of G-equivalence classes of a given weight, and Theorem 5.14.2 will tell us how this may be computed. From (5.14.6) we say the "inventory" of G on R^C is

(5.14.7) $$r^4 + r^3 b + 2 r^2 b^2 + r b^3 + b^4,$$

in accordance with the following definition.

Assume the hypotheses of Theorem 5.14.1, and let $W_1, W_2, \ldots, W_n$ be all the possible weights of functions $f \in R^C$, p_i the number of G-equivalence classes in R^C that have weight W_i. Then the **inventory** of G on R^C is the polynomial

(5.14.8) $$p_1 W_1 + p_2 W_2 + \cdots + p_n W_n.$$

Before stating Theorem 5.14.2, we need some additional definitions. Denote $|C| = k$, and assume that $\pi \in G$ has $l_i(\pi)$ cycles of length i when we write π as a product of disjoint cycles; then

(5.14.9) $$l_1(\pi) + l_2(\pi) + \cdots + l_k(\pi) = l(\pi)$$

and, since the union of the disjoint cycles is C,

(5.14.10) $$1 \cdot l_1(\pi) + 2 \cdot l_2(\pi) + \cdots + k \cdot l_k(\pi) = k.$$

Choose k symbols $x_1, x_2, \ldots, x_k$. We define the **cycle index of π** to be

(5.14.11) $$x_1^{l_1(\pi)} x_2^{l_2(\pi)} \cdots x_k^{l_k(\pi)},$$

and the **cycle index of G** to be

(5.14.12) $$P_G(x_1, x_2, \ldots, x_k) = \frac{1}{|G|} \sum_{\pi \in G} x_1^{l_1(\pi)} x_2^{l_2(\pi)} \cdots x_k^{l_k(\pi)}.$$

Theorem 5.14.2 (Redfield–Pólya). *Assume that R and C are finite sets, and assume that G is a permutation group on C; denote $k = |C|$, $R = \{r_1, r_2, \ldots, r_m\}$. Then the inventory of G on R^C is*

(5.14.13) $$P_G(r_1 + \cdots + r_m, r_1^2 + \cdots + r_m^2, \ldots, r_1^k + \cdots + r_m^k).$$

When using this theorem, we substitute $r_1^i + \cdots + r_m^i$ for x_i in the polynomial $P_G(x_1, \ldots, x_k)$ and then multiply out the resulting products, obtaining the desired polynomial (5.14.8). For example, in the problem of Figure 5.14.1, list (5.14.4) shows that the cycle indexes of the elements of G are, respectively,

(5.14.14) $$x_1^4, x_4, x_2^2, x_4, x_1^2 x_2, x_1^2 x_2, x_2^2, x_2^2$$

Sec. 5.14 Pólya Enumeration Theory

and, consequently,

(5.14.15) $\quad P_G(x_1, x_2, x_3, x_4) = \frac{1}{8}(x_1^4 + 2x_4 + 3x_2^2 + 2x_1^2 x_2).$

Theorem 5.14.2 says that the inventory of G on R^C is

$$P_G(r+b, r^2+b^2, r^3+b^3, r^4+b^4)$$
$$= \frac{1}{8}[(r+b)^4 + 2(r^4+b^4) + 3(r^2+b^2)^2 + 2(r+b)^2(r^2+b^2)]$$
$$= \frac{1}{8}(r^4 + 4r^3b + 6r^2b^2 + 4rb^3 + b^4 + 2r^4 + 2b^4 + 3r^4 + 6r^2b^2$$
$$+ 3b^4 + 2r^4 + 4r^3b + 4r^2b^2 + 4rb^3 + 2b^4)$$
$$= r^4 + r^3b + 2r^2b^2 + rb^3 + b^4,$$

in agreement with (5.14.7).

As another conceivably practical example, let us assume that we are making bead bracelets with five beads each, and we can choose three colors of beads: red = r, yellow = y, and blue = b. How many distinguishably different bracelets can we make?

The group G of symmetries of the regular pentagon (Figure 5.14.3) is dihedral of order 10 and was discussed in Section 4.18. When we consider G

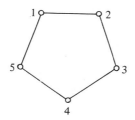

FIGURE 5.14.3. *Bead Bracelet*

as a group of permutations of the set $C = \{1, 2, 3, 4, 5\}$ of vertices we find there is the identity element, with five cycles of length 1 and cycle index x_1^5; there are four rotations of order 5 with one cycle of length 5 and cycle index x_5; and there are five reflections of order 2 in an axis through one vertex, with one cycle of length 1, two cycles of length 2, and cycle index $x_1 x_2^2$. Therefore, the cycle index polynomial of G is

(5.14.16) $\quad P_G(x_1, x_2, x_3, x_4, x_5) = \frac{1}{10}(x_1^5 + 4x_5 + 5x_1 x_2^2).$

(Reasoning like this can often be used to find the cycle index of a group without listing all its elements.)

Here $R = \{r, y, b\}$, so the inventory of G on R^C is

$$P_G(r+y+b, r^2+y^2+b^2, r^3+y^3+b^3, r^4+y^4+b^4, r^5+y^5+b^5)$$
$$= \frac{1}{10}[(r+y+b)^5 + 4(r^5+y^5+b^5) + 5(r+y+b)(r^2+y^2+b^2)^2],$$

which after simplification equals

$$r^5+r^4y+2r^3y^2+2r^2y^3+ry^4+y^5+r^4b+2r^3yb+4r^2y^2b+2ry^3b+y^4b$$
$$+2r^3b^2+4r^2yb^2+4ry^2b^2+2y^3b^2+2r^2b^3+2ryb^3+2y^2b^3$$
$$+rb^4+yb^4+b^5.$$

This is a total of 39 terms, so 39 different bracelets can be created. The term $4r^2y^2b$ says that four different bracelets have two red beads, two yellow beads, and one blue bead (what are they?).

The number 39, without the inventory information, can be computed quickly from Theorem 5.14.1(c). One element of G has five cycles, four elements have one cycle, five elements have three cycles, and $|R|=3$, so

$$\frac{1}{|G|}\sum_{\pi \in G}|R|^{l(\pi)}=\frac{1}{10}(1\cdot 3^5+4\cdot 3^1+5\cdot 3^3)=39.$$

Proof of Theorem 5.14.2. We again use the set

$$\mathcal{S}=\{(f,\pi)\in R^C \times G | f\circ \pi = f\}.$$

We denote $I=p_1W_1+\cdots+p_nW_n$, the desired inventory;

$$\sigma = \sum_{(f,\pi)\in \mathcal{S}} W(f) \quad \text{and} \quad \sigma_\pi = \sum_{\{f | f \circ \pi = f\}} W(f), \quad \text{so} \sum_{\pi \in G} \sigma_\pi = \sigma.$$

To prove the theorem, we shall prove the two equations

(5.14.17) $\sigma = |G| \cdot I$

and

(5.14.18) $\sigma_\pi = (r_1+\cdots+r_m)^{l_1(\pi)}(r_1^2+\cdots+r_m^2)^{l_2(\pi)}\cdots(r_1^k+\cdots+r_m^k)^{l_k(\pi)}.$

Then we shall have

$$I=\frac{1}{|G|}\sigma = \frac{1}{|G|}\sum_{\pi \in G}\sigma_\pi = P_G(r_1+\cdots+r_m, r_1^2+\cdots+r_m^2, \ldots, r_1^k+\cdots+r_m^k),$$

completing the proof.

To prove (5.14.17), we prove that for any i the coefficients of W_i on the left and right sides are equal. The coefficient on the right side is $|G| \cdot p_i$, where $\mathscr{C}_1, \mathscr{C}_2, \ldots, \mathscr{C}_{p_i}$ are all the equivalence classes of f's in R^C with $W(f)=W_i$. If $f \in \mathscr{C}_j$ and $G_f = \{\pi \in G | f \circ \pi = f\}$, then Lemma 4.17.1 shows that $|G_f|=|G|/|\mathscr{C}_j|$, so the $|\mathscr{C}_j|$ elements of each $\mathscr{C}_j$ contribute $|\mathscr{C}_j| \cdot |G|/|\mathscr{C}_j| = |G|$ terms W_i to σ. This means that the coefficient of W_i in σ is also $|G| \cdot p_i$.

To prove (5.14.18), note in Theorem 5.14.1(a) that $f \in R^C$ satisfies $f = f \circ \pi$ iff f is constant on each of the disjoint cycles of the permutation π of C. Hence $W(f)$ has form

(5.14.19) $W(f) = r_{11}^1 \cdots r_{1l_1(\pi)}^1 r_{21}^2 \cdots r_{2l_2(\pi)}^2 \cdots r_{k1}^k \cdots r_{kl_k(\pi)}^k,$

where r_{ij} is the image of f on the jth cycle of length i. Equation (5.14.19) is a term in the multiplied-out product

$$(r_1+\cdots+r_m)^{l_1(\pi)}(r_1^2+\cdots+r_m^2)^{l_2(\pi)}\cdots(r_1^k+\cdots+r_m^k)^{l_k(\pi)}.$$

On the other hand, again by Theorem 5.14.1(a), any such term is $W(f)$ for some f which is constant on each cycle of π, so (5.14.18) must hold. ∎

5.15 Exercises

1. Assume that Army officers decide to paint the five outside walls of the Pentagon building red, white, and blue. Two paintings are considered equivalent if one is a rotation of the other; but, unlike our bracelets, the Pentagon cannot be flipped over. In how many different (nonequivalent) ways can the painting be done, assuming that each color is to be used for at least one wall?

*2. We can think of $\mathbb{B}^6$, the set of all six-digit binary words, as the set of all functions $C \to \{0, 1\}$, where $C = \{c_1, c_2, c_3, c_4, c_5, c_6\}$ is some fixed and ordered set. In some applications, including coding theory, two six-digit words are considered equivalent if we get one from the other by applying the "cyclic permutation"

$$\pi: a_1a_2a_3a_4a_5a_6 \to a_6a_1a_2a_3a_4a_5$$

a number of times. For example,

$$\pi: 010110 \to 001011 \to 100101 \to 110010 \to 011001 \to 101100,$$

so all these words are equivalent. But 000000 is equivalent only to itself, and 010101 is equivalent only to itself and 101010.

The problem is to find the number and inventory of equivalence classes of $\mathbb{B}^6$. To do this, let G be the permutation group of C consisting of the cycle $(c_1 \ c_2 \ c_3 \ c_4 \ c_5 \ c_6)$ and its powers, $R = \{0, 1\}$, and apply Theorems 5.14.1 and 5.14.2.

3. Try Exercise 2 with $\mathbb{B}^n$ in place of $\mathbb{B}^6$, $n = 7, 8, 11$. Can you find general formulas for any $\mathbb{B}^n$? (Another approach to this type of problem will be discussed in Section 8.16.)

4. Try sphere-painting and bead-coloring problems similar to those in Section 5.14 for the regular n-gon. (Section 4.18 contains the needed information on the group of symmetries.)

5. This problem deals with the cube pictured in Figure 5.15.1, and the group M of rigid motions of the cube.

We now describe the 24 elements of M. One is the identity. If A_1, A_2, and A_3 are the axes through centers of opposite faces, we can make 90, 180, or 270° rotations around each; this gives nine more elements of M. If E_1, E_2, E_3, E_4, E_5, and E_6 are the axes through midpoints of opposite edges (edges AD and FG are opposite), we can make 180° rotations around each; this

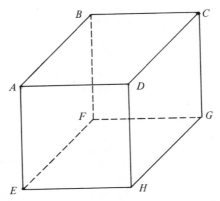

FIGURE 5.15.1. *Cube for Exercise 5*

gives six elements of M. If C_1, C_2, C_3, and C_4 are the axes through opposite vertices, then we can make 120 and 240° rotations around each; these are the remaining eight elements of M.

For example, one of the 120° rotations about the axis $C_1 = CE$ effects the following permutations:

of vertices: $(E)(A\ F\ H)(G\ D\ B)(C)$,

of faces: $(ADEH\ ABFE\ EFGH)(ABCD\ BCGF\ DCGH)$,

of edges: $(EA\ EF\ EH)(AB\ FG\ HD)(AD\ FB\ HG)(CB\ CG\ CD)$.

Each part of this problem uses the group M and Pólya theory. Similar elements of M have similar cycle structure, so you do not need to compute cycle structure for every one of the 24 elements. (However, 90 and 180° rotations are not similar. Why?)

(a) How many different ways can we paint the six faces of the cube with four colors, red, blue, green, or white?

(b) How many different ways can we label the eight vertices with two labels, 0 and 1?

(c) How many different ways can we label the 12 edges with three labels, x, y, and z?

*6. List completely the members of the six equivalence classes of the 16 possible paintings of Figure 5.14.1.

*7. Let R and C be finite sets, G a permutation group on C. For each $\pi \in G$, define $\psi_\pi : R^C \to R^C$ by $\psi_\pi(f) = f \circ \pi$, all $f \in R^C$.

(a) Prove that each ψ_π is a permutation of R^C and that we have

$$(\psi_\pi)^{-1} = \psi_{\pi^{-1}}, \qquad \psi_{1_C} = 1_{R^C}.$$

(b) Among permutations of the set R^C, let $\circ$ denote our (usual) left composition and let $\diamond$ denote right composition. Verify that if $\pi_1, \pi_2 \in G$,

Sec. 5.16 An Extension of Pólya Enumeration Theory

then

$$\psi_{\pi_1} \circ \psi_{\pi_2} = \psi_{\pi_2\pi_1} \quad \text{and} \quad \psi_{\pi_1} \diamond \psi_{\pi_2} = \psi_{\pi_1\pi_2}.$$

(We emphasize that multiplication in G is the usual left composition of functions from C to C.)

(c) Denote $\Psi = \{\psi_\pi | \pi \in G\}$. Show that Ψ is a group under either $\circ$ or $\diamond$. Show that if $|R| > 1$, then $\varphi: \pi \to \psi_\pi$ is an isomorphism $\varphi: G \to [\Psi, \diamond]$ and an anti-isomorphism $\varphi: G \to [\Psi, \circ]$, and $\bar{\varphi}: \pi \to \psi_{\pi^{-1}}$ is an isomorphism $\bar{\varphi}: G \to [\Psi, \circ]$. What happens if $|R| = 1$? (We thus think of G as a group of right permutations of R^C when $|R| > 1$, identifying G with $[\Psi, \circ]$ and writing $f \circ \pi$ instead of $(f)\psi_\pi$. In any case, G acts on R^C on the right.)

8. In this problem you are to determine how many different chemical compounds can be formed of the sort commonly pictured as in Figure 5.15.2, where each x_i is one of the radicals —H, —Cl, —CH_3, —C_2H_5. Here the correct model for this molecule is *three-dimensional*; a regular tetrahedron with C at the center, the radicals at the vertices. The group consists of 12 rigid motions of the tetrahedron. Inequivalent labelings of the vertices with the radicals correspond to distinct compounds.

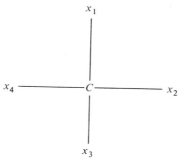

FIGURE 5.15.2. *Diagram for Exercise 8*

9. A class of eight students includes a pair of identical twins. Every day, each student wears a shirt that is red, green, blue, or yellow. How many different ways can the class be dressed (considering only shirt color), assuming that we cannot distinguish the identical twins?

5.16 An Extension of Pólya Enumeration Theory

Our goal now is a generalization of Theorem 5.14.1, in which a permutation group H also acts on the set R. We shall see in Section 5.17 how this is applied to enumerate classes of switching functions. Our result is a special case of a theorem of de Bruijn (1959).

Assume throughout this section that C and R are finite sets, G a permutation group on C, H a permutation group on R.

Lemma 5.16.1. *The function from $(G \times H) \times R^C$ to R^C defined by*

$$\psi: ((g, h), f) \to h \circ f \circ g^{-1}$$

is a left action of $G \times H$ on R^C.

Proof. We shall denote $\psi(((g, h), f))$ by $(g, h) \cdot f$, the result when (g, h) acts on f. We must prove that $(1, 1) \cdot f = f$ and $((g_1, h_1)(g_2, h_2)) \cdot f = (g_1, h_1) \cdot [(g_2, h_2) \cdot f]$; these are true because

$$(1, 1) \cdot f = 1 \circ f \circ 1^{-1} = f,$$

$$((g_1, h_1)(g_2, h_2)) \cdot f = (g_1 \circ g_2, h_1 \circ h_2) \cdot f = (h_1 \circ h_2) \circ f \circ (g_1 \circ g_2)^{-1}$$

$$= h_1 \circ h_2 \circ f \circ g_2^{-1} \circ g_1^{-1},$$

$$(g_1, h_1) \cdot [(g_2, h_2) \cdot f] = (g_1, h_1) \cdot h_2 \circ f \circ g_2^{-1} = h_1 \circ h_2 \circ f \circ g_2^{-1} \circ g_1^{-1}. \blacksquare$$

When we speak of *equivalence* of elements of R^C, we mean the $(G \times H)$-equivalence from the action of Lemma 5.16.1.

Lemma 5.16.2. *The number of equivalence classes in R^C under the action of $G \times H$ is*

$$N(R^C; G, H) = \frac{1}{|G||H|} \sum_{g \in G} \sum_{h \in H} |\{f \in R^C | f \circ g = h \circ f\}|.$$

Theorem 5.14.1(b) is the special case of this lemma when $|H| = 1$.

Proof. Let $\mathcal{S} = \{((g, h), f) \in (G \times H) \times R^C | (g, h) \cdot f = f\}$. We shall count $|\mathcal{S}|$ in two ways. For any fixed $(g, h) \in G \times H$, $(g, h) \cdot f = f$ iff $h \circ f \circ g^{-1} = f$ iff $f \circ g = h \circ f$, so indeed

$$|\mathcal{S}| = \sum_{g \in G} \sum_{h \in H} |\{f \in R^C | f \circ g = h \circ f\}|.$$

On the other hand, let $\mathcal{C}$ be an equivalence class in R^C, $f \in \mathcal{C}$. By Lemma 5.16.1, $G \times H$ acts on R^C, so Lemma 4.17.1 implies that for each f in $\mathcal{C}$,

$$|\{(g, h) \in G \times H | (g, h) \cdot f = f\}| = \frac{|G \times H|}{|\mathcal{C}|}.$$

Hence

$$\sum_{f \in \mathcal{C}} |\{(g, h) \in G \times H | (g, h) \cdot f = f\}| = |\mathcal{C}| \frac{|G \times H|}{|\mathcal{C}|} = |G||H|,$$

and therefore $|\mathcal{S}| = |G||H| \cdot N(R^C; G, H)$.

Equating the two expressions for $|\mathcal{S}|$, we have the lemma. $\blacksquare$

Lemma 5.16.3. *For a fixed $g \in G$ and $h \in H$, denote*

$$n_{g,h} = |\{f \in R^C | f \circ g = h \circ f\}|.$$

Sec. 5.16 · An Extension of Pólya Enumeration Theory

Write g and h as products of disjoint cycles (on C and R, respectively) and let $Cy(g)$ be the set of cycles of g, $Cy(h)$ the set of cycles of h. For any cycle B, denote by $\lambda(B)$ the length (number of elements) of B. Then

(5.16.1)
$$n_{g,h} = \prod_{B \in Cy(g)} \left[\sum_{\{D \in Cy(h) \mid \lambda(D) \text{ divides } \lambda(B)\}} \lambda(D) \right]$$

and

(5.16.2) $\quad N(R^C; G, H) = \dfrac{1}{|G||H|} \sum_{g \in G} \sum_{h \in H} n_{g,h}.$

Theorem 5.14.1(c) (of much simpler form) is the special case of Lemma 5.16.3 when $|H| = 1$. (Why?)

Proof. Let B be a cycle of g of length l, so $B = (b \; g(b) \; g^2(b) \cdots g^{l-1}(b))$, $g^l(b) = b$ for any $b \in B$. Suppose that $f \in R^C$ satisfies $f \circ g = h \circ f$, and set $d = f(b)$; assume that d is in the cycle D of h. Using $f \circ g = h \circ f$ repeatedly, we find that $f \circ g^t = h^t \circ f$ for any positive integer t. For any $g^t(b) \in B$ this means that $f(g^t(b)) = h^t(f(b)) = h^t(d) \in D$, so $f(B) \subseteq D$; but any $h^t(d) \in D$ satisfies $h^t(d) = h^t(f(b)) = f(g^t(b)) \in f(B)$, $D \subseteq f(B)$, so $f(B) = D$. Assume that D has length m. Then the last equation implies $l \geqslant m$; we claim $m \mid l$. Now m is the smallest positive integer such that $h^m(d) = d$. By the division algorithm, $l = qm + r$ for some integers q and r satisfying $q \geqslant 0$, $0 \leqslant r < m$; hence

$$d = f(b) = f(g^l(b)) = h^l(f(b)) = h^l(d) = h^r(h^m)^q(d) = h^r(d),$$

forcing $r = 0$, $l = qm$, and $m \mid l$.

The preceding paragraph shows that we get all functions f satisfying $f \circ g = h \circ f$ as follows: Choose some $b \in C$, suppose that b lies in the cycle B of g, and define $f(b)$ to be any $d \in R$ lying in a cycle D of h with $\lambda(D) \mid \lambda(B)$. This defines f on all of B, via $f(g^t(b)) = h^t(f(b)) = h^t(d)$. For each cycle B of g we thus have

$$\sum_{\{D \in Cy(h) \mid \lambda(D) \text{ divides } \lambda(B)\}} \lambda(D)$$

possible ways to define f. ∎

Actual use of Lemma 5.16.3 could be quite awkward. De Bruijn (1959) found the following expression, using partial derivatives and exponentials, for $N(R^C; G, H)$.

Theorem 5.16.1 (de Bruijn). *Let G be a permutation group on the finite set C, H a permutation group on the finite set R, $|C| = k$, and $|R| = r$. Let $P_G(x_1, \ldots, x_k)$ be the cycle index of G on C, $P_H(x_1, \ldots, x_r)$ the cycle index of H on R, and let $z_1, z_2, \ldots, z_k$ be real variables. Then $N(R^C; G, H)$ equals*

(5.16.3)
$$P_G\left(\frac{\partial}{\partial z_1}, \ldots, \frac{\partial}{\partial z_k}\right) P_H(e^{z_1 + \cdots + z_k}, e^{2(z_2 + z_4 + \cdots)}, \ldots, e^{r(z_r + z_{2r} + \cdots)}) \bigg|_{z_1 = \cdots = z_k = 0}.$$

Here the sums $z_i + z_{2i} + z_{3i} + \cdots$ continue as far as the terms exist in $\{z_1, z_2, \ldots, z_k\}$. If, say, $s \leq r$ but $s > k$, then $z_s, z_{2s}, \ldots$ do not exist and $e^{s(z_s + z_{2s} + \cdots)}$ is just $e^0 = 1$.

Proof. Of course,

$$(5.16.4) \quad P_G(x_1, \ldots, x_k) = \frac{1}{|G|} \sum_{g \in G} x_1^{m_1(g)} x_2^{m_2(g)} \cdots x_k^{m_k(g)},$$

where $m_i(g)$ is the number of cycles of g on C of length i. For any $h \in H$ and any positive integer l, denote

$$(5.16.5) \quad \chi_h(l) = \sum_{\{D \in \text{Cy}(h) | \lambda(D) \text{ divides } l\}} \lambda(D).$$

By (5.16.1), we have

$$(5.16.6) \quad n_{g,h} = \prod_{B \in \text{Cy}(g)} \chi_h(\lambda(B)) = \prod_{l=1}^{k} \chi_h(l)^{m_l(g)}.$$

Thinking of the numbers $\chi_h(l)$ and $m_l(g)$ as fixed nonnegative integers for the moment, we find that

$$(5.16.7) \quad \prod_{l=1}^{k} \chi_h(l)^{m_l(g)} = \left(\frac{\partial}{\partial z_1}\right)^{m_1(g)} \left(\frac{\partial}{\partial z_2}\right)^{m_2(g)} \cdots \left(\frac{\partial}{\partial z_k}\right)^{m_k(g)} e^{\sum_{l=1}^{k} \chi_h(l) z_l} \bigg|_{z_1 = \cdots = z_k = 0},$$

since each differentiation $\partial/\partial z_l$ brings out a constant $\chi_h(l)$ and setting all $z_l = 0$ afterward makes

$$e^{\sum_{l=1}^{k} \chi_h(l) z_l} = e^0 = 1.$$

Combining (5.16.2), (5.16.6), and (5.16.7), we get

$$(5.16.8) \quad N(R^C; G, H) = \frac{1}{|G||H|} \sum_{g \in G} \sum_{h \in H} \left\{ \left(\frac{\partial}{\partial z_1}\right)^{m_1(g)} \cdots \left(\frac{\partial}{\partial z_k}\right)^{m_k(g)} \right.$$
$$\left. \times e^{\sum_{l=1}^{k} \chi_h(l) z_l} \bigg|_{z_1 = \cdots = z_k = 0} \right\},$$

which, by (5.16.4), is the same as

$$(5.16.9) \quad N(R^C; G, H) = P_G\left(\frac{\partial}{\partial z_1}, \ldots, \frac{\partial}{\partial z_k}\right) \frac{1}{|H|} \sum_{h \in H} e^{\sum_{l=1}^{k} \chi_h(l) z_l} \bigg|_{z_1 = \cdots = z_k = 0}.$$

Of course,

$$(5.16.10) \quad P_H(x_1, \ldots, x_r) = \frac{1}{|H|} \sum_{h \in H} x_1^{n_1(h)} x_2^{n_2(h)} \cdots x_r^{n_r(h)},$$

where h has $n_i(h)$ cycles of length i on R. Equation (5.16.5) implies that

$$(5.16.11) \quad \chi_h(l) = \sum_{t | l} n_t(h) \cdot t,$$

Sec. 5.17 *Equivalence Classes of Switching Functions* 255

the sum running over all positive integers t such that $t|l$. Hence we have

$$P_H(e^{z_1+z_2+z_3+\cdots+z_k}, e^{2(z_2+z_4+\cdots)}, e^{3(z_3+z_6+\cdots)}, \ldots)$$

$$= \frac{1}{|H|} \sum_{h \in H} \{e^{(z_1+z_2+\cdots)n_1(h)} e^{2(z_2+z_4+\cdots)n_2(h)} e^{3(z_3+z_6+\cdots)n_3(h)} \cdots\}$$

$$= \frac{1}{|H|} \sum_{h \in H} e^{\{n_1(h)z_1 + (n_1(h)+2n_2(h))z_2 + (n_1(h)+3n_3(h))z_3 + (n_1(h)+2n_2(h)+4n_4(h))z_4 + \cdots\}}$$

$$= \frac{1}{|H|} \sum_{h \in H} e^{\sum_{l=1}^{k} \chi_h(l) z_l}.$$

Substituting this in (5.16.9), we have (5.16.3). ∎

5.17 Equivalence Classes of Switching Functions

Switching functions were discussed in Chapter 3. As before, $\mathbb{B}$ denotes $\{0, 1\}$ and $\mathbb{B}^n$ denotes the set of 2^n possible n-tuples (words of length n) of 0's and 1's. The switching functions of n variables are the $2^{(2^n)}$ functions $\mathbb{B}^n \to \mathbb{B}$. The same device may implement a number of different functions if we permute and/or complement some of the input leads; Figure 5.17.1 shows an example.

We want to determine how many essentially different devices are required to implement the switching functions of n variables, at least for n reasonably small, when we permit permutation and/or complementation of input signals. In particular, we shall show that 402 devices suffice to implement the $2^{(2^4)} = 65{,}536$ four-variable switching functions; but 222 devices are sufficient if we allow complementation of the output functions. [Harrison's book (1965) contains more information on this topic, and a catalog of the 222 devices.]

Before applying Pólya theory, we must construct and study a group G_n of permutations of $\mathbb{B}^n$, corresponding to the permutation and/or complementation of the n input leads. We first define the following notions from group theory.

If G is a group and $x, y \in G$, we say y is **conjugate** to x if $y = g^{-1}xg$, some $g \in G$. Conjugacy is an equivalence relation on the set G (see Exercise 4.15.9(b)); the equivalence classes are called **conjugacy classes** of G. (For examples, see Exercise 5.18.4.)

We now establish some notation for the rest of this section. $T = \mathbb{P}_n = \{1, 2, \ldots, n\}$. S_n, of order $n!$, is the symmetric group of all *right* permutations of T; for example, if $\sigma = (1 \quad 2 \quad 3) \in S_4$, then $1\sigma = 2, 2\sigma = 3, 3\sigma = 1, 4\sigma = 4$. We use ' to denote complementation; $0' = 1$, $1' = 0$ as usual. For any subset U of T, we denote by n_U the right permutation of $\mathbb{B}^n$ which

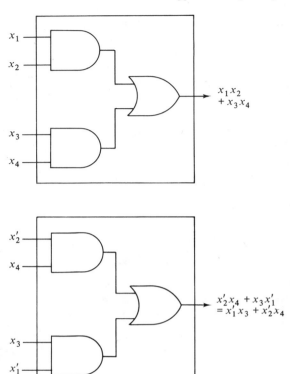

FIGURE 5.17.1. *Functions Equivalent Under G_4*

complements the ith component of a word iff $i \in U$, leaving other components alone. For example,

$$(1010)n_{\{2,3\}} = 1100, \qquad (00000)n_T = 11111, \qquad n_\varnothing = 1_{\mathbb{B}^n}.$$

We denote $N = \{n_U | U \subseteq T\}$. If $\sigma \in S_n$, we also denote by σ the right permutation of $\mathbb{B}^n$ given by $(a_1 a_2 \cdots a_n)\sigma = a_{1\sigma} a_{2\sigma} \cdots a_{n\sigma}$, all $a_1 a_2 \cdots a_n \in \mathbb{B}^n$. For example,

$$(100101)(1 \quad 2 \quad 4)(3 \quad 6) = 011100,$$

since $a_{1\sigma} = a_2 = 0$, $a_{2\sigma} = a_4 = 1$, $a_{3\sigma} = a_6 = 1$, $a_{4\sigma} = a_1 = 1$, $a_{5\sigma} = a_5 = 0$, and $a_{6\sigma} = a_3 = 0$. (Our reasons for using *right* permutations are discussed in Exercise 5.18.5.)

Lemma 5.17.1. *S_n and N are subgroups of* $\mathrm{Sym}(\mathbb{B}^n)$, *with* $|S_n| = n!$, $|N| = 2^n$. *We have the relations*
(a) $n_\varnothing = 1_{\mathbb{B}^n}$, $n_U n_V = n_{U \oplus V}$, $n_U^{-1} = n_U$.
(b) $\sigma^{-1} n_U \sigma = n_{U\sigma}$.

Sec. 5.17 *Equivalence Classes of Switching Functions*

(c) $(n_U \sigma)^{-1} = n_{U_\sigma} \sigma^{-1}$.
(d) $n_U \sigma \cdot n_V \tau = n_{U \oplus V \sigma^{-1}}(\sigma \tau)$.
(e) $(n_V \tau)^{-1} n_U \sigma (n_V \tau) = n_{(V \oplus U \oplus V \sigma^{-1})_\tau}(\tau^{-1} \sigma \tau)$.
[Of course, $U \oplus V = (U - V) \cup (V - U)$.]

Proof. (a) is trivial. For (b), note that $i\sigma^{-1} \in U$ iff $i \in U\sigma$. Hence for the word $\mathbf{a} = a_1 a_2 \cdots a_n$ with ith component a_i, the ith component of

$$\mathbf{a}\sigma^{-1} \quad \text{is} \quad a_{i\sigma^{-1}},$$

$$\mathbf{a}\sigma^{-1} n_U \quad \text{is} \quad \begin{cases} a_{i\sigma^{-1}} & \text{if } i \notin U\sigma, \\ a'_{i\sigma^{-1}} & \text{if } i \in U\sigma, \end{cases}$$

$$\mathbf{a}\sigma^{-1} n_U \sigma \quad \text{is} \quad \begin{cases} a_i & \text{if } i \notin U\sigma, \\ a'_i & \text{if } i \in U\sigma, \end{cases}$$

proving that $\sigma^{-1} n_U \sigma = n_{U\sigma}$, that is, (b). To get (c), multiply by σ on the left, getting $n_U \sigma = \sigma n_{U\sigma}$, and then apply Lemma 4.4.2 and (a).

We now prove (d), using (b) and (a):

$$n_U \sigma \cdot n_V \tau = n_U (\sigma^{-1})^{-1} n_V \sigma^{-1}(\sigma \tau) = n_U n_{V \sigma^{-1}}(\sigma \tau) = n_{U \oplus V \sigma^{-1}}(\sigma \tau).$$

Our proof of (e) uses all the previous parts:

$$(n_V \tau)^{-1} n_U \sigma n_V \tau = n_{V\tau} \tau^{-1} n_{U \oplus V \sigma^{-1}}(\sigma \tau) = n_{V\tau}(\tau^{-1} n_{U \oplus V \sigma^{-1}} \tau)(\tau^{-1} \sigma \tau)$$

$$= n_{V\tau} n_{(U \oplus V \sigma^{-1})_\tau}(\tau^{-1} \sigma \tau) = n_{(V \oplus U \oplus V \sigma^{-1})_\tau}(\tau^{-1} \sigma \tau).$$

For the last equality, we used (a) and the easy fact that $X\tau \oplus Y\tau = (X \oplus Y)\tau$. ∎

Corollary 5.17.1. $G_n = \{n_U \sigma | U \subseteq T, \sigma \in S_n\}$ *is a subgroup of* Sym$(\mathbb{B}^n)$ *of order* $2^n(n!)$, *with normal subgroup* N. *Every element of G_n has a unique expression* $n_U \sigma$, $U \subseteq T$, $\sigma \in S_n$.

Proof. By Theorem 4.4.3 and Lemma 5.17.1(d), G_n is a subgroup, and by Lemma 5.17.1(e) with $\sigma = 1_T$, N is normal in G_n.

If an element of G_n has two expressions $n_U \sigma = n_V \tau$, $U, V \subseteq T$, $\sigma, \tau \in S_n$, then $n_V^{-1} n_U = \tau \sigma^{-1}$ is an element of $N \cap S_n = \{1_{\mathbb{B}^n}\}$, so $\tau \sigma^{-1} = $ identity, $\tau = \sigma$. Also, $n_V^{-1} n_U = $ identity implies that $n_{V \oplus U} = n_\emptyset$, $V \oplus U = \emptyset$, $V = U$. Therefore, the two expressions are really the same; $|G_n| = |N||S_n| = 2^n(n!)$. ∎

Lemma 5.17.2. *A subset D of a cycle C of $\sigma \in S_n$ on T has form $B \oplus B\sigma$, B a subset of C, iff $|D|$ is even.*

Proof →. Certainly $|B| = |B\sigma|$, so if $D = B \oplus B\sigma$, then Exercise 5.18.6 shows that $|D|$ is even.

Proof ←. Let $C = (i_1 \; i_2 \cdots i_k)$ and $D = \{i_{j_1}, i_{j_2}, \ldots, i_{j_{2t}}\}$, where $1 \leq j_1 < j_2 < \cdots < j_{2t} \leq k$. Define

$$B = \{i_{j_1}, i_{j_1+1}, \ldots, i_{j_2-1}\} \cup \{i_{j_3}, i_{j_3+1}, \ldots, i_{j_4-1}\} \cup \cdots,$$

so

$$B\sigma = \{i_{j_1+1}, i_{j_1+2}, \ldots, i_{j_2}\} \cup \{i_{j_3+1}, i_{j_3+2}, \ldots, i_{j_4}\} \cup \cdots$$

and $B \oplus B\sigma = D$. ∎

Lemma 5.17.3. *Two elements of S_n are conjugate iff they have the same cycle index on T.*

Proof. Suppose that $\sigma = \rho^{-1}\tau\rho$ in S_n, so σ and τ are conjugate. If C is a cycle of τ, suppose that

$$C = \{a, a\tau, \ldots, a\tau^{t-1}\}, \qquad |C| = t, \, a\tau^t = a.$$

Then we have, using $\sigma = \rho^{-1}\tau\rho$ and $\rho\sigma = \tau\rho$,

$$C\rho = \{a\rho, a\tau\rho, \ldots, a\tau^{t-1}\rho\} = \{a\rho, a\rho\sigma, \ldots, a\rho\sigma^{t-1}\},$$

with $a\rho\sigma^t = a\tau^t\rho = a\rho$, so $C\rho$ is a cycle of σ. $|C\rho| = |C|$ since ρ is a permutation. Thus to each cycle C of τ corresponds a cycle $C\rho$ of σ, $|C\rho| = |C|$; τ and σ have the same cycle index.

If $\sigma, \tau \in S_n$ have the same cycle index, then to each cycle $(a \ a\tau \cdots a\tau^{t-1})$, $a\tau^t = a$, of τ we can assign a cycle $(b \ b\sigma \cdots b\sigma^{t-1})$, $b\sigma^t = b$, of σ of the same order t. Then by defining ρ by $b = a\rho$, $b\sigma^i = (a\tau^i)\rho$, we will indeed find that $(a\tau^s)\tau\rho = (a\tau^{s+1})\rho = (a\rho)\sigma^{s+1} = (a\rho\sigma^s)\sigma = (a\tau^s)(\rho\sigma)$, all $a\tau^s \in T$, so $\tau\rho = \rho\sigma$ and $\sigma = \rho^{-1}\tau\rho$. ∎

Since A. Young (1929) discovered the following theorem for classifying the conjugacy classes in G_n, we shall call the index that plays the role of the cycle index the *Young index*.

Assume that $n_U\sigma \in G_n$. Assume that σ (as a product of disjoint cycles on T) has α_i cycles of length i, $1 \leq i \leq n$. For each i, assume that β_i of the cycles of length i contain an *even* number of integers in U. The symbol

$$(\alpha_1, \alpha_2, \ldots, \alpha_n; \beta_1, \beta_2, \ldots, \beta_n)$$

will be called the **Young index** of $n_U\sigma$.

Note that $x_1^{\alpha_1} x_2^{\alpha_2} \cdots x_n^{\alpha_n}$ is the cycle index of σ and that we have

(5.17.1) $$\sum_{i=1}^{n} i\alpha_i = n, \qquad 0 \leq \beta_i \leq \alpha_i.$$

A little thought will reveal that all sets of nonnegative integers $\alpha_1, \ldots, \alpha_n$, $\beta_1, \ldots, \beta_n$ satisfying (5.17.1) do occur as Young indexes of elements of G_n. For example, if $n = 2$, then the possible Young indexes are

$$(2, 0; 0, 0), \quad (2, 0; 1, 0), \quad (2, 0; 2, 0), \quad (0, 1; 0, 0), \quad (0, 1; 0, 1).$$

Sec. 5.17 *Equivalence Classes of Switching Functions*   259

Elements with these Young indexes are, respectively,

$$n_{\{1,2\}}, \quad n_{\{1\}} \text{ or } n_{\{2\}}, \quad n_\varnothing,$$
$$n_{\{1\}}(1\ 2) \text{ or } n_{\{2\}}(1\ 2),$$
$$n_\varnothing(1\ 2) \text{ or } n_{\{1,2\}}(1\ 2).$$

G_2 has order $2^2(2!) = 8$, and these five indexes show (by Theorem 5.17.1) that G_2 has five conjugacy classes. All the Young indexes of elements of G_4 are listed in Table 5.17.1.

Lemma 5.17.4. *If $\sigma, \tau \in S_n$ and $V \subseteq T$, then $n_V \sigma$ and $n_{V\tau} \cdot \tau^{-1} \sigma \tau$ have the same Young index.*

Proof. Lemma 5.17.3 shows that their α's are the same, and in fact that if $C_1, \ldots, C_l$ are the cycles of σ, then $C_1\tau, \ldots, C_l\tau$ are the cycles of $\tau^{-1}\sigma\tau$. To see that the β's are the same, note that

$$|V \cap C_i| = |(V \cap C_i)\tau| = |V\tau \cap C_i\tau|,$$

so one number is even iff the other is. ∎

Theorem 5.17.1 [Young (1929)]. *Two elements of G_n are conjugate iff they have the same Young index.*

Proof ⇒. By Lemma 5.17.1(e) we must show that $n_U \sigma$ and $n_{(V \oplus U \oplus V\sigma^{-1})\tau} \cdot \tau^{-1}\sigma\tau$ have the same Young index. By Lemma 5.17.4, $n_{(V \oplus U \oplus V\sigma^{-1})\tau} \cdot \tau^{-1}\sigma\tau$ and $n_{V \oplus U \oplus V\sigma^{-1}} \cdot \sigma$ have the same Young index, so it remains to show that $n_U\sigma$ and $n_{V \oplus U \oplus V\sigma^{-1}} \cdot \sigma$ have the same Young index. If $\bar{V} = V\sigma^{-1}$, then

$$V \oplus U \oplus V\sigma^{-1} = U \oplus V \oplus V\sigma^{-1} = U \oplus \bar{V}\sigma \oplus \bar{V} = U \oplus \bar{V} \oplus \bar{V}\sigma,$$

so it suffices to prove that

(5.17.2) $n_U\sigma$ and $n_{U \oplus V \oplus V\sigma} \cdot \sigma$ have the same Young index,

for any $\sigma \in S_n$, any $U, V \subseteq T$.

By the definition of Young index, it suffices to show that

(5.17.3) $|U \cap C| \equiv |(U \oplus V \oplus V\sigma) \cap C|$ (mod 2)

for all cycles C of σ. If $U_1 = U \cap C$ and $V_1 = V \cap C$, then $(U \oplus V \oplus V\sigma) \cap C = U_1 \oplus V_1 \oplus V_1\sigma$, so it is enough to show that

(5.17.4) $|U_1| \equiv |U_1 \oplus V_1 \oplus V_1\sigma|$ (mod 2)   for all subsets U_1, V_1 of C.

By Lemma 5.17.2, this amounts to showing that $|U_1| \equiv |U_1 \oplus W_1|$ (mod 2) for all subsets U_1 and W_1 of C with $|W_1|$ even. This is *true*, because

$$|U_1 \oplus W_1| = |U_1| + |W_1| - 2|U_1 \cap W_1|.$$

Proof ⇐. Suppose that $n_U\sigma$ and $n_V\tau$ have the same Young index. To each cycle C of σ, then, we can assign a cycle $C\rho$ of τ such that $|C|=|C\rho|$ and $|U\cap C|\equiv|V\cap C\rho|$ (mod 2). If $C=(a\quad a\sigma\cdots a\sigma^{t-1})$ and $C\rho = (b\quad b\tau\cdots b\tau^{t-1})$, we can complete the definition of $\rho\in S_n$ by $a\rho=b$, $(a\sigma^i)\rho = b\tau^i$; then $\tau = \rho^{-1}\sigma\rho$, as in the proof of Lemma 5.17.3.

Define $X = U \oplus V\rho^{-1}$; for each cycle C of σ we have $X \cap C = (U\cap C) \oplus (V\rho^{-1}\cap C)$, so

$$|X\cap C| \equiv |U\cap C| + |V\rho^{-1}\cap C| = |U\cap C| + |V\cap C\rho| \equiv 0 \pmod{2}.$$

By Lemma 5.17.2, on each cycle C, $X = W \oplus W\sigma^{-1}$ for some set W. Now we have

$$V\rho^{-1} = U \oplus U \oplus V\rho^{-1} = U \oplus X = U \oplus W \oplus W\sigma^{-1}$$

and hence

$$V = (U \oplus W \oplus W\sigma^{-1})\rho.$$

We conclude that

$$n_V\tau = n_{(U\oplus W\oplus W\sigma^{-1})\rho} \cdot \rho^{-1}\sigma\rho = (n_W\rho)^{-1}(n_U\sigma)(n_W\rho)$$

by Lemma 5.17.1(e). ∎

Lemma 5.17.5. *The conjugacy class of G_n with Young index $(\alpha_1,\ldots,\alpha_n; \beta_1,\ldots,\beta_n)$ contains exactly*

(5.17.5) $$(n!)\prod_{i=1}^{n}\frac{2^{(i-1)\alpha_i}}{(\beta_i!)((\alpha_i-\beta_i)!)i^{\alpha_i}}$$

elements.

Proof. We first determine how many $\sigma \in S_n$ have cycle index $x_1^{\alpha_1}\cdots x_n^{\alpha_n}$. If we write out a blank product of cycles in which to write σ,

$$\sigma = (\underbrace{\quad)(\quad)\cdots(\quad)}_{\alpha_1 \text{ cycles of length 1}}\underbrace{(\quad)(\quad)\cdots(\quad)}_{\alpha_2 \text{ cycles of length 2}}\cdots,$$

we shall certainly have $n!$ ways we can fill it in; n choices for the first number, $n-1$ for the second, and so on. How many times will the same σ reappear? A cycle of length i can be written in i ways; for example, $(1\quad 2\quad 3) = (2\quad 3\quad 1) = (3\quad 1\quad 2)$. In addition, the α_i cycles of length i can be written in $(\alpha_i)!$ different orders; so there are $(\alpha_i)! \cdot i^{\alpha_i}$ ways of writing the product of cycles of length i within a fixed σ. The number of different σ's with cycle index $x_1^{\alpha_1}\cdots x_n^{\alpha_n}$ is therefore

(5.17.6) $$\frac{n!}{\prod_{i=1}^{n}(\alpha_i)!\cdot i^{\alpha_i}}.$$

Sec. 5.17 Equivalence Classes of Switching Functions

Now, holding σ fixed, we shall find the number of U's such that $n_U\sigma$ has the given Young index. Look at the α_i cycles of length i; β_i of them must have intersection with the set U of even order. The number

$$_{\alpha_i}C_{\beta_i} = \frac{(\alpha_i)!}{(\beta_i!)((\alpha_i - \beta_i)!)}$$

tells how many ways we can choose these β_i cycles. Exercise 5.18.7 tells us that for each cycle of length i there are 2^{i-1} ways we can choose its intersection with U of the desired order (even or odd). The number of possible U's, for fixed σ, is therefore

(5.17.7) $$\prod_{i=1}^{n} \frac{(\alpha_i)!}{(\beta_i!)((\alpha_i - \beta_i)!)} \cdot (2^{i-1})^{\alpha_i}.$$

Multiplying (5.17.6) and (5.17.7) yields (5.17.5). ∎

We now wish to use Sections 5.14 and 5.16 to study switching functions of n variables, n reasonably small. This means we must find the cycle index polynomial of the group G_n acting on the set $C = \mathbb{B}^n$. We save lots of effort with the following lemma.

Lemma 5.17.6. *Conjugate elements of G_n have the same cycle index on the set $\mathbb{B}^n$.*

Proof. This is really a special case of Lemma 5.17.3, with the set $\mathbb{B}^n$ in place of T, the group Sym $(\mathbb{B}^n)$ in place of S_n. If two elements of G_n are conjugate, then they are also conjugate in Sym $(\mathbb{B}^n)$ and hence have the same cycle index. ∎

Table 5.17.1 completely describes G_n when $n = 4$. G_4 has $2^4 \cdot (4!) = 384$ elements, but we do not have to consider them all, just one from each conjugacy class. In the Table, column I lists all the possible Young indexes of conjugacy classes in G_4, column $|I|$ gives the orders of the conjugacy classes computed from Lemma 5.17.5, column $n_U\sigma$ gives a typical element of each conjugacy class, and column "cycle index" gives the cycle index of any element of the class on $\mathbb{B}^4$. (Verification of the Table entries is Exercise 5.18.8. A systematic way to determine the cycle indexes is studied in Section 8.17.) The table implies that the cycle index polynomial of G_4 is

$$\tfrac{1}{384}(x_2^8 + 4x_2^8 + 6x_2^8 + 4x_2^8 + x_1^{16} + 12x_4^4 + 24x_4^4 + 12x_4^4 + 12x_2^8$$
$$+ 24x_2^8 + 12x_1^8 x_2^4 + 12x_4^4 + 24x_4^4 + 12x_1^4 x_2^6 + 32x_2^2 x_6^2$$
$$+ 32x_2^2 x_6^2 + 32x_2^2 x_6^2 + 32x_1^4 x_3^4 + 48x_8^2 + 48x_1^2 x_2 x_4^3),$$

TABLE 5.17.1. The group G_4.

I	$|I|$	$n_U\sigma$	cycle index
(4, 0, 0, 0; 0, 0, 0, 0)	1	$n_{\{1,2,3,4\}}$	x_2^8
(4, 0, 0, 0; 1, 0, 0, 0)	4	$n_{\{1,2,3\}}$	x_2^8
(4, 0, 0, 0; 2, 0, 0, 0)	6	$n_{\{1,2\}}$	x_2^8
(4, 0, 0, 0; 3, 0, 0, 0)	4	$n_{\{1\}}$	x_2^8
(4, 0, 0, 0; 4, 0, 0, 0)	1	$n_\varnothing = 1_{\mathbb{B}^n}$	x_1^{16}
(2, 1, 0, 0; 0, 0, 0, 0)	12	$n_{\{1,3,4\}}(1\ 2)$	x_4^4
(2, 1, 0, 0; 1, 0, 0, 0)	24	$n_{\{1,3\}}(1\ 2)$	x_4^4
(2, 1, 0, 0; 2, 0, 0, 0)	12	$n_{\{1\}}(1\ 2)$	x_4^4
(2, 1, 0, 0; 0, 1, 0, 0)	12	$n_{\{3,4\}}(1\ 2)$	x_2^8
(2, 1, 0, 0; 1, 1, 0, 0)	24	$n_{\{3\}}(1\ 2)$	x_2^8
(2, 1, 0, 0; 2, 1, 0, 0)	12	$(1\ 2)$	$x_1^8 x_2^4$
(0, 2, 0, 0; 0, 0, 0, 0)	12	$n_{\{1,3\}}(1\ 2)(3\ 4)$	x_4^4
(0, 2, 0, 0; 0, 1, 0, 0)	24	$n_{\{1\}}(1\ 2)(3\ 4)$	x_4^4
(0, 2, 0, 0; 0, 2, 0, 0)	12	$(1\ 2)(3\ 4)$	$x_1^4 x_2^6$
(1, 0, 1, 0; 0, 0, 0, 0)	32	$n_{\{1,4\}}(1\ 2\ 3)$	$x_2^2 x_6^2$
(1, 0, 1, 0; 1, 0, 0, 0)	32	$n_{\{1\}}(1\ 2\ 3)$	$x_2^2 x_6^2$
(1, 0, 1, 0; 0, 0, 1, 0)	32	$n_{\{4\}}(1\ 2\ 3)$	$x_2^2 x_6^2$
(1, 0, 1, 0; 1, 0, 1, 0)	32	$(1\ 2\ 3)$	$x_1^4 x_3^4$
(0, 0, 0, 1; 0, 0, 0, 0)	48	$n_{\{1\}}(1\ 2\ 3\ 4)$	x_8^2
(0, 0, 0, 1; 0, 0, 0, 1)	48	$(1\ 2\ 3\ 4)$	$x_1^2 x_2 x_4^3$

which simplifies to

$$(5.17.8) \quad P_{G_4}(x_1, \ldots, x_{16}) = \tfrac{1}{384}(x_1^{16} + 51x_2^8 + 84x_4^4 + 12x_1^8 x_2^4 + 12x_1^4 x_2^6$$
$$+ 96 x_2^2 x_6^2 + 32 x_1^4 x_3^4 + 48 x_8^2 + 48 x_1^2 x_2 x_4^3).$$

With $C = \mathbb{B}^4$, $R = \mathbb{B}$ and $G = G_4$, Theorem 5.14.1(c) implies that the number of equivalence classes of four-variable switching functions under permutation and complementation of inputs is

$$\tfrac{1}{384}(2^{16} + 51 \cdot 2^8 + 84 \cdot 2^4 + 12 \cdot 2^{12} + 12 \cdot 2^{10} + 96 \cdot 2^4 + 32 \cdot 2^8 + 48 \cdot 2^2$$
$$+ 48 \cdot 2^6) = 402.$$

Allowing complementation of the output function amounts to letting a group H of order 2 permute the set $R = \mathbb{B} = \{0, 1\}$ of outputs. The identity of H is $(0)(1)$ and the other element of H is $(0\ 1)$, so the cycle index polynomial of H on R is

$$P_H(x_1, x_2) = \tfrac{1}{2}(x_1^2 + x_2).$$

By Theorem 5.16.1, the number of equivalence classes of four-variable switching functions under the action of G_4 and H is

$$P_{G_4}\left(\frac{\partial}{\partial z_1}, \frac{\partial}{\partial z_2}, \ldots, \frac{\partial}{\partial z_{16}}\right) P_H(e^{z_1+z_2+\cdots+z_{16}}, e^{2(z_2+z_4+\cdots+z_{16})})\bigg|_{z_1=\cdots=z_{16}=0}$$

$$= \frac{1}{384}\left[\left(\frac{\partial}{\partial z_1}\right)^{16} + 51\left(\frac{\partial}{\partial z_2}\right)^8 + 84\left(\frac{\partial}{\partial z_4}\right)^4 + 12\left(\frac{\partial}{\partial z_1}\right)^8\left(\frac{\partial}{\partial z_2}\right)^4\right.$$

$$+ 12\left(\frac{\partial}{\partial z_1}\right)^4\left(\frac{\partial}{\partial z_2}\right)^6 + 96\left(\frac{\partial}{\partial z_2}\right)^2\left(\frac{\partial}{\partial z_6}\right)^2 + 32\left(\frac{\partial}{\partial z_1}\right)^4\left(\frac{\partial}{\partial z_3}\right)^4$$

$$\left. + 48\left(\frac{\partial}{\partial z_8}\right)^2 + 48\left(\frac{\partial}{\partial z_1}\right)^2\left(\frac{\partial}{\partial z_2}\right)\left(\frac{\partial}{\partial z_4}\right)^3\right]$$

$$\cdot \frac{1}{2}[e^{2(z_1+z_2+\cdots+z_{16})} + e^{2(z_2+z_4+\cdots+z_{16})}]\bigg|_{z_1=z_2=\cdots=z_{16}=0}$$

$$= \frac{1}{768}\{[2^{16} + 51\cdot 2^8 + 84\cdot 2^4 + 12\cdot 2^8\cdot 2^4 + 12\cdot 2^4\cdot 2^6 + 96\cdot 2^2\cdot 2^2$$

$$+ 32\cdot 2^4\cdot 2^4 + 48\cdot 2^2 + 48\cdot 2^2\cdot 2\cdot 2^3]e^{2(z_1+z_2+\cdots+z_{16})}$$

$$+ [51\cdot 2^8 + 84\cdot 2^4 + 96\cdot 2^2\cdot 2^2$$

$$+ 48\cdot 2^2]e^{2(z_2+z_4+\cdots+z_{16})}\}\bigg|_{z_1=z_2=\cdots=z_{16}=0}$$

$$= \frac{1}{768}[2^{16} + 51\cdot 2^9 + 84\cdot 2^5 + 12\cdot 2^{12} + 12\cdot 2^{10} + 96\cdot 2^5 + 32\cdot 2^8 + 48\cdot 2^3$$

$$+ 48\cdot 2^6]$$

$$= \frac{2^7}{768}[512 + 204 + 21 + 384 + 96 + 24 + 64 + 3 + 24]$$

$$= \frac{1}{6}(1332) = 222.$$

5.18 Exercises

*1. In the situation of Lemma 5.16.1, show that if $|R| > 1$ then $G \times H$ is a permutation group on R^C. (Compare with Exercise 5.15.7.)

2. Find the cycle index polynomials of the symmetric groups S_2 and S_3, and use them and Theorem 5.16.1 to find $N(R^C; \text{Sym}(C), \text{Sym}(R))$ in the cases
  (a) $|C| = 2, |R| = 2$. (b) $|C| = 3, |R| = 2$.
  (c) $|C| = 2, |R| = 3$. (d) $|C| = 3, |R| = 3$.

3. Prove that if a group G acts on a set S and x and y are conjugate elements of G, then x and y have the same number of orbits of each length.

*4. (a) Show that in any abelian group G, the conjugacy classes are the one-element subsets of G.

(b) Find the conjugacy classes in S_3.

(c) Find the conjugacy classes in the dihedral groups D_4 and D_5 (Section 4.18). Can you do this for any D_n?

(d) Show that conjugate elements in a finite group have the same order.

*5. (a) Let S_n be the group of *right* permutations of $T = \{1, 2, \ldots, n\}$. For each $\sigma \in S_n$, define $\bar{\sigma} \colon \mathbb{B}^n \to \mathbb{B}^n$ by $(a_1 a_2 \cdots a_n)\bar{\sigma} = a_{1\sigma} a_{2\sigma} \cdots a_{n\sigma}$. Show that each $\bar{\sigma}$ is a permutation of $\mathbb{B}^n$. Also, show that $\sigma \to \bar{\sigma}$ is an *isomorphism* of S_n onto a subgroup of Sym $(\mathbb{B}^n)$, the group of all *right* permutations of $\mathbb{B}^n$.

(b) Now denote by S_n the group of *left* permutations of $T = \{1, 2, \ldots, n\}$. For each $\sigma \in S_n$, define $\bar{\sigma} \colon \mathbb{B}^n \to \mathbb{B}^n$ by $\bar{\sigma}(a_1 a_2 \cdots a_n) = a_{\sigma(1)} a_{\sigma(2)} \cdots a_{\sigma(n)}$. Show that each $\bar{\sigma}$ is a permutation of $\mathbb{B}^n$. Also show that $\sigma \to \bar{\sigma}$ is an *anti-isomorphism* of S_n onto a subgroup of Sym $(\mathbb{B}^n)$, the group of all *left* permutations of $\mathbb{B}^n$. (Use the definition of $\bar{\sigma}$ carefully.)

[Parts (a) and (b) explain why we used *right* instead of *left* permutations in Section 5.17. Harrison's book (1965) uses a different strategy; use left permutations and send $\sigma \to \bar{\sigma}$, where $\bar{\sigma}(a_1 a_2 \cdots a_n) = a_{\sigma^{-1}(1)} a_{\sigma^{-1}(2)} \cdots a_{\sigma^{-1}(n)}$. One can further investigate the reason for (a) and (b) by using the fact that $\mathbb{B}^n = \mathbb{B}^T$, the set of all functions $T \to \mathbb{B} = \{0, 1\}$; $\sigma \in S_n$ then acts on $f \in \mathbb{B}^T$ on the *right*, sending $f \to f \circ \sigma$, just as G acted on R^C in Section 5.14.]

*6. Show that if X and Y are finite sets with $|X| = |Y|$, then $|X \oplus Y|$ is even.

*7. Show that a finite (nonempty) set of order i has 2^{i-1} subsets of even order and 2^{i-1} subsets of odd order.

(HINT: Use induction on i.)

*8. (a) Verify that the entries in column I of Table 5.17.1 are the only possible Young indexes with $n = 4$. [Use (5.17.1).]

(b) Verify the entries in column $|I|$ of Table 5.17.1. (Use Lemma 5.17.5.) Also verify that each element in column $n_U \sigma$ has the Young index in column I of the same row.

(c) Verify the *cycle index* column of Table 5.17.1. [This requires some computation, applying the elements $n_U \sigma$ to the 16 elements of $\mathbb{B}^4$ to actually see the disjoint cycles of $n_U \sigma$. Using more advanced theory, Harrison (1965) is able to avoid computation and obtain a (complicated!) formula for the cycle index polynomial of any G_n. Enough of this theory is given in Section 8.17 to greatly shorten this exercise.]

9. Find the cycle index polynomial of G_3. Show that the number of equivalence classes of three-variable switching functions under permutation and/or complementation of inputs is 22. Show also that this number decreases to 14 when complementation of the output function is allowed. [For $n = 5$, the corresponding numbers are 1,228,158 and 616,126; for $n = 6$, they are 400,507,806,843,728 and 200,253,952,527,184; see Harrison (1965).]

CHAPTER 6
Lattices

6.1 Definition of a Lattice

In Section 2.8 we defined a poset $P_{\leq}$ to be a set P on which is defined a relation "$\leq$" that is reflexive, antisymmetric, and transitive. Recall also that an element u of P is an *upper bound* for a subset S of $P_{\leq}$ iff $s \leq u$ for all s in S. An upper bound l of S that is $\leq$ every upper bound u of S is necessarily unique if it exists (Theorem 2.9.4); such an upper bound is called the *least upper bound* (l.u.b.) of S. Similar definitions were given for the *lower bound* and the *greatest lower bound* of S.

A poset $P_{\leq}$ is called a **lattice** iff every pair of elements of P has a l.u.b. and a g.l.b. Familiar examples are the following:

1. The set $\mathbb{R}$ of real numbers with $\leq$ as usual, l.u.b. $\{x, y\} = \max \{x, y\}$, g.l.b. $\{x, y\} = \min \{x, y\}$.
2. The set $\mathscr{P}(U)$ of all subsets of a set U with $\leq$ the inclusion relation $\subseteq$, g.l.b. $\{A, B\} = A \cap B$, l.u.b. $\{A, B\} = A \cup B$.
3. The set $\mathbb{B}^n$ of all sequences of n 0's and 1's with $\leq$ as in (2.2.6) for vectors and matrices, g.l.b. $\{\mathbf{x}, \mathbf{y}\} = \mathbf{x} \cap \mathbf{y}$ as in (2.2.3), l.u.b. $\{\mathbf{x}, \mathbf{y}\} = \mathbf{x} \cup \mathbf{y}$ as in (2.2.3).

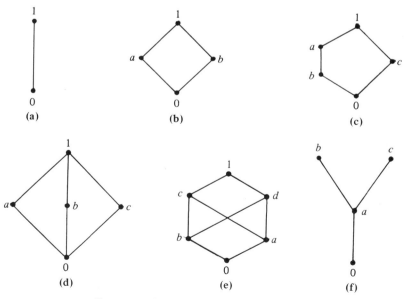

FIGURE 6.1.1. *Lattices and Nonlattices*

4. The set Π_S of all partitions of a set S where $\pi_1 \leq \pi_2$ iff every block of π_1 is a subset of some block of π_2 (see Section 2.14), and l.u.b. $\{\pi_1, \pi_2\} = \pi_1 + \pi_2$, g.l.b. $\{\pi_1, \pi_2\} = \pi_1 \pi_2$.

5. The set of all subgroups H of a given group G, where $H_1 \leq H_2$ means $H_1 \subseteq H_2$ (set inclusion), l.u.b. $\{H_1, H_2\}$ is the subgroup generated by the elements of H_1 and H_2, and g.l.b. $\{H_1, H_2\}$ is the subgroup $H_1 \cap H_2$.

Examples 2 and 3 are also examples of Boolean algebras, while 1, 4, and 5 are not. We shall see later that a Boolean algebra is a special kind of lattice.

We see from the definition that the posets with Hasse diagrams (a), (b), (c), and (d) of Figure 6.1.1 are all lattices, while the posets represented by (e) and (f) are not.

6.2 A Basic Theorem

We begin with an immediate consequence of the definition:

Theorem 6.2.1. *In every lattice $L_\leq$, every finite, nonempty subset S has a least upper bound and a greatest lower bound.*

Proof. Let $|S| = n$. If S is a subset of $L_\leq$ and if $S = \{a\}$, then g.l.b. $S =$ l.u.b. $S = a$. If $S = \{a, b\}$, then g.l.b. S and l.u.b. S exist by the definition of a lattice. Thus the theorem holds if $n = 1$ or 2. Now let n denote any integer such that the theorem holds if $|S| = n$. Let $T = \{t_1, t_2, \ldots, t_n, t_{n+1}\}$ be a subset of L and let $S = \{t_1, t_2, \ldots, t_n\}$. Then there exist $t = $ l.u.b. S and $u = $ l.u.b. $\{t, t_{n+1}\}$. Then $u \geq t$, and $u \geq t_i$, $i = 1, 2, \ldots, n+1$, so u is an upper bound for T. Let v be any upper bound for T. Then v is an upper bound for S also since $S \subseteq T$; hence $v \geq t$. But $v \geq t_{n+1}$. Therefore, v is an upper bound for $\{t, t_{n+1}\}$ so $v \geq u$ and $u = $ l.u.b. T. The result follows for the l.u.b. by induction. A dual argument proves the result for g.l.b. ∎

The proof of the theorem shows that the l.u.b. u of S, where $S = \{s_1, s_2, \ldots, s_n\}$ is any finite subset of L, may be constructed iteratively as follows:

$$u_1 = s_1,$$
$$u_2 = \text{l.u.b.} \{u_1, s_2\},$$
$$u_3 = \text{l.u.b.} \{u_2, s_3\},$$
$$\vdots$$
$$u_n = \text{l.u.b.} \{u_{n-1}, s_n\} = u.$$

Because l.u.b. S is unique, the end result is independent of the order in which the elements s_i of S are listed.

A similar construction yields g.l.b. S.

Sec. 6.3 The Operations "Cup" and "Cap"

A lattice $L_\leqslant$ is said to be **complete** iff every subset S of L, whether finite or infinite, has a g.l.b. and a l.u.b. Corollary 6.2.1 follows from the preceding theorem.

Corollary 6.2.1. *Every finite lattice is complete.* ∎

An example of an infinite, complete lattice is $\mathscr{P}(U)_\subseteq$ for an infinite set U. On the other hand, $\mathbb{R}_\leqslant$ is not complete. Why?

Recall from Section 2.9 that in a poset $P_\leqslant$, an element z such that $z \leqslant p$ for all $p \in P$ is necessarily unique if it exists, is called the *zero element* of $P_\leqslant$, and is denoted by "0." Dually, an element u of P such that $p \leqslant u$ for all $p \in P$ is necessarily unique if it exists, is called the *unit element* of $P_\leqslant$, and is denoted by "1." Whereas a given finite poset $P_\leqslant$ may have both, one, or neither of these special elements, in the case of a finite lattice we have the following corollary.

Corollary 6.2.2. *Every finite lattice $L_\leqslant$ has both a zero and a unit element.*

Proof. By Theorem 6.2.1, every subset of L, in particular L itself, has a g.l.b. and a l.u.b. In the case of L, these are zero and unit elements, respectively. ∎

Figure 6.1.1 illustrates this result.

6.3 The Operations "Cup" and "Cap"

In any lattice $L_\leqslant$, the binary operations cup ($\vee$) and cap ($\wedge$) are defined thus: For all elements a and b of L,

(6.3.1) $\quad a \vee b = \text{l.u.b. } \{a, b\}$ (the **join** or **sum** or **union** of a and b}
$\quad\quad\quad a \wedge b = \text{g.l.b. } \{a, b\}$ (the **meet** or **product** or **intersection** of a and b).

These definitions are suggested by the fact that these relations hold for union ($\cup$) and intersection ($\cap$) in $\mathscr{P}(U)_\subseteq$.

For example, in the lattice of Figure 6.1.1(c), $a \wedge b = b$, $a \vee c = 1$, $a \vee b = a$, $a \wedge c = 0$. For another illustration, recall that a poset $P_\leqslant$ (like $\mathbb{R}_\leqslant$, for example) in which $a \leqslant b$ or $b \leqslant a$ for all a and b in P is called an *ordered set* or a *chain*. In a chain, $a \vee b = \max\{a, b\}$ and $a \wedge b = \min\{a, b\}$. These examples serve also to illustrate the next lemma.

Lemma 6.3.1. *For all elements a and b of an arbitrary lattice $L_\leqslant$,*
(a) $a \leqslant a \vee b$ and $a \wedge b \leqslant a$.
(b) $a \leqslant b$ iff $a \vee b = b$.
(c) $a \leqslant b$ iff $a \wedge b = a$.

Proof. (a) Immediate from the definition of $\vee$ as l.u.b. and $\wedge$ as g.l.b.

(b) If $a \vee b = b$, then $b = $ l.u.b. $\{a, b\}$, so $a \leq b$. Conversely, if $a \leq b$, then because $b \leq b$, $a \vee b \leq b$. By (a), $b \leq a \vee b$. Hence $a \vee b = b$.

(c) Similar to the proof of (b). ∎

We now make the following two definitions for arbitrary elements $a_1, a_2, \ldots, a_n$ of $L_\leq$:

$$\bigvee_{k=1}^{n} a_k = a_1 \vee a_2 \vee \cdots \vee a_n = \text{l.u.b.}\{a_1, a_2, \ldots, a_n\},$$

$$\bigwedge_{k=1}^{n} a_k = a_1 \wedge a_2 \wedge \cdots \wedge a_n = \text{g.l.b.}\{a_1, a_2, \ldots, a_n\}.$$

The iterative procedure used in the proof of Theorem 6.2.1 shows then that

$$\bigvee_{k=1}^{n} a_k = (\cdots ((a_1 \vee a_2) \vee a_3) \vee \cdots) \vee a_n$$

and similarly for ∧, and also that the results are independent of the order of the elements in the sum or product; that is, these are *commutative* operations.

As special cases we note that by Corollary 6.2.2, if $L = \{a_1, a_2, \ldots, a_n\}$,

$$\bigvee_{k=1}^{n} a_k = \text{l.u.b.}\{a_1, a_2, \ldots, a_n\} = 1,$$

$$\bigwedge_{k=1}^{n} a_k = \text{g.l.b.}\{a_1, a_2, \ldots, a_n\} = 0.$$

6.4 Another Description of a Lattice

From what precedes, we have readily the following theorem.

Theorem 6.4.1. *In every lattice $L_\leq$, for all a, b, and c of L, these four properties hold*:

(L1) $a \vee a = a$, $a \wedge a = a$ (*laws of idempotency*).
(L2) $a \vee b = b \vee a$, $a \wedge b = b \wedge a$ (*commutative laws*).
(L3) $a \vee (b \vee c) = (a \vee b) \vee c$, $a \wedge (b \wedge c) = (a \wedge b) \wedge c$ (*associative laws*).
(L4) $a \vee (a \wedge b) = a$, $a \wedge (a \vee b) = a$ (*absorption laws*).

Proof. (L1) is immediate from the definitions of ∨ and ∧. (L2) holds because the definitions of ∨ (l.u.b.) and ∧ (g.l.b.) do not depend on the order in which elements are listed. (L3) follows from (L2) and the fact that the l.u.b. and g.l.b. may be computed iteratively, as shown in the preceding section. Indeed,

$$a \vee (b \vee c) = (a \vee b) \vee c = \text{l.u.b.}\{a, b, c\} = a \vee b \vee c$$

Sec. 6.4 Another Description of a Lattice

and
$$a \wedge (b \wedge c) = (a \wedge b) \wedge c = \text{g.l.b.} \{a, b, c\} = a \wedge b \wedge c.$$

(L4) follows from the facts that $a \wedge b \leq a$ and $a \leq a \vee b$ so, by Lemma 6.3.1 and (L2), $a \vee (a \wedge b) = a$ and $a \wedge (a \vee b) = a$. ∎

To prove the converse of the preceding result, we need the following lemma.

Lemma 6.4.1. *In every set L on which are defined two binary operations $\vee$ and $\wedge$ that satisfy (L1), (L2), (L3), and (L4), $a \vee b = a$ iff $a \wedge b = b$.*
Proof. In fact, $a \vee b = a$ implies that $a \wedge b = (a \vee b) \wedge b = b \wedge (b \vee a) = b$ and $a \wedge b = b$ implies that $a \vee b = a \vee (a \wedge b) = a$, both by (L4). ∎

An alternative definition of a lattice is implied by the next theorem.

Theorem 6.4.2. *Every set L on which are defined two binary operations $\vee$ and $\wedge$ that satisfy (L1), (L2), (L3), and (L4) is a lattice; that is, these properties characterize a lattice.*
Proof. We have to show that L is a poset in which every pair of elements has a l.u.b. and a g.l.b. To do this we first define, for all a and b in L,

(6.4.1) $\qquad\qquad\qquad a \leq b \quad \text{iff} \quad a \vee b = b.$

Then $a \leq a$ for all a in L because $a \vee a = a$ by (L1). Thus "$\leq$" is a *reflexive* relation. Next, if $a \leq b$ and $b \leq a$, then $a \vee b = b$ and $b \vee a = a$. Since $a \vee b = b \vee a$ by (L2), $a = b$ and "$\leq$" is *antisymmetric*. Finally, if $a \leq b$ and $b \leq c$, then $a \vee b = b$ and $b \vee c = c$. Hence $a \vee c = a \vee (b \vee c) = (a \vee b) \vee c = b \vee c = c$, using (L3). Thus $a \leq c$, so "$\leq$" is *transitive* and $L_\leq$ is a poset.

Now we have to show that each pair of elements of $L_\leq$ has a l.u.b. and a g.l.b. in L. Let a and b belong to L. Then $a \leq a \vee b$ because $a \vee (a \vee b) = (a \vee a) \vee b = a \vee b$, using (L3) and (L1). Similarly, $b \leq a \vee b$. Hence $a \vee b$ is an upper bound for $\{a, b\}$. Now let u be *any* upper bound for $\{a, b\}$. Then $a \leq u$, $b \leq u$, $a \vee u = u$, and $b \vee u = u$. Hence, using (L3),

$$(a \vee b) \vee u = a \vee (b \vee u) = a \vee u = u,$$

so $a \vee b \leq u$. Thus $a \vee b = \text{l.u.b.} \{a, b\}$.

Now consider $a \wedge b$. Because $a \vee (a \wedge b) = a$ by (L4), we have $a \wedge b \leq a$ and because $b \vee (b \wedge a) = b$, we have $(b \wedge a) \leq b$, so by (L2) $a \wedge b \leq b$. Thus $a \wedge b$ is a lower bound for $\{a, b\}$. Now let v be *any* lower bound for $\{a, b\}$. Then $v \vee a = a$ and $v \vee b = b$, so by Lemma 6.4.1 we have $v \wedge a = v$ and $v \wedge b = v$. Thus $v \wedge (a \wedge b) = (v \wedge a) \wedge b = v \wedge b = v$, so, again by the lemma, we have $v \vee (a \wedge b) = a \wedge b$; that is, $v \leq a \wedge b$. Thus $a \wedge b$ is the g.l.b. of $\{a, b\}$, and $L_\leq$ is a lattice. Moreover, because of the way in which "$\leq$" was defined for L, the "$\vee$" and "$\wedge$" with which we began are the same as the "$\vee$" and "$\wedge$" defined in (6.3.1) for $L_\leq$. ∎

The list (L1), (L2), (L3), (L4) is convenient but not independent, as we see in the following corollary.

Corollary 6.4.1. *Properties (L2), (L3), and (L4) suffice to define a lattice.*
Proof. For all a and b of L, (L4) used twice yields $a = a \vee (a \wedge (a \vee b)) = a \vee a$, $a = a \wedge (a \vee (a \wedge b)) = a \wedge a$. ∎

We also have, since (L2), (L3), and (L4) hold in every Boolean algebra:

Corollary 6.4.2. *Every Boolean algebra is a lattice.* ∎

Many rules of Boolean algebra are actually lattice-theoretic results, such as Lemma 6.4.2.

Lemma 6.4.2. *In every lattice $L_\leq$,*
(a) $a \leq b$ and $a \leq c$ iff $a \leq b \wedge c$.
(b) $a \leq c$ and $b \leq c$ iff $a \vee b \leq c$.
(c) If $a \leq b$ and $c \leq d$, then $a \vee c \leq b \vee d$ and $a \wedge c \leq b \wedge d$.
(d) If $a \leq b \leq c$ and $d \wedge c = a$, then $d \wedge b = a$, and dually.
Proof. The reader may provide the proof; the methods are familiar. ∎

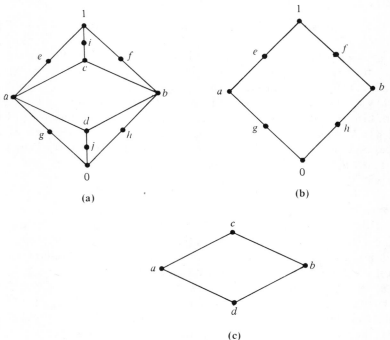

FIGURE 6.4.1. *Subset That Is a Lattice; Sublattice*

Sec. 6.5 The Modular and Distributive Laws

A frequently useful definition is the following: Let $S_\leqslant$ denote a subset of a lattice $L_\leqslant$ such that $x \leqslant y$ in $S_\leqslant$ iff $x \leqslant y$ in $L_\leqslant$. Then $S_\leqslant$ is called a **sublattice** of $L_\leqslant$ iff for all x and y in S, $x \vee y$ and $x \wedge y$, as computed in $L_\leqslant$, are also in S. That is, the operations $\vee$ and $\wedge$ must yield the same elements in the lattice $S_\leqslant$ as they do in $L_\leqslant$. For example, (c) of Figure 6.4.1 is a sublattice of (a).

A subset $S_\leqslant$ of $L_\leqslant$ may be a lattice with respect to $\leqslant$ but still not be a sublattice of $L_\leqslant$, because the element $x \vee y$ or $x \wedge y$ of $L_\leqslant$ is not in S for some x and y in S. Thus lattice (b) in Figure 6.4.1 is not a sublattice of (a) because $a \vee b = c$ in $L_\leqslant$ and c is not an element of (b).

6.5 The Modular and Distributive Laws

A lattice $L_\leqslant$ is said to be **modular** iff, for all x, y, and z in L,

(6.5.1) $\quad x \wedge (y \vee (x \wedge z)) = (x \wedge y) \vee (x \wedge z) \quad$ (*the modular law*).

Since (6.5.1) holds if any one of x, y, or z is a unit or zero element, and if any two of x, y, and z are equal (check these claims), the relation (6.5.1) needs to be verified only for elements distinct from each other and distinct from 0 and 1, in order to establish that $L_\leqslant$ is modular. Of course, only one set of values for which (6.5.1) is false need be produced to show that $L_\leqslant$ is not modular. Thus the lattice of Figure 6.1.1(c) is not modular, for $a \wedge (c \vee (a \wedge b)) = a \wedge (c \vee b) = a \wedge 1 = a$ while $(a \wedge c) \vee (a \wedge b) = 0 \vee b = b$. On the other hand, the lattice of Figure 6.1.1(d) is modular, as is proved by showing that (6.5.1) holds for the six permutations of a, b, and c (again, check the claim).

Another characterization of modularity is given in the following theorem.

Theorem 6.5.1. *A lattice $L_\leqslant$ is modular iff $x \geqslant z$ implies that for all y in L,*

(6.5.2) $\quad\quad\quad\quad x \wedge (y \vee z) = (x \wedge y) \vee z.$

Proof. Suppose first that (6.5.1) holds and that $x \geqslant z$. Then $z = x \wedge z$ so, for all y in L,

$$x \wedge (y \vee z) = x \wedge (y \vee (x \wedge z)) = (x \wedge y) \vee (x \wedge z) = (x \wedge y) \vee z,$$

using (6.5.1), so (6.5.2) holds.

Now suppose that (6.5.2) holds. Since for all x and z in L, $x \geqslant x \wedge z$, we have by (6.5.2) and for all y in L,

$$x \wedge (y \vee (x \wedge z)) = (x \wedge y) \vee (x \wedge z),$$

so (6.5.1) holds. ∎

A frequently manageable test for modularity is given in the next theorem.

Theorem 6.5.2. *A lattice $L_\leq$ is modular iff it contains no sublattice whose Hasse diagram is the pentagon of Figure* 6.1.1(c).

Proof. Suppose first that $L_\leq$ is modular. Then no sublattice $S_\leq$ can have the pentagon as its diagram, for the lattice of the pentagon is not modular and computations in $S_\leq$ yield the same results as in $L_\leq$.

Suppose now that $L_\leq$ is not modular. Then by Theorem 6.5.1 there exist in $L_\leq$ elements x, y, and z with $x > z$ such that $x \wedge (y \vee z) \neq (x \wedge y) \vee z$. Since $x > z$, $x \wedge (y \vee z) \geq z \wedge (y \vee z) = z$. Also, since $y \vee z \geq y$, we have $x \wedge (y \vee z) \geq x \wedge y$. Hence $x \wedge (y \vee z) \geq (x \wedge y) \vee z$, but since by hypothesis equality does not hold, we have

(6.5.3) $\qquad x > z, \qquad x \wedge (y \vee z) > (x \wedge y) \vee z.$

We now show that the five elements $y \vee z$, $x \wedge (y \vee z)$, $(x \wedge y) \vee z$, $x \wedge y$, and y constitute a lattice with the pentagon as Hasse diagram (Figure 6.5.1).

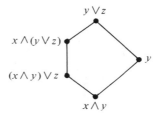

FIGURE 6.5.1. *Basic Pentagonal Lattice*

First we show that the necessary inequalities hold. We have $y \vee z \geq y$; if $y \vee z = y$, then, from (6.5.3), $x \wedge y > (x \wedge y) \vee z$, which is impossible. Hence $y \vee z > y$. In closely similar fashion, we show $y > x \wedge y$.

Next, $(x \wedge y) \vee z \geq x \wedge y$. If $(x \wedge y) \vee z = x \wedge y$, then $z \leq x \wedge y$, so $z \leq y$, or $y \vee z = y$, already shown to be false. Hence $(x \wedge y) \vee z > x \wedge y$.

If $y \vee z = x \wedge (y \vee z)$, then $y \wedge (y \vee z) = x \wedge (y \wedge (y \vee z))$, $y = x \wedge y$, which is false. Hence $y \vee z > x \wedge (y \vee z)$, and all the inequalities implied by the pentagon hold.

Now we show that y is not comparable with $x \wedge (y \vee z)$ or $(x \wedge y) \vee z$. We have that $[x \wedge (y \vee z)] \wedge y = x \wedge (y \wedge (y \vee z)) = x \wedge y$. Hence, by Lemma 6.4.2(d), $[(x \wedge y) \vee z] \wedge y = x \wedge y$. These equations show the claimed noncomparability (why?) and also show that $x \wedge y$ is indeed the zero element of the pentagon. Similarly, $[(x \wedge y) \vee z] \vee y = [(x \wedge y) \vee y] \vee z = y \vee z$, so again by the lemma, $[x \wedge (y \vee z)] \vee y = y \vee z$, and $y \vee z$ is the unit element of the pentagon. Since the computations with $\vee$ and $\wedge$ are all in $L_\leq$, the pentagon is a sublattice of $L_\leq$. ∎

A lattice $L_\leq$ is **distributive** iff, for all a, b, and c in L,

(6.5.4) $\qquad a \wedge (b \vee c) = (a \wedge b) \vee (a \wedge c) \qquad$ (the *distributive law*).

Sec. 6.5 *The Modular and Distributive Laws* 273

The lattices (c) and (d) of Figure 6.1.1 are not distributive; in the case of (c), $a \wedge (b \vee c) = a \wedge 1 = a$ while $(a \wedge b) \vee (a \wedge c) = b \vee 0 = b$, and in the case of (d), $a \wedge (b \vee c) = a$ while $(a \wedge b) \vee (a \wedge c) = 0$. Every Boolean algebra is distributive, but the lattice Π_S of partitions of S, where $|S| \geq 3$, is not. Indeed, if s_1, s_2, and s_3 are arbitrary members of S and $T = S - \{s_1, s_2, s_3\}$, the diagram represents a sublattice of Π_S that is not distributive. (Omit blocks $\bar{T}$ if $|S| = 3$.)

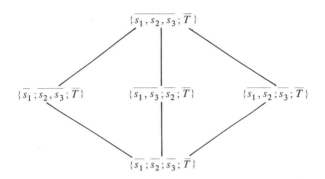

It is not hard to verify that (6.5.4) holds if any one of a, b, or c is 0 or 1 and also that it holds if any two of a, b, and c are equal. Hence to establish that a given lattice $L_\leq$ is distributive, we need only verify relation (6.5.4) in the cases where a, b, and c are distinct elements different from 0 and 1.

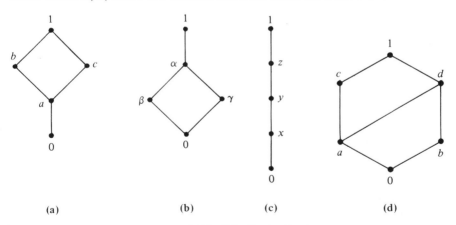

FIGURE 6.5.2. *Distributive Lattices*

For example, consider the lattice of Figure 6.5.2(a). To show that it is distributive, we need check only six cases of (6.5.4), for there are six permutations of a, b, and c. Of these six cases, we actually need to examine only three, for the equation $x \wedge (y \vee z) = (x \wedge y) \vee (x \wedge z)$ implies that $x \wedge (z \vee y) = (x \wedge z) \vee (x \wedge y)$, by (L2). We have, then, in this case

$$a \wedge (b \vee c) = a = (a \wedge b) \vee (a \wedge c),$$
$$b \wedge (a \vee c) = a = (b \wedge a) \vee (b \wedge c),$$
$$c \wedge (a \vee b) = a = (c \wedge a) \vee (c \wedge b),$$

so the lattice is distributive, as claimed; similarly for the other three lattices of Figure 6.5.2.

Lemma 6.5.1. *Each of the following three identities implies that $L_\leqslant$ is modular.*

(6.5.5)
(D1) $a \wedge (b \vee c) = (a \wedge b) \vee (a \wedge c),$
(D2) $a \vee (b \wedge c) = (a \vee b) \wedge (a \vee c),$
(D3) $(a \vee b) \wedge (b \vee c) \wedge (c \vee a) = (a \wedge b) \vee (b \wedge c) \vee (c \wedge a).$

In particular, every distributive lattice is modular.

[Note the symmetry and the self-dual character of (D3).]

Proof. Assume (D1). This implies, with the aid of (L3) and (L1), that

$$a \wedge (b \vee (a \wedge c)) = (a \wedge b) \vee (a \wedge (a \wedge c)) = (a \wedge b) \vee (a \wedge c),$$

so $L_\leqslant$ is modular.

Now assume that (D2) holds and assume also that $a \geqslant c$, so $a \vee c = a$ and $a \wedge c = c$. Then (L2) and (D2) imply that

$$a \wedge (b \vee c) = (a \vee c) \wedge (b \vee c) = (a \wedge b) \vee c,$$

so $L_\leqslant$ is modular by Theorem 6.5.1.

Finally, assume (D3) and $a \geqslant c$. Substituting for $a \vee c$ and $a \wedge c$ and using (L2) and (L3) repeatedly, we reduce (D3) to

$$[a \wedge (a \vee b)] \wedge (b \vee c) = (a \wedge b) \vee [c \vee (c \wedge b)].$$

By (L4), this reduces to

$$a \wedge (b \vee c) = (a \wedge b) \vee c. \quad \blacksquare$$

Theorem 6.5.3. *In any lattice $L_\leqslant$, the identities (D1), (D2), and (D3) are equivalent.*

Proof. (D1) $\Rightarrow$ (D2):

$$\begin{aligned}
(a \vee b) \wedge (a \vee c) &= ((a \vee b) \wedge a) \vee ((a \vee b) \wedge c) && \text{by (D1)} \\
&= a \vee ((a \vee b) \wedge c) && \text{by (L2) and (L4)} \\
&= a \vee ((a \wedge c) \vee (b \wedge c)) && \text{by (L2) and (D1)} \\
&= (a \vee (a \wedge c)) \vee (b \wedge c) && \text{by (L3)} \\
&= a \vee (b \wedge c) && \text{by (L4).}
\end{aligned}$$

Sec. 6.5 *The Modular and Distributive Laws*

(D2)$\Rightarrow$(D3):

$(a \wedge b) \vee (b \wedge c) \vee (c \wedge a)$

$\quad = [((a \wedge b) \vee b) \wedge ((a \wedge b) \vee c)] \vee (c \wedge a)$ by (L3), (D2)

$\quad = [b \wedge ((a \wedge b) \vee c)] \vee (c \wedge a)$ by (L2), (L4)

$\quad = [b \wedge ((a \vee c) \wedge (b \vee c))] \vee (c \wedge a)$ by (L2), (D2)

$\quad = [b \wedge (a \vee c)] \vee (c \wedge a)$ by (L2), (L3), (L4)

$\quad = (b \vee (c \wedge a)) \wedge ((a \vee c) \vee (c \wedge a))$ by (L2), (D2)

$\quad = ((b \vee c) \wedge (b \vee a)) \wedge [a \vee (c \vee (c \wedge a))]$ by (L3), (D2)

$\quad = ((b \vee c) \wedge (b \vee a)) \wedge (a \vee c)$ by (L4)

$\quad = (a \vee b) \wedge (b \vee c) \wedge (c \vee a)$ by (L2).

(D3)$\Rightarrow$(D1): We shall intersect both members of (D3) with a. On the left-hand side, we get

$a \wedge [(a \vee b) \wedge (b \vee c) \wedge (c \vee a)]$

$\quad = [a \wedge (a \vee b)] \wedge [(b \vee c) \wedge (c \vee a)]$ by (L3)

$\quad = a \wedge [(b \vee c) \wedge (c \vee a)]$ by (L4)

$\quad = [a \wedge (a \vee c)] \wedge (b \vee c)$ by (L2), (L3)

$\quad = a \wedge (b \vee c)$ by (L4).

In the case of the right-hand side:

$a \wedge [(a \wedge b) \vee (b \wedge c) \vee (c \wedge a)]$

$\quad = a \wedge [\{(b \wedge c) \vee (c \wedge a)\} \vee (a \wedge b)]$ by (L2), (L3)

$\quad = [a \wedge \{(b \wedge c) \vee (c \wedge a)\}] \vee (a \wedge b)$

by the modular law (6.5.2), since $a \wedge b \leq a$,

$\quad = [\{a \wedge (b \wedge c)\} \vee (c \wedge a)] \vee (a \wedge b)$

by the modular law (6.5.2), since $c \wedge a \leq a$,

$\quad = [(a \wedge c) \vee \{(a \wedge c) \wedge b\}] \vee (a \wedge b)$ by (L2), (L3)

$\quad = (a \wedge c) \vee (a \wedge b)$ by (L4)

$\quad = (a \wedge b) \vee (a \wedge c)$ by (L2). ∎

In general, one cannot conclude from $a \vee b = a \vee c$ that $b = c$ or from $a \wedge b = a \wedge c$ that $b = c$; that is, there are no cancellation laws for $\vee$ or $\wedge$. However, there is a joint cancellation law in a distributive lattice:

Theorem 6.5.4. *In a distributive lattice* $L_\leqslant$, *if* $a \vee b = a \vee c$ *and* $a \wedge b = a \wedge c$, *then* $b = c$.

Proof

$b = b \vee (b \wedge a) = b \vee (a \wedge b) = b \vee (a \wedge c) = (b \vee a) \wedge (b \vee c) = (a \vee b) \wedge (b \vee c)$

$= (a \vee c) \wedge (b \vee c) = (c \vee a) \wedge (c \vee b) = c \vee (a \wedge b) = c \vee (a \wedge c) = c \vee (c \wedge a)$

$= c$. ∎

6.6 Complements in Lattices

A lattice $L_\leqslant$ with 0 and 1 is **complemented** iff for each a of L there exists at least one element a' of L such that $a \vee a' = 1$ and $a \wedge a' = 0$. Such an element a' is called a **complement** of a. For example:

1. In every finite lattice, $0' = 1$ and $1' = 0$.
2. $L_\leqslant = \{\mathcal{P}(U), \subseteq\}$. Here $A \wedge A' = \varnothing$, $A \vee A' = U$, where A' has the usual meaning for subsets of U.
3. $L_\leqslant = \{\{\text{positive integral divisors of } 30\}, |\}$ where "$|$" means "is a divisor of." Here $a' = 30/a$; $a \vee a' = \text{l.u.b.} \{a, 30/a\} = \text{l.c.m.} \{a, 30/a\} = 30$ and $a \wedge a' = \text{g.l.b.} \{a, 30/a\} = \text{g.c.d.} \{a, 30/a\} = 1$, because 30 has no repeated prime factors. [Note that the integer 30 is the unit element here and that the integer 1 is the zero element. See Figure 6.6.1(a).]

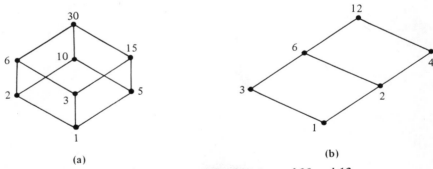

FIGURE 6.6.1. *Lattices of Divisors of 30 and 12*

4. The lattices (c) and (d) of Figure 6.1.1 are both complemented. In each case $c \vee a = 1$, $c \wedge a = 0$ and $c \vee b = 1$, $c \wedge b = 0$, so in each case c has two complements.
5. $L_\leqslant = \{\{\text{positive integral divisors of } 12\}, |\}$, where again "$|$" means "is a divisor of." Here 2 and 6 do not have complements. [See Figure 6.6.1(b).]
6. In the lattice of Figure 6.5.2(d), $b' = c$ and $c' = b$, but a and d have no complements.

Although complements need not in general be unique, we have the following theorem.

Theorem 6.6.1. *In a distributive lattice with 0 and 1, a complement is unique if it exists.*

Proof. Let $L_\leqslant$ with 0 and 1 be distributive and let $a \vee b = 1$ and $a \wedge b = 0$; also, $a \vee c = 1$ and $a \wedge c = 0$. Then $b = 1 \wedge b = (a \vee c) \wedge b = (a \wedge b) \vee (c \wedge b) = 0 \vee (c \wedge b) = (a \wedge c) \vee (c \wedge b) = c \wedge (a \vee b) = c \wedge 1 = c$. ∎

This theorem is useful for showing that a lattice is *not* distributive. For example, the lattices (c) and (d) of Figure 6.1.1 have nonunique complements, hence are not distributive, a fact that we proved earlier by producing counterexamples.

We can now show the relationship between lattice theory and Boolean algebra.

Theorem 6.6.2. *A distributive, complemented lattice is a Boolean algebra and conversely.*

Proof. We compare the list of postulates that define a distributive, complemented lattice with the list of postulates that define a Boolean algebra in Section 3.8, interpreting "+" as "$\vee$" and "·" as "$\wedge$." The conclusion follows. ∎

As preceding examples suggest, many important lattices are not Boolean algebras.

6.7 Exercises

1. Let $\mathbb{P}$ denote the set of all positive integers and let "$a \leqslant b$" mean "a divides b." Show that $\mathbb{P}$ is a lattice with $a \wedge b = $ g.c.d. $\{a, b\}$ and $a \vee b = $ l.c.m. $\{a, b\}$.

2. Give an example of an infinite lattice that has
(a) Neither zero nor unit element.
(b) A zero but not a unit element.
(c) A unit but not a zero element.
(d) Both a zero and a unit element.

3. Draw the Hasse diagram of the lattice of all partitions of $U = \{a, b, c, d\}$.

*4. Prove that in every lattice $L_\leqslant$, for all a, b, and c in L,

$$[(a \wedge b) \vee (a \wedge c)] \wedge [(a \wedge b) \vee (b \wedge c)] = a \wedge b.$$

(Show that each side is equal to or less than the other.)

5. In the lattice of positive integral divisors of $p_1 p_2 \cdots p_n$, where the p_i are distinct primes, give explicit formulas for $a \vee b$, $a \wedge b$, and a' in terms of the p_i.

6. Show that in every modular lattice $L_\leq$, if $(x \vee y) \wedge z = y \wedge z$, then $x \wedge (y \vee z) = x \wedge y$.

7. Show that if complements in a complemented lattice are not necessarily unique, then the value of $(a \wedge b') \vee (a' \wedge b)$ is not necessarily unique either, so the ring sum is not well defined in this case. (One example will suffice.)

8. State a principle of duality for lattices and apply it to Exercise 4.

*9. Show that if, in a lattice $L_\leq$, the element c covers each of a and b and if $a \neq b$, then $c = a \vee b$. Dualize. (The covering relation within any poset was discussed in Section 2.8.)

10. Let C be the ring of functions continuous on $0 \leq x \leq 1$. Define $f \leq g$ iff for all x such that $0 \leq x \leq 1$, $f(x) \leq g(x)$. Show that the corresponding l.u.b. $\{f, g\}$ and g.l.b. $\{f, g\}$ always exist in C, so $C_\leq$ is a lattice. Show also that $C_\leq$ has neither a zero element nor a unit element. Illustrate with some examples of f, g, l.u.b. $\{f, g\}$, and g.l.b. $\{f, g\}$. (For all $f \in C$, $f(x) \in \mathbb{R}$.)

11. Are these lattices distributive? Why?

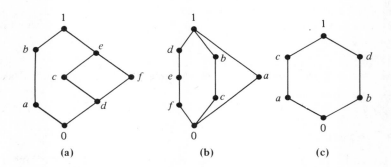

12. If a finite lattice S with $|S| = n$ is to be proved distributive by checking cases, at most how many cases actually need to be checked?

13. Prove that every lattice that is not complete necessarily contains an infinite chain. This implies that if all chains of a lattice are finite, the lattice is complete. (This problem is fairly hard.)

14. Draw (incompletely) the Hasse diagram of a lattice that has arbitrarily long but nevertheless only finite chains.

15. Show that in every chain, elements other than 0 and 1 have no complements.

16. If L_1 and L_2 denote arbitrary finite lattices, does the following diagram represent a lattice? A complemented lattice, if L_1 and L_2 are complemented? A distributive lattice, if L_1 and L_2 are distributive?

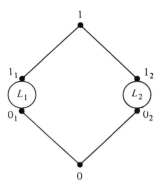

17. Show that repeated application of the operations "+" and "·" for partitions to

$$\pi_1 = \overline{\{1, 2; 3, 4; 5, 6; 7, 8\}},$$
$$\pi_2 = \overline{\{1, 2, 3, 4; 5, 6, 7, 8\}},$$
$$\pi_3 = \overline{\{\bar{1}; \bar{2}; \bar{3}; 4, 5; \bar{6}; \bar{7}; \bar{8}\}},$$
$$\pi_4 = \overline{\{\bar{1}; \bar{2}; 3, 6; \bar{4}; \bar{5}; \bar{7}; \bar{8}\}},$$

generates a total of eight distinct partitions (including the four given ones) and that these eight partitions form a lattice. Draw the Hasse diagram.

18. Show that the Hasse diagram of the lattice of subgroups of the Klein 4-group (which is generated by two elements α and β such that $\alpha^2 = \beta^2 = 1$, $\alpha\beta = \beta\alpha$) is identical with Figure 6.1.1(d).

19. Construct the Hasse diagram of the set of subgroups of the group $Q = \{\pm 1, \pm i, \pm j, \pm k\}$, where $i^2 = j^2 = k^2 = -1$ and $ij = k$, $jk = i$, $ki = j$, $ji = -k$, $kj = -i$, and $ik = -j$.

20. Prove that if $L_\leq$ has no 0, then it has an infinite descending chain and, dually, if $L_\leq$ has no 1, it has an infinite ascending chain. Provide examples of such chains in the lattice $L_\leq$, where

$$L = \left\{ 1 \pm \frac{1}{p^n} \,\middle|\, n \in \mathbb{P}, p \in \mathbb{P}, p \geq 2 \right\},$$

and "$\leq$" has its usual arithmetic meaning. (Thus the conditions "no 0" and "no 1" are not *necessary*.)

21. Is the lattice $\mathbb{R}_\leq$ complete? Is the lattice $\mathbb{Q}_\leq$ complete?

22. Prove that the lattices in Figure 6.7.1 are not distributive.

23. Prove that $L_\leq$ is distributive if the equations $x \vee y = x \vee z$ and $x \wedge y = x \wedge z$ always imply that $y = z$.

24. Prove that in every lattice $L_\leq$, for all x, y, and z in L,

$$x \wedge (y \vee z) \geq (x \wedge y) \vee (x \wedge z) \quad \text{and} \quad x \vee (y \wedge z) \leq (x \vee y) \wedge (x \vee z),$$

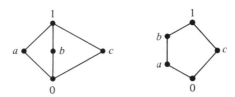

FIGURE 6.7.1. *Nondistributive Lattices*

and hence that, if $x \geq z$,

$$x \wedge (y \vee z) \geq (x \wedge y) \vee z \quad \text{and} \quad x \vee (y \wedge z) \leq (x \vee y) \wedge z.$$

25. Prove that in a distributive lattice $L_\leq$ with 0 and 1, if a has a complement $\tilde{a}$, then

$$a \vee (\tilde{a} \wedge b) = a \vee b \quad \text{and} \quad a \wedge (\tilde{a} \vee b) = a \wedge b.$$

26. Using the results of the preceding exercise, prove that in any distributive lattice $L_\leq$, the subset C of all elements that have complements is a sublattice of $L_\leq$.

*27. Show that if S is a subset of the lattice L such that $s_1 \in S$ and $s_2 \in S$ imply that $s_1 \wedge s_2 \in S$ and $s_1 \vee s_2 \in S$, then S is actually a sublattice of L.

28. Using the properties of the l.u.b. and g.l.b. (not the distributive law!), complete the proof of Lemma 6.4.2.

29. Prove that the set of all subgroups of a group G is a lattice, where $H \vee K$ is the smallest subgroup containing both H and K and $H \wedge K$ is the set-theoretic intersection of H and K. What does "$\leq$" mean here?

30. (a) Prove that the lattice of all subgroups of a commutative group is modular.

(b) Find the lattice of all subgroups of the alternating group on four letters and show that it is not modular. (See Exercise 7.14.1 for a definition of the alternating group.)

31. Construct the lattice of subgroups of the additive group $\mathbb{B}^3$. (The addition is componentwise modulo 2.)

32. Show that the lattice of normal subgroups of a group is always modular.

33. Let L be a class of subsets of a universal set U [that is, let L be a subset of $\mathcal{P}(U)$] such that for all A and B in L, $A \cup B$, $A \cap B$, and $\bar{A}$ are in L. Such a class of subsets of U is called a **field of subsets** of U.

(a) Show that every field of subsets of U is a Boolean algebra.

(b) Give a nontrivial example derived from $U = \{1, 2, 3, 4, 5, 6\}$.

(c) Show that if $A_1, A_2, \ldots, A_n$ are subsets of U, then the set of subsets consisting of all possible unions of "minterms" of the A's is a field of subsets of U.

6.8 Free Distributive Lattices

In this section, in order to make writing lengthy lattice expressions more convenient, we use $+$ and $\cdot$ in the usual way in place of $\vee$ and $\wedge$, respectively.

Given n literals $x_1, x_2, \ldots, x_n$, the set L_n of **lattice polynomials** in these literals is defined to consist of

1. $x_1, x_2, \ldots, x_n$.
2. $x_1 + x_2, x_1 x_2$.
3. Any expression obtained by substituting any lattice polynomial, appropriately enclosed in parentheses, for any literal appearing in any lattice polynomial.

For example, if $n \geq 4$, then $x_3 + x_4$, $x_{n-1} x_n$, $(x_1 + x_2) x_4$, $x_1 + x_2 + \cdots + x_n$, and $x_1(x_2 + x_1 x_3) + x_2 x_4$ are all lattice polynomials.

For arbitrary members a, b, and c of L_n, we now postulate the familiar rules:

(L1)	$a + a = a$, $aa = a$,
(L2)	$a + b = b + a$, $ab = ba$,
(L3)	$a + (b + c) = (a + b) + c$, $a(bc) = (ab)c$,
(L4)	$a + ab = a$, $a(a + b) = a$,
(D1)	$a(b + c) = ab + ac$,
(D2)	$a + bc = (a + b)(a + c)$,

thus making L_n a distributive lattice. This lattice is called the **free distributive lattice** on $x_1, x_2, \ldots, x_n$. Of course, (D3) also holds in L_n, $a \leq b$ iff $ab = a$, $a \leq b$ iff $a + b = b$, and two lattice polynomials are equal iff either can be transformed into the other with the aid of (L1) to (L4), (D1), and (D2).

The laws (D1) and (D2) permit us to expand a given lattice polynomial into a sum-of-products form in the usual way; the result may then be reduced, as is usually desirable, by (L1) to (L4), until no further reduction by (L1) or (L4) is possible. For example:

$(x_1 + x_2)(x_1 + x_3)(x_1 + x_2 + x_4)(x_3 + x_4)$

$= (x_1 + x_2 x_3)(x_4 + x_1 x_3 + x_2 x_3)$ by (D2)

$= x_1 x_4 + x_1 x_3 + x_1 x_2 x_3 + x_2 x_3 x_4 + x_1 x_2 x_3 + x_2 x_3$ by (D1)

$= x_1 x_4 + x_1 x_3 + x_2 x_3$ by (L4).

(Can you do it more quickly?)

If neither (L1) nor (L4) can be used to remove literals or terms from such a sum of products, it is said to be **minimal**. We have the following theorem.

Theorem 6.8.1. *A given lattice polynomial has a unique minimal sum-of-products representation.*

Proof. Let f and g be minimal sum-of-products representations of the same lattice polynomial, so $f = g$. Suppose that $f = h + p$, where p is a product term that does not appear in the sum g. Then $f + p = f$, so $g + p = g$, and since p is not a term of g but is necessarily absorbed by some term of g, $p = qr$, where q is a term of g. Then $g + q = g$, so $f + q = f$ and either q is a product term of f or $q = st$, where s is a product term of f. Then either q or s absorbs p and the minimality of f is contradicted. Hence f contains no term that is not in g and, of course, vice versa. ∎

If n is small, the distinct sums of this type are not hard to list. For $n = 1$, there is just the polynomial x_1. For $n = 2$, there are four; for $n = 3$, there are 18. The Hasse diagrams for $n = 2$ and $n = 3$ appear in Figure 6.8.1. The numbers of these polynomials are known for small n [p. 188 of Harrison (1965)]: $|L_4| = 166$, $|L_5| = 7579$, and $|L_6| = 7,828,352$, but no general formula for L_n appears to be known.

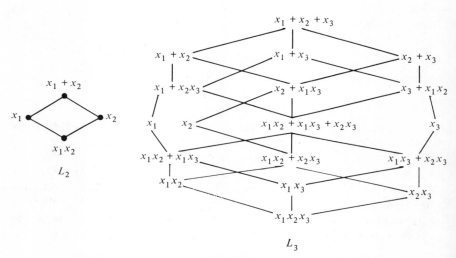

FIGURE 6.8.1. *Lattice Polynomials, $n = 2$ and $n = 3$*

An application of lattice polynomials is given in the next section.

If we interpret the literals $x_1, x_2, \ldots, x_n$ as variables over $\mathbb{B} = \{0, 1\}$, then the set of functions $L_n \cup \{0, 1\}$ may be interpreted as the set of *positive* or *monotone-increasing* switching functions. The minimal sum-of-products representations of Theorem 6.8.1 are then the minimal sums of Section 3.19, for positive switching functions. (See Exercises 3.21.4, 5, and 6.)

6.9 Compatibility Relations and Covers

Let S be an arbitrary set. A reflexive and symmetric binary relation γ on S is called a **compatibility relation** on S. If $s_1 \gamma s_2$, s_1 and s_2 are said to be

Sec. 6.9 Compatibility Relations and Covers

compatible. An equivalence relation on S is thus a compatibility relation that is also transitive. Let $\{S_\alpha\}$ be the set of maximal subsets of mutually compatible elements of S as determined by γ. Since γ is not required to be transitive, it is not necessary that $S_\alpha \cap S_\beta = \emptyset$ when $\alpha \neq \beta$. However, $\bigcup_\alpha S_\alpha = S$ since every element of S is compatible at least with itself. Conversely, every set of subsets S_α of S such that $\bigcup_\alpha S_\alpha = S$ determines a compatibility relation γ on S, for the property of belonging to a common subset S_α is reflexive and symmetric, but not necessarily transitive since the S_α may overlap. This is analogous to the fact that every partition of S determines a corresponding equivalence relation on S, and vice versa.

When $\bigcup_\alpha S_\alpha = S$, we call $\mathscr{C} = \{S_\alpha\}$ a **cover** for S, so every compatibility relation on S determines a cover for S, and conversely. The subsets S_α of a cover are called its **components**.

For example, let $S = \{s_0, s_1, \ldots, s_9\}$ and let $\mathscr{C} = \{S_1, S_2, \ldots, S_6\}$, where $S_1 = \{s_0, s_2, s_4, s_6\}$, $S_2 = \{s_0, s_1, s_3, s_4, s_9\}$, $S_3 = \{s_3, s_5, s_6, s_7\}$, $S_4 = \{s_1, s_7, s_8\}$, $S_5 = \{s_4, s_7, s_8, s_9\}$, and $S_6 = \{s_1, s_3, s_7\}$. Then $\mathscr{C}$ is a cover for S, since $\bigcup_\alpha S_\alpha = S$. The corresponding compatibility relation γ is such that $s_i \gamma s_j$ iff s_i and s_j both belong to some S_k.

In certain classes of problems in which S is *finite*, an initial cover $\mathscr{C}$ for S is determined according to criteria of compatibility that are characteristic of the class of problems at hand. Then one has to find a subset of $\mathscr{C}$ that is also a cover of S and contains a minimum number of the subsets S_α. There is a simple algorithm for determining such a minimum cover once $\mathscr{C}$ is given.

One first makes a table with a row for each element S_i of $\mathscr{C}$ and a column for each element s_j of S (Figure 6.9.1). If $s_j \in S_i$, we enter an "×" in the ij-position in the table; otherwise, no entry is made.

	s_1	s_2	$\cdots$	s_n
S_1				
S_2				
$\vdots$				
S_m				

FIGURE 6.9.1. *Covering Table*

If a column k contains precisely one ×, say in row i, the corresponding subset S_i is an **essential component** of any cover, for only by including S_i can we cover s_k. If S_i is essential, we mark its row with an asterisk to identify that fact and then delete every column that contains an × in row i, since the elements s_j associated with these columns are all covered by S_i. This process is repeated for every essential component S_i.

After the essential components have been determined and all columns corresponding to elements covered by these components have been deleted, there remain only columns each of which contains at least two ×'s, so choices must be made. If there is no basis for preferring one S_i to another except for its ability to contribute to a minimum cover, it is not hard at this point to determine all minimum covers. One treats the labels S_i of the rows that have ×'s in undeleted columns as the literals of a free distributive lattice and, for each remaining column, writes the sum of the S_i that correspond to the ×'s in that column. The product of all these sums is then expanded into a sum-of-products form σ. The initially written product of sums expresses the fact that to obtain a cover for the subset $\{s_{j_1}, s_{j_2}, \ldots, s_{j_k}\}$ of still-to-be-covered elements of S, one must choose one *or* other of the S_i that cover s_{j_1}, *and* one *or* other of the S_i that cover s_{j_2}, *and* so on. That is, addition has the meaning *or*, multiplication has the meaning *and*, in the choosing process. In the sum-of-products expansion σ, a term $S_{i_1} S_{i_2} \cdots S_{i_r}$ means that the set of subsets $\{S_{i_1}, S_{i_2}, \ldots, S_{i_r}\}$ is a cover of S.

The appropriateness of the postulates of a free distributive lattice for the process of identifying alternative choices is not hard to verify: a choice "a or a" is simply the choice "a," so $a + a = a$; the choice "a or b" is equivalent to the choice "b or a," so $a + b = b + a$; the choice "a and one of a and b" is most simply satisfied by choosing a, so $a(a+b) = a$; and so on.

Each term of the sum-of-products σ (reduced as far as possible by the lattice postulates) with a minimum number of factors identifies a choice of subsets S_i which, together with those that are essential, constitutes a minimum cover for S, and the set of all such minimum terms yields in this way the set of all minimum covers.

We illustrate the algorithm just described for the example given earlier. First we prepare the table described above (Figure 6.9.2). The table reveals

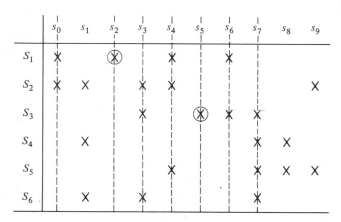

FIGURE 6.9.2. *Reduced Covering Table*

that S_1 is essential because it alone covers s_2; S_3 is essential because it alone covers s_5. These two subsets cover all elements of S except s_1, s_8, and s_9.
We now write, expand, and reduce the product of sums described earlier:

$$\sigma = (S_2 + S_4 + S_6)(S_4 + S_5)(S_2 + S_5) = (S_2 + S_4 S_5 + S_5 S_6)(S_4 + S_5)$$
$$= S_2 S_4 + S_2 S_5 + S_4 S_5 + S_5 S_6.$$

This result shows that there are four minimal covers: $\{S_1, S_3, S_2, S_4\}$, $\{S_1, S_3, S_2, S_5\}$, $\{S_1, S_3, S_4, S_5\}$, and $\{S_1, S_3, S_5, S_6\}$.

Variations of this procedure are used in Boolean algebra, automata theory, and combinatorics. We discussed one variation in Section 3.19.

6.10 Atomic Lattices and Boolean Algebras

Any element of a lattice $L_\leqslant$ which covers 0 is an **atom** of $L_\leqslant$; that is, an atom of $L_\leqslant$ is an element $a \neq 0$ of L such that $0 \leqslant b \leqslant a$ implies that $b = 0$ or $b = a$. Equivalently, $a \neq 0$ is an atom of $L_\leqslant$ if there is no b in L such that $a > b > 0$. Dually, any element of $L_\leqslant$ that is covered by 1 is a **coatom** (or an **antiatom**) of $L_\leqslant$. If every nonzero element of $L_\leqslant$ includes at least one atom, $L_\leqslant$ is an **atomic lattice**; if every element except 1 is included in at least one coatom, $L_\leqslant$ is **coatomic**.

As examples, consider

1. The lattice $\{\mathbb{P}, \leqslant\}$, where "$\leqslant$" has its usual arithmetic meaning; the single atom is the integer 2, and there are no coatoms.

2. The lattice $\{\mathbb{B}^n, \leqslant\}$, where "$\leqslant$" has the customary meaning for strings of n 0's and 1's; the atoms are the strings containing a single 1, and the coatoms are the strings containing a single 0.

3. The infinite lattice of Figure 6.10.1, with atoms $a_{11}, a_{21}, a_{31}, \ldots$ and coatoms $a_{11}, a_{22}, a_{33}, \ldots$.

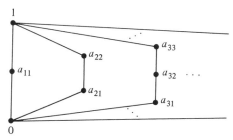

FIGURE 6.10.1. *An Infinite Atomic Lattice*

4. The lattice $\{\mathscr{P}(U), \subseteq\}$; the atoms are the singleton subsets, and the coatoms are the subsets $U - \{u\}$, where $u \in U$.

5. The lattice $\{\mathbb{Z}, \leqslant\}$, where "$\leqslant$" has its usual arithmetic meaning; there are neither atoms nor coatoms.

Theorem 6.10.1. *Every lattice $L_\leqslant$ in which every descending chain is finite is atomic. In particular, every finite lattice is atomic.*

Proof. Let b_1 be any nonzero element of L. If there exists no element b_2 such that $b_1 > b_2 > 0$, then b_1 is an atom. If such a b_2 does exist, we seek b_3 such that $b_2 > b_3 > 0$. If no such b_3 exists, b_2 is an atom. Otherwise, we seek b_4 such that $b_3 > b_4 > 0$. Continuing thus, since every descending chain is finite, we obtain a chain

$$b_1 > b_2 > \cdots > b_k > 0$$

such that there is no b_{k+1} such that $b_k > b_{k+1} > 0$. Then b_k is an atom. ∎

An immediate consequence of the theorem is that if $L_\leqslant$ with 0 is not atomic, then it contains at least one infinite descending chain. On the other hand, the converse of this observation is not true, for $\{\mathscr{P}(U), \subseteq\}$ is atomic for all U, but if U is infinite, an infinite descending chain is always possible in $\mathscr{P}(U)$ (explain how).

Some basic properties enjoyed by atoms are given in the two following lemmas.

Lemma 6.10.1. *In every lattice $L_\leqslant$, the meet of any two distinct atoms a_1 and a_2 is 0.*

Proof. If $L_\leqslant$ has at least one atom, $L_\leqslant$ has an element 0. If a_1 and a_2 are distinct atoms of $L_\leqslant$, then $a_1 \not\leqslant a_2$, so $a_1 \wedge a_2 \neq a_1$. However, $a_1 \wedge a_2 \leqslant a_1$ in any case, so we have $0 \leqslant a_1 \wedge a_2 < a_1$. Since a_1 is an atom, this implies that $a_1 \wedge a_2 = 0$. ∎

Lemma 6.10.2. *If a is an atom of a lattice $L_\leqslant$, and if x is an arbitrary element of $L_\leqslant$, then $a \wedge x = a$ or $a \wedge x = 0$.*

Proof. Either $a \leqslant x$, in which case $a \wedge x = a$, or $a \not\leqslant x$ and therefore $a \wedge x \neq a$. Since $a \wedge x \leqslant a$ in any case, we must then have $a \wedge x < a$; that is, when $a \not\leqslant x$, $0 \leqslant a \wedge x < a$. Since a is an atom, $a \wedge x = 0$ in this case. ∎

To illustrate, in the pentagonal lattice of Figure 6.1.1(c), b and c are atoms since they cover 0; $b \wedge c = 0$ and $b \wedge a = b$.

We now turn to our main purpose in this section, further study of finite Boolean algebras, which are atomic by Theorem 6.10.1 and complete by Corollary 6.2.1, for these are the most important atomic lattices in nonmathematical applications. We return to the notations "+" (sum) and "·" (product) in place of "∨" (join) and "∧" (meet), in accord with the most common usage in applications.

Lemma 6.10.3. *In every finite Boolean algebra B, the sum of all the atoms is 1.*

Sec. 6.10 Atomic Lattices and Boolean Algebras

Proof. Denote the atoms by $a_1, a_2, \ldots, a_n$ and let $\sum_{i=1}^{n} a_i = b$. Since B is complete, b is an element of B. Suppose that $\bar{b} \neq 0$. Then there exists an atom $a_j \leq \bar{b}$, since B is atomic, so $a_j\bar{b} = a_j$. But for every atom a_j, $a_jb = a_j$, since $a_j \leq b$. Hence $a_j\bar{b} = a_jb \cdot \bar{b} = 0$, which is a contradiction. Hence $\bar{b} = 0$ and $b = 1$. ∎

Lemma 6.10.4. *Given a finite Boolean algebra B, and an arbitrary nonzero element b of B, every atom a of B is included in precisely one of b and $\bar{b}$.*

Proof. In every Boolean algebra, $a\bar{b} = 0$ implies that $a \leq b$. Hence if $a \not\leq b$, then $a\bar{b} \neq 0$ and, by Lemma 6.10.2, $a\bar{b} = a$; that is, $a \leq \bar{b}$. ∎

Lemma 6.10.5. *Given elements b and c of a finite Boolean algebra B, if $b < c$, there exists an atom of B included in c but not in b.*

Proof. Since $b < c$, $\bar{b}c \neq 0$, for $\bar{b}c = 0$ implies that $c \leq b$. Hence there exists an atom $a_j \leq \bar{b}c$. This implies that $a_j \leq \bar{b}$ and $a_j \leq c$. By Lemma 6.10.4, $a_j \not\leq b$. ∎

Theorem 6.10.2. *Every nonzero element c of a finite Boolean algebra B is the sum of precisely the atoms it includes. No other sum of atoms is equal to c.*

Proof. Let b denote the sum of all the atoms included in c. Since b is the least upper bound of this set of atoms and c is an upper bound, $b \leq c$. If $b < c$, there exists an atom included in c that is not in b. This contradicts the fact that every atom included in c is also included in b. Hence $b = c$.

Now let $\tilde{b}$ be any sum of atoms different from the sum b. Then there exists at least one atom a_j such that $a_j \leq \tilde{b}$, $a_j \not\leq b$ or such that $a_j \not\leq \tilde{b}$, $a_j \leq b$. In the first case, $a_j\tilde{b} = a_j$ and $a_jb \neq a_j$, so $a_j\tilde{b} \neq a_jb$, which implies that $\tilde{b} \neq b$. In the second case, the argument is similar. Hence $\tilde{b} \neq c$, so the representation is unique. ∎

Corollary 6.10.1. *Every finite Boolean algebra B contains exactly 2^n distinct elements, where n is the number of atoms of B.*

Proof. Since every finite Boolean algebra has at least one nonzero element 1, by Theorem 6.10.1 such an algebra necessarily contains at least one atom. (The atom of $\mathbb{B} = \{0, 1\}$ is 1 and $|\mathbb{B}| = 2^1$.) Let the atoms of B be $a_1, a_2, \ldots, a_n$. Then there are 2^n distinct joins

(6.10.1) $$\alpha_1 a_1 + \alpha_2 a_2 + \cdots + \alpha_n a_n,$$

where each α_j is 0 or 1. Theorem 6.10.2 implies that these are all distinct, so $|B| = 2^n$. ∎

Corollary 6.10.2. *For each positive integer n, there exists a Boolean algebra B with precisely 2^n elements.*

Proof. It suffices to note that the algebra $\mathbb{B}^n$ provides an example for each n. ∎

Theorem 6.10.3. *Every finite Boolean algebra B is isomorphic to $\mathbb{B}^n$ for some positive integer n.*

Thus there is essentially just one Boolean algebra with 2^n elements, but it may be modeled in many ways.

Proof. If B has n atoms $a_1, a_2, \ldots, a_n$, the correspondence

(6.10.2) $\qquad \alpha_1 a_1 + \alpha_2 a_2 + \cdots + \alpha_n a_n \leftrightarrow (\alpha_1, \alpha_2, \ldots, \alpha_n)$

suggested by (6.10.1) is the required operation-preserving bijection. The reader may provide full details. ∎

6.11 Exercises

*1. Given the set $S = \{s_i | i = 1, 2, \ldots, 12\}$ and the subsets $S_1 = \{s_1, s_3, s_9\}$, $S_2 = \{s_1, s_4, s_6, s_9, s_{11}\}$, $S_3 = \{s_2, s_3, s_4\}$, $S_4 = \{s_5, s_7, s_8, s_{10}, s_{11}\}$, $S_5 = \{s_2, s_5, s_8, s_9\}$, $S_6 = \{s_6, s_7, s_{11}\}$, and $S_7 = \{s_3, s_4, s_6, s_{12}\}$, find all minimum covers.

*2. If each set S_α of a given cover $\mathscr{C}$ is assigned a real-number weight w_α, if the weight of a cover $\{S_{i_1}, S_{i_2}, \ldots, S_{i_k}\}$ is $\sum_{j=1}^{k} w_{i_j}$, and if a *preferred cover* is a minimum cover (as already defined) of minimum weight, how may all preferred covers be determined?

3. Using the definition $w_\alpha = |S_\alpha|$, find all preferred covers for the set S of Exercise 1.

*4. A switching function $f(X)$, where $X = (x_1, x_2, \ldots, x_n)$, is said to be **monotone-increasing** iff $X_i \leq X_j$ always implies that $f(X_i) \leq f(X_j)$, and to be **monotone-decreasing** iff $X_i \leq X_j$ implies that $f(X_i) \geq f(X_j)$. Show how to arrange the function values in a table with $n+1$ columns in a way that facilitates determination of the monotone character of f, whether increasing, decreasing, or neither. Illustrate by determining all monotone-increasing and all monotone-decreasing functions of two variables.

*5. Prove that the set of monotone-increasing functions of n variables is indeed $L_n \cup \{0, 1\}$ as indicated in Section 6.8. [See Harrison (1965), p. 189.]

6. Using the result of Exercise 5, list in some systematic fashion all monotone-increasing functions of x_1, x_2, and x_3. (There are 20 of them.)

7. What are the atoms and coatoms of the lattices in Figure 6.11.1?

*8. In the Boolean algebra of positive integral divisors of $N = p_1 p_2 \cdots p_n$, where the p_i are distinct primes, what are the atoms and the coatoms? (Recall that "∨" means "least common multiple," "∧" means "greatest common divisor," and "≤" means "divides," in this case.)

9. Show that the set of all positive integral divisors of $M = p_1^{\alpha_1} p_2^{\alpha_2} \cdots p_n^{\alpha_n}$, where the p_i are distinct primes and the α_i are all positive integers, is an atomic and coatomic lattice. (Here ∨, ∧, and ≤ have the meanings given in Exercise 8.)

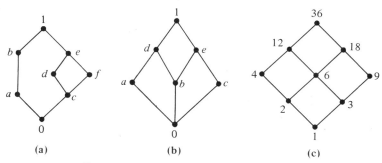

FIGURE 6.11.1. *Lattices for Exercise 7*

10. Show that in the Boolean algebra of switching functions of n variables, if $f > g$, there exists a minterm p such that $p \leq f$ and $pg = 0$. Give an example in the case $n = 3$. What theorem of Section 6.10 does this exemplify?

11. Prove that if $a_{j_1}, a_{j_2}, \ldots, a_{j_n}$ is an arbitrary permutation of the atoms $a_1, a_2, \ldots, a_n$ of a finite Boolean algebra B, then

$$\overline{\sum_{j=1}^{k} a_{ij}} = \sum_{j=k+1}^{n} a_{ij}.$$

12. Consider the set of all possible expressions that can be developed from m symbols $a_1, a_2, \ldots, a_m$ by means of a finite number of applications of the Boolean operations $+$, $\cdot$, and $^-$. Two such expressions are equal if one can be reduced to the other by means of the laws of Boolean algebra. The resulting system is called the **free Boolean algebra** with generators $a_1, a_2, \ldots, a_m$. What are the atoms of this algebra? How many distinct elements does it contain? Write a complete set of distinct expressions for the case $n = 2$.

*13. If B is the free Boolean algebra with generators $a_1, a_2, \ldots, a_m$, the set $B[x_1, x_2, \ldots, x_n]$ of all Boolean polynomials over B is also a Boolean algebra. (These polynomials look like switching functions except that coefficients come from B instead of just from $\mathbb{B} = \{0, 1\}$.) Determine the atoms and coatoms of this algebra. Then define and write expressions for the disjunctive and conjunctive normal forms of a polynomial in this algebra.

14. Using the results of Exercise 13, write the polynomial

$$\bar{a}_1 x_1 x_2 + a_2 x_2 \bar{x}_3 + (a_1 + \bar{a}_2) \bar{x}_1 \bar{x}_2 x_3 \qquad (m = 2, n = 3)$$

in both disjunctive and conjunctive normal form.

15. Regarding the Boolean polynomials $B[x_1, x_2, \ldots, x_n]$ over an arbitrary finite Boolean algebra B as functions from B^n to B, show that each such function is determined by its values at the 2^n combinations of 0's and 1's.

16. Prove that in every finite Boolean algebra B, atoms and coatoms are complements of each other.

17. Write the duals (dealing with coatoms rather than atoms) of the lemmas and theorems of Section 6.10.

18. Give full details of the proof of Theorem 6.10.3.

Exercises 19 to 21 require a knowledge of linear algebra (e.g., from Chapter 7).

19. In the set $L(\mathbb{R}^n)$ of all vector subspaces of the vector space $\mathbb{R}^n$, let $X \vee Y$ denote the smallest subspace containing each of the subspaces X and Y. Let $X \wedge Y$ denote the intersection of X and Y. Use Theorem 6.4.2 to prove that $L(\mathbb{R}^n)$ is a lattice. What is the meaning of "$\leq$" in this case? Is $L(\mathbb{R}^n)$ complete? Show by means of a counterexample that $L(\mathbb{R}^n)$ is not distributive.

20. Let $\mathbb{E}^n$ denote $\mathbb{R}^n$ with the usual definition of orthogonality added. For the zero subspace Z of $\mathbb{E}^n$, define the complement thus: $\bar{Z} = \mathbb{E}^n$. Define also $\bar{\mathbb{E}^n} = Z$. For every other subspace V of $\mathbb{E}^n$, define $\bar{V}$ to be the orthogonal complement of V. [See p. 363 of Hohn (1973) for details.] Show that this definition of complement makes $L(\mathbb{E}^n)$ a complemented lattice in the sense of Section 6.6 and that complements are unique.

21. What are the atoms and coatoms of $L(\mathbb{E}^n)$?

6.12 Closed Partitions in a Finite-State Machine

Throughout this section we deal with the finite-state machine $M = [S, X, Z, \tau, \omega]$ and certain partitions of the state set S. The set $\mathscr{L}$ of all partitions of the set S forms a lattice with operations $+$ and $\cdot$ (Section 2.14).

For any $s \in S$ and any input sequence $\mathbf{x} \in X^*$, we denote by

(6.12.1) $$s\mathbf{x} = \tau(s, \mathbf{x})$$

the final state after the machine is started in state s and sequence $\mathbf{x}$ has entered. In the example of Figure 6.12.1,

$$C1 = \tau(C, 1) = B \quad \text{and} \quad B(010) = (B0)10 = D10 = A0 = C.$$

If T is a subset of S and $\mathbf{x} \in X^*$, denote $T\mathbf{x} = \{t\mathbf{x} | t \in T\}$. If π is a partition of the set S and $s_1, s_2 \in S$, we write

$$s_1 \equiv s_2 \pmod{\pi}$$

iff s_1 and s_2 are in the same block of π; "$\equiv \pmod{\pi}$" is simply the equivalence relation determined by π on S.

Partition π of the state set S is a **closed partition (SP-partition, partition with the substitution property)** iff

(6.12.2) $$s_1 \equiv s_2 \pmod{\pi} \text{ implies } s_1\mathbf{x} \equiv s_2\mathbf{x} \pmod{\pi},$$

Sec. 6.12 Closed Partitions in a Finite-State Machine

for all $s_1, s_2 \in S$, $x \in X$. This means that the elements of a given block of π all have their x-successors in the same "next block" of π; application of input x induces in this case a well-defined mapping $\pi \to \pi$.

	τ		ω	
	0	1	0	1
A	C	A	0	1
B	D	A	1	0
C	D	B	1	1
D	C	A	0	0

FIGURE 6.12.1. Machine M_1: $S = \{A, B, C, D\}$, $X = \{0, 1\} = Z$

In the example of Figure 6.12.1, $\pi = \{\{A, B\}, \{C, D\}\}$ is closed, as we see by checking that

$$A0 \equiv B0, \quad A1 \equiv B1, \quad C0 \equiv D0, \quad C1 \equiv D1 \pmod{\pi};$$

in terms of blocks,

$$\{A, B\}0 \subseteq \{C, D\}, \quad \{A, B\}1 \subseteq \{A, B\}, \quad \{C, D\}0 \subseteq \{C, D\},$$

and

$$\{C, D\}1 \subseteq \{A, B\}.$$

For convenience, instead of $\pi = \{\{A, B\}, \{C, D\}\}$ we shall simply write $\pi = (A\ B)(C\ D)$.

In this section we shall derive some properties of closed partitions and present an algorithm for finding them. In the next section we shall see how they can often be used to decompose a given machine into two or more smaller machines.

Theorem 6.12.1. *If π_1 and π_2 are closed partitions of S, so are $\pi_1 \pi_2$ and $\pi_1 + \pi_2$.*

Proof. First we consider $\pi_1 \pi_2$. If C is a block of $\pi_1 \pi_2$, then $C = A \cap B$, A a block of π_1, B a block of π_2. For any $x \in X$, then, $Ax \subseteq A'$ and $Bx \subseteq B'$ for blocks A' of π_1 and B' of π_2, so $Cx = (A \cap B)x \subseteq Ax \cap Bx \subseteq A' \cap B'$, a block of $\pi_1 \pi_2$.

Next consider $\pi_1 + \pi_2$. If D is a block of $\pi_1 + \pi_2$, then

$$D = A_1 \cup B_1 \cup A_2 \cup B_2 \cup A_3 \cup B_3 \cup \cdots \quad \text{(finite union)},$$

where all $A_i \in \pi_1$, all $B_i \in \pi_2$, and $A_1 \cap B_1 \neq \varnothing$, $B_1 \cap A_2 \neq \varnothing$, $A_2 \cap B_2 \neq \varnothing, \ldots$. For any $x \in X$ we have $A_i x \subseteq A'_i$ and $B_i x \subseteq B'_i$ for blocks $A'_i \in \pi_1$, $B'_i \in \pi_2$, so that

(6.12.3) $\quad Dx = A_1 x \cup B_1 x \cup A_2 x \cup \cdots \subseteq A'_1 \cup B'_1 \cup A'_2 \cup \cdots.$

Since

$$\emptyset \neq (A_1 \cap B_1)x \subseteq A_1x \cap B_1x \subseteq A_1' \cap B_1', \qquad \emptyset \neq (B_1 \cap A_2)x \subseteq B_1' \cap A_2',$$

and so on, the right-hand side of (6.12.3) is contained in a single block of $\pi_1 + \pi_2$. ∎

Corollary 6.12.1. *The closed partitions of S form a sublattice $\mathscr{C}$ of the lattice $\mathscr{L}$ of all partitions of S.* (Sublattice $\mathscr{C}$ is the π-**lattice** of the machine M.)
Proof. Theorem 6.12.1 and Exercise 6.7.27. ∎

If s_i and s_j are distinct elements of S, we denote by $\tau_{s_is_j}$ the partition of S with block $(s_i\,s_j)$, all other states being in singleton blocks. (Partition $\tau_{s_is_j}$ covers the 0-partition 0 in the Hasse diagram of $\mathscr{L}$.) In the example of Figure 6.12.1, $\tau_{AC} = (AC)(B)(D)$. The **basic partition** determined by s_i and s_j, denoted by $\pi_{s_is_j}$, is the product of all closed partitions π of M with $\tau_{s_is_j} \leq \pi$; by Theorem 6.12.1, $\pi_{s_is_j}$ is closed. In fact, $\pi_{s_is_j}$ is the smallest closed partition of M in which $s_i \equiv s_j$; $\pi_{s_is_j}$ may well be the unit partition 1.

If $s_i, s_j \in S$, $x \in X$, and $s_ix \neq s_jx$, we say that $\{s_ix, s_jx\}$ is an **implied pair** of the pair $\{s_i, s_j\}$. We can find the basic partition $\pi_{s_is_j}$ by finding all implied pairs of the pair $\{s_i, s_j\}$, then all implied pairs of those, and so on, until no new implied pairs are obtained. By uniting overlapping implied pairs, we obtain the nonsingleton blocks of $\pi_{s_is_j}$. In the example, writing pairs and blocks without commas, and not bothering to record any implied pairs that have already been recorded in the same line of computations, we have

$$AB \Rightarrow CD, \quad \text{so} \quad \pi_{AB} = (A\,B)(C\,D);$$
$$AC \Rightarrow CD, AB, \quad \text{so} \quad \pi_{AC} = (A\,B\,C\,D) = 1;$$
$$AD \Rightarrow \text{no pairs}, \quad \text{so} \quad \pi_{AD} = (A\,D)(B)(C) = \tau_{AD};$$
$$BC \Rightarrow AB \Rightarrow CD, \quad \text{so} \quad \pi_{BC} = (A\,B\,C\,D) = 1;$$
$$BD \Rightarrow CD \Rightarrow AB, \quad \text{so} \quad \pi_{BD} = (A\,B\,C\,D) = 1;$$
$$CD \Rightarrow AB, \quad \text{so} \quad \pi_{CD} = (A\,B)(C\,D) = \pi_{AB}.$$

After finding the basic partitions, we can find any other closed partitions with the aid of the following lemma.

Lemma 6.12.1. *Any non-0 closed partition is a sum of basic partitions.*
Proof. If π is closed and $\pi \neq 0$, denote $\pi^* = \sum \{\pi_{s_is_j} | s_i \equiv s_j \pmod{\pi}\}$. It suffices to show that $\pi^* = \pi$. Since each $\pi_{s_is_j} \leq \pi$, also $\pi^* \leq \pi$. But if $s_i \equiv s_j \pmod{\pi}$, then by the definition of π^*, $s_i \equiv s_j \pmod{\pi^*}$, so that $\pi \leq \pi^*$. ∎

In our example, addition of closed partitions yields no new closed partitions; the π-lattice is as pictured in Figure 6.12.2.

Sec. 6.12 *Closed Partitions in a Finite-State Machine*   293

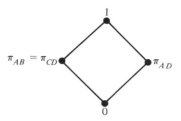

FIGURE 6.12.2. π-lattice of Machine M_1

It is not always necessary to compute all the basic partitions individually in the manner just illustrated. We shall see in the next example that early computations yield information useful in later steps.

For the example in Figure 6.12.3, we construct a stair-step, *implied-pair table*. In the cell for s_i and s_j we enter all their implied pairs and leave a space

	τ		ω	
	0	1	0	1
A	G	D	1	0
B	F	C	0	0
C	E	F	0	1
D	B	A	1	1
E	A	F	1	0
F	H	C	1	0
G	C	D	0	0
H	D	A	1	1

	A	B	C	D	E	F	G
B	FG, CD / 1						
C	EG, DF / π_2	EF, CF / 1					
D	BG / π_1	BF, AC / π_4	BE, AF / π_1				
E	AG, DF / π_3	AF, CF / π_1	AE / π_3	AB, AF / 1			
F	GH, CD / π_1	FH / π_4	EH / π_1	BH, AC / π_2	AH, CF / 1		
G	CG / π_3	CF, CD / π_1	CE, DF / π_3	BC, AD / 1	AC, DF / π_2	CH, CD / 1	
H	DG, AD / 1	DF, AC / π_2	DE, AF / 1	BD / π_4	AD, AF / π_1	DH, AC / π_4	CD, AD / π_1

FIGURE 6.12.3. Machine M_2: $S = \{A, B, C, D, E, F, G, H\}$, $X = \{0, 1\} = Z$

in the upper right corner. Other entries, including subscripts, in the figure will be determined in the following work.

We shall denote the nontrivial basic partitions, in the order we find them, by $\pi_1, \pi_2, \pi_3, \ldots$. When we determine that $\pi_{s_i s_j} = \pi_k$, we enter π_k in the upper right corner of the cell for s_i and s_j. If the pair $s_i s_j$ appears as an implied pair of some other pair $s_l s_m$, we shall also give the pair $s_i s_j$ within the cell $s_l s_m$ the subscript k, if $\pi_{s_l s_m}$ is not yet known; this information $\pi_{s_i s_j}$ will aid in the computation of $\pi_{s_l s_m}$, since $\pi_{s_l s_m} \geq \pi_{s_i s_j}$ (why?). (If we find that $\pi_{s_i s_j} = 1$ and $s_i s_j$ appears as an implied pair of the pair $s_l s_m$, we immediately know that $\pi_{s_l s_m} = 1$. Why?)

Now we work the example of Figure 6.12.3. We first note that the pair AF appears several times, so we compute π_{AF}: $AF \Rightarrow GH$, $CD \Rightarrow AD$, $BE \Rightarrow BG$, $CF \Rightarrow EH$, so that $\pi_{AF} = (A\,C\,D\,F)(B\,E\,G\,H) = \pi_1$. We enter "$\pi_1$" in cell AF, and wherever pair AF appears we add the subscript 1, showing that we already know what is implied by AF. Some other cells can be quickly finished; pair BE implies only AF and a pair congruent mod π_1, so that $\pi_{BE} = \pi_1$; similarly, $\pi_{CD} = \pi_1$ and $\pi_{EH} = \pi_1$. All occurrences of pairs BE, CD, and EH in uncompleted cells now receive subscript 1. On the other hand, pairs CH and DE imply AF and are not themselves in blocks of π_1, forcing $\pi_{CH} = \pi_{DE} = 1$. So cells CH and DE are finished, with the 1-partition 1 in the upper right corner. There is no need to subscript the occurrence of CH in cell FG, since we immediately conclude that $\pi_{FG} = 1$; similarly, AB implies FG, so $\pi_{AB} = 1$. When we enter subscript 1 on all appearances of BE, CD, and EH, to record the fact that each implies π_1, we find that $\pi_{BG} = \pi_1$, which in turn gives $\pi_{AD} = \pi_1$. We subscript as before, and then from AD we get $\pi_{AH} = 1$, $\pi_{DG} = 1$, and $\pi_{GH} = \pi_1$. These data in turn imply $\pi_{EF} = 1$, which gives $\pi_{BC} = 1$. Also, $\pi_{CF} = \pi_1$, since $\pi_{EH} = \pi_1$.

The above paragraph shows how we can quickly obtain 16 basic partitions in our example, after computing only one. Now we note that AC appears several times in uncompleted cells, so we compute π_{AC}:

$$AC \Rightarrow EG, DF \Rightarrow BH, \quad \text{so} \quad \pi_{AC} = (A\,C)(B\,H)(D\,F)(E\,G) = \pi_2.$$

We enter subscript 2 on all occurrences of AC, and note that also $\pi_{BH} = \pi_2 = \pi_{DF} = \pi_{EG}$. (We also enter subscript 2 on two occurrences of DF.)

We next compute π_{AE}:

$$AE \Rightarrow AG, DF \Rightarrow CG, \pi_2 \Rightarrow CE, \quad \text{so} \quad \pi_{AE} = (A\,C\,E\,G)(B\,H)(D\,F) = \pi_3.$$

Also $\pi_{CE} = \pi_3$, $\pi_{CG} = \pi_3$, $\pi_{AG} = \pi_3$.

The only unfinished cells are now BD, BF, DH, and FH. We compute π_{BD}:

$$BD \Rightarrow BF, \pi_2 \Rightarrow FH \Rightarrow DH, \quad \text{so} \quad \pi_{BD} = (A\,C)(B\,D\,F\,H)(E\,G) = \pi_4.$$

We also get $\pi_{DH} = \pi_4$, $\pi_{FH} = \pi_4$, $\pi_{BF} = \pi_4$.

We conclude that 1, π_1, π_2, π_3, and π_4 are the only basic partitions. Also, $\pi_3 + \pi_4 = (A\,C\,E\,G)(B\,D\,F\,H) = \pi_5$ is the only new closed partition given by Lemma 6.12.1; the π-lattice appears in Figure 6.12.4.

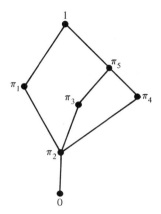

FIGURE 6.12.4. π-lattice of Machine M_2

6.13 Series and Parallel Decomposition of Machines

In some cases, a machine's purpose is simply to generate certain sequences of internal states in response to sequences of inputs. The output identifies (we may say briefly, "is") the present internal state, so such a machine is, in fact, a Moore machine. It may be represented by the symbol $M = [S, X, S, \tau, \omega]$ where for all $s \in S$, $\omega(s) = s$. The symbol for M is commonly shortened to $M = [S, X, \tau]$, where it is understood that the output is the present state. Such a machine is called a **state machine**.

For example, if $\pi = \{B_1, B_2, \ldots, B_k\}$ is a closed partition of a machine $M = [S, X, Z, \tau, \omega]$, the inputs x of M send blocks of π into blocks of π. We can construct a machine M_1 that effects these block-to-block transitions by letting the states of M_1 be the blocks of π and, if $B(s)$ is the block of π to which state s of M belongs, defining $\tau_1(B(s)) = B(\tau(s, x))$. This latter block is uniquely defined because, since π is closed, every state of the block $B(s)$ is mapped by x into the same block $B(\tau(s, x))$. The resulting state machine $M_1 = [\pi, X, \tau_1]$ is of a type frequently used as building blocks for larger machines.

The theorem to follow illustrates the last remark. In it we consider the case when there exist partitions π and θ of the state set S of a machine M such that $\pi\theta = 0$. This means that if we know the block of π and the block of θ containing some state s of M, we know s itself. The theorem specifies how to construct machines M_1 and M_2 that have blocks of π and θ as their states, and that may be interconnected in such a way as to represent precisely the specified behavior of M.

Theorem 6.13.1. Assume that π is a closed partition of the set S of states of the machine $M = [S, X, Z, \tau, \omega]$, and θ is another partition of S such that

(6.13.1) $$\pi \cdot \theta = 0.$$

Let $M_1 = [\pi, X, \pi, \tau_1, \omega_1]$ and $M_2 = [\theta, \pi \times X, Z, \tau_2, \omega_2]$ be machines such that if $s \in S$, $s \in B_s \in \pi$, and $s \in C_s \in \theta$, then

(6.13.2)
$$\tau_1(B_s, x) = B_{\tau(s,x)}, \quad \omega_1(B_s) = B_s, \quad \tau_2(C_s, (B_s, x)) = C_{\tau(s,x)},$$
$$\omega_2(C_s, (B_s, x)) = \omega(s, x).$$

Then the series composition (Figure 6.13.1) of M_1 and M_2 is a machine that covers M.

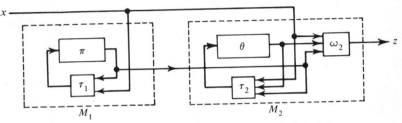

FIGURE 6.13.1. *Series Decomposition of a Machine*

Proof. We need only show that any output sequence produced by M is also produced by the composite machine. Assume that M starts in state s_0, with input sequence $x_1 x_2 x_3 \cdots$. It then produces output $z_1 = \omega(s_0, x_1)$ and goes to state $s_1 = \tau(s_0, x_1)$; produces output $z_2 = \omega(s_1, x_2)$ and goes to state $s_2 = \tau(s_1, x_2)$; produces output $z_3 = \omega(s_2, x_3)$ and goes to state $s_3 = \tau(s_2, x_3)$; and so on.

We start M_1 in that state $B_{s_0} \in \pi$ with $s_0 \in B_{s_0}$, and M_2 in that state $C_{s_0} \in \theta$ with $s_0 \in C_{s_0}$. The composite machine then produces output $\omega_2(C_{s_0}, (B_{s_0}, x_1)) = \omega(s_0, x_1) = z_1$ and goes to states

$$\tau_1(B_{s_0}, x_1) = B_{\tau(s_0, x_1)} = B_{s_1} \quad \text{and} \quad \tau_2(C_{s_0}, (B_{s_0}, x_1)) = C_{\tau(s_0, x_1)} = C_{s_1};$$

produces output $\omega_2(C_{s_1}, (B_{s_1}, x_2)) = \omega(s_1, x_2) = z_2$ and goes to states

$$\tau_1(B_{s_1}, x_2) = B_{\tau(s_1, x_2)} = B_{s_2} \quad \text{and} \quad \tau_2(C_{s_1}, (B_{s_1}, x_2)) = C_{\tau(s_1, x_2)} = C_{s_2};$$

and so on. ∎

Equations (6.13.2) do not necessarily define τ_2 and ω_2 on all state-input pairs, but they may be defined arbitrarily at other pairs, and the composite machine will cover M. For example, if $s \not\equiv t \pmod{\pi}$, then $\tau_2(C_s, (B_t, x))$ is not defined by (6.13.2), but we may define it as any $C_{s'}$ we wish, for the pair

Sec. 6.13 Series and Parallel Decomposition of Machines

$(C_s, (B_t, x))$ is never encountered when the composite machine is used to simulate M.

The importance of Theorem 6.13.1 stems from the fact that π and θ are (in all nontrivial cases) sets of smaller order than S, so each of the component machines M_1 and M_2 is simpler (has fewer states) than the original M. (However, it can happen that $|\pi|+|\theta|>|S|$.)

For example, the machine M of Figure 6.12.1 has the closed partition $\pi = \pi_{AB} = (A\,B)(C\,D)$. If we choose $\theta = (A\,C)(B\,D)$, then $\pi \cdot \theta = 0$, $|\pi| = |\theta| = 2$, and Theorem 6.13.1 says that M has a serial decomposition into two 2-state machines M_1 and M_2.

In the example of Figure 6.12.3, we can use the closed partition $\pi = \pi_2 = (A\,C)(B\,H)(D\,F)(E\,G)$. The partition $\theta = (A\,B\,D\,E)(C\,F\,G\,H)$ satisfies $\pi \cdot \theta = 0$, so we obtain a serial decomposition of the 8-state machine into a 4-state machine and a 2-state machine.

If a machine possesses two closed partitions π_1 and π_2 such that

(6.13.3) $$\pi_1 \cdot \pi_2 = 0,$$

then a stronger decomposition, the *parallel decomposition*, can be obtained. Neither component machine requires information from the other; the input set is X for both machines, as we see in Figure 6.13.2. The example of Figure 6.12.1 possesses closed partitions π_{AB} and π_{AD} with $|\pi_{AB}|=2$, $|\pi_{AD}|=3$, and $\pi_{AB} \cdot \pi_{AD} = 0$, so that it has a parallel decomposition into a two-state machine and a three-state machine.

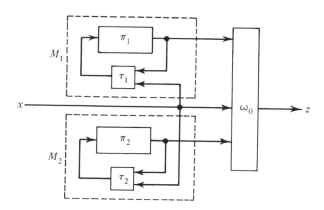

FIGURE 6.13.2. *Parallel Decomposition of a Machine*

Machines are sometimes decomposable into three or more components, and methods of composition more general than series or parallel exist; good references include Kohavi (1970), Harrison (1965), Arbib (1968), and Hartmanis and Stearns (1966).

6.14 Exercises

*1. Construct the lattices of all closed partitions of each of the following machines:

M_1:		Z		
	0	1	0	1
A	C	D	0	0
B	A	C	1	1
C	A	B	1	0
D	C	A	1	0

M_2:		Z		
	0	1	0	1
A	C	D	0	0
B	D	C	0	1
C	B	A	0	0
D	A	A	1	0

M_3:				Z		
	x_1	x_2	x_3	x_1	x_2	x_3
A	B	D	E	0	0	1
B	C	E	D	1	1	0
C	A	D	E	0	0	1
D	C	E	B	0	1	0
E	A	D	B	1	1	0

M_4:		Z		
	0	1	0	1
A	B	F	0	1
B	C	E	0	0
C	A	D	0	0
D	F	H	0	0
E	G	A	0	1
F	E	C	0	1
G	H	B	1	1
H	C	C	1	1

2. Let π_1, π_2, π_3 be closed partitions of the state set of a machine M such that $\pi_1 \pi_2 \pi_3 = 0$, but no product of any two of them is 0. Draw a schematic diagram (i.e., a diagram similar to Figures 6.13.1 and 6.13.2) of M as an interconnection of three state machines "π_1," "π_2," and "π_3," and a circuit "ω" that provides the output z.

3. Let π_1 and π_2 be closed partitions, and let τ be a partition that is not closed, of the state set of M such that $\pi_1 \pi_2 \tau = 0$, but the product of no two of these partitions is 0. In this case, the next state of τ is determined not only by the present input and the present block of τ, but also by the present blocks of π_1 and π_2. Draw the corresponding schematic diagram of M.

*4. A partition ζ of the state set of M is called **output-consistent** iff for all states s of a given block ζ_0 of ζ and for each fixed input $x \in X$, the outputs $\omega(s, x)$ are all the same. Thus the output of M depends only on the present block of ζ and the present input.

Given that π is closed, ζ is output-consistent, and $\pi \zeta = 0$, draw a schematic diagram of M with component machines "π," "ζ," and an output circuit "ω."

Sec. 6.14 *Exercises*

5. For the machine M_1 of Exercise 1, find a closed partition π and an output-consistent partition ζ such that $\pi\zeta = 0$.

6. Invent another example of a machine M with a closed partition π and an output-consistent partition ζ such that $\pi\zeta = 0$.

*****7.** A partition λ of the state set of M is called **input-consistent** iff for each given $s \in S$ and for all $x \in X$, the states $\tau(s, x)$ all belong to the same block of λ, that is, all next states in any row of the defining table of M belong to the same block of λ. If λ is both closed and input-consistent, the next block of λ is determined only by the present block of λ and we call λ **input-independent**. The component machine defined by λ is called an **autonomous clock**.

Let π be a closed partition and let λ be an input-independent partition of the state set of M such that $\pi\lambda = 0$. Draw the corresponding schematic diagram of M.

8. Given that the partitions π, λ, ζ of the state set of M are respectively closed, input-independent, and output-consistent, and given also that $\pi\lambda\zeta = 0$, draw the corresponding schematic diagram of M.

9. Find an input-independent partition λ and a partition τ of the state set of the machine M_2 of Exercise 1 such that $\lambda\tau = 0$. Then draw the corresponding schematic diagram of M_2.

*****10.** Show that

(a) if $\pi \geq \lambda$ where λ is input-consistent, then π is also input-consistent.

(b) if $\pi \leq \zeta$ where ζ is output-consistent, then π is also output-consistent.

(In each case, a single appropriate sentence will suffice.)

*****11.** The minimum input-consistent partition λ of the state set of a machine M may be found thus: Each state s of M determines a block of states consisting of the set of next states in its row of the defining table of M. Each such block must be included in some block of λ. When all the blocks determined by the individual states have been recorded, we unite blocks having a nonzero intersection in the usual way. The resulting blocks, together with any singleton states that remain, constitute the minimum input-consistent partition λ.

Find λ for the machine M_4 of Exercise 1 and show that it is also closed.

12. Find by inspection the maximum output-consistent partition ζ for the machine M_4 of Exercise 1, show that $\lambda\zeta = 0$ (where λ is as in Exercise 11), and draw the corresponding schematic diagram for M_4.

CHAPTER 7
Linear Algebra and Field Theory

BASIC facts from linear algebra are needed in Chapters 8 and 9, on linear machines and algebraic coding theory. In this chapter we aim to present a complete, although condensed, account of the necessary linear algebra. The coverage is rather concise, since this is not a textbook on linear algebra; mastery of this chapter by a student without a previous course in linear algebra will require high mathematical ability. For fuller exposition and additional examples and exercises, we refer the reader to Hohn (1973), Hoffman and Kunze (1971), and the many other available linear algebra texts. In this chapter we also present the basic facts about finite fields that are needed in Chapters 8 and 9.

7.1 Matrices

Fields were introduced in Chapter 3; matrices have been informally used in some previous chapters. Let m and n be positive integers. An $m \times n$ (read "m-by-n") **matrix over the field F** is a rectangular array

$$(7.1.1) \qquad \mathbf{A} = [a_{ij}]_{m \times n} = \begin{bmatrix} a_{11} & a_{12} & \cdots & a_{1n} \\ a_{21} & a_{22} & \cdots & a_{2n} \\ \multicolumn{4}{c}{\dotfill} \\ a_{m1} & a_{m2} & \cdots & a_{mn} \end{bmatrix}$$

of elements $a_{ij} \in F$, containing m rows and n columns; a_{ij} is the **element**, or **entry**, in the ith row and jth column. The ordered pair (m, n) is the **size**, or **order**, of the matrix $\mathbf{A}$. We may write $[a_{ij}]$ instead of $[a_{ij}]_{m \times n}$ if the size of $\mathbf{A}$ is known.

If $\mathbf{A}$ and $\mathbf{B}$ are $m \times n$ matrices with $\mathbf{A} = [a_{ij}]$, $\mathbf{B} = [b_{ij}]$, and $c \in F$, we define the **sum** $\mathbf{A} + \mathbf{B}$ and the **scalar multiple** $c\mathbf{A}$ by

$$(7.1.2) \qquad \mathbf{A} + \mathbf{B} = [a_{ij} + b_{ij}], \qquad c\mathbf{A} = [ca_{ij}].$$

The set of $m \times n$ matrices over F forms a group under addition. The additive identity is the **zero matrix** $\mathbf{0}_{m \times n} = \mathbf{0}$ with all entries the zero element 0 of F:

$$(7.1.3) \qquad \mathbf{A} + \mathbf{0} = \mathbf{0} + \mathbf{A} = \mathbf{A} \qquad \text{for all } m \times n \; \mathbf{A}.$$

Any element $a \in F$ has a *negative* (additive inverse) $-a \in F$ satisfying $a + (-a) = 0$; if we define the **negative** $-\mathbf{A}$ of a matrix $\mathbf{A} = [a_{ij}]$ to be $-\mathbf{A} = [-a_{ij}]$, we then have

$$(7.1.4) \qquad \mathbf{A} + (-\mathbf{A}) = (-\mathbf{A}) + \mathbf{A} = \mathbf{0}.$$

Sec. 7.2 *Elementary Row Operations*

It is easy to see that addition and scalar multiplication of matrices satisfy the laws

(7.1.5)
$$1 \cdot \mathbf{A} = \mathbf{A}, \quad (-1) \cdot \mathbf{A} = -\mathbf{A}, \quad (a+b)\mathbf{A} = a\mathbf{A} + b\mathbf{A},$$
$$a(\mathbf{A}+\mathbf{B}) = a\mathbf{A} + a\mathbf{B}, \quad a(b\mathbf{A}) = (ab)\mathbf{A} = b(a\mathbf{A}),$$
$$\mathbf{A} + \mathbf{B} = \mathbf{B} + \mathbf{A}.$$

If $\mathbf{A} = [a_{ij}]_{m \times n}$ and $\mathbf{B} = [b_{ij}]_{n \times p}$ are matrices over F, we define the **product** $\mathbf{AB}$ to be $[c_{ij}]_{m \times p}$, where

(7.1.6)
$$c_{ij} = \sum_{k=1}^{n} a_{ik} b_{kj};$$

we form the i,j-entry of $\mathbf{AB}$ by multiplying corresponding entries in the ith row of $\mathbf{A}$ and the jth column of $\mathbf{B}$, then adding the results. Note that we can only form the product $\mathbf{AB}$ when the number of columns of $\mathbf{A}$ equals the number of rows of $\mathbf{B}$. Multiplication of matrices satisfies the associative and distributive laws

(7.1.7)
$$\mathbf{A}(\mathbf{BC}) = (\mathbf{AB})\mathbf{C}, \quad \mathbf{A}(\mathbf{B}+\mathbf{C}) = \mathbf{AB} + \mathbf{AC},$$
$$(\mathbf{A}+\mathbf{B})\mathbf{C} = \mathbf{AC} + \mathbf{BC}, \quad a(\mathbf{AB}) = (a\mathbf{A})\mathbf{B} = \mathbf{A}(a\mathbf{B})$$

whenever the matrix products and sums appearing are all defined. However, the commutative law fails; even if $\mathbf{AB}$ and $\mathbf{BA}$ are both defined, $\mathbf{AB}$ need not equal $\mathbf{BA}$.

Here are some examples of matrix multiplication over the field $\mathbb{R}$:

$$\begin{bmatrix} 1 & -1 \\ -2 & 2 \end{bmatrix} \begin{bmatrix} 1 & 3 \\ 1 & 3 \end{bmatrix} = \begin{bmatrix} 1 \cdot 1 + (-1) \cdot 1 & 1 \cdot 3 + (-1) \cdot 3 \\ (-2) \cdot 1 + 2 \cdot 1 & (-2) \cdot 3 + 2 \cdot 3 \end{bmatrix} = \begin{bmatrix} 0 & 0 \\ 0 & 0 \end{bmatrix},$$

$$\begin{bmatrix} 1 & 3 \\ 1 & 3 \end{bmatrix} \begin{bmatrix} 1 & -1 \\ -2 & 2 \end{bmatrix} = \begin{bmatrix} -5 & 5 \\ -5 & 5 \end{bmatrix}.$$

These examples show that the commutative law fails, and the first example shows that the product of nonzero matrices may be the zero matrix.

The **main diagonal** of an $n \times n$ matrix $\mathbf{A} = [a_{ij}]$ consists of the entries $a_{11}, a_{22}, \ldots, a_{nn}$. If all other entries of $\mathbf{A}$ are 0, $\mathbf{A}$ is a **diagonal matrix**. A diagonal matrix $\mathbf{A}$ is a **scalar matrix** if $a_{11} = a_{22} = \cdots = a_{nn}$, and is the $n \times n$ **identity matrix** if $a_{11} = a_{22} = \cdots = a_{nn} = 1$. The $n \times n$ identity matrix $\mathbf{I}_n = \mathbf{I}$ satisfies $\mathbf{IB} = \mathbf{B}$, $\mathbf{CI} = \mathbf{C}$ whenever the matrix products are defined.

7.2 Elementary Row Operations

Let $\mathbf{R}_i$ denote the ith row of the $m \times n$ matrix $\mathbf{A}$ over F, and let c be any element of F. The following operations on $\mathbf{A}$ are known as **elementary row operations**:

1. Replacing $\mathbf{R}_i$ by $c\mathbf{R}_i$, if $c \neq 0$.

2. Interchanging $\mathbf{R}_i$ and $\mathbf{R}_j$.
3. Replacing $\mathbf{R}_i$ by $\mathbf{R}_i + c\mathbf{R}_j$, $j \neq i$.

These operations can be used to solve a system of m linear equations in n unknowns over any field F. The procedure, known as the **Gauss–Jordan procedure** or the **sweepout process**, is completely systematic and relatively efficient; it is often programmed for computers.

Assume that the given system of equations is

(7.2.1)
$$a_{11}x_1 + a_{12}x_2 + \cdots + a_{1n}x_n = b_1$$
$$a_{21}x_1 + a_{22}x_2 + \cdots + a_{2n}x_n = b_2$$
$$\vdots$$
$$a_{m1}x_1 + a_{m2}x_2 + \cdots + a_{mn}x_n = b_m.$$

These m equations are equivalent to the one matrix equation

(7.2.2) $\quad \mathbf{AX} = \mathbf{B}$, $\quad$ where $\mathbf{A} = [a_{ij}]_{m \times n}$, $\mathbf{X} = [x_i]_{n \times 1}$, $\mathbf{B} = [b_i]_{m \times 1}$.

System (7.2.2) is called **homogeneous** if $\mathbf{B} = \mathbf{0}_{m \times 1}$. The **augmented matrix** of (7.2.2) is the $m \times (n+1)$ matrix

(7.2.3)
$$\begin{bmatrix} a_{11} & a_{12} & \cdots & a_{1n} & b_1 \\ a_{21} & a_{22} & \cdots & a_{2n} & b_2 \\ \vdots & & & & \\ a_{m1} & a_{m2} & \cdots & a_{mn} & b_m \end{bmatrix} = [\mathbf{A} \ \ \mathbf{B}].$$

We now describe the sweepout process as it is applied to (7.2.1). We may assume that some coefficient a_{i1} of x_1 is not zero; if they are all zero, x_1 does not appear in the equations and is arbitrary. Since we may interchange the first and ith equations if necessary, we may assume that $a_{11} \neq 0$. Since we may multiply the first equation by a_{11}^{-1}, we may assume that $a_{11} = 1$. For all $i > 1$, we then add $-a_{i1}$ times the first equation to the ith equation; the coefficient of x_1 in the ith equation becomes $a_{i1} + (-a_{i1}) \cdot 1 = 0$. The column of coefficients of x_1 is now

$$\begin{bmatrix} 1 \\ 0 \\ \vdots \\ 0 \end{bmatrix};$$

we have completed *sweepout* of the first column.

Now choose the first variable x_{i_2} (it may or may not be x_2) that has a nonzero coefficient in some equation other than the first, and interchange equations to place that equation second. Multiply the new second equation $a_{2i_2}x_{i_2} + \cdots$ by $a_{2i_2}^{-1}$ to give x_{i_2} a coefficient 1; add multiples of this second equation to all others to give x_{i_2} the coefficient 0 in the others. This completes sweepout of the first, second, ..., i_2th columns. Sweepout continues until all variables are considered. The final system of equations makes the solutions (if any) obvious.

In practice, sweepout can be done by writing only the augmented matrix of the given system at each step. We illustrate this with an example over $\mathbb{C}$, writing both the equations-all-the-way solution and the augmented matrix solution.

$$\begin{array}{ll} \mathbf{R}_1: & x_1 + x_2 - x_3 + x_4 = 2 \\ \mathbf{R}_2: & -x_1 - x_2 + x_3 + 2x_4 = 1 \\ \mathbf{R}_3: & 4x_1 + 4x_2 - 4x_3 + x_4 = 5 \end{array} \qquad \begin{bmatrix} 1 & 1 & -1 & 1 & 2 \\ -1 & -1 & 1 & 2 & 1 \\ 4 & 4 & -4 & 1 & 5 \end{bmatrix}.$$

Coefficient a_{11} is already 1, so we add $\mathbf{R}_1$ to $\mathbf{R}_2$ and $-4\mathbf{R}_1$ to $\mathbf{R}_3$:

$$\begin{array}{ll} \mathbf{R}'_1: & x_1 + x_2 - x_3 + x_4 = 2 \\ \mathbf{R}'_2: & 3x_4 = 3 \\ \mathbf{R}'_3: & -3x_4 = -3 \end{array} \qquad \begin{bmatrix} 1 & 1 & -1 & 1 & 2 \\ 0 & 0 & 0 & 3 & 3 \\ 0 & 0 & 0 & -3 & -3 \end{bmatrix}.$$

We have swept the first column and we go immediately to the fourth, multiplying $\mathbf{R}'_2$ by $\frac{1}{3}$ to get $\mathbf{R}''_2$, then adding $-\mathbf{R}''_2$ to $\mathbf{R}'_1$ and $3\mathbf{R}''_2$ to $\mathbf{R}'_3$:

$$\begin{array}{ll} \mathbf{R}''_1: & x_1 + x_2 - x_3 = 1 \\ \mathbf{R}''_2: & x_4 = 1 \\ \mathbf{R}''_3: & 0 = 0 \end{array} \qquad \begin{bmatrix} 1 & 1 & -1 & 0 & 1 \\ 0 & 0 & 0 & 1 & 1 \\ 0 & 0 & 0 & 0 & 0 \end{bmatrix}.$$

These last equations make it obvious that $x_4 = 1$ and $x_1 = 1 - x_2 + x_3$, where x_2 and x_3 can be any elements of $\mathbb{C}$.

When sweepout is applied to a matrix, the final resulting matrix is always in **reduced echelon form**; that is,
1. All zero rows appear below all nonzero rows.
2. The first nonzero entry in any nonzero row is a 1 (called the **leading** 1 of that row).
3. Any column containing a leading 1 has all its other entries 0.
4. If nonzero row $\mathbf{R}_j$ is below nonzero row $\mathbf{R}_i$, then the leading 1 of $\mathbf{R}_j$ is to the right of the leading 1 of $\mathbf{R}_i$.

7.3 The Inverse of a Matrix

An $n \times n$ (square) matrix $\mathbf{A}$ is called **invertible** (or **nonsingular**) if there is an $n \times n$ matrix $\mathbf{B}$ such that $\mathbf{AB} = \mathbf{I}$. The set of all invertible $n \times n$ matrices over F is denoted $GL_n(F)$. If $\mathbf{A}_1, \mathbf{A}_2 \in GL_n(F)$ with $\mathbf{A}_1 \mathbf{B}_1 = \mathbf{I}$ and $\mathbf{A}_2 \mathbf{B}_2 = \mathbf{I}$, then $(\mathbf{A}_1 \mathbf{A}_2)(\mathbf{B}_2 \mathbf{B}_1) = \mathbf{A}_1(\mathbf{A}_2 \mathbf{B}_2)\mathbf{B}_1 = \mathbf{A}_1 \mathbf{I} \mathbf{B}_1 = \mathbf{A}_1 \mathbf{B}_1 = \mathbf{I}$, so $\mathbf{A}_1 \mathbf{A}_2 \in GL_n(F)$. Since $\mathbf{II} = \mathbf{I}$, $\mathbf{I} \in GL_n(F)$. Thus $GL_n(F)$ is a semigroup under matrix multiplication; by the right-handed analogue (Exercise 4.15.1) of Theorem 4.11.3, $GL_n(F)$ is, in fact, a group, called the **general linear group** of $n \times n$ matrices over F. [This explains the notation $GL_n(F)$.] If $\mathbf{AB} = \mathbf{I}$, this implies that the right inverse $\mathbf{B}$ of $\mathbf{A}$ is the unique two-sided inverse of $\mathbf{A}$; we denote $\mathbf{B} = \mathbf{A}^{-1}$ and have $\mathbf{A}^{-1}\mathbf{A} = \mathbf{A}\mathbf{A}^{-1} = \mathbf{I}$.

It is often important to find the inverse of an invertible matrix $\mathbf{A}$. For example, if we have to solve 100 systems of equations $\mathbf{AX}_1 = \mathbf{C}_1$, $\mathbf{AX}_2 = \mathbf{C}_2, \ldots, \mathbf{AX}_{100} = \mathbf{C}_{100}$ and we know $\mathbf{A}^{-1}$, then

$$\mathbf{AX}_i = \mathbf{C}_i \quad \text{iff} \quad \mathbf{A}^{-1}\mathbf{AX}_i = \mathbf{A}^{-1}\mathbf{C}_i \quad \text{iff} \quad \mathbf{X}_i = \mathbf{A}^{-1}\mathbf{C}_i.$$

Thus each system has a unique solution $\mathbf{X}_i$, obtainable by a single matrix multiplication $\mathbf{A}^{-1}\mathbf{C}_i$. Hence all solutions may be obtained at once as the columns of the product $\mathbf{A}^{-1}\mathbf{C}$, where $\mathbf{C} = [\mathbf{C}_1 \ \mathbf{C}_2 \cdots \mathbf{C}_{100}]$.

We shall now see that the sweepout process can be used to find the inverse $\mathbf{B}$ of an invertible matrix $\mathbf{A}$ (and reveal that $\mathbf{A}$ is not invertible, if that is the case). We wish to solve $\mathbf{AB} = \mathbf{I}$ for $\mathbf{B}$. Denoting by $\mathbf{B}_i$ the ith column of $\mathbf{B}$ and by $\mathbf{E}_i$ the ith column of $\mathbf{I}$, we see that the equation

$$\mathbf{A}[\mathbf{B}_1 \ \mathbf{B}_2 \cdots \mathbf{B}_n] = \mathbf{AB} = \mathbf{I} = [\mathbf{E}_1 \ \mathbf{E}_2 \cdots \mathbf{E}_n]$$

amounts to n systems $\mathbf{AB}_i = \mathbf{E}_i$ of simultaneous linear equations, each $\mathbf{B}_i$ a column of unknowns.

Columns $\mathbf{B}_1, \mathbf{B}_2, \ldots, \mathbf{B}_n$ can be found by a single extended sweepout process. We form the $n \times 2n$ matrix $[\mathbf{A} \ \mathbf{I}] = [\mathbf{A} \ \mathbf{E}_1 \ \mathbf{E}_2 \cdots \mathbf{E}_n]$ and perform sweepout, seeking to convert $\mathbf{A}$ (the first n columns) to $\mathbf{I}$. If we succeed, the last n columns are then $\mathbf{A}^{-1}$; if we fail, $\mathbf{A}$ is not invertible. (To see that the last sentence is true, just remember that the inverse is unique if it exists and think about the process of solving any one system $\mathbf{AB}_i = \mathbf{E}_i$ for $\mathbf{B}_i$, as in Section 7.2.)

As an example, we compute the inverse of the matrix

$$\mathbf{A} = \begin{bmatrix} 3 & 1 & 2 \\ 1 & -4 & 1 \\ 2 & 3 & 0 \end{bmatrix} \quad \text{over} \quad \mathbb{Q}.$$

The successive steps are

$$\begin{bmatrix} 3 & 1 & 2 & 1 & 0 & 0 \\ 1 & -4 & 1 & 0 & 1 & 0 \\ 2 & 3 & 0 & 0 & 0 & 1 \end{bmatrix},$$

$$\begin{bmatrix} 1 & -4 & 1 & 0 & 1 & 0 \\ 0 & 11 & -2 & 0 & -2 & 1 \\ 0 & 13 & -1 & 1 & -3 & 0 \end{bmatrix} \quad \begin{array}{l} \text{(We shifted row 1 to} \\ \text{the bottom, then} \\ \text{swept column 1.)} \end{array}$$

$$\begin{bmatrix} 1 & 0 & \frac{3}{11} & 0 & \frac{3}{11} & \frac{4}{11} \\ 0 & 1 & -\frac{2}{11} & 0 & -\frac{2}{11} & \frac{1}{11} \\ 0 & 0 & \frac{15}{11} & 1 & -\frac{7}{11} & -\frac{13}{11} \end{bmatrix},$$

$$\begin{bmatrix} 1 & 0 & 0 & -\frac{3}{15} & \frac{6}{15} & \frac{9}{15} \\ 0 & 1 & 0 & \frac{2}{15} & -\frac{4}{15} & -\frac{1}{15} \\ 0 & 0 & 1 & \frac{11}{15} & -\frac{7}{15} & -\frac{13}{15} \end{bmatrix},$$

so

$$\mathbf{A}^{-1} = \begin{bmatrix} -\frac{3}{15} & \frac{6}{15} & \frac{9}{15} \\ \frac{2}{15} & -\frac{4}{15} & -\frac{1}{15} \\ \frac{11}{15} & -\frac{7}{15} & -\frac{13}{15} \end{bmatrix}.$$

7.4 Exercises

1. (a) Show that (7.1.5) and (7.1.7) hold for the matrices

$$\mathbf{A} = \begin{bmatrix} 1 & 3 \\ 0 & 2 \end{bmatrix}, \quad \mathbf{B} = \begin{bmatrix} 7 & -3 \\ 1 & 0 \end{bmatrix}, \quad \mathbf{C} = \begin{bmatrix} 1 & -2 \\ -1 & 4 \end{bmatrix}$$

over $\mathbb{Q}$ and the values $a = 2$ and $b = -1$ in $\mathbb{Q}$.
   (b) Prove (7.1.5) and (7.1.7) in general.

2. Use sweepout to solve the systems of equations over $\mathbb{R}$.

(a) $\quad x_1 - x_2 + x_3 = 4$
$\quad\ 2x_1 + x_2 - 3x_3 = -3$
$\ -3x_1 + 2x_2 + x_3 = 6$

(b) $\quad x_1 - x_2 + x_3 = 0$
$\quad\ 2x_1 - 3x_2 + 4x_3 = 0$
$\quad\ 3x_1 + x_2 - 5x_3 = 0$

3. Solve the two systems of equations

(a) $\quad x_1 + 2x_2 - x_3 = 0$
$\quad 3x_1 + x_2 - 3x_3 + 2x_4 = 0$
$\quad -x_1 + x_3 + x_4 = 0$
$\quad x_1 + x_2 - x_3 + 4x_4 = 0$
$\quad x_1 + x_2 - x_3 + x_4 = 0$

(b) $\quad x_1 + 2x_2 - x_3 = 12$
$\quad 3x_1 + x_2 - 3x_3 + 2x_4 = -6$
$\quad -x_1 + x_3 + x_4 = 0$
$\quad x_1 + x_2 - x_3 + 4x_4 = 8$
$\quad x_1 + x_2 - x_3 + x_4 = 9$

over $\mathbb{Q}$ by using only *one* sweepout computation.

4. Show that the matrix

$$\begin{bmatrix} 1 & 2 & 0 \\ 0 & 1 & 2 \\ 2 & 0 & 1 \end{bmatrix}$$

is not invertible over the field GF(3).

5. Find the inverses, if they exist, of the following matrices over $\mathbb{Q}$:

$$\begin{bmatrix} 2 & 1 & 0 \\ -1 & 1 & -3 \\ 3 & 2 & -1 \end{bmatrix}, \quad \begin{bmatrix} 1 & 1 & 0 & -1 \\ 1 & 0 & -1 & 1 \\ 0 & -1 & 1 & 1 \\ -1 & 1 & 1 & 0 \end{bmatrix}, \quad \begin{bmatrix} 1 & 1 & 0 \\ 2 & -1 & 3 \\ -1 & 5 & -6 \end{bmatrix}.$$

6. (a) Solve the system of equations

$$x_1 = y_1$$
$$x_2 = y_1 + y_2$$
$$x_3 = y_2 + y_3$$
$$\vdots$$
$$x_{n-1} = y_{n-2} + y_{n-1}$$
$$x_n = \phantom{y_1 + \cdots + y_{n-2} +} y_{n-1} + y_n$$

over GF(2) for the y's in terms of the x's. (Do not use matrices; just manipulate equations.)

(b) Use (a) to find the inverse over GF(2) of the matrix

$$\begin{bmatrix} 1 & 0 & 0 & \cdots & 0 & 0 & 0 \\ 1 & 1 & 0 & \cdots & 0 & 0 & 0 \\ 0 & 1 & 1 & \cdots & 0 & 0 & 0 \\ \vdots & & & \ddots & & & \vdots \\ 0 & 0 & 0 & \cdots & 1 & 1 & 0 \\ 0 & 0 & 0 & \cdots & 0 & 1 & 1 \end{bmatrix}.$$

*7. This problem concerns the 2×2 matrices

$$\mathbf{A} = \begin{bmatrix} w & x \\ y & z \end{bmatrix} \quad \text{and} \quad \mathbf{D} = \begin{bmatrix} a & 0 \\ 0 & b \end{bmatrix}$$

over the field F.

(a) Compute **DA** and **AD**.

(b) Use (a) to show that if $\mathbf{DA} = \mathbf{AD}$ for all 2×2 diagonal matrices **D** over F, then **A** is diagonal.

(c) Show that if $\mathbf{MA} = \mathbf{AM}$ for all 2×2 matrices **M** over F, then **A** is a scalar matrix.

[HINT: By part (b), you already know that **A** is diagonal. Now try a nondiagonal **M**.]

(d) Show that if **A** is a 2×2 scalar matrix over F, then $\mathbf{MA} = \mathbf{AM}$ for all 2×2 matrices **M** over F.

*8. Generalize all parts of Exercise 7 to $n \times n$ matrices.

9. Put the following matrices over GF(3) in reduced echelon form:

$$\begin{bmatrix} 1 & 2 & 0 & 1 \\ 0 & 0 & 2 & 1 \\ 1 & 1 & 1 & 2 \end{bmatrix}, \quad \begin{bmatrix} 1 & 1 & 2 \\ 2 & 1 & 0 \\ 0 & 0 & 1 \\ 1 & 2 & 1 \end{bmatrix}.$$

*10. Two $m \times n$ matrices **A** and **B** over a field F are **row-equivalent** if **B** can be obtained from **A** by a sequence of elementary row operations.

(a) Show that if matrix $\mathbf{M}_2$ can be obtained from matrix $\mathbf{M}_1$ by a single elementary row operation, then $\mathbf{M}_1$ can be obtained from $\mathbf{M}_2$ by a single elementary row operation. (Consider separately the three kinds of row operations.)

(b) Show that row equivalence is an equivalence relation on the set of $m \times n$ matrices over F. [Use part (a) when proving symmetry.]

(c) Show that each equivalence class for the equivalence relation in part (b) contains a matrix in reduced echelon form.

(d) Show that if $\mathbf{R}_1$ and $\mathbf{R}_2$ are distinct $m \times n$ matrices in reduced echelon form, then sweepout will not convert $\mathbf{R}_1$ to $\mathbf{R}_2$. [This requires careful thinking. Parts (c) and (d) imply that each row-equivalence class contains exactly one matrix in reduced echelon form. Reduced echelon matrices are therefore said to be *canonical forms* for the equivalence relation.]

7.5 Vector Spaces

Assume that V is an additive abelian group and F is a field. We say that V is a **vector space** over F if there is a function "$\cdot$": $F \times V \to V$, called *scalar multiplication* and denoted by juxtaposition $(\alpha, v) \to \alpha \cdot v = \alpha v \in V$, all $\alpha \in F$, $v \in V$, such that

(7.5.1)
$$\alpha(v+w) = \alpha v + \alpha w, \quad (\alpha+\beta)v = \alpha v + \beta v,$$
$$\alpha(\beta v) = (\alpha\beta)v, \quad 1v = v,$$

for all $\alpha, \beta \in F$, $v, w \in V$. [The reader should read equations (7.5.1) carefully, making sure which operations + are in V and which in F, which juxtapositions are scalar multiplication and which are the multiplication in F.] Elements of a vector space are called **vectors**, regardless of the nature of the particular objects they represent.

Here are some important examples of vector spaces. In each case, F may be any field.

1. Denote by F^n the Cartesian product $F \times F \times \cdots \times F$ of n copies of F, so F^n is the set of all ordered n-tuples $(\alpha_1, \alpha_2, \ldots, \alpha_n)$, all $\alpha_i \in F$. F^n is an additive abelian group under componentwise addition:

(7.5.2)
$$(\alpha_1, \alpha_2, \ldots, \alpha_n) + (\beta_1, \beta_2, \ldots, \beta_n)$$
$$= (\alpha_1 + \beta_1, \alpha_2 + \beta_2, \ldots, \alpha_n + \beta_n).$$

In fact, F^n is a vector space over F, with scalar multiplication defined by

(7.5.3) $\alpha(\alpha_1, \alpha_2, \ldots, \alpha_n) = (\alpha\alpha_1, \alpha\alpha_2, \ldots, \alpha\alpha_n),$ all $\alpha \in F$,

$$(\alpha_1, \alpha_2, \ldots, \alpha_n) \in F^n.$$

2. For fixed positive integers m and n, let $V = \text{Mat}_{m \times n}(F)$ be the set of all $m \times n$ matrices over F. We know how to add elements of V: $[\alpha_{ij}] + [\beta_{ij}] = [\alpha_{ij} + \beta_{ij}]$. We also know a scalar multiplication on V: $\alpha[\alpha_{ij}] = [\alpha \alpha_{ij}]$, all $\alpha \in F$, $[\alpha_{ij}] \in V$. Here, V is a vector space over F with these operations of addition and scalar multiplication. In what sense is example 1 a special case of example 2?

3. Let $V = P_{n,F}(x)$ be the set of all polynomials of degree at most n, with coefficients in F, in the variable x:

$$P_{n,F}(x) = \{\alpha_0 + \alpha_1 x + \cdots + \alpha_n x^n | \alpha_0, \alpha_1, \ldots, \alpha_n \in F\}.$$

Addition and scalar multiplication in V are defined by

(7.5.4)
$$(\alpha_0 + \alpha_1 x + \cdots + \alpha_n x^n) + (\beta_0 + \beta_1 x + \cdots + \beta_n x^n)$$
$$= (\alpha_0 + \beta_0) + (\alpha_1 + \beta_1)x + \cdots + (\alpha_n + \beta_n)x^n,$$
$$\alpha(\alpha_0 + \alpha_1 x + \cdots + \alpha_n x^n)$$
$$= (\alpha \alpha_0) + (\alpha \alpha_1)x + \cdots + (\alpha \alpha_n)x^n.$$

With these operations, V is a vector space over F.

Lemma 7.5.1. *If V is a vector space over F, then*

$$0 \cdot x = 0 \quad \text{and} \quad (-1)x = -x \quad \text{for all } x \in V.$$

Proof. $0 \cdot x + 0 \cdot x = (0+0)x = 0 \cdot x$. Adding $-(0 \cdot x)$ to both sides, we have $0 \cdot x = 0$. The equation

$$x + (-1)x = 1x + (-1)x = (1 + (-1))x = 0 \cdot x = 0$$

implies that $(-1)x$ is the negative $-x$ of x. ∎

Let V be a vector space over F. A subset W of V is a **subspace** of V if W is itself a vector space (with the same operations as in V).

Theorem 7.5.1. *A nonempty subset W of V is a subspace iff for all $x, y \in W$ and for all $\alpha \in F$ we have $x + y \in W$ and $\alpha x \in W$.*

Proof. Certainly, if W is a subspace, then the properties stated in the theorem hold. Conversely, if they hold, choose $\alpha = -1$ to see that $x \in W$ implies that $(-1)x = -x \in W$. By Theorem 4.4.1, W is an (additive abelian) subgroup of V. The rules (7.5.1) all hold in W because they hold in V, so W is a subspace of V. ∎

Corollary 7.5.1. *Let $\mathbf{AX} = \mathbf{0}$ be a system of m homogeneous linear equations in n unknowns over F. The solutions of the system form a subspace S of the vector space $\text{Mat}_{n \times 1}(F)$.*

Sec. 7.5   Vector Spaces

Proof. We shall show that if $X_1, X_2 \in S$ and $\alpha \in F$, then $X_1 + X_2 \in S$ and $\alpha X_1 \in S$. Since $X_1, X_2 \in S$, we have $AX_1 = 0$, $AX_2 = 0$. By (7.1.7) we have

$$A(X_1 + X_2) = AX_1 + AX_2 = 0 + 0 = 0 \quad \text{and} \quad A(\alpha X_1) = \alpha(AX_1) = \alpha 0 = 0,$$

so $X_1 + X_2 \in S$, $\alpha X_1 \in S$. ∎

Corollary 7.5.2. *Let $AX = B$ be a system of m linear equations in n unknowns over F. Let $V = \text{Mat}_{n \times 1}(F)$ and let S be the subspace of V of solutions of $AX = 0$. Then the solutions of $AX = B$, if any, form a left coset of the subgroup S in the additive group V.*

Proof. If there are any solutions, let X_0 be one; $AX_0 = B$. Then $AX = B$ iff $AX - AX_0 = B - B$ iff $A(X - X_0) = 0$ iff $X - X_0 \in S$ iff $X \in X_0 + S$, so the solutions consist of the coset $X_0 + S = \{X_0 + Y | Y \in S\}$. ∎

For example, consider the system of equations over $\mathbb{Q}$

(7.5.5)
$$x_1 + x_2 - 2x_3 = 2,$$
$$2x_1 - 3x_2 + x_3 = -1.$$

One solution of this system is

$$\begin{bmatrix} x_1 \\ x_2 \\ x_3 \end{bmatrix} = \begin{bmatrix} 2 \\ 2 \\ 1 \end{bmatrix}.$$

The corresponding homogeneous system

$$x_1 + x_2 - 2x_3 = 0,$$
$$2x_1 - 3x_2 + x_3 = 0$$

has the solution set

$$S = \left\{ \alpha \begin{bmatrix} 1 \\ 1 \\ 1 \end{bmatrix} \middle| \alpha \in \mathbb{Q} \right\},$$

as determined by sweepout. S is a subspace of $\text{Mat}_{3 \times 1}(\mathbb{Q})$, and the complete set of solutions of (7.5.5) is

$$\left\{ \begin{bmatrix} 2 \\ 2 \\ 1 \end{bmatrix} + \alpha \begin{bmatrix} 1 \\ 1 \\ 1 \end{bmatrix} \middle| \alpha \in \mathbb{Q} \right\},$$

by Corollary 7.5.2. The value $\alpha = 6$ yields the particular solution

$$\begin{bmatrix} x_1 \\ x_2 \\ x_3 \end{bmatrix} = \begin{bmatrix} 2 \\ 2 \\ 1 \end{bmatrix} + 6 \begin{bmatrix} 1 \\ 1 \\ 1 \end{bmatrix} = \begin{bmatrix} 8 \\ 8 \\ 7 \end{bmatrix}.$$

If W_1 and W_2 are subspaces of the vector space V over F, it is easy to prove (see Exercise 7.7.3) that their set-theoretic intersection $W_1 \cap W_2$ is itself a subspace. For example, $W_1 = \{(a, 0, 0) | a \in \mathbb{R}\}$ and $W_2 = \{(0, b, 0) | b \in \mathbb{R}\}$ are subspaces of $\mathbb{R}^3$, with intersection the "zero subspace" $\{(0, 0, 0)\}$. The union $W_1 \cup W_2$ is *not* a subspace: $(1, 0, 0) \in W_1 \cup W_2$ and $(0, 1, 0) \in W_1 \cup W_2$, but $(1, 0, 0) + (0, 1, 0) = (1, 1, 0) \notin W_1 \cup W_2$.

The **sum** $W_1 + W_2 + \cdots + W_k$ of subspaces $W_1, W_2, \ldots, W_k$ of a vector space V is defined to be $\{w_1 + w_2 + \cdots + w_k \in V | w_i \in W_i\}$; it *is* a subspace of V. In fact, it is the smallest subspace of V containing $W_1 \cup W_2 \cup \cdots \cup W_k$ (Exercise 7.7.3). In the example above, $W_1 + W_2$ is the subspace $\{(a, b, 0) | a, b \in \mathbb{R}\}$ of $\mathbb{R}^3$.

7.6 Linear Independence and Dimension

Let S be a subset of the vector space V over F. A **linear combination** of vectors in S is a vector $x = \alpha_1 v_1 + \alpha_2 v_2 + \cdots + \alpha_k v_k$, the $\alpha_i \in F$, $v_i \in S$. The set of all linear combinations of vectors in S is denoted $\lceil S \rceil$. In $GF(3)^2$, $(2, 2)$ is a linear combination of $(2, 1)$ and $(0, 2)$ because $(2, 2) = 1 \cdot (2, 1) + 2 \cdot (0, 2)$.

Theorem 7.6.1. *If S is a nonempty subset of V, then $\lceil S \rceil$ is a subspace of V; it is the smallest subspace containing S as a subset.*

Proof. Suppose that $x, y \in \lceil S \rceil$, say $x = \alpha_1 v_1 + \alpha_2 v_2 + \cdots + \alpha_k v_k$, $y = \beta_1 w_1 + \beta_2 w_2 + \cdots + \beta_l w_l$; the $v_i, w_j \in S$; $\alpha_i, \beta_j \in F$. Then we have $x + y = \alpha_1 v_1 + \cdots + \alpha_k v_k + \beta_1 w_1 + \cdots + \beta_l w_l \in \lceil S \rceil$ and, for each $\alpha \in F$, $\alpha x = (\alpha \alpha_1) v_1 + \cdots + (\alpha \alpha_k) v_k \in \lceil S \rceil$; by Theorem 7.5.1, $\lceil S \rceil$ is a subspace. If $v \in S$, then $v = 1 \cdot v \in \lceil S \rceil$, so $S \subseteq \lceil S \rceil$. If W is any subspace with $S \subseteq W$, then for any $\alpha_i \in F$ and $v_i \in S$ we have $\alpha_1 v_1 + \cdots + \alpha_k v_k \in W$, so $\lceil S \rceil \subseteq W$. ∎

If $S = \{v_1, v_2, \ldots, v_k\}$ is a finite subset of V, we write $\lceil S \rceil$ as $\lceil v_1, v_2, \ldots, v_k \rceil$ instead of $\lceil \{v_1, v_2, \ldots, v_k\} \rceil$. $\lceil S \rceil$ is called the subspace **spanned** by S, and we say that S **spans** $\lceil S \rceil$. Thus if $\mathbf{X}_0$ and $\mathbf{X}_1$ are solutions of $\mathbf{AX} = \mathbf{0}$, $\lceil \mathbf{X}_0, \mathbf{X}_1 \rceil$ is a subspace of the solution space of $\mathbf{AX} = \mathbf{0}$.

A nonempty subset S of V is **linearly independent** if any equation $\alpha_1 v_1 + \alpha_2 v_2 + \cdots + \alpha_n v_n = 0$, $\alpha_i \in F$, distinct $v_i \in S$, implies all $\alpha_i = 0$. S is **linearly dependent** if S is not linearly independent. We say that S is a **basis** of V if S spans V and is linearly independent. A vector space V is **finite-dimensional** if some finite subset of V spans V.

For example, denote the rows of the identity matrix $\mathbf{I}_n$ over a field F by $e_1 = (1, 0, \ldots, 0)$, $e_2 = (0, 1, 0, \ldots, 0), \ldots, e_n = (0, \ldots, 0, 1)$. These n vectors span F^n because any $(\beta_1, \beta_2, \ldots, \beta_n) \in F^n$ is a linear combination

$$\beta_1(1, 0, \ldots, 0) + \beta_2(0, 1, 0, \ldots, 0) + \cdots + \beta_n(0, \ldots, 0, 1)$$

of them. This proves that F^n is finite-dimensional. They constitute a linearly independent set, because the equation $\alpha_1 e_1 + \alpha_2 e_2 + \cdots + \alpha_n e_n = 0$ implies that

$$\alpha_1(1, 0, \ldots, 0) + \alpha_2(0, 1, 0, \ldots, 0) + \cdots + \alpha_n(0, \ldots, 0, 1) = (0, 0, \ldots, 0),$$

$$(\alpha_1, \alpha_2, \ldots, \alpha_n) = (0, 0, \ldots, 0), \quad \text{all } \alpha_i = 0.$$

Therefore, $\{e_1, e_2, \ldots, e_n\}$ is a basis of F^n.

Theorem 7.6.2. *Assume that S and T are finite subsets of V such that S spans V and T is linearly independent. Then*

$$|S| \geq |T|.$$

Proof. Suppose that the inequality is false, say $S = \{v_1, \ldots, v_m\}$ and $T = \{w_1, \ldots, w_n\}$, with $m < n$. Since S spans V we can find elements $\alpha_{ij} \in F$ such that $w_i = \sum_{j=1}^{m} \alpha_{ij} v_j$. The system

$$\sum_{i=1}^{n} \alpha_{ij} x_i = 0, \quad 1 \leq j \leq m$$

of m homogeneous equations in n unknowns, $m < n$, will have a nontrivial solution $x_1 = \beta_1, \ldots, x_n = \beta_n$ with not all β_i's equal to 0. (It is clear from the sweepout process that some x's will be solved for in terms of others; we can choose these others to be nonzero.) We then have

$$\sum_{i=1}^{n} \beta_i w_i = \sum_{i=1}^{n} \beta_i \left(\sum_{j=1}^{m} \alpha_{ij} v_j \right) = \sum_{j=1}^{m} \left(\sum_{i=1}^{n} \alpha_{ij} \beta_i \right) v_j = \sum_{j=1}^{m} 0 v_j = 0,$$

a contradiction to the linear independence of T. ∎

Corollary 7.6.1. *Any non-$[0]$ finite-dimensional vector space has a basis; in fact, any linearly independent subset may be enlarged to a basis.*

Proof. If V is finite-dimensional, then some finite subset $S = \{v_1, \ldots, v_m\}$ spans V. Any nonzero element of V is a one-element linearly independent subset of V (Exercise 7.7.5), so it is enough to show that any linearly independent subset $T = \{w_1, \ldots, w_k\}$ may be enlarged to a basis of V.

If T spans V, then T is a basis; done. If not, there is some $w_{k+1} \in V - [T]$; we claim that $T_1 = T \cup \{w_{k+1}\}$ is linearly independent. If

$$\alpha_1 w_1 + \cdots + \alpha_k w_k + \alpha_{k+1} w_{k+1} = 0$$

and $\alpha_{k+1} \neq 0$, the equation can be solved for w_{k+1} to show that $w_{k+1} \in [T]$, a contradiction; hence $\alpha_{k+1} = 0$. Linear independence of T now shows $\alpha_1 = \cdots = \alpha_k = 0$, so T_1 is linearly independent. We have thus shown that a linearly independent set which is not a basis can be enlarged; by Theorem 7.6.2, no linearly independent set can contain more than $m = |S|$ vectors, so this enlargement process must end with a basis of V. ∎

Corollary 7.6.1 shows that any nontrivial finite-dimensional vector space has many different bases, but all have the same number of elements.

Corollary 7.6.2. *Any two bases of a finite-dimensional vector space have the same number of elements.*

Proof. If B_1 and B_2 are bases of V, Theorem 7.6.2 implies that $|B_1| \leq |B_2|$ and $|B_2| \leq |B_1|$. ∎

The number of vectors in any basis of a finite-dimensional V is called the **dimension** of V and is denoted by dim V or $\dim_F V$. (The dimension of $\lceil 0 \rceil$ is 0.) Since the rows $e_1, \ldots, e_n$ of $\mathbf{I}_n$ constitute a basis of F^n, we have $\dim F^n = n$.

Corollary 7.6.3. *Any subspace W of a finite-dimensional vector space V is also finite-dimensional, with $\dim W \leq \dim V$.*

Proof. Let $B = \{v_1, v_2, \ldots, v_m\}$ be a basis of V. Subsets of W are subsets of V, so Theorem 7.6.2 shows that a linearly independent subset of W can contain at most m elements. A maximal linearly independent subset of W will therefore be a basis (otherwise it could be enlarged), and will contain $\dim W \leq m = \dim V$ elements. ∎

Theorem 7.6.3. *If $B = \{v_1, \ldots, v_n\}$ is an ordered basis of V, then each $v \in V$ has a unique expression $v = \alpha_1 v_1 + \cdots + \alpha_n v_n$, where the $\alpha_i \in F$.*

The column

$$[v]_B = \begin{bmatrix} \alpha_1 \\ \alpha_2 \\ \vdots \\ \alpha_n \end{bmatrix}$$

is called the **coordinate vector** of v in the ordered basis B.

Proof. There certainly exists one such expression, since B spans V. If there are two, say

$$v = \alpha_1 v_1 + \cdots + \alpha_n v_n = \beta_1 v_1 + \cdots + \beta_n v_n, \qquad \alpha_i, \beta_i \in F,$$

then $0 = v - v = (\alpha_1 - \beta_1)v_1 + \cdots + (\alpha_n - \beta_n)v_n$. Since B is linearly independent, this means that $0 = \alpha_i - \beta_i$, $\alpha_i = \beta_i$ for all i. ∎

If $v_1, \ldots, v_m$ are vectors in F^n, we can form an $m \times n$ matrix $\mathbf{A}$ with rows $v_1, \ldots, v_m$. Then $\lceil v_1, \ldots, v_m \rceil$ is the **row space** of $\mathbf{A}$, the subspace of F^n spanned by the rows of $\mathbf{A}$. Elementary row operations on $\mathbf{A}$ do not change the row space (Exercise 7.7.8). If we perform the sweepout process on $\mathbf{A}$, the result is the reduced echelon form $\mathbf{E}$ of $\mathbf{A}$; in $\mathbf{E}$, the nonzero rows are linearly independent (Exercise 7.7.9). The number of such nonzero rows is the dimension of $\lceil v_1, \ldots, v_m \rceil$, also called the **rank** of $\mathbf{A}$. In this way we can use sweepout to study subspaces of F^n.

Sec. 7.6 Linear Independence and Dimension

If $V_1, V_2, \ldots, V_k$ are vector spaces over F, we know from Chapter 4 that the *Cartesian product* $V_1 \times V_2 \times \cdots \times V_k$ is an additive abelian group, with addition

$$(v_1, v_2, \ldots, v_k) + (w_1, w_2, \ldots, w_k) = (v_1 + w_1, v_2 + w_2, \ldots, v_k + w_k),$$

for all $v_i, w_i \in V_i$. In fact, $V_1 \times V_2 \times \cdots \times V_k$ is a vector space over F, with scalar multiplication defined by

$$\alpha(v_1, v_2, \ldots, v_k) = (\alpha v_1, \alpha v_2, \ldots, \alpha v_k)$$

for all $v_i \in V_i$ and all $\alpha \in F$. The identities (7.5.1) are all easily verified; for example, if $v = (v_1, v_2, \ldots, v_k)$ and $w = (w_1, w_2, \ldots, w_k)$ in $V_1 \times V_2 \times \cdots \times V_k$, then

$$\begin{aligned}
\alpha(v + w) &= \alpha((v_1, v_2, \ldots, v_k) + (w_1, w_2, \ldots, w_k)) \\
&= \alpha(v_1 + w_1, v_2 + w_2, \ldots, v_k + w_k) \\
&= (\alpha(v_1 + w_1), \alpha(v_2 + w_2), \ldots, \alpha(v_k + w_k)) \\
&= (\alpha v_1 + \alpha w_1, \alpha v_2 + \alpha w_2, \ldots, \alpha v_k + \alpha w_k) \\
&= (\alpha v_1, \alpha v_2, \ldots, \alpha v_k) + (\alpha w_1, \alpha w_2, \ldots, \alpha w_k) = \alpha v + \alpha w.
\end{aligned}$$

Space $V_1 \times V_2 \times \cdots \times V_k$ is called the **(external) direct sum** of $V_1, V_2, \ldots, V_k$. For example, the vector space $\mathbb{R}^3$ over $\mathbb{R}$ is really $\mathbb{R} \times \mathbb{R} \times \mathbb{R}$, where $\mathbb{R}$ is a vector space over itself; $\mathbb{R}^3 \times \mathbb{R}^2$ and $\mathbb{R}^5$ differ slightly symbolically but are used interchangeably in practice.

Assume that $V_1 \times V_2 \times \cdots \times V_k$ is the external direct sum of vector spaces $V_1, V_2, \ldots, V_k$ over F, and define, for each i, $\bar{V}_i = \{(0, 0, \ldots, 0, v_i, 0, \ldots, 0) \in V_1 \times V_2 \times \cdots \times V_k | v_i \in V_i\}$, the v_i in the ith position. Then each $\bar{V}_i$ is a subspace of $V_1 \times V_2 \times \cdots \times V_k$, and it is obvious that

(7.6.1) $\quad \bar{V}_i \cap (\bar{V}_1 + \cdots + \bar{V}_{i-1} + \bar{V}_{i+1} + \cdots + \bar{V}_k) = \lceil 0 \rceil$

for each i. For example,

$$\{(0, \gamma, 0) | \gamma \in \mathbb{R}\} \cap \{(\alpha, 0, \beta) | \alpha, \beta \in \mathbb{R}\} = \lceil 0 \rceil$$

in $\mathbb{R}^3 = \mathbb{R} \times \mathbb{R} \times \mathbb{R}$. We shall next see that this property (7.6.1) characterizes sets of subspaces of a single vector space that behave like the terms in an external direct sum.

Assume now that $V_1, V_2, \ldots, V_k$ are subspaces of the vector space V over F, with property

(7.6.2) $\quad V_i \cap (V_1 + \cdots + V_{i-1} + V_{i+1} + \cdots + V_k) = \lceil 0 \rceil$

for all $i = 1, 2, \ldots, k$. Define $\varphi \colon V_1 \times V_2 \times \cdots \times V_k \to V_1 + V_2 + \cdots + V_k$ by

$$\varphi \colon (v_1, v_2, \ldots, v_k) \to v_1 + v_2 + \cdots + v_k;$$

φ is a homomorphism of additive groups because

$$\varphi((v_1, v_2, \ldots, v_k) + (w_1, w_2, \ldots, w_k))$$
$$= \varphi((v_1 + w_1, v_2 + w_2, \ldots, v_k + w_k))$$
$$= (v_1 + w_1) + (v_2 + w_2) + \cdots + (v_k + w_k)$$
$$= (v_1 + v_2 + \cdots + v_k) + (w_1 + w_2 + \cdots + w_k)$$
$$= \varphi((v_1, v_2, \ldots, v_k)) + \varphi((w_1, w_2, \ldots, w_k)).$$

If $\varphi((v_1, v_2, \ldots, v_k)) = 0$, then $v_1 + v_2 + \cdots + v_k = 0$, $-v_1 = v_2 + \cdots + v_k \in V_1 \cap (V_2 + \cdots + V_k) = [0]$, $v_1 = 0$; similarly, all $v_i = 0$, ker $\varphi = [0]$, φ is an isomorphism. In this case we denote $V_1 + V_2 + \cdots + V_k$ by $V_1 \oplus V_2 \oplus \cdots \oplus V_k$, and we call $V_1 \oplus V_2 \oplus \cdots \oplus V_k$ an (**internal**) **direct sum** of $V_1, V_2, \ldots, V_k$.

Every external direct sum $V_1 \times V_2 \times \cdots \times V_k$ is also an internal direct sum $\bar{V}_1 \oplus \bar{V}_2 \oplus \cdots \oplus \bar{V}_k$ (notation as before). If

$$W_1 = \{(\alpha, \alpha, 0) | \alpha \in \mathbb{R}\}, \quad W_2 = \{(0, \beta, 0) | \beta \in \mathbb{R}\}, \quad W_3 = \{(0, 0, \gamma) | \gamma \in \mathbb{R}\},$$

then W_1, W_2, and W_3 are subspaces of $\mathbb{R}^3$ such that $\mathbb{R}^3 = W_1 + W_2 + W_3$ and

$$W_1 \cap (W_2 + W_3) = W_2 \cap (W_1 + W_3) = W_3 \cap (W_1 + W_2) = [0],$$

so $\mathbb{R}^3 = W_1 \oplus W_2 \oplus W_3$.

Note that condition (7.6.2) is strictly stronger than the simpler condition $V_i \cap V_j = [0]$ for $i \neq j$. Indeed, the three subspaces $V_1 = \{(\alpha, 0) | \alpha \in \mathbb{R}\}$, $V_2 = \{(\beta, \beta) | \beta \in \mathbb{R}\}$, $V_3 = \{(0, \gamma) | \gamma \in \mathbb{R}\}$ of $\mathbb{R}^2$ satisfy $V_i \cap V_j = [0]$ for $i \neq j$, but (7.6.2) fails.

In Exercise 7.7.12 you are asked to prove that

$$\dim(V_1 \oplus V_2 \oplus \cdots \oplus V_k) = \dim V_1 + \dim V_2 + \cdots + \dim V_k$$

for any internal direct sum of finite-dimensional subspaces $V_1, V_2, \ldots, V_k$ of a space V, and that each vector in a direct sum has a unique expression as a sum of elements in the summands.

7.7 Exercises

1. Verify in detail that examples 1 to 3 of Section 7.5 are examples of vector spaces.

2. Illustrate Corollaries 7.5.1 and 7.5.2 with the systems of equations

$$\begin{aligned} x_1 + x_2 - 3x_3 + x_4 &= 0 \\ 2x_1 - x_2 + x_3 - x_4 &= 0 \\ x_1 + 4x_2 - 10x_3 + 4x_4 &= 0 \end{aligned} \quad \text{and} \quad \begin{aligned} x_1 + x_2 - 3x_3 + x_4 &= 0 \\ 2x_1 - x_2 + x_3 - x_4 &= 3 \\ x_1 + 4x_2 - 10x_3 + 4x_4 &= -3. \end{aligned}$$

Sec. 7.7 Exercises

*3. Let $W_1, W_2, \ldots, W_k$ be subspaces of a vector space V. Show that $W_1 \cap W_2 \cap \cdots \cap W_k$ is a subspace of V. Show also that $W_1 + W_2 + \cdots + W_k$ equals the subspace of V spanned by the set $W_1 \cup W_2 \cup \cdots \cup W_k$.

*4. Show that the solution space of a *set* of homogeneous linear equations is the intersection of the solution spaces of the individual equations.

*5. Prove in detail that any nonzero element of a vector space is a one-element linearly independent subset, while the 0-vector is a one-element linearly dependent subset.

*6. (a) Show that if W is a subspace of the finite-dimensional vector space V and dim $W =$ dim V, then $W = V$.

(b) Show that if W is a subspace of dimension k of a finite-dimensional vector space V, then any k linearly independent vectors of W constitute a basis for W.

7. Prove that a linearly independent subset of a finite-dimensional vector space is a basis iff it is maximal (not contained in a larger linearly independent subset). In what other ways can one prove a finite subset of vectors of a finite-dimensional vector space V is a basis for V?

*8. Prove in detail that elementary row operations do not alter the row space of a matrix. [Exercise 7.4.10(a) may help.]

*9. Prove that in any matrix in reduced echelon form, or any matrix of form

$$\begin{bmatrix} 1 & a_{12} & a_{13} & \cdots & & & a_{1n} \\ 0 & 1 & a_{23} & \cdots & & & a_{2n} \\ \cdots & & & & & & \\ 0 & 0 & 0 & \cdots & 1 & a_{k,k+1} & \cdots & a_{kn} \end{bmatrix},$$

the nonzero rows constitute a linearly independent set.

*10. Show that $B = \{(1, 2, -1), (2, 3, 0), (-1, 0, 4)\}$ is an (ordered) basis of $\mathbb{R}^3$. Find the coordinate vector of $(0, 1, 0)$ in this basis.

11. Find a basis for the subspace of $\mathbb{R}^3$ spanned by $(2, 1, 3)$, $(-1, 4, 0)$, $(4, 2, 6)$, and $(7, -10, 6)$.

*12. (a) Prove that if $V_1 \oplus V_2 \oplus \cdots \oplus V_k$ is an internal direct sum of finite-dimensional subspaces of a vector space V, then $V_1 \oplus V_2 \oplus \cdots \oplus V_k$ is finite-dimensional and

$$\dim(V_1 \oplus V_2 \oplus \cdots \oplus V_k) = \dim V_1 + \dim V_2 + \cdots + \dim V_k.$$

(HINT: Consider the union of the bases of the V_i.)

(b) Show that every vector v in $V_1 \oplus V_2 \oplus \cdots \oplus V_k$ has a *unique* expression $v = v_1 + v_2 + \cdots + v_k$, $v_i \in V_i$.

(c) Given the subspaces $W_1 = \{(\alpha, 0, \alpha, 0) | \alpha \in \mathbb{R}\}$ and $W_2 = \{(0, \beta, 0, \beta) | \beta \in \mathbb{R}\}$ of $\mathbb{R}^4$, find a subspace W_3 of $\mathbb{R}^4$ such that $W_1 \oplus W_2 \oplus W_3 = \mathbb{R}^4$.

(d) Under what condition(s) on two subspaces V_1 and V_2 of a vector space V will there exist a third subspace V_3 such that $V = V_1 \oplus V_2 \oplus V_3$?

7.8 Linear Transformations

If V and W are vector spaces over F, a **linear transformation (linear operator, linear mapping, homomorphism, F-homomorphism)** from V to W is a function $f: V \to W$ such that

(7.8.1) $\quad f(u+v) = f(u) + f(v)$ and $f(\alpha v) = \alpha f(v)$, all $u, v \in V$, all $\alpha \in F$.

The set of all linear transformations from V to W is denoted by $\text{Hom}_F(V, W)$.

Theorem 7.8.1. *Let V and W be vector spaces over F, $B = \{v_1, \ldots, v_n\}$ an ordered basis of V. For any elements $w_1, \ldots, w_n$ of W, there is a unique $f \in \text{Hom}_F(V, W)$ satisfying $f(v_i) = w_i$, $1 \le i \le n$.*

Proof. By Theorem 7.6.3, for all $v \in V$, $v = \alpha_1 v_1 + \cdots + \alpha_n v_n$ for uniquely determined $\alpha_1, \ldots, \alpha_n \in F$, so we can define a function f on all of V with the equation $f(\alpha_1 v_1 + \cdots + \alpha_n v_n) = \alpha_1 w_1 + \cdots + \alpha_n w_n$, all $\alpha_i \in F$. Then, in particular,

$$f(v_i) = f(0 \cdot v_1 + \cdots + 0 \cdot v_{i-1} + 1 \cdot v_i + 0 \cdot v_{i+1} + \cdots + 0 \cdot v_n)$$
$$= 0 \cdot w_1 + \cdots + 0 \cdot w_{i-1} + 1 \cdot w_i + 0 \cdot w_{i+1} + \cdots + 0 \cdot w_n = w_i,$$

as desired. To see that f is linear, suppose that $u, v \in V$ and $\alpha \in F$, say $u = \alpha_1 v_1 + \cdots + \alpha_n v_n$, $v = \beta_1 v_1 + \cdots + \beta_n v_n$. Then

$$f(\alpha v) = f((\alpha \beta_1) v_1 + \cdots + (\alpha \beta_n) v_n) = (\alpha \beta_1) w_1 + \cdots + (\alpha \beta_n) w_n$$
$$= \alpha(\beta_1 w_1 + \cdots + \beta_n w_n) = \alpha f(v),$$

and

$$f(u + v) = f((\alpha_1 + \beta_1) v_1 + \cdots + (\alpha_n + \beta_n) v_n) = (\alpha_1 + \beta_1) w_1 + \cdots + (\alpha_n + \beta_n) w_n$$
$$= (\alpha_1 w_1 + \cdots + \alpha_n w_n) + (\beta_1 w_1 + \cdots + \beta_n w_n) = f(u) + f(v),$$

so f is linear.

As for uniqueness, any $f_1 \in \text{Hom}_F(V, W)$ with $f_1(v_i) = w_i$, all i, being linear, must satisfy

$$f_1(\alpha_1 v_1 + \cdots + \alpha_n v_n) = f_1(\alpha_1 v_1) + \cdots + f_1(\alpha_n v_n)$$
$$= \alpha_1 f_1(v_1) + \cdots + \alpha_n f_1(v_n) = \alpha_1 w_1 + \cdots + \alpha_n w_n,$$

so $f_1 = f$. ∎

Theorem 7.8.1 shows that there are always lots of linear transformations from one nontrivial vector space to another. We will frequently define a linear transformation by specifying the images of the elements of a basis; Theorem 7.8.1 shows that this always results in a uniquely determined linear transformation.

Sec. 7.8 Linear Transformations

For example, let $F = \mathrm{GF}(5)$ and use the basis $e_1 = (1, 0, 0)$, $e_2 = (0, 1, 0)$, $e_3 = (0, 0, 1)$ of F^3. Suppose that we define $f \in \mathrm{Hom}_F(F^3, F^3)$ by

(7.8.2) $f((1, 0, 0)) = (2, 3, 1)$, $f((0, 1, 0)) = (1, 0, 4)$, $f((0, 0, 1)) = (2, 1, 4)$.

Then

$$f((\alpha_1, \alpha_2, \alpha_3)) = f(\alpha_1(1, 0, 0) + \alpha_2(0, 1, 0) + \alpha_3(0, 0, 1))$$
$$= \alpha_1 f((1, 0, 0)) + \alpha_2 f((0, 1, 0)) + \alpha_3 f((0, 0, 1))$$
$$= \alpha_1(2, 3, 1) + \alpha_2(1, 0, 4) + \alpha_3(2, 1, 4);$$

in particular,

$$f((2, 0, 3)) = 2 \cdot (2, 3, 1) + 0 \cdot (1, 0, 4) + 3 \cdot (2, 1, 4)$$
$$= (4, 1, 2) + (0, 0, 0) + (1, 3, 2) = (0, 4, 4),$$
$$f((1, 4, 2)) = 1 \cdot (2, 3, 1) + 4 \cdot (1, 0, 4) + 2 \cdot (2, 1, 4)$$
$$= (2, 3, 1) + (4, 0, 1) + (4, 2, 3) = (0, 0, 0).$$

If $V_1, \ldots, V_k$ are subspaces of V, then the additive group homomorphism $\varphi: V_1 \times V_2 \times \cdots \times V_k \to V$ in Section 7.6 is always a homomorphism of vector spaces; that is, $\varphi \in \mathrm{Hom}_F(V_1 \times V_2 \times \cdots \times V_k, V)$.

If $f \in \mathrm{Hom}_F(V, W)$, we define the **kernel** (or **null space**) of f to be $\ker f = \{v \in V | f(v) = 0\}$. In the example just given, $(1, 4, 2) \in \ker f$. The range of f is, of course, $f(V) = \{f(v) \in W | v \in V\}$.

Lemma 7.8.1. *If $f \in \mathrm{Hom}_F(V, W)$, then $\ker f$ is a subspace of V and $f(V)$ is a subspace of W.*

Proof. If $u, v \in \ker f$ and $\alpha \in F$, then $f(u + v) = f(u) + f(v) = 0 + 0 = 0$ and $f(\alpha v) = \alpha f(v) = \alpha \cdot 0 = 0$. By Theorem 7.5.1, $\ker f$ is therefore a subspace. If $f(v_1), f(v_2) \in f(V)$ and $\alpha \in F$, then $f(v_1) + f(v_2) = f(v_1 + v_2) \in f(V)$ and $\alpha \cdot f(v_1) = f(\alpha v_1) \in f(V)$, so $f(V)$ is a subspace. ∎

Lemma 7.8.2. $f \in \mathrm{Hom}_F(V, W)$ *is one-one iff $\ker f = [0]$.*

Proof. Since every linear transformation of vector spaces is a homomorphism of (additive) groups, this is a special case of Exercise 4.15.10. ∎

We have seen that the homomorphism of (7.8.2) has $\ker f \neq [0]$, so that f is not one-one.

An $f \in \mathrm{Hom}_F(V, W)$ is an **isomorphism** iff f is a bijection; in this case we also say that V and W are **isomorphic**. (We shall see in Exercise 7.11.2 that if $f \in \mathrm{Hom}_F(V, W)$ is an isomorphism, then $f^{-1} \in \mathrm{Hom}_F(W, V)$.) Isomorphic finite-dimensional vector spaces have the same dimension (why?).

Now let U be a subspace of the vector space V over F. Then U is in particular a (normal) subgroup of the additive abelian group V, so we can form the quotient group V/U of cosets $v + U$, $v \in V$, with addition $(v_1 + U) + (v_2 + U) = (v_1 + v_2) + U$. We define a scalar multiplication on V/U by $\alpha(v + U) = \alpha v + U$, all $v \in V$. This definition is valid, because if $v + U = v' + U$, $v' \in V$, then $v - v' \in U$, $\alpha v - \alpha v' = \alpha(v - v') \in U$, $\alpha v + U = \alpha v' + U$; we have proved that the product is the same, regardless of which name $(v + U$ or $v' + U)$ of the coset we use. It is now trivial to verify the laws (7.5.1) for V/U; V/U is itself a vector space over F, called the **quotient space** of V by U.

Lemma 7.8.3. *If $\{v_1, \ldots, v_k\}$ is a basis of the subspace U of V and we extend it to a basis $\{v_1, \ldots, v_k, v_{k+1}, \ldots, v_n\}$ of V, then $\{v_{k+1} + U, \ldots, v_n + U\} = B$ is a basis of V/U, so that*

$$\dim(V/U) = \dim V - \dim U.$$

Proof. If $v + U \in V/U$ and if $v = \alpha_1 v_1 + \cdots + \alpha_k v_k + \alpha_{k+1} v_{k+1} + \cdots + \alpha_n v_n$, then

$$v + U = \alpha_{k+1} v_{k+1} + \cdots + \alpha_n v_n + U = \alpha_{k+1}(v_{k+1} + U) + \cdots + \alpha_n(v_n + U),$$

so B spans V/U. If $\beta_{k+1}(v_{k+1} + U) + \cdots + \beta_n(v_n + U) = 0 + U$ for some $\beta_i \in F$, then $\beta_{k+1} v_{k+1} + \cdots + \beta_n v_n \in U$, say

$$\beta_{k+1} v_{k+1} + \cdots + \beta_n v_n = \gamma_1 v_1 + \cdots + \gamma_k v_k.$$

Then

$$(-\gamma_1) v_1 + \cdots + (-\gamma_k) v_k + \beta_{k+1} v_{k+1} + \cdots + \beta_n v_n = 0,$$

and linear independence of $\{v_1, \ldots, v_k, v_{k+1}, \ldots, v_n\}$ implies that all $-\gamma_i = 0$, all $\beta_j = 0$. The fact that all $\beta_j = 0$ implies that $\{v_{k+1} + U, \ldots, v_n + U\}$ is linearly independent. ∎

Every vector space homomorphism is in particular a group homomorphism, and every subspace of a vector space is a normal subgroup of the (abelian) additive group of the vector space. Theorem 4.14.2 therefore implies

(1) For any subspace U of V, the function $f: V \to V/U$ defined by $f(v) = v + U$ is onto, is in $\text{Hom}_F(V, V/U)$, and has kernel U.

(2) If $\varphi \in \text{Hom}_F(V, W)$ has kernel U, then we obtain an isomorphism $\bar{\varphi} \in \text{Hom}_F(V/U, \varphi(V))$ by defining $\bar{\varphi}(v + U) = \varphi(v)$.

If $f \in \text{Hom}_F(V, W)$ and V and W are finite-dimensional, so are their subspaces $\ker f$ and $f(V)$. The **nullity** of f is $\dim(\ker f)$, and the **rank** of f is $\dim f(V)$.

Theorem 7.8.2. *If $f \in \text{Hom}_F(V, W)$ where V and W are finite-dimensional, then*

(7.8.3) $$\text{nullity}(f) + \text{rank}(f) = \dim V.$$

Proof. Statement (2) above and Lemma 7.8.3 imply that for $U = \ker f$, we have $\dim V - \text{nullity}(f) = \dim V - \dim U = \dim(V/U) = \dim f(V) = \text{rank}(f)$. ∎

For example, let $F = \mathbb{R}$, $V = \mathbb{R}^3$, and let $U = \{(x, x, 0) | x \in \mathbb{R}\}$ (the "45° line" in the 1,2-plane). The subspace U has basis $\{(1, 1, 0)\}$, and $\{(1, 1, 0), (1, 0, 0), (0, 0, 1)\}$ is a basis of V, so V/U is a two-dimensional vector space with basis $\{(1, 0, 0) + U, (0, 0, 1) + U\}$. [The geometrical interpretation of this space is the set of lines parallel to U. One such line intersects each point of the 1,3-plane, which has basis $(1, 0, 0)$ and $(0, 0, 1)$, so the space is indeed two-dimensional. The "sum" of the lines through $(a, 0, b)$ and $(c, 0, d)$ is the line through $(a + c, 0, b + d)$.]

Lemma 7.8.4. *If $f \in \text{Hom}_F(V, V)$ and the subspace U of V satisfies $f(U) \subseteq U$, then the relation*

$$f^*(v + U) = f(v) + U, \quad \text{all } v \in V,$$

defines an $f^* \in \text{Hom}_F(V/U, V/U)$.

Proof. We must first show that if $v + U = v' + U$, $v, v' \in V$, then $f(v) + U = f(v') + U$; this will show that the definition of f^* is independent of the name ($v + U$ or $v' + U$) of a single coset. The equation $v + U = v' + U$ implies that $v' = v + u$, some $u \in U$; then

$$f(v') + U = f(v + u) + U = [f(v) + f(u)] + U = f(v) + [f(u) + U] = f(v) + U,$$

since $f(u) \in U$ by hypothesis.

If $v_1 + U$, $v_2 + U \in V/U$ and $\alpha \in F$, then

$$f^*((v_1 + U) + (v_2 + U)) = f^*((v_1 + v_2) + U) = f(v_1 + v_2) + U$$
$$= f(v_1) + f(v_2) + U = [f(v_1) + U] + [f(v_2) + U] = f^*(v_1 + U) + f^*(v_2 + U)$$

and

$$f^*(\alpha(v_1 + U)) = f^*(\alpha v_1 + U) = f(\alpha v_1) + U = \alpha f(v_1) + U = \alpha(f(v_1) + U)$$
$$= \alpha f^*(v_1 + U),$$

proving that $f^* \in \text{Hom}_F(V/U, V/U)$. ∎

In the preceding example, if $f \in \text{Hom}_\mathbb{R}(\mathbb{R}^3, \mathbb{R}^3)$ is defined by

(7.8.4) $\qquad f((a, b, c)) = (3a - b + c, 2a + c, a - b + 3c),$

then $f((x, x, 0)) = (2x, 2x, 0)$, so $f(U) \subseteq U$ and f^* exists. We have

$$f^*((1, 0, 0) + U) = f((1, 0, 0)) + U = (3, 2, 1) + U$$

(7.8.5) $\qquad\qquad\qquad = (1, 0, 1) + U$

$$= ((1, 0, 0) + U) + ((0, 0, 1) + U)$$

and

(7.8.6)
$$f^*((0, 0, 1) + U) = f((0, 0, 1)) + U = (1, 1, 3) + U$$
$$= (0, 0, 3) + U$$
$$= 3 \cdot ((0, 0, 1) + U).$$

7.9 Matrices of Linear Transformations

If $B_1 = \{v_1, \ldots, v_m\}$ is an ordered basis of V over F, $B_2 = \{w_1, \ldots, w_n\}$ an ordered basis of W over F, $f \in \text{Hom}_F(V, W)$, and

(7.9.1)
$$f(v_j) = \sum_{i=1}^{n} \alpha_{ij} w_i,$$

then

(7.9.2)
$$[\alpha_{ij}]_{n \times m} = [f]_{B_1, B_2}$$

is the **matrix of f in the ordered bases B_1 and B_2**. By Theorem 7.8.1, any f is completely specified by its matrix. On the other hand, Theorem 7.6.3 shows that f completely determines the coefficients in (7.9.1) and hence determines a unique matrix. Therefore, $f \leftrightarrow [f]_{B_1, B_2}$ is a one-one correspondence between the sets $\text{Hom}_F(V, W)$ and $\text{Mat}_{(\dim W) \times (\dim V)}(F)$. (We see in Exercise 7.11.13 that this is actually a vector space isomorphism.)

By (7.9.1), the coordinates of $f(v_j)$ in the basis B_2 are the entries $\alpha_{1j}, \alpha_{2j}, \ldots, \alpha_{nj}$ of the jth column of $[f]_{B_1, B_2}$. For example, in the bases $B_1 = \{(1, 0), (0, 1)\}$ of $\mathbb{R}^2$ and $B_2 = \{(1, 0, 0), (0, 1, 0), (0, 0, 1)\}$ of $\mathbb{R}^3$, the equation

$$[f]_{B_1, B_2} = \begin{bmatrix} 2 & 7 \\ 1 & -3 \\ -2 & 4 \end{bmatrix}$$

defines an $f \in \text{Hom}_\mathbb{R}(\mathbb{R}^2, \mathbb{R}^3)$ such that

$$f((1, 0)) = 2 \cdot (1, 0, 0) + 1 \cdot (0, 1, 0) + (-2) \cdot (0, 0, 1) = (2, 1, -2),$$
$$f((0, 1)) = 7 \cdot (1, 0, 0) + (-3) \cdot (0, 1, 0) + 4 \cdot (0, 0, 1) = (7, -3, 4).$$

Equations (7.8.5) and (7.8.6) show that in the basis $\{(1, 0, 0) + U, (0, 0, 1) + U\}$, the f^* there has matrix

$$\mathbf{M}^* = \begin{bmatrix} 1 & 0 \\ 1 & 3 \end{bmatrix}.$$

Note that in the basis $\{(1, 1, 0), (1, 0, 0), (0, 0, 1)\}$ of $\mathbb{R}^3$, the f of (7.8.4) has matrix

$$\mathbf{M} = \begin{bmatrix} 2 & 2 & 1 \\ 0 & 1 & 0 \\ 0 & 1 & 3 \end{bmatrix} = \begin{bmatrix} 2 & 2 & 1 \\ 0 & & \mathbf{M}^* \end{bmatrix}.$$

Sec. 7.9 Matrices of Linear Transformations

Why could we have expected the block **0** and the block **M***, within **M**?

Theorem 7.9.1. If $B_1 = \{v_1, v_2, \ldots, v_m\}$ is an ordered basis of V, $B_2 = \{w_1, w_2, \ldots, w_n\}$ an ordered basis of W, $f \in \text{Hom}_F(V, W)$, and $v \in V$, then

(7.9.3) $$[f(v)]_{B_2} = [f]_{B_1, B_2}[v]_{B_1}.$$

Theorem 7.9.1 says that to find the coordinate vector of $f(v)$ in the basis B_2, we can multiply the matrix of f in the ordered bases B_1 and B_2 by the coordinate vector of v in the basis B_1. If

$$[f(v)]_{B_2} = \begin{bmatrix} \gamma_1 \\ \gamma_2 \\ \vdots \\ \gamma_n \end{bmatrix}, \quad \text{then } f(v) = \sum_{i=1}^{n} \gamma_i w_i.$$

Proof of Theorem 7.9.1. Assume that $[f]_{B_1, B_2} = [\alpha_{ij}]_{n \times m}$ and that v has coordinates $\beta_1, \ldots, \beta_m$ in B_1, so $v = \beta_1 v_1 + \cdots + \beta_m v_m$. Then

$$f(v) = f\left(\sum_{j=1}^{m} \beta_j v_j\right) = \sum_{j=1}^{m} \beta_j f(v_j) = \sum_{j=1}^{m} \sum_{i=1}^{n} \beta_j \alpha_{ij} w_i = \sum_{i=1}^{n} \left(\sum_{j=1}^{m} \alpha_{ij} \beta_j\right) w_i.$$

Thus the values $\sum_{j=1}^{m} \alpha_{ij} \beta_j$ ($1 \leq i \leq n$) are the components of $f(v)$ in the basis B_2; they are also the entries in the $n \times 1$ matrix product $[\alpha_{ij}]_{n \times m}[\beta_j]_{m \times 1}$. ∎

Note that if, in the above proof, we put $v = v_i$, then $\beta_j = \delta_{ij}$ for $j = 1, 2, \ldots, m$. Then $[\alpha_{ij}][v]_{B_1}$ is the ith column of $[\alpha_{ij}]$. Thus, again, if we know the coordinates in B_2 of the images of the basis vectors of V, we know the matrix of f.

For an example, let us use the bases $B_1 = \{(1, 0, 0), (0, 1, 0), (0, 0, 1)\}$ and $B_2 = \{(1, 2, -1), (2, 3, 0), (-1, 0, 4)\}$ (Exercise 7.7.10) of $\mathbb{R}^3$, and define $f \in \text{Hom}_\mathbb{R}(\mathbb{R}^3, \mathbb{R}^3)$ by

$$[f]_{B_1, B_2} = \begin{bmatrix} 2 & -1 & 3 \\ 0 & 4 & 1 \\ 3 & 0 & -1 \end{bmatrix}.$$

We first compute $f((3, 1, 2))$ using Theorem 7.9.1:

$$[f((3, 1, 2))]_{B_2} = [f]_{B_1, B_2}[(3, 1, 2)]_{B_1} = \begin{bmatrix} 2 & -1 & 3 \\ 0 & 4 & 1 \\ 3 & 0 & -1 \end{bmatrix} \begin{bmatrix} 3 \\ 1 \\ 2 \end{bmatrix} = \begin{bmatrix} 11 \\ 6 \\ 7 \end{bmatrix},$$

so that

$$f((3, 1, 2)) = 11 \cdot (1, 2, -1) + 6 \cdot (2, 3, 0) + 7 \cdot (-1, 0, 4) = (16, 40, 17).$$

We can check our work by using the linearity of f:

$$f((3, 1, 2)) = 3 \cdot f((1, 0, 0)) + 1 \cdot f((0, 1, 0)) + 2 \cdot f((0, 0, 1))$$
$$= 3[2(1, 2, -1) + 3(-1, 0, 4)] + 1[(-1)(1, 2, -1) + 4(2, 3, 0)]$$
$$+ 2[3(1, 2, -1) + 1(2, 3, 0) + (-1)(-1, 0, 4)]$$
$$= 3 \cdot (-1, 4, 10) + 1 \cdot (7, 10, 1) + 2 \cdot (6, 9, -7) = (16, 40, 17).$$

Theorem 7.9.2. *Assume that B_1, B_2, and B_3 are ordered bases for the finite-dimensional vector spaces V, W, and X over F. Assume also that $f \in \text{Hom}_F(V, W)$ and $g \in \text{Hom}_F(W, X)$. Then the composite function $g \circ f \colon V \to X$ satisfies $g \circ f \in \text{Hom}_F(V, X)$ and*

(7.9.4) $$[g \circ f]_{B_1, B_3} = [g]_{B_2, B_3} [f]_{B_1, B_2}.$$

Theorem 7.9.2 says that the matrix of the "product" $g \circ f$ is the product of the matrices of g and f. This is the underlying reason why the multiplication of matrices is defined as it is in (7.1.6).

Proof of Theorem 7.9.2. The proof that $g \circ f \in \text{Hom}_F(V, X)$ is Exercise 7.11.9. Let $B_1 = \{v_1, \ldots, v_m\}$, $B_2 = \{w_1, \ldots, w_n\}$, $B_3 = \{x_1, \ldots, x_p\}$, $[f]_{B_1, B_2} = [\alpha_{ij}]_{n \times m}$, and $[g]_{B_2, B_3} = [\gamma_{ij}]_{p \times n}$, so (7.9.1) implies that $f(v_j) = \sum_{i=1}^{n} \alpha_{ij} w_i$ and $g(w_j) = \sum_{i=1}^{p} \gamma_{ij} x_i$. We then have

$$(g \circ f)(v_j) = g(f(v_j)) = g\left(\sum_{k=1}^{n} \alpha_{kj} w_k\right) = \sum_{k=1}^{n} \alpha_{kj} g(w_k) = \sum_{k=1}^{n} \alpha_{kj} \sum_{i=1}^{p} \gamma_{ik} x_i$$

$$= \sum_{i=1}^{p} \left(\sum_{k=1}^{n} \gamma_{ik} \alpha_{kj}\right) x_i.$$

Therefore, $\sum_{k=1}^{n} \gamma_{ik} \alpha_{kj}$ is both the i,j-entry of $[g \circ f]_{B_1, B_3}$ and the i,j-entry of the matrix product $[\gamma_{ik}][\alpha_{kj}]$. ∎

An important special case of the matrix of a linear transformation occurs when $V = W$, so $f \in \text{Hom}_F(V, V)$. Then it is customary to use only one ordered basis $B_1 = B_2 = \{v_1, \ldots, v_n\}$ of V, and the **matrix of f in the ordered basis B_1** is $[f]_{B_1} = [f]_{B_1, B_1} = [\alpha_{ij}]_{n \times n}$, where $f(v_j) = \sum_{i=1}^{n} \alpha_{ij} v_i$, in accordance with (7.9.1) and (7.9.2). Theorem 7.9.2 implies that if $f, g \in \text{Hom}_F(V, V)$, then

(7.9.5) $$[g \circ f]_{B_1} = [g]_{B_1} [f]_{B_1}.$$

For example, let $F = \text{GF}(2)$ and let B_1 be the ordered basis $\{(1, 0), (1, 1)\}$ of F^2. Define $f, g \in \text{Hom}_F(F^2, F^2)$ by

$$[f]_{B_1} = \begin{bmatrix} 1 & 1 \\ 0 & 1 \end{bmatrix}, \quad [g]_{B_1} = \begin{bmatrix} 0 & 1 \\ 1 & 0 \end{bmatrix}.$$

Then, by (7.9.5),

$$[g \circ f]_{B_1} = \begin{bmatrix} 0 & 1 \\ 1 & 0 \end{bmatrix} \begin{bmatrix} 1 & 1 \\ 0 & 1 \end{bmatrix} = \begin{bmatrix} 0 & 1 \\ 1 & 1 \end{bmatrix}.$$

Hence, by Theorem 7.9.1,

$$[(g \circ f)((1, 1))]_{B_1} = [g \circ f]_{B_1}[(1, 1)]_{B_1} = \begin{bmatrix} 0 & 1 \\ 1 & 1 \end{bmatrix} \begin{bmatrix} 0 \\ 1 \end{bmatrix} = \begin{bmatrix} 1 \\ 1 \end{bmatrix},$$

so

$$(g \circ f)((1, 1)) = 1 \cdot (1, 0) + 1 \cdot (1, 1) = (0, 1).$$

We shall compute $(g \circ f)((1, 1))$ a second time, using the definition of $g \circ f$. We have, using $[f]_{B_1}$ and $[g]_{B_1}$,

$$(g \circ f)((1, 1)) = g(f((1, 1))) = g(1 \cdot (1, 0) + 1 \cdot (1, 1)) = g((0, 1))$$
$$= g((1, 0) + (1, 1)) = g((1, 0)) + g((1, 1))$$
$$= (1, 1) + (1, 0) = (0, 1).$$

The **direct sum** of square matrices $\mathbf{B}_1, \mathbf{B}_2, \ldots, \mathbf{B}_l$ over a field F is the matrix

(7.9.6)
$$\mathbf{M} = \begin{bmatrix} \mathbf{B}_1 & 0 & \cdots & 0 \\ 0 & \mathbf{B}_2 & \cdots & 0 \\ \multicolumn{4}{c}{\dotfill} \\ 0 & 0 & \cdots & \mathbf{B}_l \end{bmatrix},$$

the **0**'s denoting whole blocks of zeros of appropriate size.

Assume that V is a finite-dimensional vector space over F and $f \in \mathrm{Hom}_F(V, V)$. A subspace U of V is **f-invariant** if $f(U) \subseteq U$. For example, the subspace $U = \{(x_1, x_2, 0) | x_1, x_2 \in \mathbb{R}\}$ of $\mathbb{R}^3$ is f-invariant, where $f \in \mathrm{Hom}_{\mathbb{R}}(\mathbb{R}^3, \mathbb{R}^3)$ is defined by

$$f((x_1, x_2, x_3)) = (x_1 - 2x_2 + x_3, 2x_1 + x_2, 3x_3).$$

When U is an f-invariant subspace of V, the **restriction** f_U of f to U is defined by $f_U(u) = f(u) \in U$, all $u \in U$; then $f_U \in \mathrm{Hom}_F(U, U)$. In the preceding example,

$$f_U((x_1, x_2, 0)) = (x_1 - 2x_2, 2x_1 + x_2, 0).$$

The basis $B_U = \{(1, 0, 0), (0, 1, 0)\}$ of U extends to a basis $B = \{(1, 0, 0), (0, 1, 0), (0, 0, 1)\}$ of $V = \mathbb{R}^3$, and in these bases we have

$$[f_U]_{B_U} = \begin{bmatrix} 1 & -2 \\ 2 & 1 \end{bmatrix}, \quad [f]_B = \begin{bmatrix} 1 & -2 & 1 \\ 2 & 1 & 0 \\ 0 & 0 & 3 \end{bmatrix} = \begin{bmatrix} [f_U]_{B_U} & \cdots \\ \mathbf{0} & \cdots \end{bmatrix}.$$

In general, whenever U is an f-invariant subspace of V and we extend a basis B_U of U to a basis B of V, the matrix $[f]_B$ of f necessarily has this block form,

$$[f]_B = \begin{bmatrix} [f_U]_{B_U} & \cdots \\ 0 & \cdots \end{bmatrix}.$$

Lemma 7.9.1. *Assume that the vector space V over F is a direct sum $V = V_1 \oplus V_2 \oplus \cdots \oplus V_l$ of subspaces V_i, $B_i = \{v_{i1}, v_{i2}, \ldots, v_{id_i}\}$ is an ordered basis of V_i for each i, and $f \in \operatorname{Hom}_F(V, V)$ satisfies $f(V_i) \subseteq V_i$ for each i; denote by $[f_{V_i}]_{B_i}$ the matrix of the restriction f_{V_i} of f to V_i in the basis B_i. Then*

$$B = \{v_{11}, v_{12}, \ldots, v_{1d_1}, v_{21}, \ldots, v_{2d_2}, \ldots, v_{l1}, \ldots, v_{ld_l}\}$$

is an (ordered) basis of V, and the matrix of f in this basis is the direct sum

$$[f]_B = \begin{bmatrix} [f_{V_1}]_{B_1} & 0 & \cdots & 0 \\ 0 & [f_{V_2}]_{B_2} & \cdots & 0 \\ \vdots & & & \vdots \\ 0 & 0 & \cdots & [f_{V_l}]_{B_l} \end{bmatrix}.$$

Proof. The fact that B is a basis of V was verified in Exercise 7.7.12(a); the rest follows from the definition of the matrix of a linear transformation. ∎

For example, consider $F = \mathbb{R}$, $V = \mathbb{R}^5$, $V_1 = [e_1, e_2]$, and $V_2 = [e_3, e_4, e_5]$, where the e_i are the standard basis elements $e_1 = (1, 0, 0, 0, 0)$, $e_2 = (0, 1, 0, 0, 0), \ldots, e_5 = (0, 0, 0, 0, 1)$. Then $V = V_1 \oplus V_2$, and the $f \in \operatorname{Hom}_F(V, V)$ defined by $f(e_1) = e_1 + e_2$, $f(e_2) = -e_1 + 3e_2$, $f(e_3) = 2e_4$, $f(e_4) = 6e_3 - e_4 + 4e_5$, and $f(e_5) = e_4 + 2e_5$ certainly satisfies $f(V_1) \subseteq V_1$ and $f(V_2) \subseteq V_2$. The matrices of f on these subspaces are

$$[f_{V_1}] = \begin{bmatrix} 1 & -1 \\ 1 & 3 \end{bmatrix}, \quad [f_{V_2}] = \begin{bmatrix} 0 & 6 & 0 \\ 2 & -1 & 1 \\ 0 & 4 & 2 \end{bmatrix},$$

so the matrix of f on V is

$$\begin{bmatrix} [f_{V_1}] & 0 \\ 0 & [f_{V_2}] \end{bmatrix} = \begin{bmatrix} \begin{bmatrix} 1 & -1 \\ 1 & 3 \end{bmatrix} & 0 \\ 0 & \begin{bmatrix} 0 & 6 & 0 \\ 2 & -1 & 1 \\ 0 & 4 & 2 \end{bmatrix} \end{bmatrix} = \begin{bmatrix} 1 & -1 & 0 & 0 & 0 \\ 1 & 3 & 0 & 0 & 0 \\ 0 & 0 & 0 & 6 & 0 \\ 0 & 0 & 2 & -1 & 1 \\ 0 & 0 & 0 & 4 & 2 \end{bmatrix}.$$

7.10 Rank

The **transpose** of the $m \times n$ matrix $\mathbf{A} = [a_{ij}]$ is the $n \times m$ matrix $\mathbf{A}^T = [a_{ji}]$; the entry in the ith row, jth column of $\mathbf{A}$ is in the jth row, ith column of $\mathbf{A}^T$. It is easy to verify the properties

(7.10.1) $\quad (\mathbf{A}^T)^T = \mathbf{A}, \quad (\mathbf{A} + \mathbf{B})^T = \mathbf{A}^T + \mathbf{B}^T, \quad (\alpha \mathbf{A})^T = \alpha \mathbf{A}^T,$
$\quad (\mathbf{A}\mathbf{B})^T = \mathbf{B}^T \mathbf{A}^T; \quad (\mathbf{A}^{-1})^T = (\mathbf{A}^T)^{-1} \quad$ if $\mathbf{A}$ is invertible.

Note that the rows of $\mathbf{A}$ are the columns of $\mathbf{A}^T$, and the columns of $\mathbf{A}$ are the rows of $\mathbf{A}^T$; we say that $\mathbf{A}^T$ is obtained from $\mathbf{A}$ by "interchanging rows and columns."

We defined the *row space* of an $m \times n$ $\mathbf{A}$ to be the subspace of F^n spanned by the rows of $\mathbf{A}$, and the *rank* of $\mathbf{A}$ to be the dimension of its row space. Similarly, the *column space* of $\mathbf{A}$ is the subspace of $\text{Mat}_{m \times 1}(F)$ spanned by the columns of $\mathbf{A}$, and the *column rank* of $\mathbf{A}$ is the dimension of its column space. The columns of $\mathbf{A}$ are the rows of $\mathbf{A}^T$, so the column rank of $\mathbf{A}$ is the rank of $\mathbf{A}^T$.

Theorem 7.10.1. rank $\mathbf{A}$ = rank $\mathbf{A}^T$; *that is, the column rank of $\mathbf{A}$ equals the rank of $\mathbf{A}$.*

Proof. By Exercise 7.7.8, elementary row operations on $\mathbf{A}$ do not change the row space of $\mathbf{A}$. If $\mathbf{R}$ is the reduced echelon form of $\mathbf{A}$, this implies that rank $\mathbf{A}$ = rank $\mathbf{R}$. If rank $\mathbf{R} = k$, then $\mathbf{R}$ has exactly k nonzero rows, and k of the columns of $\mathbf{R}$ are columns from an identity matrix. These k "identity columns" are linearly independent and certainly span the column space of $\mathbf{R}$, so the column rank of $\mathbf{R}$ is the rank k of $\mathbf{R}$.

It is enough now to show that $\mathbf{A}$ and $\mathbf{R}$ have the same column rank, and for this it suffices to prove that an elementary row operation does not change the column rank of a matrix $\mathbf{A}$ (it usually does change the column space; give an example). Let $\mathbf{A}_1, \ldots, \mathbf{A}_n$ be the columns of $\mathbf{A}$ before the operation, $\mathbf{B}_1, \ldots, \mathbf{B}_n$ the columns afterward. By considering separately the three types of elementary row operation, it is not difficult to show that a set of columns $\{\mathbf{A}_{i_1}, \ldots, \mathbf{A}_{i_k}\}$ is linearly dependent iff the corresponding set $\{\mathbf{B}_{i_1}, \ldots, \mathbf{B}_{i_k}\}$ is linearly dependent. This implies that the column ranks are the same; further details are left as Exercise 7.11.5. ∎

Lemma 7.10.1. *If $\varphi \in \text{Hom}_F(V, W)$ is one-one and S is a subset of V, then S is linearly independent iff $\varphi(S)$ is linearly independent, so $\dim \lceil S \rceil$ and $\dim \lceil \varphi(S) \rceil$ are equal in this case.*

Proof. Set S is dependent iff there is an equation

(7.10.2) $\quad \alpha_1 s_1 + \cdots + \alpha_k s_k = 0, \quad s_i \in S, \quad$ not all $\alpha_i = 0$.

Applying φ and recalling Lemma 7.8.2, we see that (7.10.2) holds iff

$\alpha_1 \varphi(s_1) + \cdots + \alpha_k \varphi(s_k) = 0$, and such a relation holds iff $\varphi(S)$ is linearly dependent. ∎

Theorem 7.10.2. *Assume that $f \in \mathrm{Hom}_F(V, W)$, and let B_1 be an ordered basis of V, B_2 an ordered basis of W. Then*

(7.10.3) $$\mathrm{rank}\, f = \mathrm{rank}\, [f]_{B_1, B_2}.$$

Proof. Let $B_1 = \{v_1, \ldots, v_m\}$, $B_2 = \{w_1, \ldots, w_n\}$, and $[f]_{B_1, B_2} = [\alpha_{ij}]_{n \times m}$, so $f(v_j) = \sum_{i=1}^n \alpha_{ij} w_i$. The rank of f is by definition $\dim f(V) = \dim [f(v_1), \ldots, f(v_m)]$.

Let $\varepsilon_1, \ldots, \varepsilon_n$ be the *standard basis* of the vector space $\mathrm{Mat}_{n \times 1}(F)$. That is, the ith entry in ε_i is 1 and all other entries are 0. We obtain an isomorphism $\varphi \in \mathrm{Hom}_F(W, \mathrm{Mat}_{n \times 1}(F))$ by defining $\varphi(w_i) = \varepsilon_i$. Indeed, since φ is linear, φ maps $w = \sum_{i=1}^n a_i w_i$ onto $\varphi(w) = \sum_{i=1}^n a_i \varepsilon_i = \begin{bmatrix} a_1 \\ a_2 \\ \vdots \\ a_n \end{bmatrix}$ and the mapping $\sum_{i=1}^n a_i w_i \leftrightarrow \begin{bmatrix} a_1 \\ a_2 \\ \vdots \\ a_n \end{bmatrix}$ is clearly a bijection. In particular, φ maps $f(v_j) = \sum_{i=1}^n \alpha_{ij} w_i$ onto $\sum_{i=1}^n \alpha_{ij} \varepsilon_i = \begin{bmatrix} \alpha_{1j} \\ \alpha_{2j} \\ \vdots \\ \alpha_{nj} \end{bmatrix}$, the jth column of $[\alpha_{ij}]$. That is, φ sends $f(V) = [f(v_1), \ldots, f(v_m)]$ onto the column space of $[\alpha_{ij}]$. Lemma 7.10.1 implies that $\dim f(V)$ is the dimension of this column space; that is, the rank of f equals the column rank of $[f]_{B_1, B_2}$. Now apply Theorem 7.10.1 to complete the proof. ∎

Lemma 7.10.2. *Assume that $f \in \mathrm{Hom}_F(V, W)$ and $g \in \mathrm{Hom}_F(W, X)$. Then*
  (a) $\mathrm{rank}\, g \circ f \leq \min(\mathrm{rank}\, g, \mathrm{rank}\, f)$.
  (b) *If $V = W = X$ and g is an isomorphism, then $\mathrm{rank}\, g^{-1} \circ f \circ g = \mathrm{rank}\, f$.*

Proof. (a) We have first

$$\mathrm{rank}\, g \circ f = \dim g(f(V)) \leq \dim g(W) = \mathrm{rank}\, g,$$

since $f(V) \subseteq W$. By (7.8.3),

$$\mathrm{rank}\, g \circ f = \dim g(f(V)) = \dim f(V) - \{\text{nullity of } g \text{ on } f(V)\}$$
$$\leq \dim f(V) = \mathrm{rank}\, f.$$

(b) By (a), $\mathrm{rank}\, g^{-1} \circ f \circ g \leq \mathrm{rank}\, f$. But if $h = g^{-1} \circ f \circ g$, then $f = g \circ h \circ g^{-1}$, so $\mathrm{rank}\, f = \mathrm{rank}\, g \circ h \circ g^{-1} \leq \mathrm{rank}\, h = \mathrm{rank}\, g^{-1} \circ f \circ g$. ∎

Corollary 7.10.1. *If A and B are matrices over F such that AB is defined, then* rank $AB \leq$ min (rank A, rank B). *If A and B are $n \times n$ and B is invertible,* rank $B^{-1}AB =$ rank A.

Proof. Use A and B to define linear transformations on finite-dimensional vector spaces, and apply Lemma 7.10.2 and Theorem 7.10.2. ∎

Lemma 7.10.3. $f \in \text{Hom}_F(V, V)$ *is an isomorphism iff* rank $f =$ dim V.

Proof. By definition, rank $f =$ dim $f(V)$, so rank $f =$ dim V iff f is onto. By (7.8.3), dim ker $f =$ dim $V -$ rank f, so rank $f =$ dim V iff dim ker $f = 0$; by Lemma 7.8.2, iff f is one-one. ∎

Corollary 7.10.2. *An $n \times n$ matrix A over F is invertible iff* rank $A = n$.

Proof. An exercise. ∎

Corollary 7.10.2 implies that A is invertible iff its rows (columns) are linearly independent.

7.11 Exercises

1. Are the following functions linear transformations? Whenever $f \in \text{Hom}_F(V, W)$, describe ker f and $f(V)$.

(a) $F = \mathbb{R}$, $V = W =$ polynomials of degree $\leq n$ over F, $f =$ differentiation.

(b) F any field, $V = W = \text{Mat}_{2 \times 2}(F)$, $f: \mathbf{X} \to \begin{bmatrix} 1 & 0 \\ 0 & 0 \end{bmatrix} \mathbf{X}$.

(c) $F = V = W = \mathbb{R}$, $f: x \to 5x$.

(d) $F = V = W = \mathbb{R}$, $f: x \to 5x + 3$.

***2.** Show that if $f \in \text{Hom}_F(V, W)$ is an isomorphism, then

(a) $f^{-1} \in \text{Hom}_F(W, V)$.

(b) dim $V =$ dim W.

(c) If U is any subspace of V, dim $U =$ dim $f(U)$.

3. (a) Show that $B_1 = \{(2, 1), (0, 3)\}$ is a basis of $\mathbb{R}^2$ and $B_2 = \{(1, 0, 0), (1, 1, 0), (1, 1, 1)\}$ is a basis of $\mathbb{R}^3$.

(b) If $f \in \text{Hom}_\mathbb{R}(\mathbb{R}^2, \mathbb{R}^3)$ has matrix

$$[f]_{B_1, B_2} = \begin{bmatrix} 3 & 0 \\ 1 & -1 \\ 0 & 2 \end{bmatrix} \quad [B_1, B_2 \text{ as in (a)}],$$

find $[f((1, 1))]_{B_2}$ in two different ways.

(c) If also $g \in \text{Hom}_\mathbb{R}(\mathbb{R}^3, \mathbb{R}^2)$ has matrix

$$[g]_{B_2, B_1} = \begin{bmatrix} 1 & 0 & -1 \\ 0 & 3 & 2 \end{bmatrix},$$

find $[(g \circ f)((1, 1))]_{B_1}$ in two different ways. What is $[g \circ f]_{B_1}$?

4. Prove (7.10.1).
*5. Prove Theorem 7.10.1 in detail.
*6. Assume that $f \in \mathrm{Hom}_F(V, V)$, where dim $V = n$. Prove that the following are equivalent:
   (a) f is an isomorphism.
   (b) For every basis B of V, $f(B)$ is a basis of V.
   (c) For some basis B of V, $f(B)$ is a basis of V.
   (d) $\ker f = [0]$.
   (e) f is onto.
7. Find matrices $\mathbf{A}_1$, $\mathbf{A}_2$, $\mathbf{A}_3$, $\mathbf{B}_1$, $\mathbf{B}_2$, and $\mathbf{B}_3$ such that:
   (a) rank $\mathbf{A}_1 \mathbf{B}_1 =$ rank $\mathbf{A}_1$.
   (b) rank $\mathbf{A}_2 \mathbf{B}_2 =$ rank $\mathbf{B}_2$.
   (c) rank $\mathbf{A}_3 \mathbf{B}_3 <$ min (rank $\mathbf{A}_3$, rank $\mathbf{B}_3$).
*8. Use Theorem 7.8.2 to show that in Corollaries 7.5.1 and 7.5.2, dim $S = n - $ rank $\mathbf{A}$. In particular, any solution of $\mathbf{AX} = \mathbf{B}$ is unique iff rank $\mathbf{A} = n$.
*9. Show that if $f \in \mathrm{Hom}_F(V, W)$ and $g \in \mathrm{Hom}_F(W, X)$, then $g \circ f \in \mathrm{Hom}_F(V, X)$.
10. Prove Corollary 7.10.2.
*11. Let $F = \mathrm{GF}(3)$; this problem concerns the ordered bases $B_1 = \{(1, 0, 0), (1, 2, 0), (2, 0, 1)\}$ and $B_2 = \{(1, 1, 1), (0, 2, 0), (0, 0, 2)\}$ of $F^3 = V$.
   (a) Prove that the identity function $1_V \colon V \to V$ defined by $1_V(v) = v$, all $v \in V$, is a linear transformation. Find $[1_V]_{B_1}$, $[1_V]_{B_1, B_2}$, $[1_V]_{B_2, B_1}$, and $[1_V]_{B_2}$.
   (b) Prove that $[1_V]_{B_2, B_1} = ([1_V]_{B_1, B_2})^{-1}$.
   (c) Find $[(1, 0, 0)]_{B_2}$ in two ways (one way should use $f = 1_V$ in Theorem 7.9.1).
*12. Let $B_1 = \{v_1, v_2, \ldots, v_n\}$ and $B_2 = \{w_1, w_2, \ldots, w_n\}$ be two ordered bases for the vector space V over the field F.
   (a) Prove that 1_V, defined as in Exercise 11(a), is in $\mathrm{Hom}_F(V, V)$. Denote $[1_V]_{B_1} = [\alpha_{ij}]$, $[1_V]_{B_1, B_2} = [\beta_{ij}]$, and $[1_V]_{B_2, B_1} = [\gamma_{ij}]$. Find all α_{ij}'s, and write equations with the v_i's and w_j's showing where the β's and γ's occur as coefficients. (Use Theorem 7.9.1.)
   (b) Prove that $[1_V]_{B_2, B_1} = ([1_V]_{B_1, B_2})^{-1}$.
   (c) Show that $[v_i]_{B_2}$ is the ith column of $[1_V]_{B_1, B_2}$, and $[w_j]_{B_1}$ is the jth column of $[1_V]_{B_2, B_1}$.
   (d) Show that for any $v \in V$,

(7.11.1) $$[v]_{B_2} = [1_V]_{B_1, B_2} [v]_{B_1}.$$

[Because of (7.11.1), $[1_V]_{B_1, B_2}$ is called the **change-of-basis matrix** from basis B_1 to basis B_2.] Illustrate (7.11.1) with F, V, B_1, and B_2 as in Exercise 11, $v = (2, 1, 1)$.

*13. (a) Show that if V and W are finite-dimensional vector spaces over F and for each f and g in $\mathrm{Hom}_F(V, W)$ and $\alpha \in F$ we define $f + g$ and

Sec. 7.12 *Determinants* 329

αf by $(f+g)(v) = f(v)+g(v)$, $(\alpha f)(v) = \alpha f(v)$, then $f+g$ and αf are in $\mathrm{Hom}_F(V, W)$. Show that with these two operations, $\mathrm{Hom}_F(V, W)$ is a vector space over F.

(b) Show that if B_1 is an ordered basis of V and B_2 an ordered basis of W, then $f \to [f]_{B_1, B_2}$ is a vector space isomorphism from $\mathrm{Hom}_F(V, W)$ onto $\mathrm{Mat}_{(\dim W) \times (\dim V)}(F)$.

*14. Show that if $\mathbf{A}$ is invertible, then rank $\mathbf{AB}$ = rank $\mathbf{B}$.

7.12 Determinants

We need to review some aspects of the theory of determinants; we use this opportunity to present the subject as a topic in multilinear algebra. Before obtaining the basic facts about determinants, we need to know more about the group S_n of all permutations (bijections) of the set $\mathbb{P}_n = \{1, 2, \ldots, n\}$. A **transposition** in S_n is a cycle $t = (i \quad j)$ of length 2; that is, $t: i \to j$, $t: j \to i$, and $t: k \to k$ for all $k \in \mathbb{P}_n - \{i, j\}$. We collect the needed information in the following theorem.

Theorem 7.12.1. (a) *Every permutation in S_n is a product of transpositions.*
(b) *If for each $p \in S_n$ we define*

$$\mathrm{sgn}\, p = \prod_{\{i, j\} \subseteq \mathbb{P}_n} \frac{p(i) - p(j)}{i - j} \in \mathbb{Q}, \tag{7.12.1}$$

then sgn is a homomorphism from S_n onto the multiplicative group $\{\pm 1\}$.
(c) $\mathrm{sgn}\, t = -1$ *for all transpositions t.*
(d) *If $p = t_1 t_2 \cdots t_k$, the t_i being transpositions, then $\mathrm{sgn}\, p = (-1)^k$.*

We call sgn p the **sign** of p. We say that $p \in S_n$ is an **even permutation** if sgn $p = +1$, an **odd permutation** if sgn $p = -1$. By (d), an even permutation is always a product of an even number of transpositions, and an odd permutation is always a product of an odd number of transpositions.

Proof. (a) We know that every permutation is a product of (disjoint) cycles. And

$$(1 \quad 2 \quad 3 \cdots k-1 \quad k) = (1 \quad k)(1 \quad k-1) \cdots (1 \quad 3)(1 \quad 2)$$

(remember we work with the rightmost cycle first), so every cycle is a product of transpositions. The identity permutation 1 (the only one if $n = 1$) is considered an empty product of 0 transpositions, and $\mathrm{sgn}\,(1) = +1$.

(b) The product in (7.12.1) runs through all two-element subsets $\{i, j\}$ of $\mathbb{P}_n$. Note that $\{i, j\}$ and $\{j, i\}$ are the same subset, so we do not use both. However,

$$\frac{p(i) - p(j)}{i - j} = \frac{p(j) - p(i)}{j - i},$$

so the order of i and j really does not matter; the product is well defined. The function sgn is a homomorphism because

$$\operatorname{sgn} pq = \prod_{\{i,j\} \subseteq \mathbb{P}_n} \frac{p(q(i)) - p(q(j))}{i - j}$$

$$= \prod_{\{i,j\} \subseteq \mathbb{P}_n} \frac{p(q(i)) - p(q(j))}{q(i) - q(j)} \cdot \frac{q(i) - q(j)}{i - j} = (\operatorname{sgn} p)(\operatorname{sgn} q).$$

[As $\{i, j\}$ runs through all two-element subsets of $\mathbb{P}_n$, so does $\{q(i), q(j)\}$, so the first factors do form sgn p.]

We have sgn $p \in \{\pm 1\}$ because

$$|\operatorname{sgn} p| = \prod_{\{i,j\} \subseteq \mathbb{P}_n} \frac{|p(i) - p(j)|}{|i - j|} = 1,$$

the numerator and denominator being different arrangements of the same factors.

(c) From (7.12.1) we see that

$$\operatorname{sgn}((k \quad l)) = \prod_{\{i,j\} \cap \{k,l\} = \varnothing} \frac{i-j}{i-j} \cdot \prod_{j \notin \{k,l\}} \frac{l-j}{k-j} \cdot \prod_{j \notin \{k,l\}} \frac{k-j}{l-j} \cdot \frac{k-l}{l-k} = -1.$$

(d) is immediate from (b) and (c). ∎

For example, in S_3 the permutations 1, $(1 \ 2 \ 3) = (1 \ 3)(1 \ 2)$, and $(1 \ 3 \ 2) = (1 \ 2)(1 \ 3)$ are even, while $(1 \ 2)$, $(1 \ 3)$, and $(2 \ 3)$ are odd. The cycle $(1 \ 2 \ 3 \cdots k-1 \ k) = (1 \ k)(1 \ k-1) \cdots (1 \ 3)(1 \ 2)$ of "length" k is the product of $k - 1$ transpositions, and hence has sign $(-1)^{k-1}$. For example,

$$\operatorname{sgn} (1 \ 13 \ 7 \ 2 \ 6)(3 \ 14 \ 11 \ 5)(4 \ 12)(8 \ 10 \ 9)$$
$$= (-1)^4 (-1)^3 (-1)(-1)^2 = (-1)^{10} = +1 \quad \text{in} \quad S_{14}.$$

In the remainder of this section we discuss $n \times n$ matrices $\mathbf{A}$ over F; the notation $\mathbf{A} = [\mathbf{A}_1, \ldots, \mathbf{A}_i, \ldots, \mathbf{A}_n]$ means that $\mathbf{A}_i$ is the ith column of $\mathbf{A}$. Denote $V = \operatorname{Mat}_{n \times 1}(F)$ in this section, so all $\mathbf{A}_i \in V$. A function $\varphi: \operatorname{Mat}_{n \times n}(F) \to F$ is called **multilinear** if, for every $i \in \{1, \ldots, n\}$ and any fixed $\mathbf{A}_1, \ldots, \mathbf{A}_{i-1}, \mathbf{A}_{i+1}, \ldots, \mathbf{A}_n \in V$, the function $\mathbf{X} \to \varphi([\mathbf{A}_1, \ldots, \mathbf{A}_{i-1}, \mathbf{X}, \mathbf{A}_{i+1}, \ldots, \mathbf{A}_n])$ is in $\operatorname{Hom}_F(V, F)$. That is,

$$\varphi([\mathbf{A}_1, \ldots, \mathbf{A}_{i-1}, \mathbf{X}_1 + \mathbf{X}_2, \mathbf{A}_{i+1}, \ldots, \mathbf{A}_n])$$

(7.12.2)
$$= \varphi([\mathbf{A}_1, \ldots, \mathbf{A}_{i-1}, \mathbf{X}_1, \mathbf{A}_{i+1}, \ldots, \mathbf{A}_n])$$
$$+ \varphi([\mathbf{A}_1, \ldots, \mathbf{A}_{i-1}, \mathbf{X}_2, \mathbf{A}_{i+1}, \ldots, \mathbf{A}_n])$$

Sec. 7.12 *Determinants*

and

(7.12.3) $\varphi([\mathbf{A}_1, \ldots, \mathbf{A}_{i-1}, \alpha \mathbf{X}, \mathbf{A}_{i+1}, \ldots, \mathbf{A}_n])$
$= \alpha \varphi([\mathbf{A}_1, \ldots, \mathbf{A}_{i-1}, \mathbf{X}, \mathbf{A}_{i+1}, \ldots, \mathbf{A}_n])$

for all $\alpha \in F$ and for all $\mathbf{X}, \mathbf{X}_1, \mathbf{X}_2, \mathbf{A}_1, \ldots, \mathbf{A}_{i-1}, \mathbf{A}_{i+1}, \ldots, \mathbf{A}_n \in V$. A function $\varphi: \text{Mat}_{n \times n}(F) \to F$ is **alternating** if $\varphi([\mathbf{A}_1, \ldots, \mathbf{A}_n]) = 0$ whenever two of the $\mathbf{A}_i$'s are equal.

Theorem 7.12.2. *If $\varphi: \text{Mat}_{n \times n}(F) \to F$ is multilinear and alternating, then*
(a) *If any $\mathbf{A}_i$ is a column of zeros, $\varphi([\mathbf{A}_1, \ldots, \mathbf{A}_n]) = 0$.*
(b) *If $i \neq j$, then $\varphi([\mathbf{A}_1, \ldots, \mathbf{A}_i, \ldots, \mathbf{A}_j, \ldots, \mathbf{A}_n]) = \varphi([\mathbf{A}_1, \ldots, \mathbf{A}_i + \gamma \mathbf{A}_j, \ldots, \mathbf{A}_j, \ldots, \mathbf{A}_n])$, any $\gamma \in F$.*
(c) *If $i \neq j$, then $\varphi([\mathbf{A}_1, \ldots, \mathbf{A}_i, \ldots, \mathbf{A}_j, \ldots, \mathbf{A}_n]) = -\varphi([\mathbf{A}_1, \ldots, \mathbf{A}_j, \ldots, \mathbf{A}_i, \ldots, \mathbf{A}_n])$.*
(d) *If $\mathbf{B} = [\beta_{ij}]_{n \times n}$, then $\varphi(\mathbf{AB}) = \sum_{p \in S_n} (\text{sgn } p) \beta_{p(1)1} \cdots \beta_{p(n)n} \varphi(\mathbf{A})$.*

Part (b) says that the value of φ is not changed if a multiple of one column is added to another. Part (c) says that if two columns are interchanged, the value of φ changes in sign.

Proof. Part (a) follows from (7.12.3), with $\mathbf{X} = \mathbf{A}_i$, $\alpha = 0$. For (b), we have

$\varphi([\mathbf{A}_1, \ldots, \mathbf{A}_i + \gamma \mathbf{A}_j, \ldots, \mathbf{A}_j, \ldots, \mathbf{A}_n])$
$= \varphi([\mathbf{A}_1, \ldots, \mathbf{A}_i, \ldots, \mathbf{A}_j, \ldots, \mathbf{A}_n])$
$\quad + \varphi([\mathbf{A}_1, \ldots, \gamma \mathbf{A}_j, \ldots, \mathbf{A}_j, \ldots, \mathbf{A}_n])$
$= \varphi([\mathbf{A}_1, \ldots, \mathbf{A}_i, \ldots, \mathbf{A}_j, \ldots, \mathbf{A}_n])$
$\quad + \gamma \varphi([\mathbf{A}_1, \ldots, \mathbf{A}_j, \ldots, \mathbf{A}_j, \ldots, \mathbf{A}_n])$
$= \varphi([\mathbf{A}_1, \ldots, \mathbf{A}_i, \ldots, \mathbf{A}_j, \ldots, \mathbf{A}_n])$,

by (7.12.2), (7.12.3), and the fact that φ is alternating.
To prove (c), we write

$\varphi([\mathbf{A}_1, \ldots, \mathbf{A}_i, \ldots, \mathbf{A}_j, \ldots, \mathbf{A}_n])$
$= \varphi([\mathbf{A}_1, \ldots, \mathbf{A}_i + \mathbf{A}_j, \ldots, \mathbf{A}_j, \ldots, \mathbf{A}_n])$
$= \varphi([\mathbf{A}_1, \ldots, \mathbf{A}_i + \mathbf{A}_j, \ldots, -\mathbf{A}_i, \ldots, \mathbf{A}_n])$
$= \varphi([\mathbf{A}_1, \ldots, \mathbf{A}_i, \ldots, -\mathbf{A}_i, \ldots, \mathbf{A}_n])$
$\quad + \varphi([\mathbf{A}_1, \ldots, \mathbf{A}_j, \ldots, -\mathbf{A}_i, \ldots, \mathbf{A}_n])$
$= (-1) \varphi([\mathbf{A}_1, \ldots, \mathbf{A}_i, \ldots, \mathbf{A}_i, \ldots, \mathbf{A}_n])$
$\quad + (-1) \varphi([\mathbf{A}_1, \ldots, \mathbf{A}_j, \ldots, \mathbf{A}_i, \ldots, \mathbf{A}_n])$
$= -\varphi([\mathbf{A}_1, \ldots, \mathbf{A}_j, \ldots, \mathbf{A}_i, \ldots, \mathbf{A}_n])$,

using (b) twice [add $\mathbf{A}_j$ to $\mathbf{A}_i$, then add $-(\mathbf{A}_i + \mathbf{A}_j)$ to $\mathbf{A}_j$], (7.12.2), (7.12.3) twice, and the fact that φ is alternating.

For (d), first note that the entries in the kth column $\mathbf{C}_k$ of $\mathbf{C} = \mathbf{AB}$ have form $\sum_{j=1}^n \alpha_{ij}\beta_{jk}$, $1 \leq i \leq n$, where $\mathbf{A} = [\alpha_{ij}]$; in fact, $\mathbf{C}_k = \sum_{j=1}^n \mathbf{A}_j \beta_{jk}$, $\mathbf{A}_j$ the jth column of $\mathbf{A}$. We then have

$$\varphi(\mathbf{AB}) = \varphi([\mathbf{C}_1, \ldots, \mathbf{C}_n]) = \varphi\left(\left[\sum_{j=1}^n \mathbf{A}_j \beta_{j1}, \ldots, \sum_{j=1}^n \mathbf{A}_j \beta_{jn}\right]\right)$$

$$= \sum_{j_1=1}^n \cdots \sum_{j_n=1}^n \varphi([\mathbf{A}_{j_1}\beta_{j_11}, \ldots, \mathbf{A}_{j_n}\beta_{j_nn}])$$

$$= \sum_{p \in S_n} \varphi([\mathbf{A}_{p(1)}\beta_{p(1)1}, \ldots, \mathbf{A}_{p(n)}\beta_{p(n)n}])$$

$$= \sum_{p \in S_n} \beta_{p(1)1} \cdots \beta_{p(n)n} \varphi([\mathbf{A}_{p(1)}, \ldots, \mathbf{A}_{p(n)}])$$

$$= \sum_{p \in S_n} (\operatorname{sgn} p) \beta_{p(1)1} \cdots \beta_{p(n)n} \varphi(\mathbf{A}),$$

as desired. [The reasons for the third, fourth, fifth, and sixth equalities are, respectively, (7.12.2), φ is alternating, (7.12.3), and (c) in conjunction with Theorem 7.12.1(d).] ∎

A function det: $\operatorname{Mat}_{n \times n}(F) \to F$ is called a **determinant** if det is multilinear, alternating, and $\det \mathbf{I} = 1$. We can now prove the important facts about *det*.

Theorem 7.12.3. (a) *For each $n \in \mathbb{P}$ there exists a unique determinant function* det: $\operatorname{Mat}_{n \times n}(F) \to F$.
 (b) $\det [\beta_{ij}]_{n \times n} = \sum_{p \in S_n} (\operatorname{sgn} p) \beta_{p(1)1} \cdots \beta_{p(n)n}$.
 (c) $\det \mathbf{AB} = (\det \mathbf{A})(\det \mathbf{B})$ *for all* $\mathbf{A}, \mathbf{B} \in \operatorname{Mat}_{n \times n}(F)$.
 (d) $\mathbf{A}$ *is invertible iff* $\det \mathbf{A} \neq 0$. *If $\mathbf{A}$ is invertible,* $\det \mathbf{A}^{-1} = (\det \mathbf{A})^{-1}$.
 (e) $\det \mathbf{A}^T = \det \mathbf{A}$.
 (f) *det satisfies* (7.12.2), (7.12.3), *Theorem* 7.12.2(a), (b), *and* (c), *and the corresponding statements for rows instead of columns.*
 (g) *Let $\mathbf{A}_{ij}$ be the $(n-1) \times (n-1)$ matrix obtained by deleting the ith row and jth column from $\mathbf{A}$, and define the i,j-cofactor of $\mathbf{A}$ to be $\Gamma_{ij} = (-1)^{i+j} \det \mathbf{A}_{ij}$. Then if $\mathbf{A} = [\alpha_{ij}]$,*

(7.12.4) $$\det \mathbf{A} = \sum_{j=1}^n \alpha_{ij} \Gamma_{ij}, \quad \text{any } i \in \{1, \ldots, n\};$$

(7.12.5) $$\det \mathbf{A} = \sum_{i=1}^n \alpha_{ij} \Gamma_{ij}, \quad \text{any } j \in \{1, \ldots, n\}.$$

Proof. Assuming that for each n there exists a function det with the stated properties and setting $\varphi = \det$ and $\mathbf{A} = \mathbf{I}$ in Theorem 7.12.2(d), we get formula (b). The existence of specific formula (b) shows that for any n, the determinant is unique if it exists. Theorem 7.12.2(d) also yields (c), since with $\varphi = \det$ we have

$$\sum_{p \in S_n} (\operatorname{sgn} p) \beta_{p(1)1} \cdots \beta_{p(n)n} \varphi(\mathbf{A}) = (\det \mathbf{B})(\det \mathbf{A}) = (\det \mathbf{A})(\det \mathbf{B}).$$

The proof that a determinant exists for each n is by induction on n. The definition $\det[\alpha] = \alpha$ provides a determinant for $n = 1$, since $\det \mathbf{I} = \det[1] = 1$; the alternating requirement is vacuously satisfied; and $\det[\alpha + \beta] = \alpha + \beta = \det[\alpha] + \det[\beta]$, $\det[\alpha \cdot \gamma] = \alpha \cdot \gamma = \alpha \cdot \det[\gamma]$ prove multilinearity.

For $n > 1$, fix some $i \in \{1, \ldots, n\}$ and define det on $\operatorname{Mat}_{n \times n}(F)$ by (7.12.4); this is possible, since by the induction hypothesis we know that a unique determinant on $(n-1) \times (n-1)$ matrices exists. We must show det is multilinear, alternating, and $\det \mathbf{I} = 1$. Denoting $\mathbf{I} = [\delta_{ij}]_{n \times n}$, we have

$$\det \mathbf{I} = \sum_{j=1}^{n} \delta_{ij} \Gamma_{ij} = 0 + \cdots + 0 + \delta_{ii} \Gamma_{ii} + 0 + \cdots + 0 = 1 \cdot (-1)^{i+i} \det \mathbf{I}_{n-1}$$

$$= 1 \cdot 1 \cdot 1 = 1,$$

so the last requirement is satisfied.

Suppose that $\mathbf{A}_k = \mathbf{A}_l$ for some $k < l$ in $\mathbf{A} = [\mathbf{A}_1, \ldots, \mathbf{A}_k, \ldots, \mathbf{A}_l, \ldots, \mathbf{A}_n]$. Then

$$\det \mathbf{A} = \sum_{j=1}^{n} \alpha_{ij} \Gamma_{ij} = \alpha_{ik} \Gamma_{ik} + \alpha_{il} \Gamma_{il} = \alpha_{ik}(-1)^{i+k} \det \mathbf{A}_{ik} + \alpha_{il}(-1)^{i+l} \det \mathbf{A}_{il},$$

since the other submatrices $\mathbf{A}_{ij}$ have two equal columns, hence have determinant 0. $\mathbf{A}_k = \mathbf{A}_l$ implies that $\alpha_{ik} = \alpha_{il}$. If we denote $\mathbf{A}_{ik} = [\mathbf{B}_1, \ldots, \mathbf{B}_{k-1}, \mathbf{B}_k, \ldots, \mathbf{B}_{l-1}, \mathbf{B}_l, \mathbf{B}_{l+1}, \ldots, \mathbf{B}_{n-1}]$, then $\mathbf{A}_{il}$ has the same columns, rearranged in the order $\mathbf{A}_{il} = [\mathbf{B}_1, \ldots, \mathbf{B}_{k-1}, \mathbf{B}_{l-1}, \mathbf{B}_k, \ldots, \mathbf{B}_{l-2}, \mathbf{B}_l, \ldots, \mathbf{B}_{n-1}]$ (note that $\mathbf{B}_{l-1}$ is part of column $\mathbf{A}_l = \mathbf{A}_k$ of $\mathbf{A}$). There are $l - k - 1$ columns in the set $\{\mathbf{B}_k, \ldots, \mathbf{B}_{l-2}\}$, and $\mathbf{B}_{l-1}$ must be interchanged with each successively to change $\mathbf{A}_{il}$ to $\mathbf{A}_{ik}$; by Theorem 7.12.2(c) and the induction hypothesis, $\det \mathbf{A}_{il} = (-1)^{l-k-1} \det \mathbf{A}_{ik}$. Therefore,

$$\det \mathbf{A} = \alpha_{ik}(-1)^{i+k} \det \mathbf{A}_{ik} + \alpha_{ik}(-1)^{i+l}(-1)^{l-k-1} \det \mathbf{A}_{ik}$$

$$= \alpha_{ik}(-1)^{i+k} \{\det \mathbf{A}_{ik} - \det \mathbf{A}_{ik}\} = 0,$$

proving that det is alternating. The proof that det is multilinear is the subject of Exercise 7.14.6; this completes the proof of (a), and of (7.12.4) in (g).

To prove (e), let $\mathbf{A} = [\alpha_{ij}]$, so $\mathbf{A}^T = [\alpha_{ji}]$. If $p \in S_n$, then $j = p(i)$ implies that $i = p^{-1}(j)$, so we have $\alpha_{1p(1)} \cdots \alpha_{np(n)} = \alpha_{p^{-1}(1)1} \cdots \alpha_{p^{-1}(n)n}$ by a rearrangement of factors in the product. Also, if $p = t_1 t_2 \cdots t_k$, each t_i a transposition, then $p^{-1} = t_k^{-1} \cdots t_2^{-1} t_1^{-1} = t_k \cdots t_2 t_1$ since $t_i^{-1} = t_i$; this proves that $\operatorname{sgn} p =$

sgn p^{-1}. We have, therefore,

$$\det \mathbf{A}^T = \sum_{p \in S_n} (\operatorname{sgn} p) \alpha_{1p(1)} \cdots \alpha_{np(n)} = \sum_{p \in S_n} (\operatorname{sgn} p^{-1}) \alpha_{p^{-1}(1)1} \cdots \alpha_{p^{-1}(n)n}$$
$$= \det \mathbf{A},$$

the last equality holding since p^{-1} runs through all elements of S_n while p does.

Since det is multilinear and alternating, the first two parts of (f) hold. The rows of $\mathbf{A}$ are the columns of $\mathbf{A}^T$, so (e) implies that the rest of statements (f) and (g) hold.

It remains to prove (d). If $\mathbf{A}$ is invertible, then $\mathbf{AA}^{-1} = \mathbf{I}$ so, by (c), $(\det \mathbf{A})(\det \mathbf{A}^{-1}) = \det \mathbf{I} = 1$, which proves that $\det \mathbf{A} \neq 0$ and $\det \mathbf{A}^{-1} = (\det \mathbf{A})^{-1}$. If $\mathbf{A}$ is not invertible, then Corollary 7.10.2 implies that some row of $\mathbf{A}$ is a linear combination of others; adding suitable multiples of other rows to that one yields a zero row [and does not change $\det \mathbf{A}$, by Theorem 7.12.2(b)], so $\det \mathbf{A} = 0$ by Theorem 7.12.2(a) (with rows instead of columns). ∎

Computation of determinants is discussed in Exercises 7.14.3 to 7.14.5.

7.13 Similarity

Let $\mathbf{A}$ and $\mathbf{B}$ be $n \times n$ matrices over F. We say that $\mathbf{B}$ is **similar** to $\mathbf{A}$ iff there is an invertible $n \times n$ matrix $\mathbf{P}$ over F such that $\mathbf{B} = \mathbf{P}^{-1}\mathbf{AP}$.

Lemma 7.13.1. *Similarity is an equivalence relation on* $\operatorname{Mat}_{n \times n}(F)$.

Proof. $\mathbf{A} = \mathbf{I}^{-1}\mathbf{AI}$, so similarity is reflexive. If $\mathbf{B} = \mathbf{P}^{-1}\mathbf{AP}$, then $(\mathbf{P}^{-1})^{-1}\mathbf{BP}^{-1} = \mathbf{PBP}^{-1} = \mathbf{PP}^{-1}\mathbf{APP}^{-1} = \mathbf{A}$, so similarity is symmetric. If $\mathbf{B} = \mathbf{P}^{-1}\mathbf{AP}$ and $\mathbf{C} = \mathbf{Q}^{-1}\mathbf{BQ}$, then $\mathbf{C} = \mathbf{Q}^{-1}\mathbf{P}^{-1}\mathbf{APQ} = (\mathbf{PQ})^{-1}\mathbf{A}(\mathbf{PQ})$, proving transitivity. ∎

The importance of similarity stems from the following results:

Theorem 7.13.1. (a) *If B_1 and B_2 are ordered bases of V and $f \in \operatorname{Hom}_F(V, V)$, then $[f]_{B_1}$ and $[f]_{B_2}$ are similar.*

(b) *Conversely, if $f \in \operatorname{Hom}_F(V, V)$ has matrix $\mathbf{A} = [f]_{B_1}$ in the ordered basis B_1 and if $\mathbf{C}$ is a matrix similar to $\mathbf{A}$, then f has matrix $\mathbf{C} = [f]_{B_2}$ in some ordered basis B_2.*

Proof. (a) Let 1_V be the identity function $1_V: V \to V$; certainly $f = 1_V \circ f = f \circ 1_V$. By Theorem 7.9.2, this implies that

(7.13.1) $\qquad [f]_{B_1, B_2} = [1_V]_{B_1, B_2}[f]_{B_1} = [f]_{B_2}[1_V]_{B_1, B_2}.$

Denote $\mathbf{P} = [1_V]_{B_1, B_2}$; $\mathbf{P}$ is invertible since 1_V is, by Theorem 7.10.2. Our equation says $\mathbf{P}[f]_{B_1} = [f]_{B_2}\mathbf{P}$, so $[f]_{B_1} = \mathbf{P}^{-1}[f]_{B_2}\mathbf{P}$.

(b) Given **C** similar to **A**, let **P** be an invertible matrix such that $C = P^{-1}AP$, say $P = [\pi_{ij}]_{n \times n}$. If $B_1 = \{v_1, \ldots, v_n\}$, define $B_2 = \{w_1, \ldots, w_n\}$, where $w_j = 1_V(w_j) = \sum_{i=1}^{n} \pi_{ij} v_i$. Set B_2 is a basis of V (why?), and by our construction we have $P = [1_V]_{B_2, B_1}$. Using Theorem 7.9.2, we have

$$PC = AP = [f]_{B_1}[1_V]_{B_2, B_1} = [f]_{B_2, B_1} = [1_V]_{B_2, B_1}[f]_{B_2} = P[f]_{B_2},$$

so $C = [f]_{B_2}$. ∎

For example, let $B_2 = \{(1, 0, 0), (0, 1, 0), (0, 0, 1)\}$ be the standard basis of F^3, where $F = GF(2)$. Define $f \in \text{Hom}_F(F^3, F^3)$ by

$$[f]_{B_2} = \begin{bmatrix} 1 & 0 & 1 \\ 1 & 1 & 0 \\ 0 & 1 & 1 \end{bmatrix},$$

and let **P** be the invertible matrix

$$P = \begin{bmatrix} 1 & 1 & 1 \\ 0 & 1 & 1 \\ 0 & 0 & 1 \end{bmatrix}.$$

Then **P** must be $[1_V]_{B_1, B_2}$ for some basis B_1; in fact, Exercise 7.11.12(c) implies that $B_1 = \{v_1, v_2, v_3\} = \{(1, 0, 0), (1, 1, 0), (1, 1, 1)\}$. Now (7.13.1) implies that

$$[f]_{B_1} = P^{-1}[f]_{B_2}P = \begin{bmatrix} 1 & 1 & 0 \\ 0 & 1 & 1 \\ 0 & 0 & 1 \end{bmatrix}\begin{bmatrix} 1 & 0 & 1 \\ 1 & 1 & 0 \\ 0 & 1 & 1 \end{bmatrix}\begin{bmatrix} 1 & 1 & 1 \\ 0 & 1 & 1 \\ 0 & 0 & 1 \end{bmatrix} = \begin{bmatrix} 0 & 1 & 0 \\ 1 & 1 & 0 \\ 0 & 1 & 0 \end{bmatrix}.$$

After some preparatory ring theory (which is also used in our discussion of finite fields), we will discuss *rational canonical forms* of matrices. That is, we will find a "nice" set of matrices which have the property that any given matrix is similar to one and only one matrix in the set. By Theorem 7.13.1, we will then know that any $f \in \text{Hom}_F(V, V)$ has a matrix of this "nice" form, with respect to some basis.

7.14 Exercises

*1. Prove that the set of even permutations in S_n is a normal subgroup of S_n. (It is called the **alternating group** of degree n.) What is $|S_n : A_n|$?

2. Find sgn $(1\ 7\ 2\ 12)(6\ 8\ 11\ 9\ 7)$ and sgn $(1\ 7\ 2\ 6)(3\ 4) \circ (1\ 5\ 3\ 4)(2\ 5) \circ (1\ 3\ 5\ 4\ 6\ 7)$.

*3. Write out the formula of Theorem 7.12.3(b) completely in the cases $n = 2$ and $n = 3$. Show that (7.12.4) (with $i = 1$) yields these same formulas for the determinants of 2×2 and 3×3 matrices.

*4. Use Theorem 7.12.3(g) to compute the determinants of the following matrices over $\mathbb{R}$:

$$\begin{bmatrix} 1 & 2 & 7 \\ 0 & -1 & 3 \\ 1 & -2 & 0 \end{bmatrix}, \begin{bmatrix} 2 & 7 & -1 \\ 1 & 0 & 3 \\ 1 & -1 & 2 \end{bmatrix}, \begin{bmatrix} 1 & 7 & 2 & 6 \\ 0 & 3 & 1 & -3 \\ 2 & 1 & 0 & 5 \\ 1 & 4 & -1 & -2 \end{bmatrix}.$$

*5. For $n \geqslant 3$, the formula of Theorem 7.12.3(b) is rarely used for computation of determinants. Instead, one uses either the method of Exercise 4 or the method we now outline.

(a) Describe what happens to det $\mathbf{A}$ when an elementary row operation is performed on $\mathbf{A}$. (Consider three cases.)

(b) Show that if $\mathbf{A} = [\alpha_{ij}]$ is **upper-triangular** (i.e., $\alpha_{ij} = 0$ for $i > j$), then det $\mathbf{A}$ is the product of the diagonal entries of $\mathbf{A}$.

(c) Use parts (a) and (b) to describe a "sweepout" procedure for computing det $\mathbf{A}$.

(d) Use your method in part (c) to do Exercise 4.

(e) Estimate the time required for a computer to compute the determinant of an $n \times n$ matrix by the method in part (c) and by Theorem 7.12.3(g), and compare your estimates.

*6. Complete the proof that det, as defined in the proof of Theorem 7.12.3, is multilinear.

*7. Suppose that $n \times n$ matrices $\mathbf{A}$ and $\mathbf{B}$ over F are similar. Show that rank $\mathbf{A}$ = rank $\mathbf{B}$. Also show that $\mathbf{A}^k$ and $\mathbf{B}^k$ are similar, for any $k \in \mathbb{P}$, and that $\mathbf{A}^T$ and $\mathbf{B}^T$ are similar. Show that det $\mathbf{A}$ = det $\mathbf{B}$.

8. This problem concerns the field $F = GF(3)$ and the vector space $V = F^3$ over F.

(a) Show that

$$B_1 = \{(1, 0, 2), (2, 1, 1), (0, 2, 1)\}$$

and

$$B_2 = \{(2, 1, 1), (1, 2, 0), (0, 1, 2)\}$$

are (ordered) bases of V.

(b) Let $B_3 = \{(1, 0, 0), (0, 1, 0), (0, 0, 1)\}$ be the standard ordered basis of V, and write out matrices $\mathbf{P} = [1_V]_{B_1, B_3}$ and $\mathbf{Q} = [1_V]_{B_2, B_3}$. (Why is this easier than writing out $[1_V]_{B_1, B_2}$?) Then use Theorem 7.9.2 and $1_V = 1_V \circ 1_V$ to show that $[1_V]_{B_1, B_2} = \mathbf{Q}^{-1}\mathbf{P}$.

(c) Define $f \in \text{Hom}_F(V, V)$ by

$$[f]_{B_1} = \begin{bmatrix} 2 & 0 & 1 \\ 0 & 1 & 1 \\ 1 & 2 & 0 \end{bmatrix},$$

and find $[f]_{B_2}$ and $[f]_{B_3}$.

(d) Show that the matrix

$$M = \begin{bmatrix} 1 & 2 & 0 \\ 0 & 1 & 2 \\ 2 & 0 & 2 \end{bmatrix}$$

is invertible, and find an ordered basis B of V such that $[f]_B = M^{-1}[f]_{B_1}M$.

9. Determine the equivalence classes in $\text{Mat}_{2\times 2}(\text{GF}(2))$ under similarity.

*__10.__ (a) Show that if square matrices A and B are identical except that each entry in some one column (or row) of B is λ times the corresponding entry of A, then $\det B = \lambda \cdot \det A$.

(b) Show that if $\alpha_1, \alpha_2, \ldots, \alpha_n \in F$, then the determinant of the $n \times n$ **Vandermonde matrix**

$$\begin{bmatrix} 1 & 1 & \cdots & 1 \\ \alpha_1 & \alpha_2 & \cdots & \alpha_n \\ \alpha_1^2 & \alpha_2^2 & \cdots & \alpha_n^2 \\ \vdots & & & \vdots \\ \alpha_1^{n-1} & \alpha_2^{n-1} & \cdots & \alpha_n^{n-1} \end{bmatrix}$$

is $\prod_{1 \leq i < j \leq n}(\alpha_j - \alpha_i)$.

[HINT: Use induction on n. Subtract $\alpha_n \cdot ((n-1)\text{th row})$ from the nth row, ..., $\alpha_n \cdot$ (1st row) from the second row. Then expand by minors of the last column and use (a) to reduce the $n \times n$ matrix problem to the corresponding $(n-1) \times (n-1)$ matrix problem.]

7.15 Ideals and Homomorphisms in Rings

Recall from Chapter 3 that a *ring* is a set R with binary operations of addition and multiplication, which is a commutative group under addition and satisfies the identities $x(yz) = (xy)z$, $x(y+z) = xy + xz$, $(x+y)z = xz + yz$. Ring R is *commutative* if $xy = yx$ for all $x, y \in R$; R has the *identity element* 1 if $1 \in R$ satisfies $1x = x1 = x$, all $x \in R$. Most of the rings we shall encounter will be commutative, with identity element 1.

An **ideal** I in a ring R is an additive subgroup I of R such that $rx, xr \in I$ for all $r \in R$ and all $x \in I$.

Assume that R and S are rings. A function $\varphi: R \to S$ is a **homomorphism** if

(7.15.1) $\quad \varphi(x+y) = \varphi(x) + \varphi(y) \quad \text{and} \quad \varphi(xy) = \varphi(x)\varphi(y), \quad \text{all } x, y \in R.$

The **kernel** $\ker \varphi$ of φ is $\{r \in R | \varphi(r) = 0\}$; this is the kernel of φ as a homomorphism of additive groups, just as was the case with $\ker \tau$ for any $\tau \in \text{Hom}_F(V, W)$. As always, φ is an **isomorphism**, and R and S are **isomorphic**, if φ is one-one and onto.

EXAMPLE 1. If m is a positive integer, then $(m) = \{zm | z \in \mathbb{Z}\}$, the set of multiples of m, is an ideal in the ring $\mathbb{Z}$. For any integer n, $n = qm + r$,

$0 \le r < m$ for uniquely determined integers q and $r = r_m(n)$. The function $n \to r_m(n)$ is then a homomorphism from $\mathbb{Z}$ to the ring $\mathbb{Z}_m$ of integers modulo m, and the kernel is (m).

EXAMPLE 2. For any field F, we denote by $F[X]$ the set of all polynomials in the variable X with coefficients from F. Then $F[X]$ is a ring under the usual addition and multiplication of polynomials. For a fixed $\alpha \in F$, the **evaluation function** $\varphi_\alpha \colon f(X) \to f(\alpha)$ is a ring homomorphism from $F[X]$ to F, with kernel $I_\alpha = \{f(X) \in F[X] | f(\alpha) = 0\}$. Also, for a fixed $n \times n$ matrix $\mathbf{A}$ over F and any polynomial $f(X) = a_m X^m + \cdots + a_1 X + a_0$ in $F[X]$, define $f(\mathbf{A}) = a_m \mathbf{A}^m + \cdots + a_1 \mathbf{A} + a_0 \mathbf{I}$. The set $\text{Mat}_{n \times n}(F)$ is a ring under addition and multiplication of matrices, and $\varphi_\mathbf{A} \colon f(X) \to f(\mathbf{A})$ is a ring homomorphism from $F[X]$ to $\text{Mat}_{n \times n}(F)$, with kernel $I_\mathbf{A} = \{f(X) \in F[X] | f(\mathbf{A}) = \mathbf{0}\}$. (Can you prove that $I_\mathbf{A} \ne \{0\}$? HINT: Consider the vector space $[\{\mathbf{A}^p | p \in \mathbb{N}\}]$.)

Now suppose that I is an ideal in the ring R. Since I is a normal subgroup of the additive group R, we can form the quotient group $R/I = \{r + I | r \in R\}$ of all cosets of I in R, cosets added according to the rule

(7.15.2) $\qquad (r_1 + I) + (r_2 + I) = (r_1 + r_2) + I.$

We claim that the equation

(7.15.3) $\qquad (r + I)(s + I) = rs + I$

defines a multiplication under which R/I is a semigroup. By Theorem 4.6.1, to show this we must verify that $r + I = r' + I$ and $s + I = s' + I$ imply that $rs + I = r's' + I$. The first two equations do imply that $r' = r + x$ and $s' = s + y$, some $x, y \in I$; hence $r's' = (r + x)(s + y) = rs + ry + xs + xy \in rs + I$, implying that $rs + I = r's' + I$.

The following theorem is the analogue for rings of Theorem 4.14.2 for groups.

Theorem 7.15.1. (a) *If I is an ideal in R, then R/I is a ring, and the function $f \colon R \to R/I$ defined by $f(r) = r + I$ is a homomorphism with kernel I.*

(b) *Conversely, if $\varphi \colon R \to S$ is a ring homomorphism onto S, then $\ker \varphi$ is an ideal in R, and the equation $\bar{\varphi}(r + \ker \varphi) = \varphi(r)$ defines an isomorphism $\bar{\varphi} \colon R/\ker \varphi \to S$.*

Proof. (a) is an easy exercise, now that we have shown that addition and multiplication are well-defined binary operations on R/I. To prove (b), we note that φ is a homomorphism of additive groups. By Theorem 4.14.2(2), $\ker \varphi$ is a (normal) additive subgroup of R and $\bar{\varphi}$ is a well-defined isomorphism of additive groups. It remains to show that $\ker \varphi$ is an ideal in R (that is, if $x \in \ker \varphi$ and $r \in R$, then $rx, xr \in \ker \varphi$), and that $\bar{\varphi}$ is an isomorphism of multiplicative semigroups. We have $\varphi(rx) = \varphi(r)\varphi(x) = \varphi(r) \cdot 0 = 0$, so $rx \in \ker \varphi$; similarly, $xr \in \ker \varphi$. Also,

$$\bar{\varphi}((r + \ker \varphi)(s + \ker \varphi)) = \bar{\varphi}(rs + \ker \varphi) = \varphi(rs) = \varphi(r)\varphi(s)$$
$$= \bar{\varphi}(r + \ker \varphi)\bar{\varphi}(s + \ker \varphi). \blacksquare$$

Sec. 7.16   *Polynomial Rings*   339

The ring R/I is called the **quotient ring of R modulo I**. Sometimes, each coset $r+I$ is called a **residue class**, and then R/I is called a **residue class ring**.

An ideal I in a ring R is a **maximal ideal** if $I \neq R$, but $I \subseteq J \subseteq R$, $I \neq J$ for an ideal J implies that $J = R$. Thus $J = [3\mathbb{Z}, +, \cdot]$ is maximal in $\mathbb{Z}$ (why?), but $I = [6\mathbb{Z}, +, \cdot]$ is not because $I \subset J \neq \mathbb{Z}$.

Lemma 7.15.1. *Let I be an ideal in a commutative ring R with identity 1. Then the following are equivalent.*
(a) *I is maximal.*
(b) *R/I is a field.*

Proof. Suppose that I is maximal. To prove (b), it is enough to show that $R/I - \{0 + I\}$ is a multiplicative group. The identity is $1 + I$, so it is enough to find a multiplicative inverse for any $a + I \in R/I - \{0 + I\}$. The ideal $J = \{x + ra \mid x \in I, r \in R\}$ satisfies $I \subseteq J$ and $I \neq J$ (since $a = 0 + 1a \in J - I$), so, since I is maximal, $J = R$. Choose $x \in I$ and $r \in R$ satisfying $x + ra = 1$; this must be possible since $J = R$. Then $(r+I)(a+I) = 1 - x + I = 1 + I$, so $(a+I)^{-1} = r + I$.

Now suppose that R/I is a field. If J is an ideal in R satisfying $I \subseteq J \subseteq R$, $I \neq J$, $J \neq R$, then J/I is a proper ideal in R/I. But an ideal in a field containing a nonzero element a must contain every element $b = (ba^{-1})a$ of the field, hence cannot be proper. ∎

A familiar example is the field GF(3), isomorphic to $\mathbb{Z}/3\mathbb{Z}$.

7.16 Polynomial Rings

An ideal I in a commutative ring R with 1 is a **principal ideal** if $I = (a)$, the set of all multiples ra, $r \in R$, of a fixed element $a \in R$. The element a is called a **generator** of I. Again, a prime example is $3\mathbb{Z}$, with generator 3.

We now need several facts about $F[X]$, the ring of polynomials in the variable X over the field F. The **degree** $\deg f(X)$ of $f(X) = \alpha_m X^m + \cdots + \alpha_1 X + \alpha_0$, $\alpha_m \neq 0$, is m. If $m \geq 1$, $f(X)$ is a **nonconstant polynomial**; otherwise $f(X) \in F \subseteq F[X]$ is a **constant**. (The degree of the constant polynomial 0 is not defined.) A nonconstant polynomial is **irreducible** if it is not the product of other nonconstant polynomials over F. (All polynomials of degree 1 are irreducible.) The polynomial $f(X) = \alpha_m X^m + \cdots + \alpha_1 X + \alpha_0$ is **monic** if $\alpha_m = 1$. A monic polynomial $d(X)$ is the **greatest common divisor (g.c.d.)** of nonzero $f(X)$, $g(X) \in F[X]$, if (i) $d(X) | f(X)$; (ii) $d(X) | g(X)$; and (iii) $c(X) | f(X)$ and $c(X) | g(X)$ imply $c(X) | d(X)$. We then write $(f(X), g(X)) = d(X)$. In particular, $f(X)$ and $g(X)$ are **relatively prime** if $(f(X), g(X)) = 1$. For any $f(X) \in F[X]$, $\mathscr{I}(f(X))$ denotes the principal ideal in $F[X]$ consisting of all multiples of $f(X)$.

Theorem 7.16.1. (a) *If $f(X)$, $g(X) \in F[X]$ with $f(X) \neq 0$, then there are uniquely determined $q(X)$, $r(X) \in F[X]$ with*

(7.16.1) $$g(X) = q(X)f(X) + r(X),$$

where $r(X) = 0$ or $\deg r(X) < \deg f(X)$.
  (b) *Every ideal I in $F[X]$ is principal.*
  (c) *If $(f(X), g(X)) = d(X)$, then there are $a(X)$, $b(X) \in F[X]$ such that*

(7.16.2) $$a(X)f(X) + b(X)g(X) = d(X).$$

  (d) *If $p(X)$ is irreducible and $f_1(X), f_2(X), \ldots, f_r(X)$ are polynomials such that $p(X)$ divides $f_1(X)f_2(X) \cdots f_r(X)$, then $p(X)$ divides some $f_i(X)$.*
  (e) *Every nonzero $f(X) \in F[X]$ has a factorization*

(7.16.3) $$f(X) = \alpha_m p_1(X)^{t_1} \cdots p_r(X)^{t_r},$$

$\alpha_m \in F$, the $p_i(X)$ uniquely determined monic irreducible polynomials, the t_i uniquely determined positive integers.
  (f) *The following are equivalent, for an $f(X) \in F[X]$:*
    (1) *$\mathscr{I}(f(X))$ is a maximal ideal.*
    (2) *$f(X)$ is irreducible.*
    (3) *$F[X]/\mathscr{I}(f(X))$ is a field.*

Part (f) will be our basic tool for the construction of new fields from a known field F, using irreducible polynomials.

Proof. (a) This is the familiar "division algorithm," the result of dividing $g(X)$ by $f(X)$.
  (b) We may assume $I \neq (0)$, which is principal, so choose $f(X) \in I$ nonzero of lowest possible degree. Then $\mathscr{I}(f(X)) \subseteq I$. For any $g(X) \in I$ we have (7.16.1); $r(X) \neq 0$ would imply that $r(X) = g(X) - q(X)f(X) \in I$ has lower degree than $f(X)$ does, a contradiction; this proves $r(X) = 0$, $g(X) = q(X)f(X) \in \mathscr{I}(f(X))$, $I = \mathscr{I}(f(X))$.
  (c) Denote $J = \{a(X)f(X) + b(X)g(X) | a(X), b(X) \in F[X]\}$; J is certainly an ideal in $F[X]$. Hence, by (b), $J = \mathscr{I}(d_0(X))$ for some $d_0(X) \in F[X]$. If $d(X)$ is any monic polynomial satisfying the g.c.d. definition, then $d_0(X)|d(X)$ by part (iii) of that definition. However, $d_0(X) = a_0(X)f(X) + b_0(X)g(X)$ for some $a_0(X)$, $b_0(X)$, so $d(X)|d_0(X)$ by parts (i) and (ii) of the definition. Since $d_0(X)$ and $d(X)$ are monic, $d_0(X) = d(X)$, proving that $d_0(X)$, of the required form, is the unique g.c.d. of $f(X)$ and $g(X)$. (Computation of such g.c.d.'s is discussed and applied in Section 8.8.)
  (d) We use induction on r; the result holds for $r = 1$, so assume that $r > 1$ and $p(X)|f_1(X)f_2(X) \cdots f_r(X)$. If $p(X)|f_1(X) \cdots f_{r-1}(X)$, then by the induction hypothesis $p(X)|f_i(X)$ for some $i = 1, \ldots, r-1$, and we are done. Hence assume that $p(X) \nmid f_1(X) \cdots f_{r-1}(X)$. Then $1 = (p(X), f_1(X) \cdots f_{r-1}(X))$ and by (c) there are $a(X)$, $b(X)$ such that $1 =$

Sec. 7.17 *Rational Canonical Form* 341

$a(X)p(X) + b(X)f_1(X) \cdots f_{r-1}(X)$. Multiplying by $f_r(X)$, we have

$$f_r(X) = a(X)p(X)f_r(X) + b(X)f_1(X) \cdots f_{r-1}(X)f_r(X).$$

Since $p(X)$ divides both terms on the right side, $p(X)|f_r(X)$.

(e) To see that such factorizations exist, suppose that $q(X)$ is a polynomial of least degree not known to have such a factorization. Then $q(X)$ is irreducible, in which case it *is* the factorization referred to, or $q(X) = q_1(X)q_2(X)$, some $q_1(X), q_2(X)$ of lower degree. The polynomials $q_1(X)$ and $q_2(X)$, being of lower degree, have such factorizations, and the product of their factorizations is such a factorization of $q(X)$.

To prove uniqueness, suppose that $q(X)$ is a polynomial of lowest degree with two distinct factorizations

$$q(X) = \alpha_m p_1(X)^{t_1} \cdots p_k(X)^{t_k} = \alpha_m q_1(X)^{u_1} \cdots q_l(X)^{u_l}.$$

The monic irreducible polynomials $p_i(X)$ and $q_j(X)$ are *all* distinct; for otherwise we could factor out a common factor and contradict minimality of $\deg q(X)$. But by (d), $p_1(X)$ divides some $q_j(X)$, forcing $p_1(X) = q_j(X)$, a contradiction.

(f) By Lemma 7.15.1, (1) and (3) are equivalent; we shall prove (1) and (2) equivalent. If $f(X)$ is not irreducible, say $f(X) = f_1(X)f_2(X)$ for nonconstant polynomials $f_1(X)$ and $f_2(X)$, then $\mathscr{I}(f(X)) \subseteq \mathscr{I}(f_1(X)) \subseteq F[X]$ and $f_1(X) \in \mathscr{I}(f_1(X)) - \mathscr{I}(f(X))$, $1 \in F[X] - \mathscr{I}(f_1(X))$, so $\mathscr{I}(f(X)) \subset \mathscr{I}(f_1(X)) \subset F[X]$, proving that $\mathscr{I}(f(X))$ is not maximal.

On the other hand, if $\mathscr{I}(f(X))$ is not maximal, then $\mathscr{I}(f(X)) \subseteq J \subseteq F[X]$ for some ideal $J \neq F[X]$, $J \neq \mathscr{I}(f(X))$. By (b), $J = \mathscr{I}(f_0(X))$ for some $f_0(X)$, so since $f(X) \in J$, $f(X) = f_0(X)f_1(X)$ for some $f_1(X)$. $f_1(X) = 0$ or $f_1(X) \in F$ would mean that $f(X) = 0$ or $\mathscr{I}(f(X)) = J$, each a contradiction; hence $f_1(X)$ is not a constant, proving that $f(X)$ is not irreducible. ∎

7.17 Rational Canonical Form

Assume that V is a finite-dimensional vector space over F and $f \in \text{Hom}_F(V, V)$, and fix a nonzero $v \in V$. Consider the vectors v, $f(v)$, $f^2(v), \ldots$, and choose the smallest $m \in \mathbb{P}$ such that $f^m(v) \in [v, f(v), \ldots, f^{m-1}(v)]$, say

(7.17.1) $\qquad f^m(v) = \alpha_0 v + \alpha_1 f(v) + \cdots + \alpha_{m-1} f^{m-1}(v).$

Denote $W = [v, f(v), \ldots, f^{m-1}(v)]$. Then also

$$f^{m+1}(v) = f(f^m(v)) = f(\alpha_0 v + \cdots + \alpha_{m-1} f^{m-1}(v))$$
$$= \alpha_0 f(v) + \cdots + \alpha_{m-1} f^m(v) \in W,$$

and in fact all $f^i(v) \in W$. Therefore, W is f-invariant and is called the *f-cyclic* **subspace** generated by v; denote $W = \langle v \rangle_f$. In the ordered basis

$\{v, f(v), \ldots, f^{m-1}(v)\}$ of W, the restriction f_W of f to W has matrix

(7.17.2) $\quad \mathbf{C}_p = \begin{bmatrix} 0 & 0 & \cdots & 0 & \alpha_0 \\ 1 & 0 & \cdots & 0 & \alpha_1 \\ 0 & 1 & \cdots & 0 & \alpha_2 \\ \multicolumn{5}{c}{\dotfill} \\ 0 & 0 & \cdots & 1 & \alpha_{m-1} \end{bmatrix};$

all entries are 0 except for the 1's on the **subdiagonal** (just below the main diagonal) and the α_i's in the last column. The matrix $\mathbf{C}_p$ is as given, because

$f(1\text{st basis element}) = 2\text{nd basis element},$

$f(2\text{nd basis element}) = 3\text{rd basis element},$

$\vdots$

$f((m-1)\text{st basis element}) = m\text{th basis element},$

$f(m\text{th basis element}) = \alpha_0 \cdot (1\text{st basis element})$

$+ \alpha_1 \cdot (2\text{nd basis element})$

$+ \cdots + \alpha_{m-1} \cdot (m\text{th basis element}).$

For any polynomial $p(X) \in F[X]$, say $p(X) = \beta_n X^n + \cdots + \beta_1 X + \beta_0$, we define $p(f) = \beta_n f^n + \cdots + \beta_1 f + \beta_0 1_V \in \text{Hom}_F(V, V)$. If f has matrix $\mathbf{A}$ in basis B of V, then $p(f)$ has matrix $p(\mathbf{A})$ (defined in Example 2, Section 7.15) in basis B.

Equation (7.17.1) says that

$$f^m(v) - \alpha_{m-1} f^{m-1}(v) - \cdots - \alpha_1 f(v) - \alpha_0 v = 0.$$

Denoting

(7.17.3) $\quad p(X) = X^m - \alpha_{m-1} X^{m-1} - \cdots - \alpha_1 X - \alpha_0,$

we may rewrite the preceding equation as

$$p(f)(v) = 0.$$

Matrix $\mathbf{C}_p$ in (7.17.2) is the **companion matrix** of $p(X)$.

Lemma 7.17.1. *If* $f \in \text{Hom}_F(V, V)$ *and* $0 \neq v \in V$, *then*

$$I = I_{v,f} = \{g(X) \in F[X] | g(f)(v) = 0\}$$

is a principal ideal in $F[X]$, *with generator the* $p(X)$ *of* (7.17.3).

Proof. If $g_1(X), g_2(X) \in I$, and $r(X) \in F[X]$, denote $h(X) = g_1(X) + g_2(X)$, $s(X) = r(X)g_1(X)$; then $h(f) = g_1(f) + g_2(f)$ and $s(f) = r(f)g_1(f)$, so

$$h(f)(v) = g_1(f)(v) + g_2(f)(v) = 0 + 0 = 0,$$
$$s(f)(v) = r(f)g_1(f)(0) = r(f)(0) = 0,$$

Sec. 7.17 Rational Canonical Form

proving that I is an ideal. We saw in Theorem 7.16.1 that I must be principal, with generator the unique monic polynomial of lowest degree in I; since $v, f(v), \ldots, f^{m-1}(v)$ are linearly independent in V, this must be $p(X)$. ∎

The generator $p(X) = p_{v,f}(X)$ is called the **f-annihilator** of v. Given $p(X) \in F[X]$ and monic, we shall call any f-cyclic subspace $W = \langle v \rangle_f$ which determines $p(X)$ in the manner of the preceding paragraphs an **f-companion space of $p(X)$**. Clearly,

(7.17.4) $\dim W = m = \deg p(X)$, if W is an f-companion space of $p(X)$.

(We see in Exercise 7.18.7 that no subspace can be an f-companion space of two different polynomials.)

For example, consider $f \in \operatorname{Hom}_{\mathbb{R}}(\mathbb{R}^3, \mathbb{R}^3)$ defined by

$$f((a, b, c)) = (a + b + c, -a - 2b, 2a + 8b - c),$$

and $v = (1, 0, 2)$. We find that

$$f(v) = (3, -1, 0) \notin \lceil v \rceil$$

but

$$f^2(v) = f((3, -1, 0)) = (2, -1, -2) = (3, -1, 0) - (1, 0, 2) \in \lceil v, f(v) \rceil = \langle v \rangle_f = W.$$

Since $v - f(v) + f^2(v) = 0$, we have $p(X) = 1 - X + X^2$.

We shall see, in our discussion of the rational canonical form, that whenever $f \in \operatorname{Hom}_F(V, V)$, then V is the direct sum of f-companion spaces of certain polynomials. Also, any square matrix over F is similar to a direct sum of companion matrices of certain polynomials.

Lemma 7.17.2. *If $p(X) \in F[X]$ and $f \in \operatorname{Hom}_F(V, V)$, then the set*

(7.17.5) $W = W_{f,p(X)} = \{v \in V \mid p(f)^j(v) = 0, \text{ some } j = j(v) \in \mathbb{P}\}$

is an f-invariant subspace of V (which we call the $p(f)^\infty$-kernel of V).

Proof. If $v_1, v_2 \in W$ and $\alpha \in F$, say $p(f)^{j_1}(v_1) = 0$, $p(f)^{j_2}(v_2) = 0$, and $j = \max\{j_1, j_2\}$, then

$$p(f)^j(v_1 + v_2) = p(f)^j(v_1) + p(f)^j(v_2) = 0 + 0 = 0,$$
$$p(f)^j(\alpha v_1) = \alpha(p(f)^j(v_1)) = \alpha \cdot 0 = 0,$$

so $\alpha v_1, v_1 + v_2 \in W$; by Theorem 7.5.1, W is a subspace. If $v \in W$, say $p(f)^j(v) = 0$, then

$$p(f)^j(f(v)) = (p(f)^j f)(v) = (f p(f)^j)(v) = f(0) = 0,$$

so $f(v) \in W$; W is f-invariant. ∎

Theorem 7.17.1. *Assume that V is a finite-dimensional vector space over F and $f \in \operatorname{Hom}_F(V, V)$. Then there are uniquely determined distinct monic irreducible polynomials $p_1(X), p_2(X), \ldots, p_r(X)$, such that $V = V_1 \oplus V_2 \oplus \cdots \oplus V_r$, $0 \neq V_i$ the $p_i(f)^\infty$-kernel of V.*

Proof. Let $\{v_1, v_2, \ldots, v_n\}$ be a basis of V, and let $q_i(X)$ be the f-annihilator of v_i; set $q(X) = \text{l.c.m.} \{q_1(X), q_2(X), \ldots, q_n(X)\}$, and let $p_1(X), \ldots, p_r(X)$ be the distinct irreducible factors of $q(X)$, V_i the $p_i(f)^\infty$-kernel of V. For a fixed $p_i(X)$, choose $q_j(X)$ such that $p_i(X) | q_j(X)$, say $q_j(X) = p_i(X)s(X)$. Then $v_0 = s(f)(v_j)$ is not 0, as $s(X)$ is not a multiple of the f-annihilator $q_j(X)$ of v_j; but $p_i(f)(v_0) = p_i(f)s(f)(v_j) = q_j(f)(v_j) = 0$, proving that $v_0 \in V_i$; each subspace V_i is nonzero.

Since each $q_j(X)$ divides $q(X)$, $q(f)(v_j) = 0$ for all j; $\{v_1, v_2, \ldots, v_n\}$ is a basis of V, so $q(f)(v) = 0$, all $v \in V$. Let $p(X) \in F[X]$ be any irreducible polynomial other than $p_1(X), \ldots, p_r(X)$, and let w be an element of the $p(f)^\infty$-kernel W in V, say $p(f)^t(w) = 0$. Since $(p(X)^t, q(X)) = 1$, there are $a(X), b(X) \in F[X]$ such that $a(X)p(X)^t + b(X)q(X) = 1$. Substituting f for X, we have $a(f)p(f)^t + b(f)q(f) = 1_V$, so that

$$w = 1_V(w) = a(f)p(f)^t(w) + b(f)q(f)(w) = a(f)(0) + b(f)(0) = 0,$$

proving that $W = [0]$. We have now characterized the $p_i(X)$ as the only monic irreducible polynomials in $F[X]$ with nonzero $p_i(f)^\infty$-kernels.

It remains to show that $V = V_1 \oplus V_2 \oplus \cdots \oplus V_r$. This is trivial if $r = 1$, since then $q(X)$ is a power of $p_1(X)$; hence assume that $r > 1$.

Let $q(X) = p_1(X)^{t_1} p_2(X)^{t_2} \cdots p_r(X)^{t_r}$, and define $s_i(X)$ by $q(X) = p_i(X)^{t_i} s_i(X)$ for $1 \leq i \leq r$. By Exercise 7.18.4, there are $a_1(X), a_2(X), \ldots, a_r(X) \in F[X]$ such that

$$a_1(X)s_1(X) + a_2(X)s_2(X) + \cdots + a_r(X)s_r(X) = 1.$$

For any $v \in V$, this implies that

$$v = a_1(f)s_1(f)(v) + a_2(f)s_2(f)(v) + \cdots + a_r(f)s_r(f)(v).$$

Since

$$p_i(f)^{t_i} \cdot a_i(f)s_i(f)(v) = a_i(f)q(f)(v) = a_i(f)(0) = 0$$

for each i, $a_i(f)s_i(f)(v) \in V_i$, proving that $V = V_1 + V_2 + \cdots + V_r$.

We must show that $V_i \cap (V_1 + \cdots + V_{i-1} + V_{i+1} + \cdots + V_r) = [0]$ for each i. If w_i is in this space, say $w_i = w_1 + \cdots + w_{i-1} + w_{i+1} + \cdots + w_r$, all $w_j \in V_j$, choose $c_i(X), d_i(X) \in F[X]$ such that $c_i(X)p_i(X)^{t_i} + d_i(X)s_i(X) = 1$. Since the f-annihilator of w_i is a power of $p_i(X)$ and divides $q(X)$, it divides $p_i(X)^{t_i}$, and $p_i(f)^{t_i}(w_i) = 0$. Hence

$$w_i = c_i(f)p_i(f)^{t_i}(w_i) + d_i(f)s_i(f)(w_1 + \cdots + w_{i-1} + w_{i+1} + \cdots + w_r)$$
$$= c_i(f)(0) + d_i(f)(0) = 0 + 0 = 0. \blacksquare$$

Sec. 7.17 Rational Canonical Form

For example, set $F = GF(3)$, let $V = \text{Mat}_{3 \times 1}(F)$ be the space of 3×1 column vectors over F, and define $f \in \text{Hom}_F(V, V)$ by $f: \mathbf{M} \to \mathbf{AM}$, where

$$\mathbf{A} = \begin{bmatrix} 2 & 1 & 1 \\ 2 & 0 & 2 \\ 2 & 0 & 0 \end{bmatrix}.$$

The procedure of the previous proof is not often used in practice, but we shall apply it here, using the standard basis

$$v_1 = \begin{bmatrix} 1 \\ 0 \\ 0 \end{bmatrix}, \quad v_2 = \begin{bmatrix} 0 \\ 1 \\ 0 \end{bmatrix}, \quad v_3 = \begin{bmatrix} 0 \\ 0 \\ 1 \end{bmatrix}$$

of V. We get

$$f(v_1) = \mathbf{A}v_1 = \begin{bmatrix} 2 & 1 & 1 \\ 2 & 0 & 2 \\ 2 & 0 & 0 \end{bmatrix} \begin{bmatrix} 1 \\ 0 \\ 0 \end{bmatrix} = \begin{bmatrix} 2 \\ 2 \\ 2 \end{bmatrix}, \quad \text{so } f(v_1) \notin \lceil v_1 \rceil;$$

$$f^2(v_1) = \mathbf{A}^2 v_1 = \begin{bmatrix} 2 & 1 & 1 \\ 2 & 0 & 2 \\ 2 & 0 & 0 \end{bmatrix} \begin{bmatrix} 2 \\ 2 \\ 2 \end{bmatrix} = \begin{bmatrix} 2 \\ 2 \\ 1 \end{bmatrix}, \quad \text{so } f^2(v_1) \notin \lceil v_1, f(v_1) \rceil;$$

$$f^3(v_1) = \mathbf{A}^3 v_1 = \begin{bmatrix} 2 & 1 & 1 \\ 2 & 0 & 2 \\ 2 & 0 & 0 \end{bmatrix} \begin{bmatrix} 2 \\ 2 \\ 1 \end{bmatrix} = \begin{bmatrix} 1 \\ 0 \\ 1 \end{bmatrix} = 2f^2(v_1) + f(v_1) + v_1,$$

so

$$(f^3 + f^2 + 2f + 2 \cdot 1_V)(v_1) = 0,$$

implying that

$$q_1(X) = X^3 + X^2 + 2X + 2 = (X + 2)(X + 1)^2.$$

Similarly, we can calculate

$$q_2(X) = (X + 2)(X + 1)^2, \quad q_3(X) = (X + 1)^2,$$

so $q(X) = \text{l.c.m.} \{q_1(X), q_2(X), q_3(X)\} = (X + 2)(X + 1)^2$, $p_1(X) = X + 2$, $p_2(X) = X + 1$, $V_1 = \ker(f + 2 \cdot 1_V)$, $V_2 = \ker(f + 1_V)^2$.

To find V_1 explicitly, we solve

$$\begin{bmatrix} 0 \\ 0 \\ 0 \end{bmatrix} = (\mathbf{A} + 2\mathbf{I}) \begin{bmatrix} x \\ y \\ z \end{bmatrix} = \begin{bmatrix} 1 & 1 & 1 \\ 2 & 2 & 2 \\ 2 & 0 & 2 \end{bmatrix} \begin{bmatrix} x \\ y \\ z \end{bmatrix};$$

we get $y = 0$, $x + z = 0$, so $z = 2x$,

$$V_1 = \left\{ \begin{bmatrix} x \\ 0 \\ 2x \end{bmatrix} \middle| x \in F \right\}.$$

We find V_2 explicitly by solving

$$\begin{bmatrix} 0 \\ 0 \\ 0 \end{bmatrix} = (\mathbf{A}+\mathbf{I})^2 \begin{bmatrix} x \\ y \\ z \end{bmatrix} = \begin{bmatrix} 0 & 1 & 1 \\ 2 & 1 & 2 \\ 2 & 0 & 1 \end{bmatrix}^2 \begin{bmatrix} x \\ y \\ z \end{bmatrix} = \begin{bmatrix} 1 & 1 & 0 \\ 0 & 0 & 0 \\ 2 & 2 & 0 \end{bmatrix} \begin{bmatrix} x \\ y \\ z \end{bmatrix};$$

we get $x + y = 0$, $z =$ anything, so $y = 2x$,

$$V_2 = \left\{ \begin{bmatrix} x \\ 2x \\ z \end{bmatrix} \middle| x, z \in F \right\}.$$

V_1 has basis $\left\{ \begin{bmatrix} 1 \\ 0 \\ 2 \end{bmatrix} \right\}$ and V_2 has basis $\left\{ \begin{bmatrix} 1 \\ 2 \\ 0 \end{bmatrix}, \begin{bmatrix} 0 \\ 0 \\ 1 \end{bmatrix} \right\}$; since $\left\{ \begin{bmatrix} 1 \\ 0 \\ 2 \end{bmatrix}, \begin{bmatrix} 1 \\ 2 \\ 0 \end{bmatrix}, \begin{bmatrix} 0 \\ 0 \\ 1 \end{bmatrix} \right\}$ is a basis of V, we do have $V = V_1 \oplus V_2$.

In Theorem 7.17.2 we shall further decompose the V_i's of Theorem 7.17.1 into direct sums of f-companion spaces; but first we need a definition and three lemmas. If $f \in \text{Hom}_F(U, U)$ and nonzero vectors $u_1, u_2, \ldots, u_k \in U$ are such that for any $s_1(X), s_2(X), \ldots, s_k(X) \in F[X]$,

(7.17.6) $\qquad s_1(f)(u_1) + \cdots + s_k(f)(u_k) = 0$

implies that all $s_i(f)(u_i) = 0$, then we say that $u_1, u_2, \ldots, u_k$ are **f-independent** in U.

Lemma 7.17.3. *If $f \in \text{Hom}_F(U, U)$, then elements $u_1, u_2, \ldots, u_k$ are f-independent iff the subspace $\langle u_1 \rangle_f + \langle u_2 \rangle_f + \cdots + \langle u_k \rangle_f$ is a direct sum $\langle u_1 \rangle_f \oplus \langle u_2 \rangle_f \oplus \cdots \oplus \langle u_k \rangle_f$.*

Proof. Denote $W_i = \langle u_i \rangle_f$, the f-cyclic subspace generated by $u_i \in U$. Assume that $u_1, u_2, \ldots, u_k \in U$ are f-independent. To prove that the sum $W_1 + W_2 + \cdots + W_k$ is a direct sum, it suffices to show that

$$W_i \cap (W_1 + \cdots + W_{i-1} + W_{i+1} + \cdots + W_k) = [0].$$

If we assume that $w \in W_i \cap (W_1 + \cdots + W_{i-1} + W_{i+1} + \cdots + W_k)$, then since $W_i = \lceil u_i, f(u_i), \ldots \rceil$, we have $w = s_i(f)(u_i)$ for some polynomial $s_i(X) \in F[X]$. Similarly, the fact $w \in W_1 + \cdots + W_{i-1} + W_{i+1} + \cdots + W_k$ implies that $w = s_1(f)(u_1) + \cdots + s_{i-1}(f)(u_{i-1}) + s_{i+1}(f)(u_{i+1}) + \cdots + s_k(f)(u_k)$ for polynomials $s_j(X), j = 1, \ldots, i-1, i+1, \ldots, k$. The two expressions for w

Sec. 7.17 Rational Canonical Form

yield

$$s_1(f)(u_1) + \cdots + s_{i-1}(f)(u_{i-1}) + [-s_i(f)(u_i)] + s_{i+1}(f)(u_{i+1}) + \cdots + s_k(f)(u_k) = 0,$$

so f-independence implies that $-s_i(f)(u_i) = 0$, $w = 0$.

Conversely, if $W_1 + W_2 + \cdots + W_k$ is a direct sum $W_1 \oplus W_2 \oplus \cdots \oplus W_k$, suppose we have a relation

$$s_1(f)(u_1) + s_2(f)(u_2) + \cdots + s_k(f)(u_k) = 0.$$

Each $s_j(f)(u_j) \in W_j$, so for any fixed i we have

$$s_i(f)(u_i) = -s_1(f)(u_1) - \cdots - s_{i-1}(f)(u_{i-1}) - s_{i+1}(f)(u_{i+1}) - \cdots$$
$$- s_k(f)(u_k) \in W_i \cap (W_1 + \cdots + W_{i-1} + W_{i+1} + \cdots + W_k) = [0],$$

proving that $s_i(f)(u_i) = 0$; the set $\{u_1, u_2, \ldots, u_k\}$ is f-independent. ∎

For example, set $F = \mathrm{GF}(2)$, $U = F^4$, and define $f \in \mathrm{Hom}_F(U, U)$ by $f((a, b, c, d)) = (d, a + c, b + d, a)$; consider $u_1 = (1, 0, 1, 0)$, $u_2 = (0, 1, 0, 1)$. We find that $f(u_1) = (0, 0, 0, 1)$, $f^2(u_1) = u_1$, so $W_1 = \langle u_1 \rangle_f = [(1, 0, 1, 0), (0, 0, 0, 1)]$; similarly, $f(u_2) = (1, 0, 0, 0)$, $f^2(u_2) = u_2$, so $W_2 = \langle u_2 \rangle_f = [(0, 1, 0, 1), (1, 0, 0, 0)]$. Since $\{(1, 0, 1, 0), (0, 0, 0, 1), (0, 1, 0, 1), (1, 0, 0, 0)\}$ is a basis of U we have $U = W_1 \oplus W_2$, and the set $\{u_1, u_2\}$ is f-independent.

Lemma 7.17.4. *Assume that $f \in \mathrm{Hom}_F(V, V)$, and assume that $p(X)$ is an irreducible polynomial in $F[X]$ such that $p(f)^r = 0$ on all of V, some $r \in \mathbb{P}$. Assume $v_0 \in V$ is such that $p(f)^{r-1}(v_0) \neq 0$, and let W be the f-cyclic subspace of V generated by v_0. Define $f^* \in \mathrm{Hom}_F(V/W, V/W)$ by $f^*(v + W) = f(v) + W$, all $v \in V$. If $\bar{v}_1, \bar{v}_2, \ldots, \bar{v}_m$ are f^*-independent in V/W, then there are vectors $v_1, v_2, \ldots, v_m$ in V such that $\bar{v}_i = v_i + W$ and $v_0, v_1, v_2, \ldots, v_m$ are f-independent in V.*

Proof. Choose some u_i in the coset $\bar{v}_i$ for $1 \leq i \leq m$, so $\bar{v}_i = u_i + W$.

In this paragraph, fix $i \in \{1, 2, \ldots, m\}$. Pick the smallest $j \in \mathbb{P}$ such that $p(f)^j(u_i) \in W$. Since $p(f)^r(u_i) = 0 \in W$, $j \leq r$. Elements in W have form $t(f)(v_0)$ for polynomials $t(X) \in F[X]$, so we can set

(7.17.7) $$p(f)^j(u_i) = t(f)(v_0) = p(f)^k q(f)(v_0),$$

where $k \geq 0$, and $q(X) \in F[X]$ satisfies $(p(X), q(X)) = 1$. If $k \geq r$, then $p(f)^j(u_i) = 0$; in that case we define $v_i = u_i$. If $k < r$, we claim that $j \leq k$. By (7.17.7), we have

(7.17.8) $$p(f)^{r-j+k} q(f)(v_0) = p(f)^{r-j+j}(u_i) = p(f)^r(u_i) = 0.$$

If $a(X), b(X) \in F[X]$ are such that $1 = a(X)p(X)^r + b(X)q(X)$, we multiply by $p(X)^{r-j+k}$ and get

$$p(X)^{r-j+k} = a(X)p(X)^{2r-j+k} + b(X)p(X)^{r-j+k} q(X).$$

By (7.17.8), this implies that

$$p(f)^{r-j+k}(v_0) = a(f)p(f)^{2r-j+k}(v_0) + b(f)p(f)^{r-j+k}q(f)(v_0) = 0 + 0 = 0.$$

Since $p(f)^{r-1}(v_0) \neq 0$, this means that $r-j+k \geq r$, $j \leq k$. We can therefore define $v_i = u_i - p(f)^{k-j}q(f)(v_0)$; then $v_i + W = u_i + W = \bar{v}_i$. If $p(f)^l(v_i) \in W$, then $p(f)^l(u_i) \in W$, so $l \geq j$; (7.17.7) then implies that

(7.17.9)
$$\begin{aligned}p(f)^l(v_i) &= p(f)^{l-j}[p(f)^j[u_i - p(f)^{k-j}q(f)(v_0)]] \\ &= p(f)^{l-j}[p(f)^j(u_i) - p(f)^k q(f)(v_0)] = 0.\end{aligned}$$

Now we prove that $v_0, v_1, \ldots, v_m$ are f-independent. Assume that $s_0(X)$, $s_1(X), \ldots, s_m(X) \in F[X]$ are such that

(7.17.10) $\qquad s_0(f)(v_0) + s_1(f)(v_1) + \cdots + s_m(f)(v_m) = 0.$

We saw that $f^* \in \operatorname{Hom}_F(V/W, V/W)$ in Lemma 7.8.4. We have

$$\begin{aligned}(f^2)^*(v+W) &= f^2(v) + W = f(f(v)) + W = f^*(f(v) + W) \\ &= f^*(f^*(v+W)) = (f^*)^2(v+W),\end{aligned}$$

so $(f^2)^* = (f^*)^2$; in fact, $s_i(f)^* = s_i(f^*)$ for any polynomial $s_i(X)$. Therefore, we get

$$\begin{aligned}s_1(f^*)(\bar{v}_1) + \cdots + s_m(f^*)(\bar{v}_m) &= s_1(f)^*(v_1+W) + \cdots + s_m(f)^*(v_m+W) \\ &= s_1(f)(v_1) + W + \cdots + s_m(f)(v_m) + W \\ &= -s_0(f)(v_0) + W = 0 + W,\end{aligned}$$

using (7.17.10). The f^*-independence of $\bar{v}_1, \ldots, \bar{v}_m$ now implies that all $s_i(f^*)(\bar{v}_i) = 0 + W$; since

$$s_i(f^*)(\bar{v}_i) = s_i(f)^*(v_i + W) = s_i(f)(v_i) + W,$$

we have proved that all $s_i(f)(v_i) \in W$. For any fixed i, let $s_i(X) = p(X)^l q_1(X)$, $l \geq 0$, where $(p(X), q_1(X)) = 1$, and choose $a_1(X), b_1(X) \in F[X]$ with $1 = a_1(X)p(X)^r + b_1(X)q_1(X)$. We get

$$p(f)^l(v_i) = a_1(f)p(f)^{r+l}(v_i) + b_1(f)q_1(f)p(f)^l(v_i) = 0 + b_1(f)s_i(f)(v_i) \in W,$$

so, by (7.17.9), $p(f)^l(v_i) = 0$, implying that

$$s_i(f)(v_i) = q_1(f)p(f)^l(v_i) = 0,$$

and proving the f-independence. ∎

Lemma 7.17.5. *Assume that $f \in \operatorname{Hom}_F(V, V)$, where V is an f-companion space of $p(X)^t$, some $t \in \mathbb{P}$ and some irreducible $p(X) \in F[X]$. Then for any s, $0 \leq s \leq t$, the subspace $U_s = \ker p(f)^s$ has dimension $s \cdot (\text{degree of } p(X))$.*

Sec. 7.17 *Rational Canonical Form* 349

Proof. Let $V = \langle v \rangle_f$, the f-cyclic subspace generated by $v \in V$. We shall show that U_s is an f-companion space of $p(X)^s$, and in fact $U_s = \langle p(f)^{t-s}(v) \rangle_f$; this will complete the proof, by (7.17.4).

If $u \in U_s$, then $p(f)^s(u) = 0$. The fact that $u \in \langle v \rangle_f = V$ implies that $u = r(f)(v)$, some $r(X) \in F[X]$; since $0 = p(f)^s(u) = p(f)^s r(f)(v)$, the polynomial $p(X)^s r(X)$ must be a multiple of the f-annihilator $p(X)^t$ of v (Lemma 7.17.1). Hence $r(X) = p(X)^{t-s} r_0(X)$ for some $r_0(X) \in F[X]$, and we have
$$u = r(f)(v) = r_0(f) p(f)^{t-s}(v) \in \langle p(f)^{t-s}(v) \rangle_f,$$
proving that $U_s \subseteq \langle p(f)^{t-s}(v) \rangle_f$. On the other hand, $0 = p(f)^t(v) = p(f)^s p(f)^{t-s}(v)$, so $p(f)^{t-s}(v) \in U_s$, proving that $U_s = \langle p(f)^{t-s}(v) \rangle_f$. Clearly, $p(X)^s$ is the f-annihilator of $p(f)^{t-s}(v)$. ∎

Theorem 7.17.2. *Assume that $f \in \mathrm{Hom}_F(V, V)$ and that $p(X)$ is a monic irreducible polynomial in $F[X]$ such that $p(f)^r = 0$ on all of V, for some $r \in \mathbb{P}$. Then V is a direct sum $V = W_1 \oplus W_2 \oplus \cdots \oplus W_k$ of f-companion spaces of polynomials $p(X)^{n_1}, p(X)^{n_2}, \ldots, p(X)^{n_k}$, where $n_1 \geq n_2 \geq \cdots \geq n_k$ are uniquely determined positive integers.*

Proof. The proof is by induction on $\dim V$. The theorem is true for $\dim V = 1$, for if V has basis $\{v_1\}$, then $f(v_1) = \gamma v_1$, some $\gamma \in F$, so $f(v) = \gamma v$ for all $v \in V$; $p(X) = X - \gamma$, $r = n_1 = k = 1$, $V = W_1$.

Now suppose that $\dim V > 1$, and the result has been proved for spaces of smaller dimension. We may assume that r is chosen as small as possible, so there is some $v_1 \in V$ such that $p(f)^{r-1}(v_1) \neq 0$; let W be the f-cyclic subspace of V generated by v_1 ($W = \langle v_1 \rangle_f$). Then W is an f-companion space of $p(X)^{n_1}$ for $n_1 = r$. If $W = V$, then we are done, with $V = W = W_1$, $k = 1$.

It remains to consider the case when $W \neq V$, so $V/W \neq [0]$. By Lemma 7.8.4 there is an $f^* \in \mathrm{Hom}_F(V/W, V/W)$ satisfying $f^*(v + W) = f(v) + W$, all $v \in V$. We saw in Lemma 7.8.3 that V/W has smaller dimension than V. Also, $p(f^*)^r = [p(f)^r]^*$ as seen in the proof of Lemma 7.17.4, so
$$p(f^*)^r(v + W) = [p(f)^r]^*(v + W) = p(f)^r(v) + W = 0 + W,$$
proving that $p(f^*)^r = 0$ on all of V/W. By induction, the theorem therefore states that V/W is the direct sum of f^*-companion spaces of polynomials $p(X)^{n_2}, \ldots, p(X)^{n_k}$, $n_2 \geq \cdots \geq n_k$. Let the f^*-companion space $\bar{W}_i$ of $p(X)^{n_i}$ have generator $\bar{v}_i$: $\bar{W}_i = \langle \bar{v}_i \rangle_{f^*}$. By Lemma 7.17.3, the elements $\bar{v}_2, \ldots, \bar{v}_k$ are f^*-independent, so Lemma 7.17.4 asserts that the set $\{v_1, v_2, \ldots, v_k\}$ is f-independent for some vectors $v_i \in V$ satisfying $v_i + W = \bar{v}_i$, $i = 2, \ldots, k$. With W_i the f-cyclic subspace generated by v_i for $i = 1, 2, \ldots, k$ (so $W = W_1$), this means (Lemma 7.17.3) that the sum $W_1 + W_2 + \cdots + W_k$ is a direct sum $W_1 \oplus W_2 \oplus \cdots \oplus W_k$.

Fix $i \in \{2, \ldots, k\}$. Since $p(f)^r(v_i) = 0$, the f-annihilator of v_i is a power $p(X)^{t_i}$ of $p(X)$ (Lemma 7.17.1). The fact that $p(f)^{t_i}(v_i) = 0$ implies that
$$p(f^*)^{t_i}(\bar{v}_i) = [p(f)^{t_i}]^*(v_i + W) = p(f)^{t_i}(v_i) + W = 0 + W,$$

so $p(X)^{t_i}$ is a multiple of the f^*-annihilator $p(X)^{n_i}$ of $\bar{v}_i$; $t_i \geq n_i$. We therefore have

(7.17.11) $\qquad \dim W_i = \deg p(X)^{t_i} \geq \deg p(X)^{n_i} = \dim \bar{W}_i$.

From the previous paragraph, we conclude that

$$\dim (W_1 \oplus W_2 \oplus \cdots \oplus W_k) = \dim W_1 + \dim W_2 + \cdots + \dim W_k$$
$$\geq \dim W + (\dim \bar{W}_2 + \cdots + \dim \bar{W}_k) = \dim W + \dim (V/W) = \dim V,$$

forcing $V = W_1 \oplus W_2 \oplus \cdots \oplus W_k$ [and forcing equality in every case in (7.17.11)]. Note that $n_1 \geq n_2 \geq \cdots \geq n_k$, since $n_1 = r$.

We still must prove that $n_1, n_2, \ldots, n_k$ are uniquely determined by V and f. If k_j is the number of n_i's with $n_i \geq j$, it is certainly enough to prove that $k_1, k_2, \ldots, k_r$ are uniquely determined. For each j, we form the subspace $U_j = \{v \in V | p(f)^j(v) = 0\} = \ker p(f)^j$ of V, and we note that $[0] = U_0 \subseteq U_1 \subseteq \cdots \subseteq U_r = V$. The spaces U_j are uniquely determined by f, independently of any choice of v_1, W, and so on, so it is enough to show that

(7.17.12) $\qquad \dim U_j - \dim U_{j-1} = k_j \cdot (\deg p(X))$,

for $j = 1, 2, \ldots, r$.

Fix j, and denote $X_i = W_i \cap U_j$, $Y_i = W_i \cap U_{j-1}$ for $i = 1, 2, \ldots, k$. If $u \in U_j$, say $u = w_1 + w_2 + \cdots + w_k$ for $w_i \in W_i$, then $0 = p(f)^j(u) = p(f)^j(w_1) + p(f)^j(w_2) + \cdots + p(f)^j(w_k)$. Each $p(f)^j(w_i) \in W_i \cap (W_1 + \cdots + W_{i-1} + W_{i+1} + \cdots + W_k) = [0]$ by this equation, so each $w_i \in W_i \cap U_j = X_i$; we conclude that $U_j = X_1 + X_2 + \cdots + X_k = X_1 \oplus X_2 \oplus \cdots \oplus X_k$. Similarly, $U_{j-1} = Y_1 \oplus Y_2 \oplus \cdots \oplus Y_k$. The definition of k_j and the fact that $n_1 \geq n_2 \geq \cdots \geq n_k$ together imply that

$$n_1 \geq n_2 \geq \cdots \geq n_{k_j} \geq j > n_{k_j+1} \geq \cdots \geq n_k.$$

For $i > k_j$, then, $n_i \leq j - 1$, implying that $p(f)^{j-1}(W_i) = [0]$, so $W_i = X_i = Y_i$. For $i \leq k_j$, Lemma 7.17.5 applied to the restriction of f to W_i implies that $\dim X_i = j(\deg p(X))$, $\dim Y_i = (j-1)(\deg p(X))$. We conclude that

$$\dim U_j - \dim U_{j-1} = \sum_{i=1}^{k} \dim X_i - \sum_{i=1}^{k} \dim Y_i = \sum_{i=1}^{k} (\dim X_i - \dim Y_i)$$

$$= \sum_{i=1}^{k_j} (\dim X_i - \dim Y_i) + 0 + \cdots + 0$$

$$= \sum_{i=1}^{k_j} [j(\deg p(X)) - (j-1)(\deg p(X))]$$

$$= \sum_{i=1}^{k_j} \deg p(X) = k_j(\deg p(X)),$$

completing the proof of (7.17.12). ∎

Sec. 7.17 Rational Canonical Form

The hard work of this section is now complete. We still have to collect the information of Theorems 7.17.1 and 7.17.2 into the most useful general form, in Corollaries 7.17.1 and 7.17.3 for linear transformations and Corollaries 7.17.2 and 7.17.4 for matrices. We shall also give some computational hints and examples, as one need not trace through all the preceding proofs when studying a specific matrix.

Corollary 7.17.1 (Elementary Divisor Form of the Rational Canonical Form). *Assume that $f \in \text{Hom}_F(V, V)$, V finite-dimensional. Then there are uniquely determined powers*

$$p_1(X)^{n_{11}}, p_1(X)^{n_{12}}, \ldots, p_1(X)^{n_{1k_1}},$$

$$p_2(X)^{n_{21}}, \ldots, p_2(X)^{n_{2k_2}}, \ldots, p_r(X)^{n_{r1}}, \ldots, p_r(X)^{n_{rk_r}}$$

of monic irreducible polynomials in $F[X]$ such that

$$V = V_{11} \oplus V_{12} \oplus \cdots \oplus V_{1k_1} \oplus V_{21} \oplus \cdots \oplus V_{2k_2} \oplus \cdots \oplus V_{r1} \oplus \cdots \oplus V_{rk_r},$$

V_{ij} *an f-companion space of $p_i(X)^{n_{ij}}$, with*

$$n_{i1} \geq n_{i2} \geq \cdots \geq n_{ik_i} \quad \text{for each } i.$$

Proof. By Theorem 7.17.1, $V = W_1 \oplus W_2 \oplus \cdots \oplus W_r$, where W_i is an f-invariant subspace of V and $W_i = \{v \in V \mid p_i(f)^j(v) = 0, \text{ some } j = j(v)\}$. Applying powers of $p_i(f)$ to a basis of W_i, we see that there is some power $p_i(f)^{r_i}$ of $p_i(f)$ such that $p_i(f)^{r_i}(W_i) = [0]$. We then apply Theorem 7.17.2 to the restriction of f to each W_i, obtaining the direct sums $W_i = V_{i1} \oplus V_{i2} \oplus \cdots \oplus V_{ik_i}$ and the exponents $n_{i1} \geq n_{i2} \geq \cdots \geq n_{ik_i}$. The uniqueness in the corollary follows from the uniqueness in the theorems. ∎

The polynomials $p_i(X)^{n_{ij}}$ are called the **elementary divisors** of f. We can also speak of *elementary divisors of matrices*, as in the following corollary.

Corollary 7.17.2 (Elementary Divisors of Matrices). *Any square matrix $\mathbf{M}$ over a field F is similar to a direct sum of companion matrices of uniquely determined powers of monic irreducible polynomials in $F[X]$. (These powers are called the* **elementary divisors** *of $\mathbf{M}$.)*

Proof. If $\mathbf{M}$ is $n \times n$ and B is a basis of an n-dimensional vector space V over F, define $f \in \text{Hom}_F(V, V)$ by $[f]_B = \mathbf{M}$. Apply Corollary 7.17.1 to f, and for each V_{ij} choose a basis $C_{ij} = \{v_{ij}, f(v_{ij}), f^2(v_{ij}), \ldots\}$. Then $C = \bigcup_{\text{all } i,j} C_{ij}$ is a basis of V [Exercise 7.7.12(a)]. By Lemma 7.9.1, $[f]_C$ has the direct sum form described; $[f]_C$ is similar to $\mathbf{M}$ by Theorem 7.13.1. ∎

Note in Corollaries 7.17.1 and 7.17.2 that even with the specification $n_{i1} \geq n_{i2} \geq \cdots \geq n_{ik_i}$, the elementary divisor rational canonical form of a given f or $\mathbf{M}$ is not quite unique; the $p_i(X)$ may be arranged in different orders. To achieve absolute uniqueness, we must use the following corollary.

Corollary 7.17.3 (Invariant Factor Form of the Rational Canonical Form). *Assume that $f \in \text{Hom}_F(V, V)$. Then there are uniquely determined monic polynomials $q_1(X), \ldots, q_k(X)$ such that $q_{j+1}(X)|q_j(X)$, all $1 \le j < k$, and $V = V_1 \oplus \cdots \oplus V_k$, where V_j is an f-companion space of $q_j(X)$.*

Proof. We use Corollary 7.17.1. Let $k = \max\{k_1, k_2, \ldots, k_r\}$, and if $k_i < k$, define $n_{i,k_i+1} = \cdots = n_{ik} = 0$, $V_{i,k_i+1} = \cdots = V_{ik} = [0]$. Now n_{ij} and V_{ij} are defined for all $1 \le i \le r$, $1 \le j \le k$; for each $j = 1, 2, \ldots, k$, we define

$$q_j(X) = p_1(X)^{n_{1j}} p_2(X)^{n_{2j}} \cdots p_r(X)^{n_{rj}}, \quad V_j = V_{1j} + V_{2j} + \cdots + V_{rj}.$$

By our construction, each irreducible factor $p_i(X)$ divides $q_j(X)$ n_{ij} times and divides $q_{j+1}(X)$ $n_{i,j+1}$ times, where $n_{ij} \ge n_{i,j+1}$, so $q_{j+1}(X)|q_j(X)$, all $j = 1, \ldots, k-1$. Also, each V_{ij} in Corollary 7.17.1 appears exactly once as a summand of V_j, so we have $V = V_1 \oplus V_2 \oplus \cdots \oplus V_k$.

If $V_{i1} = \langle v_i \rangle_f$, consider $v = v_1 + v_2 + \cdots + v_r \in V_{11} \oplus V_{21} \oplus \cdots \oplus V_{r1} = V_1$. If $r(X) \in F[X]$ satisfies $r(f)(v) = 0$, then we have $0 = r(f)(v_1) + r(f)(v_2) + \cdots + r(f)(v_r)$; each $r(f)(v_i) \in V_{i1}$, so the directness of the sum implies that $r(f)(v_i) = 0$ for all i. The power $p_i(X)^{n_{i1}}$ is the f-annihilator of v_i, so $p_i(X)^{n_{i1}}$ divides $r(X)$ for every i. We conclude that the product $q_1(X)$ of these $p_i(X)^{n_{i1}}$ must be the f-annihilator of v; $V_1 = \langle v \rangle_f$ is an f-companion space of $q_1(X)$.

Similarly, each V_j is an f-companion space of $q_j(X)$. Since the $p_i(X)$, with their exponents $n_{i1} \ge \cdots \ge n_{ik_i}$, are unique except for order, the $q_j(X)$ are absolutely unique. ∎

The $q_i(X)$ are called the **invariant factors** of f. We can also speak of *invariant factors of a matrix*, as follows.

Corollary 7.17.4 (Invariant Factors of Matrices). *If $\mathbf{M}$ is a square matrix over a field F, then there are uniquely determined monic polynomials $q_1(X)$, $q_2(X), \ldots, q_k(X) \in F[X]$ such that $q_{j+1}(X)|q_j(X)$, all $1 \le j < k$, and $\mathbf{M}$ is similar to the direct sum of matrices $\mathbf{B}_1, \mathbf{B}_2, \ldots, \mathbf{B}_k$, $\mathbf{B}_j$ the companion matrix of $q_j(X)$.*

Proof. Apply Corollary 7.17.3 to any $f \in \text{Hom}_F(V, V)$, some V, f with matrix $\mathbf{M}$. ∎

In Corollary 7.17.4, the $q_i(X)$ are called the **invariant factors** of $\mathbf{M}$, and the matrix that is the direct sum of $\mathbf{B}_1, \mathbf{B}_2, \ldots, \mathbf{B}_k$ is called the **rational canonical form** of $\mathbf{M}$. [A matrix is said to be in *rational canonical form* if it is the direct sum of companion matrices of monic polynomials $r_1(X), r_2(X), \ldots, r_l(X)$, where $r_{j+1}(X)|r_j(X)$ for all $j < l$. We have seen that every square matrix is similar to one and only one matrix in rational canonical form.]

Sec. 7.17 Rational Canonical Form

In the notation of Corollaries 7.17.3 and 7.17.4, $q_1(X)$ is called the **minimal polynomial** of f or $\mathbf{M}$, $q_1(X)q_2(X)\cdots q_k(X)$ the **characteristic polynomial** of f or $\mathbf{M}$. The following lemma is important in actual computations with matrices.

Lemma 7.17.6. (a) *If $q_1(X)$ is the minimal polynomial of $f \in \mathrm{Hom}_F(V, V)$ and $r(X) \in F[X]$, then $r(f) = 0$ iff $q_1(X)|r(X)$.*

(b) *If $q_1(X)$ is the minimal polynomial of the square matrix $\mathbf{M}$ over F and $r(X) \in F[X]$, then $r(\mathbf{M}) = \mathbf{0}$ iff $q_1(X)|r(X)$.*

(c) *If $c(X)$ is the characteristic polynomial of the $n \times n$ matrix $\mathbf{M}$, then $c(X) = (-1)^n \det(\mathbf{M} - X\mathbf{I})$.*

Proof. (a) Assume that the V_j and $q_j(X)$ are as in Corollary 7.17.3. By Lemma 7.17.1, the f-companion space V_j of $q_j(X)$ satisfies $r(f)(V_j) = [0]$ iff $q_j(X)|r(X)$. Since all $q_j(X)$ divide $q_1(X)$, we have $r(f)(V_j) = [0]$ for *all* j iff $q_1(X)$ divides $r(X)$.

(b) We obtain part (b) from part (a) by choosing a vector space V, a basis B of V, and an $f \in \mathrm{Hom}_F(V, V)$ such that f has matrix $\mathbf{M}$ in basis B.

(c) Similar matrices certainly have the same characteristic polynomial, since they are matrices of the same linear transformation in different bases. Also, if $\mathbf{A}$ is an invertible $n \times n$ matrix, then $\det(\mathbf{A}^{-1}\mathbf{M}\mathbf{A} - X\mathbf{I}) = \det(\mathbf{A}^{-1}\mathbf{M}\mathbf{A} - \mathbf{A}^{-1}X\mathbf{I}\mathbf{A}) = \det \mathbf{A}^{-1}(\mathbf{M} - X\mathbf{I})\mathbf{A} = (\det \mathbf{A})^{-1}(\det(\mathbf{M} - X\mathbf{I}))(\det \mathbf{A}) = \det(\mathbf{M} - X\mathbf{I})$, so when proving (c) we can replace $\mathbf{M}$ by any similar matrix; we can therefore assume that $\mathbf{M}$ is in rational canonical form. The determinant of a direct sum of matrices is the product of their determinants (Exercise 7.18.8) and the characteristic polynomial $q_1(X)q_2(X)\cdots q_k(X)$ of $\mathbf{M}$ is the product of the characteristic polynomials $q_i(X)$ of the individual blocks, so we may even assume that $\mathbf{M}$ is the companion matrix of some (characteristic) polynomial $p(X) = X^m - \alpha_{m-1}X^{m-1} - \cdots - \alpha_1 X - \alpha_0$. This means that

$$\mathbf{M} - X\mathbf{I} = \begin{bmatrix} -X & 0 & \cdots & 0 & \alpha_0 \\ 1 & -X & \cdots & 0 & \alpha_1 \\ 0 & 1 & \cdots & 0 & \alpha_2 \\ \vdots & & & & \vdots \\ 0 & 0 & \cdots & -X & \alpha_{m-2} \\ 0 & 0 & \cdots & 1 & \alpha_{m-1} - X \end{bmatrix}.$$

It suffices to show, by induction on m, that

(7.17.13) $\det(\mathbf{M} - X\mathbf{I}) = (-1)^m(X^m - \alpha_{m-1}X^{m-1} - \cdots - \alpha_1 X - \alpha_0).$

Now (7.17.13) holds for $m = 1$ because $\det[\alpha_0 - X] = \alpha_0 - X = (-1)(X - \alpha_0)$. So assume that (7.17.13) holds for m; we shall prove it for $m + 1$, using (7.12.4) ("expansion by minors of the first row").

$$\det \begin{bmatrix} -X & 0 & \cdots & 0 & \alpha_0 \\ 1 & -X & \cdots & 0 & \alpha_1 \\ \hdotsfor{5} \\ 0 & 0 & \cdots & -X & \alpha_{m-1} \\ 0 & 0 & \cdots & 1 & \alpha_m - X \end{bmatrix}$$

$$= (-X) \det \begin{bmatrix} -X & 0 & \cdots & 0 & \alpha_1 \\ 1 & -X & \cdots & 0 & \alpha_2 \\ \hdotsfor{5} \\ 0 & 0 & \cdots & -X & \alpha_{m-1} \\ 0 & 0 & \cdots & 1 & \alpha_m - X \end{bmatrix}$$

$$+ (-1)^m \alpha_0 \det \begin{bmatrix} 1 & -X & \cdots & 0 & 0 \\ 0 & 1 & \cdots & 0 & 0 \\ \hdotsfor{5} \\ 0 & 0 & \cdots & 1 & -X \\ 0 & 0 & \cdots & 0 & 1 \end{bmatrix}.$$

By induction and the fact that the determinant of an upper triangular matrix is the product of its diagonal entries [Exercise 7.14.5(b)], this equals

$$(-X)(-1)^m(X^m - \alpha_m X^{m-1} - \cdots - \alpha_2 X - \alpha_1) + (-1)^m \alpha_0$$
$$= (-1)^{m+1}(X^{m+1} - \alpha_m X^m - \cdots - \alpha_1 X - \alpha_0). \blacksquare$$

EXAMPLE 1. Consider, over $\mathbb{R}$, the matrix

$$\mathbf{M}_1 = \begin{bmatrix} 1 & 1 & 1 & 1 \\ 0 & 1 & 1 & 1 \\ 0 & 0 & 1 & 1 \\ 0 & 0 & 0 & 1 \end{bmatrix}.$$

We find that

$$\det(\mathbf{M}_1 - X\mathbf{I}) = \det \begin{bmatrix} 1-X & 1 & 1 & 1 \\ 0 & 1-X & 1 & 1 \\ 0 & 0 & 1-X & 1 \\ 0 & 0 & 0 & 1-X \end{bmatrix} = (1-X)^4,$$

so $\mathbf{M}_1$ has characteristic polynomial $(-1)^4(1-X)^4 = (X-1)^4$. The minimal polynomial $q_1(X)$ must be the divisor of $(X-1)^4$ of smallest degree satisfying $q_1(\mathbf{M}_1) = \mathbf{0}$. We try successively the divisors $X-1$, $(X-1)^2$, and $(X-1)^3$ of $(X-1)^4$; we find that $\mathbf{M}_1 - \mathbf{I} \neq \mathbf{0}$ and $(\mathbf{M}_1 - \mathbf{I})^2 \neq \mathbf{0}$ but $(\mathbf{M}_1 - \mathbf{I})^3 = \mathbf{0}$, so $q_1(X) = (X-1)^3$. The fact that the characteristic polynomial is $q_1(X)q_2(X) \cdots q_k(X) = (X-1)^4$ now implies that $k = 2$ and $q_2(X) = X-1$.

Sec. 7.17 Rational Canonical Form

Since $q_1(X) = X^3 - 3X^2 + 3X - 1$, the rational canonical form of $\mathbf{M}_1$ is

$$\begin{bmatrix} 0 & 0 & 1 & | & 0 \\ 1 & 0 & -3 & | & 0 \\ 0 & 1 & 3 & | & 0 \\ \hline 0 & 0 & 0 & | & 1 \end{bmatrix}.$$

EXAMPLE 2. Consider, over GF(2), the matrix

$$\mathbf{M}_2 = \begin{bmatrix} 1 & 0 & 0 & 1 \\ 0 & 1 & 0 & 1 \\ 0 & 0 & 1 & 1 \\ 0 & 0 & 0 & 1 \end{bmatrix}.$$

As in the previous example, we find that $\det(\mathbf{M}_2 - X\mathbf{I}) = (1 - X)^4$, so $\mathbf{M}_2$ has characteristic polynomial $(-1)^4(1-X)^4 = (X-1)^4 = (X+1)^4$ [since $-1 = +1$ in GF(2)]. When we try the divisors $(X+1)$ and $(X+1)^2$ of $(X+1)^4$, we find that $\mathbf{M}_2 + \mathbf{I} \neq \mathbf{0}$ but $(\mathbf{M}_2 + \mathbf{I})^2 = \mathbf{0}$, so the minimal polynomial of $\mathbf{M}_2$ is $(X+1)^2 = q_1(X)$. This time the equations

$$q_1(X)q_2(X)\cdots q_k(X) = (X+1)^4, \qquad q_1(X) = (X+1)^2$$

do not completely determine the $q_i(X)$; we can have either $k = 2$, $q_2(X) = (X+1)^2$, or $k = 3$, $q_2(X) = q_3(X) = X+1$. Correspondingly, the rational canonical form of $\mathbf{M}_2$ must be [since $(X+1)^2 = X^2 + 1$ over GF(2)]

$$\begin{bmatrix} 0 & 1 & | & 0 & 0 \\ 1 & 0 & | & 0 & 0 \\ \hline 0 & 0 & | & 0 & 1 \\ 0 & 0 & | & 1 & 0 \end{bmatrix} \quad \text{or} \quad \begin{bmatrix} 0 & 1 & | & 0 & | & 0 \\ 1 & 0 & | & 0 & | & 0 \\ \hline 0 & 0 & | & 1 & | & 0 \\ \hline 0 & 0 & | & 0 & | & 1 \end{bmatrix}.$$

We can determine which is the correct rational canonical form by using (7.17.12), with $j = 1$. We have $U_0 = [0]$, of dimension 0, and $U_1 = \ker p(\mathbf{M}_2) = \ker(\mathbf{M}_2 + \mathbf{I})$, using $p(X) = X + 1$ and using $\mathbf{M}_2$ as a linear transformation $\mathbf{X} \to \mathbf{M}_2 \mathbf{X}$ of $\text{Mat}_{4 \times 1}(\text{GF}(2))$. When we solve the equation

$$\begin{bmatrix} 0 \\ 0 \\ 0 \\ 0 \end{bmatrix} = (\mathbf{M}_2 + \mathbf{I}) \begin{bmatrix} w \\ x \\ y \\ z \end{bmatrix} = \begin{bmatrix} 0 & 0 & 0 & 1 \\ 0 & 0 & 0 & 1 \\ 0 & 0 & 0 & 1 \\ 0 & 0 & 0 & 0 \end{bmatrix} \begin{bmatrix} w \\ x \\ y \\ z \end{bmatrix},$$

we find that $z = 0$, while w, x, and y are arbitrary. This means that U_1 has basis

$$\begin{bmatrix} 1 \\ 0 \\ 0 \\ 0 \end{bmatrix}, \begin{bmatrix} 0 \\ 1 \\ 0 \\ 0 \end{bmatrix}, \begin{bmatrix} 0 \\ 0 \\ 1 \\ 0 \end{bmatrix},$$

so dim $U_1 = 3$ and (7.17.12) yields $k_1 = 3$. Three exponents n_i are ≥ 1, so there are three elementary divisors, which must be $(X+1)^2$, $(X+1)$, and $(X+1)$. We conclude that $k = 3$, $X+1 = q_2(X) = q_3(X)$, and the correct rational canonical form of $\mathbf{M}_2$ is

$$\begin{bmatrix} 0 & 1 & 0 & 0 \\ 1 & 0 & 0 & 0 \\ \hline 0 & 0 & 1 & 0 \\ \hline 0 & 0 & 0 & 1 \end{bmatrix}.$$

EXAMPLE 3. We now seek a complete list of rational canonical forms of 4×4 matrices $\mathbf{A}$ over $\mathbb{R}$ which satisfy $\mathbf{A}^4 = \mathbf{I}$. (If $\mathbf{A}$ satisfies this equation, so does any matrix similar to $\mathbf{A}$. Why?) Such an $\mathbf{A}$ satisfies the equation $X^4 = 1$, $X^4 - 1 = 0$, so Lemma 7.17.6(b) implies that the minimal polynomial $q_1(X)$ of $\mathbf{A}$ divides $X^4 - 1$. When we factor $X^4 - 1$ into irreducible factors over $\mathbb{R}$, we get

$$X^4 - 1 = (X-1)(X+1)(X^2+1).$$

The minimal polynomial $q_1(X)$ can be any divisor of this product. The characteristic polynomial $c(X) = q_1(X)q_2(X) \cdots q_k(X)$ can be any polynomial multiple of $q_1(X)$ which has degree 4 and satisfies $q_{i+1}(X) | q_i(X)$, all $j \geq 1$. It is now easy to list all the possibilities for $q_1(X)$ and $c(X)$, and the corresponding rational canonical forms.

1. If $q_1(X) = X-1$, then $q_2(X) = q_3(X) = q_4(X) = X-1$, $c(X) = (X-1)^4$.

$$\begin{bmatrix} 1 & 0 & 0 & 0 \\ \hline 0 & 1 & 0 & 0 \\ \hline 0 & 0 & 1 & 0 \\ \hline 0 & 0 & 0 & 1 \end{bmatrix}.$$

2. If $q_1(X) = X+1$, then $q_2(X) = q_3(X) = q_4(X) = X+1$, $c(X) = (X+1)^4$.

$$\begin{bmatrix} -1 & 0 & 0 & 0 \\ \hline 0 & -1 & 0 & 0 \\ \hline 0 & 0 & -1 & 0 \\ \hline 0 & 0 & 0 & -1 \end{bmatrix}.$$

3. If $q_1(X) = X^2+1$, then $q_2(X) = X^2+1$, $c(X) = (X^2+1)^2$.

$$\begin{bmatrix} 0 & -1 & 0 & 0 \\ 1 & 0 & 0 & 0 \\ \hline 0 & 0 & 0 & -1 \\ 0 & 0 & 1 & 0 \end{bmatrix}.$$

4. If $q_1(X) = (X+1)(X-1) = X^2-1$, then $q_2(X)$ may be $X+1$, in which case $q_3(X) = X+1$ and $c(X) = (X+1)^3(X-1)$; or $q_2(X)$ may be $X-1$, in

which case $q_3(X) = X - 1$ and $c(X) = (X+1)(X-1)^3$; or $q_2(X)$ may be $q_1(X)$, in which case $c(X) = (X+1)^2(X-1)^2$.

$$\left[\begin{array}{cc|cc} 0 & 1 & 0 & 0 \\ 1 & 0 & 0 & 0 \\ \hline 0 & 0 & -1 & 0 \\ 0 & 0 & 0 & -1 \end{array}\right] \text{ or } \left[\begin{array}{cc|cc} 0 & 1 & 0 & 0 \\ 1 & 0 & 0 & 0 \\ \hline 0 & 0 & 1 & 0 \\ 0 & 0 & 0 & 1 \end{array}\right] \text{ or } \left[\begin{array}{cc|cc} 0 & 1 & 0 & 0 \\ 1 & 0 & 0 & 0 \\ \hline 0 & 0 & 0 & 1 \\ 0 & 0 & 1 & 0 \end{array}\right].$$

5. If $q_1(X) = (X+1)(X^2+1) = X^3 + X^2 + X + 1$, then $q_2(X) = X + 1$, $c(X) = (X+1)^2(X^2+1)$.

$$\left[\begin{array}{ccc|c} 0 & 0 & -1 & 0 \\ 1 & 0 & -1 & 0 \\ 0 & 1 & -1 & 0 \\ \hline 0 & 0 & 0 & -1 \end{array}\right].$$

6. If $q_1(X) = (X-1)(X^2+1) = X^3 - X^2 + X - 1$, then $q_2(X) = X - 1$, $c(X) = (X-1)^2(X^2+1)$.

$$\left[\begin{array}{ccc|c} 0 & 0 & 1 & 0 \\ 1 & 0 & -1 & 0 \\ 0 & 1 & 1 & 0 \\ \hline 0 & 0 & 0 & 1 \end{array}\right].$$

7. If $q_1(X) = (X-1)(X+1)(X^2+1) = X^4 - 1$, then $c(X) = q_1(X)$.

$$\left[\begin{array}{cccc} 0 & 0 & 0 & 1 \\ 1 & 0 & 0 & 0 \\ 0 & 1 & 0 & 0 \\ 0 & 0 & 1 & 0 \end{array}\right].$$

A 4×4 matrix $\mathbf{A}$ over $\mathbb{R}$ satisfies $\mathbf{A}^4 = \mathbf{I}$, iff $\mathbf{A}$ is similar to one of the nine listed matrices. No two of the nine listed matrices are similar to one another. In particular, if $f \in \text{Hom}_\mathbb{R}(V, V)$, $\dim_\mathbb{R} V = 4$, and $f^4 = 1_V$, then f has one of the nine listed matrices in some basis.

More general methods are known for finding the rational canonical form $\mathbf{R}$ of a given square matrix $\mathbf{A}$, and a matrix $\mathbf{P}$ such that $\mathbf{R} = \mathbf{P}^{-1}\mathbf{A}\mathbf{P}$. See, for example, Browne (1940) and Section 7.5 of Booth (1967).

7.18 Exercises

*1. Show that the only ideals in a field F are $\{0\}$ and F.

2. Write the addition and multiplication tables for the field GF(3) and the quotient ring $\mathbb{Z}/3\mathbb{Z}$, where $3\mathbb{Z} = \{3z \mid z \in \mathbb{Z}\}$. Show how the tables correspond, resulting in the isomorphism of these two rings.

3. Let **A** be a fixed $n \times n$ matrix over the field F. Show that the set $\{f(\mathbf{A}) | f(X) \in F[X]\}$ is a commutative ring (a *subring* of $\text{Mat}_{n \times n}(F)$) under addition and multiplication of matrices.

*4. For the case $r > 1$, we say that the set of r polynomials $f_1(X)$, $f_2(X), \ldots, f_r(X) \in F[X]$ is **relatively prime**, and write $(f_1(X), f_2(X), \ldots, f_r(X)) = 1$, if no nonconstant polynomial divides all of them. Prove that if $\{f_1(X), f_2(X), \ldots, f_r(X)\}$ is relatively prime, then there are polynomials $a_1(X), a_2(X), \ldots, a_r(X)$ such that

$$a_1(X)f_1(X) + a_2(X)f_2(X) + \cdots + a_r(X)f_r(X) = 1.$$

[This generalizes Theorem 7.16.1(c).]

*5. (a) Show that if $f(X) \in F[X]$ is irreducible of degree m, then the field $F[X]/\mathcal{I}(f(X))$ is a vector space of dimension m over F, with basis $\{X^i + \mathcal{I}(f(X)) | 0 \leq i \leq m-1\}$.

(b) If $F = \text{GF}(3)$ and $f(X)$ is irreducible of degree 7, how many elements are there in $F[X]/\mathcal{I}(f(X))$?

6. Define $f \in \text{Hom}_\mathbb{R}(\mathbb{R}^3, \mathbb{R}^3)$ by $f((x, y, z)) = (3x - 2y - z, x + 2y + 4z, 7x - y)$, and find the f-annihilator of $v = (2, 1, -1)$. Also, find a matrix of f_W on $W = \langle v \rangle_f$.

*7. Show that if $f \in \text{Hom}_F(V, V)$, W is a subspace of V, and $W = \langle v_1 \rangle_f = \langle v_2 \rangle_f$ where v_1 has f-annihilator $p_1(X)$ and v_2 has f-annihilator $p_2(X)$, then $p_2(f)(v_1) = 0$ and $p_1(f)(v_2) = 0$. Conclude that $p_1(X) = p_2(X)$, so that W can only be an f-companion space of one polynomial.

*8. (a) Prove that

$$\det \begin{bmatrix} \mathbf{A} & \mathbf{0} \\ \mathbf{0} & \mathbf{B} \end{bmatrix} = (\det \mathbf{A})(\det \mathbf{B})$$

for any square matrices **A** and **B** over the field F.

(b) Extend part (a) to the direct sum of k square matrices $\mathbf{B}_1, \mathbf{B}_2, \ldots, \mathbf{B}_k$.

9. Denote $F = \text{GF}(5)$, $V = F^3$, and define $f \in \text{Hom}_F(V, V)$ by $f((a, b, c)) = (3b + 4c, a + 3c, 2c)$. Set $p(X) = X^2 + 2$, and describe the $p(f)^\infty$-kernel of V.

10. Assume that the matrix **M** over $\mathbb{R}$ has elementary divisors $(X-1)^2$, $X-1$, $X+1$, $X+1$, $(X^2+1)^2$. Find a matrix that is similar to **M** and in the form given by Corollary 7.17.2. Also, find the rational canonical form of **M**.

*11. Let **M** be a given $n \times n$ matrix over the field F. Many authors define the *minimal polynomial* of **M** to be the monic polynomial $r(X) \in F[X]$ of smallest degree with $r(\mathbf{M}) = \mathbf{0}$, and the *characteristic polynomial* of **M** to be $(-1)^n \det(\mathbf{M} - X\mathbf{I}) = c(X)$.

(a) Verify that our definitions agree with these.

(b) Prove that $r(X)$ divides $c(X)$. (This result is called the *Cayley–Hamilton theorem*.)

12. Find rational canonical forms for all 3×3 matrices **A** over $\text{GF}(3)$ which satisfy $\mathbf{A}^3 = \mathbf{A}^2$.

13. Find rational canonical forms for the following matrices over $\mathbb{Q}$:

$$\begin{bmatrix} 1 & 2 \\ 0 & 3 \end{bmatrix}, \quad \begin{bmatrix} 1 & 3 & 0 \\ 0 & 1 & 6 \\ 0 & 0 & 1 \end{bmatrix}, \quad \begin{bmatrix} 1 & 1 & 3 & 1 \\ 0 & 2 & 0 & -2 \\ 0 & 0 & 1 & -1 \\ 0 & 0 & 0 & 1 \end{bmatrix}.$$

Must they have the same rational canonical forms as matrices over $\mathbb{R}$? Why?

*14. In Section 7.12 we discuss only determinants of matrices with entries in a field F. However, in Lemma 7.17.6(c) we allow expressions in the variable X to be matrix entries, and $X \notin F$. We describe here two different ways in which this usage may be justified.

(a) We define the **quotient field** $F(X)$ of $F[X]$;

$$F(X) = \left\{ \frac{f(X)}{g(X)} \,\Big|\, f(X) \in F[X], 0 \neq g(X) \in F[X] \right\},$$

where addition and multiplication are as usual for fractions, and $f(X)/g(X) = f_1(X)/g_1(X)$ iff $f(X)g_1(X) = g(X)f_1(X)$. Prove that $F(X)$ is a field, and every polynomial $f(X) \in F[X]$ is in $F(X)$. [That is, $f(X) \to f(X)/1$ is an isomorphism of $F[X]$ onto a subring of $F(X)$.] The determinant computations then take place inside the *field* $F(X)$.

(b) Verify that Theorems 7.12.2 and 7.12.3 hold for matrices over any commutative ring with unity, say $F[X]$. [In Theorem 7.12.3(d), one must verify that if $\mathbf{A}$ is invertible, then det $\mathbf{A}$ is an invertible element of the ring.]

*15. Assume that $\mathbf{M}$ is an $n \times n$ matrix over F with characteristic polynomial $c(X)$. Prove that if $\lambda \in F$, then the following are equivalent:

(a) $c(\lambda) = 0$. (Then λ is a **characteristic root** or **eigenvalue** of $\mathbf{M}$.)
(b) $\det(\mathbf{M} - \lambda \mathbf{I}) = 0$.
(c) There is a nonzero $n \times 1$ column vector $\mathbf{Y} \in \mathrm{Mat}_{n \times 1}(F)$ such that $\mathbf{MY} = \lambda \mathbf{Y}$. (Then $\mathbf{Y}$ is a **characteristic vector** or **eigenvector** of $\mathbf{M}$. Eigenvalues and eigenvectors are of central importance in many applications of linear algebra, but they are not used in the later parts of this book.)

*16. State and prove results similar to Exercise 15 for a linear transformation $f \in \mathrm{Hom}_F(V, V)$.

*17. (a) Show that if the $n \times n$ matrix $\mathbf{M}$ is similar to a diagonal matrix $\mathbf{D}$, say $\mathbf{D} = \mathbf{P}^{-1}\mathbf{MP}$, then the columns of $\mathbf{P}$ are n linearly independent characteristic vectors of $\mathbf{M}$.

(HINT: Look at the columns of $\mathbf{PD} = \mathbf{MP}$.)

(b) Conversely, show that if $\mathbf{M}$ has n linearly independent characteristic vectors, then $\mathbf{M}$ is similar to a diagonal matrix.

(c) Using Corollary 7.17.2, show that an $n \times n$ matrix $\mathbf{M}$ is similar to a diagonal matrix iff its minimal polynomial has form $(X - \lambda_1)(X - \lambda_2) \cdots (X - \lambda_r)$, the λ_i distinct elements of F.

7.19 Field Extensions

Assume that K is a subfield of the field L. (That is, K is a subset of L which is itself a field under the operations of L.) Then we say that L is an **extension** (**field extension**) of K. L is then a vector space over K, with its own additive structure and with the scalar multiplication of elements of L by elements of K being the multiplication in L. If L is finite-dimensional over K, then $\dim_K L = [L:K]$ is the **degree** of L over K and L is a **finite extension** of K; otherwise, L is an **infinite extension** of K. For example, $[\mathbb{C}:\mathbb{R}] = 2$, while $\mathbb{R}$ is an infinite extension of $\mathbb{Q}$. (Reasons for these facts will become clear as we proceed.)

If S is a subset of the field L, then the intersection of all subfields of L containing S is the unique smallest subfield of L containing S. In the case when K is a subfield of L, $u \in L$, and $S = K \cup \{u\}$, we denote the smallest subfield containing S by $K(u)$. Note that $K(u) = K$ iff $u \in K$.

For example, let $K = \mathbb{Q}$, $L = \mathbb{R}$, $u = \sqrt{2}$; it is well known that $\sqrt{2} \notin \mathbb{Q}$. The field $\mathbb{Q}(\sqrt{2})$ is therefore larger than $\mathbb{Q}$. Denote $R = \{a + b\sqrt{2} \mid a, b \in \mathbb{Q}\}$; we claim that $R = \mathbb{Q}(\sqrt{2})$. If $a, b \in \mathbb{Q}$ then $a + b\sqrt{2}$ is in any field containing $\mathbb{Q}$ and $\sqrt{2}$, so at least $R \subseteq \mathbb{Q}(\sqrt{2})$, and it is enough to show that R is a field. At least R is a ring, because if $a, b, c, d \in \mathbb{Q}$ so $a + b\sqrt{2}, c + d\sqrt{2} \in R$, then

$$(a+b\sqrt{2}) + (c+d\sqrt{2}) = (a+c) + (b+d)\sqrt{2} \in R,$$
$$(a+b\sqrt{2})(c+d\sqrt{2}) = (ac+2bd) + (ad+bc)\sqrt{2} \in R,$$
$$-(a+b\sqrt{2}) = (-a) + (-b)\sqrt{2} \in R.$$

Finally, if $a + b\sqrt{2}$ is a nonzero element of R, then $a^2 - 2b^2 \neq 0$ (why?) and

$$\frac{a}{a^2-2b^2} - \frac{b}{a^2-2b^2}\sqrt{2} = (a+b\sqrt{2})^{-1} \quad \text{in} \quad R$$

(check this). Since R contains the inverses of its nonzero elements, R is therefore a field. The fact that $\mathbb{Q}(\sqrt{2}) = \{a + b\sqrt{2} \mid a, b \in \mathbb{Q}\}$ implies that $\{1, \sqrt{2}\}$ is a basis for $\mathbb{Q}(\sqrt{2})$ over $\mathbb{Q}$ (why?), so, in fact, $[\mathbb{Q}(\sqrt{2}):\mathbb{Q}] = 2$.

One of the following two cases must occur whenever $u \in L$ and K is a subfield of L.

CASE I. There is no nonzero $f(X) \in K[X]$ such that $f(u) = 0$. In this case we say that u is **transcendental** over K. Then $K(u)$ consists of all rational functions $f(u)/g(u)$, where $f(X), g(X) \in K[X]$ and $g(X) \neq 0$. In fact, $K(u)$ is isomorphic to the quotient field $K(X)$ of $K[X]$, discussed in Exercise 7.18.14(a). Field $K(u)$ is infinite-dimensional over K. Famous nontrivial nineteenth-century theorems state that if π is the ratio of a circle's circumference to its diameter and e is the base of natural logarithms, then the real numbers π and e are transcendental over $\mathbb{Q}$.

CASE II. There does exist a nonzero $f(X) \in K[X]$ such that $f(u) = 0$. Then we say that u is **algebraic** over K, and u is a **zero**, or a **root**, of $f(X)$. In

Sec. 7.19 *Field Extensions*

the previous example, if $f(X) = X^2 - 2$, then $f(\sqrt{2}) = 0$, so $\sqrt{2}$ is algebraic over $\mathbb{Q}$. [If in fact $u \in K$, take $f(X) = X - u$ to see that u is algebraic over K.] This case is described by the following theorem.

Theorem 7.19.1. *Let K be a field, u an element of a field of which K is a proper subfield, and assume that u is algebraic over K. Let $f(X)$ be a monic polynomial in $K[X]$ of least degree such that $f(u) = 0$, and assume that $f(X)$ has degree n. Then:*

(a) $f(X)$ *is unique.*

(b) $f(X)$ *is irreducible over K.* [$f(X)$ *is called the* **irreducible polynomial for u over K**, *and we write* $f(X) = \text{Irr}(u, K)$.]

(c) $1, u, u^2, \ldots, u^{n-1}$ *form a vector space basis of $K(u)$ over K. Hence* $[K(u):K] = n$.

(d) $g(X) \in K[X]$ *satisfies* $g(u) = 0$ *iff* $f(X)|g(X)$.

(e) $K(u)$ *is isomorphic to* $K[X]/\mathcal{I}(f(X))$, *via an isomorphism in which u and $X + \mathcal{I}(f(X))$ correspond, and κ and $\kappa + \mathcal{I}(f(X))$ correspond for all $\kappa \in K$.*

Proof. (a) If $f_0(X) \in K[X]$ is also monic of degree n and $f_0(u) = 0$, then $f_1(X) = f(X) - f_0(X)$ has degree $< n$ and $f_1(u) = 0$. Therefore, $f_1(X) = 0$, proving that $f_0(X) = f(X)$.

(b) If $f(X) = g(X)h(X)$ for $g(X)$ and $h(X)$ over K and of degree smaller than that of $f(X)$, then $0 = f(u) = g(u)h(u)$, so either $g(u) = 0$ or $h(u) = 0$, contradicting the minimality of the degree of $f(X)$.

(c) By minimality of the degree of $f(X)$, these powers of u are linearly independent over K. If T is the vector space spanned by them, the fact $f(u) = 0$ enables us to express higher powers of u in terms of lower powers; hence T is a ring. (The reader should supply the missing details.)

We shall show that any nonzero $t \in T$ has an inverse in T; this will show that T is a field, so $T = K(u)$. We have $t = h(u)$, some $h(X) \in K[X]$ of degree $< n$. By (b), $(h(X), f(X)) = 1$, so there are $a(X), b(X) \in F[X]$ such that $1 = a(X)h(X) + b(X)f(X)$. Thus $1 = a(u)h(u) + b(u) \cdot 0$, proving that $a(u)$ is the inverse of $t = h(u)$. Since T is a ring and $u \in T$, $a(u) \in T$, we are done.

(d) If $g(X)$ is not a multiple of $f(X)$, there are $c(X), d(X)$ such that $1 = c(X)g(X) + d(X)f(X)$; hence $1 = c(u)g(u) + d(u) \cdot 0$, proving that $g(u) \neq 0$.

(e) Define $\varphi: K[X] \to K(u)$ by $\varphi: g(X) \to g(u)$ for each $g(X)$ in $K[X]$. Then φ is a ring homomorphism (prove this) and by (d) has kernel $\mathcal{I}(f(X))$, so (e) follows from Theorem 7.15.1(b). ∎

For example, $f(X) = X^2 - 2$ is irreducible over $\mathbb{Q}$ and is the unique monic polynomial of lowest degree with $\sqrt{2}$ as root. The field $\mathbb{Q}(\sqrt{2})$ has basis $\{1, \sqrt{2}\}$ over $\mathbb{Q}$. Polynomial $g(X)$ satisfies $g(\sqrt{2}) = 0$ iff $f(X)|g(X)$. The field $\mathbb{Q}(X)/\mathcal{I}(X^2 - 2)$ consists of $\{a + bX | a, b \in \mathbb{Q}\}$, where higher powers of X are reduced using the identity $X^2 = 2$; it is isomorphic to $\mathbb{Q}(\sqrt{2}) = \{a + b\sqrt{2} | a, b \in \mathbb{Q}\}$, discussed before.

The field $K(u)$ is called the *field obtained by adjoining u to K*.

In the remainder of this section, we consider field extensions of finite degree. For any given field K and polynomial $f(X) \in K[X]$, we shall see that there is a smallest field L, unique up to isomorphism, such that $K \subseteq L$ and $f(X)$ factors into linear (degree 1) factors in L. For example, in the case $K = \mathbb{R}$, we have $L \subseteq \mathbb{C}$, since every real polynomial $f(X)$ of degree n can be factored into a product $a(X - r_1)(X - r_2) \cdots (X - r_n)$, where the $r_i \in \mathbb{C}$ [this is the "fundamental theorem of algebra," first proved by Gauss in 1799; see Gauss (1876)].

We begin our discussion of finite extensions with the following theorem.

Theorem 7.19.2. *If K is a field and $f(X) \in K[X]$ is irreducible over K, then there exists a field L containing K and a root of $f(X)$.*

Proof. Denote $\bar{L} = K[X]/\mathscr{I}(f(X))$; by Theorem 7.16.1(f), $\bar{L}$ is a field, consisting of cosets $g(X) + \mathscr{I}(f(X))$, $g(X) \in K[X]$. Define the function $\varphi : K \to \bar{L}$ by $\varphi(\alpha) = \alpha + \mathscr{I}(f(X))$, all $\alpha \in K$. If also $\beta \in K$, then

$$\varphi(\alpha + \beta) = (\alpha + \beta) + \mathscr{I}(f(X)) = (\alpha + \mathscr{I}(f(X))) + (\beta + \mathscr{I}(f(X)))$$
$$= \varphi(\alpha) + \varphi(\beta),$$
$$\varphi(\alpha\beta) = \alpha\beta + \mathscr{I}(f(X)) = (\alpha + \mathscr{I}(f(X)))(\beta + \mathscr{I}(f(X))) = \varphi(\alpha) \cdot \varphi(\beta),$$

so φ is a ring homomorphism. Since

$$\ker \varphi = \{\alpha \in K | \alpha + \mathscr{I}(f(X)) = 0 \text{ in } \bar{L}\} = \{\alpha \in K | \alpha \in \mathscr{I}(f(X))\} = \{0\},$$

φ is one-one. This means that φ is an isomorphism from K onto the subring $\bar{K} = \{\alpha + \mathscr{I}(f(X)) | \alpha \in K\}$ of $\bar{L}$. For any nonzero $\alpha \in K$,

$$(\alpha + \mathscr{I}(f(X)))(\alpha^{-1} + \mathscr{I}(f(X))) = \alpha\alpha^{-1} + \mathscr{I}(f(X)) = 1 + \mathscr{I}(f(X)),$$

so $\bar{K}$ contains the inverses of its nonzero elements. Therefore, $\bar{K}$ is a subfield of $\bar{L}$ which is isomorphic to K. Our theorem asserts more; that L contains K rather than an isomorphic copy of K. We can satisfy this precise requirement by replacing each element $\alpha + \mathscr{I}(f(X))$ of $\bar{K}$ by $\alpha \in K$, leaving the operations $+$ and $\cdot$ unchanged, and calling the new field L instead of $\bar{L}$.

The element $u = X + \mathscr{I}(f(X))$ of $\bar{L}$ satisfies $f(u) = f(X + \mathscr{I}(f(X))) = f(X) + \mathscr{I}(f(X)) = 0$ in $\bar{L}$, so u is a root of $f(X)$. ∎

For example, let $K = \mathbb{Q}$, $f(X) = X^2 + X + 1$. The well-known quadratic formula shows that $f(X)$ has no rational roots, so $f(X)$ has no linear factors and must be irreducible over $\mathbb{Q}$. Then $\bar{L} = \mathbb{Q}[X]/\mathscr{I}(X^2 + X + 1)$ is a field. Its elements are really cosets $g(X) + \mathscr{I}(X^2 + X + 1)$. We can assume that $g(X)$ has degree ≤ 1, because otherwise we can divide $g(X)$ by $X^2 + X + 1$, obtaining

$$g(X) = (X^2 + X + 1)q(X) + r(X), \quad \deg r(X) \leq 1$$

Sec. 7.19   *Field Extensions*   363

and, in $\bar{L}$,

$$g(X)+\mathscr{I}(X^2+X+1)=q(X)(X^2+X+1)+r(X)+\mathscr{I}(X^2+X+1)$$
$$=r(X)+\mathscr{I}(X^2+X+1).$$

We can therefore replace $g(X)$ by $r(X)$.

Given elements $\tilde{a}=a_1X+a_0+\mathscr{I}(X^2+X+1)\in\bar{L}$ and $\tilde{b}=b_1X+b_0+\mathscr{I}(X^2+X+1)\in\bar{L}$, addition and multiplication satisfy

$$\tilde{a}+\tilde{b}=(a_1+b_1)X+(a_0+b_0)+\mathscr{I}(X^2+X+1)\in\bar{L},$$
$$\tilde{a}\cdot\tilde{b}=(a_1X+a_0)(b_1X+b_0)+\mathscr{I}(X^2+X+1)$$
$$=a_1b_1X^2+(a_1b_0+a_0b_1)X+a_0b_0+\mathscr{I}(X^2+X+1)$$
$$=a_1b_1(X^2+X+1-X-1)+(a_1b_0+a_0b_1)X+a_0b_0+\mathscr{I}(X^2+X+1)$$
$$=a_1b_1(-X-1)+(a_1b_0+a_0b_1)X+a_0b_0+a_1b_1(X^2+X+1)$$
$$+\mathscr{I}(X^2+X+1)$$
$$=(a_1b_0+a_0b_1-a_1b_1)X+(a_0b_0-a_1b_1)+\mathscr{I}(X^2+X+1).$$

We say that multiplication proceeds "modulo X^2+X+1"; whenever X^2 appears, we can replace it by $-X-1$. When we do arithmetic in $\bar{L}$, we customarily write only the terms a_1X+a_0, identify $\bar{L}$ with the set $\{a_1X+a_0|a_1, a_0\in\mathbb{Q}\}$, do addition as usual for polynomials, and do multiplication by replacing X^2 by $-X-1$ wherever it appears. In particular,

$$(7X+5)(4X-3)=28X^2-X-15=28(-X-1)-X-15=-29X-43$$

in $\mathbb{Q}[X]/\mathscr{I}(X^2+X+1)$.

By the quadratic formula, $(-1-i\sqrt{3})/2$ is a root of X^2+X+1 in $\mathbb{C}$, so, by Theorem 7.19.1(e), $\bar{L}$ is isomorphic to $\mathbb{Q}((-1-i\sqrt{3})/2)$.

Theorem 7.19.3. *If $f(X)\in K[X]$ is irreducible over K and u and v are roots of $f(X)$ in fields containing K, then $K(u)$ and $K(v)$ are isomorphic via an isomorphism which fixes all elements of K and sends u to v.*

Proof. By Theorem 7.19.1(e) there are isomorphisms

$$\varphi: K(u)\to K[X]/\mathscr{I}(f(X))\quad\text{and}\quad\psi: K[X]/\mathscr{I}(f(X))\to K(v)$$

such that

$$u\stackrel{\varphi}{\to}X+\mathscr{I}(f(X))\stackrel{\psi}{\to}v\quad\text{and}\quad\kappa\to\kappa+\mathscr{I}(f(X))\to\kappa,\quad\text{all }\kappa\in K;$$

therefore, $\psi\circ\varphi$ is the desired isomorphism. ∎

For example, consider again the case $F=\mathbb{Q}$, $f(X)=X^2+X+1$. Then $u=(-1+i\sqrt{3})/2$ and $v=(-1-i\sqrt{3})/2$ are the two roots of $f(X)$ in $\mathbb{C}$.

Since $u+v=-1$, the subfields $\mathbb{Q}(u)$ and $\mathbb{Q}(v) = \mathbb{Q}(-1-u)$ of $\mathbb{C}$ are identical. The isomorphism $\psi \circ \varphi$ of the theorem is an *automorphism*, an isomorphism of $\mathbb{Q}(u)$ onto itself.

Theorems 7.19.2 and 7.19.3 are existence and uniqueness theorems for the smallest field containing a given field K and *one* root of a given *irreducible* polynomial over K. Theorems 7.19.4 and 7.19.5, to follow, are similar theorems for the smallest field containing K and *all* roots of *any* given polynomial over K.

For any finite set $u_1, \ldots, u_r$ of elements in an extension field M of K, $K(u_1, \ldots, u_r) = K(u_1)(u_2) \cdots (u_r)$ is the smallest subfield of M containing $K \cup \{u_1, \ldots, u_r\}$ (Exercise 7.22.8). For any $f(X) \in K[X]$, a field extension L of K is a **splitting field** of $f(X)$ over K if $f(X)$ factors into linear factors in L and $L = K(u_1, \ldots, u_r)$, $u_1, \ldots, u_r$ the roots of $f(X)$ in L.

Theorem 7.19.4. *If $f(X) \in K[X]$, then there exists a splitting field L of $f(X)$ over K.*

Proof. We use induction on the degree of $f(X)$. The result is certainly true if $f(X)$ has degree 1; assume that it is true for all degrees $\leq r$ and $f(X)$ has degree $r+1$. The theorem is true with $L = K$ if $f(X)$ factors into linear factors over K, so we may assume that $f(X)$ has an irreducible factor $g(X) \in K[X]$ of degree >1. By Theorem 7.19.2 we can construct $L_1 = K(u)$, u a root of $g(X)$. The fact that $f(u) = 0$ implies that $(X-u) | f(X)$, so $f(X) = (X-u)h(X)$, some $h(X) \in L_1[X]$ of degree r. By the induction hypothesis, we can find a splitting field L of $h(X)$ over L_1 (L may well be L_1). Clearly, $L = K(u, \text{roots of } h(X))$, so L is also a splitting field of $f(X)$ over K. ∎

As an example, consider $K = \mathbb{Q}, f(X) = X^3 - 2$. The number 2 has a real, irrational cube root $u = \sqrt[3]{2}$, a zero of $f(X)$, so $X - u$ divides $X^3 - 2$;

$$X^3 - 2 = X^3 - u^3 = (X-u)(X^2 + uX + u^2).$$

The quadratic formula shows that the other two roots of $X^3 - 2$ are

$$v = \frac{-1+i\sqrt{3}}{2}u \quad \text{and} \quad w = \frac{-1-i\sqrt{3}}{2}u,$$

which are not real, not in $L_1 = \mathbb{Q}(u)$. Therefore, $X^2 + uX + u^2$ has no linear factors, and must be irreducible, in $L_1[X]$. We have

$$v + w = -u, \quad v = -u - w, \quad w = -v - u,$$

implying that $v \in L_1(w)$ and $w \in L_1(v)$, so $L_1(v) = L_1(w) = L$ is a splitting field of $X^3 - 2$ over $\mathbb{Q}$. By Theorem 7.19.1(c), $[L_1 : \mathbb{Q}] = 3$ and $[L : L_1] = 2$; we see in Exercise 7.22.2 that this implies that $[L : \mathbb{Q}] = 6$.

For another example, consider $K = GF(2)$, $f(X) = X^3 + X + 1$. We have $f(0) = f(1) = 1 \neq 0$ so $f(X)$ has no roots in $GF(2)$; $f(X)$ therefore has no linear factors, and must be irreducible, in $K[X]$. The field $L_1 = K(u)$, u a zero of $f(X)$, therefore satisfies $[L_1:K] = 3$; $u \in L_1$ satisfies $0 = f(u) = u^3 + u + 1$, $u^3 = u + 1$. [Remember that $2 = 0$ in $GF(2)$ or any field containing $GF(2)$, so $-w = -w + 2w = w$, all $w \in L_1$.] Now $X - u = X + u$ must divide $f(X)$; we have

$$X^3 + X + 1 = (X + u)(X^2 + uX + (u^2 + 1)).$$

We find that $h(X) = X^2 + uX + (u^2 + 1)$ satisfies

$$h(u^2) = u^4 + uu^2 + u^2 + 1 = u(u+1) + (u+1) + u^2 + 1 = 2u^2 + 2u + 2 = 0,$$

so $u^2 \in L_1$ is a zero of $h(X)$. Division of $h(X)$ by $X - u^2 = X + u^2$ yields

$$X^2 + uX + (u^2 + 1) = (X + u^2)(X + (u^2 + u)).$$

Therefore, $u^2 + u = u(u+1) = uu^3 = u^4$ is the last root of $f(X)$, and indeed

$$f(X) = X^3 + X + 1 = (X + u)(X + u^2)(X + u^4).$$

We have seen that $f(X)$ factors into linear factors in $L_1[X]$, so $L = L_1 = K(u)$ is a splitting field of $X^3 + X + 1$ over $K = GF(2)$, with $[L:K] = 3$. In the next few sections we will further study the structure of finite fields such as $K(u)$, and will in particular find that the pattern of roots u, u^2, and u^4 of $f(X)$ is no accident.

Theorem 7.19.5. *If $f(X) \in K[X]$ and M and M_0 are both splitting fields of $f(X)$ over K, then M and M_0 are isomorphic via an isomorphism which fixes all elements of K.*

We had an example of this: $\mathbb{Q}((-1+i\sqrt{3})/2)$ and $\mathbb{Q}((-1-i\sqrt{3})/2)$ are splitting fields for $X^2 + X + 1 \in \mathbb{Q}[X]$, and are in fact automorphic representations of the same field.

Theorem 7.19.5 is the special case $K = K_0$, φ = identity, of the following theorem (which is easier to prove, using induction).

Theorem 7.19.6. *Let K and K_0 be fields, φ an isomorphism of K onto K_0. For a given $f(X) \in K[X]$, let $f_0(X) \in K_0[X]$ be the polynomial obtained by applying φ to the coefficients of $f(X)$. Let M be a splitting field of $f(X)$ over K, M_0 a splitting field of $f_0(X)$ over K_0. Then φ extends to an isomorphism of M onto M_0.*

Proof. As in the proof of Theorem 7.19.4, we use induction on $\deg f(X)$. The theorem certainly holds if $\deg f(X) = 1$ or if $f(X)$ is a product of linear factors over K, for then $M = K$ and $M_0 = K_0$ and the extension of φ is trivial. Assume that the result holds for any $f(X)$ with $\deg f(X) \leq r$, and now let $f(X) \in K[X]$ have degree $r + 1$. We may assume $f(X)$ has an irreducible factor $g(X)$ of degree >1. Let $g_0(X) \in K_0[X]$ be obtained by applying φ to the coefficients of $g(X)$. Let u be one root of $g(X)$ in M, u_0 one root of $g_0(X)$ in M_0. There is a ring isomorphism $\varphi^*: K[X] \to K_0[X]$, obtained by applying φ to the coefficients of polynomials; in particular, $\varphi^*(f(X)) = f_0(X)$, $\varphi^*(g(X)) = g_0(X)$. In turn, φ^* induces an isomorphism

$$\bar\varphi: K[X]/\mathcal{I}(g(X)) \to K_0[X]/\mathcal{I}(g_0(X)),$$

since $\varphi^*(g(X)) = g_0(X)$. By Theorem 7.19.1(e) there are isomorphisms ψ and τ,

$$K(u) \stackrel{\psi}{\to} K[X]/\mathcal{I}(g(X)) \stackrel{\bar\varphi}{\to} K_0[X]/\mathcal{I}(g_0(X)) \stackrel{\tau}{\to} K_0(u_0),$$

whose composite $\varphi_1: K(u) \to K_0(u_0)$ agrees with φ on K. Certainly, M is a splitting field of $f(X)$ over $K(u)$ and M_0 is a splitting field of $f_0(X)$ over $K_0(u_0)$ [see Exercise 7.22.8(b)]. $[M:K(u)] < [M:K]$ (why?), so by the induction hypothesis φ_1 extends to an isomorphism $\varphi_2: M \to M_0$. ∎

For example, consider again the case $K = \mathbb{Q}, f(X) = X^3 - 2$. We constructed a splitting field $M = L$ of $f(X)$ over $\mathbb{Q}$ by first choosing the real cube root $u = \sqrt[3]{2}$, forming $L_1 = \mathbb{Q}(u)$ with $[L_1:\mathbb{Q}] = 3$, and then enlarging L_1 to M, $[M:L_1] = 2$. This was easy because we knew that the two nonreal zeros of $f(X)$ could not be in $\mathbb{Q}(u) \subseteq \mathbb{R}$. If we had, instead, started with a nonreal root v and formed $\mathbb{Q}(v) = L_{10}, [L_{10}:\mathbb{Q}] = 3$, we would have had difficulty in determining whether a splitting field M_0 of $f(X)$ over $\mathbb{Q}$, $M_0 \supseteq L_{10}$, was indeed strictly larger than L_{10}. But Theorem 7.19.5 implies that M and M_0 are isomorphic; in particular, $[M_0:\mathbb{Q}] = [M:\mathbb{Q}] = 6$, $M_0 \neq L_{10}$.

7.20 Finite Fields

If K is a field, $\alpha \in K$, and $n \in \mathbb{P}$, we denote

$$n\alpha = \alpha + \alpha + \cdots + \alpha \quad \text{to } n \text{ terms.}$$

We also denote $0\alpha = 0$, all $\alpha \in K$, and $(-n)\alpha = n(-\alpha)$ if $n \in \mathbb{P}$. The elements $n1$ (often denoted simply by n) are the **field integers**. For example, in the field $\mathbb{R}$ the field integers are all the ordinary integers $\mathbb{Z}$. If $k, l \in \mathbb{Z}$, then

(7.20.1) $\qquad k1 + l1 = (k+l)1 \quad \text{and} \quad (k1)(l1) = (kl)1,$

so the field integers form a subring in any field.

If $1 \in$ a field K and there is an integer $n \in \mathbb{P}$ such that $n1 = 0$, let p be the smallest such. Then $p = kl, k, l \in \mathbb{P}, k < p, l < p$, would by (7.20.1) imply that $(k1)(l1) = 0$, $k1 = 0$ or $l1 = 0$, contradicting minimality of p. Therefore, p must be a prime; p is called the **characteristic** of the field K, and we denote $p = \text{char } K$. For any $\alpha \in K$, $p = \text{char } K$ implies that

$$p\alpha = \underbrace{\alpha + \cdots + \alpha}_{p\ \alpha\text{'s}} = \alpha(\underbrace{1 + \cdots + 1}_{p\ 1\text{'s}}) = \alpha(p1) = \alpha \cdot 0 = 0.$$

If no such p exists, we say that K has **characteristic zero** and we write char $K = 0$. For example, char $\mathbb{Q}$ = char $\mathbb{R}$ = char $\mathbb{C} = 0$.

A **finite field** is a field that contains finitely many elements. We already know that for each prime p there exists a field $GF(p) = \{0, 1, \ldots, p-1\}$ of p elements. The operations are addition and multiplication modulo p. The field $GF(p)$ consists entirely of field integers, and char $GF(p) = p$.

Lemma 7.20.1. *If K is a finite field, then for some prime p, $GF(p)$ is a subfield of K. We have $|K| = p^n$ for some $n \in \mathbb{P}$.*

Proof. Since K is finite, the field integers $m1$, $m \in \mathbb{P}$, cannot all be different; $l1 = m1$, some $0 < l < m$. Adding $-1 \, l$ times to both sides, we get $0 = (m-l)1$. Therefore, char $K = p \neq 0$ for a prime p. The set of field integers $\{0, 1, \ldots, p-1\}$ forms a subfield of K isomorphic to $GF(p)$, since they are added and multiplied modulo $p = p1$. Then for some $n \in \mathbb{P}$, K is an n-dimensional vector space over $GF(p)$ (prove this), and if $\{u_1, \ldots, u_n\}$ is a basis, then

$$K = \{x_1 u_1 + \cdots + x_n u_n | \text{all } x_i \in GF(p)\}.$$

There are p choices for each x_i and hence p^n choices in all, proving that $|K| = p^n$. ∎

If $f(X) \in K[X]$, say $f(X) = \alpha_m X^m + \cdots + \alpha_2 X^2 + \alpha_1 X + \alpha_0$, we define the **derivative** $f'(X)$ of $f(X)$ by $f'(X) = m\alpha_m X^{m-1} + \cdots + 2\alpha_2 X + \alpha_1$. The differentiation rules

(7.20.2) $(f(X)g(X))' = f(X)g'(X) + f'(X)g(X)$, $(f(X)^n)' = nf(X)^{n-1}f'(X)$

are easily proved (Exercise 7.22.11).

Lemma 7.20.2. *If K is a field of characteristic $p \neq 0$, $x, y \in K$, and $n \in \mathbb{P}$, then*

(7.20.3) $$(x+y)^{p^n} = x^{p^n} + y^{p^n}.$$

Proof. By the usual binomial expansion with binomial coefficients $\binom{p^n}{i}$, we have

(7.20.4) $$(x+y)^{p^n} = \sum_{i=0}^{p^n} \binom{p^n}{i} x^{p^n - i} y^i.$$

But

$$\binom{p^n}{i} = \frac{p^n(p^n - 1) \cdots (p^n - p) \cdots (p^n - 2p) \cdots (p^n - i + 1)}{(i)(1) \cdots (p) \cdots (2p) \cdots (i-1)}$$

is divisible by p for $0 < i < p^n$, so $\binom{p^n}{i} = 0$ in K, and (7.20.4) is simply

$$(x+y)^{p^n} = x^{p^n} + 0 + \cdots + 0 + y^{p^n} = x^{p^n} + y^{p^n}. \quad \blacksquare$$

Lemma 7.20.3. *If $X^t - 1$ has a double root α in a field K of characteristic p, then $p | t$.*

Proof. We have $X^t - 1 = (X - \alpha)^2 h(X)$ for some $h(X) \in K[X]$, so

$$tX^{t-1} = (X^t - 1)' = 2(X - \alpha)h(X) + (X - \alpha)^2 h'(X).$$

Setting $X = \alpha$, we get $t\alpha^{t-1} = 0$, forcing $t = 0$ in K; that is, $p | t$. ∎

Theorem 7.20.1. *A field K has p^n elements (p a prime) iff it is a splitting field of $X^{p^n} - X$ over $GF(p)$.*

Proof. Suppose that $|K| = p^n$. The multiplicative group of nonzero elements in K has order $p^n - 1$, so these elements all satisfy $X^{p^n - 1} = 1$ (Corollary 4.12.1). Therefore,

$$X^{p^n} - X = X(X^{p^n - 1} - 1)$$

has p^n distinct roots in K; they are all of K, so K is a splitting field of $X^{p^n} - X$ over $GF(p)$.

Conversely, suppose that K is a splitting field of $f(X) = X^{p^n} - X$ over $GF(p)$. If $\alpha \in K$ is a double root of $f(X) = X(X^{p^n - 1} - 1)$, then α is a double root of $X^{p^n - 1} - 1$, a contradiction to Lemma 7.20.3. Therefore, $f(X)$ has p^n distinct roots in K. Let $S = \{u \in K | f(u) = 0\}$, so $|S| = p^n$, $S \subseteq K$. If $u, v \in S$, then $(uv)^{p^n} = u^{p^n} v^{p^n} = uv$, so $uv \in S$, and $(u + v)^{p^n} = u^{p^n} + v^{p^n} = u + v \in S$ by Lemma 7.20.2. Since K is finite-dimensional over $GF(p)$, K is finite. Therefore, Theorem 4.4.3 implies that S is an additive subgroup of K and $S - \{0\}$ is a multiplicative subgroup of $K - \{0\}$. Thus S is a subfield of K which is a splitting field of $X^{p^n} - X$, forcing $S = GF(p)(\text{roots of } f) = K$. ∎

For example, we have seen that with $F = GF(2)$, $n = 3$, a splitting field K of $X^3 + X + 1$ over F satisfies $[K : F] = 3$, implying that $|K| = 2^3 = 8$. By Theorem 7.20.1, K is also a splitting field of $f(X) = X^8 - X$ over F; in fact, all elements γ of K satisfy $0 = f(\gamma) = \gamma^8 - \gamma$. In particular, any root u of $X^3 + X + 1$ satisfies $u^8 - u = 0$, $u^8 = u$, $u^7 = 1$. Factorization shows that

$$X^8 - X = X^8 + X = X(X + 1)(X^3 + X + 1)(X^3 + X^2 + 1).$$

The factor $X(X + 1)$ has zeros 0 and 1, and we have seen that u, u^2, and u^4 are zeros of $X^3 + X + 1$. The reader can verify, using the relation $u^3 = u + 1$, that $X^3 + X^2 + 1$ has zeros u^3, u^5, and u^6. Thus $K = \{0, u, u^2, \ldots, u^6, u^7 = 1\}$, the set of 8 distinct zeros of $X^8 - X$; we will see that any finite field consists of 0 and the powers of a "primitive" element u.

Corollary 7.20.1. *For any power p^n of a prime p there exists a field $GF(p^n)$ of p^n elements. For given p and n, any two such fields are isomorphic. All elements of $GF(p^n)$ are roots of $X^{p^n} - X$, of which $GF(p^n)$ is a splitting field.*

Proof. Existence and uniqueness follow from Theorems 7.19.4, 7.20.1, and 7.19.5. The last sentence also follows from Theorem 7.20.1. ∎

We shall use the following theorem from group theory to show that the multiplicative group in $GF(p^n)$ is cyclic.

Theorem 7.20.2. *Let G be a finite abelian multiplicative group such that for any positive integer m, $x^m = 1$ has at most m solutions in G. Then G is cyclic.*

Proof. Let $x \in G$ have maximal order $n \leq |G|$. If every $y \in G$ has order dividing n, then $y^n = 1$ for all $y \in G$. By the hypothesis, $y^n = 1$ has at most n solutions; so $n \geq |G|$, $n = |G|$, $G = \langle x \rangle$.

So assume some $y \in G$ has order k, $k \nmid n$. Then we can choose a prime p such that $k = p^t k_0$, $n = p^u n_0$, $p \nmid k_0$, $p \nmid n_0$, $t > u \geq 0$. Set $x_0 = x^{p^u}$, of order n_0, and $y_0 = y^{k_0}$, of order p^t. The fact $(n_0, p^t) = 1$ implies that $\langle x_0 \rangle \cap \langle y_0 \rangle = \{1\}$. If $(x_0 y_0)^l = 1$, then $x_0^{-l} = y_0^l \in \langle x_0 \rangle \cap \langle y_0 \rangle = \{1\}$; $x_0^l = 1$ implies that $n_0 | l$ and $y_0^l = 1$ implies that $p^t | l$, so $p^t n_0 | l$. Therefore, $x_0 y_0$ has order $\geq p^t n_0 > p^u n_0 =$ order of x, contradicting the choice of x. ∎ (Compare Exercise 4.20.14.)

Corollary 7.20.2. *Any finite multiplicative subgroup of a field is cyclic; in particular, the multiplicative group of a finite field is cyclic.*

Proof. A polynomial $f(X)$ of degree m can have at most m distinct roots $\alpha_1, \ldots, \alpha_m$ in a field, because $(X - \alpha_1) \cdots (X - \alpha_m)$ must divide $f(X)$. Apply this to $f(X) = X^m - 1$ in Theorem 7.20.2. ∎

7.21 Computation in Finite Fields

In this section, p is a fixed prime, $F = GF(p)$, $K = GF(p^n)$. A **primitive element** in K is a generator of the (cyclic) multiplicative group of nonzero elements of K. By Corollary 7.20.2 and Theorem 4.16.1, K contains exactly $\varphi(p^n - 1)$ primitive elements. If $\gamma \in K$ is primitive, $f(X) = \text{Irr}(\gamma, F)$ is called a **primitive polynomial**. For example, we have seen that in $K_1 = GF(2^3)$, every nonzero element is a power of a given root u of $X^3 + X + 1$, so this polynomial is primitive.

As a second example, consider $K_2 = GF(3^2)$, which must be a splitting field of $X^9 - X$ over $GF(3)$. Its nonzero elements form a cyclic multiplicative group of order 8, so there are exactly $4 = \varphi(8)$ primitive elements (Theorem 4.16.1). When we factor $X^9 - X$, we find the incomplete factorization

$$X^9 - X = X(X^8 - 1) = X(X^4 - 1)(X^4 + 1).$$

Zeros of $X^4 - 1$ have order dividing 4 and cannot be primitive; the four primitive elements $\gamma_1, \gamma_2, \gamma_3$, and γ_4 must be the four zeros of $X^4 + 1$. Each γ_i lies in $K_2 = GF(3^2)$ and $[K_2 : GF(3)] = 2$, so each $\text{Irr}(\gamma_i, GF(3))$ has degree 2 by Theorem 7.19.1(c). Therefore, $X^4 + 1$ must be the product of two primitive factors of degree 2. Inspection of the expression

$$X^4 + 1 = (X^2 + aX + b)(X^2 + cX + d), \qquad a, b, c, d \in GF(3),$$

readily yields the factorization
$$X^4 + 1 = (X^2 + X + 2)(X^2 + 2X + 2),$$
so the latter two factors are the two primitive polynomials of degree 2 over GF(3).

Theorem 7.21.1. *There are exactly $(1/n)\varphi(p^n - 1)$ primitive polynomials of degree n over* GF(p), *where φ is the Euler φ-function of* (4.16.1). *Each such polynomial has n distinct zeros, which are primitive elements in* GF(p^n).

Proof. Let γ be one primitive element in $K = \text{GF}(p^n)$, and denote $F = \text{GF}(p)$, $f(X) = \text{Irr}(\gamma, F)$, $g(X) = X^{p^n} - X$. Since $g(\gamma) = 0$ by Corollary 7.20.1, it follows from Theorem 7.19.1(d) that $f(X)$ divides $g(X)$. Because $g(X)$ factors into distinct linear factors over K, so does $f(X)$; all roots of $f(X)$ are in K. Since γ is primitive, $F(\gamma) = K$. Since $[K:F] = n$, Theorem 7.19.1(c) implies that $f(X)$ has degree n. If γ' is another root of $f(X)$, then the uniqueness Theorem 7.19.3 implies that γ' is also primitive. We remarked at the beginning of this section that K contains $\varphi(p^n - 1)$ primitive elements; since each such $f(X)$ has n roots, the number of primitive polynomials of degree n over F is $(1/n)\varphi(p^n - 1)$. ∎

For given p and n, there is no easy way known to produce a primitive polynomial of degree n over GF(p); however, extensive tables of primitive polynomials have been computed, especially in case $p = 2$ (the case occurring most often in applications). Some primitive polynomials are listed in Table 7.21.1; more extensive tables may be found in Peterson and Weldon (1972) and Gill (1966).

Whenever $\alpha \in K$ has Irr (α, F) of degree n, then $K = F(\alpha)$, and Theorem 7.19.1(c) guarantees that every element of K has a unique "vector" expression $x_0 \cdot 1 + x_1 \cdot \alpha + \cdots + x_{n-1} \cdot \alpha^{n-1}$, where the $x_i \in \text{GF}(p)$. It is easy to add and subtract such expressions; just work componentwise modulo p. If γ is primitive, then, in addition to the unique vector representation, each nonzero element of K has a unique "exponent" expression $\gamma^i, 0 \leq i \leq p^n - 2$; it is easy to multiply and divide such expressions by adding or subtracting exponents. We can do all computations in $K = \text{GF}(p^n)$ easily if we have conversion tables between the "vector" and "exponent" expressions of elements. By definition, a logarithm is an exponent, so the table converting "vector" to "exponent" expressions is a *table of logarithms*, and the table converting "exponent" to "vector" expressions is a *table of antilogarithms*. Construction of such tables requires knowledge of Irr (γ, F), perhaps from Table 7.21.1.

EXAMPLE 1. Let γ be a root in $K = \text{GF}(3^2)$ of the primitive polynomial $X^2 + X + 2$ over $F = \text{GF}(3)$. The vector expressions of the $9 = 3^2$ elements of K are

$$0, \quad 1, \quad 2, \quad \gamma, \quad \gamma+1, \quad \gamma+2, \quad 2\gamma, \quad 2\gamma+1, \quad 2\gamma+2.$$

Sec. 7.21 Computation in Finite Fields

TABLE 7.21.1

A. Some primitive polynomials over GF(2)

$X^2 + X + 1$
$X^3 + X + 1$
$X^4 + X + 1$
$X^5 + X^2 + 1$
$X^6 + X + 1$
$X^7 + X^3 + 1$
$X^8 + X^4 + X^3 + X^2 + 1$
$X^9 + X^4 + 1$
$X^{10} + X^3 + 1$
$X^{11} + X^2 + 1$
$X^{12} + X^6 + X^4 + X + 1$
$X^{13} + X^4 + X^3 + X + 1$
$X^{14} + X^{10} + X^6 + X + 1$

$X^{15} + X + 1$
$X^{16} + X^{12} + X^3 + X + 1$
$X^{17} + X^3 + 1$
$X^{18} + X^7 + 1$
$X^{19} + X^5 + X^2 + X + 1$
$X^{20} + X^3 + 1$
$X^{21} + X^2 + 1$
$X^{22} + X + 1$
$X^{23} + X^5 + 1$
$X^{24} + X^7 + X^2 + X + 1$
$X^{60} + X + 1$
$X^{100} + X^8 + X^7 + X^2 + 1$

B. Other primitive polynomials

$X^2 + X + 2$ over GF(3)
$X^3 + 2X + 1$ over GF(3)
$X^4 + X + 2$ over GF(3)
$X^5 + 2X + 1$ over GF(3)
$X^6 + X + 2$ over GF(3)
$X^7 + X^2 + 2X + 1$ over GF(3)

$X^2 + X + 2$ over GF(5)
$X^3 + 3X + 2$ over GF(5)
$X^4 + X^2 + 2X + 2$ over GF(5)
$X^5 + 4X + 2$ over GF(5)
$X^2 + X + 3$ over GF(7)
$X^3 + 3X + 2$ over GF(7)
$X^5 + X^2 + 3X + 5$ over GF(7)

Since $\gamma^2 + \gamma + 2 = 0$, we have $\gamma^2 = -\gamma - 2 = 2\gamma + 1$. Also,
$$\gamma^3 = \gamma \cdot \gamma^2 = \gamma(2\gamma + 1) = 2\gamma^2 + \gamma = 2(2\gamma + 1) + \gamma = \gamma + 2 + \gamma = 2\gamma + 2.$$

Continuing in this manner, we can complete Table 7.21.2, describing $GF(3^2)$.

TABLE 7.21.2. $GF(3^2)$, using primitive polynomial $X^2 + X + 2$.

I. Logarithms

$\gamma = \gamma^1$ $2\gamma = \gamma^5$
$1 = \gamma^0 = \gamma^8$ $\gamma + 1 = \gamma^7$ $2\gamma + 1 = \gamma^2$
$2 = \gamma^4$ $\gamma + 2 = \gamma^6$ $2\gamma + 2 = \gamma^3$

II. Antilogarithms

$\gamma^0 = 1$ $\gamma^2 = 2\gamma + 1$ $\gamma^5 = 2\gamma$
$\gamma^1 = \gamma$ $\gamma^3 = 2\gamma + 2$ $\gamma^6 = \gamma + 2$
  $\gamma^4 = 2$ $\gamma^7 = \gamma + 1$

EXAMPLE 2. The polynomial $X^4 + X + 1$ is primitive over GF(2); with γ a root, Table 7.21.3 describes $GF(2^4)$, except for the element 0, of course. For further work with such tables, see Exercise 7.22.13.

TABLE 7.21.3. GF(2^4), using primitive polynomial $X^4 + X + 1$.

I. Logarithms

$1 = \gamma^0$
$\gamma = \gamma^1$
$\gamma + 1 = \gamma^4$
$\gamma^2 = \gamma^2$
$\gamma^2 + 1 = \gamma^8$
$\gamma^2 + \gamma = \gamma^5$
$\gamma^2 + \gamma + 1 = \gamma^{10}$
$\gamma^3 = \gamma^3$
$\gamma^3 + 1 = \gamma^{14}$
$\gamma^3 + \gamma = \gamma^9$
$\gamma^3 + \gamma + 1 = \gamma^7$
$\gamma^3 + \gamma^2 = \gamma^6$
$\gamma^3 + \gamma^2 + 1 = \gamma^{13}$
$\gamma^3 + \gamma^2 + \gamma = \gamma^{11}$
$\gamma^3 + \gamma^2 + \gamma + 1 = \gamma^{12}$

II. Antilogarithms

$\gamma^0 = 1$
$\gamma^1 = \gamma$
$\gamma^2 = \gamma^2$
$\gamma^3 = \gamma^3$
$\gamma^4 = \gamma + 1$
$\gamma^5 = \gamma^2 + \gamma$
$\gamma^6 = \gamma^3 + \gamma^2$
$\gamma^7 = \gamma^3 + \gamma + 1$
$\gamma^8 = \gamma^2 + 1$
$\gamma^9 = \gamma^3 + \gamma$
$\gamma^{10} = \gamma^2 + \gamma + 1$
$\gamma^{11} = \gamma^3 + \gamma^2 + \gamma$
$\gamma^{12} = \gamma^3 + \gamma^2 + \gamma + 1$
$\gamma^{13} = \gamma^3 + \gamma^2 + 1$
$\gamma^{14} = \gamma^3 + 1$

Sometimes such tables are not available. Also, in some cases it is important to work with an α that is not primitive, but $K = F(\alpha)$ nevertheless; that is, $f(X) = \text{Irr}(\alpha, F)$ is of degree n but not primitive. Then we must work with the "vector" expressions of elements of $K = \text{GF}(p^n)$, since a representation in powers of α does not exist. Each element of K has a unique expression $g(\alpha)$, where $g(X) \in F[X]$ has degree $< n$. Addition of these expressions is easy; as before, it is componentwise addition modulo p. To multiply $g(\alpha)$ and $h(\alpha)$ in K, we can use the fact $f(\alpha) = 0$. Multiply $g(X)$ and $h(X)$ as usual, obtaining a product whose degree may be $\geq n$. Divide $g(X)h(X)$ by $f(X)$ [see (7.16.1)], obtaining

$$g(X)h(X) = q(X)f(X) + r(X), \quad r(X) = 0 \text{ or } \deg r(X) < n.$$

Setting $f(\alpha) = 0$ shows that $g(\alpha)h(\alpha) = r(\alpha)$, and $r(\alpha)$ is the desired expression of $g(\alpha)h(\alpha)$ as a polynomial in α of degree $< n$. We often say that multiplication in K, isomorphic to $F[X]/\mathcal{I}(f(X))$, is polynomial multiplication **modulo** $f(X)$. This procedure may of course also be applied if α is primitive. The arithmetical procedures are exactly the ones we illustrated in the example $\mathbb{Q}[X]/\mathcal{I}(X^2 + X + 1)$, after Theorem 7.19.2.

Let us do a sample computation inside $\text{GF}(3^3)$. We know that $\text{GF}(3^3)$ is a splitting field of $X^{27} - X$ over $\text{GF}(3)$, and $X^{27} - X$ has the incomplete factorization

$$X^{27} - X = X(X^{13} - 1)(X^{13} + 1)$$
$$= X(X - 1)(X^{12} + X^{11} + \cdots + X + 1)(X^{13} + 1).$$

A zero α of $X^{12} + X^{11} + \cdots + X + 1$ will have multiplicative order 13, and hence $\text{Irr}(\alpha, \text{GF}(3))$ will be of degree 3 but not primitive. It turns out that

$$X^{12} + X^{11} + \cdots + X + 1 = (X^3 + X^2 + X + 2)(X^3 + 2X^2 + 2X + 2)$$
$$\cdot (X^3 + X^2 + 2)(X^3 + 2X + 2),$$

the factors irreducible. [We shall see in Exercise 7.22.16 how to systematically carry out such factorizations, if we have log and antilog tables for $GF(3^3)$.] Choosing $\text{Irr}(\alpha, GF(3)) = X^3 + X^2 + X + 2$, let us multiply $\alpha^2 + 2\alpha + 2$ and $\alpha^2 + 2\alpha + 1$ in $GF(3^3)$. We find that

$$(X^2 + 2X + 2)(X^2 + 2X + 1) = X^4 + X^3 + X^2 + 2$$

and

$$X^4 + X^3 + X^2 + 2 = X(X^3 + X^2 + X + 2) + (X + 2),$$

so

$$(\alpha^2 + 2\alpha + 2)(\alpha^2 + 2\alpha + 1) = \alpha + 2.$$

7.22 Exercises

1. Explain why $[\mathbb{C}:\mathbb{R}] = 2$ and why $\mathbb{R}$ is an infinite extension of $\mathbb{Q}$.

*2. Prove that if M is a finite extension of K and L is a subfield of M containing K, then

$$[M:K] = [M:L][L:K].$$

[HINT: If $\{x_1, \ldots, x_m\}$ is a basis of L over K and $\{y_1, \ldots, y_n\}$ a basis of M over L, prove that $\{x_i y_j \mid 1 \leq i \leq m, 1 \leq j \leq n\}$ is a basis of M over K, using the definition of "basis."]

3. Gauss first proved the "fundamental theorem of algebra": every $f(X) \in \mathbb{C}[X]$ has a root in $\mathbb{C}$. Using this, prove that $\mathbb{C}$ has no proper finite extensions. Show that the quotient field $\mathbb{C}(X)$ of $\mathbb{C}[X]$ [Exercise 7.18.14(a)] is an infinite extension of $\mathbb{C}$, and show (by construction) that $\mathbb{C}(X)$ has proper finite extensions.

*4. Assume that $n \in \mathbb{P}$, $n > 1$. An element $a \in \mathbb{C}$ is an **nth root of 1** if $a^n = 1$; it is a **primitive nth root** of 1 if also $a^k \neq 1$, all $1 \leq k < n$.

(a) Prove that the nth roots of 1 in $\mathbb{C}$ constitute a cyclic multiplicative group of order n, whose generators are the $\varphi(n)$ primitive nth roots of 1.

(b) Let a be a primitive nth root of 1. Show that $\mathbb{Q}(a)$ is a splitting field of $X^n - 1$ over $\mathbb{Q}$.

(c) With a as in (b), let $f(X) = \text{Irr}(a, \mathbb{Q})$. A nontrivial theorem in algebraic number theory says that $\deg f(X) = \varphi(n)$. [An ingenious proof appears in Landau (1928).] Show that $f(X)$ has as roots all the primitive nth roots of 1 in $\mathbb{C}$. [The polynomial $f(X)$ is called the **nth cyclotomic polynomial**.]

(d) Find the nth cyclotomic polynomials for all $n \leq 8$.

5. Let p be a prime, $n \in \mathbb{P}$, $n > 1$, and let α be a root of $X^n - p$ in $\mathbb{C}$. Given that $X^n - p$ is irreducible in $\mathbb{Q}[X]$, prove that $\mathbb{Q}(\alpha, a)$ is a splitting field for $X^n - p$ over $\mathbb{Q}$, where a is as in Exercise 4(b). What do you think $[\mathbb{Q}(\alpha, a):\mathbb{Q}]$ might be? In what cases can you prove this?

6. Show that if $f(X) \in F[X]$ has degree n and K is a splitting field of $f(X)$ over F, then $[K:F] \leq n!$.

7. Show that if φ is a (ring) isomorphism from a ring R onto a field K, then R is also a field.

*8. (a) Show that if $u_1, u_2, \ldots, u_r$ are elements of the field M and K is a subfield of M, then

$$K(u_1)(u_2) \cdots (u_r) = K(u_{\pi(1)})(u_{\pi(2)}) \cdots (u_{\pi(r)}) = K(u_1, u_2, \ldots, u_r),$$

where π is any permutation of $\{1, 2, \ldots, r\}$ and $K(u_1, u_2, \ldots, u_r)$ is the intersection of all subfields of M containing $K \cup \{u_1, u_2, \ldots, u_r\}$.

(b) Show that if u is one root of $f(X) \in K[X]$ and $u \in M$, then M is a splitting field of $f(X)$ over $K(u)$ iff M is a splitting field of $f(X)$ over K.

9. Verify in detail that the functions φ^*, $\bar{\varphi}$, and φ_1 in the proof of Theorem 7.19.6 are isomorphisms.

*10. Show that every field of characteristic $p \ne 0$ has a unique subfield isomorphic to GF(p), and every field of characteristic 0 has a unique subfield isomorphic to $\mathbb{Q}$. (In either case, the subfield is called the **prime subfield** of the given field.)

*11. Prove (7.20.2).

*12. Show that any GF(p^n) is a splitting field of some polynomial of degree n over GF(p).

*13. Given that $X^5 + X^2 + 1$ is a primitive polynomial over GF(2) and $X^3 + 2X + 1$ is a primitive polynomial over GF(3), construct log and antilog tables for GF(2^5) and GF(3^3). (Save them for use in later chapters.)

14. Which elements $\alpha \in$ GF(2^4) (Table 7.21.3) are primitive? Which elements are not primitive but still satisfy GF(2)(α) = GF(2^4)? Can you make some general conjectures?

15. Find a primitive polynomial of degree 2 over GF(5), and use it to construct log and antilog tables for GF(5^2).

*16. In this exercise we need the log and antilog tables for GF(3^3) obtained in Exercise 13; we will assume that γ is the primitive element (zero of $X^3 + 2X + 1$) on which the tables are based.

(a) Since $\gamma^{26} = 1$, show that the 12 roots of

$$X^{12} + X^{11} + \cdots + X + 1 = \frac{X^{13} - 1}{X - 1}$$

are $\gamma^2, \gamma^4, \gamma^6, \ldots, \gamma^{22}, \gamma^{24}$.

(b) Show that deg Irr (γ^{2i}, GF(3)) = 3 for $1 \le i \le 12$.

(c) Find Irr (γ^2, GF(3)) by solving the equation

$$(\gamma^2)^3 + a(\gamma^2)^2 + b\gamma^2 + c = 0, \quad a, b, c \in \text{GF}(3),$$

using the "vector" representation of powers $(\gamma^2)^i$.

(d) Similarly, find Irr (γ^4, GF(3)), Irr (γ^8, GF(3)), and Irr (γ^{14}, GF(3)), obtaining the other three irreducible factors of $X^{12} + X^{11} + \cdots + X + 1$. (See the discussion at the end of Section 7.23, to learn why γ^4, γ^8, and γ^{14} are chosen.)

*17. Generalize Lemma 7.20.3 to determine what polynomials over a field of characteristic p may have multiple roots.

7.23 Automorphisms of Finite Fields

Theorem 7.23.1. *Let p be a prime, $m, n \in \mathbb{P}$. Then $\mathrm{GF}(p^m)$ is a subfield of $\mathrm{GF}(p^n)$ iff $m \mid n$.*

Proof. If $\mathrm{GF}(p^m)$ is a subfield of $\mathrm{GF}(p^n)$, then $\mathrm{GF}(p^n)$ is a vector space, say of dimension d, over $\mathrm{GF}(p^m)$. As in the proof of Lemma 7.20.1, then $p^n = |\mathrm{GF}(p^n)| = (p^m)^d$, $n = md$, $m \mid n$.

Conversely, if $m \mid n$, then $(p^m - 1)$ divides $(p^n - 1)$, so $(X^{p^{m-1}} - 1)$ divides $(X^{p^n-1} - 1)$ and $(X^{p^m} - X)$ divides $(X^{p^n} - X)$. Therefore, $\mathrm{GF}(p^m)$, the splitting field of $X^{p^m} - X$, is contained in $\mathrm{GF}(p^n)$, the splitting field of $X^{p^n} - X$ over $\mathrm{GF}(p)$. ∎

An **automorphism** of a field L is an isomorphism from L onto L; it is an automorphism of the additive and multiplicative groups of L. The set of all automorphisms of L is a group $\mathrm{Aut}\,(L)$ under function composition (Exercise 7.25.3). If K is a subfield of L, the **Galois group** of L over K is $\{\psi \in \mathrm{Aut}\,(L) \mid \psi(\kappa) = \kappa$, all $\kappa \in K\}$, a subgroup of $\mathrm{Aut}\,(L)$. We can describe all the automorphisms of finite fields:

Theorem 7.23.2. (a) *Every automorphism of $L = \mathrm{GF}(p^n)$ is a power $\varphi^i : x \to x^{p^i}$ of the **Galois automorphism** $\varphi : x \to x^p$ of the Galois group of $\mathrm{GF}(p^n)$ over $\mathrm{GF}(p)$. Every such power is an automorphism, so $\mathrm{Aut}\,\mathrm{GF}(p^n)$ is cyclic of order n.*

(b) *The Galois group of $\mathrm{GF}(p^n)$ over $\mathrm{GF}(p^m)$ is cyclic of order $[\mathrm{GF}(p^n) : \mathrm{GF}(p^m)] = n/m$, with generator $\varphi^m : x \to x^{p^m}$. In particular, $|\mathrm{Aut}\,(L)| = |\langle \varphi \rangle| = n$.*

Proof. We have $\varphi(xy) = (xy)^p = x^p y^p = \varphi(x)\varphi(y)$ and, by Lemma 7.20.2, $\varphi(x+y) = (x+y)^p = x^p + y^p = \varphi(x) + \varphi(y)$, so φ is a homomorphism from L into L. The equation $x^p = y^p$ implies that $0 = x^p - y^p = (x^p + (-y)^p) = (x-y)^p$, so $x - y = 0$, $x = y$; therefore, φ is one-one and must be an automorphism of L. Nonzero elements of $\mathrm{GF}(p)$ lie in a multiplicative group of order $p-1$, hence satisfy the identity $1 = x^{p-1}$. Therefore, every element x of $\mathrm{GF}(p)$ satisfies $x = x^p = \varphi(x)$, and φ is in the Galois group of L over $\mathrm{GF}(p)$.

If τ is any automorphism of L and γ a primitive element of L, assume $\tau(\gamma) = \gamma^k$, where $0 < k < p^n - 1$. Since every nonzero element of L is a power of γ, $\tau(x) = x^k$ for all $x \in L$. If we show that k is a power p^i of p, then $\tau = \varphi^i$ and (a) will be proved. If k is not such a power, then $k = p^s k_0$, $p \nmid k_0$, $k_0 > 1$. Consider in this case

$$f(X) = (1+X)^k - 1 - X^k \in \mathrm{GF}(p)[X].$$

Since τ is an automorphism,

$$f(x) = \tau(1+x) - \tau(1) - \tau(x) = 0, \quad \text{all } x \in L.$$

Thus $f(X)$ has degree $< p^n$ but has p^n roots, forcing $f(X) = 0$. However,

$$f(X) = (1+X)^{p^s k_0} - 1 - X^k = (1+X^{p^s})^{k_0} - 1 - X^{p^s k_0}$$
$$= k_0 X^{p^s} + \text{other terms}$$

does not vanish identically, the contradiction that proves (a).

Now φ^i is the identity on a subfield $K = \text{GF}(p^m)$ of L iff $x = x^{p^i}$ for all $x \in K$. Elements of K satisfy $x = x^{p^m}$ by Corollary 7.20.1, so $\varphi^m =$ identity on K. If φ^t were the identity on K for some $t < m$, then x would equal x^{p^t} for all p^m elements of K, a contradiction since $X^{p^t} - X = 0$ can have at most p^t roots. The Galois group of L over K must be cyclic by Theorem 4.16.2 and, since m is minimal, must be $\langle \varphi^m \rangle$ with order n/m. ∎

Corollary 7.23.1. *If $K = \text{GF}(p^m)$ and $f(X) \in K[X]$ is irreducible of degree d with root $\alpha \in \text{GF}(p^{md})$, then all the roots of $f(X)$ are in $\text{GF}(p^{md})$ and are the set $\{\alpha, \alpha^{p^m}, \alpha^{p^{2m}}, \ldots, \alpha^{p^{(d-1)m}}\}$. In particular, all the roots of an irreducible $f(X) \in \text{GF}(p)[X]$ of degree d are in $\text{GF}(p^d)$ and are the set $\{\alpha, \alpha^p, \alpha^{p^2}, \ldots, \alpha^{p^{d-1}}\}$.*

Proof. Let $f(X) = a_d X^d + a_{d-1} X^{d-1} + \cdots + a_1 X + a_0$, so $a_d \alpha^d + a_{d-1} \alpha^{d-1} + \cdots + a_1 \alpha + a_0 = 0$. The automorphism $\psi: x \to x^{p^m}$ of $\text{GF}(p^{md})$ fixes all $a_i \in \text{GF}(p^m)$, so when we apply ψ to both sides of the equation it becomes

$$a_d (\alpha^{p^m})^d + a_{d-1} (\alpha^{p^m})^{d-1} + \cdots + a_1 (\alpha^{p^m}) + a_0 = 0,$$

showing that α^{p^m} is a root of $f(X)$. Repeatedly applying ψ shows that each $\alpha^{p^{im}}$ is a root of $f(X)$.

We claim that the d roots

$$\{\alpha, \alpha^{p^m}, \ldots, \alpha^{p^{(d-1)m}}\} = \{\alpha, \psi(\alpha), \ldots, \psi^{d-1}(\alpha)\}$$

are all distinct. Indeed, $\psi^i(\alpha) = \psi^j(\alpha)$ for $0 \leq i < j < d$ would mean that $\alpha = \psi^{j-i}(\alpha)$, $0 < j - i < d$. Every element of $\text{GF}(p^{md})$ is a polynomial in α with coefficients in $\text{GF}(p^m)$, so this would mean $\psi^{j-i} =$ identity on $\text{GF}(p^{md})$, a contradiction, since ψ has order d by Theorem 7.23.2. ∎

For example, we saw in the second example after Theorem 7.19.4 that if u is one root of the irreducible polynomial $X^3 + X + 1$ over $\text{GF}(2)$, then the other roots are u^2 and u^4; Corollary 7.23.1 would have enabled us to predict this occurrence, if we had known it then. The primitive polynomial $X^4 + X + 1$ over $\text{GF}(2)$ (Table 7.21.3) must have roots γ, γ^2, γ^4, and γ^8. If τ is one root of the primitive polynomial $X^5 + X^2 + 1$ over $\text{GF}(2)$ (Exercise 7.22.13), then the other roots are τ^2, τ^4, τ^8, and τ^{16}.

Corollary 7.23.1 also helps us determine what powers γ^i to study in Exercise 7.22.16(d). In that exercise we know γ is a root of the primitive polynomial X^3+2X+1 over $F = \mathrm{GF}(3)$, so $\gamma^{26} = 1$ and $\gamma^3 = \gamma + 2$, and we must find the four degree-3 irreducible factors of $X^{12}+X^{11}+\cdots+X+1$, which we know has roots $\gamma^2, \gamma^4, \gamma^6, \ldots, \gamma^{22}, \gamma^{24}$. By Corollary 7.23.1, Irr (γ^2, F) has roots γ^2, $\gamma^{2\cdot 3} = \gamma^6$, $\gamma^{2\cdot 3^2} = \gamma^{18}$, so another factor is Irr (γ^4, F), with roots γ^4, γ^{12}, $\gamma^{36} = \gamma^{10}$. Now Irr $(\gamma^6, F) = $ Irr (γ^2, F) has already been found, but a third factor is Irr (γ^8, F), with roots γ^8, $\gamma^{8\cdot 3} = \gamma^{24}$, and $\gamma^{8\cdot 3^2} = \gamma^{72} = \gamma^{20}$. The remaining factor must have the remaining roots γ^{14}, γ^{16}, and γ^{22}; indeed, Irr (γ^{14}, F) *will* have roots γ^{14}, $\gamma^{14\cdot 3} = \gamma^{42} = \gamma^{16}$, and $\gamma^{14\cdot 3^2} = \gamma^{126} = \gamma^{22}$.

7.24 Number of Irreducibles

The **Möbius μ-function** $\mu: \mathbb{P} \to \mathbb{Z}$ of number theory is defined as follows:

$$\mu(1) = 1;$$

(7.24.1) $\quad \mu(p_1 p_2 \cdots p_r) = (-1)^r \quad$ for distinct primes $p_1, p_2, \ldots, p_r$;

$\quad \mu(n) = 0 \quad$ if n is divisible by the square of a prime.

For example, $\mu(3) = -1$, $\mu(4) = 0$, $\mu(6) = 1$, $\mu(30) = -1$, and $\mu(50) = 0$.

Lemma 7.24.1. *For any $n > 1$ in $\mathbb{P}$,*

(7.24.2) $$\sum_{t|n} \mu(t) = 0.$$

Proof. The sum, of course, runs through all divisors $t \in \mathbb{P}$ of n. Let $n = p_1^{u_1} \cdots p_r^{u_r}$, the p_i distinct primes, the $u_i \in \mathbb{P}$. Certainly,

$$\sum_{t|n} \mu(t) = \sum_{t|p_1 \cdots p_r} \mu(t),$$

since the two sums differ only by some zero terms. For any divisor t of $p_2 \cdots p_r$, $\mu(p_1 t) = -\mu(t)$, since t and $p_1 t$ differ in the one prime factor p_1. We conclude that

$$\sum_{t|p_1\cdots p_r} \mu(t) = \sum_{t|p_2\cdots p_r} \mu(t) + \sum_{t|p_2\cdots p_r} \mu(p_1 t) = \sum_{t|p_2\cdots p_r}(\mu(t) - \mu(t)) = 0. \quad \blacksquare$$

Theorem 7.24.1. *The number of irreducible polynomials of degree d over $K = \mathrm{GF}(p^m)$ is exactly*

(7.24.3) $$\frac{1}{d}\sum_{t|d} \mu\left(\frac{d}{t}\right) p^{mt}.$$

Proof. By Theorem 7.19.1, $\alpha \in \text{GF}(p^{md}) = M$ has Irr (α, K) of degree d iff $K(\alpha) = M$, that is, iff α is not in any proper subfield of M. Each irreducible polynomial of degree d has d roots, so the number of irreducible polynomials is $(1/d)|\{\alpha \in M | K(\alpha) = M\}|$.

Fields L with $K \subseteq L \subseteq M$ have form $L = \text{GF}(p^{mt})$, some divisor t of d (Theorem 7.23.1), so the sum in (7.24.3) equals

$$\frac{1}{d} \sum_{K \subseteq L \subseteq M} \mu([M:L])|L| = \frac{1}{d} \sum_{\alpha \in M} \sum_{\substack{K \subseteq L \subseteq M \\ \text{with } \alpha \in L}} \mu([M:L]).$$

If $K(\alpha) = M$, the inner sum is simply $\mu([M:M]) = 1$. By the previous paragraph, our proof will be complete if we can show that for any $\alpha \in M$ with $K(\alpha) \neq M$,

$$\sum_{\substack{K \subseteq L \subseteq M \\ \text{with } \alpha \in L}} \mu([M:L]) = 0.$$

For such an α, let $K(\alpha) = \text{GF}(p^{mc})$, where $c|d$. Then $\{L | K \subseteq L \subseteq M$ with $\alpha \in L\} = \{\text{GF}(p^{md/t}) | t$ divides $d/c\}$ and $[M : \text{GF}(p^{md/t})] = t$, so

$$\sum_{\substack{K \subseteq L \subseteq M \\ \text{with } \alpha \in L}} \mu([M:L]) = \sum_{t|(d/c)} \mu(t) = 0,$$

by Lemma 7.24.1. ∎

For example, the number of irreducible polynomials of degree 10 over GF(2) is

(7.24.4)
$$\frac{1}{10} \sum_{t|10} \mu\left(\frac{10}{t}\right) 2^t = \frac{1}{10}[\mu(10) \cdot 2 + \mu(5) \cdot 2^2 + \mu(2) \cdot 2^5 + \mu(1) \cdot 2^{10}]$$
$$= \frac{1}{10}(2 - 4 - 32 + 1024) = \frac{1}{10}(990) = 99;$$

by Theorem 7.21.1,

$$\frac{1}{10}\varphi(2^{10} - 1) = \frac{1}{10}\varphi(3 \cdot 11 \cdot 31) = \frac{1}{10} \cdot 2 \cdot 10 \cdot 30 = 60$$

of these are primitive. The number of irreducible polynomials of degree 11 over GF(2) is

$$\frac{1}{11} \sum_{t|11} \mu\left(\frac{11}{t}\right) \cdot 2^t = \frac{1}{11}[\mu(11) \cdot 2 + \mu(1) \cdot 2^{11}] = \frac{1}{11}(-2 + 2048) = 186,$$

and

$$\frac{1}{11}\varphi(2^{11}-1) = \frac{1}{11}\varphi(2047) = \frac{1}{11}\varphi(23 \cdot 89) = \frac{1}{11} \cdot 22 \cdot 88 = 176$$

of them are primitive.

7.25 Exercises

1. List all the subfields of $GF(2^{30})$.
2. If $m, d \in \mathbb{P}$ and $n = md$, write the quotients

$$\frac{p^n - 1}{p^m - 1} \quad \text{and} \quad \frac{X^{p^n-1} - 1}{X^{p^m-1} - 1}$$

as sums, without denominators.

*3. Prove that the set $\text{Aut}(L)$ of all the automorphisms of a field L is indeed a group under function composition.

4. Divide the 16 elements of $GF(2^4)$ (Table 7.21.3) into equivalence classes, two elements being equivalent iff they are roots of the same irreducible polynomial over $GF(2)$. (Use Corollary 7.23.1.)

*5. Show that if $K = GF(p^m)$ and $f(X) \in K[X]$, say $f(X) = \alpha_r X^r + \alpha_{r-1} X^{r-1} + \cdots + \alpha_1 X + \alpha_0$, then

$$f(X)^p = \alpha_r^p X^{rp} + \alpha_{r-1}^p X^{(r-1)p} + \cdots + \alpha_1^p X^p + \alpha_0^p.$$

In particular, with $q = p^m$, show that $f(X)^q = f(X^q)$.

6. Use the methods of Exercise 7.22.16 (and the discussion at the end of Section 7.23) to factor the polynomial $X^{32} - X$ over $GF(2)$ completely into irreducible factors. Which factors are primitive?

7. Find the number of irreducible polynomials of degree 8 over $GF(3)$. How many of them are primitive?

8. Determine the number of subgroups of each possible order in the additive group of $GF(p^n)$. [This is a fairly hard problem. Think of them as vector subspaces of a vector space of dimension n over $GF(p)$.]

9. If p and q are primes, find necessary and sufficient conditions for all irreducible polynomials over $GF(p)$ of degree q to be primitive. Can this ever happen if q is not prime?

*10. Even if $m > 1$, we can still define an irreducible $f(X)$ of degree d over $GF(p^m)$ to be *primitive* if a root α of $f(X)$ in $GF(p^{md})$ is a primitive element in $GF(p^{md})$. Show that this definition is independent of the choice of α, and find a formula for the number of such primitive polynomials.

*11. Show that if $K = GF(p^m)$ and $I_K(d)$ is the number of irreducible polynomials of degree d over K, then

$$p^{md} = \sum_{t \mid d} t I_K(t).$$

Use this formula, without (7.24.3), to find $I_{GF(2)}(10)$, and compare with (7.24.4).

12. (a) When you are given a primitive polynomial of degree n over $GF(p)$ (p a prime), how would you represent, add, and multiply elements of $GF(p^n)$ in a computer program?

(b) Use the ideas in (a) and the information in Table 7.21.1 to write computer programs which will produce and print log and antilog tables for $GF(2^6)$, $GF(3^5)$, and $GF(2^9)$. What would go wrong if the chosen polynomial were not primitive?

CHAPTER 8
Linear Machines

8.1 Definition

In this chapter, if F is a field and $n \in \mathbb{P}$, we denote $F_n = \text{Mat}_{n \times 1}(F)$, the n-dimensional vector space of "column vectors" over F. Also, $F_0 = [0]$ is a 0-dimensional vector space over F. Recall from Section 1.11 that a finite-state machine is a 5-tuple $[S, X, Z, \tau, \omega]$, with S, X, and Z finite sets, $\tau: S \times X \to S$ and $\omega: S \times X \to Z$ functions. If $S = F_k$, $X = F_l$, and $Z = F_m$ for some field F and $k, l, m \in \mathbb{N}$, then $\tau: F_k \times F_l \to F_k$ can be thought of as a function $\tau: F_{k+l} \to F_k$ by identifying any

$$\left(\begin{bmatrix} y_1 \\ y_2 \\ \vdots \\ y_k \end{bmatrix}, \begin{bmatrix} x_1 \\ x_2 \\ \vdots \\ x_l \end{bmatrix} \right) \in F_k \times F_l \quad \text{with} \quad \begin{bmatrix} y_1 \\ \vdots \\ y_k \\ x_1 \\ \vdots \\ x_l \end{bmatrix} \in F_{k+l}.$$

Similarly, we can think of ω as a function $\omega: F_{k+l} \to F_m$. The machine $[F_k, F_l, F_m, \tau, \omega]$ is called a **linear machine** (**over** F) if $\tau: F_{k+l} \to F_k$ and $\omega: F_{k+l} \to F_m$ are linear transformations. The **dimension** of this machine is k. In this finite-state machine, F must be a finite field. [Some authors discuss "linear machines" for infinite fields F; see Harrison (1969) and Rabin (1960). In this chapter, F is always finite.] A **binary machine** is a linear machine over GF(2).

The idea of a linear machine arises naturally if we consider how finite-state machines are usually implemented as electronic circuits. Consider again the machine $[S, X, Z, \tau, \omega]$ of Table 1.11.1. We have $X = \{0, 1\} = F_1$ with $F = \text{GF}(2)$. That is, a device implementing this machine has one input terminal, to which values 0 (low voltage) or 1 (high voltage) may be applied. We can consider Z to be

$$\left\{ \begin{bmatrix} 0 \\ 0 \end{bmatrix}, \begin{bmatrix} 0 \\ 1 \end{bmatrix}, \begin{bmatrix} 1 \\ 0 \end{bmatrix}, \begin{bmatrix} 1 \\ 1 \end{bmatrix} \right\} = F_2.$$

That is, a circuit realizing this machine has two output terminals, and output $\begin{bmatrix} 0 \\ 1 \end{bmatrix}$ denotes 0 (low voltage) at the first terminal, 1 (high voltage) at the second, or vice versa. In general, in a machine designed to provide specified output responses to given inputs, X and Z are likely to be F_l and F_m, $l, m \in \mathbb{P}$, $F = \text{GF}(2)$. (Examples where F is naturally some other finite field will be encountered later.) Initial design of this machine to do its task (it is a "binary ring counter"; see Section 1.11) results in the set $S = \{\sigma_0, \sigma_1, \sigma_2\}$ of three states. When a circuit to perform this task is constructed, σ_0, σ_1, and σ_2 are usually assigned values in some F_m ($m = 2$ will suffice); for example, we could assign

$$\sigma_0 \to \begin{bmatrix} 0 \\ 0 \end{bmatrix}, \quad \sigma_1 \to \begin{bmatrix} 0 \\ 1 \end{bmatrix}, \quad \text{and} \quad \sigma_2 \to \begin{bmatrix} 1 \\ 0 \end{bmatrix}.$$

The circuit then contains two "memory devices," most commonly "flipflops," able to sustain a constant output voltage level for some time. For example, when the machine is in state σ_2, the first flipflop is producing signal 1 and the second is producing signal 0. If the machine shifts to state σ_1, the two flipflops change to produce signals 0 and 1, respectively.

Actual procedures and devices for designing circuits to implement specific finite-state machines are beyond the scope of this book; good discussions can be found in Kohavi (1970), Hennie (1968), Hartmanis and Stearns (1966), and Booth (1967). However, we hope that the discussion in the preceding paragraph will convince the reader that in practical situations, X and Z are often vector spaces over a finite field and S is at least a subset of such a space. We shall see that linear machines are natural tools for encoding and decoding the codes treated in Chapters 5 and 9.

8.2 Shift Registers

Linear machines over a (finite) field F are generally constructed with three types of devices: *adders*, *multipliers*, and *delays*. The **adder** is pictured as in Figure 8.2.1(a); it has some number $s > 1$ of input leads, receiving some

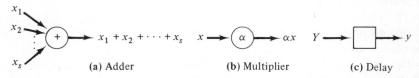

(a) Adder (b) Multiplier (c) Delay

FIGURE 8.2.1. *Components of Linear Machines*

values $x_1, x_2, \ldots, x_s \in F$, and one output lead, carrying the value $x_1 + x_2 + \cdots + x_s$ (addition in F). A (constant) **multiplier** by $\alpha \in F$ is pictured as in Figure 8.2.1(b); it has one input lead, receiving some $x \in F$, and one output lead, carrying the value $\alpha x \in F$. The response of adders and multipliers is assumed to be instantaneous; as soon as inputs $x_1, x_2, \ldots, x_s, x$ are stabilized at constant values in F, the constant values $x_1 + x_2 + \cdots + x_s$ and αx appear at the output leads.

Sec. 8.2 Shift Registers

The **delay (delay element)** is a memory device, pictured as in Figure 8.2.1(c). We assume that time is divided into equal, appropriately short intervals marked by the points $t = 0, 1, 2, \ldots$. The output of the delay at time $t+1$ is its input at time t; we denote by $Y(t)$ its input at time t and by $y(t)$ its output at time t, so $y(t+1) = Y(t)$. (There are simple algorithms for replacing these delay elements by appropriately controlled flipflops when actual circuits are being designed.) An implementation of a linear machine $M = [F_k, F_l, F_m, \tau, \omega]$ will contain k delay elements. The **state** of the machine at any time t is defined to be

$$\begin{bmatrix} y_1(t) \\ y_2(t) \\ \vdots \\ y_k(t) \end{bmatrix} \in F_k,$$

where $y_i(t)$ denotes the output of the ith delay at time t. In particular, the **initial state** of a machine starting at time $t = 0$ is the vector

$$\begin{bmatrix} y_1(0) \\ y_2(0) \\ \vdots \\ y_k(0) \end{bmatrix}.$$

A linear machine $M = [F_k, F_l, F_m, \tau, \omega]$ with $l = m = 1$ is called a **two-terminal machine**, since there is one input terminal and one output terminal in a circuit implementation of M, each carrying as signal a single element of F.

An important type of two-terminal machine is the **feedforward shift register**, with circuit pictured in Figure 8.2.2; $a_0, a_1, \ldots, a_k$ are fixed

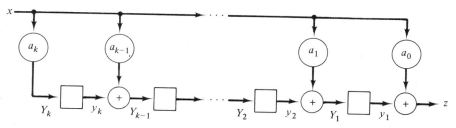

FIGURE 8.2.2. *Feedforward Shift Register*

elements of F. Here x and z stand for the input and output signals, respectively, and Y_i and y_i for the input and output of the ith delay element. That is, $x(t)$ is the input at time t, $y_2(3)$ the output of the second delay at time

$t = 3$, and so on. The equations

(8.2.1) $\quad Y_k = a_k x, \quad Y_{k-1} = y_k + a_{k-1} x, \ldots, Y_1 = y_2 + a_1 x, \quad z = y_1 + a_0 x$

express the fact, clear from the diagram, that at any time t we have

$$Y_k(t) = a_k x(t), \quad Y_i(t) = y_{i+1}(t) + a_i x(t) \quad \text{for } 1 \leq i < k, \, z(t) = y_1(t) + a_0 x(t).$$

The fact that $y_i(t) = Y_i(t-1)$ is expressed by means of a **delay operator** D; if $w: \text{time units} \to F$ is any function mapping $\mathbb{N}$ into F, then Dw is a function such that $(Dw)(t) = w(t-1)$. Powers D^k for $k > 1$ are defined inductively by $D^k w = D(D^{k-1} w)$, so $(D^k w)(t) = w(t-k)$. Algebraic properties of D are discussed in Exercise 8.4.2. Note that if w is defined for $t = 0, 1, 2, \ldots$, then $D^k w$ is defined only for $t = k, k+1, k+2, \ldots$; we will often extend $D^k w$ to $\{0, 1, 2, \ldots\}$ by defining $(D^k w)(t) = 0$ if $t < k$.

By using D, we can express concisely the output function z of a feedforward shift register in terms of its input function x. The equation $y_i(t) = Y_i(t-1)$ means that $y_i(t) = DY_i(t)$ for all $t \geq 1$, so we have

$$z = a_0 x + y_1 = a_0 x + DY_1 = a_0 x + D(a_1 x + y_2) = a_0 x + a_1 Dx + D(DY_2)$$
$$= a_0 x + a_1 Dx + D^2(a_2 x + y_3) = \cdots = a_0 x + a_1 Dx + \cdots + a_k D^k x.$$

Therefore, we have

(8.2.2) $\quad z = (a_0 + a_1 D + \cdots + a_k D^k) x,$

symbolizing the fact that

(8.2.3) $\quad z(t) = a_0 x(t) + a_1 x(t-1) + \cdots + a_k x(t-k), \quad \text{all } t \geq k.$

If $t < k$, the output $z(t)$ at time t depends on the initial state $[y_1(0) \, y_2(0) \cdots y_k(0)]^T$ of the machine, as well as the inputs received until time t. We have

$$z(t) = a_0 x(t) + y_1(t) = a_0 x(t) + Y_1(t-1) = a_0 x(t) + a_1 x(t-1)$$
$$+ y_2(t-1) = a_0 x(t) + a_1 x(t-1) + a_2 x(t-2) + y_3(t-2) = \cdots,$$

so

$$z(t) = a_0 x(t) + a_1 x(t-1) + \cdots + a_t x(0) + y_{t+1}(0) \quad \text{for } 0 \leq t < k.$$

The polynomial

(8.2.4) $\quad T(D) = a_0 + a_1 D + \cdots + a_k D^k$

is called the **transfer function** of the feedforward shift register of Figure 8.2.2.

Another important type of linear machine, the **feedback shift register**, is pictured in Figure 8.2.3. As in the case of the feedforward shift register, we

Sec. 8.2 Shift Registers

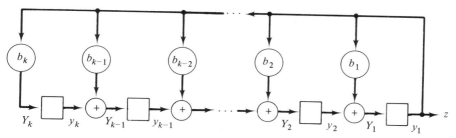

FIGURE 8.2.3. *Feedback Shift Register*

get equations

(8.2.5) $\quad Y_k = b_k z, \; Y_{k-1} = y_k + b_{k-1} z, \ldots, \; Y_1 = y_2 + b_1 z, \qquad z = y_1.$

These equations lead to

$$z = y_1 = D(y_2 + b_1 z) = b_1 Dz + D^2(y_3 + b_2 z) = \cdots$$
$$= b_1 Dz + b_2 D^2 z + \cdots + b_k D^k z.$$

Therefore, we write

(8.2.6) $\qquad z = (b_1 D + b_2 D^2 + \cdots + b_k D^k) z,$

expressing the fact that

(8.2.7) $\quad z(t) = b_1 z(t-1) + b_2 z(t-2) + \cdots + b_k z(t-k) \quad \text{for } t \geq k.$

Equation (8.2.7) expresses the present output in terms of past outputs, for $t \geq k$. For $t < k$, the output $z(t)$ depends on the initial state as well as the previous outputs. We have

$$z(t) = y_1(t) = Y_1(t-1) = b_1 z(t-1) + y_2(t-1) = b_1 z(t-1)$$
$$+ b_2 z(t-2) + y_3(t-2) = \cdots$$
$$= b_1 z(t-1) + b_2 z(t-2) + \cdots + b_t z(0) + y_{t+1}(0) \qquad \text{for } t < k.$$

[If all delays have initial output 0, the output $z(t)$ will be 0 for every t.]

Where is x, the input, in the feedback shift register? In this case $l = 0$, $m = 1$; we may assume that there is only one input symbol, appearing anew at each time interval but not applied to any of the hardware, so it is not pictured. We also do not picture the "clock pulse" or other timing mechanism used to cause the content of the delays to shift at each time interval.

Later we shall describe a type of two-terminal machine that includes both feedforward and feedback shift registers as special cases.

EXAMPLE 1. Consider $GF(2^4)$, described in Table 7.21.3. If $F = GF(2)$, then $GF(2^4) = F(\gamma)$, where $\gamma^4 = \gamma + 1$. The feedback shift register pictured in Figure 8.2.4 can be used to construct Table 7.21.3. When we are given any vector expression $\alpha = c_0 + c_1 \gamma + c_2 \gamma^2 + c_3 \gamma^3 \in GF(2^4)$, we start the shift

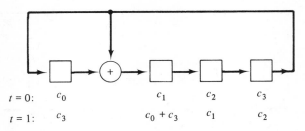

| $t = 0$: | c_0 | | c_1 | c_2 | c_3 |
| $t = 1$: | c_3 | | $c_0 + c_3$ | c_1 | c_2 |

FIGURE 8.2.4. *Shift Register for* $GF(2^4)$

register with the four delays, left to right, having initial contents (outputs) c_0, c_1, c_2, c_3. Figure 8.2.4 shows the contents of the register after one shift (time interval). Because of the equation

$$(c_0 + c_1\gamma + c_2\gamma^2 + c_3\gamma^3)\gamma = c_0\gamma + c_1\gamma^2 + c_2\gamma^3 + c_3(\gamma + 1)$$
$$= c_3 + (c_0 + c_3)\gamma + c_1\gamma^2 + c_2\gamma^3,$$

these are exactly the coefficients of $\alpha\gamma$. Therefore, starting with $\alpha = \gamma^0 = 1 + 0\gamma + 0\gamma^2 + 0\gamma^3$, we can find the vector expressions for $\gamma^1 = \gamma^0 \cdot \gamma$, $\gamma^2 = \gamma^1 \cdot \gamma, \ldots, \gamma^{14} = \gamma^{13} \cdot \gamma$ with 14 shifts of the shift register. [In Exercise 8.4.4 we generalize this discussion to show how log tables for any $GF(p^n)$ can be constructed, given a primitive polynomial of degree n over $GF(p)$.]

EXAMPLE 2. A feedforward shift register can be used to multiply polynomials in $F[X]$, F a finite field. If we wish to multiply several polynomials, including $d(X) = d_0 + d_1 X + \cdots + d_r X^r$, by $c(X) = c_0 + c_1 X + \cdots + c_s X^s$, we use the circuit

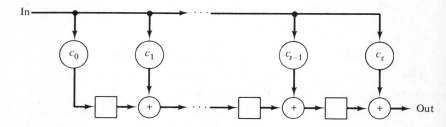

To multiply $d(X)$ by $c(X)$, we apply successively the inputs $d_r, d_{r-1}, \ldots, d_1, d_0, 0, \ldots, 0$ (s concluding 0's), and the outputs are the coefficients of the product, *high-order coefficients first*. To see that this is so, think of the register as a device that stores a polynomial. Initially, all delays store 0's. When d_r enters, the register stores the coefficients $d_r c_i$ of $d_r c(X)$, and the first output is the first coefficient $d_r c_s$ of the product. The first shift of the register, while d_{r-1} enters, puts the coefficients of $d_r X c(X) + d_{r-1} c(X) =$

Sec. 8.3 *Characterizing Matrices* 387

$(d_r X + d_{r-1})c(X)$ in the register while the second coefficient $d_{r-1}c_s + d_r c_{s-1}$ of the product is the second output. Continuing, the third shift puts all but the highest three coefficients of $(d_r X^2 + d_{r-1} X + d_{r-2})c(X)$ in the register, the highest three being the first three outputs, and so on.

EXAMPLE 3. A modified feedback shift register, with inputs, can be used to divide polynomials. If we wish to divide $d(X) = d_r X^r + \cdots + d_1 X + d_0$ by $c(X) = X^s + c_{s-1} X^{s-1} + \cdots + c_1 X + c_0$, we use the circuit

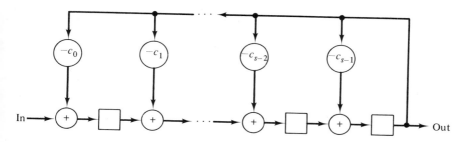

The initial outputs of the delays are all 0. The inputs are successively $d_r, d_{r-1}, \ldots, d_1, d_0$. When d_r reaches the last delay at right (after the first s shifts), the delays contain $d_{r-s+1}, d_{r-s+2}, \ldots, d_r$. We say that at this point the register stores the polynomial

$$\bar{d}_0(X) = d_{r-s+1} + d_{r-s+2} X + \cdots + d_r X^{s-1}.$$

At the next shift, d_{r-s} enters, the first coefficient $q_{r-s} = d_r$ of the quotient leaves, and the register contains

$$d_{r-s} + X\bar{d}_0(X) - d_r(c(X) - X^s) = (d_{r-s} + d_{r-s+1} X + \cdots + d_r X^s) - d_r c(X).$$

At the next shift, the next coefficient $q_{r-s-1} = d_{r-1} - d_r c_{s-1}$ of the quotient leaves, and the register contains

$$(d_{r-s-1} + d_{r-s} X + \cdots + d_r X^{s+1}) - (q_{r-s} X + q_{r-s-1}) c(X).$$

Continuing, we obtain the entire quotient $q(X)$ as the output and the remainder $d(X) - q(X)c(X)$ as the last register entry. (See Exercise 8.4.6.)

8.3 Characterizing Matrices

Now consider any linear machine $M = [F_k, F_l, F_m, \tau, \omega]$. By Theorem 7.9.1, there are matrix elements $a_{ij}, b_{ij}, c_{ij}, d_{ij} \in F$ such that for any state $\begin{bmatrix} y_1 \\ \vdots \\ y_k \end{bmatrix} \in F_k$ and input $\begin{bmatrix} x_1 \\ \vdots \\ x_l \end{bmatrix} \in F_l$, the next state $\begin{bmatrix} Y_1 \\ \vdots \\ Y_k \end{bmatrix}$ and output $\begin{bmatrix} z_1 \\ \vdots \\ z_m \end{bmatrix}$ are given by

$$\begin{bmatrix} Y_1 \\ \vdots \\ Y_k \end{bmatrix} = \begin{bmatrix} a_{11} & \cdots & a_{1k} & b_{11} & \cdots & b_{1l} \\ \cdots & \cdots & \cdots & \cdots & \cdots & \cdots \\ a_{k1} & \cdots & a_{kk} & b_{k1} & \cdots & b_{kl} \end{bmatrix} \begin{bmatrix} y_1 \\ \vdots \\ y_k \\ x_1 \\ \vdots \\ x_l \end{bmatrix}$$

$$= \begin{bmatrix} a_{11} & \cdots & a_{1k} \\ \cdots & \cdots & \cdots \\ a_{k1} & \cdots & a_{kk} \end{bmatrix} \begin{bmatrix} y_1 \\ \vdots \\ y_k \end{bmatrix} + \begin{bmatrix} b_{11} & \cdots & b_{1l} \\ \cdots & \cdots & \cdots \\ b_{k1} & \cdots & b_{kl} \end{bmatrix} \begin{bmatrix} x_1 \\ \vdots \\ x_l \end{bmatrix},$$

$$\begin{bmatrix} z_1 \\ \vdots \\ z_m \end{bmatrix} = \begin{bmatrix} c_{11} & \cdots & c_{1k} & d_{11} & \cdots & d_{1l} \\ \cdots & \cdots & \cdots & \cdots & \cdots & \cdots \\ c_{m1} & \cdots & c_{mk} & d_{m1} & \cdots & d_{ml} \end{bmatrix} \begin{bmatrix} y_1 \\ \vdots \\ y_k \\ x_1 \\ \vdots \\ x_l \end{bmatrix}$$

$$= \begin{bmatrix} c_{11} & \cdots & c_{1k} \\ \cdots & \cdots & \cdots \\ c_{m1} & \cdots & c_{mk} \end{bmatrix} \begin{bmatrix} y_1 \\ \vdots \\ y_k \end{bmatrix} + \begin{bmatrix} d_{11} & \cdots & d_{1l} \\ \cdots & \cdots & \cdots \\ d_{m1} & \cdots & d_{ml} \end{bmatrix} \begin{bmatrix} x_1 \\ \vdots \\ x_l \end{bmatrix}.$$

We denote $\mathbf{Y} = [Y_i]_{k \times 1}$, $\mathbf{z} = [z_i]_{m \times 1}$, $\mathbf{y} = [y_i]_{k \times 1}$, $\mathbf{x} = [x_i]_{l \times 1}$, $\mathbf{A} = [a_{ij}]_{k \times k}$, $\mathbf{B} = [b_{ij}]_{k \times l}$, $\mathbf{C} = [c_{ij}]_{m \times k}$, and $\mathbf{D} = [d_{ij}]_{m \times l}$, and the matrix equations become

(8.3.1)
$$\mathbf{Y} = \mathbf{A}\mathbf{y} + \mathbf{B}\mathbf{x},$$
$$\mathbf{z} = \mathbf{C}\mathbf{y} + \mathbf{D}\mathbf{x}.$$

The matrices **A**, **B**, **C**, and **D** are called **characterizing matrices** of M; they determine M completely, so we can write $M = (\mathbf{A}, \mathbf{B}, \mathbf{C}, \mathbf{D})$.

For example, in the feedforward shift register of Figure 8.2.2, (8.2.1) yields

$$\begin{bmatrix} Y_1 \\ Y_2 \\ \vdots \\ Y_{k-1} \\ Y_k \end{bmatrix} = \begin{bmatrix} 0 & 1 & 0 & \cdots & 0 & 0 \\ 0 & 0 & 1 & \cdots & 0 & 0 \\ \vdots & \vdots & & \ddots & & \vdots \\ 0 & 0 & 0 & \cdots & 0 & 1 \\ 0 & 0 & 0 & \cdots & 0 & 0 \end{bmatrix} \begin{bmatrix} y_1 \\ y_2 \\ \vdots \\ y_{k-1} \\ y_k \end{bmatrix} + \begin{bmatrix} a_1 \\ a_2 \\ \vdots \\ a_{k-1} \\ a_k \end{bmatrix} [x],$$

Sec. 8.3 *Characterizing Matrices*

$$[z] = [1 \quad 0 \quad \cdots \quad 0] \begin{bmatrix} y_1 \\ y_2 \\ \vdots \\ y_{k-1} \\ y_k \end{bmatrix} + [a_0][x],$$

so in this case all entries of $\mathbf{A}$ are 0 except for 1's on the *superdiagonal* above the main diagonal, $\mathbf{B} = [a_1 \quad a_2 \quad \cdots \quad a_k]^T$, $\mathbf{C} = [1 \quad 0 \quad \cdots \quad 0]$, $\mathbf{D} = [a_0]$.

By using the characterizing matrices, we shall express the output $\mathbf{z}(t)$ at any time t in terms of the initial state and the successive inputs. For this and other purposes, (8.3.1) can be written

(8.3.2)
$$\mathbf{y}(t+1) = \mathbf{Y}(t) = \mathbf{A}\mathbf{y}(t) + \mathbf{B}\mathbf{x}(t),$$
$$\mathbf{z}(t) = \mathbf{C}\mathbf{y}(t) + \mathbf{D}\mathbf{x}(t).$$

Theorem 8.3.1. *If the linear machine $M = (\mathbf{A}, \mathbf{B}, \mathbf{C}, \mathbf{D})$ starts at time $t = 0$ in state $\mathbf{y}(0)$, then its state at any later time t is*

(8.3.3)
$$\mathbf{y}(t) = \mathbf{A}^t \mathbf{y}(0) + \sum_{i=0}^{t-1} \mathbf{A}^{t-1-i} \mathbf{B}\mathbf{x}(i),$$

and its output at time t is

(8.3.4)
$$\mathbf{z}(t) = \mathbf{C}\mathbf{A}^t \mathbf{y}(0) + \sum_{i=0}^{t-1} \mathbf{C}\mathbf{A}^{t-1-i} \mathbf{B}\mathbf{x}(i) + \mathbf{D}\mathbf{x}(t).$$

Proof. We have $\mathbf{A}^0 = \mathbf{I}$ and the summations are vacuous for $t = 0$, so (8.3.3) and (8.3.4) hold for $t = 0$. If they hold for t, then, by (8.3.2),

$$\mathbf{y}(t+1) = \mathbf{A}\mathbf{y}(t) + \mathbf{B}\mathbf{x}(t) = \mathbf{A}\left[\mathbf{A}^t \mathbf{y}(0) + \sum_{i=0}^{t-1} \mathbf{A}^{t-1-i} \mathbf{B}\mathbf{x}(i)\right] + \mathbf{B}\mathbf{x}(t)$$

$$= \mathbf{A}^{t+1}\mathbf{y}(0) + \sum_{i=0}^{t} \mathbf{A}^{t-i} \mathbf{B}\mathbf{x}(i)$$

and

$$\mathbf{z}(t+1) = \mathbf{C}\mathbf{y}(t+1) + \mathbf{D}\mathbf{x}(t+1) = \mathbf{C}\mathbf{A}^{t+1}\mathbf{y}(0) + \sum_{i=0}^{t} \mathbf{C}\mathbf{A}^{t-i}\mathbf{B}\mathbf{x}(i) + \mathbf{D}\mathbf{x}(t+1),$$

so (8.3.3) and (8.3.4) hold for $t + 1$. ∎

In (8.3.4), $\mathbf{z}_1(t) = \mathbf{C}\mathbf{A}^t \mathbf{y}(0)$ is called the **autonomous response**, the output due to the initial state $\mathbf{y}(0)$ of the machine if all inputs are $\mathbf{0}$. The sum $\mathbf{z}_2(t) = \sum_{i=0}^{t-1} \mathbf{C}\mathbf{A}^{t-1-i} \mathbf{B}\mathbf{x}(i) + \mathbf{D}\mathbf{x}(t)$ is the **forced response**, the output due to the inputs if the initial state $\mathbf{y}(0)$ is $\mathbf{0}$.

An **autonomous linear machine** is a linear machine with $l = 0$, that is, with input space $F = F_0 = [0]$. The feedback shift register is one example. In this case no inputs are drawn in the machine diagram, no $\mathbf{x}(i)$'s appear in (8.3.1) to (8.3.4), $\mathbf{B}$ and $\mathbf{D}$ are not needed, and (8.3.3) and (8.3.4) become

(8.3.5) $\qquad \mathbf{y}(t) = \mathbf{A}^t \mathbf{y}(0), \qquad \mathbf{z}(t) = \mathbf{C}\mathbf{A}^t\mathbf{y}(0) = \mathbf{C}\mathbf{y}(t).$

Clearly, a linear machine is a Moore machine iff $\mathbf{D} = \mathbf{0}$; autonomous machines are Moore machines.

An **inert** or **quiescent** linear machine is a linear machine with initial state $\mathbf{y}(0) = \mathbf{0}$. The feedforward shift register used for polynomial multiplication in Example 2 of the previous section is an inert two-terminal machine. Equations (8.3.3) and (8.3.4) for an inert machine become

(8.3.6) $\quad \mathbf{y}(t) = \sum_{i=0}^{t-1} \mathbf{A}^{t-1-i}\mathbf{B}\mathbf{x}(i), \qquad \mathbf{z}(t) = \sum_{i=0}^{t-1} \mathbf{C}\mathbf{A}^{t-1-i}\mathbf{B}\mathbf{x}(i) + \mathbf{D}\mathbf{x}(t).$

8.4 Exercises

1. (a) No separate multipliers are needed when constructing a binary machine. Explain why. (Note that no multipliers appear in Figure 8.2.4.)

(b) The adder in a binary machine does not perform the operation "+" of the Boolean algebra $\mathbb{B} = \{0, 1\}$ of Chapter 3. Explain why.

*2. (a) Let S be any set, F a field, F^S the set of all functions $h: S \to F$. For any $h_1, h_2 \in F^S$ and $\alpha \in F$, define $h_1 + h_2 \in F^S$ and $\alpha h_1 \in F^S$ by

$$(h_1 + h_2)(s) = h_1(s) + h_2(s), \qquad (\alpha h_1)(s) = \alpha \cdot h_1(s), \qquad \text{all } s \in S.$$

Show that F^S is a vector space over F with these operations. Show that $\dim_F F^S = |S|$ if S is finite, and that F^S is infinite-dimensional if S is infinite.

(b) Denote $k = \{k, k+1, k+2, \ldots\} \subseteq \mathbb{Z}$, so $\bar{0} = \mathbb{N}$, $\bar{1} = \mathbb{P}$. For any $w \in F^{\mathbb{N}}$, define $D^1 = D: F^{\mathbb{N}} \to F^{\mathbb{P}}$ by $(D(w))(t) = w(t-1)$, and for $k > 1$ define $D^k: F^{\mathbb{N}} \to F^{\bar{k}}$ inductively by $D^k = D \circ D^{k-1}$. [If $w(t) = a_t$ for $t = 0, 1, 2, \ldots$, Table 8.4.1 illustrates the values of the functions $D^i(w)$.] Show that each D^k is a linear transformation, and that any polynomial $a_0 + a_1 D + \cdots + a_k D^k$, $a_i \in F$, is also a linear transformation from $F^{\mathbb{N}}$ to $F^{\bar{k}}$. Interpret this in terms of the delay operator D.

(c) Show that if $F = GF(2)$, then F^S is a Boolean algebra. (Be sure to distinguish OR and exclusive-OR properly.)

TABLE 8.4.1

	If $t=0$	1	2	3	4	5	$\cdots$
and	$w(t) = a_0$	a_1	a_2	a_3	a_4	a_5	$\cdots$
then	$(D(w))(t) =$	a_0	a_1	a_2	a_3	a_4	$\cdots$
	$(D^2(w))(t) =$		a_0	a_1	a_2	a_3	$\cdots$
	$(D^3(w))(t) =$			a_0	a_1	a_2	$\cdots$

Sec. 8.4 Exercises

3. Show that (8.2.3) and (8.2.7) hold for all $t \geq 0$, if the associated shift registers are inert machines.

*4. (a) Design shift registers that can be used to construct log and antilog tables for $GF(3^4)$ and $GF(2^7)$ (appropriate primitive polynomials may be found in Table 7.21.1).

(b) Generalize (a) to $GF(p^n)$, assuming that $X^n + a_{n-1}X^{n-1} + \cdots + a_1X + a_0$ is a primitive polynomial over $GF(p)$.

5. If $F = GF(5)$, design a feedforward shift register that will multiply $X^4 + 3X^3 + X^2 + 2X + 4$ by $X^3 + 2X^2 + 1$ in $F[X]$. Check that your design is correct by listing one by one the results of successive shifts.

*6. Verify that the circuit of Example 3, Section 8.2, will successfully divide $X^5 + X^3 + 2X^2 + X$ by $X^3 + X^2 + 1$ in $GF(3)[X]$. Try to prove, by induction on the degree r of $d(X)$, that it successfully divides any $d(X)$ by $c(X)$.

7. Find characterizing matrices for the feedback shift register of Figure 8.2.3 and for its modification in Example 3 of Section 8.2.

8. Show that (8.3.4) reduces to (8.2.3) for the feedforward shift register.

*9. (a) Use the characterizing matrices to show that any k-dimensional linear machine can be constructed as in Figure 8.4.1, with k delays and an unlimited number of adders and multipliers.

(b) Conversely, show that any machine constructed as in Figure 8.4.1 is a linear machine. (There are no "loops" in the combinational logic; signals flow only to the right in that region.)

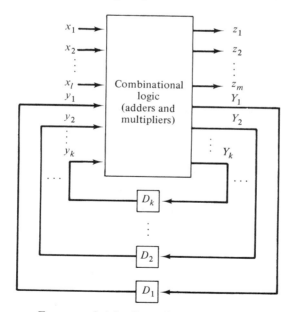

FIGURE 8.4.1. *General Linear Machine*

10. In Example 3 of Section 8.2, we assumed that we were dividing by a monic $c(X)$. What if $c(X)$ is not monic?

8.5 The Distinguishing Matrix

In this section $M = (\mathbf{A}, \mathbf{B}, \mathbf{C}, \mathbf{D})$ is any linear machine $[F_k, F_l, F_m, \tau, \omega]$ over F. We now combine the equations in (8.3.4):

$$\mathbf{z}(t) = \mathbf{CA}^t \mathbf{y}(0) + \sum_{i=0}^{t-1} \mathbf{CA}^{t-1-i} \mathbf{Bx}(i) + \mathbf{Dx}(t),$$

for $0 \leq t \leq j-1$, into one big matrix equation $(8.5.1)_j$:

$$(8.5.1)_j \quad \begin{bmatrix} \mathbf{z}(0) \\ \mathbf{z}(1) \\ \mathbf{z}(2) \\ \vdots \\ \mathbf{z}(j-1) \end{bmatrix} = \begin{bmatrix} \mathbf{C} \\ \mathbf{CA} \\ \mathbf{CA}^2 \\ \vdots \\ \mathbf{CA}^{j-1} \end{bmatrix} \mathbf{y}(0)$$

$$+ \begin{bmatrix} \mathbf{D} & 0 & 0 & \cdots & 0 & 0 \\ \mathbf{CB} & \mathbf{D} & 0 & \cdots & 0 & 0 \\ \mathbf{CAB} & \mathbf{CB} & \mathbf{D} & \cdots & 0 & 0 \\ \vdots & \vdots & \vdots & \ddots & \vdots & \vdots \\ \mathbf{CA}^{j-2}\mathbf{B} & \mathbf{CA}^{j-3}\mathbf{B} & \mathbf{CA}^{j-4}\mathbf{B} & \cdots & \mathbf{CB} & \mathbf{D} \end{bmatrix} \begin{bmatrix} \mathbf{x}(0) \\ \mathbf{x}(1) \\ \mathbf{x}(2) \\ \vdots \\ \mathbf{x}(j-1) \end{bmatrix},$$

which we abbreviate as

$$(8.5.2)_j \qquad \mathbf{Z}^{(j)} = \mathbf{G}_j \mathbf{y}(0) + \mathbf{H}_j \mathbf{X}^{(j)}.$$

[Here, $\mathbf{Z}^{(j)}$ is $mj \times 1$, $\mathbf{G}_j$ is $mj \times k$, $\mathbf{y}(0)$ is $k \times 1$, $\mathbf{H}_j$ is $mj \times lj$, and $\mathbf{X}^{(j)}$ is $lj \times 1$.]

We are mainly interested in the matrices $\mathbf{G}_j$. Certainly, rank $\mathbf{G}_j \leq$ rank $\mathbf{G}_{j+1}$ for all j, since the rows of $\mathbf{G}_j$ are some of the rows of $\mathbf{G}_{j+1}$.

Lemma 8.5.1. rank $\mathbf{G}_k =$ rank $\mathbf{G}_n$, for any $n \geq k$.

Proof. We are done if $\mathbf{C} = \mathbf{0}$, since then all $\mathbf{G}_i$ have rank 0. If $\mathbf{C} \neq \mathbf{0}$, we shall show that for some s,

$$(8.5.3) \quad 0 < \text{rank } \mathbf{G}_1 < \cdots < \text{rank } \mathbf{G}_{s-1} < \text{rank } \mathbf{G}_s = \text{rank } \mathbf{G}_{s+u}, \quad \text{all } u \geq 0.$$

This will prove the result, since then rank $\mathbf{G}_i \geq i$ for $1 \leq i \leq s$, so $s \leq$ rank $\mathbf{G}_s \leq k$, $\mathbf{G}_s$ being $ms \times k$.

To prove (8.5.3), let s be the smallest integer such that rank $\mathbf{G}_s =$ rank $\mathbf{G}_{s+1}$. Matrices $\mathbf{G}_s$ and $\mathbf{G}_{s+1}$ differ only in that $\mathbf{G}_{s+1}$ contains the rows of $\mathbf{CA}^s$. Equality of the ranks means that any row, say $\mathbf{R}$, of $\mathbf{CA}^s$ is a linear combination $\mathbf{R} = \alpha_1 \mathbf{R}_1 + \cdots + \alpha_w \mathbf{R}_w$, each $\mathbf{R}_i$ a row of some $\mathbf{CA}^{n_i}$, all $n_i < s$.

Sec. 8.5 *The Distinguishing Matrix*   393

Rows of $\mathbf{CA}^{s+1} = (\mathbf{CA}^s)\mathbf{A}$ have form $\mathbf{RA}$, $\mathbf{R}$ a row of $\mathbf{CA}^s$, so $\mathbf{RA} = \alpha_1(\mathbf{R}_1\mathbf{A}) + \cdots + \alpha_w(\mathbf{R}_w\mathbf{A})$, $\mathbf{R}_i\mathbf{A}$ a row of $\mathbf{CA}^{n_i+1}$, $n_i + 1 \leq s + 1$. Hence rank $\mathbf{G}_{s+2}$ = rank $\mathbf{G}_{s+1}$ = rank $\mathbf{G}_s$; similarly, we see that rank $\mathbf{G}_{s+u}$ = rank $\mathbf{G}_s$ for all $u \geq 0$. ∎

The matrix $\mathbf{G}_k$ is called the **distinguishing matrix** (or **diagnostic matrix**) of M.

For any finite-state machine M', whether linear or not, a **distinguishing sequence** for M' is a sequence of inputs that yields a different output sequence for each state of M'. That is, a distinguishing sequence for M' distinguishes *every* pair of states of M' (see Section 2.18). Of course, some or all *pairs* of states may have distinguishing sequences of shorter length. Clearly, a machine with a distinguishing sequence must be in reduced form (Section 2.17), but many machines in reduced form do not have distinguishing sequences.

The latter does not happen for linear machines, as the following theorem reveals.

Theorem 8.5.1. *The following are equivalent for the k-dimensional linear machine $M = (\mathbf{A}, \mathbf{B}, \mathbf{C}, \mathbf{D})$ over F:*

(a) *Some input sequence of length k is a distinguishing sequence.*
(b) *All input sequences of length k are distinguishing sequences.*
(c) rank $\mathbf{G}_k = k$.
(d) *The machine M is in reduced form.*

Proof. For any given input sequence $\mathbf{x}(0), \mathbf{x}(1), \ldots, \mathbf{x}(k-1)$ and possible output sequence $\mathbf{z}(0), \mathbf{z}(1), \ldots, \mathbf{z}(k-1)$, the system of equations $(8.5.2)_k$,

$$\mathbf{Z}^{(k)} = \mathbf{G}_k \mathbf{y}(0) + \mathbf{H}_k \mathbf{X}^{(k)},$$

is a system of mk equations over F for the k unknowns $y_1(0), \ldots, y_k(0)$ [the components of $\mathbf{y}(0)$]. By Exercise 7.11.8, if (c) holds, so that $\mathbf{G}_k$ has rank k, then the input sequence and output sequence determine the initial state uniquely; (a) and (b) hold. Conversely, the same facts about systems of equations show that either (a) or (b) implies (c). We remarked before that (a) implies (d) for any machine, linear or not.

We shall finish by showing that (d) implies (c). If (c) fails, so that rank $\mathbf{G}_k = k_0 < k$, then Lemma 8.5.1 shows that rank $\mathbf{G}_j = k_0$ for any $j \geq k$. Exercise 7.11.8 guarantees that the subspace V of F_k of solutions $\mathbf{y}(0)$ of

$$\mathbf{G}_k \mathbf{y}(0) = \mathbf{0}$$

has dimension $k - k_0 > 0$; V must also be the space of solutions of

$$\mathbf{G}_j \mathbf{y}(0) = \mathbf{0},$$

for any $j \geq k$. Equations $(8.5.2)_j$ now guarantee that all $\mathbf{y}(0) \in V$ have the same output sequence, for any given input sequence of any length j; (d) fails. ∎

8.6 Minimization of Linear Machines

If the $r \times n$ matrix $\mathbf{L}$ and the $n \times r$ matrix $\mathbf{R}$ over F satisfy $\mathbf{LR} = \mathbf{I}_r$, we say that $\mathbf{R}$ is a **right inverse** of $\mathbf{L}$ and $\mathbf{L}$ is a **left inverse** of $\mathbf{R}$. Necessarily, rank $\mathbf{L} = \text{rank } \mathbf{R} = r \leq n$, by Corollary 7.10.1. If $r < n$, then any $r \times n$ $\mathbf{L}$ of rank r will have several right inverses and any $n \times r$ $\mathbf{R}$ of rank r will have several left inverses.

If the $r \times n$ matrix $\mathbf{L}$ has rank r, we can find a right inverse $\mathbf{R}$ of $\mathbf{L}$ as follows. Choose r linearly independent columns of $\mathbf{L}$, say the i_1th, ..., i_rth, and use them to form an (invertible) $r \times r$ matrix $\mathbf{M}$. Find $\mathbf{M}^{-1}$, and use the rows of $\mathbf{M}^{-1}$ as the i_1th, ..., i_rth rows of $\mathbf{R}$, the other rows of $\mathbf{R}$ being all 0's. For example, in the 3×4 matrix

$$\mathbf{L} = \begin{bmatrix} 1 & 0 & 1 & 1 \\ 0 & 1 & 1 & 1 \\ 0 & 1 & 1 & 0 \end{bmatrix}$$

over GF(2), the first, second, and fourth columns are linearly independent. If

$$\mathbf{M} = \begin{bmatrix} 1 & 0 & 1 \\ 0 & 1 & 1 \\ 0 & 1 & 0 \end{bmatrix}, \quad \text{then} \quad \mathbf{M}^{-1} = \begin{bmatrix} 1 & 1 & 1 \\ 0 & 0 & 1 \\ 0 & 1 & 1 \end{bmatrix},$$

so take

$$\mathbf{R} = \begin{bmatrix} 1 & 1 & 1 \\ 0 & 0 & 1 \\ 0 & 0 & 0 \\ 0 & 1 & 1 \end{bmatrix}.$$

(Note the relative positioning of independent columns and rows in $\mathbf{L}$ and $\mathbf{R}$, respectively.)

Theorem 8.6.1 (Minimization Procedure). *Assume that the k-dimensional linear machine $M = (\mathbf{A}, \mathbf{B}, \mathbf{C}, \mathbf{D})$ over F has distinguishing matrix $\mathbf{G}_k$ of rank $r < k$, and let $\mathbf{L}$ be an $r \times k$ matrix whose rows are a basis for the row space of $\mathbf{G}_k$. If $\mathbf{R}$ is a right inverse of $\mathbf{L}$, then the machine $M_0 = (\mathbf{LAR}, \mathbf{LB}, \mathbf{CR}, \mathbf{D})$ is a reduced machine equivalent to M, state $\mathbf{Ls}$ of M_0 being equivalent to state $\mathbf{s}$ of M.*

(If $\mathbf{G}_k$ has rank k, then the result still holds, with $\mathbf{L} = \mathbf{R} = \mathbf{I}_k$, $M = M_0$ being the easiest choice.)

Proof. We have $M = [F_k, F_l, F_m, \tau, \omega]$ and $M_0 = [F_r, F_l, F_m, \tau_0, \omega_0]$, where $\tau(\mathbf{s}, \mathbf{x}) = \mathbf{As} + \mathbf{Bx}$, $\omega(\mathbf{s}, \mathbf{x}) = \mathbf{Cs} + \mathbf{Dx}$, $\tau_0(\mathbf{s}_0, \mathbf{x}) = \mathbf{LARs}_0 + \mathbf{LBx}$, $\omega_0(\mathbf{s}_0, \mathbf{x}) = \mathbf{CRs}_0 + \mathbf{Dx}$, all $\mathbf{s} \in F_k$, $\mathbf{s}_0 \in F_r$, $\mathbf{x} \in F_l$. Define the function $\varphi \colon F_k \to F_r$ by $\varphi(\mathbf{s}) = \mathbf{Ls}$. We will show that

(8.6.1) $$\varphi(\tau(\mathbf{s}, \mathbf{x})) = \tau_0(\varphi(\mathbf{s}), \mathbf{x})$$

and

(8.6.2) $\qquad \omega(\mathbf{s}, \mathbf{x}) = \omega_0(\varphi(\mathbf{s}), \mathbf{x}), \qquad$ all $\mathbf{s} \in F_k, \; \mathbf{x} \in F_l$.

This will show that φ is a homomorphism from M to M_0, M_0 covers M, and state $\mathbf{s}$ is equivalent to state $\varphi(\mathbf{s})$, by Exercise 2.19.12.

Assuming these results for a moment, let $F = GF(q)$, so that M_0 has $|F_r| = q^r$ states. For the input sequence $\mathbf{X}^{(k)} = \mathbf{0}$, the set of possible outputs $\{\mathbf{Z}^{(k)} = \mathbf{G}_k \mathbf{s} | \mathbf{s} \in F_k\}$ is a vector space of dimension $r = \text{rank } \mathbf{G}_k$ over F. Thus this set has order q^r, and M has q^r *distinguishable* states. The fact that M_0, which has exactly q^r states, covers M thus implies that each state in M is equivalent to only one in M_0, and that M_0 is in reduced form.

Proving (8.6.1) and (8.6.2) will now complete the proof. We have

$$\varphi(\tau(\mathbf{s}, \mathbf{x})) = \varphi(\mathbf{As} + \mathbf{Bx}) = \mathbf{LAs} + \mathbf{LBx},$$

$$\tau_0(\varphi(\mathbf{s}), \mathbf{x}) = \tau_0(\mathbf{Ls}, \mathbf{x}) = \mathbf{LARLs} + \mathbf{LBx},$$

$$\omega(\mathbf{s}, \mathbf{x}) = \mathbf{Cs} + \mathbf{Dx},$$

$$\omega_0(\varphi(\mathbf{s}), \mathbf{x}) = \omega_0(\mathbf{Ls}, \mathbf{x}) = \mathbf{CRLs} + \mathbf{Dx}.$$

For any $\mathbf{s} \in F_k$, it suffices to prove that if $\mathbf{s}_1 = \mathbf{s} - \mathbf{RLs}$, then $\mathbf{LAs}_1 = \mathbf{0}$ and $\mathbf{Cs}_1 = \mathbf{0}$. At least we do have $\mathbf{Ls}_1 = \mathbf{Ls} - \mathbf{LRLs} = \mathbf{Ls} - \mathbf{ILs} = \mathbf{0}$. All rows of $\mathbf{G}_k$ are in the row space of $\mathbf{L}$, so $\mathbf{G}_k \mathbf{s}_1 = \mathbf{0}$. By Lemma 8.5.1, $\mathbf{G}_{k+1} \mathbf{s}_1 = \mathbf{0}$, also. Rows of $\mathbf{LA}$ are linear combinations of rows of $\mathbf{G}_k \mathbf{A}$, which are rows of $\mathbf{G}_{k+1}$; so $\mathbf{LAs}_1 = \mathbf{0}$. Rows of $\mathbf{C}$ are rows of $\mathbf{G}_k$, so $\mathbf{Cs}_1 = \mathbf{0}$. ∎

Corollary 8.6.1. *Two states $\mathbf{s}_1$ and $\mathbf{s}_2$ of a linear machine with distinguishing matrix $\mathbf{G}_k$ are equivalent iff $\mathbf{G}_k \mathbf{s}_1 = \mathbf{G}_k \mathbf{s}_2$.*

Proof. By the theorem, $\mathbf{s}_1$ and $\mathbf{s}_2$ are equivalent iff $\mathbf{Ls}_1 = \mathbf{Ls}_2$, and this is true iff $\mathbf{G}_k \mathbf{s}_1 = \mathbf{G}_k \mathbf{s}_2$. ∎

EXAMPLE 1. Consider the k-dimensional feedforward shift register of Figure 8.2.2. We saw in Section 8.3 that it has characterizing matrices

$$\mathbf{A} = \begin{bmatrix} 0 & 1 & 0 & \cdots & 0 \\ 0 & 0 & 1 & \cdots & 0 \\ \vdots & \vdots & & \ddots & \\ 0 & 0 & 0 & \cdots & 1 \\ 0 & 0 & 0 & \cdots & 0 \end{bmatrix}, \quad \mathbf{B} = \begin{bmatrix} a_1 \\ a_2 \\ \vdots \\ a_{k-1} \\ a_k \end{bmatrix}, \quad \mathbf{C} = [1 \; 0 \; 0 \; \cdots \; 0], \quad \mathbf{D} = [a_0].$$

We see that

$$\mathbf{CA} = [0 \; 1 \; 0 \; \cdots \; 0]$$

and

$$\mathbf{CA}^2 = (\mathbf{CA})\mathbf{A} = [0 \; 0 \; 1 \; 0 \; \cdots \; 0], \ldots,$$

so

$$G_k = \begin{bmatrix} C \\ CA \\ \vdots \\ CA^{k-2} \\ CA^{k-1} \end{bmatrix} = \begin{bmatrix} 1 & 0 & \cdots & 0 & 0 \\ 0 & 1 & \cdots & 0 & 0 \\ \vdots & \vdots & \ddots & \vdots & \vdots \\ 0 & 0 & \cdots & 1 & 0 \\ 0 & 0 & \cdots & 0 & 1 \end{bmatrix}$$

has rank k. By Theorem 8.5.1, every feedforward shift register is already in reduced form.

EXAMPLE 2. Consider the four-dimensional binary machine with characterizing matrices

$$A = \begin{bmatrix} 1 & 0 & 0 & 1 \\ 0 & 1 & 0 & 1 \\ 1 & 0 & 1 & 0 \\ 0 & 0 & 1 & 1 \end{bmatrix}, \quad B = \begin{bmatrix} 1 & 0 & 1 \\ 0 & 1 & 1 \\ 0 & 0 & 1 \\ 1 & 1 & 0 \end{bmatrix},$$

$$C = \begin{bmatrix} 1 & 0 & 1 & 0 \\ 0 & 0 & 1 & 1 \end{bmatrix}, \quad D = \begin{bmatrix} 0 & 1 & 0 \\ 1 & 1 & 0 \end{bmatrix}.$$

We have

$$G_k = G_4 = \begin{bmatrix} C \\ CA \\ CA^2 \\ CA^3 \end{bmatrix};$$

we find that

$$CA = \begin{bmatrix} 0 & 0 & 1 & 1 \\ 1 & 0 & 0 & 1 \end{bmatrix}.$$

Since $(1, 0, 0, 1) = (1, 0, 1, 0) + (0, 0, 1, 1)$, the row space of CA is in (is, in fact, equal to) the row space of C. By (8.5.3), this means that rank G_k = rank $C = 2$, and we can take the L of Theorem 8.6.1 to be

$$L = C = \begin{bmatrix} 1 & 0 & 1 & 0 \\ 0 & 0 & 1 & 1 \end{bmatrix}.$$

Then we can take

$$R = \begin{bmatrix} 1 & 0 \\ 0 & 0 \\ 0 & 0 \\ 0 & 1 \end{bmatrix},$$

so **LR** = **I**$_2$. The given machine is equivalent to the two-dimensional machine (**LAR, LB, CR, D**), where

$$\mathbf{LAR} = \begin{bmatrix} 0 & 1 \\ 1 & 1 \end{bmatrix}, \quad \mathbf{LB} = \begin{bmatrix} 1 & 0 & 0 \\ 1 & 1 & 1 \end{bmatrix}, \quad \mathbf{CR} = \begin{bmatrix} 1 & 0 \\ 0 & 1 \end{bmatrix}, \quad \mathbf{D} = \begin{bmatrix} 0 & 1 & 0 \\ 1 & 1 & 0 \end{bmatrix}.$$

Equations (8.3.1) for this machine are

$$\begin{bmatrix} Y_1 \\ Y_2 \end{bmatrix} = \begin{bmatrix} 0 & 1 \\ 1 & 1 \end{bmatrix} \begin{bmatrix} y_1 \\ y_2 \end{bmatrix} + \begin{bmatrix} 1 & 0 & 0 \\ 1 & 1 & 1 \end{bmatrix} \begin{bmatrix} x_1 \\ x_2 \\ x_3 \end{bmatrix},$$

$$\begin{bmatrix} z_1 \\ z_2 \end{bmatrix} = \begin{bmatrix} 1 & 0 \\ 0 & 1 \end{bmatrix} \begin{bmatrix} y_1 \\ y_2 \end{bmatrix} + \begin{bmatrix} 0 & 1 & 0 \\ 1 & 1 & 0 \end{bmatrix} \begin{bmatrix} x_1 \\ x_2 \\ x_3 \end{bmatrix}.$$

A diagram of the reduced machine appears in Figure 8.6.1.

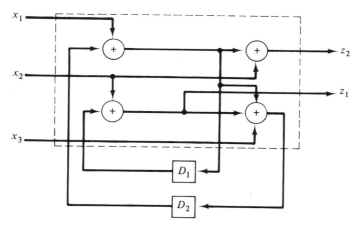

FIGURE 8.6.1. *A Reduced Machine*

A substantial amount of recent research has been devoted to the problem of determining when a given finite-state machine is equivalent to, or at least somehow covered by, a linear machine. References on this work include Davis (1968), Herman (1971), Eichner (1973), and the works cited in those papers. See also Harrison (1969) and Hartmanis and Stearns (1966).

8.7 Exercises

1. Find a finite-state machine that is in reduced form but has no distinguishing sequence.

2. Which of the following matrices have right inverses? In those cases in which a right inverse exists, find one.

$$\begin{bmatrix} 1 & 0 \\ 2 & 1 \\ 1 & -5 \end{bmatrix} \text{ over } \mathbb{R}, \quad \begin{bmatrix} 1 & 1 & 0 \\ 1 & 0 & 0 \\ 1 & 0 & 1 \end{bmatrix} \text{ over } GF(2),$$

$$\begin{bmatrix} 1 & 1 & 0 & 1 \\ 1 & 0 & 1 & 1 \\ 0 & 0 & 1 & 0 \end{bmatrix} \text{ over } GF(2).$$

3. Find *all* right inverses of the matrix

$$\begin{bmatrix} 1 & 0 & 1 \\ 0 & 1 & 1 \end{bmatrix}$$

over GF(2). Are they all obtainable, using the methods at the start of Section 8.6?

*4. Are all feedback shift registers already in reduced form?

*5. Determine whether the following linear machines are in reduced form. In each case when the given machine is not in reduced form, find a reduced machine that is equivalent to the given one.

(a) $\mathbf{A} = \begin{bmatrix} 1 & 2 & 0 \\ 0 & 1 & 1 \\ 2 & 1 & 0 \end{bmatrix}, \quad \mathbf{B} = \begin{bmatrix} 1 & 1 \\ 0 & 2 \\ 1 & 0 \end{bmatrix}, \quad \mathbf{C} = \begin{bmatrix} 1 & 1 & 0 \\ 0 & 0 & 1 \end{bmatrix}, \quad \mathbf{D} = \begin{bmatrix} 1 & 2 \\ 0 & 1 \end{bmatrix}$

over GF(3).

(b) $\mathbf{A} = \begin{bmatrix} 1 & 0 & 1 & 0 \\ 1 & 1 & 0 & 1 \\ 0 & 1 & 1 & 0 \\ 0 & 0 & 0 & 1 \end{bmatrix}, \quad \mathbf{B} = \begin{bmatrix} 1 \\ 0 \\ 0 \\ 1 \end{bmatrix}, \quad \mathbf{C} = [0 \ 1 \ 1 \ 1], \quad \mathbf{D} = [1]$

over GF(2).

(c) $\mathbf{A} = \begin{bmatrix} 0 & 1 & 1 & 0 & 1 \\ 0 & 1 & 0 & 0 & 0 \\ 1 & 0 & 1 & 0 & 0 \\ 1 & 1 & 0 & 1 & 0 \\ 1 & 1 & 0 & 0 & 1 \end{bmatrix}, \quad \mathbf{B} = \begin{bmatrix} 1 & 0 & 0 \\ 0 & 1 & 1 \\ 1 & 0 & 1 \\ 1 & 0 & 1 \\ 0 & 1 & 0 \end{bmatrix},$

$\mathbf{C} = \begin{bmatrix} 1 & 0 & 1 & 0 & 0 \\ 0 & 1 & 1 & 0 & 1 \end{bmatrix}, \quad \mathbf{D} = \begin{bmatrix} 1 & 1 & 0 \\ 0 & 1 & 1 \end{bmatrix}$ over GF(2).

*6. Show that in any linear machine, the number of inequivalent states is 1 or a power of a prime. Conclude that many finite-state machines are not equivalent to linear machines.

8.8 Rational Transfer Functions

Consider any linear machine $M = [F_k, F_l, F_m, \tau, \omega] = (\mathbf{A}, \mathbf{B}, \mathbf{C}, \mathbf{D})$, and suppose that the minimal polynomial of the matrix $\mathbf{A}$ is

$$m(X) = \sum_{j=0}^{r} m_j X^j = m_0 + m_1 X + \cdots + m_{r-1} X^{r-1} + X^r, \qquad m_r = 1.$$

In particular,

$$\mathbf{0} = m(\mathbf{A}) = \sum_{j=0}^{r} m_j \mathbf{A}^j.$$

Fix some $n \in \mathbb{N}$; with starting state $\mathbf{y}(n)$ we get, as for starting state $\mathbf{y}(0)$ in (8.3.3), the equation

$$\mathbf{y}(n+t) = \mathbf{A}^t \mathbf{y}(n) + \sum_{i=0}^{t-1} \mathbf{A}^{t-1-i} \mathbf{B} \mathbf{x}(n+i).$$

Therefore,

$$\sum_{j=0}^{r} m_j \mathbf{y}(n+j) = \sum_{j=0}^{r} m_j \left[\mathbf{A}^j \mathbf{y}(n) + \sum_{i=0}^{j-1} \mathbf{A}^{j-1-i} \mathbf{B} \mathbf{x}(n+i) \right]$$

$$= \left[\sum_{j=0}^{r} m_j \mathbf{A}^j \right] \mathbf{y}(n) + \sum_{j=0}^{r} \sum_{i=0}^{j-1} m_j \mathbf{A}^{j-1-i} \mathbf{B} \mathbf{x}(n+i)$$

$$= \mathbf{0} + \sum_{i=0}^{r-1} \left[\sum_{j=i+1}^{r} m_j \mathbf{A}^{j-1-i} \mathbf{B} \right] \mathbf{x}(n+i)$$

$$= \sum_{i=0}^{r-1} \mathbf{M}_i \mathbf{x}(n+i)$$

for matrices

$$\mathbf{M}_i = \sum_{j=i+1}^{r} m_j \mathbf{A}^{j-1-i} \mathbf{B}, \qquad 0 \leq i \leq r-1.$$

By defining $\mathbf{M}_r = \mathbf{0}$, we have

(8.8.1) $$\sum_{j=0}^{r} m_j \mathbf{y}(n+j) = \sum_{i=0}^{r} \mathbf{M}_i \mathbf{x}(n+i).$$

Using (8.3.2), we can write

$$\sum_{j=0}^{r} m_j \mathbf{z}(n+j) = \sum_{j=0}^{r} m_j [\mathbf{C} \mathbf{y}(n+j) + \mathbf{D} \mathbf{x}(n+j)]$$

$$= \mathbf{C} \cdot \sum_{j=0}^{r} m_j \mathbf{y}(n+j) + \mathbf{D} \cdot \sum_{j=0}^{r} m_j \mathbf{x}(n+j)$$

$$= \mathbf{C} \cdot \sum_{j=0}^{r} \mathbf{M}_j \mathbf{x}(n+j) + \mathbf{D} \cdot \sum_{j=0}^{r} m_j \mathbf{x}(n+j)$$

$$= \sum_{j=0}^{r} [\mathbf{C} \mathbf{M}_j + m_j \mathbf{D}] \mathbf{x}(n+j).$$

We have proved that

$$\sum_{j=0}^{r} m_j \mathbf{z}(n+j) = \sum_{j=0}^{r} \mathbf{L}_j \mathbf{x}(n+j), \tag{8.8.2}$$

for matrices $\mathbf{L}_j = \mathbf{CM}_j + m_j \mathbf{D}$. We can substitute $t = n + r$ in (8.8.2) and get

$$\sum_{j=0}^{r} m_j \mathbf{z}(t-r+j) = \sum_{j=0}^{r} \mathbf{L}_j \mathbf{x}(t-r+j), \qquad \text{all } t \geq r. \tag{8.8.3}$$

Since $m_r = 1$, (8.8.3) can be solved for $\mathbf{z}(t) = \mathbf{z}(t-r+r)$, obtaining an expression for the "present output" $\mathbf{z}(t)$ in terms of "past outputs," "past inputs," and "present input," if $t \geq r$.

In this section we study two-terminal machines, where $l = m = 1$. In this case, $\mathbf{L}_j = [l_j]$ is a 1×1 matrix. The matrices $\mathbf{z}(t-r+j) = z(t-r+j)$ and $\mathbf{x}(t-r+j) = x(t-r+j)$ are single field elements and $m_r = 1$, so (8.8.3) amounts to

$$z(t) = \sum_{j=0}^{r} l_j x(t-r+j) - \sum_{j=0}^{r-1} m_j z(t-r+j), \qquad \text{all } t \geq r. \tag{8.8.4}$$

Now we see that the present output (at time t) is a *linear combination* of the present input and the past inputs and outputs at time $t-r, \ldots, t-1$.

Equation (8.8.4) can be written using the delay operator D:

$$\bar{m}(D)z = h(D)x \qquad \text{at all times } t \geq r, \tag{8.8.5}$$

where $\bar{m}(X) = m_0 X^r + m_1 X^{r-1} + \cdots + m_{r-1} X + m_r = X^r m(1/X)$ is the *reciprocal polynomial* of the rth-degree polynomial $m(X) \in F[X]$. (For properties of reciprocal polynomials, see Exercise 8.10.2; we shall encounter them several times in this chapter and in Chapter 9.) The polynomial $h(X)$ has coefficients l_j, so it depends on $\mathbf{A}$, $\mathbf{B}$, $\mathbf{C}$, and $\mathbf{D}$.

We wish to interpret (8.8.5) as

$$z = \frac{h(D)}{\bar{m}(D)} x; \tag{8.8.6}$$

that is, we want to introduce rational functions (quotients of two polynomials) in the delay operator, and hence have an expression for output in terms of input in any two-terminal machine.

Recall that the delay operator D is in $\text{Hom}_F(F^{\mathbb{N}}, F^{\mathbb{N}})$; that is, for any function $f \in F^{\mathbb{N}}$ (so $f: \mathbb{N} \to F$), $Df \in F^{\mathbb{N}}$ is defined by

$$(Df)(t) = \begin{cases} f(t-1) & \text{if } t \geq 1, \\ 0 & \text{if } t = 0. \end{cases} \tag{8.8.7}$$

(See Exercise 8.4.2 for a discussion of $F^{\mathbb{N}}$ as a vector space. Note also that it is possible to deal with f's applied to a long finite sequence $t = 0, 1, 2, \ldots, w$ of discrete times, but there is no harm and considerable notational conveni-

Sec. 8.8 Rational Transfer Functions

ence if we think of f's defined on all of $\mathbb{N}$.) Powers D^j of $D, j > 1$, are defined by the rule

$$(D^j f)(t) = \begin{cases} f(t-j) & \text{if } t \geq j \\ 0 & \text{if } t < j. \end{cases}$$

Polynomials $p(D) = p_0 + p_1 D + \cdots + p_r D^r \in \text{Hom}_F(F^{\mathbb{N}}, F^{\mathbb{N}})$, the $p_i \in F$, satisfy

(8.8.8) $\quad (p(D)f)(t) = p_0 f(t) + p_1 (Df)(t) + \cdots + p_r (D^r f)(t),$

which equals $p_0 f(t) + p_1 f(t-1) + \cdots + p_r f(t-r)$ if $t \geq r$.

If $a(X), b(X) \in F[X]$ and $a(D) \in \text{Hom}_F(F^{\mathbb{N}}, F^{\mathbb{N}})$ is invertible, then the equations

$$a(D)z = b(D)x \quad \text{and} \quad z = a(D)^{-1} b(D) x$$

are truly equivalent; in this way we hope to justify passage from (8.8.5) to (8.8.6).

Lemma 8.8.1. $a(D) = a_0 + a_1 D + \cdots + a_r D^r \in \text{Hom}_F(F^{\mathbb{N}}, F^{\mathbb{N}})$ *is invertible iff* $a_0 \neq 0$.

Proof. If $a(D)$ is invertible it is onto, so there is a function $f \in F^{\mathbb{N}}$ such that $[a(D)f](0) = 1$. Since $[a(D)f](t) = a_0 f(t) + a_1 (Df)(t) + \cdots + a_r (D^r f)(t)$, we get $1 = [a(D)(f)](0) = a_0 f(0) + 0 + \cdots + 0$, proving that $a_0 \neq 0$.

Conversely, if $a_0 \neq 0$ we shall show that $a(D)$ is one-one and onto. Denote $a_i = 0$ for $i > r$. If $a(D)f_1 = a(D)f_2$ for $f_1 \neq f_2$, choose $i \in \mathbb{N}$ smallest such that $f_1(i) \neq f_2(i)$. Then we have

$$a_0 f_1(i) + a_1 f_1(i-1) + \cdots + a_i f_1(0) = (a(D)f_1)(i)$$
$$= (a(D)f_2)(i) = a_0 f_2(i) + a_1 f_2(i-1) + \cdots + a_i f_2(0),$$

forcing $a_0 f_1(i) = a_0 f_2(i)$, $f_1(i) = f_2(i)$, a contradiction. This proves that $a(D)$ is one-one.

To prove that $a(D)$ is onto, assume $g \in F^{\mathbb{N}}$ given; we shall define $f \in F^{\mathbb{N}}$ with $a(D)f = g$. Pick $f(0) = a_0^{-1} g(0)$ to assure that $(a(D)f)(0) = g(0)$. If we already have defined $f(0), f(1), \ldots, f(j)$ to assure that $(a(D)f)(i) = g(i)$ for $i \leq j$, we wish to define $f(j+1)$ so that

$$(a(D)f)(j+1) = a_0 f(j+1) + a_1 f(j) + \cdots + a_{j+1} f(0)$$

equals $g(j+1)$. Clearly, we are done by defining

$$f(j+1) = a_0^{-1} [g(j+1) - a_1 f(j) - \cdots - a_{j+1} f(0)]. \quad \blacksquare$$

In (8.8.5) the constant term in $\bar{m}(D)$ is $m_r = 1 \neq 0$, so $\bar{m}(D)$ is invertible. Denote

(8.8.9) $\quad F(D) = \{a(D)^{-1} \circ b(D) \in \text{Hom}_F(F^{\mathbb{N}}, F^{\mathbb{N}}) | a(X), b(X) \in F[X],$

$a(X)$ with nonzero constant term$\}$.

We see in Exercise 8.10.3 that $F(D)$ is a commutative subring of $\text{Hom}_F(F^N, F^N)$. Denoting $a(D)^{-1} \circ b(D)$ by $b(D)/a(D)$, addition and multiplication in $F(D)$ obey the familiar rules

(8.8.10)
$$\frac{b(D)}{a(D)} + \frac{d(D)}{c(D)} = \frac{b(D)c(D) + a(D)d(D)}{a(D)c(D)},$$
$$\frac{b(D)}{a(D)} \cdot \frac{d(D)}{c(D)} = \frac{b(D)d(D)}{a(D)c(D)}.$$

We conclude that each rational function $h(D)/\bar{m}(D)$, $\bar{m}(D)$ with nonzero constant term, does have a unique meaning

$$\frac{h(D)}{\bar{m}(D)} : F^N \to F^N,$$

say from input sequence x to output sequence z, so that $\bar{m}(D)z = h(D)x$.

Figure 8.8.1 depicts a circuit that realizes the rational transfer function

$$\frac{b(D)}{a(D)} = \frac{b_0 + b_1 D + \cdots + b_k D^k}{1 + a_1 D + \cdots + a_k D^k}.$$

To see that Figure 8.8.1 does realize $b(D)/a(D)$, we compute:

$Y_k = b_k x - a_k z,$

$Y_i = y_{i+1} + b_i x - a_i z = D Y_{i+1} + b_i x - a_i z \quad$ for $1 \leq i \leq k-1,$

$z = y_1 + b_0 x = D Y_1 + b_0 x = D(D Y_2 + b_1 x - a_1 z) + b_0 x$

$\quad = D^2(D Y_3 + b_2 x - a_2 z) + b_1 D x - a_1 D z + b_0 x = \cdots$

$\quad = D^{k-1}(D Y_k + b_{k-1} x - a_{k-1} z) + b_{k-2} D^{k-2} x - a_{k-2} D^{k-2} z + \cdots + b_1 D x$

$\quad - a_1 D z + b_0 x$

$\quad = (b_0 + b_1 D + \cdots + b_k D^k)x - (a_1 D + a_2 D^2 + \cdots + a_k D^k)z,$

$(1 + a_1 D + a_2 D^2 + \cdots + a_k D^k)z = (b_0 + b_1 D + \cdots + b_k D^k)x,$

so $a(D)z = b(D)x$, as desired.

Given any rational transfer function $b(D)/a(D)$, we can multiply numerator and denominator by a nonzero field element; hence we can assume that the nonzero constant term in the denominator is 1. In Figure 8.8.1, a_k and b_k need not both be nonzero, so numerator and denominator need not have the same degree. Therefore, Figure 8.8.1 gives us a design pattern for any rational transfer function, and hence for any two-terminal linear machine.

Before constructing a circuit to implement a given transfer function $b(D)/a(D)$, one should always divide away any common factors of $a(X)$ and $b(X)$; the new fraction has numerator and denominator of lower

Sec. 8.8 Rational Transfer Functions

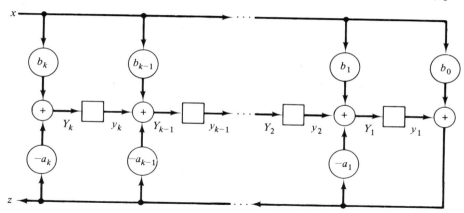

FIGURE 8.8.1. Transfer Function $(b_0 + b_1 D + \cdots + b_k D^k)/(1 + a_1 D + \cdots + a_k D^k)$

degree and can be implemented with fewer delay devices. The requisite algebra is as follows

The greatest common divisor (g.c.d.) of any nonzero polynomials $a(X)$ and $b(X)$ in $F[X]$ is, of course, the monic polynomial $d(X)$ of highest degree such that $d(X)$ divides $a(X)$ and $b(X)$. We write $(a(X), b(X)) = d(X)$; $d(X)$ is found with the following algorithm.

(8.8.11) **Euclidean Algorithm.** *Divide $a(X)$ by $b(X) = r_0(X)$, obtaining*

$$a(X) = q_1(X) r_0(X) + r_1(X), \text{ where } r_1(X) = 0 \text{ or } \deg r_1(X) < \deg r_0(X).$$

Whenever $r_i(X) \neq 0$, divide $r_{i-1}(X)$ by $r_i(X)$, obtaining

$$r_{i-1}(X) = q_{i+1}(X) r_i(X) + r_{i+1}(X), \quad r_{i+1}(X) = 0 \quad \text{or} \quad \deg r_{i+1}(X) < \deg r_i(X).$$

When finally $r_{k-1}(X) \neq 0$ but $r_k(X) = 0$, $d(X)$ is the unique monic scalar multiple of $r_{k-1}(X)$.

[Proof of existence and uniqueness of the g.c.d., and of the fact that this algorithm works, is Exercise 8.10.4 and Theorem 7.16.1(c).]

For example, suppose that we wish to implement the transfer function

(8.8.12) $$T(D) = \frac{b(D)}{a(D)} = \frac{1 + 2D^4 + D^5 + D^6}{2 + 2D + D^2 + D^4 + 2D^5}$$

over $F = GF(3)$. With $a(X) = 2X^5 + X^4 + X^2 + 2X + 2$ and $b(X) = X^6 + X^5 + 2X^4 + 1$, we start by dividing the one, $b(X)$, of higher degree by the one, $a(X) = r_0(X)$, of lower degree:

$$X^6 + X^5 + 2X^4 + 1 = (2X+1)(2X^5 + X^4 + X^2 + 2X + 2)$$
$$+ (X^4 + X^3 + X^2 + 2),$$
$$r_1(X) = X^4 + X^3 + X^2 + 2;$$

$$2X^5 + X^4 + X^2 + 2X + 2 = (2X+2)(X^4 + X^3 + X^2 + 2)$$
$$+ (2X^3 + 2X^2 + X + 1),$$
$$r_2(X) = 2X^3 + 2X^2 + X + 1;$$

$$X^4 + X^3 + X^2 + 2 = (2X)(2X^3 + 2X^2 + X + 1) + (2X^2 + X + 2),$$
$$r_3(X) = 2X^2 + X + 2;$$

$$2X^3 + 2X^2 + X + 1 = (X+2)(2X^2 + X + 2) + 0,$$
$$r_4(X) = 0.$$

Hence $(a(X), b(X))$ is the monic multiple $d(X) = 2(2X^2 + X + 2) = X^2 + 2X + 1$ of $r_3(X)$.

Dividing by $d(X)$, we have

$$a(X) = (X^2 + 2X + 1)(2X^3 + X + 2),$$
$$b(X) = (X^2 + 2X + 1)(X^4 + 2X^3 + X + 1),$$

so the reduced transfer function is

$$(8.8.13) \qquad T_r(D) = \frac{D^4 + 2D^3 + D + 1}{2D^3 + D + 2} = \frac{2 + 2D + D^3 + 2D^4}{1 + 2D + D^3};$$

an implementation is depicted in Figure 8.8.2. Note that if $a_i = b_i = 0$, the corresponding adder is not needed.

In Exercise 8.10.13 you are asked to show that, given the same input sequences, the original and reduced transfer function implementations produce the same outputs for all $t \geq 6$, the dimension of the original machine; if both start with all delay outputs initially 0 (are inert machines), then they produce the same outputs for all t.

In Figure 8.8.1, the dimension (number of delays) of the machine is k, the degree of the polynomial $[a(X)$ or $b(X)]$ of higher degree. The diagram uses two two-input adders and $k-1$ three-input adders. We shall show that a

Sec. 8.8   Rational Transfer Functions

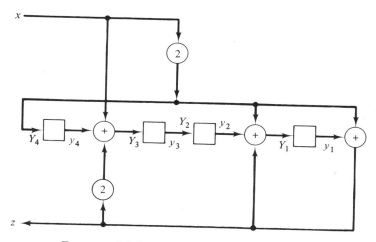

FIGURE 8.8.2. *Transfer Function* (8.8.13)

binary $[F = \text{GF}(2)]$ two-terminal machine of dimension k requires at most k two-input adders and no three-input adders.

We can see this directly from Figure 8.8.1, as follows. Since each a_i or b_i is 0 or 1, the input combination $b_i x - a_i z = b_i x + a_i z$ to the adder with output Y_i, $i < k$, is either 0, x, z, or $x + z$. It requires no adder to add 0 to y_{i+1} and form Y_i, and only a two-input adder to add x, z, or $x + z$ to y_{i+1}, provided that $x + z$ is available. If $b_0 = 1$, the final two-input adder in Figure 8.8.1 has one input x and output z, so the other input is $z - x = x + z$, meaning that $x + z$ is available for use at other adders. If $b_0 = 0$, the final adder in Figure 8.8.1 is not needed; therefore, the adder to create $x + z$, if needed, does not increase the total number of adders. With x, z, and $x + z$ available, the first adder is not needed at all.

For example, consider, over GF(2), the rational transfer function

$$\frac{1 + D^2 + D^3 + D^5 + D^6}{1 + D + D^3 + D^4 + D^7}.$$

(The reader can verify that in GF(2)$[X]$ the polynomials $1 + X^2 + X^3 + X^5 + X^6$ and $1 + X + X^3 + X^4 + X^7$ have no nontrivial common factor, so the machine dimension cannot be reduced by removing such common factors.) We can design a circuit for this transfer function using only two-input adders, using Figure 8.8.1 and the considerations of the last paragraph. The final adder must have input $x + z$, so $x + z$ is available as input to the adder creating Y_3 (needed, since D^3 appears in both numerator and denominator). We obtain the circuit of Figure 8.8.3.

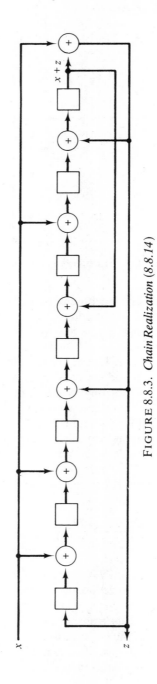

FIGURE 8.8.3. *Chain Realization (8.8.14)*

Sec. 8.8 *Rational Transfer Functions*   407

Figure 8.8.3 can be obtained in another way, with computation instead of memorization of Figure 8.8.1. The transfer function implies the equation

$$(1+D+D^3+D^4+D^7)z = (1+D^2+D^3+D^5+D^6)x,$$

and hence

$$\begin{aligned}
x+z &= (D+D^3+D^4+D^7)z + (D^2+D^3+D^5+D^6)x \\
&= D[(1+D^2+D^3+D^6)z + (D+D^2+D^4+D^5)x] \\
&= D[z + D[(D+D^2+D^5)z + (1+D+D^3+D^4)x]] \\
(8.8.14) \quad &= D[z + D[x + D\{(1+D+D^4)z + (1+D^2+D^3)x\}]] \\
&= D[z + D[x + D\{x+z+D\{(1+D^3)z + (D+D^2)x\}\}]] \\
&= D[z + D[x + D\{x+z+D\{z+D(D^2z+(1+D)x)\}\}]] \\
&= D[z + D[x + D\{x+z+D\{z+D(x+D(x+Dz))\}\}]].
\end{aligned}$$

We can draw Figure 8.8.3 (see p. 406) using the last line in (8.8.14), by starting on the left of the drawing with the innermost set of parentheses and working toward the right. Because of the chain of nested parentheses, we say that we have a **chain realization** of the given transfer function.

We shall conclude this section by pointing out that any inert linear machine $M = [F_k, F_l, F_m, \tau, \omega]$ can be constructed by using lm two-terminal machines.

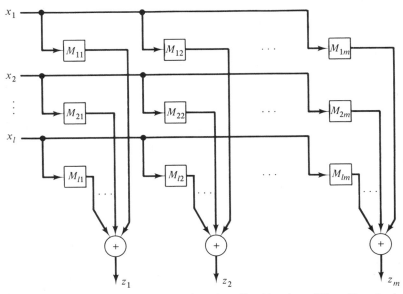

FIGURE 8.8.4. *Any Linear Machine Is a Combination of Two-Terminal Machines*

Lemma 8.8.2. *Any inert linear machine* $M = [F_k, F_l, F_m, \tau, \omega] = (\mathbf{A}, \mathbf{B}, \mathbf{C}, \mathbf{D})$ *is equivalent to a series–parallel combination (Figure 8.8.4) of* lm *two-terminal machines* $\{M_{ij} | 1 \leq i \leq l, 1 \leq j \leq m\}$*, where* $M_{ij} = (\mathbf{A}, \mathbf{B}_i, \mathbf{C}_j, \mathbf{D}_{ji})$*,* $\mathbf{B}_i$ *the* ith *column of* $\mathbf{B}$*,* $\mathbf{C}_j$ *the* jth *row of* $\mathbf{C}$*, and* $\mathbf{D}_{ji}$ *the* 1×1 *matrix whose entry is in the* jth *row,* ith *column of* $\mathbf{D}$*.*

Proof. Using (8.3.6), the jth output z_j in Figure 8.8.4 satisfies

$$z_j(t) = \sum_{i=1}^{l} \left\{ \sum_{s=0}^{t-1} \mathbf{C}_j \mathbf{A}^{t-1-s} \mathbf{B}_i x_i(s) + d_{ji} x_i(t) \right\}$$

$$= \sum_{s=0}^{t-1} \mathbf{C}_j \mathbf{A}^{t-1-s} \mathbf{B} \begin{bmatrix} x_1(s) \\ x_2(s) \\ \vdots \\ x_l(s) \end{bmatrix} + \{j\text{th row of } \mathbf{D}\} \begin{bmatrix} x_1(t) \\ x_2(t) \\ \vdots \\ x_l(t) \end{bmatrix};$$

by (8.3.6), this is also the jth output of M. ∎

8.9 Impulse Response

In this section we discuss *inert two-terminal machines*. We assume that an input sequence $x(0)x(1) \cdots x(w-1)$ of large finite length w is applied; some output sequence $z(0)z(1) \cdots z(w-1)$ is obtained.

Theorem 8.9.1. *If M is an inert two-terminal machine and we define* $T_M: F_w \to F_w$ *by*

$$T_M(\text{input sequence}) = \text{output sequence},$$

then $T_M \in \text{Hom}_F(F_w, F_w)$.

Proof. By $(8.5.2)_j$ for $j = w$, T_M is effected by a matrix multiplication by $\mathbf{H}_w$:

$$\begin{bmatrix} z(0) \\ z(1) \\ \vdots \\ z(w-1) \end{bmatrix} = \mathbf{H}_w \begin{bmatrix} x(0) \\ x(1) \\ \vdots \\ x(w-1) \end{bmatrix}. \blacksquare$$

In the notation of $(8.5.2)_j$ for $j = w$, the input and output sequences are column vectors; but we shall write them as rows [e.g., $x(0)x(1) \cdots x(w-1)$] whenever convenient.

The **impulse response** (of length w) is the output sequence when the input sequence is

$$\mathbf{i}_0 = 10000 \cdots 00 \qquad (\text{length } w).$$

Sec. 8.9 *Impulse Response*

When we work with sequences of fixed length w in an inert machine, we modify the definition of the delay operator D, defining positive powers D^i of D as follows:

(8.9.1) $\quad D^i(\alpha_1\alpha_2\cdots\alpha_w) = \underbrace{000\cdots 0}_{i\ 0\text{'s}}\alpha_1\alpha_2\cdots\alpha_{w-i}.$

Because of Theorem 8.9.1, we can compute the response of the machine to any input sequence $x(0)x(1)\cdots x(t)00\cdots 0$ if we know the impulse response $\mathbf{v}$:

(8.9.2)
$$\begin{aligned}
T_M(x(0)x(1)\cdots x(t)00\cdots 0) \\
= T_M(x(0)\mathbf{i}_0 + x(1)D\mathbf{i}_0 + \cdots + x(t)D^t\mathbf{i}_0) \\
= x(0)T_M(\mathbf{i}_0) + x(1)T_M(D\mathbf{i}_0) + \cdots + x(t)T_M(D^t\mathbf{i}_0) \\
= x(0)\mathbf{v} + x(1)D\mathbf{v} + \cdots + x(t)D^t\mathbf{v}.
\end{aligned}$$

For example, if $w = 9$, $F = GF(5)$, the impulse response is 103210000, and the input sequence is 314000000, then the output is

$3 \cdot 103210000 + 1 \cdot 010321000 + 4 \cdot 001032100$

$= 304130000 + 010321000 + 004023400 = 313424400.$

The sequence $1000\cdots 00 = \mathbf{i}_0$ is constant (after the initial 1), so by Exercise 1.13.17 the impulse response $\mathbf{v}$ is periodic after an initial nonperiodic portion. Conversely, we can, given any eventually periodic sequence $\mathbf{v}$, find an inert two-terminal machine with impulse response $\mathbf{v}$. We illustrate with an example. The sequence

$\mathbf{v} = 10212012120121201212\cdots 01212\cdots$

over $F = GF(3)$ is eventually periodic, so we can write

(8.9.3) $\quad \mathbf{v} = 01212012120121201212\cdots + 1200000\cdots = \mathbf{v}_p + \mathbf{v}_t,$

the sum of a *periodic part* $\mathbf{v}_p$ and a *transient part* $\mathbf{v}_t$, $\mathbf{v}_t$ eventually 0. The input sequence $\mathbf{x}$ that produces output $\mathbf{v}$ is $\mathbf{x} = \mathbf{i}_0 = 10000\cdots 00\cdots$, and we see that

(8.9.4) $\quad\quad\quad\quad \mathbf{v}_t = \mathbf{x} + 2D\mathbf{x} = (1 + 2D)\mathbf{x}$

while

(8.9.5)
$$\begin{aligned}
\mathbf{v}_p &= (D\mathbf{x} + 2D^2\mathbf{x} + D^3\mathbf{x} + 2D^4\mathbf{x}) + D^5(D\mathbf{x} + 2D^2\mathbf{x} + D^3\mathbf{x} + 2D^4\mathbf{x}) \\
&\quad + \cdots \\
&= (1 + D^5 + D^{10} + \cdots)(D + 2D^2 + D^3 + 2D^4)\mathbf{x}.
\end{aligned}$$

The series $1+D^5+D^{10}+\cdots$ is actually finite because $D^u = 0$ whenever $u > w$. Consequently, $1+D^5+D^{10}+\cdots$ is invertible; in fact,
$$(1+D^5+D^{10}+\cdots)^{-1} = 1-D^5 = 1+2D^5$$
because
$$(1-D^5)(1+D^5+D^{10}+\cdots) = 1-D^5+D^5-D^{10}+D^{10}-\cdots = 1.$$
Equation (8.9.5) is equivalent to

(8.9.6) $$\mathbf{v}_p = \frac{D+2D^2+D^3+2D^4}{1+2D^5}\mathbf{x}.$$

Combining (8.9.4) and (8.9.6) gives

(8.9.7)
$$\mathbf{v} = \frac{(1+2D)(1+2D^5)+D+2D^2+D^3+2D^4}{1+2D^5}\mathbf{x}$$
$$= \frac{1+2D^2+D^3+2D^4+2D^5+D^6}{1+2D^5}\mathbf{x}.$$

The Euclidean algorithm reveals that the greatest common divisor of $1+2X^2+X^3+2X^4+2X^5+X^6$ and $1+2X^5$ is $X+2$; after dividing it out of numerator and denominator, we have

(8.9.8) $$\mathbf{v} = \frac{1+D+D^3+2D^5}{1+D+D^2+D^3+D^4}\mathbf{x}.$$

The five-dimensional circuit of Figure 8.9.1 will therefore have the desired impulse response.

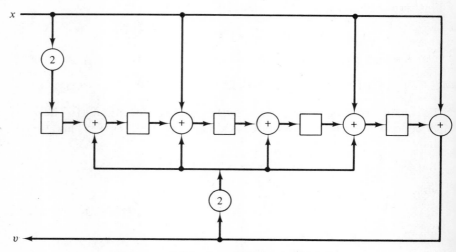

FIGURE 8.9.1. *Impulse Response (8.9.3)*

8.10 Exercises

1. Find the values m_j and L_j, and write out (8.8.2), for the specific machine of Exercise 8.7.5(a).

*2. For any $f(X) \in F[X]$ of degree r, define the **reciprocal polynomial** of $f(X)$ to be $\bar{f}(X) = X^r f(1/X)$.

(a) Show that if γ is a nonzero root of $f(X)$ in an extension field of F, then $1/\gamma$ is a root of $\bar{f}(X)$. Show that all roots of $\bar{f}(X)$ are obtained in this way.

(b) Show that if $f(X)$ is not constant and $f(X)$ has nonzero constant term, then

(1) $f(X)$ is irreducible iff $\bar{f}(X)$ is irreducible;
(2) $f(X)$ is primitive iff $\bar{f}(X)$ is primitive.

(c) Show that if $f(X)$ has nonzero constant term, then $\bar{\bar{f}}(X) = f(X)$.

*3. (a) Let V be a vector space over the field F, and consider $A = \text{Hom}_F(V, V)$. If $f_1, f_2 \in A$ and $\alpha \in F$, define $f_1 + f_2$, $f_1 f_2$, and $\alpha f_1 \in A$ by $(f_1 + f_2)(v) = f_1(v) + f_2(v)$, $(\alpha f_1)(v) = \alpha f_1(v)$, $(f_1 f_2)(v) = f_1(f_2(v))$, all $v \in V$. Show that

(1) A is a ring under this addition and multiplication.
(2) A is a vector space over F with this addition and scalar multiplication.
(3) $\alpha(f_1 f_2) = (\alpha f_1) f_2 = f_1(\alpha f_2)$, all $\alpha \in F$, $f_1, f_2 \in A$.

[Since (1), (2), and (3) hold, we say that A is an **algebra** over F.]

(b) In (a), set $V = F^\mathbb{N}$ (a vector space by Exercise 8.4.2), define $D: V \to V$ by (8.8.7), and show that $D \in \text{Hom}_F(V, V)$. Also show that (8.8.8) holds for any $p(X) = p_0 + p_1(X) + \cdots + p_r X^r$ in $F[X]$.

(c) Finally, define $F(D) \subseteq \text{Hom}_F(V, V)$ as in (8.8.9). Denote $a(D)^{-1} \circ b(D)$ in $F(D)$ by $b(D)/a(D)$ and show that addition and multiplication in $F(D)$ satisfy (8.8.10). Verify that $F(D)$ is both a commutative ring and a subspace (hence, a *subalgebra*) of $\text{Hom}_F(V, V)$.

*4. (a) In the Euclidean algorithm (8.8.11) for the g.c.d. of $a(X)$ and $b(X)$, show that if $c(X)$ is any polynomial that divides $a(X)$ and $b(X)$, then $c(X)$ divides $r_1(X)$, hence $c(X)$ divides $r_2(X), \ldots$, hence $c(X)$ divides $r_{k-1}(X)$.

(b) In (8.8.11), show that $r_{k-1}(X)$ divides $r_{k-2}(X)$, hence $r_{k-1}(X)$ divides $r_{k-3}(X), \ldots$, hence $r_{k-1}(X)$ divides $r_0(X)$, hence $r_{k-1}(X)$ divides $b(X)$ and $a(X)$.

(c) Conclude, using (a) and (b), that the g.c.d. of any two nonzero polynomials exists and may be found by using (8.8.11).

5. Design two-terminal machines, with smallest possible dimension, to realize the following transfer functions:

(a) $\dfrac{1 + D^2 + D^3 + D^5}{1 + D + D^2 + D^4}$ over GF(2).

(b) $\dfrac{1 + D^3 + D^4 + D^5 + D^7}{1 + D + D^2 + D^4 + D^5 + D^7}$ over GF(2).

(c) $\dfrac{1+2D+D^2+2D^4}{2+2D^2+D^3+2D^4+D^5}$ over GF(3).

(d) $\dfrac{1+3D+D^2+2D^4}{4+2D+3D^3+D^4}$ over GF(5).

(e) $\dfrac{1+\gamma D+\gamma^2 D^2+\gamma D^3+D^4}{\gamma^2+D^2+\gamma D^3+D^5}$ over GF(4) = \{0, 1, γ, γ^2\}, where $\gamma^2 = \gamma+1$.

*6. (a) Find the characterizing matrices **A**, **B**, **C**, and **D** of the machine in Figure 8.8.1, with transfer function $b(D)/a(D)$.

(b) If $(a(X), b(X)) = 1$, can you show that the machine of Figure 8.8.1 is minimal?

7. Find minimal chain realizations of the following transfer functions over GF(2):

$$\dfrac{1+D^2+D^4+D^5}{1+D+D^3+D^4+D^5}, \quad \dfrac{D^2+D^3+D^6}{1+D^2+D^4+D^6+D^7},$$

$$\dfrac{D+D^3+D^4+D^5+D^8}{1+D+D^4+D^5+D^6+D^7}.$$

*8. Assume that the machine of Exercise 8.7.5(a) is inert. Construct a realization of it in accordance with Lemma 8.8.2, minimizing each machine M_{ij} as much as possible. Compare this realization with a realization constructed directly from the given matrices **A**, **B**, **C**, and **D**.

9. Construct inert two-terminal machines with the following impulse responses:
   (a) 10101001100110011001 $\cdots$ over GF(2).
   (b) 100110111110111110111110111110 $\cdots$ over GF(2).
   (c) 231401210121012101210121 $\cdots$ over GF(5).

10. Show that $T_M \circ D = D \circ T_M$ in (8.9.2).

11. Can you describe the impulse response of an inert two-terminal machine, given the transfer function $b(D)/a(D)$?

12. Find the transfer functions and impulse responses of the machines in Figure 8.10.1. Simplify those machines whose dimension can be reduced.

*13. Assume that we use Figure 8.8.1 to implement the two transfer functions $a(D)d(D)/b(D)d(D)$ and $a(D)/b(D)$.

(a) Show that, given the same input sequence, the two implementations produce the same outputs for all

$$t \geq \max\{\deg a(D)d(D), \deg b(D)d(D)\}.$$

(b) Show that if the two implementations are inert, then, given the same input sequence, the two implementations produce the same outputs for all t.

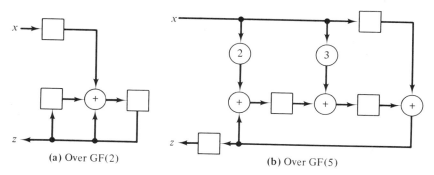

FIGURE 8.10.1. Exercise 8.10.12

8.11 Autonomous Linear Machines

Autonomous linear machines are extensively used as "sequence generators"; they are often important special parts of more complicated devices. We saw in (8.3.5) that the state and output of an autonomous machine satisfy the equations

(8.11.1) $\quad \mathbf{y}(t) = \mathbf{A}^t \mathbf{y}(0), \quad \mathbf{z}(t) = \mathbf{C}\mathbf{A}^t\mathbf{y}(0) = \mathbf{C}\mathbf{y}(t).$

From these equations, we see that the properties of our machine depend on the matrix $\mathbf{A}$; it alone determines the next-state behavior, and the output can be obtained from the state at any time t by a fixed circuit of adders and multipliers (according to matrix $\mathbf{C}$).

We start with a definition and a lemma, applicable to any linear machine. If $\mathbf{P}$ is an invertible $k \times k$ matrix, the linear machines $(\mathbf{A}, \mathbf{B}, \mathbf{C}, \mathbf{D})$ and $(\mathbf{P}^{-1}\mathbf{AP}, \mathbf{P}^{-1}\mathbf{B}, \mathbf{CP}, \mathbf{D})$ are called **similar**.

Lemma 8.11.1. *Any two similar linear machines* $M_1 = (\mathbf{A}, \mathbf{B}, \mathbf{C}, \mathbf{D})$ *and* $M_2 = (\mathbf{P}^{-1}\mathbf{AP}, \mathbf{P}^{-1}\mathbf{B}, \mathbf{CP}, \mathbf{D})$ *are equivalent. State* $\mathbf{y}(0)$ *of* M_1 *is equivalent to state* $\mathbf{P}^{-1}\mathbf{y}(0)$ *of* M_2.

Proof. Assume that M_1 starts in state $\mathbf{y}(0)$ and M_2 starts in state $\mathbf{P}^{-1}\mathbf{y}(0)$, and the input sequence is $\mathbf{x}(0), \mathbf{x}(1), \ldots, \mathbf{x}(t)$. By (8.3.4) the output of M_1 at time t is

$$\mathbf{z}_1(t) = \mathbf{C}\mathbf{A}^t\mathbf{y}(0) + \sum_{i=0}^{t-1} \mathbf{C}\mathbf{A}^{t-1-i}\mathbf{B}\mathbf{x}(i) + \mathbf{D}\mathbf{x}(t)$$

and the output of M_2 is

$$\mathbf{z}_2(t) = (\mathbf{CP})(\mathbf{P}^{-1}\mathbf{AP})^t(\mathbf{P}^{-1}\mathbf{y}(0)) + \sum_{i=0}^{t-1} (\mathbf{CP})(\mathbf{P}^{-1}\mathbf{AP})^{t-1-i}(\mathbf{P}^{-1}\mathbf{B})\mathbf{x}(i) + \mathbf{D}\mathbf{x}(t),$$

so $\mathbf{z}_1(t) = \mathbf{z}_2(t)$. ∎

Because of Lemma 8.11.1, we may assume that in a given autonomous machine the matrix **A** is in rational canonical form; we can use the elementary divisor form of Corollary 7.17.2 or the invariant factor form of Corollary 7.17.4, whichever we prefer.

Suppose first that **A** is the companion matrix

(8.11.2)
$$\mathbf{A} = \begin{bmatrix} 0 & 0 & \cdots & 0 & 0 & \alpha_0 \\ 1 & 0 & \cdots & 0 & 0 & \alpha_1 \\ 0 & 1 & & 0 & 0 & \alpha_2 \\ & & \ddots & & & \\ 0 & 0 & & 1 & 0 & \alpha_{k-2} \\ 0 & 0 & \cdots & 0 & 1 & \alpha_{k-1} \end{bmatrix}$$

of the polynomial

(8.11.3) $\quad c(X) = X^k - \alpha_{k-1} X^{k-1} - \cdots - \alpha_1 X - \alpha_0 \in F[X].$

Since

$$\begin{bmatrix} Y_1 \\ Y_2 \\ \vdots \\ Y_k \end{bmatrix} = \mathbf{A} \begin{bmatrix} y_1 \\ y_2 \\ \vdots \\ y_k \end{bmatrix},$$

we have next-state equations

(8.11.4) $\quad Y_1 = \alpha_0 y_k, \quad Y_i = y_{i-1} + \alpha_{i-1} y_k \quad$ for $2 \leq i \leq k$.

Equations (8.11.4) mean that the internal circuit is the feedback shift register of Figure 8.11.1. Of course, the output variables z_i are linear combinations, depending on **C**, of the y_j's, and need not be the output z of Figure 8.11.1. However, we shall see in Section 8.13 that any such z_i has the form $p(D)z$, $p(D)$ a polynomial in the delay operator.

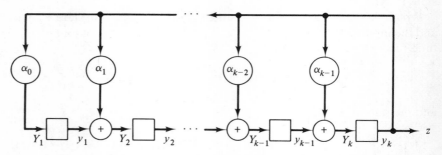

FIGURE 8.11.1. Feedback Shift Register with Characteristic Polynomial (8.11.3)

Sec. 8.11 Autonomous Linear Machines

Clearly, any monic polynomial (8.11.3) determines a feedback shift register (Figure 8.11.1), and conversely, any feedback shift register determines a monic polynomial $c(X)$. We say that (8.11.3) is the **characteristic polynomial** of Figure 8.11.1, and Figure 8.11.1 is the feedback shift register *defined* by (8.11.3).

If **A** is a direct sum

$$\begin{bmatrix} \mathbf{A}_1 & 0 & \cdots & 0 \\ 0 & \mathbf{A}_2 & \cdots & 0 \\ \vdots & \vdots & \ddots & \vdots \\ 0 & 0 & \cdots & \mathbf{A}_r \end{bmatrix}$$

of r companion matrices $\mathbf{A}_i$ of polynomials $c_i(X)$, the internal circuit of the autonomous machine is a union of r unconnected feedback shift registers, the ith one defined by $c_i(X)$.

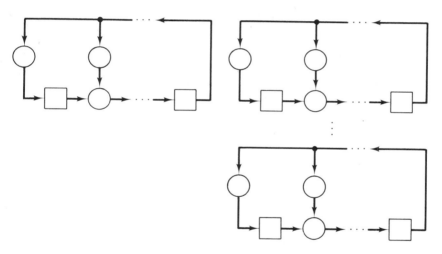

A square matrix **M** over F is called **nonderogatory** if its characteristic and minimal polynomials are equal. By Corollary 7.17.4, this will be true iff the invariant factor rational canonical form of **M** is the companion matrix of a single polynomial, not the direct sum of several. A *one-terminal autonomous machine* is an autonomous linear machine with only one output variable z; that is, $m = 1$. The following theorem relates nonderogatory matrices and one-terminal linear machines.

Theorem 8.11.1. *If the one-terminal autonomous linear machine M is minimal, its characterizing matrix* **A** *is nonderogatory.*

Proof. By passing to a similar machine, we may assume that $\mathbf{A}$ is in invariant factor rational canonical form;

$$\mathbf{A} = \begin{bmatrix} \mathbf{A}_1 & \mathbf{0} & \cdots & \mathbf{0} \\ \mathbf{0} & \mathbf{A}_2 & \cdots & \mathbf{0} \\ \vdots & \vdots & \ddots & \\ \mathbf{0} & \mathbf{0} & & \mathbf{A}_r \end{bmatrix},$$

$\mathbf{A}_i$ the companion matrix of $c_i(X)$, $c_{i+1}(X)|c_i(X)$. It will suffice to assume that $r > 1$ and to prove that the machine is not minimal.

If $r > 1$, then $c_1(X) = X^b - \beta_{b-1} X^{b-1} - \cdots - \beta_1 X - \beta_0$ for some $b < k$ and $\beta_i \in F$. Since $c_1(\mathbf{A}_1) = \mathbf{0}$, we have

$$\mathbf{A}_1^b = \beta_{b-1} \mathbf{A}_1^{b-1} + \cdots + \beta_1 \mathbf{A}_1 + \beta_0 \mathbf{I}.$$

In fact, $\mathbf{0} = c_i(\mathbf{A}_i)$ for every i and $c_i(X)|c_1(X)$, so $\mathbf{0} = c_1(\mathbf{A}_i)$ for every i, implying that

$$\mathbf{A}_i^b = \beta_{b-1} \mathbf{A}_i^{b-1} + \cdots + \beta_1 \mathbf{A}_i + \beta_0 \mathbf{I}$$

for every i. Taken together, these equations imply that

(8.11.5) $$\mathbf{A}^b = \beta_{b-1} \mathbf{A}^{b-1} + \cdots + \beta_1 \mathbf{A} + \beta_0 \mathbf{I}.$$

The diagnostic matrix of the machine is

$$\mathbf{G}_k = \begin{bmatrix} \mathbf{C} \\ \mathbf{CA} \\ \vdots \\ \mathbf{CA}^b \\ \vdots \\ \mathbf{CA}^{k-1} \end{bmatrix},$$

a $k \times k$ matrix since $\mathbf{C}$ is a $1 \times k$ matrix. Equation (8.11.5) implies that

$$\mathbf{CA}^b = \beta_{b-1} \mathbf{CA}^{b-1} + \cdots + \beta_1 \mathbf{CA} + \beta_0 \mathbf{C}.$$

In the notation of (8.5.3) this means that rank $\mathbf{G}_{b+1}$ = rank $\mathbf{G}_b$; hence rank $\mathbf{G}_k$ = rank $\mathbf{G}_b \leq b < k$, implying that the machine is not minimal. ∎

As a partial converse to Theorem 8.11.1, note that we saw in Exercise 8.7.4 that feedback shift registers, which are one-terminal autonomous machines with nonderogatory characterizing matrix $\mathbf{A}$, are always minimal. If we study a minimal one-terminal autonomous machine using the elementary divisor rational canonical form, Theorem 8.11.1 implies that the internal circuit of the machine is a union of unconnected feedback shift

8.12 Cycle Structure of Feedback Shift Registers

We saw in the previous section that feedback shift registers are basic building blocks for all autonomous linear machines. In this section we study the next-state behavior of the FSR of Figure 8.11.1, with characteristic polynomial (8.11.3); in Section 8.13, we shall study its output behavior.

Over $F = GF(q)$ there are q^k states

$$\mathbf{y}(t) = \begin{bmatrix} y_1 \\ y_2 \\ \vdots \\ y_k \end{bmatrix} \in F_k,$$

and the characterizing matrix $\mathbf{A}$ determines the next-state function $\mathbf{y}(t) \to \mathbf{A}\mathbf{y}(t)$. By Lemma 7.17.6(c),

$$c(X) = (-1)^k \det(\mathbf{A} - X\mathbf{I});$$

in particular, setting $X = 0$, we see that $\mathbf{A}$ is invertible iff the constant term $-\alpha_0$ of $c(X)$ is nonzero. Certainly, $\mathbf{A}$ is invertible iff the next-state function $\mathbf{y}(t) \to \mathbf{A}\mathbf{y}(t)$ is a *permutation* of the q^k states.

If $\alpha_0 = 0$, we say that the FSR of Figure 8.11.1 is **singular**; otherwise, it is **nonsingular**. If $\alpha_0 = 0$, the leftmost delay in Figure 8.11.1 will clearly contain 0 in every state following the initial state. Therefore it can be deleted, yielding a smaller-dimensional machine, without affecting the performance of the machine after the first few outputs. Consequently, we shall assume that the FSRs we deal with are nonsingular, whenever it is convenient to do so.

By the **cycle structure** of a nonsingular FSR, we mean the cycle index of the permutation of F_k defined by $\mathbf{y}(t) \to \mathbf{A}\mathbf{y}(t)$; the number of cycles of each length. Certainly,

$$\left\{ \begin{bmatrix} 0 \\ 0 \\ \vdots \\ 0 \end{bmatrix} \right\}$$

is always a cycle of length 1.

While studying the FSR with characteristic polynomial $c(X)$, we use the quotient ring $R = F[X]/\mathscr{I}(c(X))$, $\mathscr{I}(c(X))$ the principal ideal generated by $c(X)$. (See Sections 7.15 and 7.16 for basic properties of this quotient ring.) Since $c(X)$ has degree k, we can think of the elements of R as polynomials over F of degree less than k; the operations in R are addition and multiplication modulo $c(X)$, meaning that to find $a(X)b(X)$ in R we multiply $a(X)$ and $b(X)$ as usual and then divide the product by $c(X)$; the resulting remainder is the product in R. This is true because if

$$a(X)b(X) = q(X)c(X) + r(X), \quad r(X) = 0 \text{ or } \deg r(X) < \deg c(X),$$

then

$$[a(X) + \mathcal{I}(c(X))][b(X) + \mathcal{I}(c(X))] = r(X) + \mathcal{I}(c(X)) \quad \text{in } R.$$

Lemma 8.12.1. *If we think of the present state*

$$\begin{bmatrix} s_0 \\ s_1 \\ \vdots \\ s_{k-1} \end{bmatrix}$$

of the FSR *in Figure 8.11.1 as a polynomial* $s(X) = s_0 + s_1 X + \cdots + s_{k-1} X^{k-1}$ *in* $F[X]/\mathcal{I}(c(X))$, *then shifting the register once yields next state* $Xs(X)$, *modulo* $c(X)$.

Proof. In $F[X]/\mathcal{I}(c(X))$ we have $0 = c(X) = X^k - \alpha_{k-1} X^{k-1} - \cdots - \alpha_1 X - \alpha_0$, so we can replace X^k by $\alpha_{k-1} X^{k-1} + \cdots + \alpha_1 X + \alpha_0$. Hence, in $F[X]/\mathcal{I}(c(X))$, we have

$$\begin{aligned}
Xs(X) &= s_0 X + s_1 X^2 + \cdots + s_{k-2} X^{k-1} + s_{k-1}(\alpha_{k-1} X^{k-1} + \cdots + \alpha_0) \\
&= (\alpha_0 s_{k-1}) + (s_0 + \alpha_1 s_{k-1})X + \cdots + (s_{k-2} + \alpha_{k-1} s_{k-1})X^{k-1},
\end{aligned}$$
(8.12.1)

exactly the register contents after one shift. ∎

Corollary 8.12.1. *Assume that the* FSR *M of Figure 8.11.1 over* $F = \mathrm{GF}(q)$ *has an irreducible characteristic polynomial* $c(X)$, *and assume that a root* γ *of* $c(X)$ *in* $\mathrm{GF}(q^k)$ *has multiplicative order* b. *Then* M *has* 1 *cycle of length* 1 *and* $(q^k - 1)/b$ *cycles of length* b *on its set of* q^k *states. In particular, M has* **maximal period** *(one cycle of length* $q^k - 1$*) iff* $c(X) \in F[X]$ *is primitive.*

Proof. Since $c(X)$ is irreducible, $\mathrm{GF}(q^k) = F(\gamma) = \{s_0 + s_1 \gamma + \cdots + s_{k-1} \gamma^{k-1} | s_i \in F\}$. Since $c(\gamma) = 0$, multiplication by γ in $\mathrm{GF}(q^k)$ proceeds exactly as multiplication by X in $F[X]/\mathcal{I}(c(X))$; (8.12.1) holds with γ in place of X. Multiplication by γ certainly has one cycle of length 1, all other cycles of length b on $F(\gamma)$, since $0 \cdot \gamma = 0$, $\tau \cdot \gamma^i = \tau$ iff $\gamma^i = 1$ for $\tau \neq 0$. By definition, $c(X)$ is primitive iff its roots have multiplicative order $q^k - 1$. [If $c(X)$ is not irreducible, we shall see in Exercise 8.14.3 that M cannot have a cycle of length $q^k - 1$.] ∎

We saw in Theorem 7.21.1 that there always exist primitive polynomials of any degree k over any $\mathrm{GF}(q)$, so maximal-period FSRs always exist. They

Sec. 8.12 Cycle Structure of Feedback Shift Registers

have applications in random-number generators and radar range-measuring experiments; these are discussed in Section 9.5 of Stone (1973). They are also used in "secure" data-transmission systems, for protection from unfriendly eavesdroppers; some digits from the maximal-period FSR are added successively to the digits from the transmitter (and then subtracted at the receiver), making it difficult for the eavesdropper to read the garbled message. If, say, $q = 2$ and $k = 60$, we can construct a maximal-period FSR that cycles 10^7 times per second; since $(2^{60}-1)/10^7$ seconds is approximately 3650 years, the pattern of added digits will not repeat in time to help any eavesdropper. There are $\varphi(2^{60}-1)/60 > 10^{15}$ different primitive polynomials of degree 60 over GF(2), so it may take an eavesdropper some time to find the one being used.

We wish to discuss the cycle structure of the FSR of Figure 8.11.1 when its characteristic polynomial $c(X)$ is not irreducible; a little algebra is needed. If $F = \mathrm{GF}(q)$ and $f(X) \in F[X]$ has nonzero constant term, then the **period** of $f(X)$ is the smallest $t \in \mathbb{P}$ such that $f(X)$ divides $X^t - 1$. If $f(X)$ is *irreducible* of degree k and γ is a root of $f(X)$ in $\mathrm{GF}(q^k)$ of multiplicative order t, then γ is a root of $X^t - 1$ and t is the period of $f(X)$. If $g(X)|(X^t - 1)$ and $h(X)|(X^u - 1)$, then $g(X)h(X)$ divides $(X^t - 1)(X^u - 1)$, which divides $(X^{tu} - 1)^2$, which divides $(X^{tu} - 1)^q = X^{tuq} - 1$ (by Exercise 7.25.5). Using this repeatedly, we see that every $f(X)$ with nonzero constant term has some period.

Lemma 8.12.2. *If $f(X) \in F[X]$ has period t and $f(X)$ divides $X^u - 1$, then t divides u.*

Proof. If $t \nmid u$, then $u = qt + r$, $q, r \in \mathbb{P}$ with $0 < r < t$. Hence $X^u - 1 = X^r(X^{qt} - 1) + (X^r - 1)$. Since $f(X)$ divides $X^u - 1$ and $X^{qt} - 1$, we find that $f(X)$ divides $X^r - 1$, a contradiction. ∎

Lemma 8.12.3. *Assume that $f(X) \in F[X]$ is irreducible and has period t, where $F = \mathrm{GF}(q)$, $q = p^r$, p a prime. Then for any $m \in \mathbb{P}$, $f(X)^m$ has period tp^l, where $l \in \mathbb{N}$ satisfies $p^{l-1} < m \le p^l$.*

Proof. Since $f(X)^m$ divides $(X^t - 1)^m$, which in turn divides $(X^t - 1)^{p^l} = X^{tp^l} - 1$, the period t_m of $f(X)^m$ satisfies $t_m | tp^l$, by Lemma 8.12.2. Also, $f(X)$ divides $f(X)^m$, which divides $X^{t_m} - 1$, so t divides t_m, by Lemma 8.12.2. We conclude that $t_m = tp^j$, some j with $0 \le j \le l$.

Since $f(X)^m$ divides $X^{tp^j} - 1 = (X^t - 1)^{p^j}$, $j < l$ would mean that $p^j < m$, so $f(X)^2$ divides $X^t - 1$. If α is a root of $f(X)$ in some extension field of F, Lemma 7.20.3 now implies that $p|t$, say $t = pt_0$. Hence $X^t - 1 = (X^{t_0} - 1)^p$, and the fact that $f(X)$ divides $X^t - 1$ implies that $f(X)$ divides $X^{t_0} - 1$, contradicting the fact that t is the period of $f(X)$. We conclude that $j = l$. ∎

Theorem 8.12.1. *Assume that the FSR M of Figure 8.11.1 over $F = \mathrm{GF}(q)$, $q = p^r$, p a prime, has characteristic polynomial $c(X) = f(X)^m$, $f(X) \in F[X]$ irreducible of degree d and period t. Choose integers $l_1 = 0, l_2 = 1, \ldots, l_m$*

such that

$$p^{l_i-1} < i \leq p^{l_i} \quad \text{for } 1 \leq i \leq m.$$

Then M has 1 *cycle of length* 1 *and*

$$\frac{q^{di} - q^{d(i-1)}}{tp^{l_i}}$$

cycles of length tp^{l_i}, $1 \leq i \leq m$, *on its set of* q^{dm} *states.*

For example, suppose that $q = 2$, $d = 4$, $m = 4$, and $f(X)$ is primitive, so $t = 15$. Then $l_1 = 0$, $l_2 = 1$, and $l_3 = l_4 = 2$, so we get one cycle of length 1, $(2^4 - 1)/15 = 1$ cycle of length 15, $(2^8 - 2^4)/30 = 8$ cycles of length 30, $(2^{12} - 2^8)/60 = 64$ cycles of length 60, and $(2^{16} - 2^{12})/60 = 1024$ more cycles of length 60.

Proof. As in Lemma 8.12.1, we think of the states as polynomials in $F[X]/\mathcal{I}(c(X))$, of degree less than $dm = k$. Choose a state $s(X)$, say

$$s(X) = a(X)f(X)^j, f(X) \nmid a(X), \quad 0 \leq j < m, \quad \deg a(X) < d(m-j).$$

Then $s(X)$ is in a cycle of length r, where r is the smallest positive integer such that

$$X^r s(X) \equiv s(X) \quad (\text{modulo } c(X)).$$

We have

$$X^r s(X) \equiv s(X)$$
$$\text{iff} \quad (X^r - 1)s(X) \equiv 0$$
$$\text{iff} \quad c(X) | (X^r - 1)s(X)$$
$$\text{iff} \quad f(X)^{m-j} | (X^r - 1),$$

so r is the period of $f(X)^{m-j}$; by Lemma 8.12.3, $r = tp^{l_{m-j}}$.

For any i, $1 \leq i \leq m$, there are q^{di} polynomials $a(X)$ of degree less than di; but $q^{d(i-1)}$ of them are multiples of $f(X)$, so the previous paragraph, with $j = m - i$, applies to $q^{di} - q^{d(i-1)}$ states $s(X)$, yielding $(q^{di} - q^{d(i-1)})/tp^{l_i}$ cycles of length tp^{l_i}. ∎

By the elementary divisor rational canonical form, every FSR is similar to a union of unconnected FSRs of the type studied in Theorem 8.12.1, so we now need only study the cycle-length behavior of a union of unconnected FSRs. To do this, we construct a ring W which we shall call **cycle-set algebra**.

Choose symbols $w_1, w_2, \ldots, w_n, \ldots$ for all $n \in \mathbb{P}$, and let W be the infinite-dimensional vector space over $\mathbb{Q}$ with basis $\{w_1, w_2, \ldots, w_n, \ldots\}$. (Recall that elements of W are *finite* sums $a_1 w_{i_1} + a_2 w_{i_2} + \cdots + a_s w_{i_s}$, the $a_j \in \mathbb{Q}$.) Multiplication in W is defined by the rules

(8.12.2) $$w_a w_b = (a, b) w_{[a,b]}$$

Sec. 8.12 Cycle Structure of Feedback Shift Registers

and

(8.12.3) $$\left(\sum_{j=1}^{r} c_j w_{i_j}\right)\left(\sum_{k=1}^{s} d_k w_{l_k}\right) = \sum_{j=1}^{r} \sum_{k=1}^{s} (c_j d_k)(w_{i_j} w_{l_k}),$$

where (a, b) is the greatest common divisor and $[a, b]$ the least common multiple of a and b (see Section 3.3). For example,

$$(3w_2 + 2w_5)(w_3 + 4w_6 + w_{10})$$
$$= 3(w_2 w_3) + 12(w_2 w_6) + 3(w_2 w_{10}) + 2(w_5 w_3) + 8(w_5 w_6) + 2(w_5 w_{10})$$
$$= 3 \cdot 1 w_6 + 12 \cdot 2 w_6 + 3 \cdot 2 w_{10} + 2 \cdot 1 w_{15} + 8 \cdot 1 w_{30} + 2 \cdot 5 w_{10}$$
$$= 27 w_6 + 16 w_{10} + 2 w_{15} + 8 w_{30}.$$

In Exercise 8.14.7 the reader is asked to verify that W is a commutative ring with identity (in fact, an *algebra* over $\mathbb{Q}$). If an FSR has a_1 cycles of length t_1, a_2 cycles of length $t_2, \ldots,$ and a_s cycles of length t_s on its set of states, then we say that its **cycle sum** is

$$a_1 w_{t_1} + a_2 w_{t_2} + \cdots + a_s w_{t_s} \in W.$$

Theorem 8.12.2. *The cycle sum of a union of unconnected FSRs is the product in the cycle-set algebra W of their individual cycle sums.*

Proof. Let C_1 of length t_1 be a cycle of the first FSR, C_2 of length t_2 a cycle of the second FSR, $\ldots, C_r$ of length t_r a cycle of the rth FSR, and assume that the composite machine M starts with the ith FSR in a state of C_i, each i. Machine M will clearly return to this starting state only after

$$[t_1, t_2, \ldots, t_r] = \text{l.c.m.} \{t_1, t_2, \ldots, t_r\}$$

shifts, so the state of M is in a cycle of length $[t_1, t_2, \ldots, t_r]$. The given cycles represent $t_1 t_2 \cdots t_r$ states of M, and hence contribute the term

(8.12.4) $$\frac{t_1 t_2 \cdots t_r}{[t_1, t_2, \ldots, t_r]} w_{[t_1, t_2, \ldots, t_r]}$$

to the cycle sum of M.

In W, since

$$(a, b) = \frac{ab}{[a, b]}, \qquad \text{[see Exercise 8.14.8(a)]}$$

we have

$$w_{t_1} w_{t_2} \cdots w_{t_r} = \frac{t_1 t_2}{[t_1, t_2]} w_{[t_1, t_2]} w_{t_3} \cdots w_{t_r}$$

$$= \frac{t_1 t_2 [t_1, t_2] t_3}{[t_1, t_2][t_1, t_2, t_3]} w_{[t_1, t_2, t_3]} w_{t_4} \cdots w_{t_r}$$

$$= \cdots = \frac{t_1 t_2 \cdots t_r}{[t_1, t_2, \ldots, t_r]} w_{[t_1, t_2, \ldots, t_r]},$$

so the multiplication in W mimics exactly the performance (8.12.4) of the composite machine. ∎

Using Theorems 8.12.1 and 8.12.2, we can compute the cycle sum of any FSR (or union of unconnected FSRs). For example, consider the FSR over GF(2) with characteristic polynomial

$$(8.12.5) \qquad c(X) = (X^2+X+1)^4(X^3+X+1)^3(X^4+X+1)^2$$

[the factors are primitive over GF(2)]. Its cycle sum is the same as that of a (similar) machine which is the union of three unconnected FSRs, with characteristic polynomials $(X^2+X+1)^4$, $(X^3+X+1)^3$, and $(X^4+X+1)^2$. By Theorem 8.12.1, the cycle sum of a component machine has form

$$(8.12.6) \qquad w_1 + \sum_{i=1}^{m} \frac{q^{di} - q^{d(i-1)}}{tp^{l_i}} w_{tp^{l_i}},$$

where the values in our example are: for $(X^2+X+1)^4$, $m=4$, $q=p=2$, $d=2$, $t=3$, $l_1=0$, $l_2=1$, and $l_3=l_4=2$, giving cycle sum $w_1+w_3+2w_6+20w_{12}$; for $(X^3+X+1)^3$, $m=3$, $q=p=2$, $d=3$, $t=7$, $l_1=0$, $l_2=1$, and $l_3=2$, giving cycle sum $w_1+w_7+4w_{14}+16w_{28}$; and for $(X^4+X+1)^2$, $m=2$, $q=p=2$, $d=4$, $t=15$, $l_1=0$, and $l_2=1$, giving cycle sum $w_1+w_{15}+8w_{30}$.

By Theorem 8.12.2, then, the cycle sum of the FSR with characteristic polynomial (8.12.5) must be

$$(w_1+w_3+2w_6+20w_{12})(w_1+w_7+4w_{14}+16w_{28})(w_1+w_{15}+8w_{30});$$

when we multiply out this product in W, the final result is

$$w_1 + w_3 + 2w_6 + w_7 + 20w_{12} + 4w_{14} + 4w_{15} + w_{21} + 16w_{28} + 134w_{30}$$
$$+ 22w_{42} + 1020w_{60} + 1540w_{84} + 4w_{105} + 1222w_{210} + 78812w_{420}.$$

We see that $78812 \cdot 420$, or 98.65 per cent, of all the 2^{25} states are in cycles of length 420; only $4 \cdot 105$ of the states are in cycles of length 105. We can find a state of the composite machine in a cycle of desired length, if we need to, by retracing the details of the computation.

8.13 Recursive Equations

Now, using the delay operator D, we discuss the output behavior of the FSR of Figure 8.11.1. We have

$$Y_1 = \alpha_0 y_k, \qquad Y_i = y_{i-1} + \alpha_{i-1} y_k \qquad \text{for } 2 \leq i \leq k,$$

Sec. 8.13 Recursive Equations

and $DY_i = y_i$, so we can write

$$y_k = DY_k = D(y_{k-1} + \alpha_{k-1}y_k) = D^2 Y_{k-1} + \alpha_{k-1} Dy_k$$
$$= D^2(y_{k-2} + \alpha_{k-2}y_k) + \alpha_{k-1} Dy_k = \cdots$$
$$= D^{k-1} y_1 + \alpha_1 D^{k-1} y_k + \cdots + \alpha_{k-2} D^2 y_k + \alpha_{k-1} Dy_k$$
$$= \alpha_0 D^k y_k + \alpha_1 D^{k-1} y_k + \cdots + \alpha_{k-2} D^2 y_k + \alpha_{k-1} Dy_k.$$

Since $z = y_k$, this last equation says that

(8.13.1) $\qquad z = \alpha_0 D^k z + \alpha_1 D^{k-1} z + \cdots + \alpha_{k-2} D^2 z + \alpha_{k-1} Dz.$

That is,

(8.13.2)
$$z(t) = \alpha_0 z(t-k) + \alpha_1 z(t-k+1) + \cdots$$
$$+ \alpha_{k-2} z(t-2) + \alpha_{k-1} z(t-1),$$

a *recursive equation* expressing the present output in terms of past outputs, for $t \geq k$.

Equation (8.13.1) can be rewritten

$$(1 - \alpha_{k-1}D - \alpha_{k-2}D^2 - \cdots - \alpha_1 D^{k-1} - \alpha_0 D^k)z = 0;$$

that is,

(8.13.3) $\qquad\qquad\qquad \bar{c}(D)z = 0,$

where $\bar{c}(X)$ is the reciprocal polynomial of the characteristic polynomial $c(X)$ of the FSR.

Conversely, given any recursive equation

(8.13.4)
$$z(t) = \beta_0 z(t-k) + \beta_1 z(t-k+1) + \cdots$$
$$+ \beta_{k-2} z(t-2) + \beta_{k-1} z(t-1), \quad \beta_0 \neq 0,$$

it is clear that the FSR with characteristic polynomial

$$X^k - \beta_{k-1} X^{k-1} - \beta_{k-2} X^{k-2} - \cdots - \beta_1 X - \beta_0$$

will produce outputs satisfying the given equation. In fact, *every* sequence satisfying (8.13.4) will be an output, as we see in the following lemma.

Lemma 8.13.1. *In Figure 8.11.1, for any given set of values*

$$\begin{bmatrix} z(0) \\ z(1) \\ \vdots \\ z(k-1) \end{bmatrix} \in F_k, \text{ there is an initial state } \begin{bmatrix} y_1(0) \\ y_2(0) \\ \vdots \\ y_k(0) \end{bmatrix} = \mathbf{y}(0)$$

Proof. By $(8.5.1)_k$ we have

(8.13.5)
$$\begin{bmatrix} z(0) \\ z(1) \\ \vdots \\ z(k-1) \end{bmatrix} = \begin{bmatrix} \mathbf{C} \\ \mathbf{CA} \\ \vdots \\ \mathbf{CA}^{k-1} \end{bmatrix} \begin{bmatrix} y_1(0) \\ y_2(0) \\ \vdots \\ y_k(0) \end{bmatrix}.$$

Here $\mathbf{C} = [0 \ 0 \cdots 0 \ 0 \ 1]$ and, by (8.11.2),

$$\mathbf{A} = \begin{bmatrix} 0 & 0 & \cdots & 0 & 0 & \alpha_0 \\ 1 & 0 & & 0 & 0 & \alpha_1 \\ & \ddots & & & \vdots & \vdots \\ 0 & 0 & & 1 & 0 & \alpha_{k-2} \\ 0 & 0 & \cdots & 0 & 1 & \alpha_{k-1} \end{bmatrix},$$

so

$$\begin{bmatrix} \mathbf{C} \\ \mathbf{CA} \\ \mathbf{CA}^2 \\ \vdots \\ \mathbf{CA}^{k-1} \end{bmatrix} = \begin{bmatrix} 0 & 0 & 0 & \cdots & 0 & 0 & 1 \\ 0 & 0 & 0 & & & 0 & 1 & \alpha_{k-1} \\ \vdots & & & & \cdot^{\cdot^{\cdot}} & & \\ 0 & 1 & \cdot^{\cdot^{\cdot}} & & & & etc. \\ 1 & & & & & & \end{bmatrix}$$

is invertible, meaning that the system (8.13.5) can always be solved for the unknowns $y_i(0)$ in terms of the knowns $z(j)$. ∎

Thus, nonsingular FSRs provide an exact tool for producing solutions to recursive equations over $GF(q)$.

As remarked in Section 8.11, any autonomous linear machine is similar to one with an unconnected union of FSRs as its internal circuit. An output variable z_i of an FSR, via the ith row $[c_1 \ c_2 \cdots c_k]$ of matrix $\mathbf{C}$, is some linear combination $z_i = c_1 y_1 + c_2 y_2 + \cdots + c_k y_k$ of the y_i's, not necessarily $z = y_k$. However, we find that

$$y_1 = DY_1 = \alpha_0 D y_k,$$
$$y_2 = DY_2 = D(y_1 + \alpha_1 y_k) = \alpha_0 D^2 y_k + \alpha_1 D y_k,$$
(8.13.6) $\quad y_3 = \alpha_0 D^3 y_k + \alpha_1 D^2 y_k + \alpha_2 D y_k,$
$$\vdots$$
$$y_{k-1} = \alpha_0 D^{k-1} y_k + \alpha_1 D^{k-2} y_k + \cdots + \alpha_{k-2} D y_k,$$

and, consequently,

$$z_i = c_1 y_1 + c_2 y_2 + \cdots + c_k y_k = p(D) y_k = p(D) z$$

Sec. 8.14 *Exercises*

for some $p(X) \in F[X]$ of degree at most $k-1$. In particular, $\bar{c}(D)z_i = \bar{c}(D)p(D)z = p(D)\bar{c}(D)z = 0$, so z_i satisfies the same recursive equation as z and could be obtained as z, for some choice of starting state.

If $\alpha_1\alpha_2\cdots\alpha_m\alpha_1\alpha_2\cdots\alpha_m\alpha_1\alpha_2\cdots\alpha_m\alpha_1\alpha_2\cdots\alpha_m\cdots$ is any periodic sequence of elements of the field F, then the sequence $\alpha_{k+1}\alpha_{k+2}\cdots\alpha_m\alpha_1\alpha_2 \cdots \alpha_k\alpha_{k+1}\alpha_{k+2}\cdots\alpha_m\alpha_1\alpha_2\cdots\alpha_m\alpha_1\alpha_2\cdots\alpha_m\cdots$ is a *periodic shift* of the first. For example, the sequence $10211021102110211021\cdots$ is a periodic shift of $2110211021102110\cdots$ over GF(3). We now discuss the periodic behavior of the output sequence z of the FSR of Figure 8.11.1.

Theorem 8.13.1. *The output sequences of the nonsingular FSR of Figure 8.11.1, with characteristic polynomial $c(X)$, form a k-dimensional vector space over GF(q). Suppose the FSR has a_j cycles of length t_j on its set of states. All output sequences are periodic; declare two output sequences equivalent if one is a periodic shift of the other. Then there are exactly a_j equivalence classes of output sequences of period t_j, each j.*

Proof. We have seen in Lemma 8.13.1 that the first k outputs may be any elements of GF(q), and in (8.13.2) that the remaining outputs are then determined, so there are q^k output sequences. If two sequences satisfy (8.13.2), then their sum, and any scalar multiple of one of them, also satisfies (8.13.2), so the output sequences do form a vector space. (We can think of them either as a subspace of the vector space of sequences of large finite length w, or as a subspace of the vector space of infinite sequences.)

If the FSR starts in a state lying in a cycle of length t_j, then the output will certainly be periodic, of period dividing t_j. On the other hand, (8.13.6) shows that every state variable is a fixed polynomial in D of the state variable $y_k = z$, so the length of the state cycle can be no longer than the length of the period of the output sequence; if the output repeated, the state would, also. ∎

8.14 Exercises

1. Does Theorem 8.11.1 hold for autonomous linear machines with more than one output terminal ($m > 1$)?

2. Describe the next-state function (is it a permutation?) for the singular binary FSR with characteristic polynomial $X^3 + X$.

*3. Show that if the k-dimensional FSR over GF(q) has a cycle of maximal length $q^k - 1$ on its set of q^k states, then its characteristic polynomial $c(X)$ is irreducible (and hence primitive).

*4. Investigate whether an FSR with only two different cycle lengths (one of which is 1) must have an irreducible characteristic polynomial.

5. Suppose that $b \in \mathbb{P}$ divides $q^k - 1$ but does not divide $q^j - 1$ for any $j < k$. Show that GF(q^k) contains an element γ of order b such that

$f(X) = \text{Irr}(\gamma, \text{GF}(q))$ is irreducible. Conclude that some k-dimensional FSR over $\text{GF}(q)$ has all its cycles of length 1 or b on its set of states.

6. Each polynomial in parentheses below is known to be primitive. Find the number of cycles of each length of FSRs with the given characteristic polynomials.
  (a) $c_1(X) = (X^4 + X^3 + 1)^5$ over $\text{GF}(2)$.
  (b) $c_2(X) = (X + 1)^{17}$ over $\text{GF}(2)$.
  (c) $c_3(X) = (X^2 + 2X + 1)^5$ over $\text{GF}(3)$.
  (d) $c_4(X) = (X^5 + X^2 + 1)^6$ over $\text{GF}(2)$.

*7. (a) Show that cycle-set algebra W (Section 8.12) is a commutative ring with identity w_1.
  (b) Show that $a(\omega_1\omega_2) = (a\omega_1)\omega_2 = \omega_1(a\omega_2)$ for all $a \in \mathbb{Q}$, $\omega_1, \omega_2 \in W$, so W is an algebra over $\mathbb{Q}$. [See Exercise 8.10.3(a).]

8. Show that the greatest common divisor (a, b) and least common multiple $[a, b]$ of two positive integers a and b exist and are unique, and prove that $(a, b)[a, b] = ab$.

Also extend this discussion, where possible, to the g.c.d. and l.c.m. of r integers $a_1, a_2, \ldots, a_r \in \mathbb{P}$.

9. Find the cycle sums of the FSRs with the following characteristic polynomials:
  (a) $c_5(X) = (X+1)^4(X^2+X+1)^3$ over $\text{GF}(2)$.
  (b) $c_6(X) = (X^2+1)^2(X+2)^5$ over $\text{GF}(3)$.
  (c) $c_7(X) = (X+1)(X+2)^2(X+3)^3(X+4)^4$ over $\text{GF}(5)$.
  (d) $c_8(X) = (X^3+X+1)^3(X+1)^2$ over $\text{GF}(4)$.

*10. Show that any output sequence from an FSR is periodic, and discuss the length of the period in terms of what was learned in Section 8.12.

8.15 Null Sequences

In this section we return to inert two-terminal machines [starting state $\mathbf{y}(0) = \mathbf{0}$, one input terminal and one output terminal, $l = m = 1$]. We shall use what we know about FSRs to reach some conclusions about these other machines. By a **null sequence** of length w in a k-dimensional inert two-terminal machine M, we mean an input sequence of length w for which all outputs (after the first k) are 0's.

EXAMPLE 1. Consider the binary feedforward shift register (Figure 8.15.1) with transfer polynomial $T(D) = 1 + D^2 + D^3$. We have $z =$

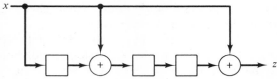

FIGURE 8.15.1. *Transfer Function* $1 + D^2 + D^3$

Sec. 8.15 Null Sequences

$(1+D^2+D^3)x$, so $z(t) = x(t) + x(t-2) + x(t-3)$ for $t \geq 3$. If all $z(t) = 0$ for $t \geq 3$, this says that

(8.15.1) $\qquad x(t) = x(t-2) + x(t-3)$,

a *recursive equation* for $x(t)$ in terms of previous inputs. If the input sequence $x(0)x(1) \cdots x(w-1)$ happens to start with $x(0) = 1$, $x(1) = 0$, $x(2) = 1$, then (8.15.1) enables us to compute $x(3) = x(1) + x(0) = 1$, $x(4) = x(2) + x(1) = 1, \ldots$. We find that

$x(0)x(1)x(2) \cdots = 1011100101110010111100 \cdots 1011100 \cdots$

is periodic with period 7. This is a nontrivial null sequence for the machine of Figure 8.15.1.

EXAMPLE 2. Consider the inert three-dimensional machine over GF(3) in Figure 8.15.2, with transfer function

$$\mathbf{z} = \frac{2+D^2}{1+2D+D^3}\mathbf{x}.$$

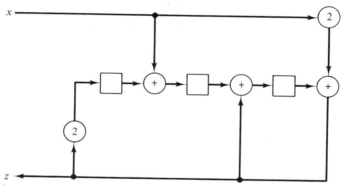

FIGURE 8.15.2. *Transfer Function* $((2+D^2)/(1+2D+D^3))\mathbf{x}$

For $t \geq 3$ we have $(1+2D+D^3)\mathbf{z} = (2+D^2)\mathbf{x}$; hence

$$z(t) + 2z(t-1) + z(t-3) = 2x(t) + x(t-2),$$

implying that

(8.15.2) $\qquad 2z(t) + z(t-1) + 2z(t-3) + x(t-2) = x(t)$.

Equation (8.15.2) says that we achieve $z(3) = 0$ by defining $x(3) = z(2) + 2z(0) + x(1)$. Then we achieve $z(4) = 0$ by defining $x(4) = 2z(1) + x(2)$. Then we achieve $z(5) = 0$ by defining $x(5) = 2z(2) + x(3)$. Having done these, we get $z(t) = 0$ for all $t \geq 6$ by choosing $x(t) = x(t-2)$.

From these two examples, we see that a null sequence for a machine with a polynomial transfer function will be periodic, while for a rational transfer function it will be ultimately periodic, the periodic part depending only on the numerator polynomial of the transfer function.

To study the periodic part, then, we need only study equations

(8.15.3) $$0 = (b_0 + b_1 D + \cdots + b_k D^k)x$$

for $t \geq k$, where $b(X) = b_0 + b_1 X + \cdots + b_k X^k \in F[X]$; we then have the recursive equation

(8.15.4) $$b_0 x(t) = -b_1 x(t-1) - \cdots - b_k x(t-k)$$

for $x(t)$ in terms of past inputs. We shall assume that $b_0 \neq 0$ and $b_k \neq 0$; if either is 0, we can reduce to a similar situation for smaller k, as we see in Exercise 8.18.2. Hence (8.15.4) can be written

(8.15.5) $$x(t) = -(b_1 b_0^{-1})x(t-1) - \cdots - (b_k b_0^{-1})x(t-k),$$

a recursive equation of the type discussed in Section 8.13. We saw there that a sequence satisfies (8.15.5) iff it is an *output* sequence of the FSR with characteristic polynomial

$$X^k + (b_1 b_0^{-1})X^{k-1} + (b_2 b_0^{-1})X^{k-2} + \cdots + (b_{k-1} b_0^{-1})X + b_k b_0^{-1} = b_0^{-1}\bar{b}(X).$$

Consequently, we may state the following theorem, describing the period of (the periodic part of) a null sequence of an inert two-terminal machine with transfer function $b(D)/a(D)$.

Theorem 8.15.1. *Let M be an inert two-terminal machine over $F = \mathrm{GF}(q)$ with transfer function $b(D)/a(D)$, where $b(X) = b_0 + b_1 X + \cdots + b_k X^k \in F[X]$, $b_0 \neq 0$, $b_k \neq 0$. Then the null sequences of M form a k-dimensional vector space over F. Suppose that the FSR with characteristic polynomial $b_0^{-1}\bar{b}(X)$ has a_j cycles of length t_j on its set of states. Declare two null sequences of M equivalent if one is (eventually) a periodic shift of the other. Then there are exactly a_j equivalence classes of null sequences of M of period t_j. In particular, a null sequence of M has maximal period $q^k - 1$ iff $b(X)$ is primitive; in that case, the only other nonequivalent null sequence is the all-0 sequence.*

Proof. By the remarks immediately preceding this theorem, almost everything follows from Theorem 8.13.1. The last sentence also requires Corollary 8.12.1 and Exercise 8.10.2(b)(2). ∎

For example, consider a binary machine with transfer function $b(D) = 1 + D + D^3 + D^4$. The polynomial $b(X) = X^4 + X^3 + X + 1$ factors as $b(X) = (X^2 + X + 1)(X + 1)^2$.

The FSR with characteristic polynomial $c(X) = \bar{b}(X) = b(X)$ has, by Section 8.12, the cycle sum $(w_1 + w_3)(w_1 + w_1 + w_2) = 2w_1 + w_2 + 2w_3 + w_6$, so there should be, up to equivalence, two null sequences of period 1, one of period 2, two of period 3, and one of period 6. The recursive equation $0 = b(D)x$ yields

$$x(t) = x(t-1) + x(t-3) + x(t-4),$$

and trial with various possibilities for $x(0)$, $x(1)$, $x(2)$, $x(3)$ yields the null sequences

00000000··· and 11111111··· of period 1,

0101010101010101··· of period 2,

100100100100100··· and 110110110110110··· of period 3,

100011100011100011100011100011··· of period 6.

(One can systematically find a null sequence of prescribed period, by tracing the FSR theory.)

8.16 Circulating Shift Registers

A *k-stage circulating shift register* (CSR) over GF(q) is an FSR over GF(q) with characteristic polynomial $X^k - 1$. Such a machine (Figure 8.16.1) certainly returns to its initial state after k shifts, so its cycles on its set of q^k states must all have length dividing k. We denote by $N(q, k)$ the number of cycles of length exactly k of a k-stage CSR over GF(q).

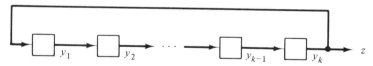

FIGURE 8.16.1. *Circulating Shift Register*

Theorem 8.16.1. *If $d|k$, then a k-stage CSR has exactly $N(q, d)$ cycles of length d on its set of q^k states. Consequently,*

(8.16.1) $$\sum_{d|k} dN(q, d) = q^k.$$

Proof. If a state (k-tuple of register entries) lies in a cycle of length d, it must have the form

$$(s_1, s_2, \ldots, s_d, s_1, s_2, \ldots, s_d, \ldots, s_1, s_2, \ldots, s_d),$$

so that after d shifts the starting state is restored. The d-tuple $(s_1, s_2, \ldots, s_d)$ cannot similarly be a union of repeated segments, or the cycle length will be less than d. We conclude that

(8.16.2) $$(s_1, s_2, \ldots, s_d, s_1, s_2, \ldots, s_d, \ldots, s_1, s_2, \ldots, s_d)$$
$$\leftrightarrow (s_1, s_2, \ldots, s_d)$$

is a one-one correspondence between states of a k-stage CSR in cycles of length d, and states of a d-stage CSR in cycles of length d. Equation (8.16.1) follows by counting all q^k states in the cycles in which they occur. ∎

Corollary 8.16.1. *With $\mu(n)$ the Möbius μ-function of Section 7.24, we have*

$$(8.16.3) \qquad N(q,k) = \frac{1}{k} \sum_{d|k} \mu(k/d) q^d.$$

Proof. Equation (8.16.1) was also obtained in Exercise 7.25.11 for the number of irreducible polynomials of degree k over $\mathrm{GF}(q)$, $q = p^m$. Since (8.16.1) can be used repeatedly to compute $N(q,k)$ for larger and larger k, we conclude that $N(q,k)$ must be the number of irreducible polynomials of degree k over $\mathrm{GF}(q)$, for any k. Equation (8.16.3) then follows from Theorem 7.24.1. ∎

The "coincidence" that $N(q,k)$ is the number of irreducible polynomials of degree k over $\mathrm{GF}(q)$ is explained by Exercise 8.18.10.

8.17 Section 5.17 Revisited

In Section 5.17 one matter was left to brute-force computation; if $n_U \sigma \in G_n$, where $U \subseteq T = \{1, 2, \ldots, n\}$, $\sigma \in S_n$, and n_U and σ act on $c_1 c_2 \cdots c_n \in \mathbb{B}^n$ by

$$(c_1 c_2 \cdots c_n) n_U = d_1 d_2 \cdots d_n, \qquad d_i = \begin{cases} c_i & \text{if } i \notin U \\ c_i' & \text{if } i \in U, \end{cases}$$

$$(c_1 c_2 \cdots c_n) \sigma = c_{1\sigma} c_{2\sigma} \cdots c_{n\sigma},$$

one must compute the cycle index of $n_U \sigma$ on $\mathbb{B}^n$ by actually applying $n_U \sigma$ to elements of $\mathbb{B}^n$.

We shall use FSRs to obtain this cycle index systematically, using ideas of Harrison (1971). First consider the case when $n_U = n_\varnothing = 1$ and $\sigma = (1 \ 2 \ 3 \cdots k)$ is a single cycle of length k.

Theorem 8.17.1. *The element $\sigma = (1 \ 2 \ 3 \cdots k)$ has exactly $N(2, d)$ cycles of length d on $\mathbb{B}^k$, for every divisor d of k.*

Proof. It is enough to show that the set of cycle lengths of σ on $\mathbb{B}^k$ is the same as the set of cycle lengths of a k-stage binary CSR on its set of states. For $\sigma = (1 \ 2 \ 3 \cdots k)$ and $c_1 c_2 c_3 \cdots c_k \in \mathbb{B}^k$ we have

$$(8.17.1) \qquad (c_1 c_2 c_3 \cdots c_k) \sigma = c_2 c_3 \cdots c_k c_1;$$

if we draw the CSR in the form

it is clear that the CSR acts in exactly the same way on its states [k-tuples of elements of GF(2)] as σ acts on $\mathbb{B}^k$ [k-tuples of elements of GF(2)]. ∎

For convenience, Table 8.17.1 gives the first few values of $N(2, d)$. More can be computed from either (8.16.3) or (8.16.1).

TABLE 8.17.1

d	1	2	3	4	5	6	7	8	9	10
$N(2, d)$	2	1	2	3	6	9	18	30	56	99

Next we consider the case when $n_U = n_{\{1\}}$ and $\sigma = (1 \ 2 \ 3 \cdots k)$ is a single cycle of length k. Then we have, for $c_1 c_2 c_3 \cdots c_k \in \mathbb{B}^k$,

(8.17.2) $\quad (c_1 c_2 c_3 \cdots c_k) n_U \sigma = (c_1' c_2 c_3 \cdots c_k)\sigma = c_2 c_3 \cdots c_k c_1'$.

We cannot imitate the complementation operation with an FSR over GF(2); indeed, if an FSR is in the all-0 state, it remains there, never complementing a 0 to get a 1. We shall, however, succeed over GF(3), using the following lemma.

Lemma 8.17.1. *Consider the k-dimensional FSR of Figure 8.17.1 over GF(3) as a permutation of the set S^* of its 2^k states with all components nonzero. Its cycle lengths d satisfy $d | 2k$, $d \nmid k$. For such a d it has T_d cycles of length d, where the sequence of integers $T_2, T_4, T_6, \ldots, T_d, \ldots$ satisfies*

(8.17.3) $\qquad \sum_{\substack{d|2k \\ d \nmid k}} dT_d = 2^k, \quad \text{all } k > 0.$

Proof. After $2k$ shifts, the FSR *always* returns to its original state; each state variable has been multiplied by 2 twice, and $2^2 = 1$ in GF(3). After k

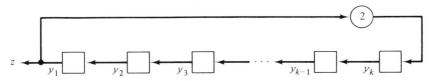

FIGURE 8.17.1. *FSR of Lemma 8.17.1*

shifts, it *never* returns to an initial state from S^*; all (nonzero) components of a state have been multiplied by 2. We conclude that the cycle lengths d on S^* do satisfy $d|2k$, $d \nmid k$. Denote by T_{2k} the number of cycles of Figure 8.17.1 on S^* of length $2k$, for any $k \geq 1$. By the same correspondence as (8.16.2), T_d is the number of cycles of Figure 8.17.1 on S^* of length d, for any d such that $d|2k$, $d \nmid k$. Equation (8.17.3) follows, by counting all the 2^k states in S^*. ∎

Theorem 8.17.2. *Element* $g = n_{\{1\}}(1\ 2\ 3 \cdots k) \in G_k$ *has exactly* T_d *cycles of length d on* $\mathbb{B}^k$, *for every d such that* $d|2k$, $d \nmid k$.

Proof. The FSR of the previous lemma, in one shift, changes from state $(y_1, y_2, y_3, \ldots, y_k)$ to state $(y_2, y_3, \ldots, y_k, 2y_1)$, exactly the behavior (8.17.2) of g on $\mathbb{B}^k$ (replacing 0 by 1, 1 by 2, say, when going from $\mathbb{B}^k$ to S^*). Consequently, the cycle-length information in Lemma 8.17.1 holds for g. ∎

For convenience, the first few values of T_d are listed in Table 8.17.2. Further values can be found by using (8.17.3).

TABLE 8.17.2

d	2	4	6	8	10	12	14	16	18	20
T_d	1	1	1	2	3	5	9	16	28	51

Now we show that, using Theorems 8.17.1 and 8.17.2, we can quickly compute the cycle index on $\mathbb{B}^n$ of a representative of any conjugacy class in G_n. By the **cycle sum** of $g \in G_n$ on $\mathbb{B}^n$ we mean, of course, $a_1 w_{t_1} + a_2 w_{t_2} + \cdots + a_s w_{t_s} \in W$, where g has a_s cycles of length t_s, each s. (The *cycle index* of g is then $x_{t_1}^{a_1} x_{t_2}^{a_2} \cdots x_{t_s}^{a_s}$.)

Theorem 8.17.3. *Given* $g = n_U \sigma \in G_n$, *write* $\sigma = \sigma_1 \sigma_2 \cdots \sigma_r$, *the* σ_i *disjoint cycles on domains* D_i *(so* $\{1, 2, \ldots, n\} = \bigcup_{i=1}^r D_i$, $D_i \cap D_j = \emptyset$ *for* $i \neq j$), *and correspondingly* $U = \bigcup_{i=1}^r U_i$, $U_i = U \cap D_i$ *for each i. Then* $g = (n_{U_1} \sigma_1)(n_{U_2} \sigma_2) \cdots (n_{U_r} \sigma_r)$, *and the cycle sum of g on $\mathbb{B}^n$ is the product in the algebra W of the cycle sums of the factors* $n_{U_i} \sigma_i$ *on* $\mathbb{B}^{D_i}$.

Proof. That g is such a product is clear, since the factors operate on disjoint sets of components of a word in $\mathbb{B}^n$. The rest is proved in the same way that Theorem 8.12.2 was proved. ∎

For example, if $g_1 = (1\ 2\ 3\ 4)$ in the last row of Table 5.17.1, Theorem 8.17.1 says there are $N(2, d)$ cycles of length d for $d = 1, 2, 4$, so g_1 has cycle sum $2w_1 + w_2 + 3w_4$, cycle index $x_1^2 x_2 x_4^3$. If $g_2 = n_{\{1,4\}}(1\ 2\ 3)$ in the fifteenth row of Table 5.17.1, we write $g_2 = n_{\{1\}}(1\ 2\ 3) \cdot n_{\{4\}}(4)$. Using Theorems 8.17.3 and 8.17.2, the cycle sum of g_2 must be

$$(T_2 w_2 + T_6 w_6) w_2 = (w_2 + w_6) w_2 = 2w_2 + 2w_6, \quad \text{cycle index } x_2^2 x_6^2.$$

For a larger example, consider

$$g_3 = n_{\{1,3,4,7,8,10\}}(2\ 3\ 4)(5\ 6\ 7\ 8)(9\ 10) \in G_{10}.$$

We have

$$g_3 = (n_{\{1\}}(1))(n_{\{3,4\}}(2\ 3\ 4))(n_{\{7,8\}}(5\ 6\ 7\ 8))(n_{\{10\}}(9\ 10)).$$

Element g_3 will have the same Young index, and hence be in the same conjugacy class, as

$$n_{\{1\}} \cdot (2\ 3\ 4) \cdot (5\ 6\ 7\ 8) \cdot n_{\{9\}}(9\ 10),$$

so the cycle sum of g_3 is

$$w_2(2w_1 + 2w_3)(2w_1 + w_2 + 3w_4)(T_4 w_4) = (2w_2 + 2w_6)(2w_4 + 2w_4 + 12w_4)$$
$$= 32(w_2 + w_6)w_4 = 64w_4 + 64w_{12}$$

and the cycle index of g_3 is $x_4^{64} x_{12}^{64}$.

8.18 Exercises

1. Check that equivalence of null sequences (defined in Theorem 8.15.1) is an equivalence relation.

*__2.__ Describe how you would study sequences **x** satisfying (8.15.3), in case (a) $b_0 = 0$; or (b) $b_k = 0$.

3. List (up to equivalence) all null sequences of the binary machine with transfer function $T(D) = (1 + D + D^3)(1 + D + D^2)$, and verify that Theorem 8.15.1 gives the correct information about periods of null sequences in this case.

*__4.__ Is every periodic sequence over GF(2) the output sequence of some FSR over GF(2)?

5. Try to find a general cycle-length theory for FSRs with characteristic polynomial $X^k - \alpha$ over GF(q). (Theorem 8.16.1 and Lemma 8.17.1 are special cases.)

6. Extend Table 8.17.1 to all $d \leq 20$. Extend Table 8.17.2 to all $d \leq 40$. (Use a computer, if you like.)

7. Check the other cycle indexes in Table 5.17.1, using methods of Section 8.17.

8. Find the cycle index on $\mathbb{B}^{12}$ of the element

$$n_{\{1,3,4,6,8,11,12\}}(1\ 3\ 9\ 4\ 12)(2\ 5\ 10\ 8\ 7) \in G_{12}.$$

9. (For an energetic student, or a whole class working together, or a skillful programmer.) Construct a table for G_5 similar to Table 5.17.1, and verify that there are 1,228,158 equivalence classes of five-variable switching functions under G_5 (only 616,126 if we use Theorem 5.16.1 to allow complementation of the output function).

10. Denote $F = \mathrm{GF}(q)$, $K = \mathrm{GF}(q^k)$, $I_F(k) =$ number of irreducible polynomials of degree k over F, and $N(q, k) =$ number of cycles of length k of a k-stage CSR over F. We seek to explain *why* $I_F(k) = N(q, k)$ (Corollary 8.16.1). Define $\varphi: K \to K$ by $\varphi(x) = x^q$. We know that the Galois group of K over F is $\langle \varphi \rangle = \{1, \varphi, \varphi^2, \ldots, \varphi^{k-1}\}$.

(a) Consider φ as a permutation of K, and show that each cycle of φ is the set of roots of some irreducible $f(X) \in F[X]$; conclude that $I_F(k)$ is the number of cycles of φ of length k in K.

(b) By the *normal basis theorem* [van der Waerden (1970, Section 8.11)], there is a $\sigma \in K$ such that $\{\sigma, \varphi(\sigma), \varphi^2(\sigma), \ldots, \varphi^{k-1}(\sigma)\}$ is a basis for the vector space K over F. Show that for all $a_i \in F$, $\varphi(a_0\sigma + a_1\varphi(\sigma) + \cdots + a_{k-1}\varphi^{k-1}(\sigma)) = a_{k-1}\sigma + a_0\varphi(\sigma) + \cdots + a_{k-2}\varphi^{k-1}(\sigma)$. Conclude that $N(q, k)$ is also the number of cycles of φ of length k in K.

CHAPTER 9
Algebraic Coding Theory

9.1 Codes over GF(q)

In this section we shall use the linear algebra of Chapter 7, and the existence of a finite field GF(q) for each prime power q, to generalize the theory of binary block codes in Sections 5.1 to 5.10. In the remainder of this chapter we use the properties of GF(q) to construct powerful new families of multiple-error-correcting and burst-error-correcting codes. Important reference texts on coding include Berlekamp (1968), Blake and Mullin (1975), van Lint (1973), and Peterson and Weldon (1972).

A *q*-ary **symmetric channel** consists of a *transmitter* sending signals $\alpha_i \in$ GF(q), a *receiver*, and a probability p of incorrect transmission of a single digit. (The case $q = 3$ is pictured in Figure 9.1.1.) We denote by $V(n, q)$ the vector space over GF(q) of all n-tuples $(\alpha_1, \alpha_2, \ldots, \alpha_n)$, all

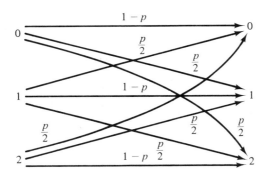

FIGURE 9.1.1. *Ternary Symmetric Channel*

$\alpha_i \in$ GF(q). We usually call the elements of $V(n, q)$ **words**, and write them solidly:

$$(0, 4, 2, 1, 3, 2) = 042132 \quad \text{in} \quad V(6, 5).$$

If word $\mathbf{c} = c_1 c_2 \cdots c_n$ is sent but word $\mathbf{r} = r_1 r_2 \cdots r_n$ is received, we say the channel has added the error pattern $\mathbf{e}$ to $\mathbf{c}$; $\mathbf{r} = \mathbf{c} + \mathbf{e}$, where $\mathbf{e} = e_1 e_2 \cdots e_n$, $e_i = r_i - c_i$ in GF(q). The number of nonzero components in a word $\mathbf{c}$ is its **weight** $w(\mathbf{c})$. We shall assume that all error patterns of the same weight are equally likely. [This is not the only reasonable assumption; some research has been done on a channel in which, say, $3 \in$ GF(7) is more likely to be received as 2 or 4 than as 1 or 6; see Berlekamp (1968, Chaps. 8 and 9).]

With this assumption, the analogue of Theorem 5.1.1 is the following theorem.

Theorem 9.1.1. (a) *If* **e** *is an n-digit error pattern of weight k over* GF(q), *then the probability of* **e** *occurring is*

$$\left(\frac{p}{q-1}\right)^k (1-p)^{n-k}.$$

(b) *There are* $(q-1)^k \binom{n}{k}$ *error patterns of weight k, so the probability that exactly k errors will be made in transmission of a word is*

$$\binom{n}{k} p^k (1-p)^{n-k}.$$

Proof. Exercise 9.2.1. ∎

An **(n, m) block code over GF(q)** consists of an **encoding function** $E: V(m, q) \to V(n, q)$ and a **decoding function** $D: V(n, q) \to V(m, q)$. Elements of im E, the image of the function E, are called **code words**. As in Chapter 5, we have $m < n$, and we define the **rate** of the code to be m/n. As in Section 5.2, a good code must correct errors within the desired tolerance, be efficient (have a high rate), and be reasonably easy to implement. [We should point out to the reader that not all important codes are block codes. *Convolutional codes*, in which check digits in the latter part of a long coded message depend on message digits farther back toward the start of the given message, are extensively studied and often used. See Peterson and Weldon (1972).]

If $\mathbf{a} = a_1 a_2 \cdots a_n$ and $\mathbf{b} = b_1 b_2 \cdots b_n$ are in $V(n, q)$, then $w(\mathbf{a} - \mathbf{b})$ is the number of locations i with $a_i \ne b_i$. The **distance** $d(\mathbf{a}, \mathbf{b})$ between $\mathbf{a}, \mathbf{b} \in V(n, q)$ is $w(\mathbf{a} - \mathbf{b})$. Lemma 5.3.1 and Theorems 5.3.1 and 5.3.2 hold over GF(q). The set $V(n, q)$ is a metric space with distance function d. If, for an encoding function E, distinct code words are distance $\ge 2k+1$ apart [i.e., $\mathbf{w}_1 \ne \mathbf{w}_2$ in $V(m, q)$ implies that $d(E(\mathbf{w}_1), E(\mathbf{w}_2)) \ge 2k+1$], then there exists a decoding function D that corrects all error patterns of weight $\le k$.

Any subspace C of $V(n, q)$ is called a **linear code** over GF(q) of **length** n, or an **(n, m) linear code over GF(q)** if dim $C = m$. An $m \times n$ matrix $\mathbf{G}$ over GF(q) whose rows form a basis of C is called a **generator matrix** for the code C. Code C will in general have many different ordered bases, and hence many different generator matrices. On the other hand, if $\mathbf{G}$ is any $m \times n$ matrix over GF(q) of rank m, then its rows form a basis for some m-dimensional subspace C of $V(n, q)$ and G is a generator matrix for C. (In Chapter 5 we artificially forced an $m \times n$ generator matrix to have rank m by demanding that its first m columns constitute the matrix $\mathbf{I}_m$.)

Sec. 9.1 *Codes over* GF(q)   437

If **G** is a generator matrix for the *m*-dimensional linear code *C* of length *n*, then the linear transformation $E: V(m, q) \to C$ defined by $E: \mathbf{w} \to \mathbf{w}\mathbf{G}$ is a vector-space isomorphism which, as an encoding function, has *C* as its exact set of code words. Let

$$C^d = \{\mathbf{y} \in V(n, q) | \mathbf{x}\mathbf{y}^T = 0, \quad \text{all } \mathbf{x} \in C\};$$

then C^d is called the **dual code** of *C*. The set C^d is a subspace of $V(n, q)$, and if **G** is any $m \times n$ generator matrix for *C*, $C^d = \{\mathbf{y} \in V(n, q) | \mathbf{G} \cdot \mathbf{y}^T = \mathbf{0}\}$; by Exercise 7.11.8, $\dim C^d = n - m = n - \dim C$.

If $\mathbf{x} \in C$, then $\mathbf{x}\mathbf{y}^T = 0$, all $\mathbf{y} \in C^d$, so $0 = (\mathbf{x}\mathbf{y}^T)^T = \mathbf{y}\mathbf{x}^T$, all $\mathbf{y} \in C^d$; this means that $\mathbf{x} \in (C^d)^d$, proving that $C \subseteq (C^d)^d$. But

$$\dim (C^d)^d = n - \dim C^d = n - (n - \dim C) = \dim C,$$

so $C = (C^d)^d$. If **H** is any generator matrix for C^d, then $\mathbf{x} \in V(n, q)$ satisfies $\mathbf{H}\mathbf{x}^T = \mathbf{0}$ iff $\mathbf{y}\mathbf{x}^T = 0$, all $\mathbf{y} \in C^d$ iff $\mathbf{x} \in (C^d)^d = C$, so $C = \{\mathbf{x} \in V(n, q) | \mathbf{H}\mathbf{x}^T = \mathbf{0}\}$; **H** is called a **parity-check matrix** for *C*. [Note that in some books the "dual code" is given a reverse-order definition:

$$C^d = \{(y_1, y_2, \ldots, y_n) \in V(n, q) | \mathbf{x}(y_n, \ldots, y_2, y_1)^T = 0, \quad \text{all } \mathbf{x} \in C\},$$

and results in this paragraph must then be suitably modified.]

Linear codes are subgroups of the additive group $V(n, q)$, and consequently Theorem 5.6.1 holds for them; the minimum distance between nonzero code words is the minimum weight of a nonzero code word. A linear code will thus have a decoding function that can correct all single errors, iff all nonzero code words have weight ≥ 3. Any parity-check matrix then affords a method of decoding, as follows:

Theorem 9.1.2. *Let C be an (n, m) linear code over* GF(q), *and let* **H** *be a parity-check matrix for C. Then the following are equivalent*:
(a) *All nonzero code words have weight* ≥ 3.
(b) *The columns of* **H** *are nonzero, and no one is a scalar multiple of another.*

If (a) and (b) hold, we can correctly decode any single error in a received word **r** *as follows*: *First, compute the syndrome* $\mathbf{H} \cdot \mathbf{r}^T$.
(1) *If* $\mathbf{H} \cdot \mathbf{r}^T = \mathbf{0}$, *assume no error.*
(2) *If* $\mathbf{H} \cdot \mathbf{r}^T = \gamma(i$th *column of* **H**), *assume error pattern* $\mathbf{e} = 00 \cdots 0\gamma 0 \cdots 0$ *of weight 1 ($\gamma \in$* GF(q) *the i*th *entry in* **e**).
(3) *If* $\mathbf{H} \cdot \mathbf{r}^T$ *is not a multiple of a column of* **H**, *at least two errors occurred.*

Partial Proof. Assume that (a) holds, and denote by $\mathbf{e}_i$ the vector with *i*th entry 1, all other entries 0. If the *j*th column of **H** is **0** then $\mathbf{H} \cdot \mathbf{e}_j^T = j$th column of $\mathbf{H} = \mathbf{0}$, proving that $\mathbf{e}_j \in C$, a contradiction to (a). If the *j*th column of $\mathbf{H} = \alpha(k$th column of **H**) for $\alpha \in$ GF(q) and $j \neq k$, then $\mathbf{H} \cdot (\mathbf{e}_j - \alpha \mathbf{e}_k)^T = \mathbf{0}$, $\mathbf{e}_j - \alpha \mathbf{e}_k \in C$, contradicting (a).

We have shown that (a) implies (b). The proof that (b) implies (a), and the decoding procedure, are Exercise 9.2.3. ∎

We should point out now that for binary codes [GF(q) = GF(2)], the codes of Chapter 5 are equivalent in error-correcting ability to any that we have here, even though we have many more generator matrices of a given size than we did then. If **G** is any $m \times n$ generator matrix (i.e., rank m) over GF(q) for a linear code C and **A** is an invertible $m \times m$ matrix over GF(q), then **AG** is an $m \times n$ matrix of rank m (Exercise 7.11.14) whose rows are linear combinations of the rows of **G**, so **AG** is also a generator matrix for C. If **P** is any $n \times n$ *permutation matrix* (one 1 in each row and each column, other entries 0) and $\mathbf{x} \in C$, then **xP** is a rearrangement (*permutation*) of the entries of **x**; in particular, $w(\mathbf{x}) = w(\mathbf{xP})$. Consequently, **AGP** is a generator matrix for an m-dimensional code C' (and the minimum weight of a nonzero code word in C is the same as the minimum weight of a nonzero code word in C'); we say that C and C' are **equivalent codes**. Since **G** has rank m, some m columns of **G** are linearly independent. Therefore, there is a permutation matrix **P** such that **GP** has its first m columns linearly independent: **GP** = [**B** **C**], **B** invertible and **C** $m \times (n-m)$. With $\mathbf{A} = \mathbf{B}^{-1}$, we have

(9.1.1) $$\mathbf{AGP} = [\mathbf{I}_m \quad \mathbf{D}], \quad \mathbf{D} = \mathbf{AC}.$$

This shows that any linear code C is equivalent to one of the type discussed in Chapter 5; the first m columns of the generator matrix form an identity matrix.

We say that an (n, m) block code is **systematic** if each m-digit message word $\mathbf{w} = w_1 w_2 \cdots w_m$ is itself the first m digits of its code word: $E(\mathbf{w}) = w_1 w_2 \cdots w_m c_1 c_2 \cdots c_{n-m}$ for the encoding function E, with check digits $c_1, c_2, \ldots, c_{n-m}$ depending on **w**. Since a generator matrix of form (9.1.1) provides such an E, we proved in the last paragraph the following theorem.

Theorem 9.1.3. *Every linear code is equivalent to a linear systematic code.*

Coset decoding can be done for any linear code over GF(q), just as it is done over GF(2) in Section 5.7. Again the code will correct (correctly) those error patterns which are coset leaders, so the code will be t-error-correcting iff all error patterns of weight $\leq t$ are coset leaders. For a fixed parity-check matrix, one can again decode using only a two-column decoding table, containing the syndrome and coset leader. Many of the powerful multiple-error-correcting codes to be constructed later in this chapter are too long (say $n > 1000$) for any coset decoding or parity-check matrix decoding to be practical; ingenious new decoding (and encoding) methods will be described.

We construct *Hamming codes* over any GF(q) as follows. For a fixed number $k \geq 2$, there are $q^k - 1$ nonzero column vectors of length k over GF(q); but any one has $q-1$ nonzero scalar multiples, so that with $n = (q^k - 1)/(q-1)$ we can form a $k \times n$ matrix **H**, all columns nonzero, no column a scalar multiple of another. The $(n, n-k)$ linear code C with parity-check matrix **H** will be single-error-correcting, by Theorem 9.1.2; it is called a **Hamming code**. It is a *perfect* 1-error-correcting code; perfect codes are defined just as in Section 5.8.

A perfect t-error-correcting (n, m) linear code over GF(q) must have as its q^{n-m} coset leaders the error patterns of weight $\leq t$; by Theorem 9.1.1, this means that

$$\binom{n}{0} + (q-1)\binom{n}{1} + (q-1)^2\binom{n}{2} + \cdots + (q-1)^t\binom{n}{t} = q^{n-m}.$$

In addition to the (23, 12) perfect 3-error-correcting binary code Golay found in 1949, he also found an (11, 6) perfect 2-error-correcting linear code over GF(3). Its existence is possible because

$$\binom{11}{0} + (3-1)\binom{11}{1} + (3-1)^2\binom{11}{2} = 1 + 2 \cdot 11 + 4 \cdot 55 = 243 = 3^{11-6}.$$

It has now been proved [van Lint (1971, 1973); Tietaväinen (1973, 1974); Zinov'ev and Leont'ev (1972)] that no more nontrivial perfect multiple-error-correcting codes exist over any GF(q).

9.2. Exercises

*1. Prove Theorem 9.1.1.

2. Verify that Lemma 5.3.1 and Theorems 5.3.1 and 5.3.2 hold over any GF(q).

*3. Complete the proof of Theorem 9.1.2.

*4. (a) Show that

$$\begin{bmatrix} 1 & 2 & 4 & 0 & 3 \\ 0 & 2 & 1 & 4 & 1 \\ 2 & 0 & 3 & 1 & 4 \end{bmatrix}$$

is a generator matrix for a (5, 3) linear code C over GF(5).

(b) Find a parity-check matrix for the same code. Does the code correct all single errors?

(c) State a general procedure for finding a parity-check matrix for a given code, if a generator matrix is given.

(d) Find the generator matrix of a systematic code that is equivalent to C.

5. Answer questions (a), (b), and (d) of Exercise 4 for the code over GF(3) with generator matrix

$$\begin{bmatrix} 1 & 2 & 0 & 2 & 1 & 0 \\ 2 & 0 & 1 & 2 & 0 & 1 \\ 1 & 1 & 1 & 2 & 1 & 2 \end{bmatrix}.$$

*6. Let $\mathcal{S}$ be the set of all nonzero column vectors $\mathbf{x}$ of length k over GF(q) with the property that the first nonzero entry in any $\mathbf{x} \in \mathcal{S}$ is 1. Show that if the vectors in $\mathcal{S}$ are the columns of a matrix $\mathbf{H}$, then $\mathbf{H}$ is a parity-check matrix for an $(n, n-k)$ Hamming code ($n = (q^k - 1)/(q-1)$) over GF(q). What is a natural way in which to order the elements of $\mathcal{S}$?

7. Use Exercise 6 to write parity-check matrices for the (8, 6) Hamming code over GF(7), the (13, 10) Hamming code over GF(3), and the (5, 3) Hamming code over GF(4). [In the last case, you must first describe GF(4).]

Use the first of these matrices to decode the received words 35234106 and 10521360.

8. Betting is legal in many countries where soccer (called *football* outside the United States) is played. Each game, or *match*, has one of three outcomes: win, loss, or tie. A popular form of wager in the national pool in some countries is as follows. Assume there are t matches on a given weekend. A *bet* consists of picking an outcome in each of the t matches. The *t-match football pool problem* is then as follows: How many bets must a bettor make to guarantee himself second prize (some bet has at most one miss)? Solve the football pool problem for $t = 4$ and $t = 13$, using Hamming codes over GF(3). [The only other $t > 3$ for which the football pool problem has been solved is $t = 5$, and that took a lot of work: Kamps and van Lint (1967).]

*9. Prove that Hamming codes over any GF(q) are perfect single-error-correcting codes.

10. Write computer programs which will do problems like Exercise 4 over GF(p) (p a prime). What about other finite fields?

9.3 Cyclic Codes

A **linear cyclic code** over GF(q) is a linear code with the property that if $c_0 c_1 \cdots c_{n-2} c_{n-1}$ is a code word, so is $c_{n-1} c_0 c_1 \cdots c_{n-2}$. (Hence $c_{n-2} c_{n-1} c_0 c_1 \cdots c_{n-3}, c_{n-3} c_{n-2} c_{n-1} c_0 \cdots c_{n-4}, \ldots, c_1 c_2 \cdots c_{n-1} c_0$ are also code words.) Most of the most useful and powerful block codes discovered recently are linear cyclic codes.

Denote $F = \text{GF}(q)$ and $A(n, q) = F[X]/\mathcal{I}(X^n - 1)$, the quotient ring of $F[X]$ modulo the ideal $\mathcal{I}(X^n - 1)$. Elements of $A(n, q)$ are actually cosets $f(X) + \mathcal{I}(X^n - 1)$. Each coset contains one polynomial of degree $< n$, since modulo $X^n - 1$ we can replace X^n by 1 wherever it appears. No coset can

Sec. 9.3 Cyclic Codes

contain two polynomials of degree $< n$, as their difference would be nonzero of degree $< n$, in the coset $0 + \mathscr{I}(X^n - 1)$, which is an impossibility. We shall therefore think of elements $f(X) + \mathscr{I}(X^n - 1) \in A(n, q)$ as polynomials $f(X)$ of degree $< n$, where multiplication in $A(n, q)$ is done with the usual multiplication of polynomials, taken modulo $X^n - 1$ (set X^n equal to 1 wherever it appears).

For example, let's multiply $X^4 + X^3 + 4X^2 + X + 2$ and $X^3 + 3X^2 + 1$ in $A(5, 5)$. We get

$$(X^3 + 3X^2 + 1)(X^4 + X^3 + 4X^2 + X + 2)$$
$$= X^7 + 4X^6 + 2X^5 + 4X^4 + X^3 + X + 2$$
$$\equiv X^2 + 4X + 2 + 4X^4 + X^3 + X + 2$$
$$= 4X^4 + X^3 + X^2 + 4 \quad \text{(modulo } X^5 - 1\text{)},$$

so the product is $4X^4 + X^3 + X^2 + 4 \in A(5, 5)$.

The ring $A(n, q)$ is a vector space over F with its own addition and the usual multiplication of polynomials by scalars; in fact, we have a vector space isomorphism

(9.3.1) $\qquad c_0 c_1 \cdots c_{n-1} \leftrightarrow c_0 + c_1 X + \cdots + c_{n-1} X^{n-1}$

between $V(n, q)$ and $A(n, q)$. Consequently, we may identify $V(n, q)$ and $A(n, q)$; we think of our code words as polynomials of degree $< n$, and our linear codes as subspaces of $A(n, q)$. Cyclic codes are usually studied in $A(n, q)$, because of the following theorem.

Theorem 9.3.1. *A linear code in $A(n, q)$ is cyclic iff it is an ideal.*

Proof. A subspace W of $A(n, q)$ is an ideal iff $X \cdot W \subseteq W$, because if $X \cdot W \subseteq W$ then $X^2 \cdot W = X \cdot (X \cdot W) \subseteq X \cdot W \subseteq W, \ldots$, any $X^i \cdot W \subseteq W$,

$$(a_0 + a_1 X + \cdots + a_{n-1} X^{n-1}) \cdot W \subseteq a_0 W + a_1 X \cdot W + \cdots + a_{n-1} X^{n-1} \cdot W$$
$$\subseteq W + W + \cdots + W = W.$$

(Certainly, if W is an ideal then $X \cdot W \subseteq W$.)

If C is cyclic and $c_0 + c_1 X + \cdots + c_{n-1} X^{n-1} \in C$, then $c_{n-1} + c_0 X + c_1 X^2 + \cdots + c_{n-2} X^{n-1} \in C$ by the definition of cyclic code; but

$$X(c_0 + c_1 X + \cdots + c_{n-1} X^{n-1}) = c_0 X + c_1 X^2 + \cdots + c_{n-1} X^n$$
$$= c_{n-1} + c_0 X + \cdots + c_{n-2} X^{n-1}$$

in $A(n, q)$, so C is an ideal. Conversely, if C is an ideal and $c_0 + c_1 X + \cdots + c_{n-1} X^{n-1} \in C$, then

$$c_{n-1} + c_0 X + \cdots + c_{n-2} X^{n-1} = X(c_0 + c_1 X + \cdots + c_{n-1} X^{n-1}) \in C,$$

proving that C is cyclic. ∎

We easily find the structure of ideals in $A(n, q)$:

Theorem 9.3.2. (a) *Let I be an ideal in $A(n, q)$, $g(X)$ the monic polynomial of smallest degree in I. Then I consists of all multiples of $g(X)$, and $g(X)$ divides $X^n - 1$. $g(X)$ is uniquely determined by I and called the **generator** of I.*
(b) *Conversely, any monic divisor of $X^n - 1$ generates an ideal in $A(n, q)$.*

Proof. (a) Let $J = \{f(X) \in F[X] | f(X) + \mathscr{I}(X^n - 1) \in I\}$. Clearly, J is an ideal in $F[X]$, so $J = \mathscr{I}(g(X))$, some uniquely determined monic $g(X)$; $X^n - 1 \in J$, so $g(X)$ divides $X^n - 1$.

(b) Such a divisor $g(X)$ generates an ideal J of $F[X]$ containing $X^n - 1$, so $g(X) \in A(n, q)$ generates the ideal

$$J/\mathscr{I}(X^n - 1) = \{h(X) + \mathscr{I}(X^n - 1) | h(X) \in J\}$$

of $A(n, q)$. ∎

Note that in the statement of Theorem 9.3.2(a), we can think of I as either all multiples of $g(X)$ of degree less than n; or all multiples of $g(X)$ whatever, reduced modulo $X^n - 1$. Since $g(X)$ divides $X^n - 1$, we obtain the same set of elements of $A(n, q)$ in either case.

This shows that $|I| = q^{n - \deg g(X)}$ since, if $\deg g(X) = r$,

$$I = \{(a_0 + a_1 X + \cdots + a_{n-r-1} X^{n-r-1}) g(X) | \text{ all } a_i \in F\}.$$

Hence the dimension of the code I is $m = n - \deg g(X)$.

9.4 BCH Codes

Let $n \in \mathbb{P}$. An element γ of a field is a **primitive nth root of 1** if $\gamma^n = 1$, but $\gamma^m \neq 1$ for $1 \leq m < n$.

Lemma 9.4.1. *Denote $F = \mathrm{GF}(q)$ for a prime power q, and fix $n \in \mathbb{P}$. Then $(n, q) = 1$ iff there is a primitive nth root γ of 1 in an extension field of F.*

Proof. Let K be a splitting field of $X^n - 1$ over F. By Corollary 7.20.2, the roots of $X^n - 1$ form a cyclic multiplicative group G. If $(n, q) = 1$ and we denote $f(X) = X^n - 1$, then $f'(X) = nX^{n-1}$, and any root α of $X^n - 1$ has $f'(\alpha) \neq 0$. By Lemma 7.20.3, α is not a double root, $f(X)$ has n distinct roots, and G has order n; any generator γ of G is a primitive nth root of 1 in K.

If $(n, q) \neq 1$, say $q = p^r$, $n = pn_0$ for a prime $p, r, n_0 \in \mathbb{P}$, then $X^n - 1 = (X^{n_0} - 1)^p$ by Lemma 7.20.2, so any root ε of $X^n = 1$ also satisfies $X^{n_0} = 1$; there are no primitive nth roots of 1 in K. ∎

Theorem 9.4.1 (Bose–Chaudhuri–Hoquenghem). *Assume that $(n, q) = 1$ and let γ be a primitive nth root of 1 in an extension field of $\mathrm{GF}(q) = F$. Assume that the generator polynomial $g(X)$ of a cyclic code of length n over $\mathrm{GF}(q)$ has $\gamma, \gamma^2, \ldots, \gamma^{d-1}$ among its roots. Then the minimum distance of this code is at least d.*

Sec. 9.4 BCH Codes

Proof. Let $K = F(\gamma) = \text{GF}(q^k)$, and form the $(d-1) \times n$ matrix

$$\mathbf{H} = \begin{bmatrix} 1 & \gamma & \gamma^2 & \cdots & \gamma^{n-1} \\ 1 & \gamma^2 & \gamma^4 & \cdots & \gamma^{2(n-1)} \\ \cdots & \cdots & \cdots & \cdots & \cdots \\ 1 & \gamma^{d-1} & \gamma^{2(d-1)} & \cdots & \gamma^{(d-1)(n-1)} \end{bmatrix}$$

over K. A vector $\mathbf{c} = c_0 c_1 \cdots c_{n-1} = c_0 + c_1 X + \cdots + c_{n-1} X^{n-1} = c(X)$ is a code word iff $g(X) | c(X)$. If this holds, then $c(\gamma^i) = 0$ for $1 \leq i \leq d-1$; these $d-1$ equations amount to the one matrix equation $\mathbf{Hc}^T = \mathbf{0}$.

Take any $d-1$ columns of $\mathbf{H}$ headed by $\beta_1 = \gamma^{i_1}, \ldots, \beta_{d-1} = \gamma^{i_{d-1}}$; then, by Exercise 7.14.10(a) and (b),

$$\det \begin{bmatrix} \beta_1 & \beta_2 & \cdots & \beta_{d-1} \\ \beta_1^2 & \beta_2^2 & \cdots & \beta_{d-1}^2 \\ \cdots & \cdots & \cdots & \cdots \\ \beta_1^{d-1} & \beta_2^{d-1} & \cdots & \beta_{d-1}^{d-1} \end{bmatrix} = \beta_1 \beta_2 \cdots \beta_{d-1} \prod_{i > j} (\beta_i - \beta_j) \neq 0.$$

(Since γ is a primitive nth root of 1, $\gamma^0, \gamma^1, \ldots, \gamma^{n-1}$ are all different; therefore, all differences $\beta_i - \beta_j$ are nonzero.) Hence any $d-1$ columns of $\mathbf{H}$ are linearly independent. If a nonzero code word $\mathbf{c}$ had less than d nonzero entries, the equation $\mathbf{Hc}^T = \mathbf{0}$ would make some $d-1$ columns linearly dependent, a contradiction. ∎

This theorem, discovered independently by Hoquenghem (1959) and Bose and Ray-Chaudhuri (1960), is actually a recipe for constructing multiple-error-correcting codes. If we want a t-error-correcting code of length n over $F = \text{GF}(q)$ [with $(n, q) = 1$], we let γ be a primitive nth root of 1 in an extension field of F and choose the generator polynomial

(9.4.1) $g(X) = \text{l.c.m.} \{\text{Irr}(\gamma, F), \text{Irr}(\gamma^2, F), \ldots, \text{Irr}(\gamma^{2t}, F)\},$

Irr (α, F) the irreducible polynomial for α over F. By Theorem 9.4.1, the cyclic code of length n over F with generator polynomial $g(X)$ will have minimum distance at least $2t + 1$, so by Theorem 5.3.2 it will correct t errors. With (9.4.1), we say we have a **t-error-correcting BCH code of length n over GF(q)**.

Let Irr (γ, F) have degree k, so $F(\gamma) = \text{GF}(q^k)$. All γ^i are in $F(\gamma)$, so all Irr (γ^i, F) have degree at most k; the $g(X)$ in (9.4.1) has degree at most $2kt$. Our code will have

$$\text{rate} = \frac{n - \deg g(X)}{n} \geq \frac{n - 2kt}{n},$$

so we would like to have k small compared to n, for an efficient code. Element γ has order n in the multiplicative group of $F(\gamma) = \text{GF}(q^k)$, of order $q^k - 1$, so n divides $q^k - 1$. The number n will be largest, for given k and q, if

$n = q^k - 1$; we then say that we have a **primitive BCH code**, as γ is a primitive element in $GF(q^k)$.

Some parameters of typical primitive (n, m) t-error-correcting BCH codes over $GF(2)$ are given in Table 9.4.1. Extensive tables are available, yielding $g(X)$ for given requirements n and t; for example, see the Appendix of Peterson and Weldon (1972).

TABLE 9.4.1. t-error-correcting (n, m) primitive binary BCH codes.

k	5	5	5	7	7	10	10	10
$n = 2^k - 1$	31	31	31	127	127	1023	1023	1023
t	2	3	5	4	9	4	10	34
m	21	16	11	99	71	983	923	708

9.5 Reed–Solomon Codes; Burst Error Correction

We keep the notation of the preceding section. A **Reed–Solomon code** [Reed and Solomon (1960)] is a (primitive) BCH code with $n = q - 1$, so $k = 1$ and $\gamma \in GF(q) = F$. [Some authors merely require that $k = 1$ and $n | (q - 1)$.] Hence

$$\text{Irr}(\gamma^i, F) = X - \gamma^i$$

has degree 1 for all i. For a t-error-correcting code, take

(9.5.1) $\qquad g(X) = (X - \gamma)(X - \gamma^2) \cdots (X - \gamma^{2t})$.

For example, suppose that $q = 2^8$, so $n = 255$. A 5-error-correcting code has $g(X)$ of degree 10 and rate

$$\frac{255 - 10}{255} = \frac{245}{255} = \frac{49}{51} > .96.$$

Elements of $GF(2^8)$ are 8-dimensional vectors over $GF(2)$, so this code can be used as a code of length $8 \cdot 255 = 2040$ over $GF(2)$, with the same rate. Any 33 consecutive error locations over $GF(2)$ will affect at most five of the elements of $GF(2^8)$ (see Figure 9.5.1), and errors there will therefore be

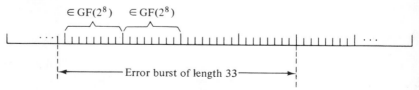

FIGURE 9.5.1. *Burst Error Correction*

corrected. We say that the code corrects any "burst" of 33 consecutive errors. For comparison, a primitive 33-error-correcting BCH code over GF(2) with $n = 2^{11} - 1 = 2047$, $k = 11$ will have $g(X)$ of degree 352,

$$\text{rate } \frac{2047 - 352}{2047} = \frac{1695}{2047} < .83.$$

The Reed–Solomon code of comparable length has as much burst-error-correcting ability, and a higher rate.

Reed–Solomon codes are among the best known for channels with bursts of errors, or bursts of errors together with occasional random errors, or "multiple bursts." The Reed–Solomon code of length 63 over $GF(2^6)$ was found especially useful in correcting errors in binary data stored in a *photodigital mass memory* [see Chien (1973) and Chien, Oldham, and Tang (1968)]. In this memory system, many thousands of lines of data are stored on a memory chip of only a few square inches, each line containing $6 \cdot 63 = 378$ binary bits of data. Each line is precisely one word in the Reed–Solomon code.

Reed–Solomon and other BCH codes became more important when efficient encoding and decoding methods were discovered. We shall discuss these, and consider some actual examples, in the next few sections.

9.6 Encoding Cyclic Codes

Consider now a cyclic code in $A(n, q)$ with generator polynomial $g(X)$. We think of the code words as polynomials of degree $< n$ which are multiples of $g(X)$. If $g(X)$ has degree r, then the code has dimension $m = n - r$, so we seek an encoding function

(9.6.1) $\qquad E: A(m, q) \to \mathscr{I}(g(X)) \subset A(n, q).$

The usual E is not multiplication by $g(X)$ [although $f(X) \to f(X)g(X)$ is an acceptable encoding function] but rather division, in a way that leaves the information digits systematically visible in the code word.

If $w(X)$, of degree less than m, is a message word, form $X^{n-m}w(X)$ and divide by $g(X)$:

$$X^{n-m}w(X) = q(X)g(X) + r(X), \quad r(X) = 0 \quad \text{or} \quad \deg r(X) < r = n - m.$$

Then

$$X^{n-m}w(X) - r(X) \in \mathscr{I}(g(X)).$$

We encode

(9.6.2) $\qquad E: w(X) \to X^{n-m}w(X) - r(X);$

since $r(X)$ has degree less than $n - m$ or $r(X) = 0$, the information digits are the higher-degree coefficients in the code word $E(w(X))$. (The reader is asked to show in Exercise 9.7.8 that this E is a linear transformation.)

We saw in Section 8.2 that a modified FSR can be used to divide one polynomial by another. Assume that $g(X) = g_0 + g_1 X + \cdots + g_{r-1} X^{r-1} + X^r$; the type of modified FSR used for encoding [via (9.6.2)] is pictured in Figure 9.6.1.

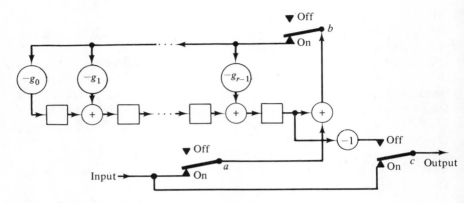

FIGURE 9.6.1. *Encoder for Cyclic Codes*

The circuit in Figure 9.6.1 differs from the division circuit of Example 3, Section 8.2, in two ways. First, the input $w(X)$ is inserted into the *rightmost adder*, high-order coefficients first, instead of the left side of the register. This shift of r positions to the right can be interpreted as the multiplication of $w(X)$ by $X^{n-m} = X^r$ to obtain $X^{n-m} w(X)$, which we are dividing by $g(X)$.

Second, three switches, a, b, and c, are provided. While the input $w(X)$ is received, all switches are in the ON position; the circuit receives $w(X)$, performs the division, and stops with the coefficients of the remainder $r(X)$ stored in the delays, as did the circuit of Example 3, Section 8.2. At the same time, the input feeds directly through switch c to the output, becoming the first (high-order) coefficients of the output (code word). After $w(X)$ is received, all switches go to the OFF position. Since a and b are OFF, the $r(X)$ stored in the register is unaffected by input or feedback; it merely feeds out through the -1 multiplier and switch c to the output, yielding the correct final digits $-r(X)$ of the code word $X^{n-m} w(X) - r(X)$. Then all switches return to the ON position, to receive the next message word.

For example, consider the cyclic code over GF(2) with length $n = 7$ and generator polynomial $g(X) = X^3 + X + 1$. The encoder is pictured in Figure 9.6.2.

If the input is the four-digit message word $c_3 c_4 c_5 c_6$ (that is, message $c_3 + c_4 X + c_5 X^2 + c_6 X^3$, c_6 the highest-order coefficient), the values of $y_1, y_2,$ and y_3 at times $t = 1, 2, 3, 4$ are given in Table 9.6.1. (At time $t = 0$, they all have value 0.) We conclude from the table that the code word produced is

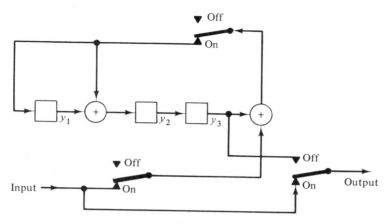

FIGURE 9.6.2. *Hamming Code Encoder*

$c_0 c_1 c_2 c_3 c_4 c_5 c_6$, where $c_0 = c_3 + c_5 + c_6$, $c_1 = c_3 + c_4 + c_5$, $c_2 = c_4 + c_5 + c_6$. That is,

$$\begin{bmatrix} 1 & 0 & 0 & 1 & 0 & 1 & 1 \\ 0 & 1 & 0 & 1 & 1 & 1 & 0 \\ 0 & 0 & 1 & 0 & 1 & 1 & 1 \end{bmatrix} \begin{bmatrix} c_0 \\ c_1 \\ c_2 \\ c_3 \\ c_4 \\ c_5 \\ c_6 \end{bmatrix} = \begin{bmatrix} 0 \\ 0 \\ 0 \end{bmatrix}.$$

Since the left matrix is the parity-check matrix of a Hamming code, this identifies the code as a Hamming (7, 4) code.

TABLE 9.6.1

t	1	2	3	4
y_1	c_6	c_5	$c_4 + c_6$	$c_3 + c_5 + c_6$
y_2	c_6	$c_5 + c_6$	$c_4 + c_5 + c_6$	$c_3 + c_4 + c_5$
y_3	0	c_6	$c_5 + c_6$	$c_4 + c_5 + c_6$

9.7 Exercises

*1. (a) Assume that $(n, q) = 1$, $F = GF(q)$, and show that $X^n - 1 = f_1(X) f_2(X) \cdots f_r(X)$, the $f_i(X)$ *distinct* irreducible polynomials in $F[X]$.

(b) Show that, in (a), $A(n, q)$ is isomorphic to

(9.7.1) $F[X]/\mathscr{I}(f_1(X)) \oplus F[X]/\mathscr{I}(f_2(X)) \oplus \cdots \oplus F[X]/\mathscr{I}(f_r(X))$,

a direct sum of fields.

(c) Describe the ideals in $A(n, q)$ in terms of (9.7.1), and compare with Theorem 9.3.2.

2. Perform the following multiplications:
 (a) $X^5 + X^3 + X^2 + 1$ and $X^6 + X^5 + X^3 + X + 1$ in $A(8, 2)$.
 (b) $2X^5 + X^4 + 2X^3 + X + 2$ and $X^4 + 2X^3 + X^2 + 2X$ in $A(6, 3)$.
 (c) $X^5 + \gamma^6 X^4 + \gamma^2 X^3 + \gamma X^2 + X + \gamma^4$ and $X^4 + \gamma^5 X^3 + \gamma^7 X^2 + \gamma X + \gamma^2$ in $A(7, 9)$, where GF(9) = GF(3)(γ) is as described in Table 7.21.2.

*3. Show that GF(q) contains $\varphi(n)$ primitive nth roots of 1 if $n | (q-1)$, and no primitive nth roots of 1 if $n \nmid (q-1)$.

*4. In this problem we will find a generator polynomial for a (primitive) 5-error-correcting BCH code of length $n = 63 = 2^6 - 1$ over $F = $ GF(2). We take as given the fact that $f(X) = X^6 + X + 1$ is a primitive polynomial over GF(2); let γ be a root of $f(X)$ in GF(2^6).

(a) Find the order of each γ^i, $1 \leq i \leq 10$, in the multiplicative group of GF(2^6). Conclude that $\gamma^9 \in $ GF(2^3) but no other γ^i, $1 \leq i \leq 10$, is in a proper subfield of GF(2^6). Find the degree of Irr (γ^i, F) for $1 \leq i \leq 10$.

(b) Since γ^3 is the root of an irreducible polynomial $X^6 + aX^5 + bX^4 + cX^3 + dX^2 + eX + f$, $a, b, \ldots, f \in $ GF(2), use the facts that $\gamma^6 = \gamma + 1$ and $1, \gamma, \ldots, \gamma^5$ are linearly independent over GF(2) to find Irr (γ^3, F).

(c) Use Corollary 7.23.1 to show that γ, γ^2, γ^4, and γ^8 are all roots of $f(X)$ and γ^3 and γ^6 are both roots of Irr (γ^3, F). With this short cut, find Irr (γ^i, F) for $1 \leq i \leq 10$ and find $g(X) = $ l.c.m. {Irr $(\gamma, F), \ldots,$ Irr (γ^{10}, F)}.

*5. Use the methods of Exercise 4 to find generator polynomials for a 3-error-correcting BCH code of length 26 over GF(3) and a 2-error-correcting BCH code of length 21 over GF(2), given that $X^3 + 2X + 1$ is primitive over GF(3) and $X^6 + X + 1$ is primitive over GF(2).

6. For each of the following examples of Reed–Solomon codes, find the largest integer b such that the code will correct all errors in b consecutive error locations, using the code as a code over GF(2). Also find the rate of each code.

 (a) A 7-error-correcting code of length $2^7 - 1$ over GF(2^7).
 (b) A 13-error-correcting code of length $2^8 - 1$ over GF(2^8).

7. Diagram encoders for the codes considered in Exercise 5.

*8. Show that the encoding function (9.6.2) is a linear transformation.

*9. Generalize Theorem 9.4.1 by replacing $\{\gamma, \gamma^2, \ldots, \gamma^{d-1}\}$ by $\{\gamma^{m_0}, \gamma^{m_0+1}, \ldots, \gamma^{m_0+d-2}\}$, any integer m_0.

10. Write a computer program which will do encoding for binary BCH codes, by imitating Figure 9.6.1. What about Reed–Solomon codes over GF(q), q a power of 2?

Sec. 9.8   BCH Decoding as an FSR Problem

*11. Assume that $g(X) = g_0 + g_1 X + \cdots + g_{r-1} X^{r-1} + X^r$ is the generator of the cyclic code C in $A(n, q)$, and $X^n - 1 = g(X)h(X)$, where $h(X) = h_0 + h_1 X + \cdots + h_{m-1} X^{m-1} + X^m$, $m + r = n$. Show that

$$\begin{bmatrix} g_0 & g_1 & g_2 & \cdots & g_{r-1} & 1 & 0 & 0 & \cdots & 0 & 0 \\ 0 & g_0 & g_1 & \cdots & g_{r-2} & g_{r-1} & 1 & 0 & \cdots & 0 & 0 \\ \vdots & & & & & & & & & & \\ 0 & 0 & 0 & \cdots & 0 & 0 & 0 & g_0 & \cdots & g_{r-1} & 1 \end{bmatrix}$$

is a generator matrix and

$$\begin{bmatrix} 1 & h_{m-1} & \cdots & h_1 & h_0 & 0 & 0 & \cdots & 0 & 0 \\ \vdots & & & & & & & & & \\ 0 & 0 & \cdots & 0 & 1 & h_{m-1} & h_{m-2} & \cdots & h_0 & 0 \\ 0 & 0 & \cdots & 0 & 0 & 1 & h_{m-1} & \cdots & h_1 & h_0 \end{bmatrix}$$

is a parity-check matrix for C.

12. (a) Assume F is a field, $\mathbf{T} \in \mathrm{Mat}_{n \times n}(F)$ and $\mathbf{x} \in \mathrm{Mat}_{n \times 1}(F)$. Show that $\{p(X) \in F[X] \mid p(\mathbf{T})\mathbf{x} = \mathbf{0}\} = I$ is an ideal in $F[X]$ (see Lemma 7.17.1). The generator of I is called the **T-*annihilator*** of $\mathbf{x}$.

(b) Suppose that the state $\mathbf{y}$ of the FSR M over the field $F = \mathrm{GF}(q)$ lies in a state cycle of length n. With $\mathbf{A}$ as in (8.11.2), then, the cycle consists of states $\mathbf{y}, \mathbf{Ay}, \mathbf{A}^2\mathbf{y}, \ldots, \mathbf{A}^{n-1}\mathbf{y}$, where $\mathbf{A}^n\mathbf{y} = \mathbf{y}$. (Why?) Denote $\mathbf{H} = [\mathbf{y} \ \mathbf{Ay} \cdots \mathbf{A}^{n-1}\mathbf{y}]$, a $k \times n$ matrix, and $C = \{\mathbf{c} \in F^n \mid \mathbf{Hc}^T = \mathbf{0}\}$ (so $\mathbf{H}$ is a parity-check matrix for code C). Show that C is a cyclic code whose generator is the $\mathbf{A}$-annihilator of $\mathbf{y}$, by writing the equation $\mathbf{Hc}^T = \mathbf{0}$ in terms of powers of $\mathbf{A}$ and components of $\mathbf{c}$.

(c) [Ash (1965, Chap. 5)] Show conversely that if C is any cyclic code generated by $g(X) \in A(n, q)$, $\mathbf{A}$ the companion matrix of $g(X)$,

$$\mathbf{y} = \begin{bmatrix} 1 \\ 0 \\ \vdots \\ 0 \end{bmatrix}, \quad \text{and} \quad \mathbf{H} = [\mathbf{y} \ \mathbf{Ay} \cdots \mathbf{A}^{n-1}\mathbf{y}],$$

then $\mathbf{H}$ is a parity-check matrix for C. (Show by induction that the columns of $\mathbf{H}$ are the columns of the parity-check matrix in Exercise 11. Ash thus says that cyclic codes are precisely the codes generated by FSRs.)

9.8. BCH Decoding as an FSR Problem

The discovery by Berlekamp of an efficient procedure for decoding BCH code words [Berlekamp (1968, 1973)] was a truly major advance in coding. Soon afterward, Massey (1969) was able to interpret the decoding problem as a problem concerning FSRs, and thus slightly simplify the statement and proof of the decoding algorithm. We shall follow Massey (1969) in the next two sections. [See also Tzeng (1969) and Burton (1971).]

Assume that $(n, q) = 1$, and let γ be a primitive nth root of 1 in an extension field of $F = \mathrm{GF}(q)$. Let

(9.8.1) $\quad g(X) = \mathrm{l.c.m.} \{\mathrm{Irr}\,(\gamma, F), \mathrm{Irr}\,(\gamma^2, F), \ldots, \mathrm{Irr}\,(\gamma^{2t}, F)\}$,

so $g(X)$ generates a t-error-correcting BCH code of length n over $\mathrm{GF}(q)$.

If code word $\mathbf{f} = f_0 f_1 \cdots f_{n-1}$, identified as $f_0 + f_1 X + \cdots + f_{n-1} X^{n-1}$, is transmitted and word $\mathbf{w} = w_0 w_1 \cdots w_{n-1} \leftrightarrow w_0 + w_1 X + \cdots + w_{n-1} X^{n-1}$ received, the error pattern $\mathbf{e} = \mathbf{w} - \mathbf{f}$ may be thought of as an *error polynomial* $e(X) = e_0 + e_1 X + \cdots + e_{n-1} X^{n-1}$, $e_i = w_i - f_i$ in F.

Now $g(\gamma^i) = 0$ for $1 \leq i \leq 2t$; since $g(X)$ divides $f(X)$, we also have $f(\gamma^i) = 0$ for $1 \leq i \leq 2t$. The equation $w(X) = f(X) + e(X)$ implies that $w(\gamma^i) = e(\gamma^i)$ for $1 \leq i \leq 2t$. Therefore, the values

(9.8.2) $\quad\quad\quad S_i = e(\gamma^i) = w(\gamma^i), \quad 1 \leq i \leq 2t,$

can be computed at the receiver from the (known) $w(X)$. The S_i play the role of the syndrome in **H**-matrix decoding (Section 5.4); they are the usual tool in determining the error pattern $e(X)$ from the received word $w(X)$.

Assume that ν errors occurred, so $e(X)$ has ν nonzero coefficients $e_{j_1}, \ldots, e_{j_\nu}$;

$$S_i = e(\gamma^i) = e_{j_1}(\gamma^i)^{j_1} + e_{j_2}(\gamma^i)^{j_2} + \cdots + e_{j_\nu}(\gamma^i)^{j_\nu}.$$

We say that $\gamma^{j_1} = x_1, \gamma^{j_2} = x_2, \ldots, \gamma^{j_\nu} = x_\nu$ are the **error locators** and $e_{j_1} = y_1$, $e_{j_2} = y_2, \ldots, e_{j_\nu} = y_\nu$ are the **error values**; we have $x_1, \ldots, x_\nu \in F(\gamma)$, $y_1, \ldots, y_\nu \in F$, and

(9.8.3) $\quad\quad\quad S_i = \sum_{j=1}^{\nu} y_j x_j^i, \quad 1 \leq i \leq 2t.$

We wish to solve the $2t$ equations (9.8.3) for ν and the y_j's and x_j's; this will completely determine $e(X)$. (Later we shall describe an efficient procedure, the *Chien search*, for actually correcting the errors after the y_j's and x_j's are found.) Sections 9.8 and 9.9 and Theorem 9.10.1 are devoted to the determination of ν and $x_1, \ldots, x_\nu$; Theorem 9.10.2 will then explain how to find $y_1, \ldots, y_\nu$.

Set $K = F(\gamma)$ and define polynomials $r(X), p(X) \in K[X]$ by

(9.8.4) $\quad\quad\quad\quad r(X) = \prod_{j=1}^{\nu} (1 - x_j X),$

(9.8.5) $\quad p(X) = \sum_{j=1}^{\nu} \left[y_j x_j \prod_{\substack{k=1 \\ k \neq j}}^{\nu} (1 - x_k X) \right] = \sum_{j=1}^{\nu} y_j x_j \frac{r(X)}{1 - x_j X}.$

Sec. 9.8 BCH Decoding as an FSR Problem

In the formal power series ring $K[[X]]$ (see Exercise 9.11.1) we have

$$\sum_{j=1}^{\nu} \frac{y_j x_j}{1-x_j X} = \sum_{j=1}^{\nu} y_j x_j \left(\sum_{i=0}^{\infty} x_j^i X^i \right) = \sum_{i=0}^{\infty} \left(\sum_{j=1}^{\nu} y_j x_j^{i+1} \right) X^i$$

$$= S_1 + S_2 X + S_3 X^2 + \cdots,$$

so $p(X) = r(X)s(X)$ by (9.8.5), where $s(X)$ is the power series $S_1 + S_2 X + S_3 X^2 + \cdots$; we have

(9.8.6) $$\frac{p(X)}{r(X)} = s(X), \qquad \text{where } (p(X), r(X)) = 1.$$

To see that $(p(X), r(X)) = 1$, note that otherwise $p(X)$ and $r(X)$ have a common root α in an extension field of K; $0 = r(\alpha)$ implies that $0 = 1 - x_j \alpha$ for some j, so $\alpha = x_j^{-1}$; but

$$p(x_j^{-1}) = y_j x_j \prod_{\substack{k=1 \\ k \neq j}}^{\nu} (1 - x_k x_j^{-1}) \neq 0,$$

a contradiction. Certainly, $r(X)$ has degree ν and $p(X)$ has degree less than ν.

We saw in Section 8.13 that the output sequences of a k-dimensional FSR with characteristic polynomial

(9.8.7) $$c(X) = X^k - \alpha_{k-1} X^{k-1} - \cdots - \alpha_1 X - \alpha_0$$

are the solutions of the recursive equation

(9.8.8) $\quad z(t) = \alpha_0 z(t-k) + \alpha_1 z(t-k+1) + \cdots + \alpha_{k-1} z(t-1)$

for $t \geq k$; that is,

$$z = (\alpha_0 D^k + \alpha_1 D^{k-1} + \cdots + \alpha_{k-1} D)z,$$
$$0 = (1 - \alpha_{k-1} D - \cdots - \alpha_1 D^{k-1} - \alpha_0 D^k)z = \bar{c}(D)z.$$

Here $\bar{c}(X)$ is the reciprocal polynomial of $c(X)$. In this chapter we shall call $\tilde{r}(X) = \bar{c}(X)$ the **recursion polynomial** of the FSR. Thus $\tilde{r}(X)$ has degree k, $\tilde{r}(0) = 1$, and the output sequences z are the solutions of $\tilde{r}(D)z = 0$. We denote

(9.8.9) $$\tilde{r}(X) = 1 + \rho_1 X + \cdots + \rho_{k-1} X^{k-1} + \rho_k X^k = \sum_{i=0}^{k} \rho_i X^i,$$

where $\rho_i = -\alpha_{k-i}$ and $\rho_0 = 1$. If the output sequence is $s_0, s_1, \ldots, s_{k-1}, s_k, \ldots$, then $s_0, s_1, \ldots, s_{k-1}$ are arbitrary elements of K. By (9.8.8) with $s_j = z(j)$,

(9.8.10) $$s_l = -\sum_{i=1}^{k} \rho_i s_{l-i}, \qquad l \geq k.$$

Since $\rho_0 = 1$, (9.8.10) can also be written

$$\sum_{i=0}^{k} \rho_i s_{l-i} = 0, \quad l \geq k. \tag{9.8.11}$$

If $s(X)$ is the power series

$$s(X) = s_0 + s_1 X + s_2 X^2 + \cdots + s_n X^n + \cdots, \tag{9.8.12}$$

then $\sum_{i=0}^{k} \rho_i s_{l-i}$ is the coefficient of X^l in $\tilde{r}(X)s(X)$, so (9.8.11) says that $\tilde{r}(X)s(X)$ is a polynomial $p(X)$ of degree less than k:

$$\tilde{r}(X)s(X) = p(X), \quad \frac{p(X)}{\tilde{r}(X)} = s(X). \tag{9.8.13}$$

If $p(X) = p_0 + p_1 X + \cdots + p_{k-1} X^{k-1}$, $p_i \in K$, then by (9.8.13) we have

$$\sum_{i=0}^{l} \rho_i s_{l-i} = p_l, \quad l = 0, 1, \ldots, k-1. \tag{9.8.14}$$

These k equations amount to the single matrix equation

$$\begin{bmatrix} 1 & 0 & 0 & \cdots & 0 & 0 \\ \rho_1 & 1 & 0 & \cdots & 0 & 0 \\ \rho_2 & \rho_1 & 1 & \cdots & 0 & 0 \\ \cdots & \cdots & \cdots & \cdots & \cdots & \cdots \\ \rho_{k-2} & \rho_{k-3} & \rho_{k-4} & \cdots & 1 & 0 \\ \rho_{k-1} & \rho_{k-2} & \rho_{k-3} & \cdots & \rho_1 & 1 \end{bmatrix} \begin{bmatrix} s_0 \\ s_1 \\ s_2 \\ \vdots \\ s_{k-2} \\ s_{k-1} \end{bmatrix} = \begin{bmatrix} p_0 \\ p_1 \\ p_2 \\ \vdots \\ p_{k-2} \\ p_{k-1} \end{bmatrix}, \tag{9.8.15}$$

since $\rho_0 = 1$. Since this coefficient matrix is nonsingular, every $p(X)$ of degree less than k occurs for some choice of output sequence.

Lemma 9.8.1 (Massey). *A sequence* $\mathbf{s} = s_0 s_1 s_2 \cdots s_n \cdots$ *is an output sequence of the k-dimensional FSR with recursion polynomial $\tilde{r}(X)$ over the finite field K, iff the power series $s(X) = s_0 + s_1 X + \cdots + s_n X^n + \cdots$ has form $p(X)/\tilde{r}(X)$, $p(X) \in F[X]$ of degree less than k. Also, $(p(X), \tilde{r}(X)) = 1$ iff no smaller-dimensional FSR produces output sequence $\mathbf{s}$; the FSR of dimension k producing $\mathbf{s}$ is then unique.*

Proof. It remains only to prove the last sentence. If $(p(X), \tilde{r}(X)) = m(X)$, say $p(X) = m(X)p_0(X)$, $\tilde{r}(X) = m(X)\tilde{r}_0(X)$, then

$$s(X) = \frac{p(X)}{\tilde{r}(X)} = \frac{p_0(X)}{\tilde{r}_0(X)},$$

so a smaller FSR of dimension $\deg \tilde{r}_0(X)$ produces $\mathbf{s}$. Conversely, if $s(X) = p_1(X)/\tilde{r}_1(X)$ for some other $\tilde{r}_1(X)$ of degree $\leq k$ and $p_1(X)$ with $\deg p_1(X) <$

deg $\tilde{r}_1(X)$, then $p(X)\tilde{r}_1(X) = p_1(X)\tilde{r}(X)$. We have

$$\deg p(X) = \deg p_1(X) + (\deg \tilde{r}(X) - \deg \tilde{r}_1(X)) \geq \deg p_1(X),$$

so either $p(X)$ and $\tilde{r}(X)$ have an irreducible factor in common or $p(X) = p_1(X)$, $\tilde{r}(X) = \tilde{r}_1(X)$, uniqueness. ∎

Theorem 9.8.1. *Assume that $(n, q) = 1$ and let γ be a primitive nth root of 1 in the extension field $K = F(\gamma)$ of $F = GF(q)$. Assume that we use the t-error-correcting BCH code of length n over F with generator $g(X) = $ l.c.m. $\{Irr\,(\gamma, F), \ldots, Irr\,(\gamma^{2t}, F)\}$, receive word $w(X)$ with error pattern $e(X)$ of weight $\nu \leq t$, and define $S_i = w(\gamma^i)$, all $i > 0$. Then the unique smallest-dimensional FSR over K with output sequence $S_1, S_2, \ldots, S_{2t}$ has dimension ν, and its characteristic polynomial has as its roots the ν error locators $x_1, \ldots, x_\nu$.*

Proof. Denote by $s(X)$ the power series $S_1 + S_2 X + S_3 X^2 + \cdots$. We saw in (9.8.6) that

$$s(X) = \frac{p(X)}{r(X)},$$

where $r(X) = \prod_{j=1}^{\nu}(1 - x_j X)$, $(p(X), r(X)) = 1$, and $\deg p(X) < \nu$. By Lemma 9.8.1, the unique smallest-dimensional FSR with output sequence $S_1, S_2, \ldots, S_{2t}$ has dimension ν and recursion polynomial $r(X)$. From (9.8.4) we have $r(0) = 1$ and $r(x_j^{-1}) = 0$, $1 \leq j \leq \nu$; the roots of $r(X)$ are $x_1^{-1}, \ldots, x_\nu^{-1}$, so the roots of the characteristic polynomial $\tilde{r}(X)$ are $x_1, \ldots, x_\nu$. ∎

9.9 Shortest FSR with Given Output

Now we work over the finite field K. Assume that we have a sequence $\mathbf{s} = s_0 s_1 \cdots s_{n-1} s_n$ of elements of K. We seek an algorithm for finding the smallest-dimensional FSR with output sequence $\mathbf{s}$. We follow Massey (1969).

Lemma 9.9.1. *If the k-dimensional FSR M over K has output sequence $s_0, s_1, \ldots, s_{n-1}$ but not $s_0, s_1, \ldots, s_{n-1}, s_n$, then any FSR M' over K with output sequence $s_0, s_1, \ldots, s_{n-1}, s_n$ has dimension $k' \geq n + 1 - k$.*

Proof. The result is trivial if $k \geq n$ ($k' \geq 1$ is trivial), so we may assume that $k < n$. Assume that M has recursion polynomial

$$r(X) = 1 + \rho_1 X + \cdots + \rho_k X^k = \sum_{i=0}^{k} \rho_i X^i,$$

M' has recursion polynomial

$$r'(X) = 1 + \rho_1' X + \cdots + \rho_{k'}' X^{k'} = \sum_{j=0}^{k'} \rho_j' X^j,$$

and $k' \leq n-k$; we seek a contradiction. By (9.8.10) we have

$$s_l = -\sum_{i=1}^{k} \rho_i s_{l-i} \quad \text{for } l = k, \ldots, n-1,$$

(9.9.1) $$s_n \neq -\sum_{i=1}^{k} \rho_i s_{n-i},$$

$$s_l = -\sum_{j=1}^{k'} \rho'_j s_{l-j} \quad \text{for } l = k', \ldots, n.$$

Then

$$s_n \neq -\sum_{i=1}^{k} \rho_i s_{n-i} = -\sum_{i=1}^{k} \rho_i \left(-\sum_{j=1}^{k'} \rho'_j s_{n-i-j}\right)$$

$$= -\sum_{j=1}^{k'} \rho'_j \left(-\sum_{i=1}^{k} \rho_i s_{n-j-i}\right) = -\sum_{j=1}^{k'} \rho'_j s_{n-j} = s_n,$$

the desired contradiction. (For the first and third equalities we used the containments $\{n-k, \ldots, n-1\} \subseteq \{k', \ldots, n\}$, $\{n-k', \ldots, n-1\} \subseteq \{k, \ldots, n-1\}$, true because $n-k \geq k'$, $n-k' \geq k$.) ∎

By convention, we say a "0-dimensional FSR" has recursion polynomial $r(X) = 1$ and the all-0 output sequence $00 \cdots 0 \cdots$.

Theorem 9.9.1. *Let* $\mathbf{s} = s_0 s_1 s_2 \cdots s_n \cdots$ *be a given sequence of elements of* K. *We can find, for each* n, *the recursion polynomial* $r_n(X)$ *of an FSR* M_n *of minimal dimension* k_n *and output sequence* $s_0, s_1, \ldots, s_{n-1}$ *as follows.*

I. Initial Assignments. If all $s_i = 0$, *then* $r_n(X) = 1$ *for all* n. *If* $s_{h-1} \neq 0$ *but* $s_j = 0$ *for* $j < h-1$, *set* $k_j = 0$, $r_j(X) = 1$ *for* $0 \leq j < h$, $d_0 = s_0$, $k_h = h$, $r_h(X)$ *of degree* h *with* $r_h(0) = 1$.

II. Inductive Assignments. Assume now that $r_1(X), \ldots, r_n(X)$ *have been found, satisfying*

(9.9.2) $$\text{either } r_{j+1}(X) = r_j(X) \quad \text{or} \quad k_{j+1} = \max\{k_j, j+1-k_j\},$$
$$\text{all } j = 1, \ldots, n-1;$$

we shall find $r_{n+1}(X)$. *For each* $j = 1, \ldots, n$, *let* $r_j(X) = \sum_{i=0}^{k_j} \rho_i^{(j)} X^i$, *all* $\rho_0^{(j)} = r_j(0) = 1$, *and denote*

$$d_j = \sum_{i=0}^{k_j} \rho_i^{(j)} s_{j-i}.$$

(*The quantity* d_j *is called the* j*th* **discrepancy**.)
   (a) *If* $d_n = 0$, *take*

(9.9.3) $$k_{n+1} = k_n, \quad r_{n+1}(X) = r_n(X).$$

Sec. 9.9   *Shortest FSR with Given Output*   455

(b) *If* $d_n \neq 0$, *choose* $m < n$ *such that* $k_m < k_{m+1} = k_n$ *and take*

(9.9.4) $\quad k_{n+1} = \max\{k_n, n+1-k_n\}, \quad r_{n+1}(X) = r_n(X) - d_n d_m^{-1} X^{n-m} r_m(X).$

Proof. It is clear that only an FSR of length $\geq h$ can have its hth output s_{h-1} nonzero, all previous outputs 0; (9.8.15) with $k = h$ shows that our initial conditions are right, and we need only find $r_{n+1}(X), n \geq h$, given $r_1(X), \ldots, r_n(X)$ satisfying (9.9.2). If $d_n = 0$, then $0 = \sum_{i=0}^{k_n} \rho_i^{(n)} s_{n-i}$ so M_n defined by $r_n(X)$ has output sequence $s_0, s_1, \ldots, s_{n-1}, s_n$, not just $s_0, s_1, \ldots, s_{n-1}$; $k_{n+1} \geq k_n$ always, so we may take (9.9.3).

In the remainder of the proof we may assume $d_n \neq 0$. Since $n \geq h$, there is an $m < n$ with $k_m < k_{m+1} = k_{m+2} = \cdots = k_n$. Since $k_m < k_{m+1}$, we must have

(9.9.5) $\qquad k_{m+1} = m + 1 - k_m;$

also,

(9.9.6)
$$d_m = \sum_{i=0}^{k_m} \rho_i^{(m)} s_{m-i} \neq 0,$$
$$\sum_{i=0}^{k_m} \rho_i^{(m)} s_{l-i} = 0 \qquad \text{for } l = k_m, \ldots, m-1,$$

(9.9.7)
$$d_n = \sum_{i=0}^{k_n} \rho_i^{(n)} s_{n-i} \neq 0,$$
$$\sum_{i=0}^{k_n} \rho_i^{(n)} s_{l-i} = 0 \qquad \text{for } l = k_n, \ldots, n-1,$$

as in (9.8.11).

Define $r_{n+1}(X)$ by (9.9.4) and denote $r_{n+1}(X) = 1 + \rho_1 X + \cdots + \rho_k X^k$, abbreviating $k = k_{n+1}$, $\rho_i = \rho_i^{(n+1)}$; to show that M_{n+1} has output sequence $s_0, s_1, \ldots, s_{n-1}, s_n$, we must show

(9.9.8) $\qquad \sum_{i=0}^{k} \rho_i s_{l-i} = 0, \qquad l = k, k+1, \ldots, n.$

If we prove this we are finished, since Lemma 9.9.1 implies that no machine of dimension smaller than k_{n+1} can produce outputs $s_0, s_1, \ldots, s_{n-1}, s_n$. The machine defined by $r_{n+1}(X)$ has dimension

$\max\{k_n, n - m + k_m\} = \max\{k_n, n + 1 - (m + 1 - k_m)\}$

$\qquad\qquad\qquad\qquad = \max\{k_n, n + 1 - k_{m+1}\} = \max\{k_n, n + 1 - k_n\} = k_{n+1}.$

By (9.9.4) we have

(9.9.9)
$$\sum_{i=0}^{k} \rho_i s_{l-i} = \sum_{i=0}^{k_n} \rho_i^{(n)} s_{l-i} - d_n d_m^{-1} \sum_{i=n-m}^{k_m+n-m} \rho_{i-(n-m)}^{(m)} s_{l-i}$$
$$= \sum_{i=0}^{k_n} \rho_i^{(n)} s_{l-i} - d_n d_m^{-1} \sum_{j=0}^{k_m} \rho_j^{(m)} s_{l-n+m-j}.$$

Since $k \leq l \leq n$ and $k_n \leq k$, we have, by (9.9.7),

$$\sum_{i=0}^{k_n} \rho_i^{(n)} s_{l-i} = \begin{cases} 0 & \text{if } l < n, \\ d_n & \text{if } l = n. \end{cases}$$

Since $l \geq k \geq n+1-k_n$, we have $n+1-k_n \leq l \leq n$, so $m+1-k_n \leq l-n+m \leq m$. But $k_n = k_{m+1} = m+1-k_m$, so $m+1-k_n = k_m$, implying that $k_m \leq l-n+m \leq m$. Thus, by (9.9.6),

$$\sum_{j=0}^{k_m} \rho_j^{(m)} s_{l-n+m-j} = \begin{cases} 0 & \text{if } l < n, \\ d_m & \text{if } l = n. \end{cases}$$

Therefore (9.9.9) simplifies to

$$\sum_{i=0}^{k_n} \rho_i^{(n)} s_{l-i} - d_n d_m^{-1} \sum_{j=0}^{k_m} \rho_j^{(m)} s_{l-n+m-j} = \begin{cases} 0-0 = 0 & \text{if } l < n, \\ d_n - d_n d_m^{-1} d_m = 0 & \text{if } l = n. \end{cases} \blacksquare$$

We are mainly interested in Theorem 9.9.1 for its use in BCH decoding in the next section; but we do two examples now. Let us find the shortest FSR over GF(2) with output sequence $1101001 = s_0 s_1 s_2 s_3 s_4 s_5 s_6$. Because $s_0 \neq 0$, we have initial assignments $k_0 = 0$, $r_0(X) = 1$, $d_0 = 1$, $k_1 = 1$, $r_1(X) = 1+X$. For the inductive work we use a table of values (Table 9.9.1).

TABLE 9.9.1

m	n	k_n	$r_n(X)$	d_n
	0	0	1	$s_0 = 1$
	1	1	$1+X$	$1 \cdot s_1 + 1 \cdot s_0 = 1+1 = 0$
0	2	1	$1+X$	$1 \cdot s_2 + 1 \cdot s_1 = 0+1 = 1$
	3	2	$1+X+X^2$	$1 \cdot s_3 + 1 \cdot s_2 + 1 \cdot s_1 = 0$
2	4	2	$1+X+X^2$	$1 \cdot s_4 + 1 \cdot s_3 + 1 \cdot s_2 = 1$
	5	3	$1+X+X^3$	$1 \cdot s_5 + 1 \cdot s_4 + 1 \cdot s_2 = 0$
	6	3	$1+X+X^3$	$1 \cdot s_6 + 1 \cdot s_5 + 1 \cdot s_3 = 0$
	7	3	$1+X+X^3$	

Note in Table 9.9.1 that each d_n is easily computed as a sort of "dot product"; the first term is $1 \cdot s_n$, and succeeding terms multiply coefficients of higher powers of X in $r_n(X)$ times earlier terms in the sequence. Whenever $d_n = 0$, $r_{n+1}(X) = r_n(X)$ by (9.9.3), so it remains to show how $r_3(X)$ and $r_5(X)$ are computed.

Polynomial $r_3(X)$ is computed when $n = 2$. $d_n = d_2 \neq 0$, and $k_n = k_2 = 1$ so m must be 0; $k_0 = 0 < k_1 = k_2 = 1$ and $d_0 = s_0 = 1$. (9.9.4) gives

$$k_3 = \max\{k_2, 2+1-k_2\} = \max\{1, 2\} = 2$$

Sec. 9.9 Shortest FSR with Given Output

and

$$r_3(X) = r_2(X) - d_2 d_0^{-1} X^{2-0} r_0(X) = 1 + X - 1 \cdot 1^{-1} \cdot X^2 \cdot 1 = 1 + X + X^2.$$

Polynomial $r_5(X) = r_{n+1}(X)$ is computed when $n = 4$. We have $d_4 = 1$, $k_4 = 2$, so $m = 2$. Equation (9.9.4) gives

$$k_5 = \max\{k_4, 4 + 1 - k_4\} = \max\{2, 4 + 1 - 2\} = 3$$

and

$$\begin{aligned} r_5(X) &= r_4(X) - d_4 d_2^{-1} X^{4-2} r_2(X) \\ &= (1 + X + X^2) - 1 \cdot 1^{-1} \cdot X^2 \cdot (1 + X) = 1 + X + X^3. \end{aligned}$$

The desired FSR has recursion polynomial $r_7(X) = 1 + X + X^3$ and characteristic polynomial $\bar{r}_7(X) = X^3 + X^2 + 1$. It is pictured in Figure 9.9.1.

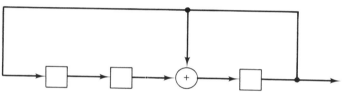

FIGURE 9.9.1. *Shortest FSR with Output Sequence 1101001*

We shall also find the shortest FSR over GF(2) with output sequence 0111010. The table obtained is Table 9.9.2. Note especially row 4, which

TABLE 9.9.2

m	n	k_n	$r_n(x)$	d_n
	1	0	1	$s_1 = 1$
1	2	2	$1 + X^2$	$s_2 + s_0 = 1$
1	3	2	$1 + X + X^2$	$s_3 + s_2 + s_1 = 1$
1	4	2	$1 + X + 0X^2$	$s_4 + s_3 = 1$
	5	3	$1 + X + X^3$	$s_5 + s_4 + s_2 = 0$
	6	3	$1 + X + X^3$	$s_6 + s_5 + s_3 = 0$
	7	3	$1 + X + X^3$	

says that $1 + X + 0X^2$, of "degree" 2, is the recursion polynomial of a shortest FSR with output $s_0 s_1 s_2 s_3 = 0111$. It is pictured in Figure 9.9.2, with initial state

$$\begin{bmatrix} y_1 \\ y_2 \end{bmatrix} = \begin{bmatrix} 1 \\ 0 \end{bmatrix}.$$

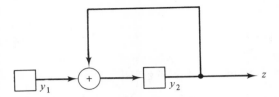

FIGURE 9.9.2. *Shortest FSR with Output Sequence 0111*

So we see that we must allow the actual degree of a recursion polynomial to be less than its "degree" = dimension of the FSR. One can easily show, using Theorem 9.12.1, that this FSR never occurs in BCH decoding.

9.10 Algorithmic BCH Decoding

The following theorem is an immediate corollary of Theorems 9.8.1 and 9.9.1.

Theorem 9.10.1. *Assume that* $(n_0, q) = 1$ *and let* γ *be a primitive* n_0*th root of 1 in the extension field* $K = F(\gamma)$ *of* $F = \mathrm{GF}(q)$. *Assume we use the t-error-correcting BCH code of length* n_0 *over F with generator* $g(X) = $ l.c.m. $\{\mathrm{Irr}\,(\gamma, F), \ldots, \mathrm{Irr}\,(\gamma^{2t}, F)\}$, *receive word* $r(X)$ *with error pattern* $e(X)$ *of weight* $\nu \leq t$, *and compute* $S_i = r(\gamma^i)$ *for* $1 \leq i \leq 2t$. *Then the error locators* $x_1, \ldots, x_\nu$ *are the roots of* $\bar{r}_{2t}(X)$, *where* $r_1(X), \ldots, r_{2t}(X)$ *are obtained as follows*:

I. Initial assignments. If all $S_i = 0$, take all $r_i(X) = 1$. If $S_h \neq 0$ but $S_1 = \cdots = S_{h-1} = 0$, set $k_j = 0$, $r_j(X) = 1$ for $0 \leq j < h$, $d_0 = S_1$, $k_h = h$, $r_h(X)$ of degree h with $r_h(0) = 1$.

II. Inductive assignments. Assume now that $r_1(X), \ldots, r_n(X)$ have been found, satisfying

(9.10.1) *either* $r_{j+1}(X) = r_j(X)$ *or* $k_{j+1} = \max\{k_j, j+1-k_j\}$,

all $j = 1, \ldots, n-1$; we will find $r_{n+1}(X)$. For each j, let $r_j(X) = \sum_{i=0}^{k_j} \rho_i^{(j)} X^i$, all $\rho_0^{(j)} = r_j(0) = 1$, and denote

$$d_j = \sum_{i=0}^{k_j} \rho_i^{(j)} S_{j-i+1}.$$

(a) If $d_n = 0$, take

(9.10.2) $\qquad k_{n+1} = k_n, \qquad r_{n+1}(X) = r_n(X).$

(b) If $d_n \neq 0$, choose $m < n$ such that $k_m < k_{m+1} = k_n$ and take

$$k_{n+1} = \max\{k_n, n+1-k_n\},$$

(9.10.3)

$$r_{n+1}(X) = r_n(X) - d_n d_m^{-1} X^{n-m} r_m(X). \blacksquare$$

Sec. 9.10 Algorithmic BCH Decoding

After we have found $x_1, \ldots, x_\nu$ from Theorem 9.10.1, equations (9.8.3) are *linear* equations for the error values $y_1, \ldots, y_\nu$. They can be solved using standard methods for linear equations, but Forney (1965) found a faster solution, using recursion. Using his method, we can find the error value y_j corresponding to an error location x_j without needing all the other error locations; computation of error values is almost simultaneous with computation of error locations.

Theorem 9.10.2 (Forney). *In Theorem 9.10.1, denote*

(9.10.4) $$\bar{r}_{2t}(X) = c(X) = X^\nu + \rho_1 X^{\nu-1} + \cdots + \rho_{\nu-1} X + \rho_\nu$$
$$= \sum_{i=0}^{\nu} \rho_i X^{\nu-i}, \qquad \rho_0 = 1.$$

For each j, $1 \leq j \leq \nu$, define $\rho_{j0}, \rho_{j1}, \ldots, \rho_{j,\nu-1}$ by

(9.10.5) $$\rho_{j0} = 1, \qquad \rho_{ji} = x_j \rho_{j,i-1} + \rho_i \qquad \text{for } i \geq 1.$$

Then

(9.10.6) $$\rho_{ji} = \sum_{k=0}^{i} \rho_k x_j^{i-k}$$

and the error values y_j can be computed from the formula

(9.10.7) $$y_j = \frac{\sum_{i=0}^{\nu-1} \rho_{ji} S_{\nu-i}}{\sum_{i=0}^{\nu-1} \rho_{ji} x_j^{\nu-i}}.$$

Proof. $c(X) = \prod_{i=1}^{\nu}(X - x_i)$, so

(9.10.8) $$\frac{c(X)}{X - x_j} = \prod_{\substack{i=1 \\ i \neq j}}^{\nu}(X - x_i) = \sum_{i=0}^{\nu-1} \rho_{ji} X^{\nu-1-i}$$

for some coefficients ρ_{ji}. Therefore,

$$\sum_{i=0}^{\nu} \rho_i X^{\nu-i} = c(X) = (X - x_j) \sum_{i=0}^{\nu-1} \rho_{ji} X^{\nu-1-i} = \sum_{i=0}^{\nu-1} \rho_{ji} X^{\nu-i} - \sum_{i=1}^{\nu} \rho_{j,i-1} x_j X^{\nu-i}.$$

Equating coefficients of $X^{\nu-i}$, we see that $\rho_{j0} = \rho_0 = 1$ and, for $i > 0$, $\rho_i = \rho_{ji} - x_j \rho_{j,i-1}$, proving (9.10.5). Equation (9.10.6) is easily proved by induction from (9.10.5).

Now, by (9.8.3) and (9.10.8),

$$\sum_{i=0}^{\nu-1} \rho_{ji} S_{\nu-i} = \sum_{i=0}^{\nu-1} \rho_{ji} \sum_{k=1}^{\nu} y_k x_k^{\nu-i} = \sum_{k=1}^{\nu} y_k x_k \sum_{i=0}^{\nu-1} \rho_{ji} x_k^{\nu-1-i}$$

$$= \sum_{k=1}^{\nu} y_k x_k \prod_{\substack{l=1 \\ l \neq j}}^{\nu} (x_k - x_l) = y_j x_j \prod_{\substack{l=1 \\ l \neq j}}^{\nu} (x_j - x_l)$$

$$= y_j x_j \sum_{i=0}^{\nu-1} \rho_{ji} x_j^{\nu-1-i} = y_j \sum_{i=0}^{\nu-1} \rho_{ji} x_j^{\nu-i},$$

so (9.10.7) is proved, using

$$\sum_{i=0}^{\nu-1} \rho_{ji} x_j^{\nu-i} = x_j \prod_{\substack{l=1 \\ l \neq j}}^{\nu} (x_j - x_l) \neq 0. \blacksquare$$

EXAMPLE 1. For our first decoding example, we shall use the primitive 3-error-correcting BCH code of length $3^3 - 1 = 26$ over GF(3). The polynomial $X^3 + 2X + 1$ is primitive over $F = \text{GF}(3)$; let γ be one of its roots in $\text{GF}(3^3) = K$. Log and antilog tables for K appear in Table 9.10.1; they were computed in Exercise 7.22.13. Assume that the error pattern in the transmission of some 26-digit code word $f(X)$ is

(9.10.9) $$e(X) = X^4 + 2X^{11} + X^{21}.$$

TABLE 9.10.1. GF(3^3).

$\gamma^0 = 1$	$\gamma^9 = \gamma + 1$	$\gamma^{18} = \gamma^2 + 2\gamma + 1$
$\gamma^1 = \gamma$	$\gamma^{10} = \gamma^2 + \gamma$	$\gamma^{19} = 2\gamma^2 + 2\gamma + 2$
$\gamma^2 = \gamma^2$	$\gamma^{11} = \gamma^2 + \gamma + 2$	$\gamma^{20} = 2\gamma^2 + \gamma + 1$
$\gamma^3 = \gamma + 2$	$\gamma^{12} = \gamma^2 + 2$	$\gamma^{21} = \gamma^2 + 1$
$\gamma^4 = \gamma^2 + 2\gamma$	$\gamma^{13} = 2$	$\gamma^{22} = 2\gamma + 2$
$\gamma^5 = 2\gamma^2 + \gamma + 2$	$\gamma^{14} = 2\gamma$	$\gamma^{23} = 2\gamma^2 + 2\gamma$
$\gamma^6 = \gamma^2 + \gamma + 1$	$\gamma^{15} = 2\gamma^2$	$\gamma^{24} = 2\gamma^2 + 2\gamma + 1$
$\gamma^7 = \gamma^2 + 2\gamma + 2$	$\gamma^{16} = 2\gamma + 1$	$\gamma^{25} = 2\gamma^2 + 1$
$\gamma^8 = 2\gamma^2 + 2$	$\gamma^{17} = 2\gamma^2 + \gamma$	$\gamma^{26} = 1$
	$\gamma^2 = \gamma^2$	$2\gamma^2 = \gamma^{15}$
$1 = \gamma^0$	$\gamma^2 + 1 = \gamma^{21}$	$2\gamma^2 + 1 = \gamma^{25}$
$2 = \gamma^{13}$	$\gamma^2 + 2 = \gamma^{12}$	$2\gamma^2 + 2 = \gamma^8$
$\gamma = \gamma^1$	$\gamma^2 + \gamma = \gamma^{10}$	$2\gamma^2 + \gamma = \gamma^{17}$
$\gamma + 1 = \gamma^9$	$\gamma^2 + \gamma + 1 = \gamma^6$	$2\gamma^2 + \gamma + 1 = \gamma^{20}$
$\gamma + 2 = \gamma^3$	$\gamma^2 + \gamma + 2 = \gamma^{11}$	$2\gamma^2 + \gamma + 2 = \gamma^5$
$2\gamma = \gamma^{14}$	$\gamma^2 + 2\gamma = \gamma^4$	$2\gamma^2 + 2\gamma = \gamma^{23}$
$2\gamma + 1 = \gamma^{16}$	$\gamma^2 + 2\gamma + 1 = \gamma^{18}$	$2\gamma^2 + 2\gamma + 1 = \gamma^{24}$
$2\gamma + 2 = \gamma^{22}$	$\gamma^2 + 2\gamma + 2 = \gamma^7$	$2\gamma^2 + 2\gamma + 2 = \gamma^{19}$

The receiver sees $r(X) = f(X) + e(X)$. Since $f(X)$ is a multiple of

$$g(X) = \text{l.c.m.} \{\text{Irr}(\gamma, F), \text{Irr}(\gamma^2, F), \ldots, \text{Irr}(\gamma^6, F)\},$$

Sec. 9.10   Algorithmic BCH Decoding

$f(\gamma^i) = 0$ for $1 \leq i \leq 6$ and the receiver computes $S_i = r(\gamma^i) = e(\gamma^i)$, $1 \leq i \leq 6$. The following lemma is helpful.

Lemma 9.10.1. $S_{iq} = (S_i)^q$ in $K = F(\gamma)$, where $F = \mathrm{GF}(q)$.

Proof. The function $\alpha \to \alpha^q$ is an automorphism of K that is the identity on F (Theorem 7.23.2). Since $S_i = y_1 x_1^i + \cdots + y_\nu x_\nu^i$, $y_j \in F$, $x_j \in K$, we have

$$(S_i)^q = y_1^q (x_1^i)^q + \cdots + y_\nu^q (x_\nu^i)^q = y_1 x_1^{iq} + \cdots + y_\nu x_\nu^{iq} = S_{iq}. \quad \blacksquare$$

In our example, with the aid of Table 9.10.1 we compute $S_1 = \gamma^{11}$, $S_2 = \gamma^{25}$, $S_3 = (S_1)^3 = \gamma^7$, $S_4 = \gamma^9$, $S_5 = \gamma^5$, $S_6 = (S_2)^3 = \gamma^{23}$. From $S_1, \ldots, S_6$, the receiver computes x_1, x_2, and x_3. In Table 9.10.2 we show the computation of $r_6(X)$, using Theorem 9.10.1.

TABLE 9.10.2

m	n	k_n	$r_n(X)$	d_n
	0	0	1	$S_1 = \gamma^{11}$
0	1	1	$1 + X$	$S_2 + S_1 = \gamma$
0	2	1	$1 + \gamma X$	$S_3 + \gamma S_2 = \gamma^4$
2	3	2	$1 + \gamma X + \gamma^6 X^2$	$S_4 + \gamma S_3 + \gamma^6 S_2 = \gamma^7$
2	4	2	$1 + X + \gamma^{16} X^2$	$S_5 + S_4 + \gamma^{16} S_3 = \gamma^{10}$
4	5	3	$1 + X + \gamma^{17} X^2 + \gamma^{20} X^3$	$S_6 + S_5 + \gamma^{17} S_4 + \gamma^{20} S_3 = \gamma^{10}$
	6	3	$1 + \gamma^5 X^2 + \gamma^{23} X^3$	

In Table 9.10.2, $r_2(X), \ldots, r_6(X)$ are computed using (9.10.3), which yields the following equations:

$$r_2(X) = (1 + X) - \gamma \cdot \gamma^{-11} \cdot X \cdot 1 = 1 + (1 - \gamma^{16}) X = 1 + \gamma X,$$

$$r_3(X) = (1 + \gamma X) - \gamma^4 \cdot \gamma^{-11} \cdot X^2 \cdot 1 = 1 + \gamma X + \gamma^6 X^2,$$

$$r_4(X) = (1 + \gamma X + \gamma^6 X^2) - \gamma^7 \gamma^{-4} X(1 + \gamma X) = 1 + X + \gamma^{16} X^2,$$

$$r_5(X) = (1 + X + \gamma^{16} X^2) - \gamma^{10} \gamma^{-4} X^2 (1 + \gamma X) = 1 + X + \gamma^{17} X^2 + \gamma^{20} X^3,$$

$$r_6(X) = (1 + X + \gamma^{17} X^2 + \gamma^{20} X^3) - \gamma^{10} \gamma^{-10} X(1 + X + \gamma^{16} X^2)$$
$$= 1 + \gamma^5 X^2 + \gamma^{23} X^3.$$

The receiver knows that the error locations are the zeros of $\bar{r}_6(X) = X^3 + \gamma^5 X + \gamma^{23}$, and finds the zeros by trying *all* powers γ^i of γ; for each i, $0 \leq i \leq 25$, it tests whether $\bar{r}_6(\gamma^i) = \gamma^{3i} + \gamma^{5+i} + \gamma^{23}$ is 0. Chien (1964) described an efficient procedure for doing this, now called the *Chien search*. A "multiplying-by-γ" circuit, as in Example 1, Section 8.2, successively produces γ^{5+i} in its ith shift, and simultaneously a "multiplying-by-γ^3" circuit produces γ^{3i} in its ith shift. These results can be added to γ^{23} (also written in form $a_2 \gamma^2 + a_1 \gamma + a_0$) and the sum tested to see if it is 0, in a single shift time. Of course, it turns out that $\bar{r}_6(\gamma^4) = \bar{r}_6(\gamma^{11}) = \bar{r}_6(\gamma^{21}) = 0$, so the receiver determines the error locators $x_1 = \gamma^4$, $x_2 = \gamma^{11}$, and $x_3 = \gamma^{21}$.

Finally, we compute the error values from Theorem 9.10.2, using $\rho_1 = 0$, $\rho_2 = \gamma^5$, and $\rho_3 = \gamma^{23}$ from $\bar{r}_6(X)$.

For y_1, $\rho_{10} = 1$, $\rho_{11} = x_1\rho_{10} + \rho_1 = \gamma^4$, and $\rho_{12} = x_1\rho_{11} + \rho_2 = \gamma^4 \cdot \gamma^4 + \gamma^5 = \gamma^6$, so

$$y_1 = \frac{\rho_{10}S_3 + \rho_{11}S_2 + \rho_{12}S_1}{\rho_{10}x_1^3 + \rho_{11}x_1^2 + \rho_{12}x_1} = \frac{\gamma^7 + \gamma^3 + \gamma^{17}}{\gamma^{12} + \gamma^{12} + \gamma^{10}} = \frac{\gamma^9}{\gamma^9} = 1.$$

For y_2, $\rho_{20} = 1$, $\rho_{21} = x_2\rho_{20} + \rho_1 = \gamma^{11}$, and $\rho_{22} = x_2\rho_{21} + \rho_2 = \gamma^{11} \cdot \gamma^{11} + \gamma^5 = \gamma^{25}$, so

$$y_2 = \frac{\rho_{20}S_3 + \rho_{21}S_2 + \rho_{22}S_1}{\rho_{20}x_2^3 + \rho_{21}x_2^2 + \rho_{22}x_2} = \frac{\gamma^7 + \gamma^{10} + \gamma^{10}}{\gamma^7 + \gamma^7 + \gamma^{10}} = \frac{\gamma^7 + 2\gamma^{10}}{2\gamma^7 + \gamma^{10}} = 2.$$

For y_3, $\rho_{30} = 1$, $\rho_{31} = x_3\rho_{30} + \rho_1 = \gamma^{21}$, and $\rho_{32} = x_3\rho_{31} + \rho_2 = \gamma^{21} \cdot \gamma^{21} + \gamma^5 = \gamma^{15}$, so

$$y_3 = \frac{\rho_{30}S_3 + \rho_{31}S_2 + \rho_{32}S_1}{\rho_{30}x_3^3 + \rho_{31}x_3^2 + \rho_{32}x_3} = \frac{\gamma^7 + \gamma^{20} + 1}{\gamma^{11} + \gamma^{11} + \gamma^{10}} = \frac{1}{1} = 1.$$

The receiver concludes that the error pattern was

$$y_1X^4 + y_2X^{11} + y_3X^{21} = X^4 + 2X^{11} + X^{21} = e(X),$$

in agreement with (9.10.9), and subtracts $e(X)$ from $r(X)$ to obtain the correct code word transmitted.

EXAMPLE 2. For the same code as in Example 1, assume that the error pattern is

(9.10.10) $$e(X) = 2X^{17}.$$

Then $S_1 = \gamma^4$, $S_2 = \gamma^{21}$, $S_3 = \gamma^{12}$, $S_4 = \gamma^3$, $S_5 = \gamma^{20}$, and $S_6 = \gamma^{11}$. From these values of S_j the decoder computes the values in Table 9.10.3. The only

TABLE 9.10.3

m	n	k_n	$r_n(X)$	d_n
	0	0	1	$S_1 = \gamma^4$
0	1	1	$1 + X$	$S_2 + S_1 = \gamma^{24}$
	2	1	$1 + \gamma^4 X$	$S_3 + \gamma^4 S_2 = 0$
	3	1	$1 + \gamma^4 X$	$S_4 + \gamma^4 S_3 = 0$
	4	1	$1 + \gamma^4 X$	$S_5 + \gamma^4 S_4 = 0$
	5	1	$1 + \gamma^4 X$	$S_6 + \gamma^4 S_5 = 0$
	6	1	$1 + \gamma^4 X$	

Sec. 9.10 Algorithmic BCH Decoding

computation involving (9.10.3) is

$$r_2(X) = (1+X) - \gamma^{24}\gamma^{-4}X(1) = 1 + (1+\gamma^7)X = 1 + \gamma^4 X.$$

Here $\bar{r}_6(X) = X + \gamma^4$ has one zero $x_1 = \gamma^{17}$, the correct error locator. Since $\nu = 1$ is the number of errors and $\rho_{10} = 1$, Theorem 9.10.2 yields the (correct) error value

$$y_1 = \frac{\rho_{10}S_1}{\rho_{10}x_1} = \frac{\gamma^4}{\gamma^{17}} = 2.$$

EXAMPLE 3. Now we consider the Reed–Solomon code of length $n = 15 = 2^4 - 1$ over $F = GF(q)$, $q = 2^4$. If γ is a primitive element in F and a root of $X^4 + X + 1$, log and antilog tables for F are given in Table 7.21.3. Suppose the error pattern is

(9.10.11) $$e(X) = \gamma^4 X^5 + \gamma^9 X^8 + \gamma X^{12}.$$

The values of the $S_i = e(\gamma^i)$ are

$$S_1 = \gamma^4, \quad S_2 = \gamma^{14}, \quad S_3 = 0, \quad S_4 = \gamma^{10}, \quad S_5 = \gamma^3, \quad S_6 = 1.$$

Table 9.10.4 shows the computation of $\bar{r}_6(X)$.

TABLE 9.10.4

m	n	k_n	$r_n(X)$	d_n
	0	0	1	$S_1 = \gamma^4$
0	1	1	$1+X$	$S_2 + S_1 = \gamma^9$
0	2	1	$1 + \gamma^{10}X$	$S_3 + \gamma^{10}S_2 = \gamma^9$
2	3	2	$1 + \gamma^{10}X + \gamma^5 X^2$	$S_4 + \gamma^{10}S_3 + \gamma^5 S_2 = \gamma^2$
2	4	2	$1 + \gamma X + \gamma^{11}X^2$	$S_5 + \gamma S_4 + \gamma^{11}S_3 = \gamma^5$
4	5	3	$1 + \gamma X + \gamma^6 X^3$	$S_6 + \gamma S_5 + \gamma^6 S_3 = \gamma$
	6	3	$1 + \gamma^6 X + \gamma^{12}X^2 + \gamma^{10}X^3$	

The polynomial $\bar{r}_6(X) = X^3 + \gamma^6 X^2 + \gamma^{12} X + \gamma^{10}$ does have zeros $x_1 = \gamma^5, x_2 = \gamma^8, x_3 = \gamma^{12}$, the correct error locations. We can also compute the error values y_1, y_2, and y_3, using $\rho_1 = \gamma^6$ and $\rho_2 = \gamma^{12}$ in Theorem 9.10.2.

For y_1, $\rho_{10} = 1$, $\rho_{11} = x_1\rho_{10} + \rho_1 = \gamma^9$, and $\rho_{12} = x_1\rho_{11} + \rho_2 = \gamma^{14} + \gamma^{12} = \gamma^5$, so

$$y_1 = \frac{\rho_{10}S_3 + \rho_{11}S_2 + \rho_{12}S_1}{\rho_{10}x_1^3 + \rho_{11}x_1^2 + \rho_{12}x_1} = \frac{0 + \gamma^8 + \gamma^9}{1 + \gamma^4 + \gamma^{10}} = \frac{\gamma^{12}}{\gamma^8} = \gamma^4.$$

For y_2, $\rho_{20} = 1$, $\rho_{21} = x_2\rho_{20} + \rho_1 = \gamma^{14}$, and $\rho_{22} = x_2\rho_{21} + \rho_2 = \gamma^7 + \gamma^{12} = \gamma^2$, so

$$y_2 = \frac{\rho_{20}S_3 + \rho_{21}S_2 + \rho_{22}S_1}{\rho_{20}x_2^3 + \rho_{21}x_2^2 + \rho_{22}x_2} = \frac{0 + \gamma^{13} + \gamma^6}{\gamma^9 + 1 + \gamma^{10}} = \frac{1}{\gamma^6} = \gamma^9.$$

For y_3, $\rho_{30} = 1$, $\rho_{31} = x_3\rho_{30} + \rho_1 = \gamma^4$, and $\rho_{32} = x_3\rho_{31} + \rho_2 = \gamma + \gamma^{12} = \gamma^{13}$, so

$$y_3 = \frac{\rho_{30}S_3 + \rho_{31}S_2 + \rho_{32}S_1}{\rho_{30}x_3^3 + \rho_{31}x_3^2 + \rho_{32}x_3} = \frac{0 + \gamma^3 + \gamma^2}{\gamma^6 + \gamma^{13} + \gamma^{10}} = \frac{\gamma^6}{\gamma^5} = \gamma.$$

We get error pattern $\gamma^4 X^5 + \gamma^9 X^8 + \gamma X^{12}$, in agreement with (9.10.11).

We should remark here that recently Sugiyama, Kasahara, Hirasawa, and Namekawa (1975) found a new algorithm for decoding BCH codes (even the more general recently discovered *Goppa codes*), using the Euclidean algorithm for the g.c.d. of polynomials. A new paper by Sarwate (1977) describes how their work can be combined with a fast Euclidean algorithm due to Moenck [see Section 8.9 of Aho, Hopcroft, and Ullman (1974)] and the "fast Fourier transform" [Pollard (1971)] over finite fields, to obtain a decoding algorithm faster than that of Berlekamp and Massey for codes of very large block length.

9.11 Exercises

*1. The **formal power series ring** $K[[X]]$ in the indeterminate X over the field K is defined to be the set of all power series

$$\sum_{i=0}^{\infty} a_i X^i = a_0 + a_1 X + a_2 X^2 + \cdots + a_n X^n + \cdots,$$

where

$$\left(\sum_{i=0}^{\infty} a_i X^i\right) + \left(\sum_{i=0}^{\infty} b_i X^i\right) = \sum_{i=0}^{\infty} (a_i + b_i) X^i,$$

$$\left(\sum_{i=0}^{\infty} a_i X^i\right)\left(\sum_{i=0}^{\infty} b_i X^i\right) = \sum_{i=0}^{\infty} \left(\sum_{j=0}^{i} a_j b_{i-j}\right) X^i$$
$$= a_0 b_0 + (a_0 b_1 + a_1 b_0) X + (a_0 b_2 + a_1 b_1 + a_2 b_0) X^2 + \cdots.$$

(a) Show that $K[[X]]$ is a commutative ring with identity under the given operations, and is also a vector space over K. Show that $K[X]$ is a subring of $K[[X]]$.

(b) Show that we can also interpret $K[[X]]$ as the ring of all functions $f: \mathbb{N} \to K$ ($f: i \to a_i$ corresponding to $\sum_{i=0}^{\infty} a_i X^i$), where $f + g$ is defined by $(f+g)(i) = f(i) + g(i)$ and $f \cdot g$ is the "convolution" defined by $(f \cdot g)(t) = \sum_{i=0}^{t} f(i)g(t-i)$.

(c) Show that $(1-\alpha X)^{-1} = \sum_{i=0}^{\infty} \alpha^i X^i$ in $K[[X]]$, for any $\alpha \in K$.

2. Find the shortest FSR over GF(2) with output sequence 100111010011.

3. Find the shortest FSR over GF(7) with output sequence 31650511.

4. Find the shortest FSR over GF(3^3) with output sequence $\gamma^6 \gamma^{19} \gamma^{18} \gamma^3 \gamma^{11} \gamma^5$, where Irr $(\gamma, \text{GF}(3)) = X^3 + 2X + 1$.

*5. For the following codes and error patterns, compute S_j, $1 \le j \le 6$. [Use Tables 9.10.1 and 7.21.3, and the table you obtained in Exercise 7.22.13 for GF(2^5).]

(a) 3-error-correcting BCH code of length $n = 26$ over GF(3), $e(X) = X^9 + 2X^{19}$.

(b) 3-error-correcting Reed–Solomon code of length $n = 15$ over GF(2^4), $e(X) = \gamma^5 X^2 + \gamma^4 X^6 + \gamma^9 X^{12}$, where Irr $(\gamma, \text{GF}(2)) = X^4 + X + 1$.

(c) 3-error-correcting Reed–Solomon code of length $n = 31$ over GF(2^5), $e(X) = \gamma^3 X^2 + \gamma^{16} X^9 + \gamma^{18} X^{18}$, where Irr $(\gamma, \text{GF}(2)) = X^5 + X^2 + 1$.

*6. (a), (b), (c) Using Theorems 9.10.1 and 9.10.2 and the values of S_j obtained in Exercise 5(a) to (c), decode and see if you can find the correct error patterns $e(X)$.

7. (a) Write a computer program which will determine the shortest FSR over GF(2) with given output.

(b) Extend (a) to other finite fields (refer to Exercise 7.25.12 for a discussion of the necessary computations within finite fields).

(c) Ambitious students may wish to actually program a BCH decoder [see Paschburg (1974) and Michaelson (1969)], for one or a family of BCH codes. [This is easier over GF(2) after the simplifications described in the next section.]

9.12 Binary BCH Decoding

We shall see in this section that for binary codes ($q = 2$), half the steps in the algorithm of Theorem 9.10.1 are trivial. We keep the notation of Section 9.10, with $q = 2$, and we shall use the fact (Lemma 9.10.1) that $S_{2i} = S_i^2$. Our first lemma concerns some equations known as *Newton's identities* [see van der Waerden (1970), Exercise 5.18, p. 101].

Lemma 9.12.1. *Assume that the elements* $S_1, S_2, \ldots, S_{2t}$ *of* GF(2^m) *satisfy* $S_{2i} = (S_i)^2$, *and the elements* $\rho_1, \rho_2, \ldots, \rho_{2t-1} \in$ GF(2^m) *satisfy the t equations*

(9.12.1$_1$) $0 = S_1 + \rho_1$

(9.12.1$_2$) $0 = S_3 + S_2 \rho_1 + S_1 \rho_2 + \rho_3$

$\vdots$

(9.12.1$_t$) $0 = S_{2t-1} + S_{2t-2}\rho_1 + S_{2t-3}\rho_2 + S_{2t-4}\rho_3 + \cdots + S_1 \rho_{2t-2} + \rho_{2t-1}$.

Then the equations

(9.12.2$_1$) $\quad 0 = S_2 + S_1\rho_1$

(9.12.2$_2$) $\quad 0 = S_4 + S_3\rho_1 + S_2\rho_2 + S_1\rho_3$

$$\vdots$$

(9.12.2$_t$) $\quad 0 = S_{2t} + S_{2t-1}\rho_1 + S_{2t-2}\rho_2 + S_{2t-3}\rho_3 + \cdots + S_2\rho_{2t-2} + S_1\rho_{2t-1}$

also hold.

Proof. When $t=1$ and (9.12.1$_1$) holds, then $S_2 + S_1\rho_1 = S_1^2 + S_1 \cdot S_1 = 0$, that is, (9.12.2$_1$) holds. So by induction on t we may assume that the result holds for $t-1$ and prove it for t. We know that equations (9.12.1$_i$) hold for $1 \le i \le t-1$, so by induction, equations (9.12.2$_i$) hold for $1 \le i \le t-1$. It remains only to prove (9.12.2$_t$), using (9.12.1$_t$). To do this, we simply add

$S_1 \cdot$ [equation (9.12.1$_t$)] + $S_2 \cdot$ [equation (9.12.2$_{t-1}$)]

$\quad + S_3 \cdot$ [equation (9.12.1$_{t-1}$)] + $S_4 \cdot$ [equation (9.12.2$_{t-2}$)] + $\cdots$

$\quad + S_{2t-3} \cdot$ [equation (9.12.1$_2$)] + $S_{2t-2} \cdot$ [equation (9.12.2$_1$)].

Most terms $S_iS_j\rho_k$ cancel out, and the final equation is

$$0 = (S_t)^2 + S_{2t-1}S_1 + (S_{t-1})^2\rho_2 + S_{2t-3}\rho_3 + \cdots + (S_1)^2\rho_{2t-2} + S_1\rho_{2t-1}$$
$$= S_{2t} + S_{2t-1}\rho_1 + S_{2t-2}\rho_2 + S_{2t-3}\rho_3 + \cdots + S_2\rho_{2t-2} + S_1\rho_{2t-1}. \quad \blacksquare$$

Theorem 9.12.1. *If $q=2$ in Theorem 9.10.1, then h is odd. We can take $r_{h+1}(X) = r_h(X) = 1 + S_h X^h$. When we do so, all odd-numbered discrepancies are zero: $d_1 = d_3 = d_5 = \cdots = d_{2t-1} = 0$.*

Theorem 9.12.1 guarantees that in the algorithm of Theorem 9.10.1, $r_{n+1}(X) = r_n(X)$ whenever n is odd, provided that $q=2$ and $r_h(X)$ is chosen as specified.

Proof. Since $S_{2i} = (S_i)^2$, the first nonzero S_h certainly has h odd. Any polynomial of degree h and constant term 1 will do for $r_h(X)$, so take $r_h(X) = 1 + S_h X^h$. If $h=1$, then $d_h = d_1 = S_2 + S_1 \cdot S_1 = (S_1)^2 + S_1 \cdot S_1 = 0$, so $r_{h+1}(X) = r_h(X)$. If $h > 1$, then

$$d_h = S_{h+1} + S_h S_1 = (S_{(h+1)/2})^2 + S_h S_1 = 0^2 + S_h \cdot 0 = 0,$$

so again $r_{h+1}(X) = r_h(X)$.

We now prove inductively that the coefficients $\rho_i^{(2l)}$ of each $r_{2l}(X) = \sum_{i=0}^{k_{2l}} \rho_i^{(2l)} X^i$ satisfy (9.12.1$_1$), ..., (9.12.1$_l$), (9.12.2$_1$), ..., (9.12.2$_l$), and each $d_{2l+1} = 0$. To start the induction, we have shown that $d_h = 0$. If $h=1$, then (9.12.1$_1$) is merely $0 = S_1 + S_1$, true; and if $h > 1$, then

Sec. 9.12 Binary BCH Decoding

(9.12.1$_1$), ..., (9.12.1$_{(h+1)/2}$) amount to $0=0$, $0=0$, ..., $0=S_h+0+\cdots+0+S_h$, again true. By Lemma 9.12.1, (9.12.2$_1$), ..., (9.12.2$_{(h+1)/2}$) also hold.

Continuing the induction, we may now assume that for some l, coefficients of $r_{2l}(X)$ satisfy the equations, and $d_{2l-1}=0$; we must prove that $d_{2l+1}=0$, and coefficients of $r_{2(l+1)}(X)$ satisfy appropriate equations.

If $d_{2l}=0$, then $r_{2l+1}(X)=r_{2l}(X)=r_{2l-1}(X)$ and

$$0 = d_{2l} = S_{2l+1} + S_{2l}\rho_1^{(2l)} + S_{2l-1}\rho_2^{(2l)} + \cdots + S_3\rho_{2l-2}^{(2l)} + S_2\rho_{2l-1}^{(2l)}.$$

Since $\rho_i^{(2l+1)} = \rho_i^{(2l)}$, and $r_{2l-1}(X)$ has degree at most $2l-1$, so higher coefficients $\rho_i^{(2l)} = \rho_i^{(2l-1)}$ for $i > 2l-1$ are 0, this is exactly equation (9.12.1$_{l+1}$). Thus $r_{2l+1}(X)$ satisfies (9.12.1$_1$), ..., (9.12.1$_{l+1}$), so that by Lemma 9.12.1 it also satisfies (9.12.2$_1$), ..., (9.12.2$_{l+1}$). Equation (9.12.2$_{l+1}$) amounts to the desired equation $0=d_{2l+1}$, so $r_{2l+2}(X)=r_{2l+1}(X)$ satisfies appropriate equations. If $d_{2l}=0$, the induction is complete.

If $d_{2l} \neq 0$, then

$$(9.12.3) \qquad r_{2l+1}(X) = r_{2l}(X) + d_{2l}d_m^{-1}X^{2l-m}r_m(X),$$

where $m < 2l$, $k_m < k_{m+1} = k_{2l}$. For m' odd and $m' < 2l$, $d_{m'} = 0$ has already been proved, so $k_{m'} = k_{m'+1}$; hence $m = 2r$ is even. We shall show that $r_{2l+1}(X) = \sum \rho_i^{(2l+1)} X^i$ satisfies (9.12.1$_1$), ..., (9.12.1$_{l+1}$); that is, we must show that

$$(9.12.4) \qquad \sum_{i=1}^{2l+1} S_i \rho_{j-i}^{(2l+1)} + \rho_j^{(2l+1)} = 0, \qquad j=1, 3, \ldots, 2l+1.$$

(Of course, $\rho_0 = 1$; denote $\rho_s = 0$ for $s < 0$.) We do know, by induction and definition of d_n, that

$$(9.12.5) \qquad \sum_{i=1}^{2l+1} S_i \rho_{j-i}^{(2l)} + \rho_j^{(2l)} = \begin{cases} 0 & \text{if } j=1, 3, \ldots, 2l-1, \\ d_{2l} & \text{if } j=2l+1; \end{cases}$$

$$(9.12.6) \qquad \sum_{i=1}^{2r+1} S_i \rho_{j'-i}^{(2r)} + \rho_{j'}^{(2r)} = \begin{cases} 0 & \text{if } j'=1, 3, \ldots, 2r-1, \\ d_{2r} & \text{if } j'=2r+1. \end{cases}$$

These and (9.12.3) imply that

$$\sum_{i=1}^{2l+1} S_i \rho_{j-i}^{(2l+1)} + \rho_j^{(2l+1)}$$

$$= \sum_{i=1}^{2l+1} S_i \rho_{j-i}^{(2l)} + \rho_j^{(2l)} + d_{2l}d_{2r}^{-1}\left(\sum_{i=1}^{2l+1} S_i \rho_{j-i-2(l-r)}^{(2r)} + \rho_{j-2(l-r)}^{(2r)}\right);$$

with $j' = j - 2(l-r)$, we have $j = 2l+1$ iff $j' = 2r+1$, so this equals $0 + d_{2l}d_{2r}^{-1} \cdot 0 = 0$ for odd $j < 2l+1$, $d_{2l} + d_{2l}d_{2r}^{-1}d_{2r} = d_{2l} + d_{2l} = 0$ for $j = 2l+1$, proving (9.12.4). By the Lemma 9.12.1, then, $r_{2l+1}(X)$ satisfies (9.12.2$_i$), $1 \le i \le l+1$; equation (9.12.2$_{l+1}$) amounts to $0 = d_{2l+1}$, so $r_{2l+2}(X) = r_{2l+1}(X)$ satisfies appropriate equations. ∎

For example, consider a 3-error-correcting primitive BCH code of length $n = 31$ over GF(2). The error pattern

$$(9.12.7) \qquad e(X) = X^3 + X^{17} + X^{20}$$

yields the values $S_1 = e(\gamma) = \gamma^{26}$, $S_2 = S_1^2 = \gamma^{21}$, $S_3 = e(\gamma^3) = \gamma^{15}$, $S_4 = S_2^2 = \gamma^{11}$, $S_5 = e(\gamma^5) = \gamma^2$, and $S_6 = S_3^2 = \gamma^{30}$ (where Irr $(\gamma, \text{GF}(2)) = X^5 + X^2 + 1$).

We have $h = 1$, $r_2(X) = r_1(X) = 1 + S_1X = 1 + \gamma^{26}X$. From (9.12.3) we have the rule (when $d_{2l} \ne 0$)

$$(9.12.8) \qquad r_{2l+2}(X) = r_{2l}(X) + d_{2l}d_{2r}^{-1}X^{2(l-r)}r_{2r}(X)$$

for computing other $r_j(X)$'s. Also,

$$k_{2l+2} = k_{2l+1} = \max\{k_{2l}, 2l+1-k_{2l}\} = \max\{k_{2l}, 2l+1-k_{2r+1}\}$$
$$= \max\{k_{2l}, 2l+1-(2r+1-k_{2r})\} = \max\{k_{2l}, 2(l-r)+k_{2r}\},$$

so

$$(9.12.9) \qquad k_{2l+2} = \max\{k_{2l}, 2(l-r)+k_{2r}\}.$$

The computation results appear in Table 9.12.1. The equations used to compute $r_4(X)$ and $r_6(X)$ were

$$r_4(X) = r_2(X) + d_2d_0^{-1}X^{2-0}r_0(X) = 1 + \gamma^{26}X + \gamma^2\gamma^5X^2 \cdot 1,$$
$$r_6(X) = r_4(X) + d_4d_2^{-1}X^{4-2}r_2(X) = (1 + \gamma^{26}X + \gamma^7X^2) + \gamma^{14}X^2(1 + \gamma^{26}X).$$

The reciprocal $\bar{r}_6(X) = X^3 + \gamma^{26}X^2 + \gamma^{29}X + \gamma^9$ of $r_6(X)$ does indeed have roots γ^3, γ^{17}, and γ^{20}, revealing to the receiver that $e(X) = X^3 + X^{17} + X^{20}$.

See Burton (1971) for another simplification of BCH decoding in the binary case.

TABLE 9.12.1

$2r$	$n = 2l$	k_{2l}	$r_{2l}(X)$	d_{2l}
0	2	1	$1 + \gamma^{26}X$	$S_3 + \gamma^{26}S_2 = \gamma^2$
2	4	2	$1 + \gamma^{26}X + \gamma^7X^2$	$S_5 + \gamma^{26}S_4 + \gamma^7S_3 = \gamma^{16}$
	6	3	$1 + \gamma^{26}X + \gamma^{29}X^2 + \gamma^9X^3$	

9.13 The Cyclic Redundancy Check

We discuss here a cyclic-code-based strategy used to correct errors in recorded data. We refer here to standard $\frac{1}{2}$-inch magnetic tape, and specifically to the 2400 tape series of IBM/System 360. Nine tracks of binary data are recorded lengthwise along the tape. Customarily, eight tracks represent desired recorded data and the ninth track consists of a single parity-check on the other eight. That is, in Figure 9.13.1, each vertical column has sum 0 (mod 2).†

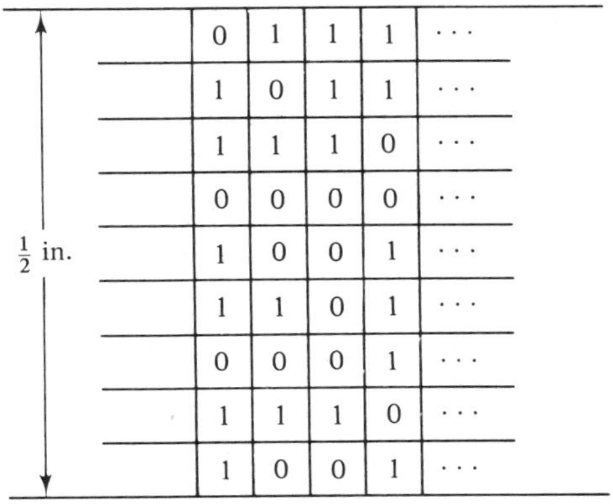

0	1	1	1	...
1	0	1	1	...
1	1	1	0	...
0	0	0	0	...
1	0	0	1	...
1	1	0	1	...
0	0	0	1	...
1	1	1	0	...
1	0	0	1	...

$\frac{1}{2}$ in.

FIGURE 9.13.1. *Nine-Track Magnetic Tape*

In this section, all data are binary and all work is over GF(2). In the 2400 tape series, data are recorded densely along each track: 800 bpi (bits per inch). Consequently, any small flaw in the tape or its magnetic coating will probably affect only one track (width approximately $\frac{1}{18}$ inch) but a whole burst of digits in that track (each digit recorded in a space $\frac{1}{800}$ inch wide). Our goal is a strategy for correcting such errors, described in Brown and Sellers (1970) and Chien (1973). [A more recent, more sophisticated, highly algebraic code used with IBM 6250-bpi 3420 series tape units is described in Patel and Hong (1974).]

A burst of errors within a track will be detected by the single-parity-check track, as the sum of digits in some columns (mod 2) will not be 0. But it does not identify the track in which the errors lie; the "cyclic redundancy check" (CRC) procedure will accomplish this and keep the code rate high.

Each vertical nine-digit column of values across the tape is called a *character*. The system uses two modified FSRs, called the CRC register

† Sometimes odd parity-check is used; vertical column sums are 1 (mod 2).

(CRCR) and the error pattern register (EPR), pictured in Figure 9.13.2 (see p. 471). Both the CRCR and EPR are FSRs over GF(2) with characteristic polynomial $c(X) = 1 + X^3 + X^4 + X^5 + X^6 + X^9$, but the CRCR admits a nine-digit word (character) $a_0 a_1 a_2 \cdots a_8$ as input at each shift, while the EPR admits a single input bit e at each shift.

We first describe the encoding and decoding procedure, without proof. Number the nine tracks $0, 1, 2, \ldots, 8$, track 0 being the parity-check track. For each eight-digit block of data $a_{1i} a_{2i} \cdots a_{8i}$, the encoder computes $a_{0i} = a_{1i} + a_{2i} + \cdots + a_{8i}$ and records $a_{0i} a_{1i} \cdots a_{8i}$ as the ith column (character) on the tape. The word $\mathbf{a}_i = a_{0i} a_{1i} \cdots a_{8i}$ is simultaneously entered into the CRCR as it is shifted. After a preassigned number n of characters $\mathbf{a}_n, \mathbf{a}_{n-1}, \ldots, \mathbf{a}_1$ have been recorded on tape and entered into the CRCR, the CRCR is shifted once more and its final contents, the *CRC character*, is written on the tape as an $(n+1)$st character $\mathbf{a}_0$. [Note that our code thus has rate $8n/9(n+1)$.]

The decoder reads the received characters $\tilde{\mathbf{a}}_n, \tilde{\mathbf{a}}_{n-1}, \ldots, \tilde{\mathbf{a}}_1$, entering them into the CRCR and shifting at each entry. At the same time, the EPR receives input $\tilde{e}_i = \tilde{a}_{0i} + \tilde{a}_{1i} + \cdots + \tilde{a}_{8i}$† at the ith character, shifting at each entry. If there is no error, all $\mathbf{a}_i = \tilde{\mathbf{a}}_i$, so the CRCR again generates the CRC character, and comparison with $\mathbf{a}_0$ on the tape leads to the report that no error occurred. (The contents of the EPR in this case are always 0's.)

If there are errors in a single track, say the jth, the CRCR receives erroneous digits $\tilde{a}_{ji} = a_{ji} + 1$ for some characters $\tilde{\mathbf{a}}_i$. The EPR receives $\tilde{e}_i = 1$ at just such $\tilde{\mathbf{a}}_i$. If $j = 8$, the contents of the CRCR will eventually be the CRC character $\mathbf{a}_0$ plus the contents of the EPR (the inputs to the CRCR have been the inputs generating $\mathbf{a}_0$, plus the inputs to the EPR). If $j < 8$, the erroneous digits entered the CRCR in a different track. After subtracting the CRC character $\mathbf{a}_0$ from the CRCR contents, we shift the CRCR (with **0** input) k times, until its contents match those of the EPR. Then $k = 8 - j$, revealing j, the track with the erroneous digits. (This decoding procedure fails in a small fraction of cases, as we shall see later.) The tape is then reread and digit $\tilde{a}_{ji}$ is corrected wherever $\tilde{e}_i$ is 1. If no CRCR–EPR match is obtained for $0 \leq k \leq 8$, then errors occurred in more than one track and are uncorrectable.

For an example, we shall assume that $n = 4$ and the four data characters $\mathbf{a}_4, \mathbf{a}_3, \mathbf{a}_2,$ and $\mathbf{a}_1$ are as in Table 9.13.1. The computation of $\mathbf{a}_0$ at the encoder, using the CRCR of Figure 9.13.2, is given in Table 9.13.2.

Assume now that errors occur in track 6, changing all four 1's to 0's. Table 9.13.3 shows how the CRCR and EPR work in tandem to identify the track errors. The CRCR, after two extra shifts, matched the EPR, revealing that the errors were in track $8 - 2 = 6$.

† $\tilde{e}_i = 1 + \tilde{a}_{0i} + \cdots + \tilde{a}_{8i}$ in case of odd parity-check.

Sec. 9.13 The Cyclic Redundancy Check

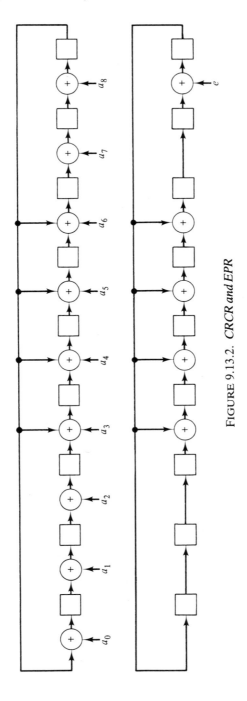

FIGURE 9.13.2. *CRCR and EPR*

TABLE 9.13.1. Example of the CRC.

Track	a_4	a_3	a_2	a_1	a_0
0	0	0	1	1	1
1	1	0	1	1	0
2	0	0	0	1	1
3	0	1	0	1	0
4	1	0	0	1	1
5	0	1	1	0	1
6	1	1	0	1	1
7	0	0	1	0	1
8	1	1	0	0	0

TABLE 9.13.2. Computation of a_0 in Table 9.13.1.

```
enter a₄:           01001010①
shift, adding a₃:   001001010  ) (a₄ shifted)
                    100111100 ⸺ (feedback)
                    000101101    (a₃)
                    ─────────
                    10101101①
shift, adding a₂:   010101101  ) (shift)
                    100111100 ⸺ (feedback)
                    110001010    (a₂)
                    ─────────
                    00001101①
shift, adding a₁:   000001101  ) (shift)
                    100111100 ⸺ (feedback)
                    111110100    (a₁)
                    ─────────
                    01100010①
final shift:        001100010  ) (shift)
                    100111100 ⸺ (feedback)
                    ─────────
                    101011110    (a₀)
```

Now we shall discuss the mathematics that justifies the procedure. Write the character $\mathbf{a}_i = a_{0i}a_{1i}\cdots a_{8i}$ as $a_i(X) = a_{0i} + a_{1i}X + \cdots + a_{8i}X^8$. The contents of the CRCR after each entry $\mathbf{a}_n, \mathbf{a}_{n-1}, \ldots, \mathbf{a}_1$ are successively

$$a_n(X),$$
$$a_{n-1}(X) + Xa_n(X) \ (\mathrm{mod}\ c(X)),$$
$$a_{n-2}(X) + Xa_{n-1}(X) + X^2 a_n(X) \ (\mathrm{mod}\ c(X)),$$
$$\vdots$$
$$\sum_{i=1}^{n} X^{i-1} a_i(X) \ (\mathrm{mod}\ c(X)).$$

Sec. 9.13 The Cyclic Redundancy Check

TABLE 9.13.3. Decoding with the cyclic redundancy check.

	CRCR:	EPR:
enter $\tilde{\mathbf{a}}_4$:	01001000①	00000000①
shift, adding $\tilde{\mathbf{a}}_3$:	001001000	000000000
	100111100	100111100
	000101001	000000001
	$\overline{10101110①}$	$\overline{10011110①}$
shift, adding $\tilde{\mathbf{a}}_2$:	010101110	010011110
	100111100	100111100
	110001010	000000000
	$\overline{00001100①}$	$\overline{11010001①}$
shift, adding $\tilde{\mathbf{a}}_1$:	000001100	011010001
	000000000	000000000
	111110000	000000001
	$\overline{11111110①}$	$\overline{01101000①}$
shift, adding $\tilde{\mathbf{a}}_0$:	011111110	001101000
	000000000	000000000
	101011010	000000001
	$\overline{110100100}$	$\overline{001101001}$
shift CRCR twice:	001101001 ←	match

After the final shift, the CRCR contains

$$\sum_{i=1}^{n} X^i a_i(X) \pmod{c(X)},$$

so the CRC character $a_0(X)$ satisfies

(9.13.1) $\qquad a_0(X) \equiv \sum_{i=1}^{n} X^i a_i(X) \pmod{c(X)}.$

Consequently, when the decoder reads into the CRCR the characters $\mathbf{a}_n, \mathbf{a}_{n-1}, \ldots, \mathbf{a}_1, \mathbf{a}_0$, the contents are

$$a_0(X) + \sum_{i=1}^{n} X^i a_i(X) \equiv 0 \pmod{c(X)};$$

this is what happens when no errors occur.

If there are errors in the jth track in characters $\tilde{\mathbf{a}}_{i_1}, \ldots, \tilde{\mathbf{a}}_{i_\nu}$, and no other errors, then the CRCR at the decoder has final contents

$$a_0(X) + \sum_{i=1}^{n} X^i a_i(X) + X^{j+i_1} + \cdots + X^{j+i_\nu} \equiv X^j E(X) \pmod{c(X)},$$

where $E(X) = X^{i_1} + \cdots + X^{i_\nu}$ is the "error pattern." The EPR receives input X^8 (a 1 in the high-order position) when reading $\tilde{\mathbf{a}}_{i_1}, \ldots, \tilde{\mathbf{a}}_{i_\nu}$, and receives 0 otherwise, so its final contents are

$$X^8 E(X) \pmod{c(X)}.$$

Clearly, then, shifting the CRCR $k = 8-j$ times will make its contents $X^{8-j} \cdot X^j E(X) \pmod{c(X)}$ match those of the EPR. The decoding will fail, however, if there is some $k < 8-j$ with

$$X^{j+k} E(X) \equiv X^8 E(X) \pmod{c(X)};$$

that is,

(9.13.2) $\qquad X^{j+k}(1 + X^{8-j-k}) E(X) \equiv 0 \pmod{c(X)},$

$8-j-k > 0$. Can this happen? $c(X) = (1+X)c_0(X)$, where $c_0(X) = 1 + X + X^2 + X^4 + X^6 + X^7 + X^8$ is irreducible over GF(2). Consequently, (9.13.2) can only happen if $c_0(X)$ divides $E(X)$; this will never happen if $E(X)$ has degree less than 8, and will happen for only $2^{-8} = \frac{1}{256}$ of all possible longer error patterns.

Thus the correction procedure succeeds for all single-track error patterns, except for less than $\frac{1}{2}$ per cent of the longer error patterns. When the procedure fails, $c_0(X)$ divides $E(X)$, so the CRCR contents is a multiple of $c_0(X)$ modulo $c(X)$; the only such multiples are 0 and $c_0(X)$. The CRCR contents is tested for these two values, and if one is found, the error pattern is declared uncorrectable. (If such an error pattern occurs with CRCR contents 0, we can still know that some errors occurred, because some e_i's will be nonzero.)

If $c(X)$ were chosen irreducible, there would be even fewer uncorrectable error patterns. However, we need the factor $1+X$ in $c(X)$, for two reasons.

First, we want to know the parity of the CRC character $a_0(X)$, to correct any errors in it. By (9.13.1),

$$a_0(X) = \sum_{i=1}^{n} X^i a_i(X) + f(X) c(X)$$

for some $f(X)$. The parity (number of nonzero coefficients) of the $a_i(X)$ is predetermined by the parity-check digit a_{0i}. If $c(X)$ has even parity, so does $f(X)c(X)$, but if $c(X)$ has odd parity, the parity of $f(X)c(X)$ is unknown (why?). Therefore, to know the parity of $a_0(X)$ from that of the $a_i(X)$ ($i > 0$), we must have $c(X)$ of even parity; $c(X)$ has even parity iff $c(1) = 0$ iff $X+1$ divides $c(X)$. [A polynomial over GF(2) has root 1 iff it is divisible by $X-1 = X+1$.]

Second, after finding that errors are present while reading the tape forward, it is desirable to find the erroneous track while reading backward. (Then correct the errors while going forward again, and proceed.) When we do this, all polynomials in the previous discussion change to their reciprocals; high-order coefficients play the role of low-order coefficients, and conversely. It is therefore desirable that $c(X)$ be *symmetric*: $c(X) = \bar{c}(X)$.

We can then use the same CRCR and EPR forward and backward, because the characteristic polynomial does not change. $c(X)$ symmetric of degree 9 means

$$c(X) = 1 + c_1 X + c_2 X^2 + c_3 X^3 + c_4 X^4 + c_4 X^5 + c_3 X^6 + c_2 X^7 + c_1 X^8 + X^9,$$

so $c(1) = 1 + c_1 + c_2 + c_3 + c_4 + c_4 + c_3 + c_2 + c_1 + 1 = 0$ in GF(2), implying that $X+1$ divides $c(X)$.

9.14 Fire Codes

A channel in which errors tend to gather in lumps is called a *burst channel*. Many practical data-transmission channels are of this type. A **burst of length** l is a vector whose nonzero digits occur in l consecutive positions, the first and last of which are nonzero. For example, 00010110000 is a burst of length 4, 11111 is a burst of length 5, and 10000000011 is a burst of length 3. (Since our codes are usually cyclic, we usually go "around the end" in finding the length of a burst.)

We first establish some necessary conditions for burst-correcting codes.

Lemma 9.14.1. *If an (n, m) linear code V has every nonzero code word a burst of length $>l$, then $n - m \geq l$.*

Proof. V is a subspace of $V(n, q)$, some q. Let W be the subspace of $V(n, q)$ of words having all their nonzero entries in the first l components. By the hypothesis, $V \cap W = \{0\}$, so $V + W = V \oplus W$; by Exercise 7.7.12,

$$m + l = \dim V + \dim W = \dim (V + W) \leq \dim V(n, q) = n,$$

implying that $l \leq n - m$. ∎

Lemma 9.14.2. *If an (n, m) linear code V corrects all bursts of length $\leq b$, then $n - m \geq 2b$.*

Proof. "Corrects all bursts of length $\leq b$" means, of course, "corrects all error patterns which are bursts of length $\leq b$." If all such bursts are corrected, they are all coset leaders, and hence no two are in the same coset of V. Any burst of length $\leq 2b$ is a difference of two bursts of length $\leq b$, and hence not a code word; by Lemma 9.14.1, $n - m \geq 2b$. ∎

The quantity $n - m - 2b$ is often denoted z. Lemma 9.14.2 shows that $z \geq 0$ in any code; this condition is called the **Rieger bound**. The rate of the code is

$$(9.14.1) \qquad R = \frac{m}{n} = \left(1 - \frac{2b}{n}\right) - \frac{z}{n} \leq 1 - \frac{2b}{n}.$$

The best burst-error-correcting codes have z very close to 0, rate nearly $1 - (2b/n)$; see Exercise 9.17.3.

One special family of burst-error-correcting codes, important because error correction is easy, is the following, discovered by P. Fire in 1959.

If $b > 0$ is given, choose an irreducible polynomial $p(X)$ of degree $c \geq b$ over $GF(q)$ and let e be the smallest positive integer such that $p(X)$ divides $X^e - 1$. [Then e is the *period* of $p(X)$; see Section 8.12.] Assume that

(9.14.2) $\qquad\qquad\qquad (e, 2b - 1) = 1,$

and define

(9.14.3) $\quad n = e(2b-1), \quad m = n - 2b - c + 1, \quad g(X) = (X^{2b-1} - 1)p(X).$

The (n, m) cyclic code over $GF(q)$ with generator polynomial $g(X)$ is called a **Fire code**.

Before proving that this Fire code corrects all bursts of $\leq b$ errors and seeing how this correction can be done, we give some examples.

EXAMPLE 1. Suppose that we want a Fire code to correct all bursts of ≤ 5 errors, over $GF(2)$. Then $b = 5$, and we try $c = 5$ also; $p(X)$ is irreducible of degree 5 over $GF(2)$, so a root α is in $GF(2^5)$ and satisfies $1 = \alpha^{2^5-1} = \alpha^{31}$. Since $p(X)$ divides $X^{31} - 1$, $e | 31$ by Lemma 8.12.2. But 31 is prime, so we have to choose $e = 31$, getting $n = e(2b-1) = 31 \cdot 9 = 279$, $m = n - 2b - c + 1 = 265$. $p(X)$ can be $X^5 + X^2 + 1$; $g(X) = (X^9 - 1)(X^5 + X^2 + 1)$ generates a (279, 265) Fire code which corrects any burst of 5 errors or less.

EXAMPLE 2. We seek a Fire code to correct all bursts of at most 13 errors over $GF(2)$. So $b = 13$, and we try $c = 13$; $p(X)$ is irreducible of degree 13 over $GF(2)$, so a root α is in $GF(2^{13})$ and satisfies $1 = \alpha^{2^{13}-1} = \alpha^{8191}$. So $p(X) | (X^{8191} - 1)$, $e | 8191$. The number 8191 is prime, forcing $e = 8191$. Then $n = e(2b-1) = 204{,}775$. We probably do not want n that large, as the risk of a second burst of errors in the same n-digit word would be too great. So we reconsider and try $c = 14$. Now a root α of $p(X)$ satisfies $1 = \alpha^{2^{14}-1} = \alpha^{16383}$. $16383 = 3 \cdot 43 \cdot 127$, where 3, 43, and 127 are primes. We require that e divide 16383, so $e \in \{3, 43, 127, 3 \cdot 43, 3 \cdot 127, 43 \cdot 127, 16383\}$. We cannot choose e as 3 or 127, as roots of $X^3 - 1$ and $X^{127} - 1$ are in $GF(2^2)$ and $GF(2^7)$, respectively, so a divisor $p(X)$ cannot be irreducible of degree 14. But we can choose e as 43, $3 \cdot 43$, $3 \cdot 127$, $43 \cdot 127$, or 16383. If we pick $e = 43$, we get a (1075, 1036) Fire code that corrects all error bursts of length ≤ 13.

Theorem 9.14.1. *The Fire code defined in (9.14.2) and (9.14.3) corrects any error burst of length $\leq b$.*

Proof. Assume that we send the code word $c(X)$ and receive $r(X) = c(X) + e(X)$, $e(X)$ the error burst of length $\leq b$. Let $r(X) = \sum_{i=0}^{n-1} r_i X^i$, and consider what happens if we divide $r(X)$ by $X^{2b-1} - 1$. Since $n = e(2b-1)$, we can break $\mathbf{r} = r_{n-1} r_{n-2} \cdots r_1 r_0$ into e blocks of $2b - 1$ digits each:

$r_{n-1} \qquad\qquad r_{n-2b+1} \quad r_{n-2b} \qquad\qquad\qquad r_{2b-2} \qquad\qquad r_0$

Sec. 9.14 *Fire Codes* 477

We claim that the remainder when $r(X)$ is divided by $X^{2b-1}-1$ is the sum (in $V(2b-1, q)$) of these blocks:

$$(r_{n-1}+r_{n-2b}+\cdots+r_{2b-2})X^{2b-2}+(r_{n-2}+r_{n-2b-1}+\cdots+r_{2b-3})X^{2b-3}$$
$$+\cdots+(r_{n-2b+1}+r_{n-4b+2}+\cdots+r_0).$$

One can see this either by carrying out a few steps of the ordinary division "by hand" or by looking at Figure 9.14.1, the FSR that divides $r(X)$ by $X^{2b-1}-1$. Polynomial $r(X)$ is entered high-order coefficients first, and after division the remainder is left in the register. Clearly, after each $2b-1$ shifts this FSR begins adding the next block of $2b-1$ digits in **r** to the sum of blocks already entered.

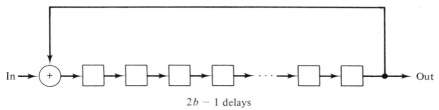

FIGURE 9.14.1. *Division by* $X^{2b-1}-1$

Let

$$e(X)=X^ib(X), \quad \text{where } b(X)=\sum_{i=0}^{b-1}b_iX^i, \quad b_0\neq 0.$$

Since $X^{2b-1}-1$ divides $c(X)$ exactly, the remainder upon dividing $r(X)$ by $X^{2b-1}-1$ is the same as the remainder upon dividing $e(X)$ by $X^{2b-1}-1$. If the error burst appears entirely in one $(2b-1)$-length block:

it is, by the above description of division, reproduced exactly in the remainder

If it overlaps two blocks:

it appears in the remainder as

In either case we know the error burst $b(X)$; we do not know j, the location of it in the received word. We do know j modulo $2b-1$ [its location within the $(2b-1)$-block in which it occurs]; we just do not know in which of the e blocks that jth entry occurs.

We need only check that if j_0 is such that

$$0 \leq j_0 < 2b-1 \quad \text{and} \quad j \equiv j_0 \pmod{2b-1},$$

so we know j_0, then the e polynomials

$$X^{j_0}b(X), X^{j_0+(2b-1)}b(X), X^{j_0+(2b-1)2}b(X), \ldots, X^{j_0+(2b-1)(e-1)}b(X)$$

all have different remainders upon division by $p(X)$. Then we know that $j = j_0 + (2b-1)i$, where $X^{j_0+(2b-1)i}b(X)$ has the same remainder as $r(X)$ upon division by $p(X)$.

If $p(X)$ has the same remainder when divided into two of the polynomials, then it divides their difference. Here $p(X)$ has degree $\geq b > \deg b(X)$ and $p(X)$ is irreducible, so $(p(X), b(X)) = 1$ and it is enough to show that

$$p(X) \nmid (X^{j+(2b-1)e'} - X^j);$$

that is,

(9.14.4) $$p(X) \nmid (X^{(2b-1)e'} - 1)$$

for any $e' < e$. But $p(X)|(X^d - 1)$ iff d is a multiple of e (Lemma 8.12.2), and $(2b-1, e) = 1$, so (9.14.4) is *true*. ∎

The ideas in the preceding proof can actually be used for decoding Fire codes. Another possible procedure, using the notation of the above proof, is as follows. Use an FSR to divide $r(X)$ by $g(X)$. After division, the register contains the remainder from dividing $e(X) = X^j b(X)$ by $g(X)$. If we now shift the FSR $n-j$ times, the register contains the remainder from dividing $X^{n-j} \cdot X^j b(X) = X^n b(X)$ by $g(X)$. Since $g(X)|(X^n - 1)$, $X^n b(X) \equiv b(X) \pmod{g(X)}$, so this remainder is $b(X)$. We can test for this, with a high probability of accuracy (see Exercise 9.17.6), by testing when the high-order coefficients of the remainder (the last $n-m-b$ entries in the register) are all zero. Then we know not only $b(X)$, but also $n-j$ and hence j.

9.15 Some Coding Tricks

We first discuss a powerful and much-used method of increasing burst-error-correcting ability. Assume that we have an (n, m) block code capable of correcting all burst error patterns of length $\leq b$. Choose an integer i, called the **degree of interleaving**. Take i k-digit message words, encode them into i n-digit code words, and store them in a "stack" (Table 9.15.1). Then transmit the digits *by columns*; send them in the order

$$c_{10}, c_{20}, \ldots, c_{i0}, c_{11}, c_{21}, \ldots, c_{i1}, \ldots, c_{1,n-1}, c_{2,n-1}, \ldots, c_{i,n-1}.$$

Sec. 9.15 *Some Coding Tricks* 479

The receiver reforms the stack, treats each row as a received word, and decodes as usual. *But now we can correct an error burst of length bi*; the errors are scattered among i code words, each with an error burst of length $\leq b$, and all are corrected. By this method of interleaving, we have constructed an (ni, mi) code with burst-correcting ability bi. In actual applications, i may be 100 or greater. For some numerical examples, see Exercises 9.17.7 and 9.17.8.

TABLE 9.15.1

$$\begin{aligned}
\mathbf{c}_1 &= c_{10} \quad c_{11} \quad \cdots \quad c_{1,n-1} \\
\mathbf{c}_2 &= c_{20} \quad c_{21} \quad \cdots \quad c_{2,n-1} \\
&\cdots\cdots\cdots\cdots\cdots\cdots\cdots\cdots \\
\mathbf{c}_i &= c_{i0} \quad c_{i1} \quad \cdots \quad c_{i,n-1}
\end{aligned}$$

Next, we shall discuss concatenated codes, using ideas of Forney (1966). Perhaps the most important example of a burst channel is the encoder–channel–decoder system itself. For the decoder gets most words right; we have long periods without an error. But when the decoder chooses the wrong code word, it gets lots of digits wrong in a burst.

Consequently, we can form a **concatenated code**, as in Figure 9.15.1, from inner and outer codes. If the inner code uses m-digit message words, we choose the outer code to correct bursts of m errors. For example, if the inner code is an (n, m) code over GF(2) and the outer code is an (N, M) code over GF(2^m), then the code takes mM information bits and partitions them into M "bytes" of m bits each. The outer encoder produces N bytes of m bits each, and the inner encoder then changes each byte into n bits. So the whole code is an (Nn, Mm) code over GF(2). For some numerical examples, see Exercise 9.17.9.

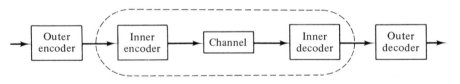

FIGURE 9.15.1. *Concatenated Code*

Another way to form a large code is to start with an (n_1, m_1) linear code A and an (n_2, m_2) linear code B, both systematic over the same GF(q). We shall form their **product code** $A \times B$. Take $m_1 m_2$ information symbols all at once and arrange them in an $m_1 \times m_2$ matrix. Extend the matrix to $n_1 \times n_2$ as follows (Figure 9.15.2).

1. Use code A to make each of the first m_2 columns an n_1-digit code word in code A.

2. Use code B to make each of the first m_1 rows an n_2-digit code word in code B.

3. Use either code A or code B to make the bottom rows code words in code B, the right columns code words in code A.

It is initially surprising that the "checks on checks" give code words in both rows and columns, so the choice between A and B does not matter in step 3. To see that this is so, let the array in Figure 9.15.2 be

$$\begin{bmatrix} [i_{st}] & [r_{st}] \\ [c_{st}] & [p_{st}] \end{bmatrix},$$

and consider the two ways of defining the entries p_{st}.

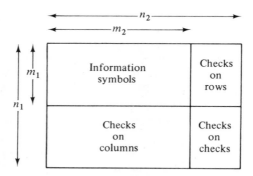

FIGURE 9.15.2. *Product Code Word*

If p_{st} is defined with code B, there are certain constants $b_1, \ldots, b_{m_2}$ (in the parity-check matrix) used to define the tth parity-check digit:

$$p_{st} = \sum_{j=1}^{m_2} b_j c_{sj}.$$

The c_{sj} are defined from code A, using certain constants $a_1, \ldots, a_{m_1}$ that define the sth parity-check digit:

$$c_{sj} = \sum_{l=1}^{m_1} a_l i_{lj}.$$

Hence

$$p_{st} = \sum_{j=1}^{m_2} b_j \sum_{l=1}^{m_1} a_l i_{lj} = \sum_{l=1}^{m_1} a_l \sum_{j=1}^{m_2} b_j i_{lj}.$$

If we define p_{st} with code A we get

$$p_{st} = \sum_{l=1}^{m_1} a_l r_{lt},$$

and the r_{lt} were defined, using code B, with

$$r_{lt} = \sum_{j=1}^{m_2} b_j i_{lj},$$

so we get the same expression for p_{st}:

$$p_{st} = \sum_{l=1}^{m_1} a_l \sum_{j=1}^{m_2} b_j i_{lj}.$$

We show in Exercise 9.17.11 that if A has minimum distance d_1 and B has minimum distance d_2, then $A \times B$ has minimum distance $d_1 d_2$. Exercise 9.17.10 shows that if A and B are single parity-check codes, merely error-detecting, then $A \times B$ is single-error-correcting.

Tricks and modifications of known codes are important in many applications. Codes are lengthened, shortened, augmented, expurgated, punctured, reordered, and so on, to fit given tasks. For examples, see Berlekamp (1968).

9.16 How Good Can a Code Be?

Suppose that we have a code of length n over $GF(q)$. Denote the number of code words by M. If the code is linear, then $M = q^m$ where the code is a subspace of dimension m in $V(n, q)$. The *rate* of the code is $(\log_q M)/n$, which is m/n if the code is linear. Denote by d the minimum distance and t the error-correcting capability, so d is $2t+1$ or $2t+2$. The following theorem gives an upper bound which the best code cannot exceed; it is achieved only in the case of perfect t-error-correcting codes.

Theorem 9.16.1 (Hamming Upper Bound) [Hamming (1950)]

(9.16.1)
$$\sum_{i=0}^{t} \binom{n}{i}(q-1)^i \leq \frac{q^n}{M},$$

and the rate R satisfies

(9.16.2)
$$R \leq 1 - \frac{1}{n} \log_q \left\{ \sum_{i=0}^{t} \binom{n}{i}(q-1)^i \right\}.$$

Proof. The minimum distance d is at least $2t+1$, so the "spheres"

$$S_t(\mathbf{c}) = \{\mathbf{a} \in V(n, q) \mid d(\mathbf{a}, \mathbf{c}) \leq t\},$$

for code words $\mathbf{c}$, do not overlap. We have

$$|S_t(\mathbf{c})| = 1 + \binom{n}{1}(q-1) + \binom{n}{2}(q-1)^2 + \cdots + \binom{n}{t}(q-1)^t$$

$$= \sum_{i=0}^{t} \binom{n}{i}(q-1)^i.$$

The M spheres $S_t(\mathbf{c})$ about code words $\mathbf{c}$ cannot contain more than the q^n words in $V(n, q)$, so

$$M \sum_{i=0}^{t} \binom{n}{i}(q-1)^i \leq q^n,$$

$$\log_q M + \log_q \left\{ \sum_{i=0}^{t} \binom{n}{i}(q-1)^i \right\} \leq n,$$

$$R = \frac{\log_q M}{n} \leq 1 - \frac{1}{n} \log_q \left\{ \sum_{i=0}^{t} \binom{n}{i}(q-1)^i \right\}. \blacksquare$$

The following lower bound [Gilbert (1952) and Varsharmov (1957)] looks like the Hamming upper bound, except that the inequality is reversed and t is replaced by $d-1$, approximately $2t$.

Theorem 9.16.2 (Gilbert Lower Bound). *There does exist a code of length n over $GF(q)$ with M code words and minimum distance d, such that*

(9.16.3) $$\left[\sum_{i=0}^{d-1} \binom{n}{i}(q-1)^i \right] \cdot M \geq q^n.$$

Proof. The number of words at distance less than d from a given code word $\mathbf{c}$ is

$$N_d = |S_{d-1}(\mathbf{c})| = \sum_{i=0}^{d-1} \binom{n}{i}(q-1)^i.$$

If M' code words in our code have been chosen so far and $N_d \cdot M' < q^n = |V(n, q)|$, then some word in $V(n, q)$ is at distance at least d from all the code words and can be chosen as a new code word. So we can increase M' until the Gilbert bound is attained. $\blacksquare$

Bounds that are more restrictive than those above, especially upper bounds for relatively low-rate codes, are discussed in Chapter 4 of Peterson and Weldon (1972). Some numerical examples are discussed in Exercise 9.17.12. It is unfortunate that the code given by the Gilbert bound is not linear and not reasonably encoded and decoded, because (for n large, t/n reasonably small) its rate is nearly the maximum permitted by the Hamming bound.

Shannon's fundamental 1948 paper not only inspired the growth of algebraic coding theory, with work of Hamming (1950), Golay (1949), and others following soon afterward, but also marked the birth of *information theory*, more closely related to probability theory than to algebra. In information theory, it is shown that many communication channels of the encoder–channel–decoder form have associated with them specific numbers, called their *capacities*. If C is the capacity of a particular channel and R

a transmission rate with $R < C$, it is often possible to show that for any $\varepsilon > 0$ there is a code (with large word length n) with rate R and probability of error per digit less than ε. For $R > C$ and some $\varepsilon > 0$, no such code exists.

We shall state the situation precisely for the binary symmetric channel.

Theorem 9.16.3 (Shannon's Fundamental Theorem for the Binary Symmetric Channel). *Assume that the probability p of error per digit in a binary symmetric channel satisfies $0 < p < \frac{1}{2}$, and define the* **capacity** *of the channel to be*

(9.16.4) $$C = 1 + p \log_2 p + (1-p) \log_2 (1-p).$$

Assume R and ε given, $0 < R < C$, $\varepsilon > 0$. Then for some n there is a code in $V(n, 2)$ with rate at least R, such that the probability of error per digit after decoding is less than ε.

We shall not prove this theorem here; a nice discussion and proof appear in Chapter I of van Lint (1973). Other discussions and generalizations can be found in Bobrow and Arbib (1974), Ash (1965), and Wolfowitz (1964).

Shannon's theorem has been not only an inspiration to coding theory, but also a source of frustration; for some time, no one was able to construct a family of codes with error probability approaching zero and rate bounded away from zero, let alone with rate at least R for a given $R < C$. Then partial success was achieved by Elias (1954), using repeated products of codes, and major success by Justesen (1972), using repeatedly concatenated codes. Finally, Goppa (1970) constructed a huge new class of algebraic codes, including some that (for n large) approach the Gilbert bound and hence fulfill the major prophecy of Shannon's theorem. For a discussion of basic properties of Goppa codes, see Berlekamp (1973). There is, however, no nice decoding scheme yet known for Goppa codes that will decode all errors they are capable of correcting. Some algorithms for decoding a reasonable number of errors have recently been published: see Chien and Choy (1975), Patterson (1975), Retter (1975), and our remarks at the end of Section 9.10.

9.17 Exercises

1. For the following binary BCH codes and S_i-values, find the error patterns, using the shortened algorithm of Section 9.12.

(a) 2-error-correcting of length $n = 15$, Irr $(\gamma, \mathrm{GF}(2)) = X^4 + X + 1$, $S_1 = \gamma^9$, $S_2 = \gamma^3$, $S_3 = \gamma^2$, and $S_4 = \gamma^6$.

(b) 3-error-correcting of length $n = 31$, Irr $(\gamma, \mathrm{GF}(2)) = X^5 + X^2 + 1$, $S_1 = \gamma^6$, $S_2 = \gamma^{12}$, $S_3 = \gamma^{27}$, $S_4 = \gamma^{24}$, $S_5 = \gamma^{20}$, and $S_6 = \gamma^{23}$.

(c) 4-error-correcting of length $n = 31$, Irr $(\gamma, \mathrm{GF}(2)) = X^5 + X^2 + 1$, $S_1 = \gamma^{26}$, $S_2 = \gamma^{21}$, $S_3 = 1$, $S_4 = \gamma^{11}$, $S_5 = \gamma^{12}$, $S_6 = 1$, $S_7 = \gamma^{30}$, and $S_8 = \gamma^{22}$.

(d) 5-error-correcting of length $n = 31$, Irr $(\gamma, GF(2)) = X^5 + X^2 + 1$, $S_1 = S_2 = 0$, $S_3 = 1$, $S_4 = 0$, $S_5 = S_6 = 1$, $S_7 = S_8 = 0$, and $S_9 = S_{10} = 1$.

2. Assume that during a use of the cyclic redundancy check, the four data characters are as in Figure 9.13.1. Determine the CRC character. Then assume that an error changes all 1's in track 5 to 0's, and show how these errors would be corrected.

*3. Compare the Rieger bound $z = n - m - 2b$ for the following codes:

(a) A t-error-correcting Reed–Solomon code of length $n = q - 1$ over $GF(q)$, where $q = 2^m$, considered as a code correcting error bursts of length $\leq m(t-1) + 1$ over $GF(2)$.

(b) The BCH codes of Table 9.4.1, considered as burst-error-correcting codes.

(c) The Fire code of (9.14.3) with $c = \deg p(X) \geq b$, which corrects error bursts of length $\leq b$.

*4. Use the results of Exercise 3 to compare Reed–Solomon, BCH, and Fire codes constructed to correct error bursts of length ≤ 34, with n approximately 1000.

5. Describe Fire codes to correct error bursts of length at most

(a) 20 errors, over $GF(2)$.

(b) 15 errors, over $GF(5)$.

*6. Show that, for an (n, m) Fire code over $GF(q)$ with data (9.14.3), $\deg p(X) = c$, the probability that one of $n - 1$ given polynomials of degree $< n - m$ actually has degree $< b$ is less than $(2b - 1)/q^{b-1}$. Interpret this as a measure of accuracy of the last Fire code decoding method described in Section 9.14.

*7. Compute the rate and Rieger bound for the codes of Exercises 3 and 4, when the degree of interleaving is i.

*8. Assume that we wish to use a code with word length approximately 5000 over $GF(2)$, to correct any error burst of length at most 200. Find several examples of long Reed–Solomon and Fire codes and interleaved short Reed–Solomon, BCH, and Fire codes to accomplish this task, and compare them with respect to rate, decoding ease, Rieger bound, and so on.

*9. Form a concatenated code by using as the inner code an $(n, m) = (2^k - 1, 2^k - 1 - k)$ Hamming code and as the outer code a t-error-correcting Reed–Solomon code over $GF(2^m)$. Considered as a code over $GF(2)$, show that this code corrects all error bursts of length at most $t(n - 1) + 1$ and also corrects many single errors. State the situation more specifically when $t = 4$ and $k \in \{3, 4\}$.

*10. In this exercise we consider the product code $A \times B$, where both A and B are $(6, 5)$ single-parity-check codes.

(a) Find the 6×6 code word for the 5×5 message word in Table 9.17.1.

(b) Assume that the 1 in the second row and third column is erroneously received as 0. Show that by merely checking the parity of all the rows and columns, the decoder can locate the error.

Sec. 9.17 *Exercises* 485

TABLE 9.17.1

0	1	0	0	1
1	1	1	0	0
1	0	0	1	1
0	1	1	0	0
1	0	1	1	1

*11. In this exercise, A is an (n_1, m_1) linear code over GF(q) with minimum distance d_1, and B is an (n_2, m_2) linear code over GF(q) with minimum distance d_2.

(a) Show that code words in $A \times B$ form a subspace of the vector space of $n_1 \times n_2$ matrices over GF(q), and hence that $A \times B$ is an $(n_1 n_2, m_1 m_2)$ linear code over GF(q).

(b) Show that a nonzero code word in $A \times B$ has at least d_1 nonzero rows, each with at least d_2 nonzero entries, and hence that $A \times B$ has minimum distance at least $d_1 d_2$.

(c) Let $\mathbf{c} = c_1 c_2 \cdots c_{n_1}$ be a word in code A of weight d_1, $\mathbf{r} = r_1 r_2 \cdots r_{n_2}$ a word in code B of weight d_2, and define the $n_1 \times n_2$ matrix $M = [m_{ij}]$ by $m_{ij} = c_i r_j$, $1 \le i \le n_1$, $1 \le j \le n_2$. Show that M is a code word in code $A \times B$ of weight $d_1 d_2$, so $A \times B$ has minimum weight exactly $d_1 d_2$.

*12. For the following values of n, t, and q, use Theorems 9.16.1 and 9.16.2 to estimate:

(a) The maximum possible number of code words in a t-error-correcting code of length n over GF(q).

(b) The minimum possible number of code words in the highest-rate t-error-correcting code of length n over GF(q).

n	8	10	12	15
t	2	3	4	4
q	5	3	2	2

If you have the use of a computer, try some larger examples and look for trends.

Bibliography

AHO, A. V., J. E. HOPCROFT, and J. D. ULLMAN, *The Design and Analysis of Computer Algorithms*, Reading, Mass.: Addison-Wesley Publishing Co., Inc., 1974.

APPEL, K., and W. HAKEN, Every planar map is four colorable, *Bull. Amer. Math. Soc. 82* (1976), 711–712.

ARBIB, M. A., ed., *Algebraic Theory of Machines, Languages, and Semigroups*, New York: Academic Press, Inc., 1968.

ASH, R., *Information Theory*, New York: Wiley–Interscience, 1965.

BERLEKAMP, E. R., *Algebraic Coding Theory*, New York: McGraw-Hill Book Company, 1968.

———, Goppa codes, *IEEE Trans. Inform. Theory IT-19* (1973), 590–592.

BLAKE, A., Canonical expressions in Boolean algebra, Ph.D. dissertation, Department of Mathematics, University of Chicago, 1937.

BLAKE, I. F., and R. C. MULLIN, *The Mathematical Theory of Coding*, New York: Academic Press, Inc., 1975.

BOBROW, L. S., and M. A. ARBIB, *Discrete Mathematics*, Philadelphia: W. B. Saunders Company, 1974.

BOOTH, T. L., *Sequential Machines and Automata Theory*, New York: John Wiley & Sons, Inc., 1967.

BOSE, R. C., and D. K. RAY-CHAUDHURI, On a class of error correcting binary group codes, *Information and Control 3* (1960), 68–79.

BROWN, D. T., and F. F. SELLERS, JR., Error correction for IBM 800-bit-per-inch magnetic tape, *IBM J. Res. Develop. 14* (1970), 384–389.

BROWNE, E. T., On the reduction of a matrix to a canonical form, *Amer. Math. Monthly 47* (1940), 437–450.

BURTON, H. O., Inversionless decoding of binary BCH codes, *IEEE Trans. Inform. Theory IT-17* (1971), 464–466.

BUSACKER, R. G., and T. L. SAATY, *Finite Graphs and Networks*, New York: McGraw-Hill Book Company, 1965.

CHIEN, R. T., Cyclic decoding procedures for Bose–Chaudhuri–Hoquenghem codes, *IEEE Trans. Inform. Theory IT-10* (1964), 357–363.

———, Memory error control: Beyond parity, *IEEE Spectrum 10* (1973), 18–23.

CHIEN, R. T., and D. M. CHOY, Algebraic generalization of BCH–Goppa–Helgert codes, *IEEE Trans. Inform. Theory IT-21* (1975), 70–79.

CHIEN, R. T., I. B. OLDHAM, and D. T. TANG, Error detection and correction in a photo-digital memory system, *IBM J. Res. Develop. 12* (1968), 422–430.

DAVIS, W. A., The linearity of sequential machines: A critical review, *1968 IEEE Conference Record, 9th Annual Symposium on Switching and Automata Theory*, Schenectady, N.Y. (1968), 427–430.

DE BRUIJN, N. G., Generalization of Pólya's fundamental theorem in enumerative combinatorial analysis, *Proc. Koninklijke Nederlandse Akad. Wetenschappen*, Ser. A., *62* (1959), 59–69.

DEO, N., *Graph Theory with Applications to Engineering and Computer Science*, Englewood Cliffs, N.J.: Prentice-Hall, Inc., 1974.

EICHNER, L., Lineare Realisierbarkeit endlicher Automaten über endlichen Körpern, *Acta Inform. 3* (1973), 75–100.

ELIAS, P., Error-free coding, *IRE Trans. Inform. Theory PGIT-4* (1954), 29–37.

FIRE, P., A class of multiple-error-correcting binary codes for nonindependent errors, Sylvania Report RSL-E-2, Sylvania Reconnaissance Systems Laboratory, Mountain View, Calif., 1959.

FORNEY, G. D., JR., On decoding BCH codes, *IEEE Trans. Inform. Theory IT-11* (1965), 549–557.

———, *Concatenated Codes*, Cambridge, Mass.: MIT Press Research Monograph 37, 1966.

GAUSS, C. F., *Werke*, Band III, Göttingen: Königliche Gesellschaft der Wissenschaften, 1876, 1–30.

GHAZALA, M. J., Irredundant disjunctive and conjunctive forms of a Boolean function, *IBM J. Res. Develop. 1* (1956), 171–176.

GILBERT, E. N., A comparison of signalling alphabets, *Bell System Tech. J. 31* (1952), 504–522.

GILL, A., *Introduction to the Theory of Finite State Machines*, New York: McGraw-Hill Book Company, 1962.

———, *Linear Sequential Circuits*, New York: McGraw-Hill Book Company, 1966.

GINZBURG, A., *Algebraic Theory of Automata*, New York: Academic Press, Inc., 1968.

GOLAY, M. J. E., Notes on digital coding, *Proc. IRE 37* (1949), 657.

GOPPA, V. D., A new class of linear error-correcting codes, *Probl. Peredach. Inform. 6* (1970), 24–30. (In Russian.)

HAMMING, R. W., Error detecting and error correcting codes, *Bell System Tech. J. 29* (1950), 147–160.

HARARY, F., *Graph Theory*, Reading, Mass.: Addison-Wesley Publishing Co., Inc., 1969.

HARRISON, M. A., Combinatorial problems in Boolean algebras and applications to the theory of switching, Ph.D. thesis, University of Michigan, Ann Arbor, 1963.

HARRISON, M. A., *Introduction to Switching and Automata Theory*, New York: McGraw-Hill Book Company, 1965.
———, *Lectures on Linear Sequential Machines*, New York: Academic Press, Inc., 1969.
———, Counting theorems and their applications to switching functions, in *Recent Developments in Switching Theory*, ed. by A. Mukhopadhyay, New York: Academic Press, Inc., 1971, 85–120.
HARTMANIS, J., and R. E. STEARNS, *Algebraic Structure Theory of Sequential Machines*, Englewood Cliffs, N.J.: Prentice-Hall, Inc., 1966.
HENNIE, F. C., *Finite-State Models for Logical Machines*, New York: John Wiley & Sons, Inc., 1968.
HERMAN, G. T., When is a sequential machine the realization of another?, *Math. Systems Theory 5* (1971), 115–121.
HOFFMAN, K., and R. KUNZE, *Linear Algebra*, 2nd ed., Englewood Cliffs, N.J.: Prentice-Hall, Inc., 1971.
HOHN, F. E., *Applied Boolean Algebra*, 2nd ed., New York: Macmillan Publishing Co., Inc., 1966.
———, *Elementary Matrix Algebra*, 3rd ed., New York: Macmillan Publishing Co., Inc., 1973.
HOQUENGHEM, A., Codes correcteurs d'erreurs, *Chiffres 2* (1959), 147–156.
HUFFMAN, D. A., The synthesis of sequential switching circuits, *J. Franklin Inst. 257* (1954), 161–190, 275–303.
HUNTINGTON, E. V., Sets of independent postulates for the algebra of logic, *Trans. Amer. Math. Soc. 5* (1904), 288–309; see also *35* (1933), 274–304, 557–558, 971.
JUSTESEN, J., A class of constructive asymptotically good algebraic codes, *IEEE Trans. Inform. Theory IT-18* (1972), 652–656.
KAMPS, H. J. L., and J. H. VAN LINT, The football pool problem for five matches, *J. Comb. Theory 3* (1967), 315–325.
KAPUR, G. K., *IBM 360 Assembler Language Programming*, New York: John Wiley & Sons, Inc., 1970.
KOHAVI, Z., *Switching and Finite Automata Theory*, New York: McGraw-Hill Book Company, 1970.
LANDAU,, E., Über die Irreduzibilität der Kreisteilungsgleichung, *Math. Zeitschr. 29* (1928), 462.
MASSEY, J. L., Shift register synthesis and BCH decoding, *IEEE Trans. Inform. Theory IT-15* (1969), 122–127.
MAYEDA, W., *Graph Theory*, New York: Wiley-Interscience, 1972.
MEALY, G. H., A method for synthesizing sequential circuits, *Bell System Tech. J. 34* (1955), 1045–1079.
MICHAELSON, A. M., Computer implementation of decoders for several BCH codes, *Proc. of the Symp. on Computer Processing in Comm.*,

Polytechnic Press of the Polytechnic Institute of Brooklyn, Brooklyn, N.Y. (1969), 401–413. (Microwave Research Institute Symposia Series, J. Fox, ed., vol. 19.)

MOORE, E. F., Gedanken experiments on sequential machines, in *Automata Studies*, ed. by C. E. Shannon and J. McCarthy, 129–153, Princeton, N.J.: Princeton University Press, 1956.

MOORE, R. E., *Interval Analysis*, Englewood Cliffs, N.J.: Prentice-Hall, Inc., 1966.

PASCHBURG, R. H., Software implementation of error-correcting codes, *Coord. Sci. Lab. Rept. R-659*, University of Illinois at Urbana, 1974.

PATEL, A. M., and S. J. HONG, Optimal rectangular code for high density magnetic tapes, *IBM J. Res. and Develop. 18* (1974), 579–588.

PATTERSON, N. J., The algebraic decoding of Goppa codes, *IEEE Trans. Inform. Theory IT-21* (1975), 203–207.

PETERSON, W. W., and E. J. WELDON, JR., *Error-correcting Codes*, 2nd ed., Cambridge, Mass.: The MIT Press, 1972.

POLLARD, J. M., The fast Fourier transform in a finite field, *Math. of Comp. 25* (1971), 365–374.

PÓLYA, G., Kombinatorische Anzahlbestimmungen für Gruppen, Graphen und chemische Verbindungen, *Acta Math. 68* (1937), 143–254.

QUINE, W. V., A way to simplify truth functions, *Amer. Math. Monthly 62* (1955), 627–631.

RABIN, M. O., Computable algebra, general theory, and theory of computable fields, *Trans. Amer. Math. Soc. 87* (1960), 341–360.

REDFIELD, J. J., The theory of group reduced distributions, *Amer. J. Math. 49* (1927), 433–455.

REED, I. S., and G. SOLOMON, Polynomial codes over certain finite fields, *J. Soc. Ind. Appl. Math. 8* (1960), 300–304.

RETTER, C. T., Decoding Goppa codes with a BCH decoder, *IEEE Trans. Inform. Theory IT-21* (1975), 112.

REUSCH, B., Generation of prime implicants from subfunctions and a unifying approach to the covering problem, *IEEE Trans. Computers C-24* (1975), 924–930.

ROSENBLOOM, P. C., *The Elements of Mathematical Logic*, New York: Dover Publications, 1950.

SARWATE, D., On the complexity of decoders for Goppa codes, to appear in *IEEE Trans. Inform. Theory, IT-23*, July 1977.

SHANNON, C. E., The mathematical theory of communication, *Bell System Tech. J. 27* (1948), 379–423, 623–656. Reprinted in C. E. Shannon and W. Weaver, *The Mathematical Theory of Communication*, Urbana, Ill.: University of Illinois Press, 1949.

STONE, H. S., *Discrete Mathematical Structures and Their Applications*, Chicago: Science Research Associates, 1973.

SUGIYAMA, Y., M. KASAHARA, S. HIRASAWA, and T. NAMEKAWA, A method for solving key equation for decoding Goppa codes, *Information and Control 27* (1975), 87–99.

TIETAVÄINEN, A., On the nonexistence of perfect codes over finite fields, *SIAM J. Appl. Math. 24* (1973), 88–96.

———, A short proof for the nonexistence of unknown perfect codes, *Ann. Acad. Sci. Fenn.*, Ser. A, No. 580 (1974).

TZENG, K. K. M., On iterative decoding of BCH codes and decoding beyond the BCH bound, Ph.D. dissertation, University of Illinois, 1969.

VAN DER WAERDEN, B. L., *Modern Algebra*, Vol. 1 (7th ed.), New York: Frederick Ungar Publishing Co., Inc., 1970.

VARSHARMOV, R. R., Estimate of the number of signals in error correcting codes, *Dokl. Akad. Nauk SSSR 117* (1957), 739–741. (In Russian.)

VICKERS, F. D., *Introduction to Machine and Assembly Language: Systems/360/370*, New York: Holt, Rinehart and Winston, 1971.

VAN LINT, J. H., Nonexistence theorems for perfect error correcting codes, in *Computers in Algebra and Number Theory*, SIAM-AMS Proceedings IV, American Mathematical Society, Providence, R.I., 1971, 89–95.

———, *Coding Theory*, Berlin: Springer-Verlag, 1973.

WINOGRAD, S., On the time required to perform addition, *J. Assoc. Comp. Mach. 12* (1965), 277–285.

WOLFOWITZ, J., *Coding Theorems of Information Theory*, 2nd ed., Berlin: Springer-Verlag, 1964.

YOUNG, A., On quantitative substitutional analysis (fifth paper), *Proc. London Math. Soc. 31*(2) (1929), 273–288.

ZINOV'EV, V. A., and V. K. LEONT'EV, On perfect codes, *Probl. Peredach. Inform. 8* (1972), 26–35. (In Russian.)

Index

Abelian binary algebra, 166
Abelian group, additive, 207–208
Abelian group, elementary, 207
Abelian groups, fundamental theorem on finite, 208
Absorption laws, 31, 33, 69, 129, 268
Accepted sequence, 21
Accessible state, 19
Action, 204–205, 210, 243, 252
Adder, 26, 382
   carry-lookahead, 241
   fast, 235–240
   full binary, 241
   ripple-carry, 241
Addition modulo p, 180
   for a representation h, 237
Additive abelian group, 207–208
Additive notation, 172
Adjunction, 362
Aho, A. V., 98, 464, 486
Algebra, 102, 411
   binary, 165
   cycle-set, 420
   fundamental theorem of, 362, 373
   linear, 300–337, 341–359
   matrix, 105
   universal, 103
Algebraic coding theory, 435–485
Algebraic element, 360–361
Alphabet, 1
Alternating function, 330–332
Alternating group, 280, 335
Alternation, 118
AND-gate (AND-element), 124–125
Annihilator, 343, 449
Antiatom, 58, 285
Anti-homomorphism, 183
Anti-isomorphic posets, 56
Anti-isomorphism, 193, 264
Antilogarithms, table of, 370
Antisymmetric property, 27, 48, 119
Appel, K., 98, 486
Arbib, M. A., 190, 297, 483, 486
Argument, 50
Arithmetic modulo 2^n, 111–114
Ash, R., 449, 483, 486
Associative law, 10, 29, 32, 46, 69, 102, 103, 116, 117, 129, 140, 165, 166, 268
Atom, 58, 285–287
Atomic lattice, 285
Augmented matrix, 302
Automaton, finite, 16
Automorphism, 191, 364, 375–376
   Galois, 375
   inner, 200
Automorphism group, 191
Autonomous clock, 299
Autonomous linear machine, 390, 413–417
Autonomous machine, 184
Autonomous response, 389

Basic partition, 292–295
Basis, 310–312
   standard, 326
BCH code, 442–445, 450, 453
   binary decoding, 465–468
   decoding, 458–461, 464
   primitive, 444
Bead bracelet, 247–248
Berlekamp, E. R., 435, 449, 464, 481, 483, 486
Bijection, 9, 10, 12
Binary adder, 26
Binary algebra, 165
Binary counter, 23
Binary machine, 381
Binary operation, 49, 165
Binary pulse divider, 23
Binary relation, 41–68
Binary ring counter, 17, 19, 382
Binary symmetric channel, 211
Binary word, 212
Bipartite graph, 85
Blake, A., 163, 486
Blake, I. F., 435, 486
Block, 64
Block code, 213–214, 436
Bobrow, L. S., 483, 486
Boole, G., 32
Boolean algebra, 32–33, 118–123, 266, 270, 277, 285–290
   free, 289
   independent postulates for, 119–120, 122
Boolean function, 128
Boolean polynomial, 289
Boolean ring, 104, 116–119
Booth, T. L., 357, 382, 486
Bose, R. C., 442–443, 486
Bound, 4, 58–59
   greatest lower, 59, 265–266
   least upper, 59, 265–266
   lower, 58, 265, 482
   universal, 119
   upper, 4, 58, 265, 481
Bracelet, 247–248
Branch, 81, 87
Brown, D. T., 469, 486
Browne, E. T., 357, 486
Burnside, W., 244
Burst, 475
Burst channel, 475
Burst error correction, 444–445, 475–479
Burton, H. O., 450, 468, 486
Busacker, R. G., 99, 486

Cancellation, laws of, 108, 115, 140, 193
   left and right, 175
Canonical form, 307, 335
   rational, 335, 341–357, 414–416
   elementary divisors, 351
   invariant factors, 352
Cap ($\wedge$), 267–268

Capacity, 98, 99, 482–483
Cardinality, 4
Carroll, L., 38
Carry-lookahead adder, 241
Cartesian product, 3, 4
Cayley, A., 91, 203–204
Cayley–Hamilton theorem, 358
Ceiling, 236
Center, 87–88, 115, 178
Chain, 53, 59–60, 62, 267
Chain realization, 407
Change-of-basis matrix, 328
Channel, 211–213, 435
Character, 469
   CRC, 470
Characteristic, 115, 366–367
Characteristic function, 35
Characteristic polynomial, 353–354, 358, 415
Characteristic root, 359
Characteristic vector, 35, 359
Characteristic zero, 367
Characterizing matrices, 388–390, 412
Chemistry, 91, 251
Chessboard, 100
Chien, R. T., 445, 461, 469, 483, 486, 487
Chien search, 450, 461–462
Chinese remainder theorem, 239
Chord, 92
Choy, D. M., 483, 486
Circuit, 84
   combinational, 125
   fundamental, 94–95, 100
   logic or switching, 124–127
   relay, 164
Circuits, equivalent, 131
Circulating shift register, 429–430
   number of cycles, 429–430, 434
Clock, autonomous, 299
Closed operation, 165, 166
Closed partition, 290–292, 296–299
Closed walk, 84
Closure, 120
Coatom, 58, 285–287
Coatomic lattice, 285
Code, 214–235, 435–485
   BCH, 442–445, 450, 453
      decoding, 458–461, 464, 465–468
      primitive, 444
   block, 213–214, 435–485
   concatenated, 479
   convolutional, 436
   cyclic, 440–445
      encoding, 445–447
   dual, 225, 437
   error-correcting, 214–235, 435–485
   error-detecting, 214–215, 217
   five-times-repetition, 216
   Golay, 232–233, 438
      extended, 235
   Goppa, 464, 483
   group, 226–227
   Hamming, 230–234, 438, 440
      extended, 233–234
   linear, 227, 436–437
   parity-check, 214–215
   perfect, 231, 438, 481

Code [cont.]
   product, 479–481
   Reed–Solomon, 444–445
   single-error-correcting, 214–215, 220–223, 230–231
   systematic 438, 479
   triple-repetition, 215–216
Code word, 213, 436
Codomain, 5
Cofactor, 332–334
Column factor, 159–161
Column rank, 325
Column space, 325
Combinational circuit, 15, 125
Commutative binary algebra, 166
Commutative law, 29, 32, 69, 102–104, 116, 117, 120, 129, 140, 165, 268
Commutative ring, 104, 337
Companion matrix, 342
Companion space, 343, 348–350, 358
Comparable elements, 53, 143
Comparator, 79
Compatibility relation, 282–283
Complement, 27, 35, 118, 276
Complement, one's and two's, 111
Complementarity, laws of, 28, 32, 120, 129, 140
Complementation, 27–28
   of input leads, 255–263
Complemented lattice, 276–277, 290
Complete graph, 97
Complete lattice, 267
Complete polygon, 2
Complete sum, 151, 153
Component, 85, 283
Composite, right, 183–184
Composite function, 9–12, 322
Composite relation, 45–46
Composition, 45
Concatenated code, 479
Concatenation, 166
Conditionally transient state, 19
Congruence modulo p_λ 63, 66, 107–109, 180
Congruence relation, 179
Conjugacy class, 200, 255
   in G_n, 259–260
   in the symmetric group, 258
Conjugacy relation, 200
Conjugate, 197, 200, 255
Conjunctive normal form (c.n.f.), 138–139
Connectedness, 85
Connection relation, 41, 47
Connectives, 124
Consensus, laws of, 31, 33, 129
Consensus method, 163–164
Consistency principle, 27
Constant, 339
Constant function, 169
Construction of machines, 20
Containment, 3, 26
Converse, 42, 56
Convolutional code, 436
Coordinate vector, 312, 321
Coset, left, 194, 197
Coset, right, 197, 199
Coset decoding, 227–230, 438
Coset leader, 228–230

Cotree, 92
Countably infinite set, 2
Counter, binary, 17, 19, 23, 24, 382
Counter, up-down, 24
Counterimage, 5
Cover, 283–285, 288
Covering relation, 53, 75, 278
Cube, 63, 249–250
Cup (∨), 267–268
Cut, 99
Cut point, 85
Cut-set, 94, 95, 100
Cycle, 202–203
Cycle index, 246, 258, 261, 417, 430–432
Cycle-set algebra, 420
Cycle structure, 209, 417–422
Cycle sum, 421–422, 432
Cyclic code, 440–447
Cyclic group, 170–171, 177, 190, 201–202, 238, 240
  subgroup of, 202
Cyclic redundancy check (CRC), 469–475
Cyclic semigroup, 170–172
Cyclic subspace, 343, 347
Cyclotomic polynomial, 373

Davis, W. A., 397, 487
DeBruijn, N. G., 251, 253, 487
Decoding, coset, 227–230, 438
Decoding function, 213, 436
Decoding table, 228
Decomposition of a machine, 296–297
Defining equations, 206
Defining table, 17
Degree, 82, 203, 339, 360, 478
  in-degree and out-degree, 82
Delay (element), 382–383
Delay machine, 24
Delay operator, 384–385, 400–402
Delta, Kronecker, 219
De Morgan's laws, 28, 32, 119
Denumerable set, 2
Deo, N., 94, 95, 99, 487
Derivative, 367
Descartes, R., 3
Determinant, 329–336
Diagnostic matrix, 393
Diagonal matrix, 301, 359
Diagram, Hasse, 53, 60
Dichotomy, law of, 53
Difference of sets, 3
Digraph, 46–47
  associated undirected graph, 83
  symmetric, 82
Dihedral group, 205–207, 240, 247
Dimension, 312, 381
Direct product, 60–61, 196
Direct sum, 313–315, 323–324, 346
Directed graph, 46–47
Discrepancy, 454, 466
Disjoint cycle notation, 203
Disjoint sets, 26
Disjunctive normal form (d.n.f.), 138–139
Distance, 87, 216–218, 436
  minimum, 217–218, 227, 481
Distinguishing matrix, 392–393

Distinguishing sequence, 77–78, 393
Distributive lattice, 272–277
  free, 281–282, 284
Distributive law, 29, 32, 102, 104, 117, 120, 129, 140, 272
Divisibility, 54–55, 106
  tests for, 109
Division algorithm, 201, 340
Division of polynomials, 387
Division ring, 104
Divisor, 48, 54–55
Divisor of zero, 104, 115
Divisors, elementary, 351
Domain, 5
Domain, integral, 104
Dominant element, 30, 32, 129, 169
Dual, 30
Dual code, 225, 437
Dual poset, 56–57
Dual theorem, 31
Duality, principle of, 31, 33, 132, 278

Eccentricity, 87
Echelon form, 303
Edge, 81, 82
Edge-disjoint subgraphs, 85
Eichner, L., 397, 487
Eigenvalue and eigenvector, 359
Element, 1
  dominant, 129, 169
  greatest or least, 58
  idempotent, 115
  identity or unit or unity, 58, 102–104, 117, 120, 129, 165, 168, 337
  invertible, 166
  maximal or minimal, 57
  ubiquitous, 237–238, 240–241
  zero, 58, 103, 116, 168, 267
Element of a matrix, 300
Elementary divisors, 351
Elementary row operations, 301–302, 312, 336
Elias, P., 483, 487
Empty product, 144
Empty relation, 41
Empty sequence, 18
Empty set, 3
Encoder for cyclic codes, 445–447
Encoding function, 213, 226, 436, 445
Endomorphism, 191
Endpoint, 59
Entry, 42, 300
Enumerable set, 2
Enumeration theory of Pólya, 242–255
Epimorphism, 181
Equation, parity-check, 219
Equations, system of, 302–303, 308–309
Equations in semigroups, 192–193
Equivalence class, 64–65, 242
Equivalence relation, 63–67
  under group action, 205, 243
    number of equivalence classes, 244, 252
Equivalent circuits, 131
Equivalent codes, 438
Equivalent machines, 21, 75
Equivalent statements, 8
Equivalent states, 18, 71–74, 80

Error-correcting code, 214–235, 435–485
Error-detecting code, 214–215, 217
Error locators, 450
Error pattern, 212–213, 435
Error polynomial, 450
Error values, 450, 459
Essential component, 283
Essential prime implicant, 156–157
Euclidean algorithm, 403–404, 411, 464
Euler, L., 95, 100
Euler phi-function, 201, 370
Euler–Venn diagram, 28
Euler walk, 100
Evaluation function, 338
Even permutation, 329
Exclusive-OR, 38, 49, 140–143
Exclusive-OR element, 124
Existence theorem, 8
Exponent expression, 370
Exponents, laws of, 105, 176
Extended Hamming code, 233–234
Extension, field, 360–366, 373
External direct sum, 313

Face, k-dimensional, 63
Factor group, 198–199
Factor semigroup, 179, 190
Factors, invariant, 352
Fast adders, 235–240
Fast Fourier transform, 464
Feedback shift register (FSR), 384–387, 414–425, 430–432, 451–458
   cycle structure, 417–422
   outputs, 425, 452
   singular or nonsingular, 417
Feedforward shift register, 383–384, 386
Fermat, P., 108
Field, 104, 339, 340, 357, 360–380
   finite, 367–372, 381
   Galois, 110–111
   splitting, 364–366, 368
Field extension, 360–366
Field integers, 366
Field of subsets, 280
Final state, 18
Finitary operation, 50
Finite abelian group, 208, 240
Finite-dimensional vector space, 310–312
Finite extension, 360, 373
Finite field, 367–372, 381
Finite-state machine, 16, 17–26
Fire, P., 475, 487
Fire code, 476–478
First homomorphism theorem, 198
Five-times-repetition code, 216
Fixed point, 7
Flipflop, 24, 382
Football pool problem, 440
Forced response, 389
Forest, 87
Formal power series, 115–116
   ring of, 464
Forney, G. D., Jr., 459, 479, 487
Four-color problem, 97–98
Four-group, Klein, 200, 279
Fourier transform, fast, 464

Free Boolean algebra, 289
Free distributive lattice, 281–282, 284
Free monoid or semigroup, 166
Frobenius, G., 244
Frontal switching function, 164
Full binary adder, 241
Function, 4–6
   as relation, 49
   Boolean, 128
   characteristic, 35
   constant, 169
   decoding, 213
   encoding, 213, 226
   exclusive-OR, 49
   identity, 7, 11, 167
   incompletely specified, 163
   invertible, 11, 12, 15
   next-state or output, 16, 128, 184
   Petrick, 157
   rational, 105
   switching, 128
   transition, 16
Function table, 127, 129–130
Functions, equality of, 5
Fundamental circuit or cut-set, 94–95, 100
Fundamental theorem of algebra, 362, 373

Galois automorphism, 375
Galois field, 110–111
Galois group, 375–376
Gate, 124, 236
Gauss, C. F., 362, 373, 487
Gauss–Jordan procedure, 302–303
General linear group $GL_n(F)$, 303
Generated subgroup or subsemigroup, 174, 176
Generator, 170, 201, 339, 442
Generator matrix, 219–223, 436–437
Ghazala, M. J., 158, 487
Gilbert, E. N., 482–483, 487
Gilbert lower bound, 482
Gill, A., 74, 370, 487
Ginzburg, A., 190, 487
Golay, M. J. E., 232, 438, 482, 487
Golay code, 232–233, 235, 438
Goppa, V. D., 483, 487
Goppa code, 464, 483
Graph, 19, 46–47, 81–101
   bipartite, 85
   complete, 97
   connected, 85
   directed, 41, 81, 82
   planar or nonplanar, 96–97
   simple, 82
   undirected, 81–101
   weighted, 92
Graph of a function, 41
Graphs, homeomorphic, 96–97
Graphs, isomorphic, 83
Greatest common divisor (g.c.d), 106, 339–341, 403, 411, 426, 464
Greatest lower bound (g.l.b.), 59, 265, 266
Group, 166, 193–210
   additive abelian, 207–208
   alternating, 280, 335
   cyclic, 170, 177, 190, 201–202, 238, 240
      subgroup of, 202

Group [*cont.*]
  dihedral, 205–207, 240, 247
  elementary abelian, 207
  factor, 198
  finite abelian, 208, 240
  Galois, 375–376
  matrix, 210
  permutation, 203–204
  quaternion, 209, 241
  quotient, 198
  symmetric, 167, 202
Group code, 226–227
Group-induced equivalence relation, 205, 243
Group of permutations and complementations of n input leads (G_n), 257–261, 430–432
  case $n = 4$, 262
  conjugacy classes, 259, 260
Group properties, 166

Haken, W., 98, 486
Hamiltonian path, 100
Hamming, R. W., 481, 482, 487
Hamming code, 230–234, 438, 440
Hamming upper bound, 481
Harary, F., 97, 99, 487
Harrison, M. A., 255, 264, 282, 288, 297, 381, 397, 430, 487–488
Hartmanis, J., 190, 297, 382, 397, 488
Hasse diagram, 53, 60
Hennie, F. C., 81, 382, 488
Herman, G. T., 397, 488
Hexagon, 206
Hirasawa, S., 464, 490
Hoffman, K., 300, 488
Hohn, F. E., 164, 290, 300, 488
Homeomorphic graphs, 96
Homogeneous system of equations, 302, 308–309
Homomorphic image, 181
Homomorphism, 81, 181–182, 226, 316, 337–338
Homomorphism theorem, first, 198
Hong, S. J., 469, 489
Hopcroft, J. E., 98, 464, 486
Hoquenghem, A., 442–443, 488
Huffman, D. A., 16, 488
Huntington, E. V., 120–122, 488
Huntington postulates, 120

Ideal, 337–340, 441–442
Idempotency, law of, 30, 32, 69, 104, 129, 268
Idempotent element, 115, 175–176
Idempotent ring, 104
Identity element, 30, 32, 102–103, 120, 129, 140, 165, 166, 168, 337
  left, 168, 193
  right, 168, 199
  for a semigroup, 169–170
Identity function, 7, 11, 167
Identity matrix, 218, 301
Identity relation, 48, 63
Iff, 8
Image, 5
Implementation, 128
Implicant, 144
  essential prime, 156–157
  prime, 144, 150–155
Implicants, poset of, 150–151

Implication, 143
Implied pair, 292–293
Impulse response, 408–410, 412
Inclusion, 3, 26–27, 33, 44, 68–69, 119
Incompletely specified function, 163
Independent postulates, 119–120, 122
Independent vectors, 346–347
Index, 194
  cycle, 246, 258, 261, 417, 430–432
  Young, 258–262
Indexing set, 2
Induction, mathematical, 33–34, 40
Induction hypothesis, 34
Inert linear machine, 390, 408
Infinite extension, 360
Infinite set, 2
Information theory, 482–483
Initial (or initialized) machine, 41, 80
Initial state, 383
Injection, 9, 10, 196
Inner automorphism, 200
Input, 124, 184
Input alphabet, 16
Input consistent partition, 299
Input independent partition, 299
Input leads, 15, 124
Input monoid, 181, 184, 187–189
Input sequence, 18
Input string, 18
Input symbols, 16
Input tape, 18, 174, 181
Input variables, 124
Input word, 18
Integers modulo p, 110
Integral domain, 104
Interleaving, 478
Internal direct sum, 314, 315
Internal state, 15, 16
International code, 7
Intersection, 26, 36, 42, 43, 66–69, 267, 310
Invariant factors, 352
Invariant subspace, 343
Inventory, 246
Inverse, 11, 12, 103, 104, 166
  left or right, 115, 176, 193, 199, 394
Inverse of a matrix, 304–305
Inverse of a product, 15, 173
Inverter, 124
Invertible element, 103, 166
Invertible function, 11, 12, 15
Invertible matrix, 303, 327, 332
Involution, law of, 28, 32, 129
Involvement, 144, 149
Irreducible polynomial, 111, 339–341, 363
  for u over K, 361
  roots of, 376–378, 434
Irredundant sum of products, 149, 158–161
Isolated vertex, 82
Isomorphism, 43, 68
  of binary algebras, 181
  of Boolean algebras, 36–37
  of graphs, 83
  of machines, 80
  of posets, 56
  of rings, 337
  of vector spaces, 317

Iterated consensus method, 163–164

JK-flipflop, 24
Join, 118, 267
Jordan–Dedekind chain condition, 60
Justesen, J., 483, 488

Kamps, H. J. L., 440, 488
Kapur, G. K., 111, 488
Kasahara, M., 464, 490
Kernel, 198, 317, 337, 343
Kirchhoff, G., 95
Klein four-group, 200, 279
Kohavi, Z., 297, 382, 488
Königsberg bridge problem, 95–96, 100
Kronecker delta, 219
Kunze, R., 300, 488
Kuratowski, K., 97

Lagrange's theorem, 193–195
Landau, E., 373, 488
Lattice, 69, 265–286
  atomic or coatomic, 285
  complemented, 276–277, 290
  complete, 267
  distributive, 272–276, 277
    free, 281–282, 284
  modular, 271–272, 274
  pi-lattice, 292
Lattice of partitions of a set, 273
Lattice polynomial, 281–282
Leading, 1, 303
Leaf, 87
Least common multiple (l.c.m.), 106, 426
Least upper bound (l.u.b.), 59, 265–266
Left action, 204–205, 210, 252
Left cancellation law, 175
Left coset, 194, 197
Left identity, 168, 193
Left inverse, 115, 193, 199, 394
Left permutation, 264
Left zero, 168
Length, 59, 84, 144, 437, 479
Leont'ev, V. K., 439, 490
Letter, 1
Line, 81
Linear algebra, 300–337, 341–359
Linear code, 227, 436–437
Linear combination, 310
Linear dependence, 310
Linear equation, solution, 140, 302–303
Linear independence, 310–311, 325
Linear machine, 381–434
  autonomous, 390, 413–417
  inert or quiescent, 390, 408
Linear mapping or operator, 316
Linear ordering, 53
Linear transformation, 316–324
  matrix of, 320–324, 334–335
Literal, 132, 143
Logarithms, table of, 370
Logic circuits, 124–127
Logic elements, 124
Loop, 46, 82
Lower bound, 58, 265
  greatest, 59, 265, 266

McCluskey, E., 164
Machine, 16–25, 41–42, 75–80
  autonomous, 184
  binary, 381
  delay, 24
  finite-state, 16
  initial(ized), 41, 80
  linear, 381–434
  Mealy, 16, 184–186
  minimal-state, 75–77
  Moore, 16, 184–187, 295, 390
  parity-check, 23
  pi-lattice of, 292
  reduced form, 75–77
  semigroup of, 184–187
  sequential, 16
  simply minimal, 25
  state, 184, 295
  state-output, 184–187
  strongly connected, 19
  two-terminal, 383, 402, 405, 408
  unit-delay, 20–21
Machine of a semigroup, 189–190
Machines, equivalent, 21, 42, 75
Machines, isomorphic or state-isomorphic, 80
Map method, 164
Mapping, 5
Massey, J. L., 449, 450, 452, 453, 464, 488
Mathematical induction, 33–34, 40
Matrix, 42, 218, 300
  augmented, 302
  companion, 342
  diagonal, 301, 359
  generator, 219–223, 436–437
  identity, 218, 301
  invertible, 303, 327, 332
  nonsingular, 303
  parity-check, 220–223, 437
  permutation, 438
  scalar, 301
  simple transition, 52
  zero, 300
Matrix group, 210
Matrix of a linear transformation, 320–324, 334–335
Matrix of a relation, 42–44, 46, 105
Max-flow min-cut theorem, 98–99
Maximal chain, 59–60
Maximal element, 57
Maximal ideal, 339, 340
Maximal period feedback shift regster, 418–419
Maximal polynomial, 136–139
Maxterm, 136–139
Mayeda, W., 99, 488
Mealy, G. H., 16, 184, 488
Mealy machine, 16, 184–186
Meet, 267
Member, 1
Michaelson, A. M., 465, 488
Minimal cover, 283–285, 288
Minimal element, 57
Minimal polynomial, 136–139, 353, 358
Minimal spanning tree, 92–94
Minimal-state machine, 75–77
Minimal sum of products, 150, 155–161, 281–282
Minimization procedure, 77, 394–397

Index

Minimum distance, 217–218, 227
Minimum weight, 227
Minterm, 136–139, 289
Möbius mu-function, 377, 430
Modular arithmetic, 63, 66, 107–114, 180, 237
Modular lattice, 271–272, 274
Modular law, 271
Moenck, R., 464
Monic polynomial, 339
Monoform variable, 159
Monoid, 165, 166, 169
Monoid, input, 181, 184, 187–189
Monomorphism, 181
Monotone switching function, 282, 288
Moore, E. F., 16, 184, 489
Moore, R. E., 178, 489
Moore machine, 16, 184–187, 295, 390
Moore state, 185
Morphism, 181–183
Morse code, 7
Mullin, R. C., 435, 486
Multigraph, 82
Multilinear function, 330–332
Multiplication modulo p, 180, 242
Multiplication of polynomials, 386
Multiplication table, 167–168
Multiplier, 382

Namekawa, T., 464, 490
Necessary and sufficient condition (n.a.s.c. or n.s.c.), 8
Necessity, 8
Negative, 103, 116, 140, 300
Newton's identities, 465
Next-block vector, 74
Next-state function, 16, 184
Node, 81
Noisy channel, 211
Nonderogatory matrix, 415–416
Nonsingular feedback shift register, 417, 425
Nonsingular matrix, 303
Normal basis theorem, 434
Normal form, conjunctive or disjunctive, 138–139
Normal form, ring, 142–143
Normal subgroup, 196–199, 280
NOT-element, 124
Null sequence, 426–428
Null set, 3
Null space, 317
Nullity, 318

Odd permutation, 329
Oldham, I. B., 445, 487
One-one (one-to-one), 8–9
One-one correspondence, 9, 12
One's complement, 111
Onto function, 8–9
Open walk, 84
Operation or operator, 50
  binary, 49, 165
Orbit, 202, 205
Order, 4, 165, 171, 300
Ordered pair, 3
Ordered set, 53, 59, 267
Ordering, linear or partial, 53
OR-gate (OR-element), 124–125

Orientation, 83
Orthogonal functions, 141
Orthogonality, 290
Output, 124, 184
Output alphabet or symbols, 16
Output consistent partition, 298–299
Output expression, 125
Output function, 16, 128, 184
Output leads, 15, 127
Output sequence or string or tape or word, 18
Output variables, 124
Output vector, 64
Overall check symbol, 233

Pair, implied, 292
Pairwise disjoint sets, 64
Parallel decomposition, 297
Parallel edges, 82
Parity-check code, 214–215
Parity-check equations, 219
Parity-check machine, 23
Parity-check matrix, 220–223, 437
Parity checker 141–143
Partial ordering, 53–61
Partially ordered set, 53
Partition (partitioning), 64–67, 179, 266
  basic, 292–295
  closed, 290–292, 296–299
  input-consistent or input-independent, 299
  output-consistent, 298–299
  unit or zero, 64, 69
Partition inclusion or intersection or union, 68–69
Partition with substitution property, 290
Paschburg, R. H., 465, 489
Patel, A. M., 469, 489
Path, 47, 84
  Hamiltonian, 100
Patterson, N. J., 483, 489
Pendant vertex, 82, 87
Pentane, 91
Pentagon, 206
  in a nonmodular lattice, 272
  symmetries of, 247
Pentagon building, 249
Perfect code, 231, 438, 481
Period, maximal, 418–419, 476
Periodic input tape, 174
Periodic part, 409
Permutation, 167, 329
  even or odd, 329
  left or right, 264
Permutation group, 203–204
Permutation matrix, 438
Permutation of input leads, 255–263
Peterson, W. W., 370, 435, 436, 444, 482, 489
Petrick function, 157
Photodigital mass memory, 445
Pigeonhole principle, 12
Pivotal variable, 160
Planar graph, 96
Pollard, J. M., 464, 489
Pólya, G., 242, 246, 489
Pólya enumeration theory, 242–255
Polygon, complete, 2
Polynomial, 105, 111
  Boolean, 289

Polynomial [*cont.*]
  characteristic, 353–354, 358, 415
  cyclotomic, 373
  error, 450
  irreducible, 111, 339–341, 361, 363, 376–378, 434
  maximal, 136–139
  minimal, 136–139, 353, 358
  monic, 339
  nonconstant, 339
  primitive, 369–371, 379
  reciprocal, 400, 411
  recursion, 451–452
  symmetric, 474
Polynomial multiplication and division, 386–387
Polynomial ring, 339–341
Poset, 53–61, 119, 150–151, 265
Positive switching function, 164, 282
Power series, formal, 115–116, 464
Power set, 3, 105
Powers of an element, 170
Powers of a function, 11
Preferred cover, 288
Preimage, 5, 15
Present input, 16
Prime, 59, 207
Prime implicant, 144, 150–157
Prime power, 208, 238
Prime subfield, 374
Primitive BCH code, 444
Primitive element, 368–369
Primitive nth root of 1, 373, 442
Primitive polynomial, 369–371, 379
Principal ideal, 339, 340
Probability, 211
Product, 267, 301
  direct, 60–61, 196
Product code, 479–481
Product term, 144, 149
Projection, 196
Proper subgroup, 174
Propositional logic, 164
Pulse divider, binary, 23
Pythagorean triad, 7

Quaternion group, 209, 241
Quiescent linear machine, 390
Quine, W. V., 163, 164, 489
Quotient field, 359
Quotient group, 198
Quotient ring, 338–339
Quotient space, 318–319

Rabin, M. O., 381, 489
Range, 5, 317
Rank, 312, 318, 325–327, 392–393
Rate, 214, 436, 481
Rational canonical for, 335, 341–357, 414–416
  elementary divisor form, 351
  invariant factor form, 352
Rational function, 105
Rational transfer function, 400–402
Ray-Chaudhuri, D. K., 442–443, 486
Reachability relation, 41
Reachable state, 19
Reciprocal polynomial, 400, 411
Recognized sequence, 21

Recursion polynomial, 451, 452
Recursive equation, 423–424, 427–428, 451
Redfield, J. J., 242, 246, 489
Reduced echelon form, 303
Reduced form, 75–77, 393
Redundancy, laws of, 31, 33, 129
Reed, I. S., 444, 489
Reed–Solomon code, 444–445
Refinement, 69, 70, 72
Reflexive property, 27, 48, 119
Regular n-gon, 205–206
Relation, 41–68
  antisymmetric, 48
  binary, 41
  compatibility, 282–283
  complementary, 42
  congruence, 179
  connection, 41, 47
  covering, 53, 278
  empty, 41
  equality, 63
  equivalence, 63–67
  finite, 42
  identity, 48, 63
  reachability, 41
  reflexive, 48
  similarity, 63
  symmetric, 48
  transitive, 48
  universal, 41, 63
Relation, digraph of, 46
Relation matrix, 42–44, 46, 48, 105
Relation of a digraph, 47
Relation on a set, 48
Relatively prime, 106, 201, 339, 358
Relay switching circuit, 164
Representation, 237
Reset, 169
Residue class, 339
Response, autonomous or forced, 389
Response, impulse, 408–410, 412
Restriction, 172, 323
Retter, C. T., 483, 489
Reusch, B., 151, 155, 158, 163, 489
Rieger bound, 479
Right action, 204–205
Right cancellation law, 175
Right composite, 183–184
Right coset, 197, 199
Right identity, 168
Right inverse, 115, 199, 394
Right permutation, 264
Right zero, 168
Rigid motions, 206, 249–250
Ring, 103–105, 337–339
  Boolean, 104, 116–119
  commutative, 104, 337
  division, 104
  idempotent, 104
Ring normal form, 142–143
Ring of polynomials, 339–341
Ring sum, 49, 95, 117, 124, 140–143
Ring with unity, 104, 337
Ripple-carry addition, 241
Root, 88, 360, 376
Root, characteristic, 359

Index

Root of 1, 373, 442
Rooted tree, 88
Rosenbloom, P. C., 123, 489
Rotations of the square, 171
Row equivalent matrices, 307
Row operations, 301–302, 312, 336
Row space, 312, 325–327

Saaty, T. L., 99, 486
Sarwate, D., 464, 489
Scalar matrix, 301
Scalar multiplication, 300, 307
Schematic diagram, 298–299
Self-conjugate subgroup, 197
Sellers, F. F., Jr., 469, 486
Semigroup, 165, 168–193
   cyclic, 170
   factor, 179, 190
   free, 166
   machine of, 189–190
Semigroup, equations in, 192–193
Semigroup of a machine, 184–187
Sequence, distinguishing, 77–78, 393
Sequence, synchronizing, 81
Sequence recognizer, 22, 24
Sequential circuit, 15
Sequential machine, 16
Serial binary adder, 26
Series decomposition, 295–297
Set, 1–4
Set-builder notation, 2
Shannon, C. E., 482–483, 489
Shannon's fundamental theorem, 483
Sheep-pen principle, 14
Shift register, 382–387
   circulating, 429–430, 434
   feedback, 384–387, 414–425, 430–432, 451–458
   feedforward, 383–384, 386
Shortening, 144
Sign (sgn), 329–330
Similar linear machines, 413
Similar matrices, 334–335
Simple graph, 82
Simple walk, 84
Simply minimal machine, 25
Single-error-correcting code, 214–215, 220–223, 230–231
Singleton, 4
Singular feedback shift register, 417
Sink, 98
Sink state, 19
Size, 300
Skeleton, 91
Slepian's standard array, 228
Soccer, 440
Solomon, G., 444, 489
Source, 98
Spanning of a subspace, 310
Spanning tree, 91–95
Splitting field, 364–366, 368
Square, rotations of, 171
Square, symmetries of, 199, 242
Stabilizer, 174, 205
Stage, 241
Standard basis, 326
State, 184, 383

State diagram, 18, 185
State-isomorphic machines, 80
State machine, 184, 295
State-output machine, 184–187
State table, 17
State transition map, 181
Stearns, R. E., 190, 297, 382, 397, 488
Stone, H. S., 419, 489
Strictly into mapping, 9
Strongly connected machine, 19
Subalgebra, 411
Subcube, 63
Subdiagonal, 342
Subfield, 360, 374, 375
Subgraph, 84, 85
Subgroup, 172–174, 176, 266, 280
   normal or self-conjugate, 196–199, 280
Subgroup of cyclic group, 202
Sublattice, 271, 280
Submachine, 19
Subring, 114, 358
Subsemigroup, 172, 174, 176
Subset, 3
Subspace, 290, 308, 310, 317–319
   f-cyclic, 341
   f-invariant, 323–324
Substitution principle, 107, 120
Substitution property, 290
Subsuming, 144
Successor function, 15
Sufficiency, 8
Sugiyama, Y., 464, 490
Sum, 118, 267, 300, 310
Sum, complete, 151, 153
Sum-of-products representation, 144
   irredundant, 149, 158–161
   minimal, 150, 155–161, 281–282
   shorter, 149
Superdiagonal, 389
Superset, 3
Surjection, 9, 10
Sweepout process, 302–303, 304–305, 312, 336
Switching circuits, 124–127
   relay, 164
Switching function, 128
   frontal or positive, 164, 282
   monotone, 282, 288
Symmetric channel, 435
Symmetric difference, 38
Symmetric digraph, 82
Symmetric group, 167, 202, 258
Symmetric polynomial, 474
Symmetric relation, 48
Symmetries, 177–178, 195, 199, 206, 242, 247
Synchronizing sequence, 81
Syndrome, 221–222, 229–230, 233–234
System of equations, 302–303, 308–309
Systematic code, 438, 479

Table of values, 50
Tang, D. T., 445, 487
Term, 144, 149
Terminal, 82
Terminal function, 83
Tetrahedron, 251
Tietäväinen, A., 439, 490

Tison, P., 163–164
Total state, 185
Totally ordered set, 53, 59
Transcendental element, 360
Transfer function, 384, 412
    rational, 400–402
Transformation, linear, 316–324
    matrix of, 320–324, 334–335
Transient part, 409
Transient state, 19
Transition function, 16
Transition matrix, simple, 52
Transition table, 17
Transitive permutation goup, 205
Transitive property, 27, 48, 119
Transitive union, 67
Transmission, 128
Transportation problem, 98, 101
Transpose, 325
Transposition, 329
Traveling salesperson problem, 98, 101
Tree, 85–95
    binary, 88, 89
    decision, 88
    degenerate or nondegenerate, 87
    rooted, 88
    sorting, 88
    spanning, 91–95
Triangle, symmetries of, 177–178, 195
Triple-repetition code, 215–216
Truth table, 129
Two-sided identity or zero, 168
Two-terminal machine, 383, 402, 405, 408
Two-s complement, 111
Tzeng, K. K. M., 450, 490

Ubiquitous element, 237–238, 240–241
Ullman, J. D., 98, 464, 486
Undirected graph, 81–101
Union, 26, 36, 42–43, 66–69, 267
Union, transitive, 67
Unit delay machine, 20–21
Unit (unity) element, 102, 104, 117, 168, 267
Unit partition, 64, 69
Universal algebra, 103
Universal bounds, 119
Universal element, 32
Universal relation, 41, 63
Universal set, 3, 27
Universe of discourse, 3

Upper bound, 4, 58, 265
    least, 59, 265, 266
Upper-triangular matrix, 336

Vandermonde matrix, 337
Van der Waerden, B. L., 434, 465, 490
Van Lint, J. H., 435, 438, 440, 483, 488, 490
Varsharmov, R. R., 482, 490
Vector, 307
    characteristic, 35–37, 359
    next-block, 74
    output, 64
Vector expression, 370
Vector space, 307–319
Vectors, complement or intersection or union of, 35–36
Venn diagram, 28–29, 64
Vertex, 81
    isolated, 82
    pendant, 82, 87
Vertex-disjoint subgraphs, 85
Vertices, adjacent, 82
Vickers, F. D., 111, 490

Walk, 47, 84
    closed or open, 84
    Euler, 100
    simple, 84
Weight, 50, 92, 214, 216–218, 243, 435
    minimum, 227, 288
Weighted graph, 92
Weldon, E. J., Jr., 370, 435, 436, 444, 482, 489
Well-defined operation, 179
Winograd, S., 237, 490
Wolfowitz, J., 483, 490
Word, 435
    binary, 212, 218
    code, 213, 436

XOR, 49

Young, A., 258, 259, 490
Young index, 258–262

Zero (element), 30, 32, 103, 116, 168, 267, 360
    left or right, 168
Zero, divisor of, 104, 115
Zero matrix, 300
Zero partition, 64, 69
Zinov'ev, V. A., 439, 490

DATE DUE			
DEC 17 1980			
JAN 21 1981			
DEC 3 1982			
JAN 30 1984			
MAY 25 1989			

DEMCO 38-297